PRINCIPLES OF ORGANIC CHEMISTRY

Principles of Organic Chemistry

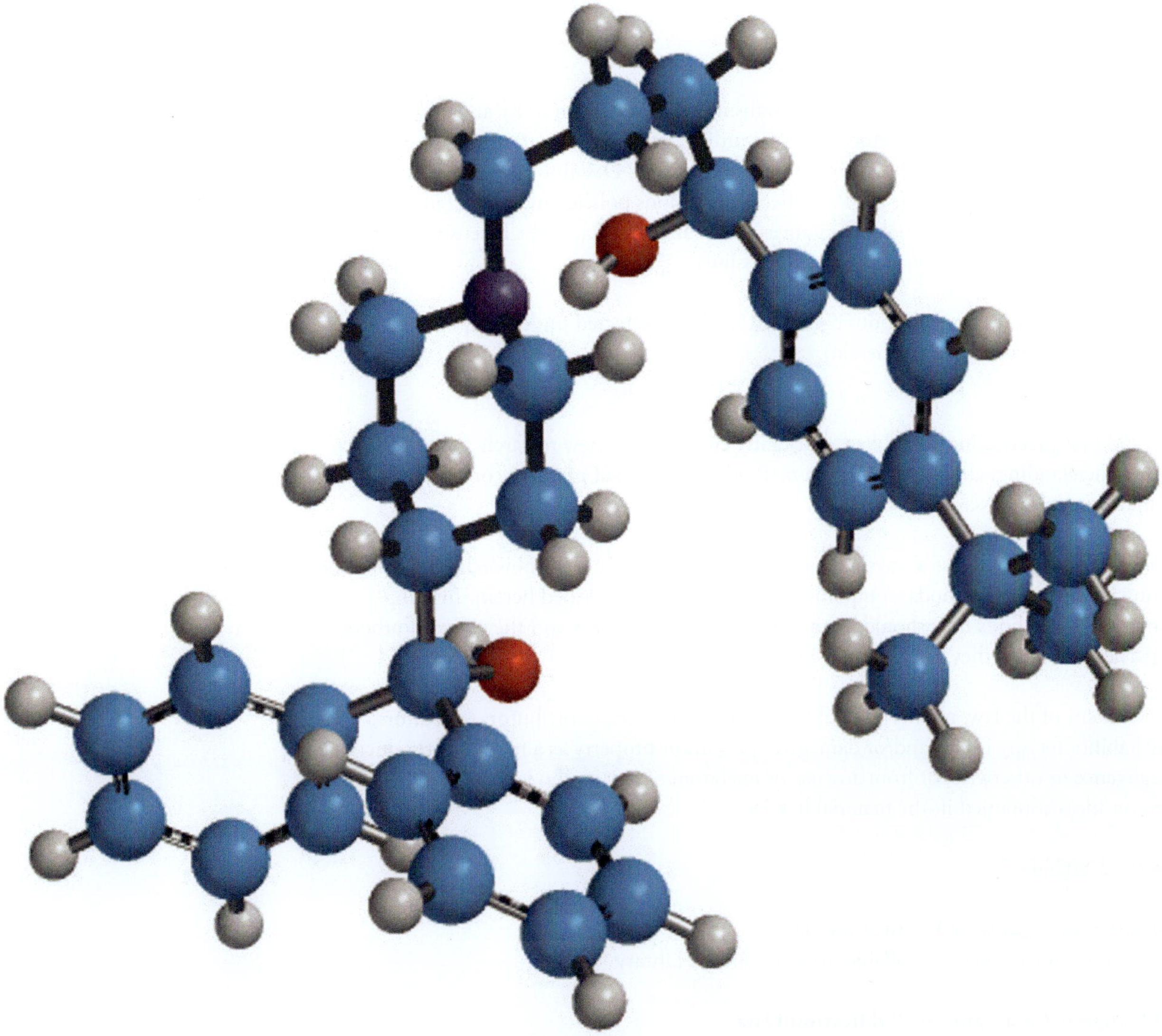

Robert J. Ouellette
Professor Emeritus
Department of Chemistry
The Ohio State University

J. David Rawn
Professor Emeritus
Department of Chemistry
Towson University

ELSEVIER AMSTERDAM · BOSTON · HEIDELBERG · LONDON · NEW YORK · OXFORD
PARIS · SAN DIEGO · SAN FRANCISCO · SINGAPORE · SYDNEY · TOKYO

Elsevier
Radarweg 29, PO Box 211, 1000 AE Amsterdam, Netherlands
The Boulevard, Langford Lane, Kidlington, Oxford OX5 1GB, UK
225 Wyman Street, Waltham, MA 02451, USA

Notices
Knowledge and best practice in this field are constantly changing. As new research and experience
broaden our understanding, changes in research methods, professional practices, or medical
treatment may become necessary.

Practitioners and researchers must always rely on their own experience and knowledge in evaluating
and using any information, methods, compounds, or experiments described herein. In using
such information or methods they should be mindful of their own safety and the safety of others,
including parties for whom they have a professional responsibility.

To the fullest extent of the law, neither the Publisher nor the authors, contributors, or editors,
assume any liability for any injury and/or damage to persons or property as a matter of products
liability, negligence or otherwise, or from any use or operation of any methods, products,
instructions, or ideas contained in the material herein.

ISBN: 978-0-12-802444-7

British Library Cataloguing in Publication Data
A catalogue record for this book is available from the British Library

Library of Congress Cataloging-in-Publication Data
A catalog record for this book is available from the Library of Congress

For Information on all Elsevier publications
visit our website at http://store.elsevier.com/

Working together
to grow libraries in
developing countries

www.elsevier.com • www.bookaid.org

TABLE OF CONTENTS

CHAPTER 1 STRUCTURE OF ORGANIC COMPOUNDS 1
1.1 ORGANIC AND INORGANIC COMPOUNDS 1
1.2 ATOMIC STRUCTURE 1
1.3 TYPES OF BONDS 4
1.4 FORMAL CHARGE 7
1.5 RESONANCE STRUCTURES 8
1.6 PREDICTING THE SHAPES OF SIMPLE MOLECULES 10
1.7 ORBITALS AND MOLECULAR SHAPES 11
1.8 FUNCTIONAL GROUPS 15
1.9 STRUCTURAL FORMULAS 18
1.10 ISOMERS 23
1.11 NOMENCLATURE 25
    EXERCISES 27

CHAPTER 2 PROPERTIES OF ORGANIC COMPOUNDS 33
2.1 STRUCTURE AND PHYSICAL PROPERTIES 33
2.2 CHEMICAL REACTIONS 38
2.3 ACID-BASE REACTIONS 39
2.4 OXIDATION-REDUCTION REACTIONS 41
2.5 CLASSIFICATION OF ORGANIC REACTIONS 44
2.6 CHEMICAL EQUILIBRIUM AND EQUILIBRIUM CONSTANTS 45
2.7 EQUILIBRIA IN ACID-BASE REACTIONS 47
2.8 EFFECT OF STRUCTURE ON ACIDITY 49
2.9 INTRODUCTION TO REACTION MECHANISMS 51
2.10 REACTION RATES 54
    EXERCISES 58

CHAPTER 3 ALKANES AND CYCLOALKANES 65
3.1 CLASSES OF HYDROCARBONS 65
3.2 ALKANES 65
3.3 NOMENCLATURE OF ALKANES 68
3.4 CONFORMATIONS OF ALKANES 72
3.5 CYCLOALKANES 75
3.6 CONFORMATIONS OF CYCLOALKANES 78
3.7 PHYSICAL PROPERTIES OF ALKANES 81
3.8 OXIDATION OF ALKANES AND CYCLOALKANES 83
3.9 HALOGENATION OF SATURATED ALKANES 84

3.10   NOMENCLATURE OF HALOALKANES   87
        SUMMARY OF REACTIONS   89
        EXERCISES   90

CHAPTER 4 ALKENES AND ALKYNES   95

4.1   UNSATURATED HYDROCARBONS   95
4.2   GEOMETRIC ISOMERISM   99
4.3   E,Z NOMENCLATURE OF GEOMETRICAL ISOMERS   101
4.4   NOMENCLATURE OF ALKENES AND ALKYNES   103
4.5   ACIDITY OF ALKENES AND ALKYNES   106
4.6   HYDROGENATION OF ALKENES AND ALKYNES   107
4.7   OXIDATION OF ALKENES AND ALKYNES   110
4.8   ADDITION REACTIONS OF ALKENES AND ALKYNES   111
4.9   MECHANISM OF ADDITION REACTIONS   113
4.10   HYDRATION OF ALKENES AND ALKYNES   115
4.11   PREPARATION OF ALKENES AND ALKYNES   116
4.12   ALKADIENES (DIENES)   119
4.13   TERPENES   120
        SUMMARY OF REACTIONS   124
        EXERCISES   126

CHAPTER 5 AROMATIC COMPOUNDS   133

5.1   AROMATIC COMPOUNDS   133
5.2   AROMATICITY   134
5.3   NOMENCLATURE OF AROMATIC COMPOUNDS   137
5.4   ELECTROPHILIC AROMATIC SUBSTITUTION   139
5.5   STRUCTURAL EFFECTS IN ELECTROPHILIC AROMATIC SUBSTITUTION 143
5.6   INTERPRETATION OF RATE EFFECTS   145
5.7   INTERPRETATION OF DIRECTING EFFECTS   148
5.8   REACTIONS OF SIDE CHAINS   150
5.9   FUNCTIONAL GROUP MODIFICATION   152
5.10   SYNTHESIS OF SUBSTITUTED AROMATIC COMPOUNDS   154
        SUMMARY OF REACTIONS   156
        EXERCISES   158

CHAPTER 6 STEREOCHEMISTRY   163

6.1   CONFIGURATION OF MOLECULES   163
6.2   MIRROR IMAGES AND CHIRALITY   163
6.3   OPTICAL ACTIVITY   167

6.4   FISCHER PROJECTION FORMULAS   168
6.5   ABSOLUTE CONFIGURATION   170
6.6   MOLECULES WITH MULTIPLE STEREOGENIC CENTERS   173
6.7   SYNTHESIS OF STEREOISOMERS   178
6.8   REACTIONS THAT PRODUCE STEREOGENIC CENTERS   179
6.9   REACTIONS THAT FORM DIASTEREOMERS   182
        EXERCISES   184

CHAPTER 7 NUCLEOPHILIC SUBSTITUTION
AND ELIMINATION REACTIONS   189
7.1   REACTION MECHANISMS AND HALOALKANES   189
7.2   NUCLEOPHILIC SUBSTITUTION REACTIONS   192
7.3   NUCLEOPHILICITY VERSUS BASICITY   194
7.4   MECHANISMS OF SUBSTITUTION REACTIONS   197
7.5 S_N2 VERSUS S_N1 REACTIONS   200
7.6   MECHANISMS OF ELIMINATION REACTIONS   201
7.7   EFFECT OF STRUCTURE ON COMPETING REACTIONS   203
        SUMMARY OF REACTIONS   206
        EXERCISES   206

CHAPTER 8 ALCOHOLS AND PHENOLS   209
8.1   THE HYDROXYL GROUP   209
8.2   PHYSICAL PROPERTIES OF ALCOHOLS   212
8.3   ACID-BASE REACTIONS OF ALCOHOLS   214
8.4   SUBSTITUTION REACTIONS OF ALCOHOLS   215
8.5   DEHYDRATION OF ALCOHOLS   216
8.6   OXIDATION OF ALCOHOLS   218
8.7   SYNTHESIS OF ALCOHOLS   221
8.8   PHENOLS   226
8.9   SULFUR COMPOUNDS: THIOLS AND THIOETHERS   229
        SUMMARY OF REACTIONS   231
        EXERCISES   232

CHAPTER 9 ETHERS AND EPOXIDES   239
9.1   STRUCTURE OF ETHERS   239
9.2   NOMENCLATURE OF ETHERS   240
9.3   PHYSICAL PROPERTIES OF ETHERS   241
9.4   THE GRIGNARD REAGENT AND ETHERS   242
9.5   SYNTHESIS OF ETHERS   244

9.6   REACTIONS OF ETHERS   245
9.7   SYNTHESIS OF EPOXIDES   246
9.8   REACTIONS OF EPOXIDES   246
       SUMMARY OF REACTIONS   254
       EXERCISES   255

CHAPTER 10 ALDEHYDES AND KETONES   259

10.1   THE CARBONYL GROUP   259
10.2   NOMENCLATURE OF ALDEHYDES AND KETONES   261
10.3   PHYSICAL PROPERTIES OF ALDEHYDES AND KETONES   263
10.4   OXIDATION-REDUCTION REACTIONS OF CARBONYL COMPOUNDS   265
10.5   ADDITION REACTIONS OF CARBONYL COMPOUNDS   267
10.6   SYNTHESIS OF ALCOHOLS FROM CARBONYL COMPOUNDS   269
10.7   ADDITION REACTIONS OF OXYGEN COMPOUNDS   272
10.8   FORMATION OF ACETALS AND KETALS   274
10.9   ADDITION OF NITROGEN COMPOUNDS   275
10.10 REACTIVITY OF THE α-CARBON ATOM   278
10.11 THE ALDOL CONDENSATION   279
       SUMMARY OF REACTIONS   282
       EXERCISES   284

CHAPTER 11 CARBOXYLIC ACIDS AND ESTERS   287

11.1   CARBOXYLIC ACIDS AND ACYL GROUPS   287
11.2   NOMENCLATURE OF CARBOXYLIC ACIDS   289
11.3   PHYSICAL PROPERTIES OF CARBOXYLIC ACIDS   292
11.4   ACIDITY OF CARBOXYLIC ACIDS   294
11.5   SYNTHESIS OF CARBOXYLIC ACIDS   297
11.6   NUCLEOPHILIC ACYL SUBSTITUTION   300
11.7   REDUCTION OF ACYL DERIVATIVES   304
11.8   ESTERS AND ANHYDRIDES OF PHOSPHORIC ACID   305
11.9   THE CLAISEN CONDENSATION   308
       SUMMARY OF REACTIONS   309
       EXERCISES   311

CHAPTER 12 AMINES AND AMIDES   315

12.1   ORGANIC NITROGEN COMPOUNDS   315
12.2   BONDING AND STRUCTURE OF AMINES   316
12.3   STRUCTURE AND CLASSIFICATION OF AMINES AND AMIDES   317
12.4   NOMENCLATURE OF AMINES AND AMIDES   319

12.5   PHYSICAL PROPERTIES OF AMINES   322
12.6   BASICITY OF NITROGEN COMPOUNDS   325
12.7   SOLUBILITY OF AMMONIUM SALTS   328
12.8   NUCLEOPHILIC REACTIONS OF AMINES   328
12.9   SYNTHESIS OF AMINES   331
12.10 HYDROLYSIS OF AMIDES   333
12.11 SYNTHESIS OF AMIDES   334
       SUMMARY OF REACTIONS   334
       EXERCISES   336

CHAPTER 13 CARBOHYDRATES   343
13.1   CLASSIFICATION OF CARBOHYDRATES   343
13.2   CHIRALITY OF CARBOHYDRATES   344
13.3   HEMIACETALS AND HEMIKETALS   349
13.4   CONFORMATIONS OF MONOSACCHARIDES   353
13.5   REDUCTION OF MONOSACCHARIDES   354
13.6   OXIDATION OF MONOSACCHARIDES   354
13.7   GLYCOSIDES   356
13.8   DISACCHARIDES   358
13.9   POLYSACCHARIDES   362
       SUMMARY OF REACTIONS   365
       EXERCISES   366

CHAPTER 14 AMINO ACIDS, PEPTIDES, AND PROTEINS   371
14.1   PROTEINS AND POLYPEPTIDES   371
14.2   AMINO ACIDS   371
14.3   ACID-BASE PROPERTIES OF α-AMINO ACIDS   372
14.4   ISOIONIC POINT   376
14.5   PEPTIDES   377
14.6   PEPTIDE SYNTHESIS   380
14.7   DETERMINATION OF PROTEIN STRUCTURE   382
14.8   PROTEIN STRUCTURE   386
       EXERCISES   393

CHAPTER 15 SYNTHETIC POLYMERS   397
15.1   NATURAL AND SYNTHETIC MACROMOLECULES   397
15.2   STRUCTURE AND PROPERTIES OF POLYMERS   397
15.3   CLASSIFICATION OF POLYMERS   399
15.4   METHODS OF POLYMERIZATION   401

15.5	ADDITION POLYMERIZATION	404
15.6	COPOLYMERIZATION OF ALKENES	405
15.7	CROSS-LINKED POLYMERS	406
15.8	STEREOCHEMISTRY OF ADDITION POLYMERIZATION	408
15.9	CONDENSATION POLYMERS	410
15.10	POLYESTERS	411
15.11	POLYCARBONATES	413
15.12	POLYAMIDES	414
15.13	POLYURETHANES	415
	EXERCISES	416

CHAPTER 16 SPECTROSCOPY — 421

16.1	SPECTROSCOPIC STRUCTURE DETERMINATION	421
16.2	SPECTROSCOPIC PRINCIPLES	422
16.3	ULTRAVIOLET SPECTROSCOPY	424
16.4	INFRARED SPECTROSCOPY	425
16.5	NUCLEAR MAGNETIC RESONANCE SPECTROSCOPY	431
16.6	SPIN-SPIN SPLITTING	435
16.7	^{13}C NMR SPECTROSCOPY	439
	EXERCISES	442

| *Solutions to In-Chapter Problems* | 447 |
| *Index* | 477 |

Please find the companion website at http://booksite.elsevier.com/9780128024447.

1

STRUCTURE
OF
ORGANIC COMPOUNDS

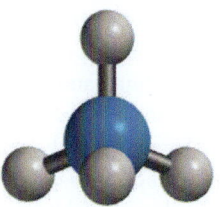

1.1 ORGANIC AND INORGANIC COMPOUNDS

Organic chemistry began to emerge as a science about 200 years ago. By the late eighteenth century, substances were divided into two classes called inorganic and organic compounds. Inorganic compounds were derived from mineral sources, whereas organic compounds were obtained only from plants or animals. Organic compounds were more difficult to work with in the laboratory, and decomposed more easily, than inorganic compounds. The differences between inorganic and organic compounds were attributed to a "vital force" associated with organic compounds. This unusual attribute was thought to exist only in living matter. It was believed that without the vital force, organic compounds could not be synthesized in the laboratory. However, by the mid-nineteenth century, chemists had learned both how to work with organic compounds and how to synthesize them.

Organic compounds always contain carbon and a limited number of other elements, such as hydrogen, oxygen, and nitrogen. Compounds containing sulfur, phosphorus, and halogens are known but are less prevalent. Most organic compounds contain many more atoms per structural unit than inorganic compounds and have more complex structures. Common examples of organic compounds include the sugar sucrose ($C_{12}H_{22}O_{11}$), vitamin B2 ($C_{117}H_{120}N_4O_6$), cholesterol ($C_{27}H_{46}O$), and the fat glycerol tripalmitate ($C_{51}H_{98}O_6$). Some organic molecules are gigantic. DNA, which stores genetic information, has molecular weights that range from 3 million in *Escherichia coli* to 2 billion for mammals.

Based on the physical characteristics of compounds, such as solubility, melting point, and boiling point, chemists have proposed that the atoms of the elements are bonded in compounds in two principal ways—ionic bonds and covalent bonds. Both types of bonds result from a change in the electronic structure of atoms as they associate with each other. Thus, the number and type of bonds formed and the resultant shape of the molecule depend on the electron configuration of the atoms. Therefore, we will review some of the electronic features of atoms and the periodic properties of the elements before describing the structures of organic compounds.

1.2 ATOMIC STRUCTURE

Each atom has a central, small, dense nucleus that contains protons and neutrons; electrons are located outside the nucleus. Protons have a +1 charge; electrons have a –1 charge. The number of protons, which determines the identity of an atom, is given as its **atomic number**. Since atoms have an equal number of protons and electrons and are electrically neutral, the atomic number also indicates the number of electrons in the atom. The number of electrons in the hydrogen, carbon, nitrogen, and oxygen atoms are one, six, seven, and eight, respectively.

The periodic table of the elements is arranged by atomic number. The elements are arrayed in horizontal rows called **periods** and vertical columns called **groups**. In this text, we will emphasize hydrogen in the first period and the elements carbon, nitrogen, and oxygen in the second period. The electronic structure of these atoms is the basis for their chemical reactivity.

Principles of Organic Chemistry. http://dx.doi.org/10.1016/B978-0-12-802444-7.00001-X

Atomic Orbitals

Electrons around the nucleus of an atom are found in **atomic orbitals**. Each orbital can contain a maximum of two electrons. The orbitals, designated by the letters s, p, d, and f, differ in energy, shape, and orientation. We need to consider only the s and p orbitals for elements such as carbon, oxygen, and nitrogen.

Orbitals are grouped in shells of increasing energy designated by the integers $n = 1, 2, 3, 4, \cdots, n$. These integers are called **principal quantum numbers**. With few exceptions, we need consider only the orbitals of the first three shells for the common elements found in organic compounds.

Each shell contains a unique number and type of orbitals. The first shell contains only one orbital—the s orbital. It is designated 1s. The second shell contains two types of orbitals—one s orbital and three p orbitals.

An s orbital is a spherical region of space centered around the nucleus (Figure 1.1). The electrons in a 2s orbital are higher in energy than those in a 1s orbital. The 2s orbital is larger than the 1s orbital, and its electrons on average are farther from the nucleus. The three p orbitals in a shell are shaped like "dumbbells." However, they have different orientations with respect to the nucleus (Figure 1.1). The orbitals are often designated p_x, p_y, and p_z to emphasize that they are mutually perpendicular to one another. Although the orientations of the p orbitals are different, the electrons in each p orbital have equal energies.

Orbitals of the same type within a shell are often considered as a group called a **subshell**. There is only one orbital in an s subshell. An s subshell can contain only two electrons, but a p subshell can contain a total of six electrons within its p_x, p_y, and p_z orbitals. Electrons are located in subshells of successively higher energies so that the total energy of all electrons is as low as possible. The order of increasing energy of subshells is 1s< 2s < 2p < 3s < 3p for elements of low atomic number. If there is more than one orbital in a subshell, one electron occupies each with parallel spins until all are half full. A single electron within an orbital is unpaired; two electrons with opposite spins within an orbital are paired and constitute an electron pair. The number and location of electrons for the first 18 elements are given in Table 1.1. The location of electrons in atomic orbitals is the **electron configuration** of an atom.

Figure 1.1
Shapes of 2s and 2p Orbitals
Electrons are pictured within a volume called an orbital. A "cloud" of negative charge surrounds the nucleus, which is located at the origin of the intersecting axes. (a) The s orbital is pictured as a sphere. (b) The three orbitals of the p subshell are arranged perpendicular to one another. Each orbital may contain two electrons. (c) Molecular model of a $2p_z$ orbital.

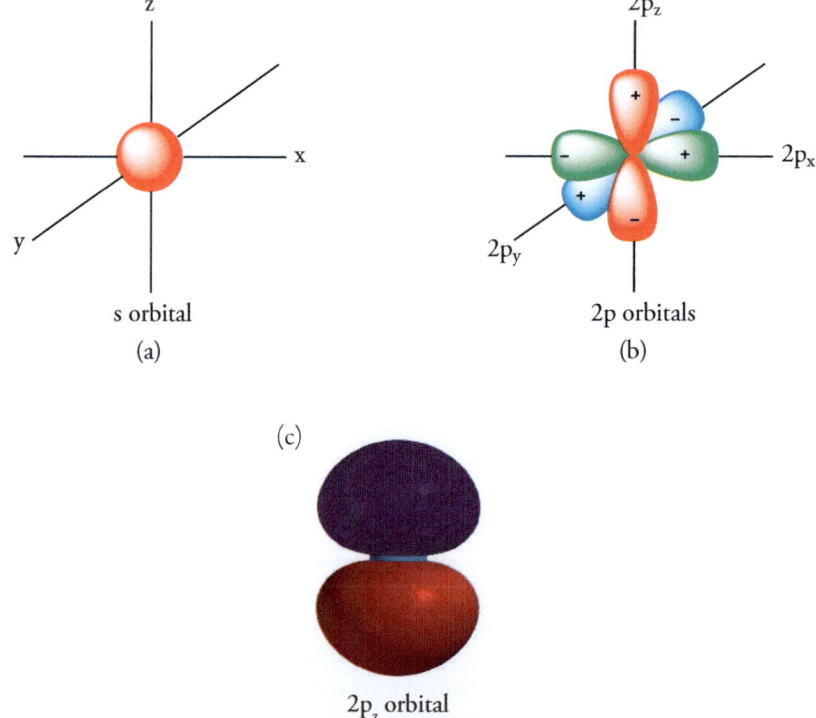

s orbital
(a)

2p orbitals
(b)

(c)

$2p_z$ orbital
(molecular model)

Table 1.1
Electron Configurations of First and Second Period Elements

Element	Atomic Number	1s	2s	$2p_x$	$2p_y$	$2p_z$	Electron Configuration
H	1	1					$1s^1$
He	2	2					$1s^2$
Li	3	2	1				$1s^2\,2s^1$
Be	4	2	2				$1s^2\,2s^2$
B	5	2	2	1(↑)			$1s^2\,2s^2\,2p^1$
C	6	2	2	1 (↑)	1 (↑)		$1s^2\,2s^2\,2p^2$
N	7	2	2	1 (↑)	1 (↑)	1 (↑)	$1s^2\,2s^2\,2p^3$
O	8	2	2	2(↑↓)	1 (↑)	1 (↑)	$1s^2\,2s^2\,2p^4$
F	9	2	2	2 (↑↓)	2 (↑↓)	1 (↑)	$1s^2\,2s^2\,2p^5$
Ne	10	2	2	2 (↑↓)	2 (↑↓)	2 (↑↓)	$1s^2\,2s^2\,2p^6$

Valence Shell Electrons

Electrons in filled, lower energy shells of atoms have no role in determining the structure of molecules, nor do they participate in chemical reactions. Only the higher energy electrons located in the outermost shell, the **valence shell**, participate in chemical reactions. Electrons in the valence shell are valence electrons. For example, the single electron of the hydrogen atom is a **valence electron**. The number of valence electrons for the common atoms contained in organic molecules is given by their group number in the periodic table. Thus carbon, nitrogen, and oxygen atoms have four, five, and six valence electrons, respectively. With this information we can understand how these elements combine to form the structure of organic compounds.

The physical and chemical properties of an element may be estimated from its position in the periodic table. Two principles that help us to explain the properties of organic compounds are atomic radius and electronegativity. The overall shape of an isolated atom is spherical, and the volume of the atom depends on the number of electrons and the energies of the electrons in occupied orbitals. The sizes of some atoms expressed as the **atomic radius**, in picometers, are given in Figure 1.2. The atomic radius for an atom does not vary significantly from one compound to another. Atomic radii increase from top to bottom in a group of the periodic table. Each successive member of a group has one additional energy level containing electrons located at larger distances from the nucleus. Thus, the atomic radius of sulfur is greater than that of oxygen, and the radii of the halogens increase in the order F < Cl < Br.

The atomic radius decreases from left to right across a period. Although electrons are located in the same energy level within the s and p orbitals of the elements, the nuclear charge increases from left to right within a period. As a result, the nucleus draws the electrons inward and the radius decreases. The radii of the common elements in organic compounds are in the order C > N > 0.

Figure 1.2
Atomic radii in picometers, pm (10^{-12} m)

H						
37						
Li	**Be**	**B**	**C**	**N**	**O**	**F**
152	111	88	77	70	66	64
Na	**Mg**	**Al**	**Si**	**P**	**S**	**Cl**
186	160	143	117	110	104	99
						Br
						114
						I
						133

Electronegativity

Electronegativity is a measure of the attraction of an atom for bonding electrons in molecules compared to that of other atoms. The electronegativity values devised by Linus Pauling, an American chemist, are dimensionless quantities that range from slightly less than one for the alkali metals to a maximum of four for fluorine. Large electronegativity values indicate a stronger attraction for electrons than small electronegativity values.

Electronegativities increase from left to right across the periodic table (Figure 1.3). Elements on the left of the periodic table have low electronegativities and are often called electropositive elements. The order of electronegativities F > O > N > C is an important property that we will use to explain the chemical properties of organic compounds. Electronegativities decrease from top to bottom within a group of elements. The order of decreasing electronegativities F > Cl > Br > I is another sequence that we will use to interpret the chemical and physical properties of organic compounds.

Figure 1.3
Electronegativity

H
2.1

Li	Be	B	C	N	O	F
1.0	1.5	2.0	2.5	3.0	3.5	4.0

Na	Mg	Al	Si	P	S	Cl
0.9	1.2	1.5	1.8	2.1	2.5	3.0

Br
2.8

I
2.5

1.3 TYPES OF BONDS

In 1916, the American chemist G.N. Lewis proposed that second period elements tend to react to obtain an electron configuration of eight electrons so that they electronically resemble the inert gases. This hypothesis is summarized in the **Lewis octet rule**: Second period atoms tend to combine and form bonds by transferring or sharing electrons until each atom is surrounded by eight electrons in its highest energy shell. Note that hydrogen requires only two electrons to complete its valence shell.

Ionic Bonds

Ionic bonds form between two or more atoms by the transfer of one or more electrons between atoms. Electron transfer produces negative ions called **anions** and positive ions called **cations**. These ions attract each other.

Let's examine the ionic bond in sodium chloride. A sodium atom, which has 11 protons and 11 electrons, has a single valence electron in its 3s subshell. A chlorine atom, which has 17 protons and 17 electrons, has seven valence electrons in its third shell, represented as $3s^23p^5$. In forming an ionic bond, the sodium atom, which is electropositive, loses its valence electron to chlorine. The resulting sodium ion has the same electron configuration as neon ($1s^22s^22p^6$) and has a +1 charge, because there are 11 protons in the nucleus, but only 10 electrons about the nucleus of the ion.

The chlorine atom, which has a high electronegativity, gains an electron and is converted into a chloride ion that has the same electron configuration as argon ($1s^22s^22p^63s^23p^6$). The chloride ion has a –1 charge because there are 17 protons in the nucleus, but there are 18 electrons about the nucleus of the ion. The formation of sodium chloride from the sodium and chlorine atoms can be shown by Lewis structures. Lewis structures represent only the valence electrons; electron pairs are shown as pairs of dots.

$$Na\cdot \ + \ :\overset{..}{\underset{..}{Cl}}\cdot \ \longrightarrow \ Na^+ \ + \ :\overset{..}{\underset{..}{Cl}}:^-$$

Note that by convention, the complete octet is shown for anions formed from electronegative elements. However, the filled outer shell of cations that results from loss of electrons by electropositive elements is not shown.

Metals are electropositive and tend to lose electrons, whereas nonmetals are electronegative and tend to gain electrons. A metal atom loses one or more electrons to form a cation with an octet. The same number of electrons are accepted by the appropriate number of atoms of a nonmetal to form an octet in the anion, producing an ionic compound. In general, ionic compounds result from combinations of metallic elements, located on the left side of the periodic table, with nonmetals, located on the upper right side of the periodic table.

Covalent Bonds

A **covalent bond** consists of the mutual sharing of one or more pairs of electrons between two atoms. These electrons are simultaneously attracted by the two atomic nuclei. A covalent bond forms when the difference between the electronegativities of two atoms is too small for an electron transfer to occur to form ions. Shared electrons located in the space between the two nuclei are called **bonding electrons.** The bonded pair is the "glue" that holds the atoms together in molecular units.

The hydrogen molecule is the simplest substance having a covalent bond. It forms from two hydrogen atoms, each with one electron in a 1s orbital. Both hydrogen atoms share the two electrons in the covalent bond, and each acquires a helium-like electron configuration.

$$H\cdot + H\cdot \longrightarrow H-H$$

A similar bond forms in Cl_2. The two chlorine atoms in the chlorine molecule are joined by a shared pair of electrons. Each chlorine atom has seven valence electrons in the third energy level and requires one more electron to form an argon-like electron configuration. Each chlorine atom contributes one electron to the bonding pair shared by the two atoms. The remaining six valence electrons of each chlorine atom are not involved in bonding and are concentrated around their respective atoms. These valence electrons, customarily shown as pairs of electrons, are variously called **nonbonding electrons, lone pair electrons**, or **unshared electron pairs**.

The covalent bond is drawn as a dash in a **Lewis structure** to distinguish the bonding pair from the lone pair electrons. Lewis structures show the nonbonding electrons as pairs of dots located about the atomic symbols for the atoms. The Lewis structures of four simple organic compounds—methane, methylamine, methanol, and chloromethane—are drawn here to show both bonding and nonbonding electrons. In these compounds carbon, nitrogen, oxygen, and chlorine atoms have four, three, two, and one bonds, respectively.

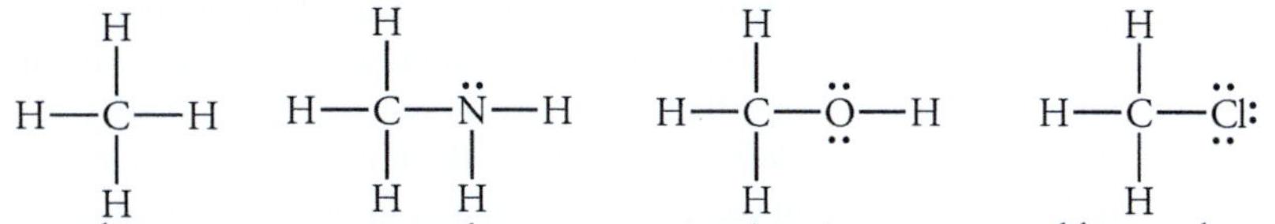

The hydrogen atom and the halogen atoms form only one covalent bond to other atoms in most stable neutral compounds. However, the carbon, oxygen, and nitrogen atoms can simultaneously bond to more than one atom. The number of such bonds is the **valence** of the atom. The valences of carbon, nitrogen, and oxygen are four, three, and two, respectively.

Multiple Covalent Bonds

In some molecules more than one pair of electrons is shared between pairs of atom. If four electrons (two pairs) or six electrons (three pairs) are shared, the bonds are called **double** and **triple bonds**, respectively. A carbon atom can form single, double, or triple bonds with other carbon atoms as well as with atoms of some other elements. Single, double, and triple covalent bonds link two carbon atoms in ethane, ethylene, and acetylene, respectively. Each carbon atom in these compounds shares one, two, and three electrons, respectively, with the other. The remaining valence electrons of the carbon atoms are contained in the single bonds with hydrogen atoms.

Polar Covalent Bonds

A polar covalent bond exists when atoms with different electronegativities share electrons in a covalent bond. Consider the hydrogen chloride (HCl) molecule. Each atom in HCl requires one more electron to form an inert gas electron configuration. Chlorine has a higher electronegativity than hydrogen, but the chlorine atom's attraction for electrons is not sufficient to remove an electron from hydrogen. Consequently, the bonding electrons in hydrogen chloride are shared unequally in a polar covalent bond. The molecule is represented by the conventional Lewis structure, even though the shared electron pair is associated to a larger extent with chlorine than with hydrogen. The unequal sharing of the bonding pair results in a partial negative charge on the chlorine atom and a partial positive charge on the hydrogen atom. The symbol δ (Greek lowercase delta) denotes these fractional charges.

$$\overset{\delta^+}{H}\!\!-\!\!\overset{\delta^-}{\ddot{\underset{\cdot\cdot}{Cl}}}$$

The hydrogen chloride molecule has a **dipole** (two poles), which consists of a pair of opposite charges separated from each other. The dipole is shown by an arrow with a cross at one end. The cross is near the end of the molecule that is partially positive, and the arrowhead is near the partially negative end of the molecule.

Single or multiple bonds between carbon atoms are nonpolar. Hydrogen and carbon have similar electronegativity values, so the C—H bond is not normally considered a polar covalent bond. Thus ethane, ethylene, and acetylene have nonpolar covalent bonds, and the compounds are nonpolar.

Bonds between carbon and other elements such as oxygen and nitrogen are polar. The polarity of a bond depends on the electronegativities of the bonded atoms. Large differences between the electronegativities of the bonded atoms increase the polarity of bonds. The direction of the polarity of common bonds found in organic molecules is easily predicted. The common nonmetals are more electronegative than carbon. Therefore, when a carbon atom is bonded to common nonmetal atoms, it has a partial positive charge.

Hydrogen is also less electronegative than the common nonmetals. Therefore, when a hydrogen atom is bonded to common nonmetals, the resulting polar bond has a partial positive charge on the hydrogen atom.

**Table 1.2
Average Dipole Moments (D)**

Structural Unit[1]	Bond Moments (D)
H—C	0.4
H—N	1.3
H—O	1.5
H—F	1.7
H—S	0.7
H—Cl	1.1
H—Br	0.8
H—I	0.4
C—C	0.0
C—N	0.2
C—O	0.7
C—F	1.4
C—Cl	1.5
C—Br	1.4
C—I	1.2
C=O	2.3
C≡N	3.5

1. The more negative element is on the right.

The magnitude of the polarity of a bond is the **dipole moment**, (D). The dipole moments of several bond types are given in Table 1.2. The dipole moment of a specific bond is relatively constant from compound to compound. When carbon forms multiple bonds to other elements, these bonds are polar. Both the carbon-oxygen double bond in formaldehyde (methanal) and the carbon–nitrogen triple bond in acetonitrile (cyanomethane) are polar.

methanal   cyanomethane

1.4 FORMAL CHARGE

Although most organic molecules are represented by Lewis structures containing the "normal" number of bonds, some organic ions and even some molecules contain less than or more than the customary number of bonds. First let's review the structures of some "inorganic" ions. The valence of the oxygen atom is two—it normally forms two bonds. However, there are three bonds in the hydronium ion and one in the hydroxide ion.

How do we predict the charge of the ions? Second, what atoms bear the charge? There is a useful formalism for answering both of these question. Each atom is assigned a formal charge by a book-keeping method that involves counting electrons. The method is also used for neutral molecules that have unusual numbers of bonds. In such cases, centers of both positive and negative charge are located at specific atoms.

The **formal charge** of an atom is equal to the number of its valence electrons as a free atom minus the number of electrons that it "owns" in the Lewis structure.

$$\text{formal charge} = \left(\begin{array}{c} \text{number of valence} \\ \text{electrons in free atom} \end{array} \right) - \left(\begin{array}{c} \text{number of valence} \\ \text{electrons in bonded atom} \end{array} \right)$$

The question of ownership is decided by two simple rules. Unshared electrons belong exclusively to the parent atom. One-half of the bonded electrons between a pair of atoms is assigned to each atom. Thus, the total number of electrons "owned" by an atom in the Lewis structure equals the number of nonbonding electrons plus half the number of bonding electrons. Therefore, we write the following:

$$\text{formal charge} = \left[\left(\begin{array}{c} \text{number of valence} \\ \text{electrons in free atom} \end{array} \right) - \left(\begin{array}{c} \text{number of valence} \\ \text{electrons in bonded atom} \end{array} \right) \right] - 1/2 \left(\text{number of bonded electrons} \right)$$

The formal charge of each atom is zero in most organic molecules. However, the formal charge may also be negative or positive. The sum of the formal charges of each atom in a molecule equals zero; the sum of the formal charges of each atom in an ion equals the charge of the ion. Let's consider the molecule hydrogen cyanide, HCN, and calculate the formal charges of the carbon and nitrogen atoms bonded in a triple bond.

H—C≡N: lone pair electrons: assign both to nitrogen

two bonding electrons: 6 bonding electrons:
assign 1 to hydrogen   assign 3 to carbon
assign 1 to carbon   assign 3 to nitrogen

The formal charge of each atom is calculated by substitution into the formula shown below:

$$\text{Formal charge of hydrogen} = 1 - 0 - 1/2(2) = 0$$
$$\text{Formal charge of carbon} = 4 - 2 - 1/2(6) = -1$$
$$\text{Formal charge of nitrogen} = 5 - 0 - 1/2(8) = +1$$

The formal charge of carbon is −1 and the formal charge of nitrogen is +1. However, the *sum* of the formal charges of these atoms equals the net charge of the species, which in this case is zero.

There are often important chemical consequences when a neutral molecule contains centers whose formal charges are not zero. It is important to be able to recognize these situations, which allow us to understand the chemical reactivity of such molecules.

1.5 RESONANCE STRUCTURES

In the Lewis structures for the molecules shown to this point, the electrons have been pictured as either between two nuclei or about a specific atom. These electrons are **localized**. The electronic structures of molecules are written to be consistent with their physical properties. However, the electronic structures of some molecules cannot be represented adequately by a single Lewis structure. For example, the Lewis structure of the acetate ion has one double bond and one single bond to oxygen atoms. Note that the formal charge of the single-bonded oxygen atom is −1 whereas that of the double-bonded oxygen atom is zero.

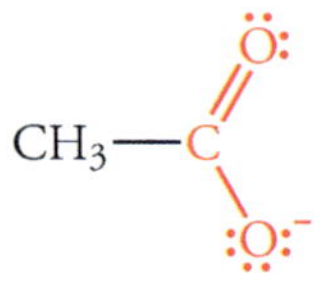

However, single and double bonds are known to have different bond lengths—a double bond between two atoms is shorter than a single bond. The Lewis structure shown implies that there is one "long" C—O bond and a "short" C=O bond in the acetate ion. But both carbon–oxygen bond lengths in the acetate ion have been shown experimentally to be equal. Moreover, both oxygen atoms bear equal amounts of negative charge. Therefore, the preceding Lewis structure with single and double bonds does not accurately describe the acetate ion. Under these circumstances, the concept of **resonance** is used. We say that a molecule is **resonance stabilized** if two or more Lewis structures can be written that have identical arrangements of atoms but different arrangements of electrons. The real structure of the acetate ion can be represented better as a **hybrid** of two Lewis structures, neither of which is completely correct.

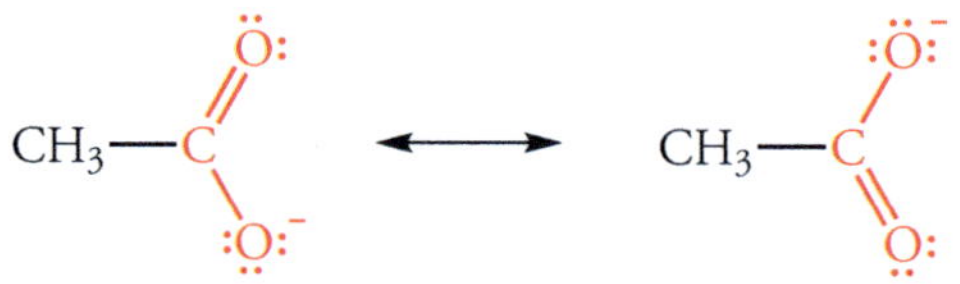

A double-headed arrow between two Lewis structures indicates that the actual structure is similar in part to the two simple structures but lies somewhere between them. The individual Lewis structures are called contributing structures or resonance structures.

Curved arrows can be used to keep track of the electrons when writing resonance structures. The tail of the arrow is located near the bonding or nonbonding pair of electrons to be "moved" or "pushed," and the arrowhead shows the "final destination" of the electron pair in the Lewis structure.

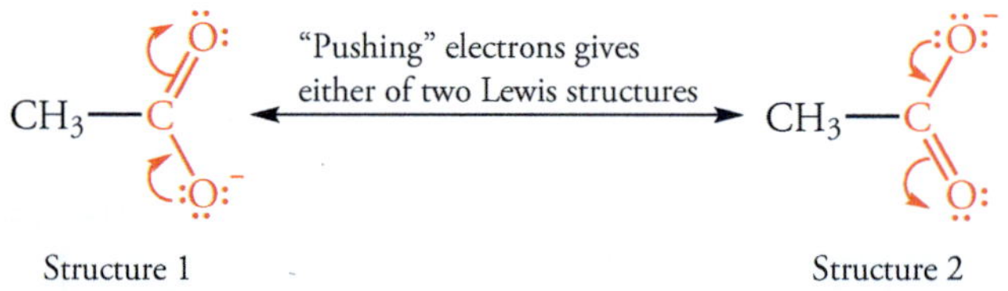

"Pushing" electrons gives either of two Lewis structures

Structure 1   Structure 2

In resonance structure 1, the nonbonding pair of electrons on the bottom oxygen atom is moved to form a double bond with the carbon atom. A bonding pair of electrons of the carbon–oxygen double bond is also moved to form a nonbonding pair of electrons on the top oxygen atom. The result is resonance structure 2. This procedure of "pushing" electrons from one position to another is only a bookkeeping formalism. *Electrons do not really move this way!* The actual ion has **delocalized** electrons distributed over three atoms—a phenomenon that cannot be shown by a single Lewis structure.

Electrons can be delocalized over many atoms. For example, benzene, C_6H_6, consists of six equivalent carbon atoms contained in a ring in which all carbon–carbon bonds are identical. Each carbon atom is bonded to a hydrogen atom. A single Lewis structure containing alternating single and double bonds can be written to satisfy the Lewis octet requirements.

benzene

However, single and double bonds have different bond lengths. In benzene, all carbon–carbon bonds have been shown to be the same length. Like the acetate ion, benzene is represented by two contributing resonance structures separated by a double-headed arrow. The positions of the alternating single and double bonds are interchanged in the two resonance structures.

equivalent contributing structures for the resonance hybrid of benzene

The electrons in benzene are delocalized over the six carbon atoms in the ring, resulting in a unique structure. There are no carbon–carbon single or double bonds in benzene; its bonds are of an intermediate type that cannot be represented with a single structure.

Problem 1.1

Consider the structure of nitromethane, a compound used to increase the power in some specialized race car engines. A nitrogen-oxygen single bond length is 136 pm; a nitrogen-oxygen double bond length is 114 pm. The nitrogen-oxygen bonds in nitromethane are equal and are 122 pm. Explain.

Solution

The actual nitrogen–oxygen bonds are neither single nor double bonds. Two resonance forms can be written to represent nitromethane. They result from "moving" a nonbonding pair of electrons from the single-bonded oxygen atom to form a double bond with the nitrogen atom. One of the bonding pairs of electrons from the nitrogen-oxygen double bond is moved to the other oxygen atom. The structures differ only in the location of the single and double bonds.

Problem 1.2
Nitrites (NO_2^-) are added as antioxidants in some processed meats. Write resonance structures for the nitrite ion.

1.6 PREDICTING THE SHAPES OF SIMPLE MOLECULES

Up to this point, we have considered the distribution of bonding electrons and nonbonding electrons within molecules without regard to their location in three-dimensional space. But molecules have characteristic shapes that reflect the spatial arrangement of electrons in bonds. For example, the shapes of carbon dioxide, formaldehyde, and methane are linear, trigonal planar, and tetrahedral, respectively. (Note that wedge-shaped bonds are used to show the location of atoms above the plane of the page and dashed lines to indicate the location of atoms behind the plane of the page.)

carbon dioxide   formaldehyde   methane

We can "predict" the geometry of these simple molecules and approximate the bond angles using valence-shell electron-pair repulsion (VSEPR) theory. This theory is based on the idea that bonding and nonbonding electron pairs about a central atom repel each other. VSEPR theory predicts that electron pairs in molecules should be arranged as far apart as possible. Thus, two electron pairs should be arranged at 180° to each other; three pairs should be at 120° in a common plane; four electron pairs should have a tetrahedral arrangement with angles of 109.5°.

All of the valence electrons about the central carbon atom in carbon dioxide, formaldehyde, and methane are in bonds. Each type of bond may be regarded as a region that contains electrons that should be arranged as far apart as possible. Carbon dioxide has two double bonds; the double bonds are separated by the maximum distance, and the resulting angle between the bonds is 180°. Formaldehyde has a double bond and two single bonds to the central carbon atom; these bonds correspond to three regions containing electrons. They are separated by the maximum distance in a trigonal planar arrangement with bond angles of 120°. Methane has four bonding electron pairs. They are best located in a tetrahedral arrangement. Each H—C—H bond angle is predicted to be 109.5° in agreement with the experimental value.

Now let's consider molecules that have both bonding and nonbonding pairs of electrons in the valence shell of the central atom. Water and ammonia have experimentally determined shapes described as angular and trigonal pyramidal, respectively. Both have four electron pairs about the central atom, as does methane. They both have central atom bonded to hydrogen atoms, but there are also unshared electron pairs.

anglular molecule   trigonal pyramidal molecule

VSEPR theory describes the distribution of electron pairs, including the nonbonding pairs. However, molecular structure is defined by the positions of the nuclei. Although the four pairs of electrons in both water and ammonia are tetrahedrally arranged, water and ammonia are angular and pyramidal molecules, respectively (Figure 1.4).

The arrangement of bonds to the oxygen atom and the nitrogen atom in organic molecule are similar to those in water and ammonia, respectively. The groups bonded to the oxygen atom of an alcohol or an ether (Section 1.9) are arranged to form angular molecules. The groups bonded to the nitrogen atom of an amine (Section 1.9) are arranged to form a pyramid.

Figure 1.4
VSEPR Model Predicts Molecular Geometry

All electron pairs in methane, ammonia, and methanol are directed to the corners of a tetrahedron. However, the geometry around the nitrogen atom in ammonia is described as trigonal pyramidal; the geometry around the oxygen atom in a water molecule is angular. There is one lone pair in ammonia and two lone pairs in water.

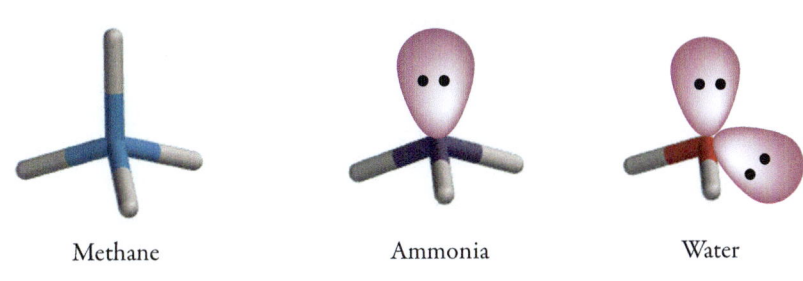

Methane   Ammonia   Water

Problem 1.3

The electronic structure of allyl isothiocyanate, a flavor ingredient in horseradish, is shown below. What are the C—N=C and N=C=S bond angles?

$$CH_2\!\!=\!\!CH\!-\!CH_2\!-\!\ddot{N}\!\!=\!\!C\!\!=\!\!\ddot{S}\!:$$

Solution

The C—N=C bond angle depends on the electrons associated with the nitrogen atom. This atom has a single bond, a double bond, and a nonbonding pair of electrons. These three electron-containing regions have trigonal planar geometry. Only two of the electron-containing regions are bonding, but the C—N=C bond angle must still be 120°.

Problem 1.4

Using one of the resonance forms for the nitrite ion (NO_2^-) determine the shape of this ion.

1.7 ORBITALS AND MOLECULAR SHAPES

Because electrons form bonds between atoms, the shapes of molecules depend on the location of the electrons in the orbitals of the various atoms. Two electrons in a covalent bond are shared in a region of space common to the bonding atoms. This region of space is pictured as an overlap or merging of two atomic orbitals. For example, the covalent bond in H_2 results from the overlap of two s orbitals to give a sigma (σ) bond (Figure 1.5). This bond is symmetrical around an axis joining the two nuclei. Viewed along the interatomic axis, the σ bond looks like an s orbital. All single bonds are also σ bonds regardless of the component orbitals used in forming the bond.

The simple picture of bonding described for H_2 has to be modified somewhat for carbon-containing compounds. Carbon has the electronic configuration $1s^2 2s^2 2p^2$, which suggests that only the two electrons in the 2p orbitals would be available to form two covalent bonds. If this were so, the molecular formula for a compound of carbon and hydrogen would be CH_2 and the carbon atom would not have four bonds.

$$H\!-\!\ddot{C}_{\diagdown H}$$

However, there are four equivalent C—H bonds in methane, CH_4. All carbon compounds presented in this chapter have a Lewis octet about the carbon atoms, and each carbon atom has four bonds. The difference between these structural facts and predictions based on the atomic orbitals of carbon is explained using the concept of hybrid orbitals, which result from the "mixing" of two or more orbitals in the bonded atoms. This mixing process, called **orbital hybridization**, was proposed by Pauling to account for the formation of bonds by using orbitals having the geometry of the actual molecule. As a result of hybridization, two or more hybrid orbitals can be formed from the appropriate number of atomic orbitals. The number of hybrid orbitals created equals the number of atomic orbitals used in hybridization.

Figure 1.5
The Sigma Bond of the Hydrogen Molecule
The region occupied by the electron pair is symmetrical about both hydrogen nuclei. Although the two electrons may be located anywhere within the volume shown, it is most probable that they are between the two nuclei.

sp³ Hybridization of Carbon

Pauling suggested that the tetrahedral geometry of methane results from hybridization of the 2s and the three 2p orbitals of carbon, which combine to form four equivalent hybrid orbitals. Each hybrid orbital contains one electron. These orbitals extend toward the corners of a tetrahedron so there is maximum separation of the electrons. Each hybrid atomic orbital then overlaps with a hydrogen 1s orbital to form a σ bond. The formation of the hybrid orbitals is illustrated in Figure 1.6. The four new orbitals are called **sp³ hybrid orbitals** because they result from the combination of one 2s and three 2p orbitals. Each sp³ orbital has the same shape, and the electrons in each orbital have the same energy. The orbitals differ only in their position in space.

Figure 1.6
sp³-Hybridized Carbon Atom
(a) The original set of four atomic orbitals on carbon are mixed, or hybridized to give four new sp³-hybridized atomic orbitals.
(b) We have represented the new hybrid orbitals with a new color to emphasize the notion that the hybrid orbitals replace the original unhybridized orbitals.

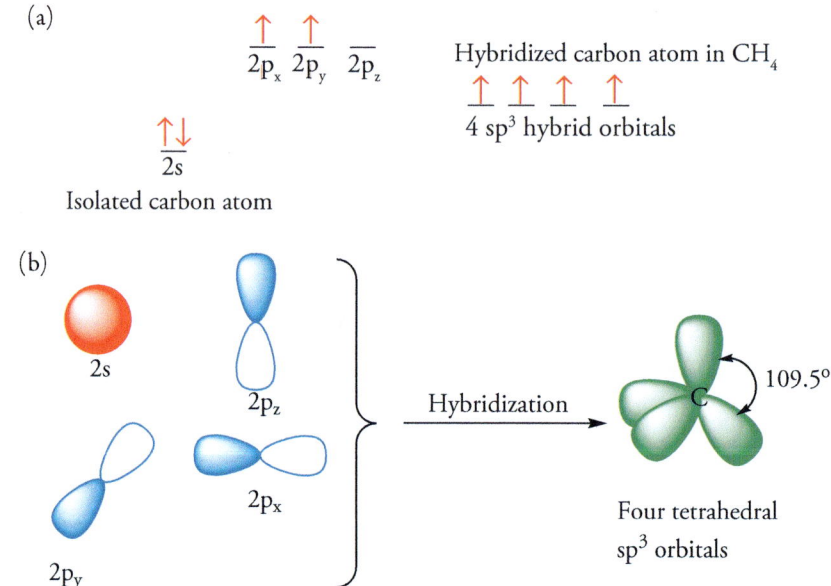

sp² Hybridization of Carbon

Now let's consider the bonding electrons in the double bond of ethylene in which each carbon atom is bonded to three atoms. All six nuclei lie in a plane, and all the bond angles are close to 120°. Each carbon atom in ethylene is pictured with three **sp² hybrid orbitals** and one remaining 2p orbital. The three sp² hybrid orbitals result from "mixing" a single 2s orbital and two 2p orbitals. Each sp² orbital has the same shape, and the electrons in each orbital have the same energy. The orbitals differ only in their position in space. They are separated by 120° and are directed to the corners of a triangle to have maximum separation of the electrons. The four valence electrons are distributed as indicated in Figure 1.7. The three sp^2 hybridized orbitals are used to make σ bonds. Two of the sp^2 orbitals, containing one electron each, form σ bonds with hydrogen. The third sp^2 orbital, which also contains one electron, forms a σ bond with the other carbon atom in ethylene.

The second bond of the double bond in ethylene results from a lateral or side-by-side overlap of the p orbitals of each carbon atom. Each p orbital is perpendicular to the plane containing the sp^2 orbitals. The 2p orbital of each atom provide one electron to the electron pair for the second bond. A bond formed by sideways overlap of p orbitals is a π (**pi**) bond. Viewed along the carbon–carbon internuclear axis, a π bond resembles a p orbital. Note that the electrons in the π bond are not concentrated along an axis between the two atoms but are shared in regions of space both above and below the plane defined by the sp^2 orbitals. Nevertheless, it is only one bond.

Figure 1.7
Hybridization and the
Double Bond of Ethylene

(a)

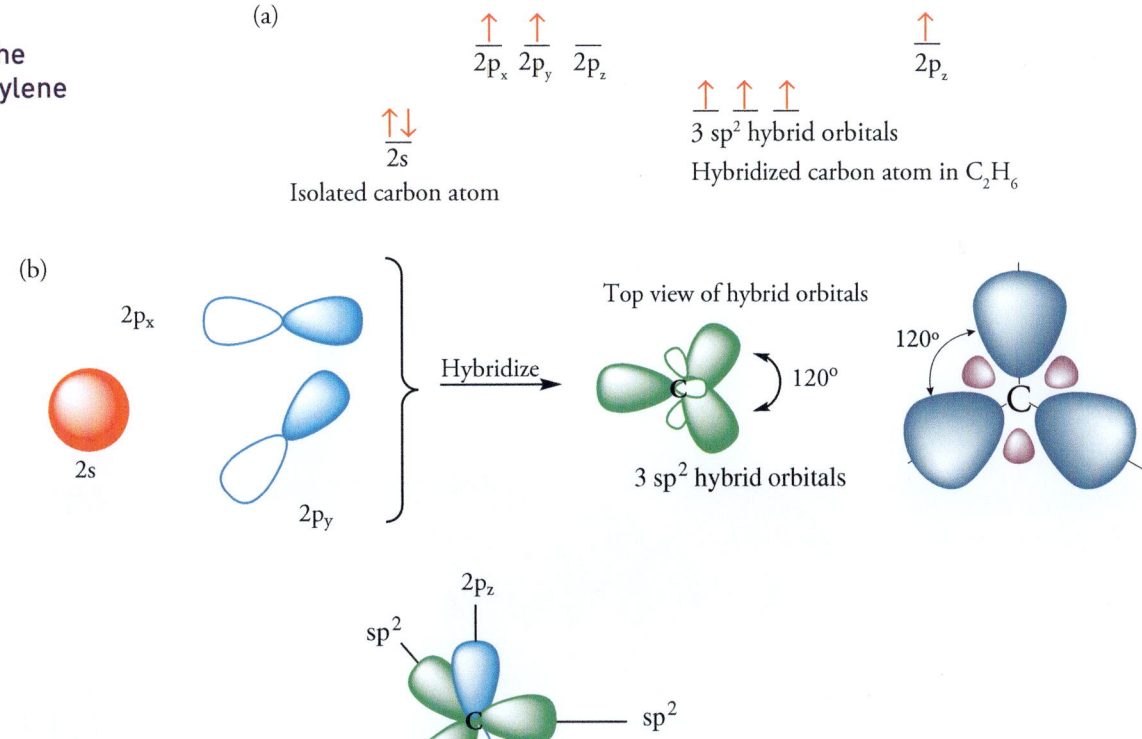

$\underset{2p_x}{\uparrow} \underset{2p_y}{\uparrow} \underset{2p_z}{\overline{\uparrow}}$

$\underset{2s}{\uparrow\downarrow}$

Isolated carbon atom

$\underset{2p_z}{\uparrow}$

$\uparrow \uparrow \uparrow$
3 sp^2 hybrid orbitals

Hybridized carbon atom in C$_2$H$_6$

(b)

2p$_x$

2s

2p$_y$

Hybridize →

Top view of hybrid orbitals

3 sp^2 hybrid orbitals

120°

120°

C

sp^2

2p$_z$

sp^2

sp^2

Side view: three sp^2 hybrid orbitals and one 2p orbital

(c)

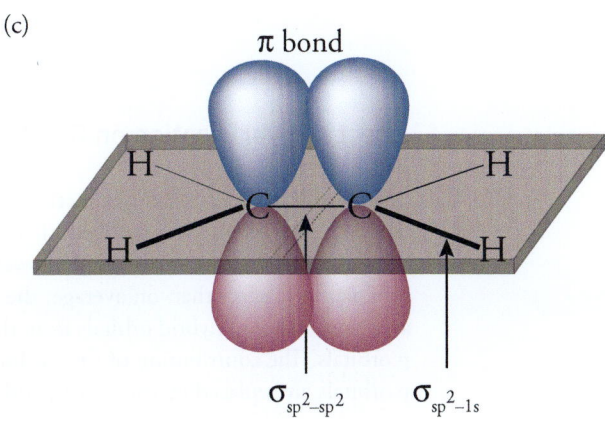

π bond

H   H

C — C

H   H

$\sigma_{sp^2-sp^2}$ σ_{sp^2-1s}

sp Hybridization of Carbon

Now let's consider the triple bond of acetylene (ethyne), in which each carbon atom is bonded to two atoms. All four nuclei are collinear, and all the bond angles are 180°. In acetylene, we mix a 2s orbital with a 2p orbital to give two **sp hybrid orbitals** of equal energy. The remaining two 2p orbitals do not change (Figure 1.8). The sp orbitals have the same shape, and the electrons in each orbital have the same energy. The orbitals differ only in their position in space; they are at 180° angles to each other—again to provide for maximum separation of the electrons. Each carbon atom in acetylene has four valence electrons. The two sp hybrid orbitals of each carbon atom contain one electron each, and the two 2p orbitals of each carbon atom contain one electron each. The carbon atoms in acetylene are linked by one σ bond and two π bonds to give a triple bond. One sp orbital and its electron form a bond with hydrogen; the other sp orbital forms a σ bond with the second carbon atom. The second and third bonds between carbon atoms result from sideways overlap of 2p orbitals. One set of 2p orbitals overlaps in front and back of the molecule to form one π bond. The second set of 2p orbitals overlaps above and below the molecule to form the second π bond.

**Figure 1.8
Structure and Bonding in Ethyne**

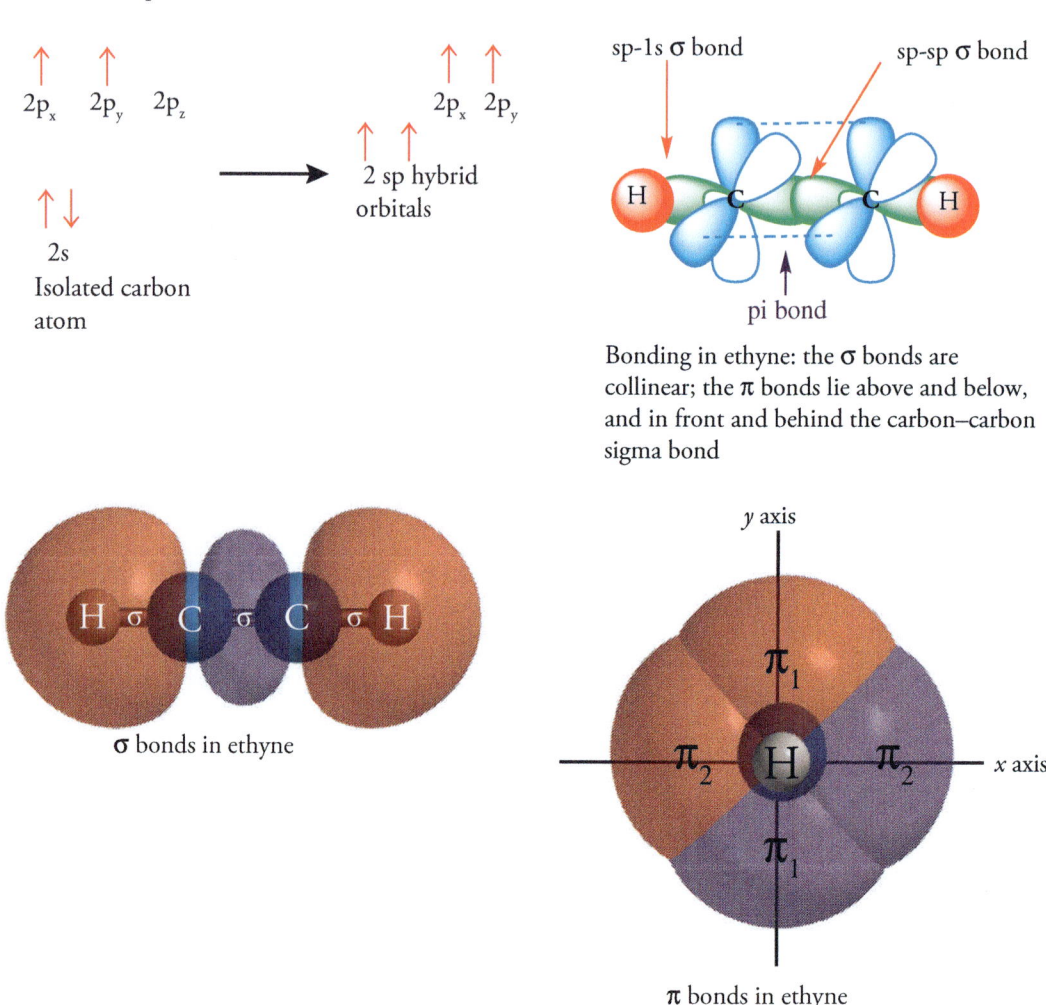

Bonding in ethyne: the σ bonds are collinear; the π bonds lie above and below, and in front and behind the carbon–carbon sigma bond

σ bonds in ethyne

π bonds in ethyne

Effect of Hybridization on Bond Length

The hybridization of carbon in methane, ethylene, and acetylene affects the C—H and C—C bond lengths (Table 1.3). Note in Table 1.3 that the length of the C—H bond decreases in the order $sp^3 > sp^2 > sp$. This order reflects the lower energy of the 2s orbital compared to the energy of the 2p orbital and the fact that, on average, the 2s orbital is closer to the nucleus than the 2p orbital. The average distance of hybrid orbitals from the nucleus depends on the percent contribution of the s and p orbitals. The contribution of the s orbital is 25% in an sp^3 hybrid orbital, because one s and three p orbitals are replaced by the four hybrid orbitals. Similarly, the contribution of the s orbital is 33%

and 50% for the sp² and sp hybrid orbitals, respectively. Because an sp³ hybrid orbital has a smaller s character than an sp² or sp hybrid orbital, the electrons in an sp³ orbital are in general farther from the nucleus. As a consequence, a bond formed with an sp³ orbital is longer than bonds involving sp² or sp hybrid orbitals.

The length of the carbon–carbon bond also decreases in the order sp³ > sp² > sp. This trend partly reflects the effect of the closer approach to the nucleus of the σ bonding electrons as the percent s character increases. However, the substantial decrease in the carbon to carbon bond length of ethane > ethylene > acetylene is also a consequence of the increased number of bonds joining the carbon atoms. Two shared pairs of electrons draw the carbon atoms closer together than a single bond. Three shared pairs move the carbon atoms still closer.

Table 1.3
Average Bond Lengths (pm)

H—C (sp³)	109
H—C (sp²)	107
H—C (sp)	105
C—C (sp³)	154
C=C (sp²)	133
C≡C (sp)	120

1.8 FUNCTIONAL GROUPS

The sheer numbers of organic compounds can make the study of organic compounds and their related physical and chemical properties a daunting objective. Fortunately, organic chemists have found ways to handle the immense number of facts by using techniques similar to those used in inorganic chemistry. Just as the elements are organized by groups in a periodic table, organic compounds are organized into families of compounds. Families of organic compounds are organized by **functional groups**. Atoms or groups of bonded atoms responsible for similar physical and chemical properties in a family of compounds are functional groups. Thus, the study of 10 million compounds is organized into more manageable groups of compounds whose reactivity is predictable. A summary of the more common functional groups is given in Table 1.4.

Some functional groups are a part of the carbon skeleton. These include the carbon–carbon double bond in compounds called alkenes, such as ethene, and the carbon–carbon triple bond in compounds called alkynes, such as ethyne. Although benzene, a member of a class called arenes, is represented as a series of alternating carbon–carbon single and double bonds, it reacts differently from ethene (Chapter 5).

ethane   ethene   ethyne   benzene

Functional groups can contain a variety of elements. The most common elements in functional groups are oxygen and nitrogen, although sulfur or the halogens may also be present. **Alcohols** and **ethers** are two classes of compounds that contain carbon–oxygen single bonds. The —OH unit in alcohols is the **hydroxyl group**.

ethanol
(an alcohol)

dimethyl ether
(an ether)

Aldehydes and ketones contain double bonds to oxygen. The unit C=O is called the **carbonyl group**. The carbon atom of the carbonyl group is called the **carbonyl carbon** atom, and the oxygen atom is called the **carbonyl oxygen atom**. Note that an aldehyde has at least one hydrogen atom bonded to the carbonyl carbon atom. In ketones, the carbonyl carbon atom is bonded to two other carbon atoms.

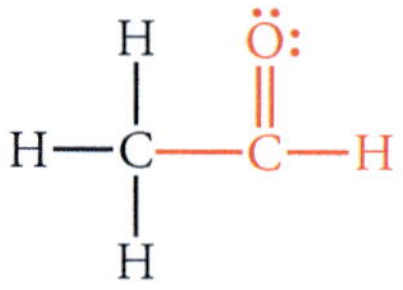

acetaldehyde
(an aldehyde)

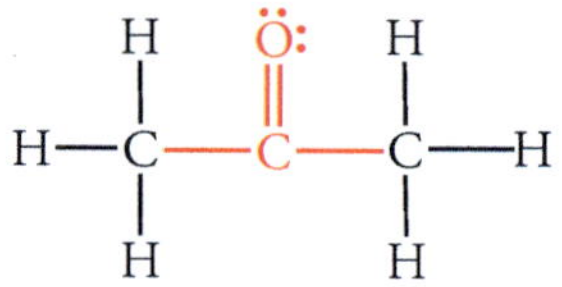

acetone
(a ketone)

Carboxylic acids and **esters** contain both single and double bonds from a carbon atom to oxygen atoms. In a carboxylic acid, the carbonyl group is bonded to a hydroxyl group and either a hydrogen or a carbon atom. In an ester, the carbonyl group is bonded to an "O—R" group, where "R" contains one or more carbon atoms, and to either a hydrogen or a carbon atom.

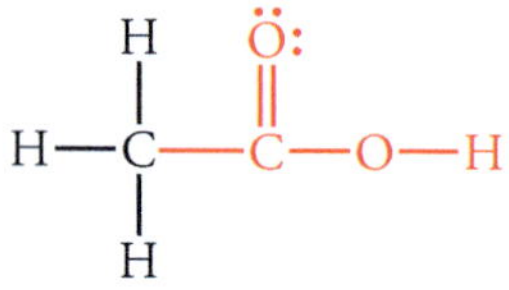

acetic acid
(a carboxylic acid)

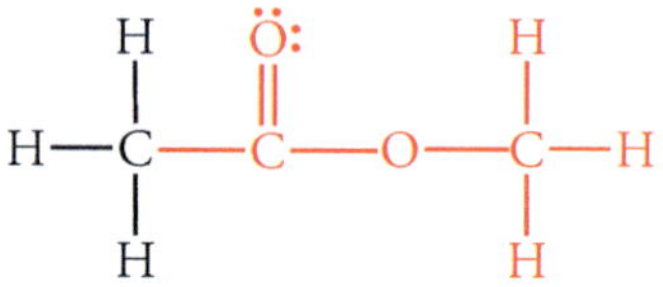

methyl acetate
(an ester)

A nitrogen atom can form single, double, or triple bonds to a carbon atom. Compounds with one or more carbon–nitrogen single bonds are **amines**. The remaining bonds to nitrogen can be to hydrogen or carbon atoms. Compounds with carbon–nitrogen double and triple bonds are **imines** and **nitriles** respectively.

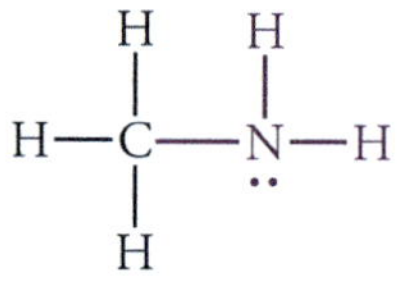

methylamine
(an amine)

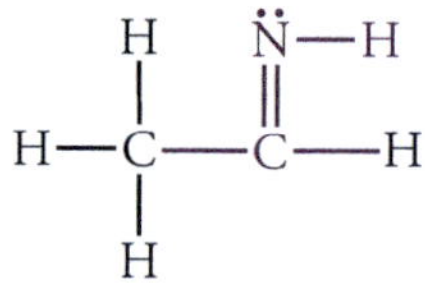

ethylimine
(an imine)

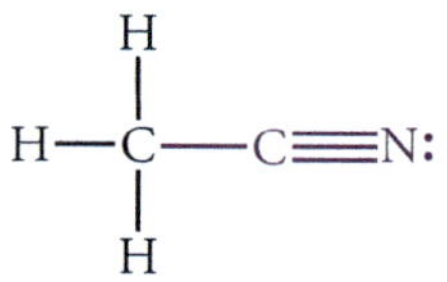

acetonitrile
(a nitrile)

Amides are functional groups in which a carbonyl carbon atom is linked by a single bond to a nitrogen atom and either a hydrogen or a carbon atom. The remaining two bonds to the nitrogen atom may be to either hydrogen or carbon atoms.

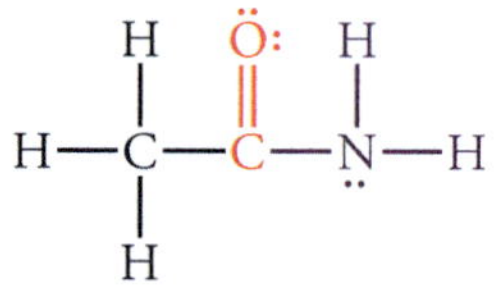

acetamide
(an amide)

Sulfur forms single bonds to carbon in two classes of compounds. **Thiols** (also called mercaptans) and **thioethers** (also called sulfides) structurally resemble alcohols and ethers, which contain oxygen, another element in the same group of the periodic table as sulfur.

Table 1.4
Functional Groups of Organic Compounds

Class	Functional Group	Expanded Structural Formula
Alkene	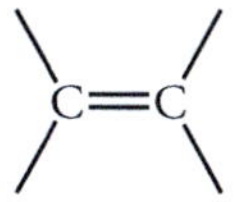	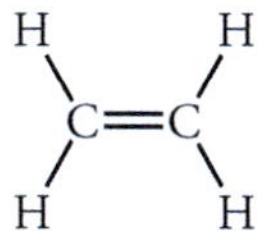
Alkyne		
Alcohol	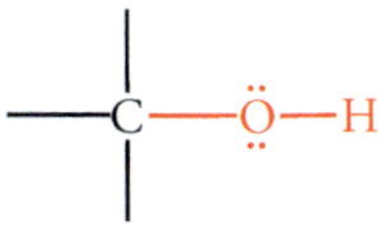	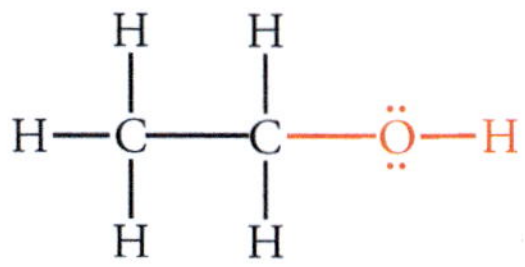
Ether		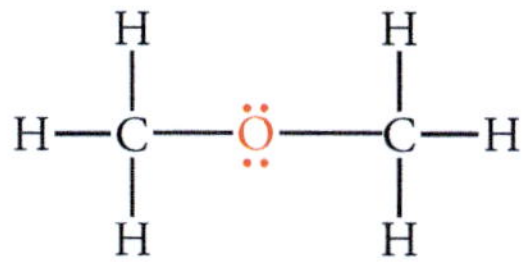
Aldehyde	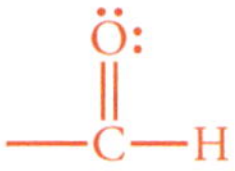	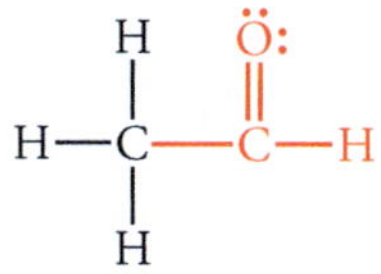
Ketone		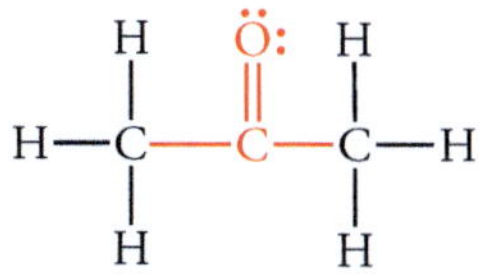
Carboxylic acid	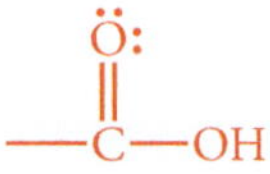	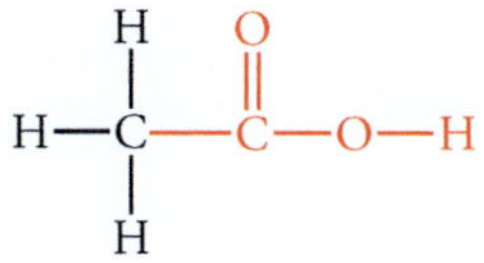
Ester	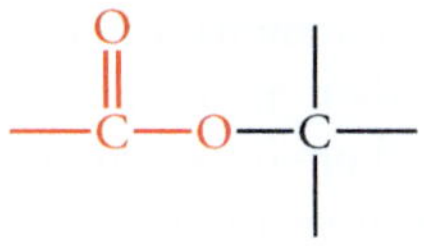	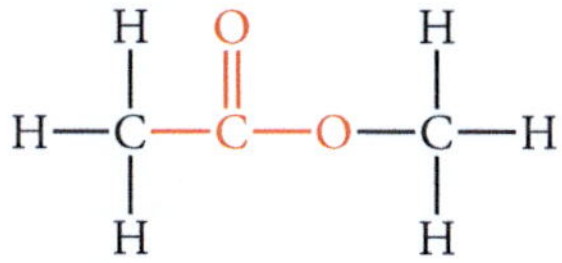
Amine		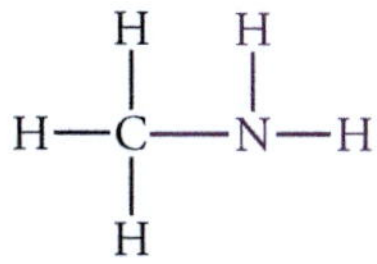
Amide		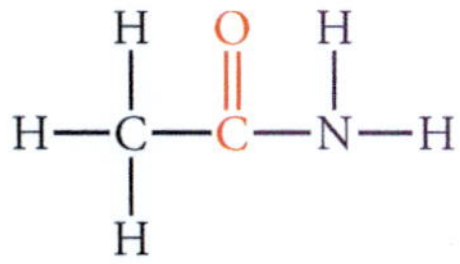
Halide	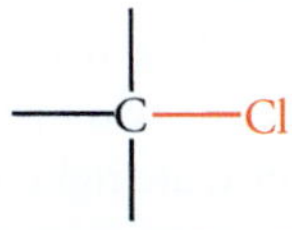	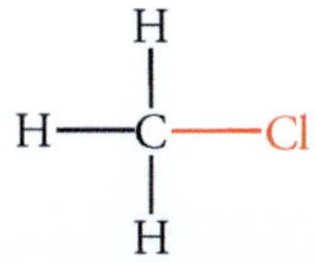
Mercaptan or thiol	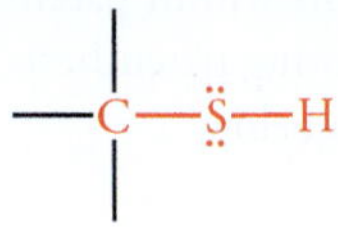	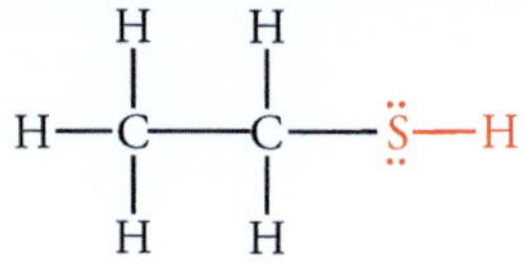

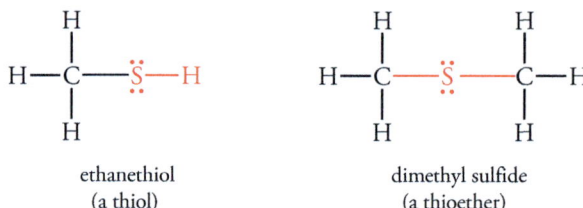

ethanethiol
(a thiol)

dimethyl sulfide
(a thioether)

The halogens form single bonds to carbon. Chlorine and bromine are the more common halogens in organic compounds. Compounds with halogens bonded to a carbon atom containing only single bonds to carbon or hydrogen are **haloalkanes** (alkyl halides). Compounds with halogens bonded to a carbonyl carbon atom are **acyl halides**.

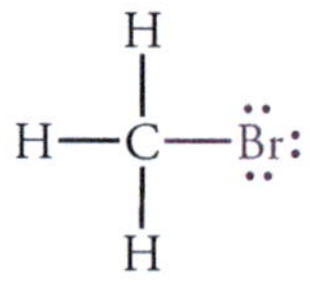

bromoethane
(a haloalkane)

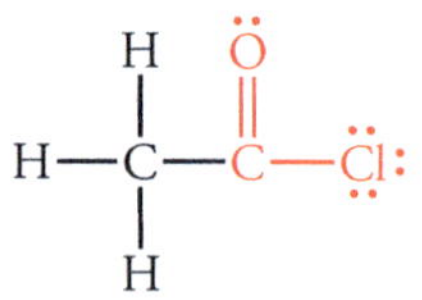

acetyl chloride
(an acyl halide)

1.9 STRUCTURAL FORMULAS

The molecular formula of a compound indicates its atomic composition. For example, the molecular formula of butane is C_4H_{10}. However, to understand the chemistry of organic compounds, it is necessary to represent the structure of a molecule by a **structural formula** that shows the arrangement of atoms and bonds.

To save time and space, we draw abbreviated or condensed versions of structural formulas. Condensed structural formulas show only specific bonds; other bonds are omitted, but implied. The degree of condensation depends on which bonds are shown and which are implied. For example, because hydrogen forms only a single bond to carbon, the C—H bond need not be shown in the condensed structure of a molecule such as butane.

$$CH_3 - CH_2 - CH_2 - CH_3$$
butane

One carbon–carbon bond is shown between a terminal carbon atom and an internal carbon atom. The terminal carbon atoms are understood to have single bonds to three hydrogen atoms. Each carbon atom in the interior of the molecule has the two carbon–carbon bonds shown; the two carbon–hydrogen bonds are implied but not shown. Note that by convention the symbol for the hydrogen atom is written to the right of the symbol for the carbon atom. In a further condensation of a structural formula, the C—C bonds are omitted.

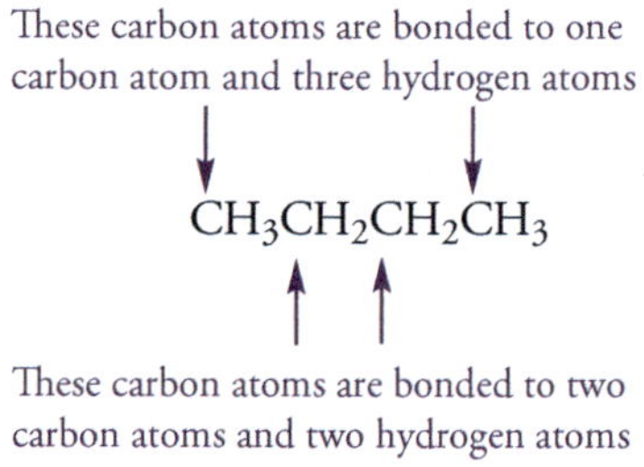

These carbon atoms are bonded to one carbon atom and three hydrogen atoms

$$CH_3CH_2CH_2CH_3$$

These carbon atoms are bonded to two carbon atoms and two hydrogen atoms

In the structural formula shown above, the carbon atom on the left is understood to be bonded to the three hydrogen atoms and the carbon atom to the right. The second carbon atom from the left is bonded to the two hydrogen atoms to the right. That carbon atom is also bonded to a carbon atom to its immediate right and a carbon atom to its left.

Large structures may have repeated structural subunits that are represented by grouping the subunits within parentheses. The number of times the unit is repeated is given by a subscript after the closing parenthesis. For example, butane is represented by an even more condensed formula, as shown below.

$$CH_3(CH_2)_2CH_3$$

The —CH_2— unit is a **methylene group**. It occurs twice in butane. Because the methylene groups are linked in a repeating chain, they are placed within the parentheses.

Two or more identical groups of atoms bonded to a common central atom may also be represented within parentheses with an appropriate subscript in a condensed formula. The parentheses may be placed to the right or left in the condensed structure depending on the way in which the molecule is drawn.

$$CH_3-\overset{\overset{\displaystyle CH_3}{|}}{CH}-CH_2-CH_2-CH_2-CH_3 \quad \text{is} \quad (CH_3)_2CHCH_2CH_2CH_2CH_3$$

$$CH_3-CH_2-CH_2-\overset{\overset{\displaystyle CH_3}{|}}{\underset{\underset{\displaystyle CH_3}{|}}{C}}-CH_3 \quad \text{is} \quad CH_3CH_2CH_2C(CH_3)_3$$

Bond-Line Structures

Condensed structural formulas are convenient but still require considerable time to draw compared to yet another shorthand method using bond-line structures. The bond-line structure method also results in a less cluttered drawing. However, we have to mentally add many more features to understand the structure. The rules for drawing bond-line structures are as follows:

1. Carbon and hydrogen atoms are not shown unless needed for special emphasis or clarity.
2. All other atoms are shown.
3. A carbon atom is assumed to be at the end of each line segment or at the intersection of lines.
4. Multiple bonds are shown with multiple lines.
5. The proper number of hydrogen atoms to provide four bonds to each carbon atom is implied. Hydrogen atoms on other atoms such as oxygen and nitrogen are explicitly indicated.

For a bond-line structure, it is best to start by drawing a zigzag arrangement of the carbon atoms and then mentally remove them.

$$CH_3-CH_2-CHBr-CH_3 \quad \text{is} \quad \text{(structure)} \quad \text{is} \quad \text{(structure)}$$

Bond-line formulas are also used to show cyclic structures. Rings of carbon atoms are shown by polygons such as an equilateral triangle, square, pentagon, and hexagon.

$$\begin{array}{cc} CH_2 & \!\!\!\!-CH_2 \\ | & | \\ CH_2 & \!\!\!\!-CH_2 \end{array} \quad \text{is} \quad \square$$

There are differences in the representation of multiple bonds. Atoms such as oxygen must be shown in carbonyl groups, but a double-bonded carbon atom is not shown.

It is important to remember the normal number of bonds formed by each common atom in an organic compound. Carbon, nitrogen, and oxygen form four, three, and two bonds, respectively.

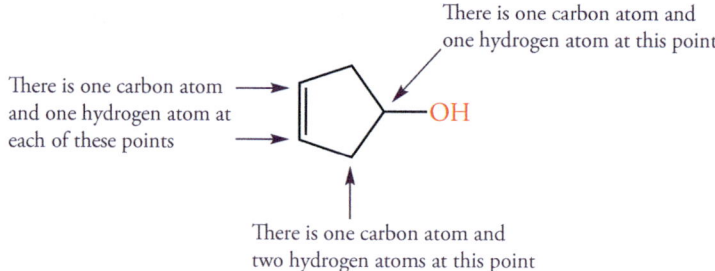

There is one carbon atom and one hydrogen atom at this point

There is one carbon atom and one hydrogen atom at each of these points

There is one carbon atom and two hydrogen atoms at this point

Three-Dimensional Structures and Molecular Models

Because structure is so important to understanding chemical reactions, we construct molecular models that can be viewed from many angles. Molecular model kits are a great help in understanding the structures of organic molecules. Molecular modeling program, such as Spartan Model, are available in many chemistry departments.

Ball-and-stick models and space-filling models are two types of molecular models; each has certain advantages and disadvantages. Ball-and-stick models show the molecular framework and bond angles: the balls represent the nuclei of the atoms, and the sticks represent the bonds (Figure 1.9). The actual volume occupied by the molecule is not shown realistically. Space-filling models show the entire volume occupied by the electrons surrounding each atom, but, as a consequence, the carbon skeleton and its bond angles are obscured, although transparent surfaces can be used to eliminate this problem.

On paper, the three-dimensional shape of molecules is shown by a wedge and a dashed line (Figure 1.9). The wedge is viewed as a bond extending out of the plane of the page toward the reader. The dashed lines represent a bond directed behind the plane of the page. The other line is a bond in the plane of the page. Three-dimensional representations of molecules using wedges and dashed lines are perspective structural formulas.

Figure 1.9
Three Views of Methane

(a) Perspective structure   (b) Ball-and-stick model   (c) Space-filling model

Recognizing Structural Features

The structural features that allow chemists to predict the physical and chemical properties of naturally occurring molecules are often only a small part of a larger structure. These large structures are written in condensed forms that are meaningful because certain conventions are used. Regardless of the size and complexity of a molecule, we examine the entire molecule, ignore the many lines indicating the carbon-carbon bonds, and focus on the important parts. Are there multiple bonds? Are there atoms, such as oxygen and nitrogen, that are part of functional groups? How are these atoms bonded, and what other atoms are nearby? For example, if a carbonyl group is present, it may be part of an aldehyde, ketone, acid, ester, or amide. The distinction between these functional groups can be decided by looking at the atoms bonded to the carbonyl carbon atom.

Consider the structure for nonactin, an antibiotic that forms pores in biological membranes. It binds potassium ions through the many oxygen atoms in the large ring of atoms. It transports potassium ions across bacterial cell membranes, and the cells die. What are the oxygen-containing functional groups in this complex structure? Concentrate on one oxygen atom at a time. Some oxygen atoms are part of a carbonyl group. There are four carbonyl groups in nonactin. Now look at the atoms bonded to the carbonyl carbon atom of the carbonyl groups. One bond is to carbon and the other to oxygen. Both carboxylic acids and esters have such features. The oxygen atom of carboxylic acids is in an —OH group, whereas the oxygen atom of esters is bonded to another carbon atom. There are four ester groups in nonactin.

Now concentrate on the second type of oxygen-containing functional group in the molecule. There are four oxygen atoms contained as part of a five-membered ring. Each of the five-membered rings contains an oxygen atom. These functional groups are ethers.

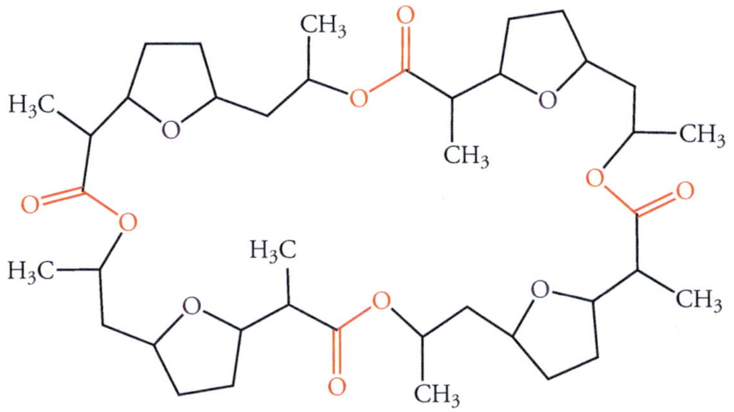

nonactin

Pheromones: Chemical Communications in the Insect World

The scope of organic chemistry is rapidly changing and contributes to many fields. For example, we cannot understand modern biology without a foundation in organic chemistry and indirectly without an understanding of functional groups. Organic chemistry underlies all life forms. As an example, we will consider the structure and functional groups of some pheromones. Pheromones (Greek pherein, "to transfer," + hormon, "to excite") are compounds (occasionally mixtures of compounds) that insects use to communicate. Higher animals, including mammals, also emit pheromones.

Pheromones are used to mark trails, warn of dangers, cause aggregation of species, defend against danger, and attract members of the opposite sex. The whip scorpion ejects a defensive spray that it uses to ward off predators. Some species of ants warn other ants of danger by an alarm pheromone. Bark beetles responsible for Dutch Elm disease emit an aggregation pheromone that results in the gathering of a large number of beetles on the trees. This species carries and transmits a fungus that kills the tree. The sex attractants, usually emitted by the female of the species, attract members of the opposite sex. They are signals that the female is ready to mate. They also aid the male in locating the female, often from great distances.

All moths that have been studied have sex attractants that are species specific. The compounds are usually derived from long chains of carbon atoms. However, the functional groups in the pheromones vary considerably. Two examples are the sex attractants of the gypsy moth and the grape berry moth. Their structures are shown below. The oxygen atom in the three-membered ring of the sex attractant of the gypsy moth is part of an ether functional group. The oxygen atoms in the sex attractant of the grape berry moth are part of an ester functional group. Note that this compound also contains a carbon-carbon double bond.

When the structures of sex attractants were determined, some scientists predicted that it might be possible to bait trap with the compound and, by removal of one sex, break the reproductive cycle. This "ideal" way to control insects and eliminate the use of pesticides has not proved effective for most species. The ultimate goal of replacing pesticides with pheromones has remained elusive.

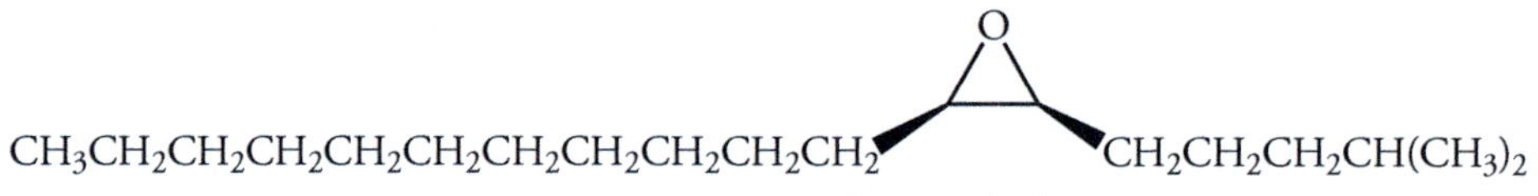

$$CH_3CH_2CH_2CH_2CH_2CH_2CH_2CH_2CH_2CH_2CH_2 \overset{\displaystyle O}{\overset{\displaystyle \triangle}{}} CH_2CH_2CH_2CH(CH_3)_2$$

Gypsy moth pheromone

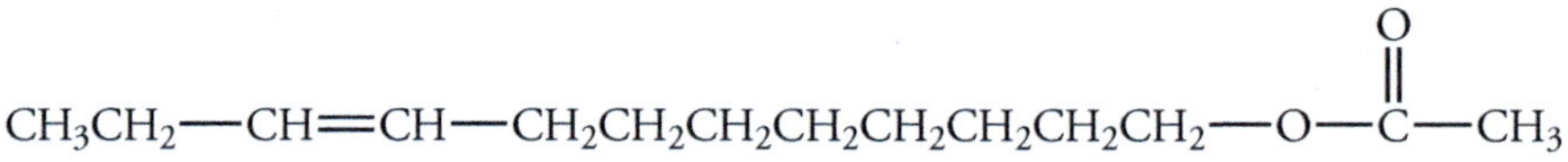

$$CH_3CH_2-CH=CH-CH_2CH_2CH_2CH_2CH_2CH_2CH_2CH_2-O-\overset{\displaystyle O}{\overset{\displaystyle \|}{C}}-CH_3$$

European grape vine moth pheromone

Problem 1.5

A species of cockroach secretes the substance shown below, which attracts other cockroaches. Write three condensed structural formulas for the substance.

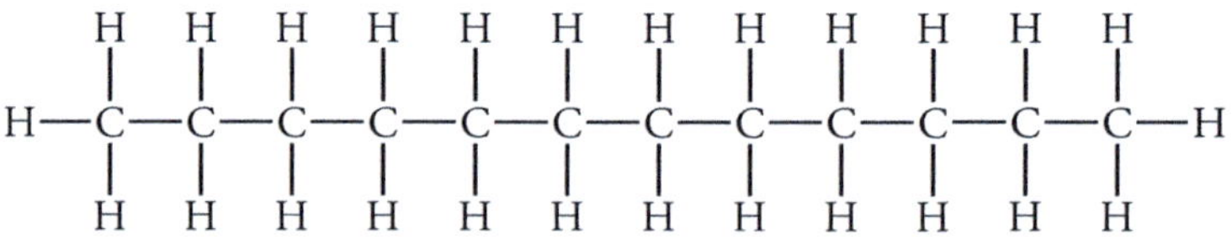

Solution

With the C—H bonds understood, we write only the C—C bonds.

$$CH_3-CH_2-CH_2-CH_2-CH_2-CH_2-CH_2-CH_2-CH_2-CH_2-CH_2-CH_3$$

With both the C—H and C—C bonds understood, we write,

$$CH_3CH_2CH_2CH_2CH_2CH_2CH_2CH_2CH_2CH_2CH_2CH_3$$

The above structure can be condensed still further to give the fully condensed formula.

$$CH_3(CH_2)_{10}CH_3$$

Problem 1.6

Hexamethylenediamine, a compound used to produce nylon, has the following structural formula. Write three condensed structural formulas for it.

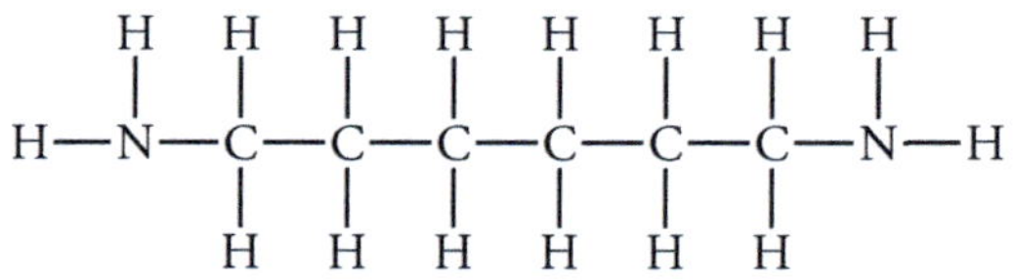

hexamethylenediamine

Problem 1.7

What is the molecular formula of carvone, which is found in oil of caraway?

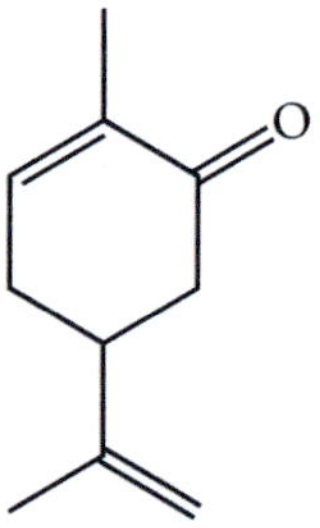

carvone

Solution

There are 10 carbon atoms in the structure, located at the ends or intersections of line segments. An oxygen atom is located at the end of a segment representing the double bond of a carbonyl group. Hydrogen atoms are counted by determining the number of bonds from each carbon atom to other atoms. Note that three carbon atoms have no hydrogen atoms. The molecular formula is $C_{10}H_{14}O$.

carvone

Problem 1.8
What is the molecular formula of indoleacetic acid, a plant growth hormone that promotes shoot growth?

indoleacetic acid

1.10 ISOMERS

Compounds that have the same molecular formula but different structures are isomers. Structure refers to the linkage of the atoms. As we examine the structure of organic compounds in increasing detail, you will learn how subtle structural differences in isomers affect the physical and chemical properties of compounds. There are several types of isomers. Isomers that differ in their bonded connectivity are **skeletal isomers**. Consider the structural differences in the two isomers of C_4H_{10}, butane and isobutane. Butane has an uninterrupted chain of four carbon atoms (Figure 1.10), but isobutane has only three carbon atoms connected in sequence and a fourth carbon atom appended to the chain. The boiling points (bp) of butane and isobutane are –1°C and –12°C, respectively; the chemical properties of the two compounds are similar but different.

Isomers that have different functional groups are **functional group isomers**. The molecular formula for both ethyl alcohol and dimethyl ether is C_2H_6O (Figure 1.10). Although the compositions of the two compounds are identical, their functional groups differ. The atomic sequence is C—C—O in ethyl alcohol, and the oxygen atom is present as an alcohol. The C—O—C sequence in the isomer corresponds to an ether.

$$CH_3-CH_2-OH \qquad\qquad CH_3-O-CH_3$$
ethyl alcohol (bp 78.5°C) \qquad\qquad dimethyl ether (bp –24°C)

The physical properties of these two functional group isomers, as exemplified by their boiling points, are very different. These substances also have different chemical properties because their functional groups differ.

Positional isomers are compounds that have the same functional groups in different positions on the carbon skeleton. For example, the isomeric alcohols 1-propanol and 2-propanol differ in the location of the hydroxyl group. The chemical properties of these two compounds are similar because they both contain the same type of functional group and have identical molecular weights.

Figure 1.10
Structure of Isomers

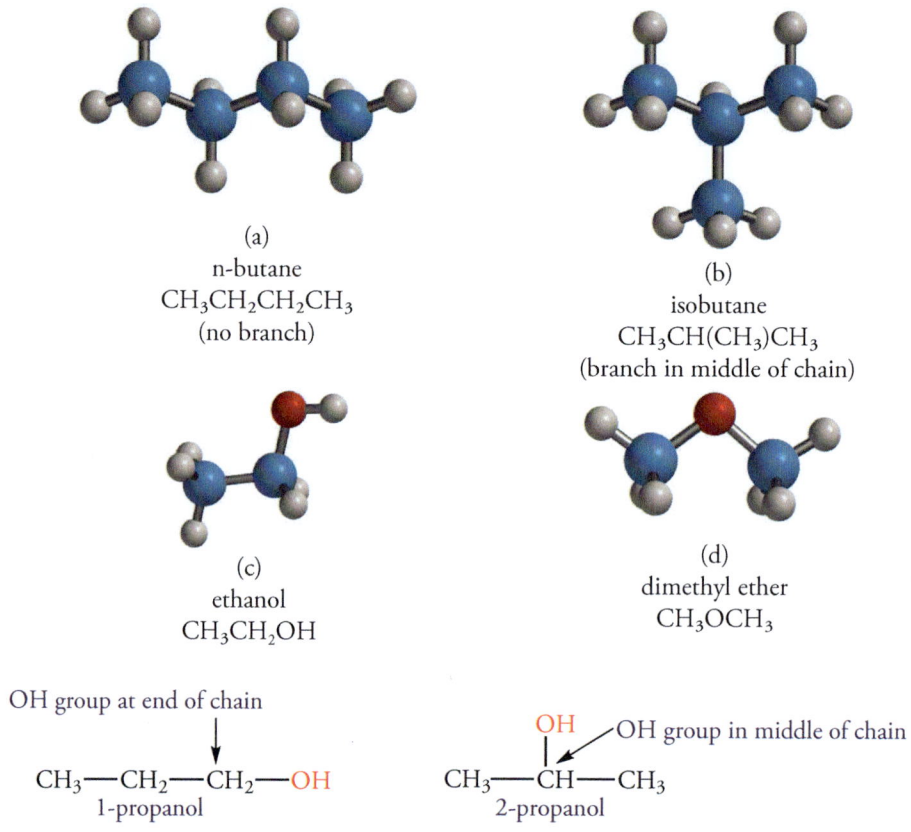

(a)
n-butane
$CH_3CH_2CH_2CH_3$
(no branch)

(b)
isobutane
$CH_3CH(CH_3)CH_3$
(branch in middle of chain)

(c)
ethanol
CH_3CH_2OH

(d)
dimethyl ether
CH_3OCH_3

OH group at end of chain

$CH_3—CH_2—CH_2—OH$
1-propanol

OH group in middle of chain

$CH_3—CH—CH_3$
$\quad\quad\;\; OH$
2-propanol

Isomerism is not always immediately obvious. Sometimes two structures appear to be isomers when in fact the structures are the same compound written in slightly different ways. It is important to be able to recognize isomers and distinguish them from equivalent representations of the same compound. For example, 1,2-dichloroethane can be written in several ways. In each formula, the bonding sequence is Cl—C—C—Cl.

1,2-dichloroethane (CH_2ClCH_2Cl)

The isomer of 1,2-dichloroethane is 1,1-dichloroethane. In 1,1-dichloroethane, the two chlorine atoms are bonded to the same carbon atom, but in 1,2-dichloroethane, the two chlorine atoms are bonded to different carbon atoms. The different condensed structural formulas, $CHCl_2CH_3$ and CH_2ClCH_2Cl, also tell us that in the first case two chlorine atoms are bound to the same carbon and that in the second case the two chlorine atoms are bound to adjacent carbons.

1,1-dichloroethane (CH_2ClCH_2Cl)

Problem 1.9

The structural formulas for two compounds used as general anesthetics are shown below. Are they isomers? How do they differ?

$$\text{Cl}-\underset{\underset{H}{|}}{\overset{\overset{F}{|}}{C}}-\underset{\underset{F}{|}}{\overset{\overset{F}{|}}{C}}-O-\underset{\underset{H}{|}}{\overset{\overset{F}{|}}{C}}-F \qquad F-\underset{\underset{F}{|}}{\overset{\overset{F}{|}}{C}}-\underset{\underset{H}{|}}{\overset{\overset{Cl}{|}}{C}}-O-\underset{\underset{F}{|}}{\overset{\overset{H}{|}}{C}}-F$$

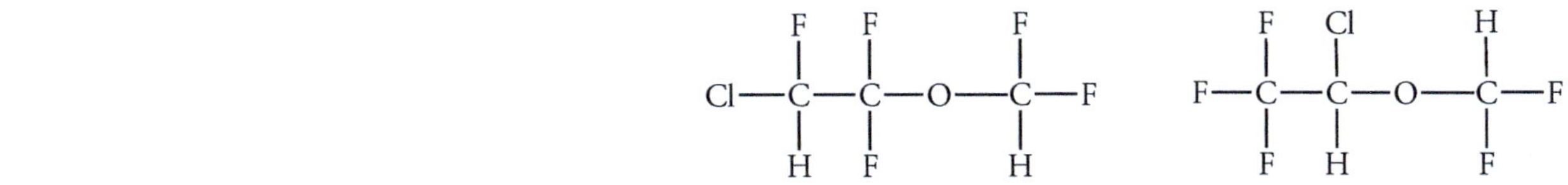

Solution

The atomic compositions of these structural formulas are identical; the molecular formula is $C_3H_2F_5ClO$. Therefore, the compounds are isomers. The carbon skeletons are identical and the compounds are both ethers.

Both isomers have a CHF_2 unit on the right side of the ether oxygen atom in spite of the different ways in which the fluorine and hydrogen are written—this is not the basis for isomerism. The two-carbon unit on the left of the oxygen atom has the halogen atoms distributed in two different ways. That is, they are positional isomers. The structure on the left has two fluorine atoms bonded to the carbon atom bonded to the oxygen atom. The carbon atom on the left has a fluorine and a chlorine atom bonded to it. The structure on the right has one chlorine atom bonded to the carbon atom appended to the oxygen atom. The carbon atom on the left has three fluorine atoms bonded to it.

Problem 1.10

Compare the following structures of two intermediates in the metabolism of glucose. Are they isomers? How do they differ?

$$H-\underset{\underset{H-C-OH}{|}}{\overset{\overset{H}{|}}{C}}-OH \qquad H-\overset{\overset{O}{\diagup\!\!\diagup}}{C}$$

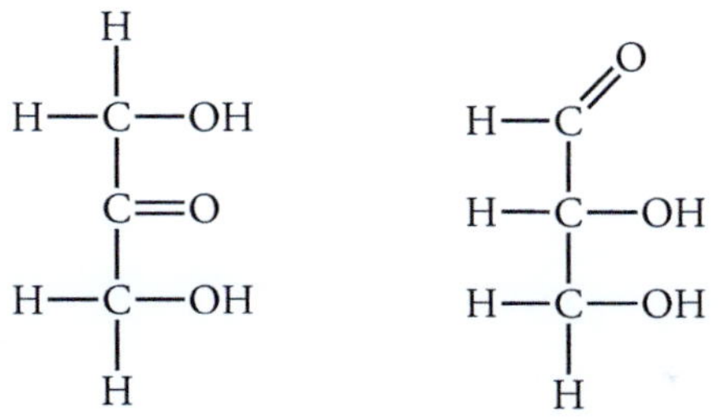

1.11 NOMENCLATURE

Nomenclature refers to a systematic method of naming materials. In chemistry, the nomenclature of compounds is exceedingly important. The existence of isomers illustrates this point. The common names *butane* and *isobutane* of the two isomeric C_4H_{10} compounds are easy to learn. However, there are 75 isomers of $C_{10}H_{21}$ and 62,491,178,805,831 isomers of $C_{40}H_{82}$. Without a system of naming compounds, organic chemistry would be difficult, if not impossible, to comprehend.

At a meeting in Geneva, Switzerland, in 1892 a systematic nomenclature was devised for all compounds, including organic compounds. Compounds are now named by rules developed by the International Union of Pure and Applied Chemistry (IUPAC). The rules result in a clear and definitive name for each compound. A universal system for naming organic compounds was needed because different names had often been given to the same compound. For example, CH_3CH_2OH had been called not only alcohol but also spirits, grain alcohol, ethyl alcohol, methyl carbinol, and ethanol. Furthermore, a variety of names developed in each language.

A chemical name consists of three parts: prefix, parent, and suffix. The parent indicates how many carbon atoms are in the main carbon skeleton. The suffix identifies most of the functional groups present in the molecule. Examples of suffixes are *-ol* for alcohols, *-al* for aldehydes, and *-one* for ketones. The prefix specifies the location of the functional group designated in the suffix as well as some other types of substituents on the main parent chain.

Once the rules are applied, there is only one name for each structure, and one structure for each name. For example, a compound that is partly responsible for the odor of a skunk is 3-methyl-1-butanethiol.

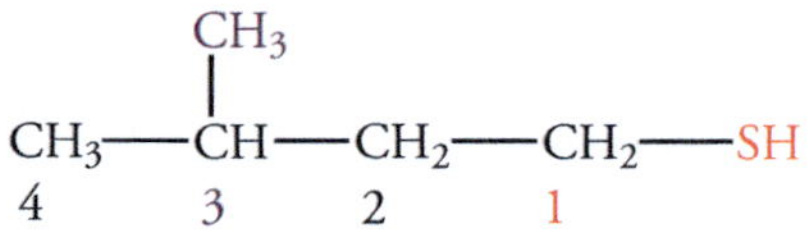

$$CH_3\text{—}\overset{\overset{\displaystyle CH_3}{|}}{\underset{3}{CH}}\text{—}\underset{2}{CH_2}\text{—}\underset{1}{CH_2}\text{—}SH$$

$$\underset{4}{}$$

3-methyl-1-butanethiol

Butane is the parent name of the four-carbon unit that is written horizontally. The prefix 3-methyl refers to the −CH$_3$ written above the chain of carbon atoms. The prefix 1- and the suffix *-thiol* refer to the position and identity of the −SH group. This method of assigning numbers to the carbon chain and other features of the IUPAC system will be discussed further in subsequent chapters.

In spite of the IUPAC system, many common names are so well established that both common and IUPAC names must be recognized. The IUPAC name for CH$_3$CH$_2$OH is ethanol, but the common name ethyl alcohol is still used. In addition, many complex compounds, particularly ones of biological origin, retain common names partly due to the unwieldy nature of their IUPAC names.

EXERCISES

Atomic Properties

1.1 How many valence shell electrons are in each of the following elements?
(a) N (b) F (c) C (d) O
(e) Cl (f) Br (g) S (h) P

1.2 Which of the following atoms has the higher electronegativity? Which has the larger atomic radius?
(a) Cl or Br (b) O or S (c) C or N (d) N or O (e) C or O

Lewis Structures of Covalent Compounds

1.3 Write a Lewis structure for each of the following compounds:
(a) NH_2OH (b) CH_3CH_3 (c) CH_3OH (d) CH_3NH_2 (e) CH_3Br (f) CH_3SH

1.4 Write a Lewis structure for each of the following compounds:
(a) HCN (b) HNNH (c) CH_2NH (d) CH_3NO (e) CH_2NOH (f) CH_2NNH_2

1.5 Add any required unshared pairs of electrons that are missing from the following formulas:

(a) $CH_3-\overset{\overset{\displaystyle O}{\|}}{C}-OH$ (b) $CH_3-\overset{\overset{\displaystyle O}{\|}}{C}-O-CH_3$ (c) $H-\overset{\overset{\displaystyle O}{\|}}{C}-NH-CH_3$ (d) $CH_3-S-CH=CH_2$

1.6 Add any required unshared pairs of electrons that are missing from the following formulas:

(a) $CH_3-\overset{\overset{\displaystyle O}{\|}}{C}-Cl$ (b) $CH_3-O-CH=CH_2$ (c) $CH_3-\overset{\overset{\displaystyle O}{\|}}{C}-SH$ (d) $NH_2-\overset{\overset{\displaystyle O}{\|}}{C}-O-CH_3$

1.7 Using the number of valence electrons in the constituent atoms and the given arrangement of atoms in the compound, write the Lewis structure for each of the following molecules:

(a) $Cl-\overset{\overset{\displaystyle O}{|}}{C}-Cl$ (b) $H-\overset{\overset{\displaystyle \,}{|}}{\underset{\underset{H}{|}}{N}}-\overset{\overset{\displaystyle O}{|}}{C}-\overset{\overset{\displaystyle \,}{|}}{\underset{\underset{H}{|}}{N}}-H$ (c) $H-\overset{\overset{\displaystyle H}{|}}{\underset{\underset{H}{|}}{C}}-\overset{\overset{\displaystyle O}{|}}{C}-S-H$

1.8 Using the number of valence electrons in the constituent atoms and the given arrangement of atoms in the compound, write the Lewis structure for each of the following molecules:

(a) $H-\overset{\overset{\displaystyle H}{|}}{\underset{\underset{H}{|}}{C}}-C-S-S-\overset{\overset{\displaystyle H}{|}}{\underset{\underset{H}{|}}{C}}-H$ (b) $Cl-\overset{\overset{\displaystyle H}{|}}{\underset{\underset{H}{|}}{C}}-\overset{\overset{\displaystyle O}{|}}{C}-O-H$ (c) $H-\overset{\overset{\displaystyle H}{|}}{\underset{\underset{H}{|}}{C}}-\overset{\overset{\displaystyle O}{|}}{\underset{\underset{\,}{}}{C}}-\overset{\overset{\displaystyle Cl}{|}}{N}-H$

1.9 Two compounds used as dry cleaning agents have the molecular formulas C_2Cl_4 and C_2HCl_3. Write the Lewis structures for each compound.

1.10 Acrylonitrile, a compound used to produce fibers for rugs, has the molecular formula CH_2CHCN. Write the Lewis structure for the compound.

Formal Charge

1.11 Assign the formal charges for the atoms other than carbon and hydrogen in each of the following species:

(a) H—Ö—C≡N: (b) H—Ö—N≡C: (c) CH₃—N̈=N=N̈:

1.12 All of the following species are isoelectronic, that is, they have the same number of electrons bonding the same number of atoms. Determine which atoms have a formal charge. Calculate the net charge for each species.

(a) :C≡O: (b) :N≡O: (c) :C≡N: (d) :C≡C: (e) :N≡N:

1.13 Acetylcholine, a compound involved in the transfer of nerve impulses, has the following structure. What is the formal charge on the nitrogen atom? What is the net charge of acetylcholine?

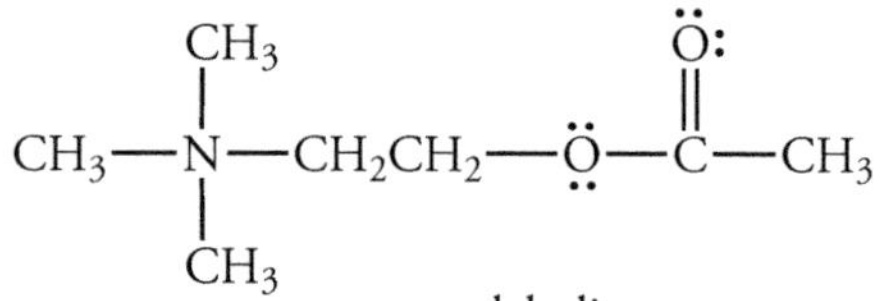

acetylcholine

1.14 Sarin, a nerve gas, has the following structure. What is the formal charge of the phosphorus atom?

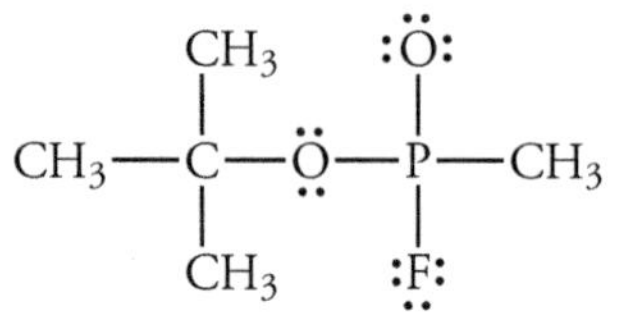

Sarin

Resonance

1.15 The small amounts of cyanide ion contained in the seeds of some fruits are eliminated from the body as SCN⁻. Draw two possible resonance forms for the ion. Which atom has the formal negative charge in each form?

1.16 Are the following pairs contributing resonance forms of a single species? Formal charges are not shown and have to be added. Explain.

(a) :N̈=N=N̈: and :N̈—N≡N: (b) H—C≡N—Ö: and H—C̈=N=Ö:

1.17 Write the resonance structure that results when electrons are moved in the direction indicated by the curved arrows for the following amide. Calculate any formal charges that result.

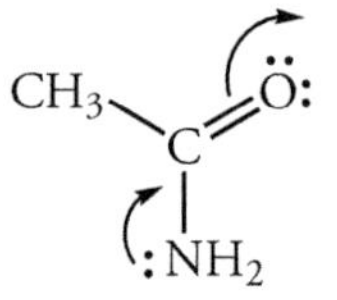

1.18 Write the resonance structure that results when electrons are moved in the direction indicated by the curved arrows for the following amide. Calculate any formal charges that result.

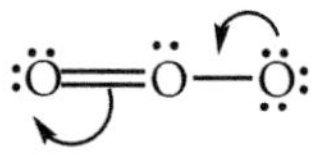

Molecular Shapes

1.19 Based on VSEPR theory, what is the expected value of the indicated bond angle in each of the following compounds?
(a) C—C—N in CH₃—C≡N
(b) C—O—C in CH₃—O—CH₃
(c) C—N—C in CH₃—NH—CH₃
(d) C—C—C in CH₃—C≡C—H

1.20 Based on VSEPR theory, what is the expected value of the indicated bond angle in each of the following ions?

(a) C—O—H in CH_3—OH_2^+

(b) C—N—H in CH_3—NH_3^+

(c) O—C—O in $CH_3CO_2^-$

(d) C—O—C in $(CH_3)_2OH^+$

1.21 Based on VSEPR theory, what is the expected value of the C—N=N bond angle in the following compound?

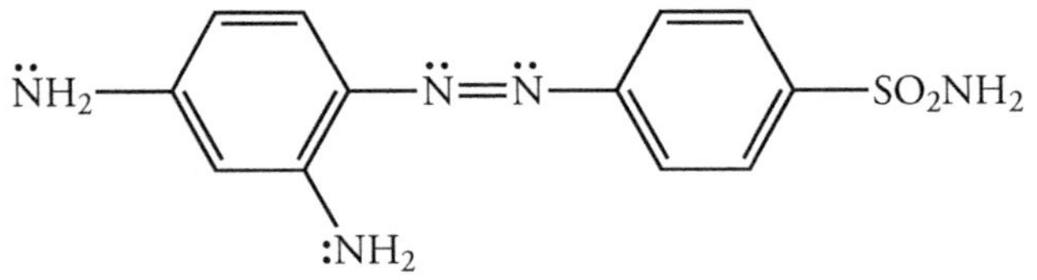

1.22 Based on VSEPR theory, what is the expected value of the S—C—S bond angle in dibenzthiozole disulfide, a catalyst used in the vulcanization of rubber?

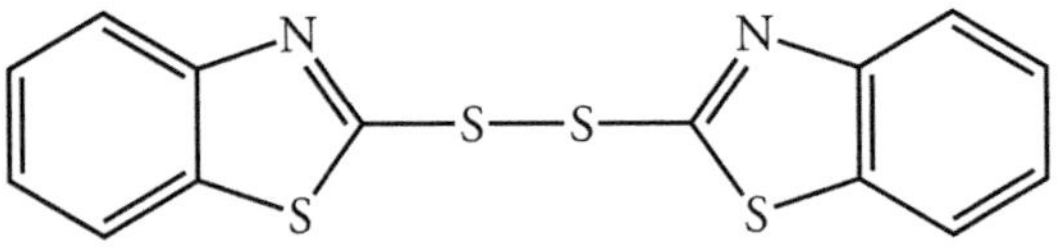

Hybridization

1.23 What is the hybridization of each carbon atom in each of the following compounds?

(a) CH_3—$\overset{\overset{\displaystyle O}{\|}}{C}$—H
(b) CH_3—O—CH=CH_2
(c) CH_3—$\overset{\overset{\displaystyle O}{\|}}{C}$—SH

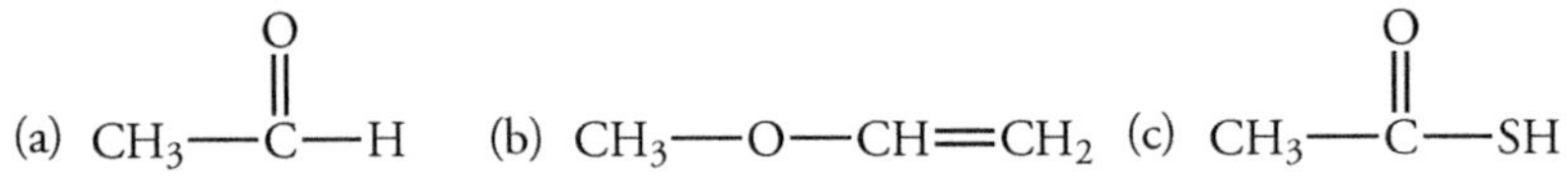

1.24 What is the hybridization of each carbon atom in each of the following compounds?

(a) CH_3—$\overset{\overset{\displaystyle O}{\|}}{C}$—NH—$CH_3$
(b) CH_3—S—CH=CH_2
(c) CH_3—$\overset{\overset{\displaystyle NH}{\|}}{C}$—$CH_3$

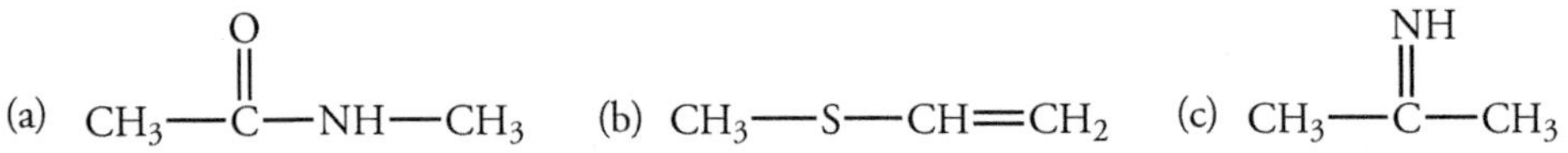

1.25 What is the hybridization of each of the carbon atoms bonded to two oxygen atoms in aspirin?

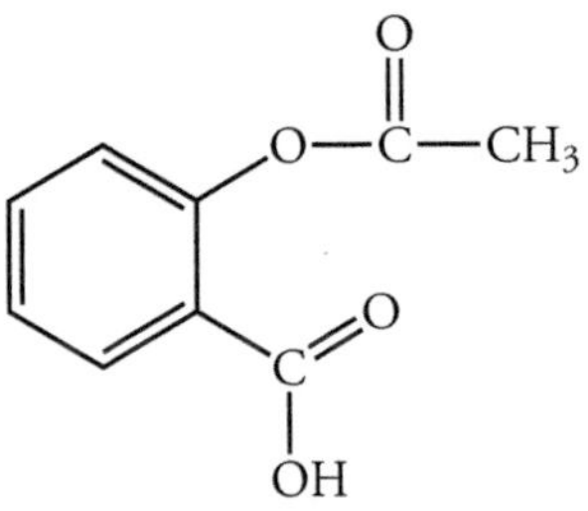

acetyl salicylic acid (aspirin)

1.26 What is the hybridization of the carbon atom bonded to the nitrogen atom and of the carbon atom bonded to two oxygen atoms in L-dopa, a drug that is used in the treatment of Parkinson's disease?

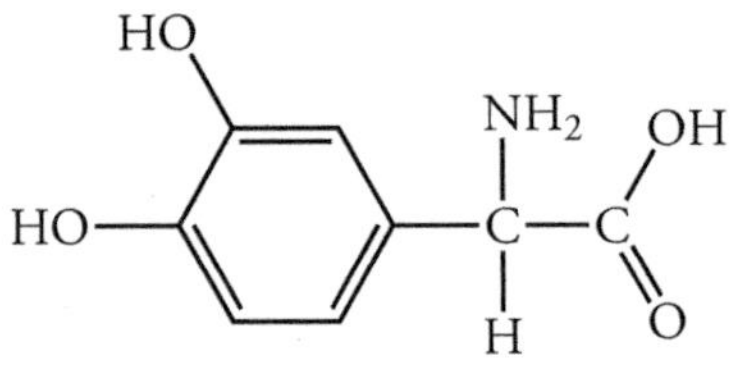

L-dopa

Molecular Formulas

1.27 Write the molecular formula for each of the following:

(a) CH_3—CH_2—CH_2—CH_2—CH_3 (b) CH_3—CH_2—CH_2—CH_3 (c) CH_2=CH—CH_2—CH_3 (d) CH_3—CH_2—C≡C—H

1.28 Write the molecular formula for each of the following:
(a) $CH_3CH_2CH_2CH_2CH_2CH_3$　　(b) $CH_3CH=CHCH_3$　　(c) $CH_3CH_2C\equiv CCH_3$ (d) $CH_3C\equiv CCH_2CH=CHCH_3$

1.29 Write the molecular formula for each of the following:
(a) $CH_3CH_2CHCl_2$　　　　　　　　　(b) $CH_3CCl_2CH_3$　(c) $BrCH_2CH_2Br$

1.30 Write the molecular formula for each of the following:
(a) $CH_3CH_2CH_2OH$　　(b) CH_3CH_2SH　(c) $CH_3CH_2CH_2NH_2$

Structural Formulas

1.31 For each of the following, write a condensed structural formula in which only C—C and C—S bonds are shown.

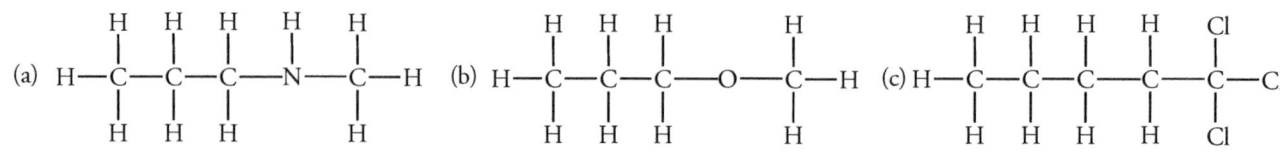

1.32 For each of the following, write a condensed structural formula in which only the bonds to hydrogen and chlorine are not shown.

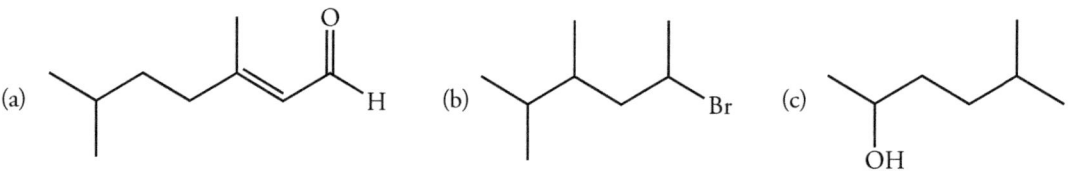

1.33 Write a condensed structural formula in which no bonds are shown for each of the structures in Problem 1.31.

1.34 Write a condensed structural formula in which no bonds are shown for each of the structures in Problem 1.32.

1.35 Write a complete structural formula, showing all bonds, for each of the following condensed formulas:
(a) $CH_3CH_2CH_2CH_3$　　(b) $CH_3CH_2CH_2Cl$　　(c) $CH_3CHClCH_2CH_3$

(d) $CH_3CH_2CHBrCH_3$　　(e) $CH_3CH_2CHBr_2$　　(f) $CH_3CBr_2CH_2CH_2CH_3$

1.36 Write a complete structural formula, showing all bonds, for each of the following condensed formulas:
(a) $CH_3CH_2CH_3$　　　　(b) $CH_3CH_2CHCl_2$　　(c) $CH_3CH_2CH_2CH_2SH$

(d) $CH_3CH_2C\equiv CCH_3$　　(e) $CH_3CH_2OCH_2CH_3$　(f) $CH_3CH_2CH_2C\equiv CH$

1.37 What is the molecular formula for each of the following bond-line structures?

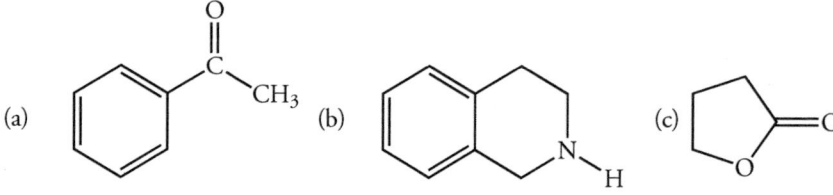

1.38 What is the molecular formula for each of the following bond-line structures?

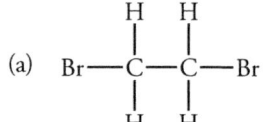

1.39   What is the molecular formula for each of the following bond-line structures?

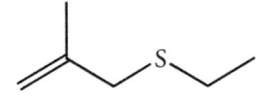

(a) Scent marker of the red fox

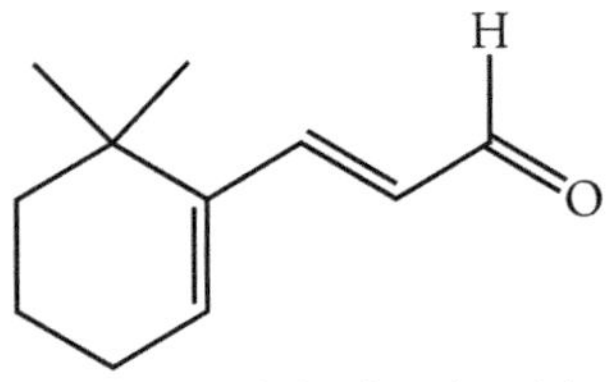

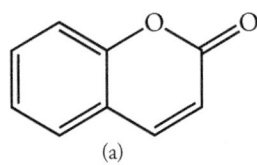

(b) Responsible for the odor of the iris

1.40   What is the molecular formula for each of the following bond-line structures?

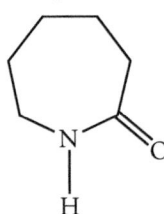

(a)

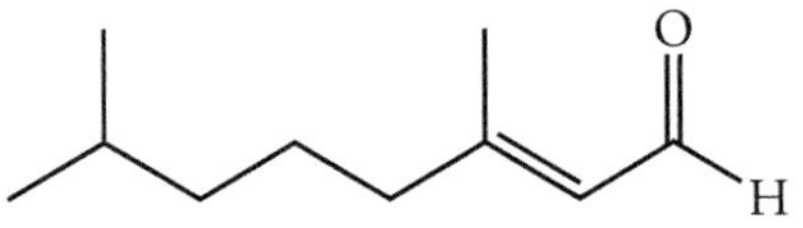

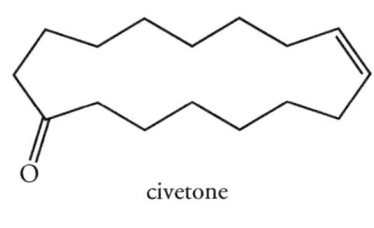

(b) An oil found in citrus fruits

Functional Groups

1.41   Identify the functional groups contained in each of the following structures:
(a) caprolactam, a compound used to produce a type of nylon; (b) civetone, a compound in the scent gland of the civet cat.

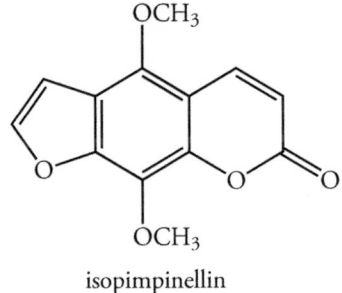

caprolactam

civetone

1.42   Identify the oxygen-containing functional groups in each of the following compounds:
(a) isopimpinellin, a carcinogen found in diseased celery;   (b) aflatoxin B_1, a carcinogen found in moldy foods.

OCH$_3$

OCH$_3$

isopimpinellin

O

O

O

O

OCH$_3$

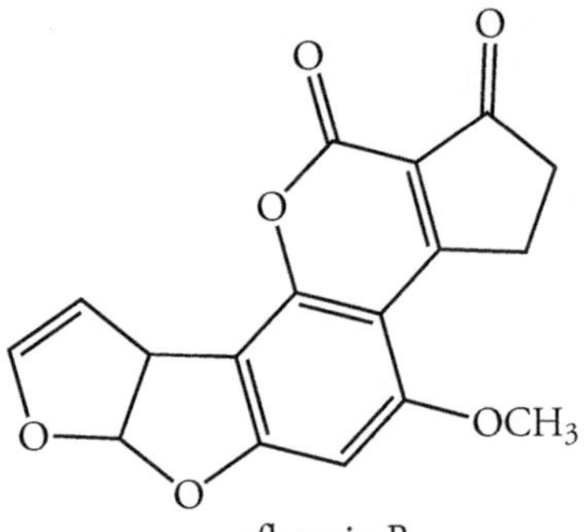

aflatoxin B_1

Isomerism

1.43 Indicate whether the following pairs of structures are isomers or different representations of the same compound:

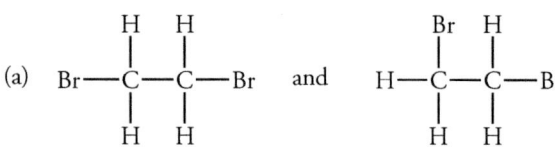

(a)   Br—C—C—Br   and   H—C—C—Br
  (with H, H above and H, H below the first; Br, H above and H, H below the second)

(b) CH₃—CH₂   and   CH₃—CH₂—CH₂—Cl
          |
         CH₂—Cl

(c) CH₃—CH—Cl   and   CH₃—CH₂—CH₂—Cl
          |
         CH₃

1.44 Indicate whether the following pairs of structures are isomers or different representations of the same compound:

(a)   H—C—C—Br   and   H—C—C—Br
  (with H, Cl above and H, H below the first; Cl, H above and H, H below the second)

(b) CH₃—CH₂   and   CH₃—CH—CH₃
          | |
         CH₂—Cl   Cl

(c) CH₃—CH—CH₂—Cl   and   CH₃—CH——CH₃
          | |
         CH₃   CH₂—Cl

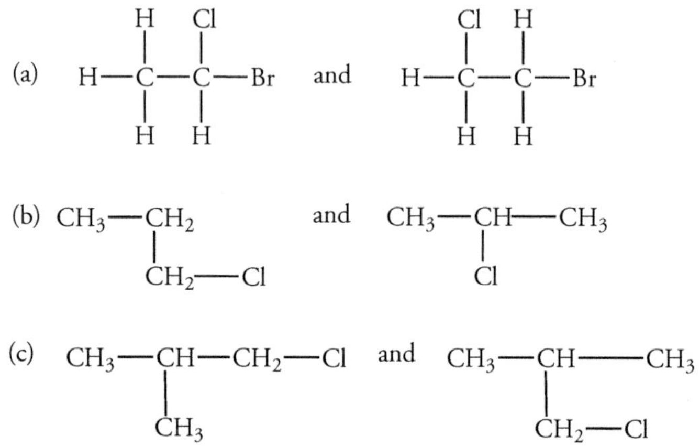

1.45 There are two isomers for each of the following molecular formulas. Draw their structural formulas.
(a) C₂H₂Br₂   (b) C₂H₆O   (c) C₂H₄BrCl   (d) C₂H₇Cl   (e) C₂H₇N

1.46 There are three isomers for each of the following molecular formulas. Draw their structural formulas.
(a) C₂H₃Br₂Cl   (b) C₃H₈O   (c) C₃H₈S

2

PROPERTIES

OF

ORGANIC COMPOUNDS

CHOLESTEROL

2.1 STRUCTURE AND PHYSICAL PROPERTIES

Each of the millions of organic compounds has unique physical and chemical properties. Thus, we might expect that understanding the relationships between the structure of compounds and their physical properties, such as melting point, boiling point, and solubility, would be a difficult task. Yet we can make reasonable guesses about the physical properties of a compound based on its structure, because organic compounds belong to a small number of classes of substances characterized by their functional groups. These structural units within a molecule are largely responsible for its properties. These properties reflect the attractive intermolecular (between molecules) forces attributable to the functional groups. Intermolecular forces are of three types: **dipole-dipole forces**, **London forces**, and **hydrogen-bonding forces**.

Dipole-Dipole Forces

The bonding electrons in polar covalent bonds are not shared equally, and a bond moment results. However, a molecule may be polar or nonpolar depending on its geometry. For example, tetrachloromethane (carbon tetrachloride, CCl_4) has polar C—Cl bonds, but the tetrahedral arrangement of the four bonds about the central carbon atom causes the individual bond moments to cancel. In contrast, dichloromethane (methylene chloride, CH_2Cl_2) is a polar molecule with a net polarity away from the partially positive carbon atom toward the partially negative chlorine atoms.

Blue arrow: direction of net dipole moment

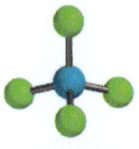

Tetrachloromethane

The bond moments cancel and there is no net polarity

The bond moments do not cancel and there is a net polarity

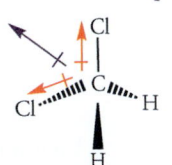

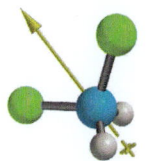

Dichloromethane

Polar molecules have a negative "end" and a positive "end." They tend to associate because the positive end of one molecule attracts the negative end of another molecule. The physical properties of polar molecules reflect this association. An increased association between molecules decreases their vapor pressure, which in turn results in a higher boiling point, because more energy is required to vaporize the molecules. The molecular weights and molecular shapes of acetone and isobutane are similar (Figure 2.1), but acetone boils at a higher temperature than isobutane. Acetone contains a polar carbonyl group, whereas isobutane is a nonpolar molecule. The higher boiling point of acetone results from strong the dipole-dipole interaction of the polar carbonyl group.

Principles of Organic Chemistry. http://dx.doi.org/10.1016/B978-0-12-802444-7.00002-1

Figure 2.1 Physical Properties of Isobutane and Acetone

The physical properties of these two molecules reflect their dipole moments. Isobutane, which has a dipole moment near zero, has a low boiling point of –11.7 °C. Acetone, however, has a large dipole moment of 2.91 D and a boiling point of 56-57 °C.

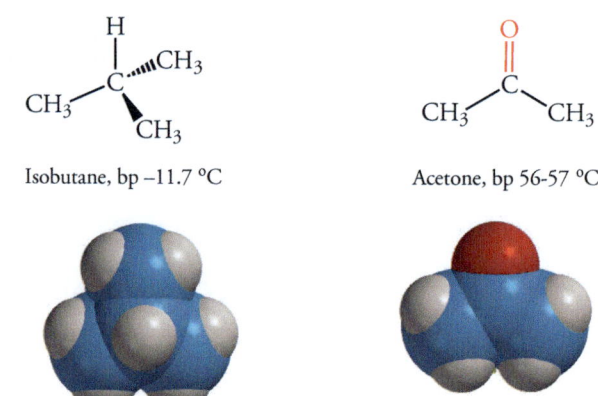

Isobutane, bp –11.7 °C

Acetone, bp 56-57 °C

London Forces

In a nonpolar molecule, the electrons, on average, are distributed uniformly in the molecule. However, the electrons at some instant may be distributed closer to one atom in a molecule or toward one side of a molecule. At that instant, a temporary dipole is present (Figure 2.2). A temporary dipole exerts an influence on nearby molecules; it polarizes neighboring molecules and results in an **induced dipole**. The resultant attractive forces between a temporary dipole and an induced dipole are called **London forces**. The ease with which an electron cloud is distorted by nearby charges or dipoles is called **polarizability**. The attractive forces between the temporary dipoles in otherwise nonpolar molecules are small and have a short lifetime at any given site in the sample. However, the cumulative effect of these attractive forces holds a collection of molecules together in the condensed state. The strength of London forces depends on the number of electrons in a molecule and on the types of atoms containing those electrons. Electrons that are far from atomic nuclei are more easily distorted or polarizable than electrons that are closer to atomic nuclei. For example, the polarizability of the halogens increases in the order F < Cl < Br < I. London forces also depend on the size and shape of a molecule. The boiling point of bromoethane is higher than the boiling point of chloroethane. Because a C—Cl bond is more polar than a C—Br bond, we might have expected the more polar chloroethane to have a higher boiling point than bromoethane. However, polarity isn't the only factor that determines molecular properties. The molecular weights of the two compounds are substantially different, and the electrons of the bromine atom are more polarizable than the electrons of the chlorine atom. Thus, the order of boiling points reflects the polarizability of the molecules and the larger London attractive forces of bromoethane.

	CH_3CH_2Br	CH_3CH_2Cl
Boiling point	38.4°C	12.3°C
Molecular weight	109 amu	64.5 amu

Even when the types of atoms in molecules are the same, London forces differ when the molecular weights are different. For example, the boiling points of pentane and hexane are 36 °C and 69 °C, respectively. These two nonpolar molecules contain the same types of atoms, but the numbers of atoms differ. Hexane is a larger molecule whose chain has more surface area to interact with neighboring molecules. As a result, the London forces are stronger in hexane than in pentane. This increased attraction between molecules decreases the vapor pressure of hexane and its boiling point is higher than the boiling point of pentane.

$CH_3H_2CH_2CH_2CH_3$
pentane
(bp 36 °C)

$CH_3CH_2CH_2CHvCH_2CH_3$
hexane
(bp 69 °C)

Figure 2.2
London Forces

(a) The approach of one nonpolar molecule induces a transient dipole in its neighbor "end-to-end." (b) Several nonpolar molecules interacting side-by-side by London interactions.

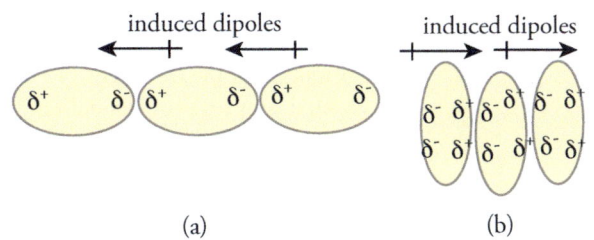

(a)   (b)

London forces also depend on molecular shape. For example, the boiling point of 2,2-dimethyl-propane is lower than that of pentane. 2,2-Dimethylpropane is more spherical, and it therefore has a smaller surface area than the more ellipsoidal-shaped pentane molecule (Figure 2.3). As a result, there is less effective contact between 2,2-dimethylpropane molecules, and the London forces are weaker.

2,2-dimethylpropane, bp 10 °C

CH_3—CH_2—CH_2—CH_2—CH_3

pentane, bp 36 °C

Figure 2.3
London Forces

Molecular models show how the difference in surface contact depends on molecular shapes.

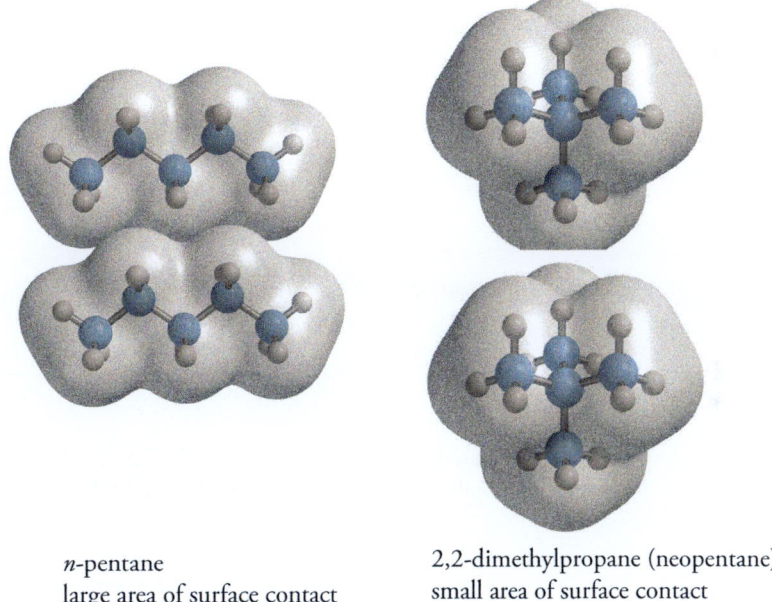

n-pentane
large area of surface contact

2,2-dimethylpropane (neopentane)
small area of surface contact

Hydrogen-Bonding Forces

Compounds that contain hydrogen bonded to oxygen or nitrogen, such as water or ammonia, interact by very strong intermolecular forces. This interaction is called a **hydrogen bond**.

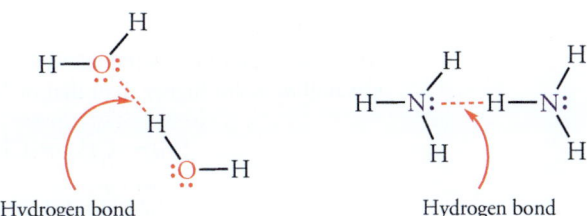

Hydrogen bond   Hydrogen bond

The hydrogen atom in a polar covalent bond to an electronegative element has a partial positive charge. As a result, there is an attraction between the hydrogen atom and the lone pair electrons of another molecule. The O—H or N—H groups in organic compounds can form hydrogen bonds. The physical properties of alcohols and amines are strongly affected by hydrogen bonds. For example, the boiling point of ethanol, an alcohol, is substantially higher than the boiling point of dimethyl ether, which has the same molecular weight.

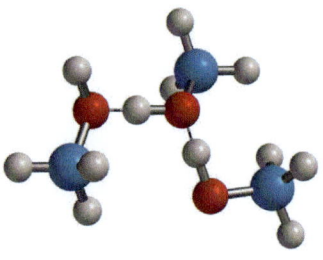

$$CH_3\text{—}CH_2\text{—}OH \qquad\qquad CH_3\text{—}O\text{—}CH_3$$

ethanol, bp 78.5 °C   dimethyl ether, bp -24 °C

Since the numbers of atoms are the same and the shapes of the molecules are similar, the boiling point difference cannot be due to differences in London forces. Both molecules have polar bonds, and the dipole-dipole forces should be similar. The higher boiling point of the ethanol results from the hydrogen bonding between the hydroxyl groups of neighboring molecules, which are much stronger interactions than London forces (Figure 2.4).

Figure 2.4
Hydrogen Bonding in Ethanol

Problem 2.1
Based on the difference in the boiling points of pentane (36 °C) and hexane (69 °C), predict the boiling point of heptane, $CH_3(CH_2)_5CH_3$.

Solution
The boiling points of pentane and hexane are 36 °C and 69 °C, respectively, a difference of 33 °C. The boiling point of heptane should be higher than that of hexane. The effect of the extra methylene group (—CH_2—) on the boiling point could be predicted to be an additional 33 °C, by assuming a linear relationship between molecular weight and boiling point. The predicted boiling point would be 102 °C. The actual boiling point is 98 °C.

Problem 2.2
The boiling points of CCl_4 and $CHCl_3$ are 77 °C and 62 °C, respectively. Which compound is more polar? Is the polarity consistent with the boiling points? Why or why not?

Problem 2.3
The boiling point of 1,2-ethanediol (ethylene glycol), which is used as antifreeze, is 190 °C. Why is the boiling point higher than that of 1-propanol (97 °C)?

$$HO\text{—}CH_2\text{—}CH_2\text{—}OH \qquad\qquad CH_3\text{—}CH_2\text{—}OH$$

1,2-ethanediol, bp 190 °C   1-propanol, bp 97 °C

Solution

Both molecules have similar molecular weights and should have comparable London forces. However, 1,2-ethanediol has two hydroxyl groups per molecule, compared to only one per molecule in 1-propanol. As a consequence, liquid 1,2-ethanediol can form twice as many hydrogen bonds. The increased number of hydrogen bonds decreases the vapor pressure of 1,2-ethanediol, which leads to a higher boiling point.

Problem 2.4

Explain why the boiling points of ethanethiol and dimethyl sulfide are very similar.

$$CH_3-CH_2-SH \qquad\qquad CH_3-O-CH_3$$

ethanethiol, bp 35 °C   dimethyl sullfide, bp 37 °C

Solubility

A maxim of the chemistry laboratory is that "like dissolves like." This generalization is reasonable because molecules of solute that are similar to molecules of solvent should interact by similar intermolecular attractive forces. Carbon tetrachloride, CCl_4, a nonpolar substance, does not dissolve ionic compounds such as sodium chloride. However, this nonpolar compound is a good solvent for nonpolar compounds such as fats and waxes. Water, which is quite polar, is a good solvent for ionic compounds and substances that dissolve to produce ions in water. Water dissolves a limited number of low molecular weight organic compounds if they are sufficiently polar or can form hydrogen bonds with water.

Liquids that dissolve in each other in all proportions are said to be **miscible**. Liquids that do not dissolve in each other are **immiscible**. Immiscible liquids form separate layers in a container. For example, ethyl alcohol is miscible with water, but carbon tetrachloride and water are immiscible. The solubility of ethyl alcohol in water is explained by its structure.

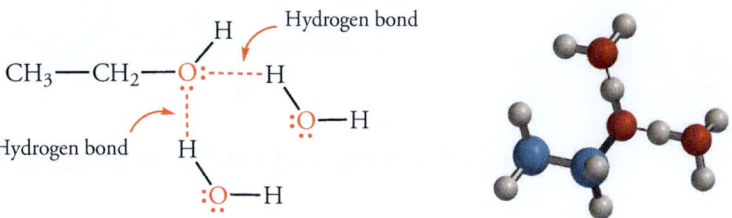

Ethyl alcohol, like water, has an —OH group. Ethyl alcohol is polar, as is water. The nonbonding electron pairs on the oxygen atom in ethyl alcohol and the hydroxyl hydrogen atom form hydrogen bonds with water, and thus make it soluble.

Water-Soluble and Fat-Soluble Vitamins

The different solubilities of vitamins, characterized as water soluble and fat soluble, illustrate the maxim that "like dissolves like." Water-soluble vitamins have large numbers of functional groups that can hydrogen bond with water. Water-insoluble vitamins are essentially nonpolar structures that are soluble in the nonpolar fatty tissue of the body. Water-soluble vitamins are not stored in the body and should be part of one's daily diet. Unneeded water-soluble vitamins are excreted. Fat-soluble vitamins are stored by the body. If excessive quantities are consumed in vitamin supplements, illness can result. The condition is known as hypervitaminosis.

The structures of several water-soluble vitamins are shown below. Note that the relatively small vitamin C molecule has a high proportion of —OH groups that can form hydrogen bonds to water. In contrast, vitamin A is not "like" water. It contains an —OH group, but that single functional group is insufficient to allow the relatively large nonpolar portion of the molecule to be accommodated within water. Vitamin B_6 and riboflavin contain not only —OH groups but also nitrogen-containing functional groups that can also hydrogen bond to water. Vitamins E and D_3 are nonpolar compounds that are not soluble in water (Figure 2.5).

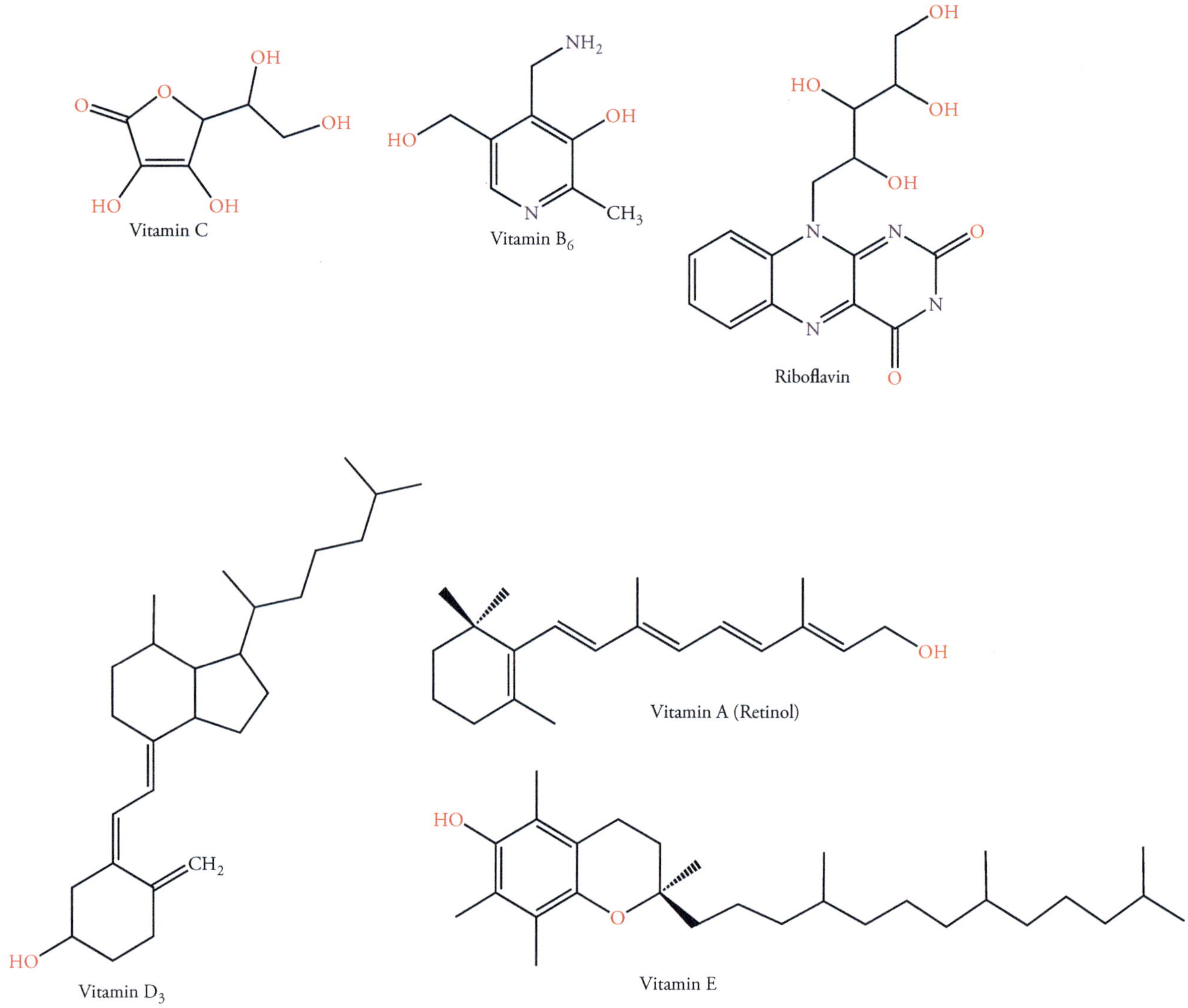

Figure 2.5 Water-and Fat-Soluble Vitamins

2.2 CHEMICAL REACTIONS

The number of known and potential reactions among the dozens of functional groups in the millions of organic compounds is astronomically large. However, we can understand these myriad reactions by learning the fundamental concepts that underlie all organic chemical reactions. In other words, we can discern patterns of chemical behavior that unify many facts into a few classes of chemical reactions.

We will review acid-base and oxidation-reduction reactions in Sections 2.3 and 2.4 and illustrate how these concepts apply to organic chemical reactions. We will briefly discuss several other classes of organic reactions in Section 2.5.

All chemical reactions are reversible to some degree, and some reactions result in an equilibrium mixture containing substantial amounts of reactants as well as products. It is important to determine the conditions that control chemical equilibria. For example, industrial processes must convert as much reactant to product as possible. Not only is the inefficient conversion of chemicals costly, but the unwanted material must be removed to purify the product. Impure materials cannot be tolerated for many products, especially those for human consumption.

The study of the kinetics of chemical reactions is also of concern in industry, where it is important to understand how to form a desired chemical product rapidly and in preference to other products. A reaction may be too slow to be economically practical, or a reaction may be so fast that it is dangerous and difficult to control. In biological systems, virtually all reactions produce only a single product, and the reactions are very fast.

The study of the kinetics of a reaction helps us determine the **mechanism** of the process. A mechanism details the order of bond cleavage and bond formation that occurs during the reaction. This information establishes general guidelines, which allow us to extrapolate a few observations on selected reactions to many other reactions.

2.3 ACID-BASE REACTIONS

According to the *Brønsted-Lowry* theory, an **acid** is a proton (H^+) donor; a **base** is a proton acceptor. For example, when gaseous hydrogen chloride dissolves in water, virtually all of the HCl molecules transfer a proton to water, and a solution of hydronium ions and chloride ions results.

Brønsted-Lowry Acids and Bases

A Brønsted-Lowry acid is a proton donor; a Brønsted-Lowry base is a proton acceptor. When an acid transfers a proton to a base, another base and acid are produced. The acid loses a proton and becomes a *conjugate base.* When a base accepts a proton, it becomes a *conjugate acid.* For example, consider the example shown below for the reaction that occurs when the acid HCl transfers a proton to water. The base, water, becomes the conjugate acid, hydronium ion; chloride ion is the conjugate base. There are no HCl molecules in an aqueous solution.

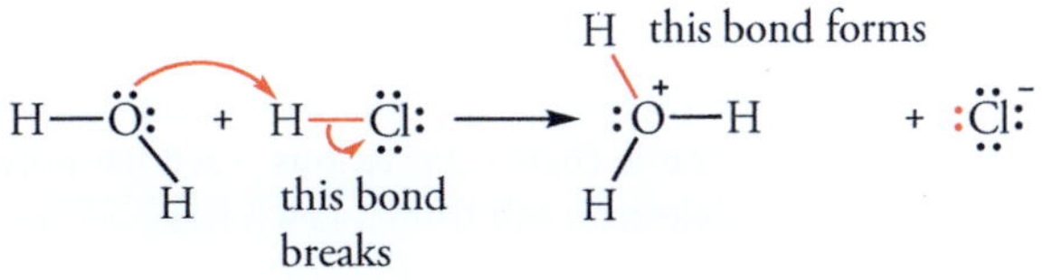

This reaction is illustrated by curved arrows. Electrons are pictured as flowing from the start of the arrow toward the arrowhead. The nonbonding pair of electrons of the oxygen atom forms a bond to the hydrogen atom, and the bonding pair of electrons in HCl is transferred to the chlorine atom.

The hydroxide ion, which exists as an ion in compounds such as NaOH, KOH, and $Ca(OH)_2$, is a base because it has nonbonding electron pairs that can accept a proton from an acid such as a hydronium ion. Ammonia is also a base because it has a nonbonding pair of electrons on its nitrogen atom, which can form a bond to a hydrogen atom of H_3O^+. A curved arrow shows the movement of the pair of electrons from the nitrogen atom toward a hydrogen atom.

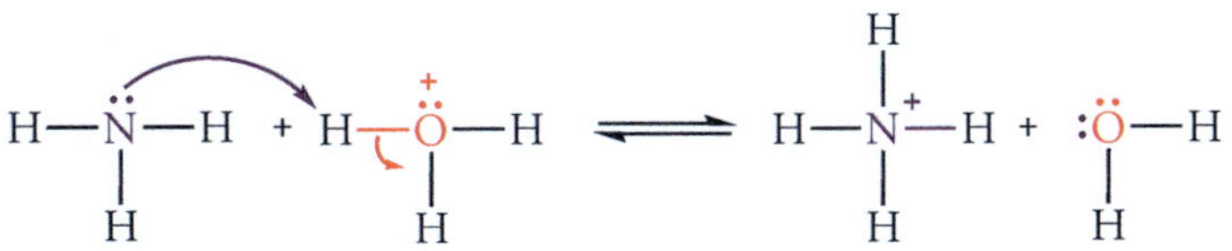

Organic acids and bases behave similarly. Carboxylic acids contain a carboxyl group, which can donate a proton to a base such as water. The curved arrow formalism that depicts the movement of electrons shows how the nonbonding electron pair of the water molecule forms a bond with the hydrogen atom of the carboxyl group.

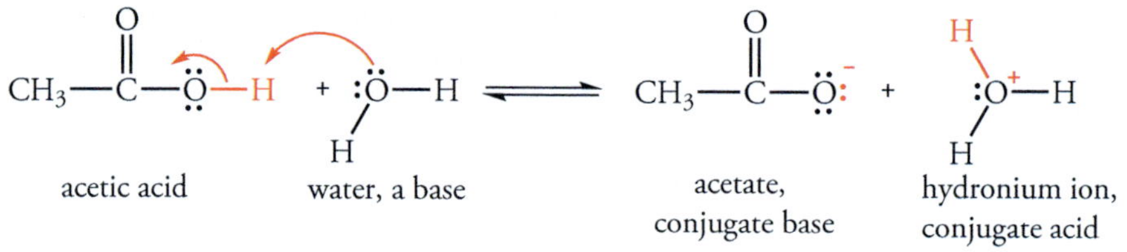

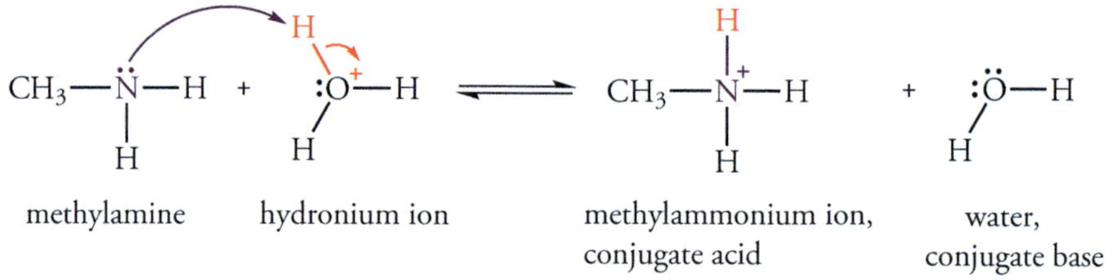

acetic acid   water, a base   acetate, conjugate base   hydronium ion, conjugate acid

methylamine   hydronium ion   methylammonium ion, conjugate acid   water, conjugate base

When an acid transfers a proton to a base, another base and acid are produced. The acid loses a proton and becomes a **conjugate base**. For example, the conjugate base of acetic acid is acetate ion. When a base accepts a proton, the substance formed is a **conjugate acid**. Thus, the conjugate acid of methylamine is the methylammonium ion.

Lewis Acids and Bases

Some chemical reactions that do not occur with proton transfer are also regarded as acid-base reactions. These reactions can be explained in terms of Lewis acids and Lewis bases, which are defined based on electron pairs. A Lewis acid is a substance that accepts an electron pair; a Lewis base is a substance that donates an electron pair. Thus HCl, which is an acid in the Brønsted-Lowry sense, is also a Lewis acid because it contains a proton that can "accept" an electron pair. Similarly, ammonia is a Lewis base as well as a Brønsted-Lowry base because it can donate an electron pair. However, the Lewis classification of acids and bases is more extensive because it is not restricted to protons. Boron trifluoride (BF_3) and aluminum trichloride ($AlCl_3$) are two common Lewis acids encountered in organic chemical reactions. Each has only six electrons in its valence shell, and each can thus accept an electron pair from a Lewis base.

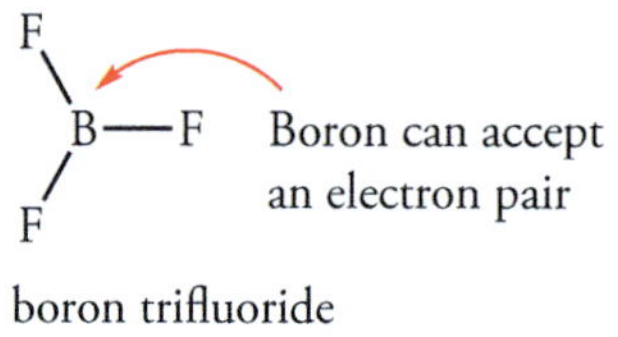

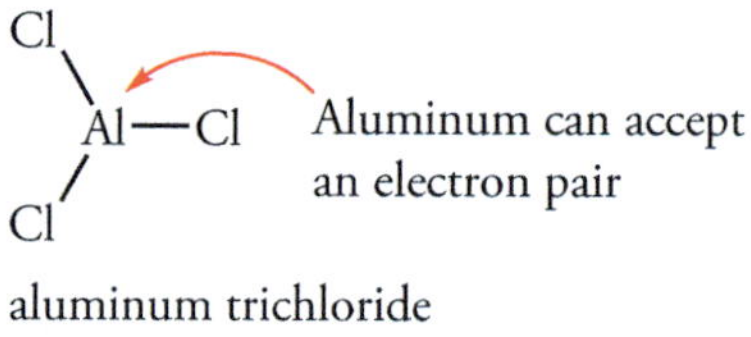

boron trifluoride   aluminum trichloride

Other Lewis acids include transition metal compounds, such as $FeBr_3$, that react by accepting a pair of electrons. For example, $FeBr_3$ reacts with molecular bromine to accept a bromide ion via a pair of electrons. In this reaction, $FeBr_3$ behaves as a Lewis acid and bromine behaves as a Lewis base.

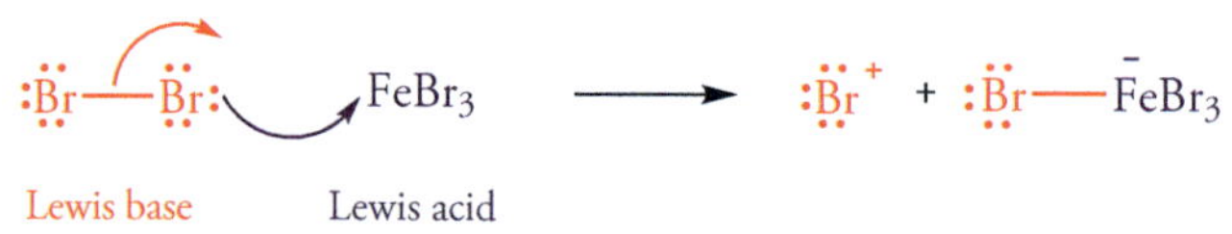

Lewis base   Lewis acid

Many organic compounds that contain oxygen and nitrogen atoms can act as Lewis bases because these atoms have nonbonding electrons that can react with Lewis acids. For example, ethers react with boron trifluoride to give a product with a bond between boron and oxygen.

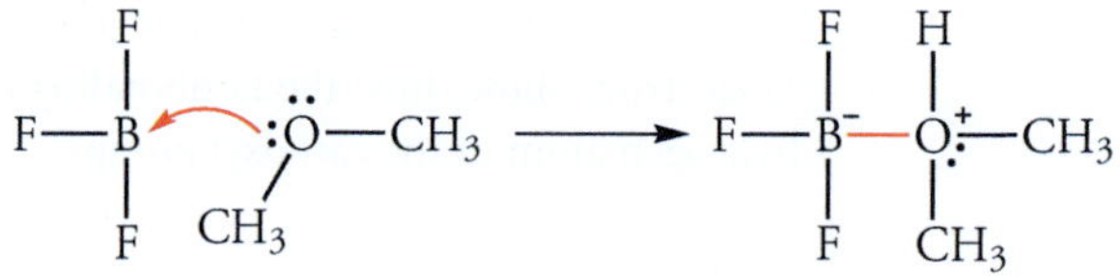

Problem 2.5

Assume that ethanol behaves as a Brønsted-Lowry acid in a reaction. What is the conjugate base of ethanol?

$$CH_3-CH_2-OH$$

Solution

Loss of a proton (H^+) from an electrically neutral acid results in a conjugate base with a negative charge. The electron pair of the O—H bond remains with the oxygen atom.

$$CH_3-CH_2-\overset{..}{\underset{..}{O}}-H \quad :B \longrightarrow CH_3-CH_2-\overset{..}{\underset{..}{O}}:^- + H-B^+$$

Problem 2.6

Consider the reaction of Br^- with ethene to give a charged intermediate called a carbocation. Which reactant is the Lewis acid, which one is the Lewis base?

$$\begin{matrix} H & & H \\ & C=C & \\ H & & H \end{matrix} \quad + \quad :\overset{..}{\underset{..}{Br}}^+ \quad \longrightarrow \quad \begin{matrix} H & & H \\ H-C & - & C^+ \\ :Br & & H \end{matrix}$$

a carbocation

2.4 OXIDATION-REDUCTION REACTIONS

Oxidation is the loss of electrons by a substance or an increase in its oxidation number. **Reduction** is the gain of electrons by a substance or a decrease in its oxidation number. From a slightly different point of view, it follows that when a substance is reduced, it gains the electrons from a substance that becomes oxidized.

The relationship between oxidation and reduction is emphasized further in the terms oxidizing agent and reducing agent. In an oxidation-reduction reaction, the substance that is reduced is the **oxidizing agent** because, by gaining electrons, it causes oxidation of another substance. The substance that is oxidized is called the **reducing agent** because, by losing its electrons, it causes the reduction of another substance.

In organic chemistry, oxidation numbers are not as easily assigned as in inorganic chemistry. However, we can decide on the change in the oxidation state of a compound in many chemical reactions by accounting for the number of hydrogen atoms or oxygen atoms gained or lost. The oxidation state of a molecule increases (oxidation) if its hydrogen content decreases or its oxygen content increases. Conversely, the oxidation state of a molecule decreases (reduction) if its hydrogen content increases or its oxygen content decreases. For example, the reaction of methanol (CH_3OH) to produce methanal (formaldehyde, CH_2O) is an oxidation because methanol loses two hydrogen atoms. Further reaction of methanal to produce methanoic acid (formic acid, HCO_2H) occurs with an increase in the oxygen content and is also an oxidation process.

$$\begin{matrix} H & \\ | & \\ H-C-OH & \xrightarrow{[O]} \\ | & \\ H & \end{matrix} \quad \begin{matrix} O \\ \parallel \\ H-C \\ \backslash \\ H \end{matrix} \quad \xrightarrow{[O]} \quad \begin{matrix} O \\ \parallel \\ H-C \\ \backslash \\ OH \end{matrix}$$

methanol   methanal   methanoic acid
(formaldehyde)   (formic acid)

The symbol [O] represents an unspecified oxidizing agent. Note also that the equations are not balanced. The focus in organic chemistry is on the conversion of one organic compound to another. Oxidizing agents such as potassium dichromate, which might be used in an oxidation reaction, are seldom balanced in writing an equation but instead are written above the reaction arrow. The following conversion of an alcohol into an acid is an oxidation, and the substance above the arrow is the oxidizing agent.

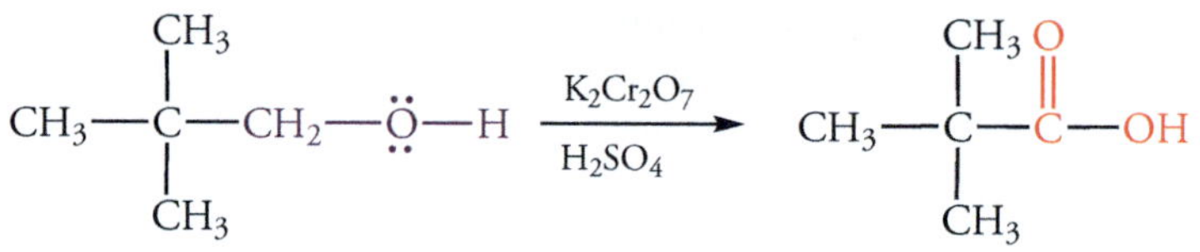

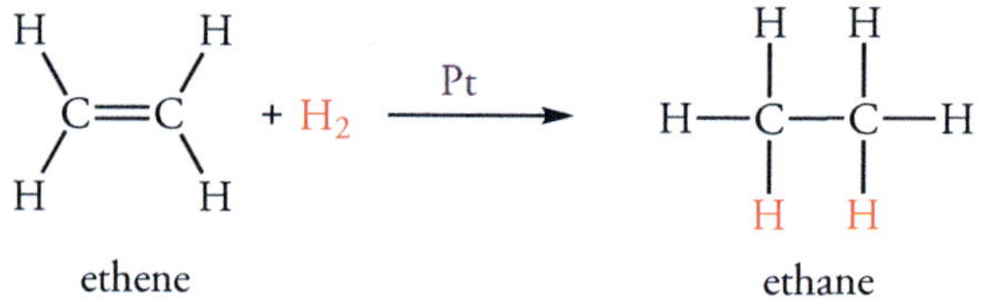

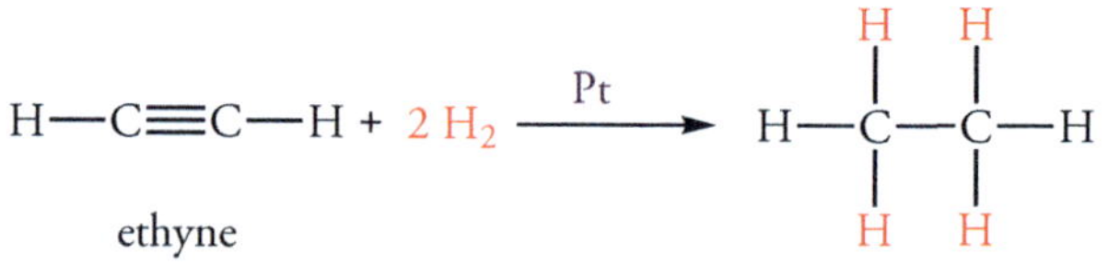

$$CH_3-\underset{\underset{CH_3}{|}}{\overset{\overset{CH_3}{|}}{C}}-CH_2-\ddot{\underset{\cdot\cdot}{O}}-H \xrightarrow[\text{H}_2\text{SO}_4]{\text{K}_2\text{Cr}_2\text{O}_7} CH_3-\underset{\underset{CH_3}{|}}{\overset{\overset{CH_3}{|}}{C}}-\overset{\overset{O}{\|}}{C}-OH$$

$$\underset{\text{ethene}}{\underset{H}{\overset{H}{>}}C=C\underset{H}{\overset{H}{<}}} + H_2 \xrightarrow{\text{Pt}} \underset{\text{ethane}}{H-\overset{\overset{H}{|}}{\underset{\underset{H}{|}}{C}}-\overset{\overset{H}{|}}{\underset{\underset{H}{|}}{C}}-H}$$

$$\underset{\text{ethyne}}{H-C\equiv C-H} + 2\,H_2 \xrightarrow{\text{Pt}} H-\overset{\overset{H}{|}}{\underset{\underset{H}{|}}{C}}-\overset{\overset{H}{|}}{\underset{\underset{H}{|}}{C}}-H$$

The simultaneous increase or decrease of two hydrogen atoms and one oxygen atom in a reactant is neither reduction nor oxidation. Thus, the conversion of ethene to ethanol is not an oxidation-reduction reaction.

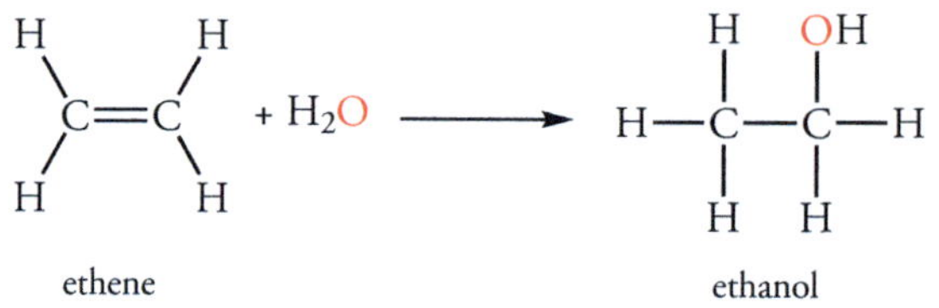

$$\underset{\text{ethene}}{\underset{H}{\overset{H}{>}}C=C\underset{H}{\overset{H}{<}}} + H_2O \longrightarrow \underset{\text{ethanol}}{H-\overset{\overset{H}{|}}{\underset{\underset{H}{|}}{C}}-\overset{\overset{OH}{|}}{\underset{\underset{H}{|}}{C}}-H}$$

Problem 2.7

Ethylene oxide is used to sterilize medical equipment that is temperature sensitive and cannot be heated in an autoclave. It is produced from ethylene by the following process. Classify the type of reaction. Is an oxidizing or reducing agent required?

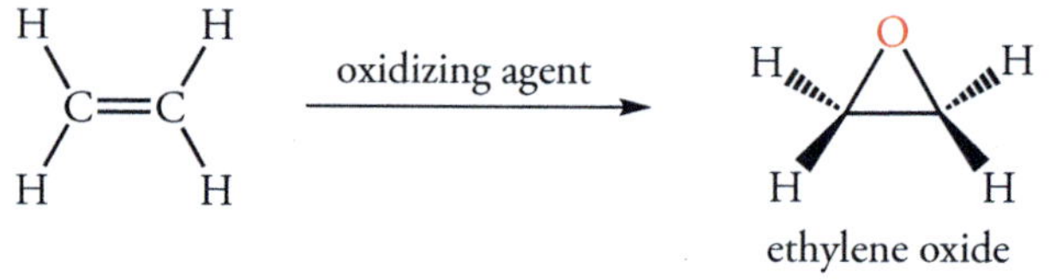

$$\underset{H}{\overset{H}{>}}C=C\underset{H}{\overset{H}{<}} \xrightarrow{\text{oxidizing agent}} \underset{\text{ethylene oxide}}{\text{ethylene oxide}}$$

Solution

The oxygen content of the ethylene increases. Thus the reaction is an oxidation. An oxidizing agent is required for the reaction.

Problem 2.8

Is the following process an oxidation or a reduction reaction?

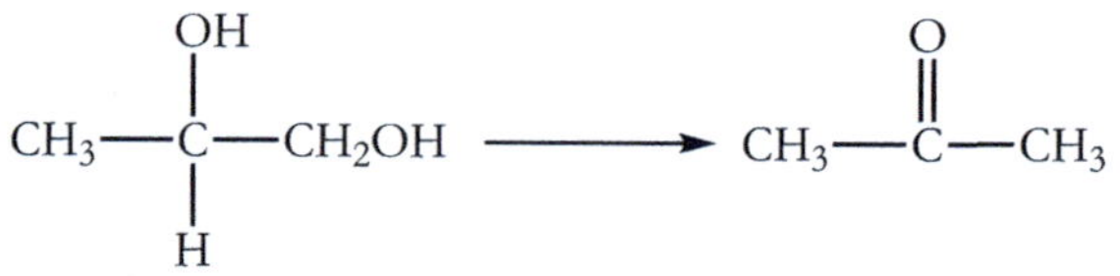

$$CH_3-\underset{\underset{H}{|}}{\overset{\overset{OH}{|}}{C}}-CH_2OH \longrightarrow CH_3-\overset{\overset{O}{\|}}{C}-CH_3$$

Biochemical Redox Reactions

Metabolic reactions provide energy in multiple-step processes in which metabolites are oxidized. Biosynthetic reactions, which build the necessary compounds to maintain organisms, are often reduction reactions. Thus, organisms require both oxidizing and reducing agents. Furthermore, because so many diverse reactions occur in oxidative degradation and reductive biosynthesis, these oxidizing and reducing agents must have a wide range of reactivity. These reactions are catalyzed by enzymes. These redox reactions require a substance called a **coenzyme** that serves as the oxidizing or reducing agent. Two of these compounds, **nicotinamide adenine dinucleotide** and **flavin adenine dinucleotide**, are represented by the shorthand **NAD⁺** and **FAD**, respectively. Each of these coenzymes oxidizes biological molecules by removing two hydrogen atoms from covalent bonds. The net result of these reactions is the formation of the reduced forms of NAD and FAD; **NADH** and **FADH$_2$**, respectively.

$$NAD^+ + 2H^+ + 2e^- \longrightarrow NADH + H^+$$

oxidized form · · · · · · · · · · · · · · · reduced form

$$FAD + 2H^+ + 2e^- \longrightarrow FADH_2$$

oxidized form · · · · · · · · · · · · · · · reduced form

The reduced form of each coenzyme can serve as a reducing agent and be oxidized to regenerate the oxidized form. NAD⁺ and FAD oxidize different classes of compounds. NAD⁺ removes hydrogen atoms from a C—H and an O—H bond, whereas FAD removes hydrogen atoms from two C—H bonds. For example, NAD⁺ oxidizes malic acid to oxaloacetic acid in one of the steps of the citric acid cycle, which is ultimately responsible for the oxidation of metabolites to carbon dioxide and water.

malic acid · · · · · · · · · · · · · · · · · · · oxaloacetic acid

FAD oxidizes fatty acids such as stearic acid, which is derived from the fats that provide mammals with long-term sources of energy.

stearic acid · · · · · · · · · · · · · · · · · · · oleic acid

Foreign compounds (xenobiotics) and many common drugs are eliminated from the body by oxidative reactions. Water-soluble substances are easily excreted, but most organic compounds are nonpolar and are **lipid soluble** (i.e., they dissolve in the fatty components of cells). If **lipophilic** (lipid loving) xenobiotics or drugs were not eliminated, they would accumulate, and an organism would eventually die. Organisms ordinarily transform lipophilic substances into more polar water-soluble products that can be excreted. The liver is the most important organ for the oxidation of xenobiotics and drugs. The oxidation of compounds, represented as R—H, in the liver requires molecular oxygen and the coenzyme nicotinamide adenosine dinucleotide phosphate, NADPH. One of the oxygen atoms is incorporated in the reactant and the other oxygen atom is incorporated in water.

$$R\text{—}H + NADPH + O_2 + H^+ \xrightarrow{\text{Cyt P450}} NADP^+ + R\text{—}OH + H_2O$$

reduced form · · · · · · · · · · · · · · · oxidized form

The enzyme responsible for catalyzing the reaction is cytochrome P-450. It contains iron surrounded by heme, a complex nitrogen-containing compound, and the amino acid lysine, which is part of the protein structure. Substrates bind the enzyme at a site within the protein—the active site—where they are oxidized with the aid of the coenzyme. The oxidation of the oral hypoglycemic drug tolbutamide (Orenase) to an alcohol is one example of the metabolism of drugs by the liver.

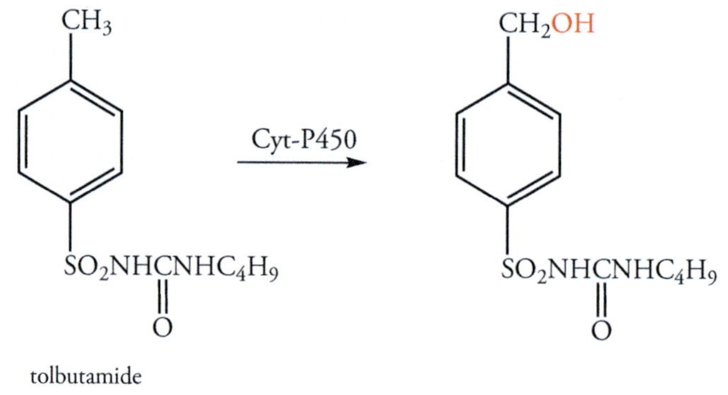

tolbutamide

2.5 CLASSIFICATION OF ORGANIC REACTIONS

In the preceding two sections we reviewed two classes of reactions that are discussed in general chemistry courses. Now we will look at several common examples of organic reactions. We will discuss these reactions in greater detail in subsequent chapters.

Addition reactions occur when two reactants combine to give a single product. An example of an addition reaction is the reaction of ethene with HBr to form bromoethane. The hydrogen and bromine atoms are added to adjacent atoms, a common characteristic of addition reactions.

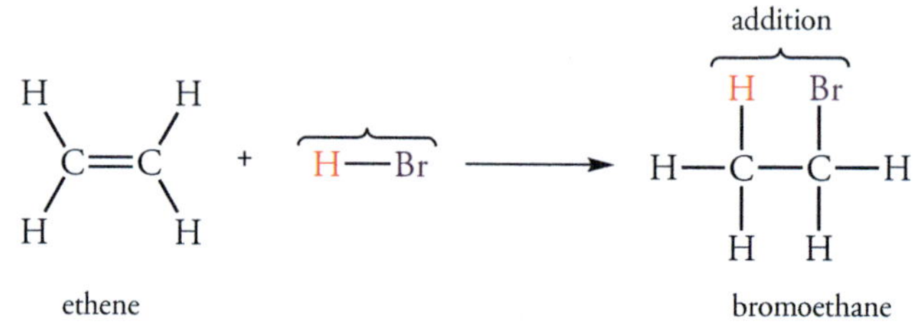

In an **elimination reaction** a single compound splits apart to give two compounds. Most elimination reactions form a product with a double bond containing most of the atoms in the reactant and a second smaller molecule such as H_2O or HCl. The atoms eliminated to form the smaller molecule are usually located on adjacent carbon atoms in the reactant. For example, 2-propanol reacts with concentrated sulfuric acid to produce propene; water is eliminated in this reaction.

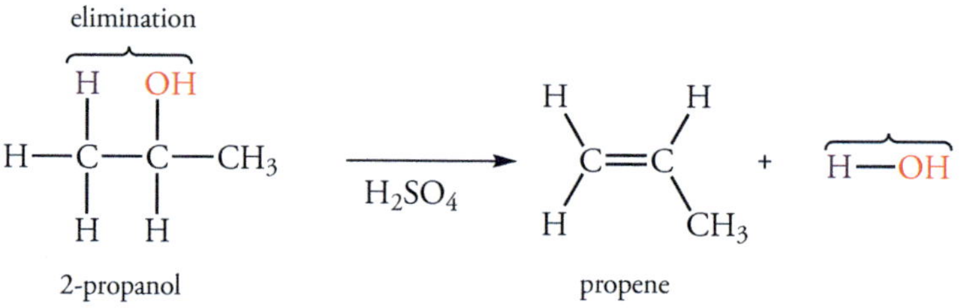

In a **substitution reaction** one atom or group of atoms (Y) replaces a second atom or group of atoms (X).

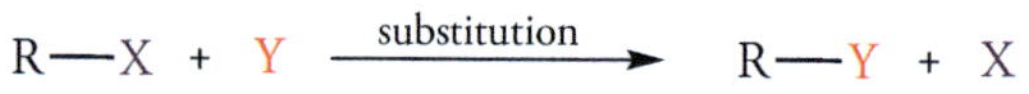

$$R\!-\!X \;+\; Y \xrightarrow{\text{substitution}} R\!-\!Y \;+\; X$$

An example of a substitution reaction is the conversion of bromomethane into methanol.

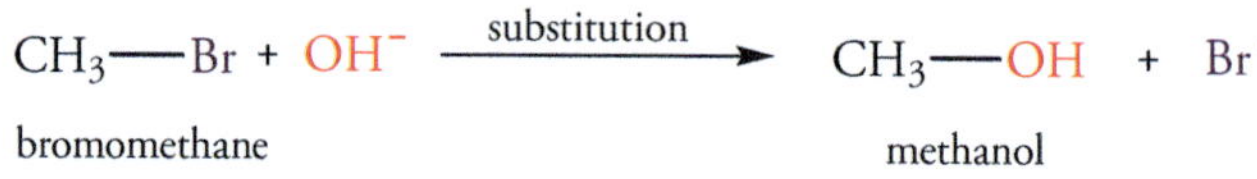

$$CH_3\!-\!Br \;+\; OH^- \xrightarrow{\text{substitution}} CH_3\!-\!OH \;+\; Br^-$$

bromomethane   methanol

In a **hydrolysis reaction** (Greek *hydro*, "water" + *lysis*, "splitting") water splits a large reactant molecule into two smaller product molecules. One product molecule is bonded to a hydrogen atom derived from water. The other product is bonded to an —OH group derived from water. The hydrolysis of an ester to produce a carboxylic acid and an alcohol is an example of this process.

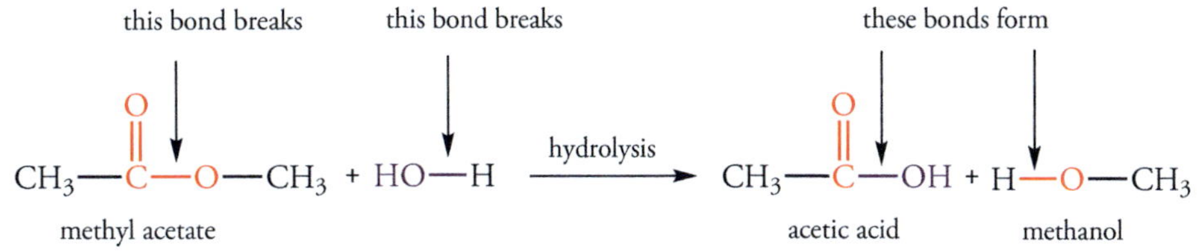

this bond breaks   this bond breaks   these bonds form

$$CH_3-\overset{\overset{\displaystyle O}{\|}}{C}-O-CH_3 \;+\; HO-H \quad \xrightarrow{\text{hydrolysis}} \quad CH_3-\overset{\overset{\displaystyle O}{\|}}{C}-OH \;+\; H-O-CH_3$$

methyl acetate   acetic acid   methanol

In a **condensation reaction**, two reactants combine to form one larger product with the simultaneous formation of a second, smaller product such as water. When the second product is water, the reaction is the reverse of a hydrolysis reaction. The formation of an ester from an alcohol and a carboxylic acid is an example of this process.

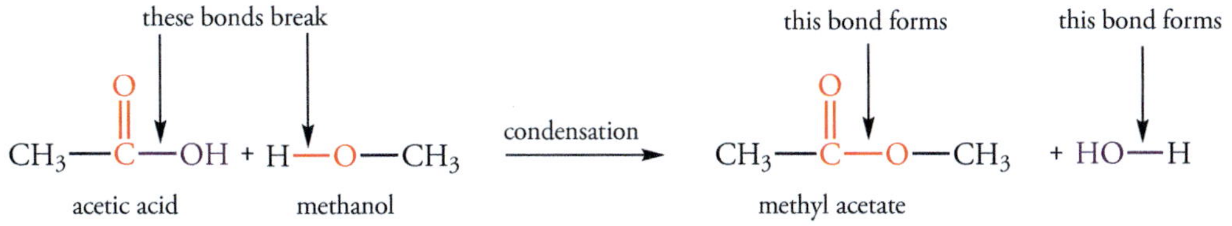

these bonds break   this bond forms   this bond forms

$$CH_3-\overset{\overset{\displaystyle O}{\|}}{C}-OH \;+\; H-O-CH_3 \quad \xrightarrow{\text{condensation}} \quad CH_3-\overset{\overset{\displaystyle O}{\|}}{C}-O-CH_3 \;+\; HO-H$$

acetic acid   methanol   methyl acetate

In a **rearrangement reaction** the bonds within a single reactant "reorganize" to give an isomeric product. One example is a rearrangement in which the location of a double bond changes to give an isomer. Note that the bromine atom in the reaction shown below also is relocated in the product. We will only discuss a few rearrangement reactions in this text.

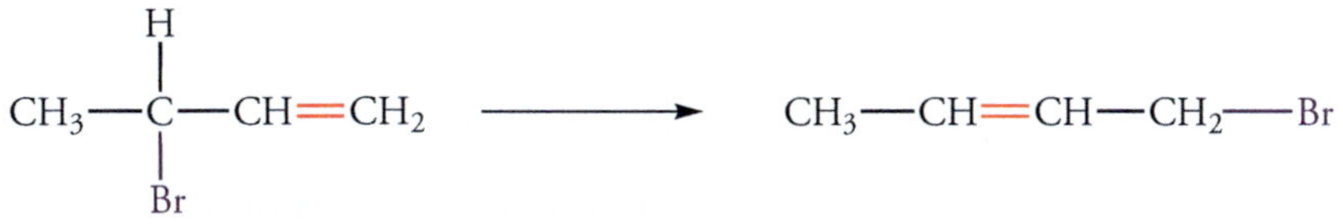

$$CH_3-\overset{\overset{\displaystyle H}{|}}{\underset{\underset{\displaystyle Br}{|}}{C}}-CH=CH_2 \quad \longrightarrow \quad CH_3-CH=CH-CH_2-Br$$

Problem 2.9
Classify the following reaction.

$$CH_3-\overset{\overset{\displaystyle CH_2Br}{|}}{\underset{\underset{\displaystyle H}{|}}{C}}-CH_2Br \;+\; Zn \quad \longrightarrow \quad CH_3-CH=CH_2 \;+\; ZnBr_2$$

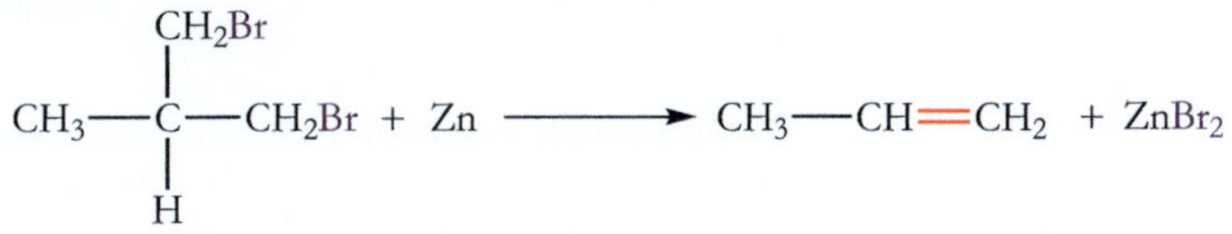

Solution
This is an elimination reaction: the two bromine atoms located on adjacent carbon atoms are transferred to Zn. The organic product contains most of the atoms of the reactant, and the by-product is $ZnBr_2$.

Problem 2.10
Classify the following reaction.

$$\underset{\underset{\displaystyle H}{\diagup}}{\overset{\overset{\displaystyle H}{\diagdown}}{C}}=\underset{\underset{\displaystyle H}{\diagdown}}{\overset{\overset{\displaystyle OH}{\diagup}}{C}} \quad \longrightarrow \quad H-\overset{\overset{\displaystyle H}{|}}{\underset{\underset{\displaystyle H}{|}}{C}}-\overset{\overset{\displaystyle O}{\diagup\diagdown}}{C}-H$$

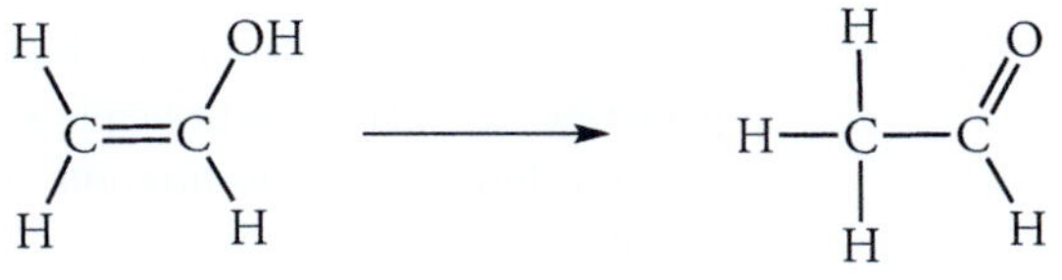

2.6 CHEMICAL EQUILIBRIUM AND EQUILIBRIUM CONSTANTS

Chemical reactions do not proceed in only one direction. As a reaction occurs, product molecules can revert to reactant molecules. Thus, two opposing reactions occur. When the rate of product formation is equal to the rate of reactant formation, an equilibrium is established.

Consider the general equation for a reaction at equilibrium, where A and B are reactants, X and Y are products, and m, n, p, and q are coefficients.

$$mA + nB \underset{\text{reverse reaction}}{\overset{\text{forward reaction}}{\rightleftharpoons}} pX + qY$$

$$K_{equilibrium} = \frac{[X]^p [Y]^q}{[A]^m [B]^n}$$

Consider the equilibrium constant expression for the addition reaction of gaseous ethylene with gaseous hydrogen bromide.

$$CH_2{=}CH_2 + HBr \overset{K_{eq}}{\rightleftharpoons} CH_3CH_2Br$$
$$\text{ethene}$$

$$K_{equilibrium} = \frac{[CH_3CH_2Br]}{[CH_2{=}CH_2][HBr]} = 10^8$$

Because the equilibrium constant is very large, the reaction "goes to completion." That is, for all practical purposes, no reactant remains at equilibrium.

Now let's consider the condensation reaction of ethanoic acid and ethanol to produce ethyl ethanoate and the related equilibrium constant expression.

$$K_{equilibrium} = \frac{[CH_3CO_2CH_2CH_3] [H_2O]}{[CH_3CO_2H] [CH_3CH_2OH]} = 4.0$$

In this reaction, significant concentrations of reactants are present at equilibrium. Thus, the product yield is less than 100% based on the balanced equation.

The position of a chemical equilibrium and the value of the equilibrium constant are not affected by catalysts. A catalyst (see Section 2.10) increases the rates of the forward and reverse reactions equally, and K does not change. In the reaction of ethanol and ethanoic acid, the reaction is acid catalyzed. Thus, the equilibrium is established in a shorter time period at the same temperature in the presence of an acid such as HCl.

Le Châtelier's Principle

Most of the reactions we will discuss have large equilibrium constants. For those that do not, reaction conditions are selected to "force" the reaction in the direction favoring product based on **Le Châtelier's principle**. Le Châtelier's principle states that a change in the conditions of a chemical equilibrium causes a shift in the concentration of reactants and products to result in a new equilibrium system. If additional reactant is added to a chemical system at equilibrium, the concentrations of both reactants and products change to establish a new equilibrium system, but the equilibrium constant is unchanged. After adding reactant, the total concentration of reactant is initially increased, but then decreases to establish a new equilibrium. As a result, the concentration of the products increases. In short, the change imposed on the system by adding reactants is offset when added reactants are converted to product. If a product is removed from a chemical system at equilibrium, the forward reaction occurs to give more product. Regardless of the condition imposed on the system at equilibrium, the concentrations change to maintain the same value of the equilibrium constant. Consider the equilibrium in the formation of ethyl ethanoate.

If water is removed from the system by some means, the equilibrium is disturbed and the equilibrium position of the reaction would shift to the right to produce more water and ethyl acetate. If the amount of alcohol is increased, a larger amount of the carboxylic acid will be converted into product.

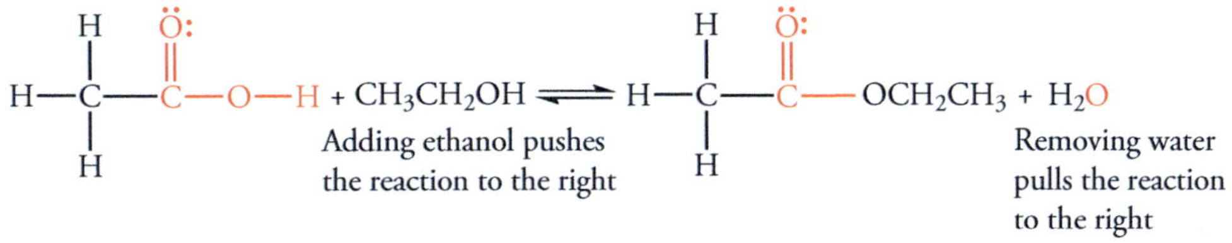

Adding ethanol pushes the reaction to the right

Removing water pulls the reaction to the right

2.7 EQUILIBRIA IN ACID-BASE REACTIONS

Water is the reference solvent commonly used to compare the strengths of acids or bases. The strengths of acids are measured by their tendencies to transfer protons to water.

$$HA + H_2O \xrightleftharpoons{K_{eq}} A^- + H_3O^+$$

A quantitative measure of the acidity of an acid with the general formula HA is given by the equilibrium constant for ionization, which is obtained from the equation for ionization.

$$K_{equilibrium} = \frac{[H_3O^+][A^-]}{[HA][H_2O]}$$

The concentration of water, about 55 M, is so large compared to that of the other components of the equilibrium system that its value changes very little when the acid HA is added. Therefore, the concentration of water is included in the acid ionization constant K_a.

$$K_a = K_{eq}[H_2O] = \frac{[H_3O^+][A^-]}{[HA]}$$

Table 2.1
K_a and pK_a Values of Common Acids

Acid	K_a	pK_a
HBr	10^9	−9
HCl	10^7	−7
H_2SO_4	10^5	−5
HNO_3	10^1	−1
HF	6×10^{-4}	3.2
CH_3CO_2H	2×10^{-5}	4.7
$(CF_3)_3COH$	2×10^{-5}	4.7
CH_3CH_2SH	3×10^{-11}	10.6
CF_3CH_2OH	4×10^{-13}	12.4
CH_3OH	3×10^{-16}	15.5
$(CH_3)_3COH$	1×10^{-18}	18
CCl_3H	10^{-25}	25
$HC{\equiv}CH$	10^{-25}	25
NH_3	10^{-36}	36
$CH_2{=}CH_2$	10^{-44}	44
CH_4	10^{-49}	49

Acids with $K_a > 10$ are strong acids. Most organic acids have $K_a < 10^{-4}$, and are weak acids. Acid dissociation constants are often expressed as pK_a values.

$$pK_a = -\log K_a.$$

Note that *pK_a values increase as K_a decreases.* Table 2.1 lists the acid ionization constants of some common acids.

Weak acids do not completely transfer their protons to water, and few ions are produced. An example of a weak acid is acetic acid, which ionizes in water to give acetate ions and hydronium ions.

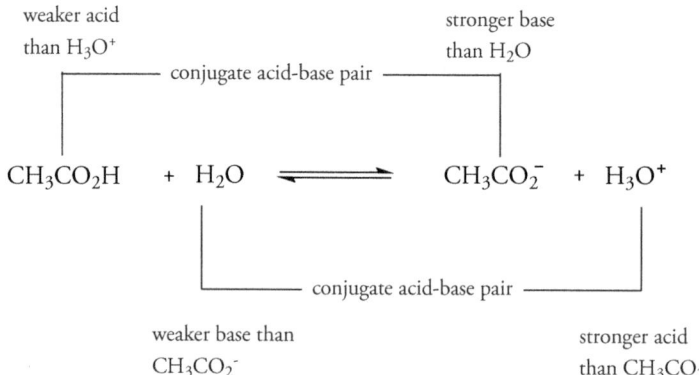

The equilibria between acids and bases and their conjugate bases and acids can be viewed as a "contest" for protons. The equilibrium position favors the side containing the weaker acid and weaker base. Acetic acid is a weaker acid than H_3O^+, and $CH_3CO_2^-$ is a stronger base than H_2O. Note that a strong acid, with its great tendency to lose protons, is paired with a weak conjugate base that has a low affinity for protons. Thus, as the tendency of an acid to lose a proton increases, the tendency of its conjugate base to accept a proton decreases.

There is a close relationship between acidity and basicity. When an acid dissociates, a base is formed that can react in the reverse direction by accepting a proton. Thus, we can discuss the acidity of the acid HA or the basicity of the base A^-. As in the case of acids, the basicity of bases is both qualitatively and quantitatively compared to the properties of water. A base, A^-, removes a proton from water to form hydroxide ion and the conjugate acid HA. The base dissociation constant, K_b, for the reaction is given by the following expression.

$$A^- + H_2O \rightleftharpoons HA + OH^-$$

$$K_b = K_{eq}[H_2O] = \frac{[HA][OH^-]}{[A^-]}$$

The K_b values of bases are conveniently expressed as pK_b values. The pK_b is defined as

$$pK_b = -\log K_b.$$

pK_b values increase with decreasing basicity. The pK_b values of some organic bases are listed in Table 2.2. A strong base has a large K_b (small pK_b) and completely removes the proton of an acid. The most common strong base is hydroxide ion, which will remove and accept protons from even weak acids such as acetic acid.

$$CH_3CO_2H + OH^- \xrightarrow{K_b} CH_3CO_2^- + H_2O$$
strong base   weak base

Table 2.2
K_b and pK_b Values of Common Bases

Acid	K_b	pK_b
⬡—NH$_2$	4×10^{-10}	9.4
$CH_3CO_2^-$	5×10^{-10}	9.3
$C{\equiv}N^-$	1.6×10^{-5}	4.8
NH_3	1.7×10^{-5}	4.8
CH_3NH_2	4.3×10^{-4}	3.4
CH_3O^-	3×10^{-16}	−1.5

Weak bases do not have a large attraction for the protons of an acid. A small fraction of the molecules of a weak base are protonated at equilibrium. For example, methylamine is a weak base. When it dissolves in water, a low concentration of methylammonium ions forms.

$$CH_3NH_2 + H_2O \overset{K_b}{\rightleftharpoons} CH_3NH_3^+ + OH^-$$

weak base   strong base

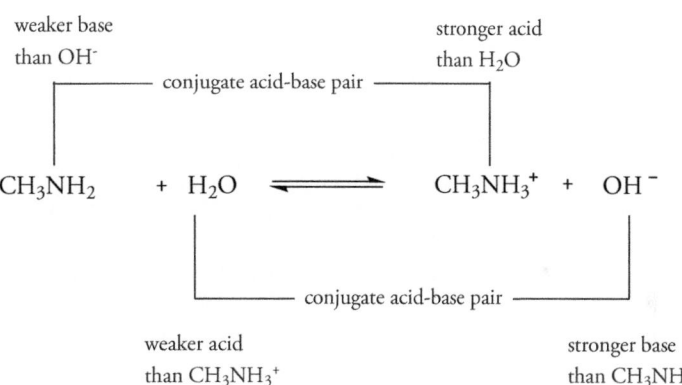

weaker base than OH⁻   stronger acid than H₂O

conjugate acid-base pair

$$CH_3NH_2 \;\; + \;\; H_2O \;\; \rightleftharpoons \;\; CH_3NH_3^+ \;\; + \;\; OH^-$$

conjugate acid-base pair

weaker acid than $CH_3NH_3^+$ stronger base than CH_3NH_2

2.8 EFFECT OF STRUCTURE ON ACIDITY

Removing a proton from an electrically neutral acid in a solvent requires breaking a bond to hydrogen, which generates a negative charge on the resulting conjugate base. Thus, K_a values depend on both the strength of the H—A bond and the stability of A⁻ in the solvent. The acidity of simple inorganic acids is related to the position in the periodic table of the atom bonded to hydrogen. The acidities of acids, HA, increase as we move down a column of the periodic table. For example, the acidities of the halogen acids increase in the order HF < HCl < HBr < HI. Similarly, for the same reasons, H_2O is also a weaker acid than H_2S. In a given row of the periodic table, acidity increases from left to right. The order of increasing acidity is CH_4 < NH_3 < H_2O < HF. This trend reflects the stability of the negative charge on the electronegative element of the conjugate base. That is, the order of increasing strength of conjugate bases is F⁻ < OH⁻ < NH_2^- < CH_3^-.

Many organic compounds are structurally related to inorganic acids and bases. As a consequence, we can predict the acid-base properties by making an appropriate comparison. For example, methanesulfonic acid has an O—H bond that is structurally similar to the O—H bond in sulfuric acid. Because sulfuric acid is a strong acid, it is reasonable to expect methanesulfonic acid to be a strong acid, and it is.

sulfuric acid,
a strong acid

methane sulfonic acid,
a strong acid

Ethylamine, $CH_3CH_2NH_2$, is structurally related to ammonia, which is a weak base. Thus, ethylamine and other amines are weak bases. As we examine various functional groups in detail, we will find that the acid-base properties do vary somewhat with structure.

ammonia

pK_b 4.74

(weak base)

ethylamine

pK_b 3.25

(weak base)

A reaction in which relatively unstable reactants are converted to more stable products has a large equilibrium constant. Thus, stabilizing the negative charge in the conjugate base formed from an acid increases K_a. When an anion produced by ionization of an acid has resonance stabilization, acid strength increases by a substantial amount. For example, both methanol and acetic acid ionize to form conjugate bases that have a negative charge on oxygen. However, acetic acid is about **ten billion** (10^{10}) times more acidic than methanol.

$$CH_3OH + H_2O \rightleftharpoons CH_3O^- + H_3O^+ \quad K_a = 10^{-16}$$

$$CH_3CO_2H + H_2O \rightleftharpoons CH_3CO_2^- + H_3O^+ \quad K_a = 1.8 \times 10^{-5}$$

The greater acidity of acetic acid is the result of **resonance stabilization** of the negative charge in the conjugate base, acetate ion. The methoxide ion (CH_3O^-) has its negative charge concentrated on a single oxygen atom.

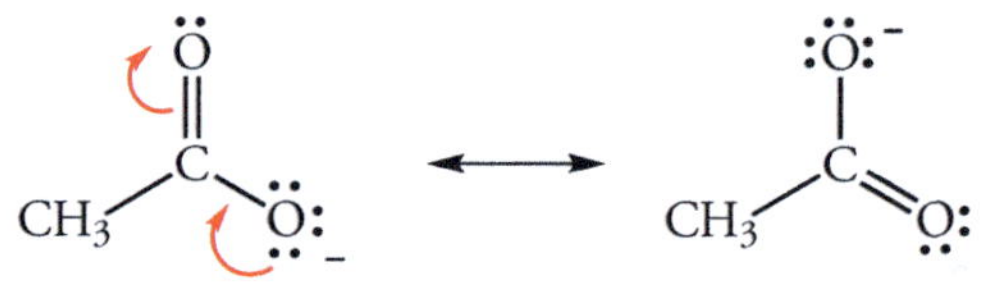

Acidity also reflects the ability of an atom to polarize neighboring bonds by an **inductive effect**. For example, chloroacetic acid is a stronger acid than acetic acid.

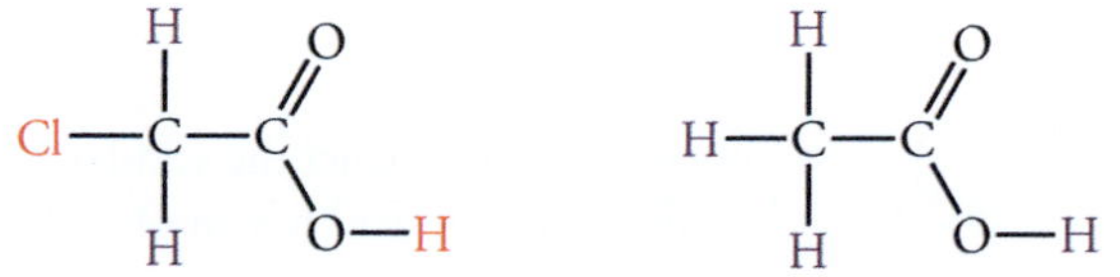

$pK_a = 2.9$

$pK_a = 4.7$

The electrons in the C—Cl bond are "pulled" toward the more electronegative chlorine atom and away from the carbon skeleton. As a result, the electrons of the oxygen atom are drawn away from the O—H bond, and the proton can therefore ionize more easily. This shifting of electrons through a network of sigma bonds is called **inductive withdrawal** of electrons.

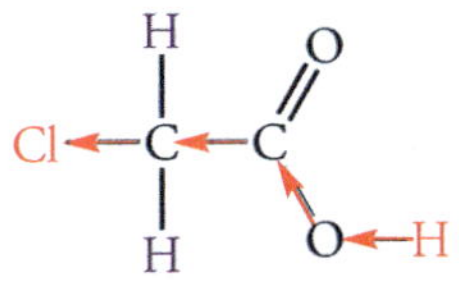

Problem 2.11

The pKa values of ethanol and 2,2,2-trifluoroethanol are 15.9 and 12.4, respectively. What is responsible for this difference?

$$CH_3\text{—}CH_2\text{—}OH \qquad\qquad CF_3\text{—}CH_2\text{—}OH$$

ethanol   2,2,2-trifluoroethanol

pKa = 15.9   pKa = 12.4

Solution

2,2,2-Trifluoroethanol is a stronger acid than ethanol because the carbon atom with three fluorine atoms has a partial positive charge as a consequence of inductive electron withdrawal by the three polar covalent C—F bonds. This carbon atom, in turn, inductively attracts electrons from the other carbon atom and, indirectly, from the oxygen atom. Thus, the oxygen-hydrogen bond is more strongly polarized, and the compound is more acidic.

Problem 2.12

The pKa value of the C—H bond of nitromethane is 10.2, whereas the pKa value of methane is approximately 49. Explain why nitromethane is so much more acidic.

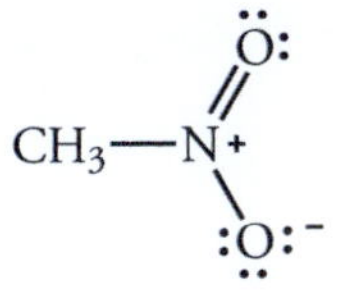

nitromethane

2.9 INTRODUCTION TO REACTION MECHANISMS

The description of the individual steps of a reaction, showing the order in which bonds are broken in the reactant and formed in the product, is called the **reaction mechanism**. Some reactions occur in a single step, and bonds form and break simultaneously. Such processes are **concerted reactions**. The reaction mechanism thus resembles that of an ordinary chemical equation.

$$A \longrightarrow P$$

reactant   product

Many reactions occur in a series of steps. For example, the conversion of reactant A into product B may occur in two steps, in which an intermediate M is formed and then reacts.

$$A \xrightarrow{\text{Step 1}} M$$

reactant   intermediate

$$\text{M} \xrightarrow{\text{Step 2}} \text{P}$$

intermediate   product

The rate of the slowest step in a sequence of reactions is called the rate-determining step because the overall rate of conversion of reactant into product can occur no faster than this slowest step.

Types of Bond Cleavage and Formation

When a bond is broken so that one electron remains with each of the two fragments, the process is **homolytic** cleavage. Consider R—Y, where R usually represents the major portion of an organic molecule and Y is an atom or group of atoms that may or may not contain carbon atoms. Homolysis of the bond produces a **carbon radical**.

$$X\text{—}Y \longrightarrow X\cdot + Y\cdot$$

general reaction for homolytic bond cleavage

Homolytic cleavage of a bond to carbon produces a carbon radical that is highly reactive because it has only seven electrons in its valence shell.

When a bond is broken so that one fragment gains both bonding electrons, the process is called **heterolytic** cleavage. The fragment produced by heterolysis that gains electrons has a negative charge. The second fragment is electron deficient and has a positive charge. If the bond to carbon breaks so that its electrons remain with the carbon atom, a negatively charged **carbanion** forms. The carbanion has an octet of electrons around the carbon atom. If the bond breaks so that its electrons are lost by the carbon atom, a positively charged **carbocation** forms. The carbocation has a sextet of electrons around the carbon atom and is an electron-deficient species.

$$X\text{—}Y \longrightarrow X^+ + :Y^-$$

general reaction for heterolytic bond cleavage

$$R\text{—}Y \longrightarrow R:^- + Y^+$$
(a carbanion)

$$R\text{—}Y \longrightarrow R^+ + :Y^-$$
(a carbocation)

The mode of heterolytic cleavage of a C—Y bond depends on the electronegativity of Y. If Y is a more electronegative element than carbon, such as a halogen atom, the carbon atom bears a partial positive charge, and the carbon bond tends to break heterolytically to form a carbocation. Conversely, if Y is a less electronegative element, such as a metal, the bond has the opposite polarity and tends to break heterolytically to form a carbanion.

$$\overset{\delta^-}{R}\text{—}\overset{\delta^+}{Li} \longrightarrow R:^- + Li^+$$
(a carbanion)

$$\overset{\delta^+}{R}\text{—}\overset{\delta^-}{Br} \longrightarrow R^+ + :Br^-$$
(a carbocation)

Heterogenic reactions are more common than homogenic reactions in organic chemistry. In organic reactions, a carbocation behaves as an **electrophile** (electron-loving species). It seeks a negatively charged center to neutralize its positive charge and to obtain a stable octet of electrons. On the other hand, a carbanion has an electron pair that causes it to react as a **nucleophile** (nucleus-loving species). It seeks a positively charged center to neutralize its negative charge.

Many organic reactions can be depicted by the following equation, in which E^+ represents an electrophile and Nu^- represents a nucleophile.

$$Nu:^- \quad E^+ \longrightarrow E\!-\!Nu$$
$$\text{nucleophile} \quad \text{electrophile}$$

The curved arrow notation shows the movement of a pair of electrons from the nucleophile to the electrophile. This notation is exactly like that used to show the reaction between a Lewis base and a Lewis acid.

Free Radical Substitution Reactions

Methane reacts with chlorine gas at elevated temperatures or in the presence of ultraviolet light as an energy source. In this reaction, a chlorine atom replaces a hydrogen atom.

$$CH_3\!-\!H \;+\; Cl\!-\!Cl \longrightarrow CH_3\!-\!Cl \;+\; H\!-\!Cl$$

The mechanism of this reaction occurs by homolytic bond cleavage and homogenic bond formation. In the first step, a chlorine molecule absorbs either heat or light energy and the Cl—Cl bond is broken to give two chlorine atoms. They are electron-deficient radicals and are highly reactive. This step starts the reaction and is called the **initiation step**.

Step 1 $$:\!\ddot{C}l\!-\!\ddot{C}l\!: \longrightarrow :\!\ddot{C}l\cdot \;+\; \cdot\ddot{C}l\!:$$

Two steps, collectively known as **propagation steps**, that involve radicals as reactants and products, then occur.

Step 2 $$CH_3\!-\!H \;+\; \cdot\ddot{C}l\!: \longrightarrow CH_3\cdot \;+\; H\!-\!\ddot{C}l\!:$$

Step 3 $$CH_3\cdot \;+\; :\!\ddot{C}l\!-\!\ddot{C}l\!: \longrightarrow CH_3\!-\!Cl \;+\; \cdot\ddot{C}l\!:$$

In step 2, a C—H bond is broken and an H—Cl bond is produced; in step 3, a Cl—Cl bond is broken and a C—Cl bond forms. Furthermore, in each step a radical reacts and a radical is produced. One radical generates another in this chain propagation sequence. The process continues as long as radicals and a supply of both reactants are present.

Nucleophilic Substitution Reactions

Reactions in which a nucleophile "attacks" a carbon atom and replaces another group are very common. The "leaving group" displaced from the carbon center is symbolized by L. A leaving group is invariably an electronegative atom or a group that can exist as a stable anion.

$$Nu:^- \quad R\!-\!L \xrightarrow{\text{nucleophilic substitution}} R\!-\!Nu + :\!L^-$$
$$\text{nucleophile} \qquad\qquad\qquad\qquad \text{leaving group}$$

Note that the nucleophile has an unshared pair of electrons that bonds to the carbon residue. Thus, bond formation is a heterogenic process. The leaving group departs with an electron pair, and cleavage of the bond between the leaving group and carbon is a heterolytic process. An example of this type of nucleophilic substitution process is the reaction of bromomethane with iodide ion.

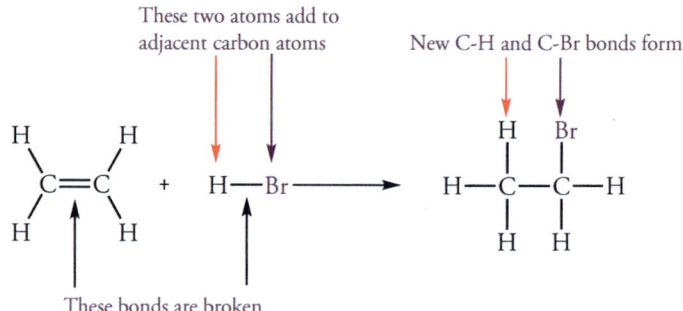

In this reaction, the nucleophile approaches the carbon atom, which is made somewhat positive by the electronegative chlorine atom. The nucleophile has a nonbonding pair of electrons that begins to bond to the carbon atom. As the nucleophile approaches the carbon, the bond between carbon and the iodide ion, a leaving group, weakens. The entire process is **concerted**—that is, both bond breaking and bond formation occur simultaneously. The mechanism of nucleophilic substitution depends on many factors to be discussed in Chapter 7.

2.10 REACTION RATES

Factors that Affect Reaction Rates

In a chemical reaction, the reactant molecules collide with each other, some bonds break, and others form. Several factors affect the rate of a reaction.

1. The nature of the reactants.
2. The concentration of the reactants.
3. Temperature.
4. The presence of substances called catalysts.

The nature of the reactants is the most important feature controlling a chemical reaction. For example, in the addition reaction of ethylene (C_2H_4) with HBr, a bond must be broken between the hydrogen and bromine atoms. Bonds must be formed between a carbon and hydrogen atom and between a carbon and bromine atom.

The addition reaction of HCl to ethylene occurs at a rate different from that of the analogous reaction with HBr. The bonding that occurs now involves carbon and chlorine atoms, and the bond broken is between hydrogen and chlorine atoms. The energy requirements associated with the reorganization of these atoms and bonds must be different.

As the concentration of reactants is increased, the reaction velocity increases because reactant molecules are more likely to collide. The rates of chemical reactions increase with a rise in temperature because the reactant molecules collide more frequently and with greater energy. As a rule of thumb, the rate approximately doubles for a 10 °C rise in temperature.

A **catalyst** is a substance that increases a reaction rate, and the process is known as catalysis. Catalysts are usually required only in small amounts. The catalyst is present in the same amount before and after the reaction takes place. Although a catalyst increases the rate of a reaction, it does not change the equilibrium constant for the reaction.

Reaction Rate Theory

Collisions between molecules that cause a chemical reaction are called **effective collisions**, and the minimum energy required for an effective collision is the **activation energy**. The activation energy for a given reaction depends on the type of bonds broken and formed in the reaction. During a reaction, the arrangement of the atoms changes as bonds are distorted and eventually broken while new bonds form. During this process, some repulsion occurs when reactant atoms move close together. This repulsion results from the proximity of the electrons surrounding each atom. During a reaction, each specific arrangement of atoms has an associated energy. All such arrangements of atoms have an energy higher than the initial energy of the reactants. The atomic arrangement lying along the lowest energy pathway connecting reactants and products whose structure has the maximum energy is the **transition state**.

Let's consider the transition state in the nucleophilic substitution of bromide ion with bromomethane that has both the hydroxide and chloride ions bonded to some degree to the carbon atom.

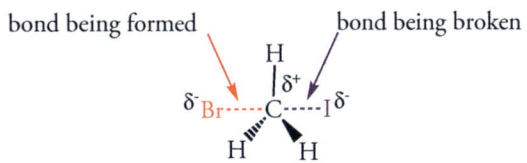

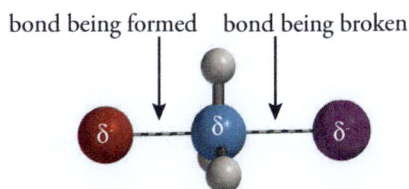

For example, in the transition state of the reaction in which bromide ion reacts with iodomethane, both the bromide and iodide ions bonded to some degree to the carbon atom. The carbon-iodine bond breaks on one side of the transition state structure, while the carbon-bromine bond forms on the other side. Transition states exist for periods of time as short as femtoseconds (10^{-15} s) during a reaction. However, the structure of the transition state for a reaction can be inferred from various kinds of experimental data.

Reaction Coordinate Diagrams

Reaction coordinate diagrams represent the progress of a reaction versus the energy changes that occur during the reaction. The vertical axis gives the total energy of the reacting system; the horizontal axis qualitatively represents the progress of a reaction from reactants (left) to the products (right). Figure 2.6 shows the energy changes that occur in an exothermic reaction ($\Delta H° < 0$). The difference between the energy of the reactants and the transition state is the activation energy ($E_a > 0$). In the transition state for the nucleophilic substitution of chloromethane by hydroxide ion, both hydroxide and chloride are partially bonded to the carbon atom.

A large activation energy results in a slow reaction, because only a small fraction of the molecules collide with sufficient energy to reach the transition state. At this point, energy is released as the reaction proceeds to form products. The energy released is equal to the activation energy originally added plus an amount equal to that characteristic for the exothermic reaction.

The kinetic energy of molecules increases with increasing temperature. As the kinetic energy increases, the chances increase for molecular collisions equal in energy to the activation energy, so the rate of reaction increases.

Figure 2.6
Reaction Coordinate Diagram for a Substitution Reaction

The reaction of hydroxide ion with chloromethane occurs in a single step. The activation energy, E_a, reflects the stability of the transition state, which depends upon the structure of the substrate, the nucleophile, and the leaving group.

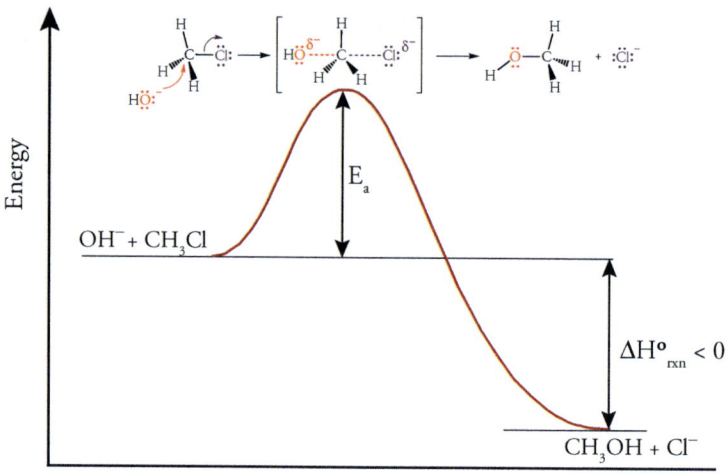

Some reactions occur in two or more steps, as in the case of the two-step addition reaction of HBr with ethene that we discussed earlier. In the first step, a proton acts an electrophile. It forms a bond to carbon using the electrons of the double bond. As a consequence, an intermediate carbocation is formed. It then reacts in a second step with the bromide ion, which is a nucleophile. Each step is shown in the reaction coordinate diagram in Figure 2.7.

In the first transition state, a hydrogen ion begins to bond to carbon as π electrons are removed from the double bond. The energy then decreases until an intermediate carbocation is formed. In the second step, which has its own activation energy, the carbocation starts to bond to the nucleophilic bromide ion. Note that the energy of the carbocation is lower than the energy of the two transition states. Finally, as the carbon-bromine bond becomes fully formed, the reaction coordinate diagram shows that the energy of the product is lower than the energy of the reactants; that is, the reaction is exothermic.

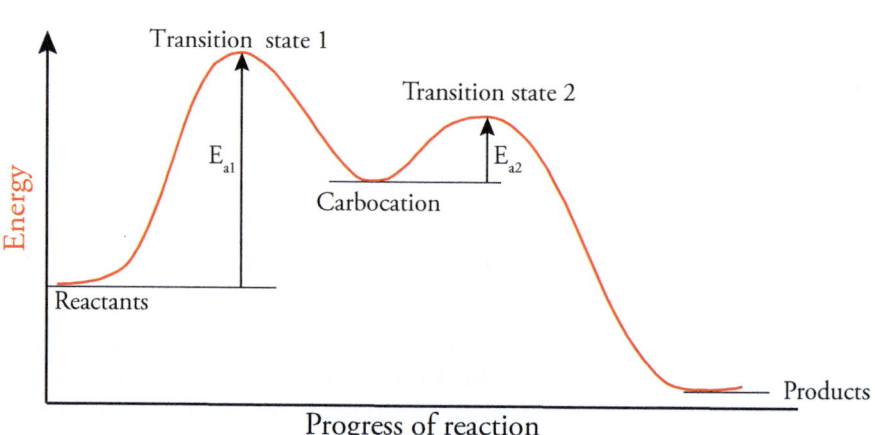

Figure 2.7 Energy Diagram for the Addition of HBr to an Alkene

The first, rate-determining step in the addition of HBr to ethene is the attack of the electrons of the double bond on a proton to give a carbocation. The second step occurs at a faster rate because the activation energy of the second step is lower than for the first step.

The Function of Catalysts

A catalyst provides a path for the reaction that is different from the path of the uncatalyzed reaction. The reaction path starts with the same reactants and concludes with the same products. However, the path for the catalyzed reaction has a different, lower activation energy (Figure 2.8). To illustrate the effect of a catalyst on the path of a reaction, consider the hypothetical concerted reaction of A and B.

$$A + B \longrightarrow X$$

The activation energy required for this reaction is available in a small fraction of high-energy molecular collisions. However, in the presence of a catalyst, represented by C, the following reactions may occur.

Step 1 $A + C \longrightarrow A—C$

Step 2 $B + A—C \longrightarrow A—B + C$

The catalyst may combine with A in a reaction with a lower activation energy. Similarly, the reaction of A—C with B may have a low activation energy. If the activation energy of each step is low, a larger fraction of molecules will be able to react faster via this catalyzed pathway than could react without the catalyst at the same temperature.

Figure 2.8 Energy Diagram for a Catalyzed and an Uncatalyzed Reaction

The activation energy for a catalyzed reaction is smaller than the activation energy for reaction in the absence of a catalyst. The catalyzed reaction may require a different number of steps than the uncatalyzed reaction.

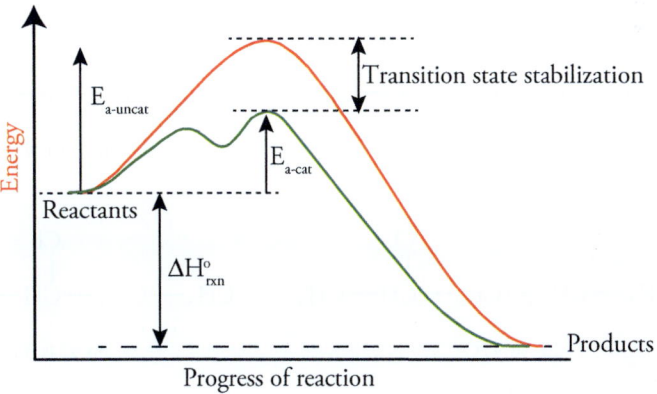

EXERCISES

Physical Properties

2.1 Suggest a reason for the difference in boiling points between the following pairs of isomeric compounds. (Two or more structural features may be responsible.)

(a) CH_3—CH_2—CH_2—O—CH_2—CH_2—CH_3

bp 90.5 °C

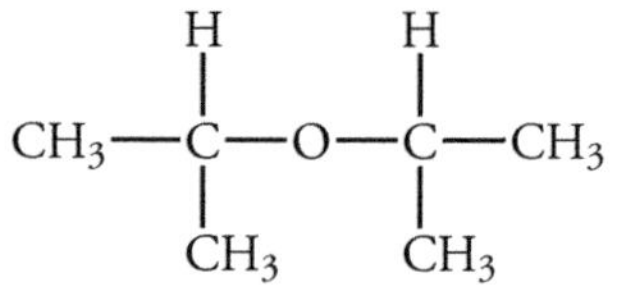

$$CH_3-\underset{\underset{CH_3}{|}}{\overset{\overset{H}{|}}{C}}-O-\underset{\underset{CH_3}{|}}{\overset{\overset{H}{|}}{C}}-CH_3$$

bp 68 °C

(b) CH_3—CH_2—CH_2—NH_2

bp 49 °C

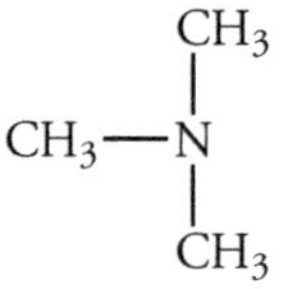

$$CH_3-\underset{\underset{CH_3}{|}}{\overset{\overset{CH_3}{|}}{N}}$$

bp 3 °C

(c) $$CH_3-\underset{\underset{CH_3}{|}}{\overset{\overset{CH_3}{|}}{C}}-CH_2-OH$$

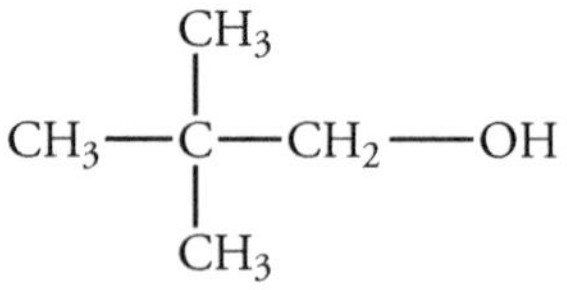

bp 113 °C

$$CH_3-\underset{\underset{CH_3}{|}}{\overset{\overset{CH_3}{|}}{C}}-O-CH_3$$

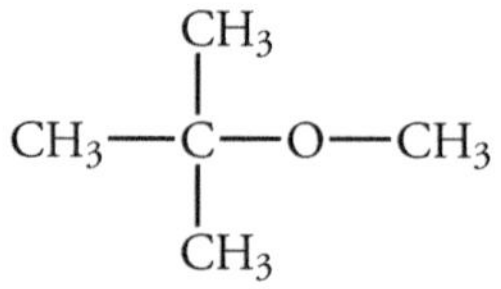

bp 55 °C

2.2 Explain why the boiling points of the following pairs of isomeric compounds do not differ very much.

(a) CH_3—CH_2—CH_2—S—CH_3

bp 95.5 °C

CH_3—CH_2—S—CH_2—CH_3

bp 92.1 °C

(b) $$CH_3-CH_2-CH_2-\underset{}{\overset{\overset{Cl}{|}}{CH}}-CH_3$$

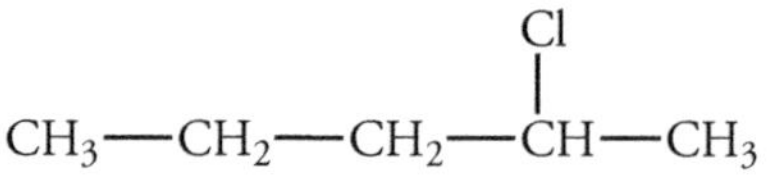

bp 96.9 °C

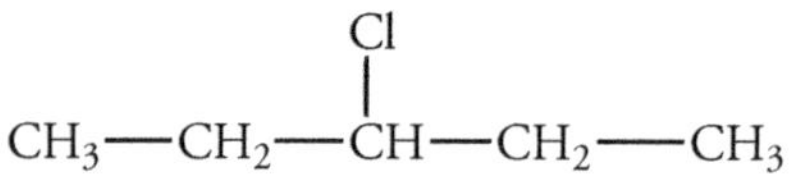

$$CH_3-CH_2-\underset{}{\overset{\overset{Cl}{|}}{CH}}-CH_2-CH_3$$

bp 97.8 °C

(c) $$CH_3-CH_2-\underset{}{\overset{\overset{H}{|}}{N}}-CH_2-CH_3$$

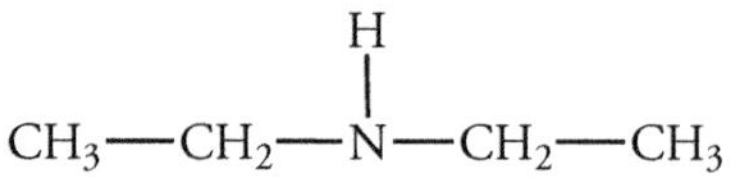

bp 56 °C

$$CH_3-CH_2-CH_2-\underset{}{\overset{\overset{H}{|}}{N}}-CH_3$$

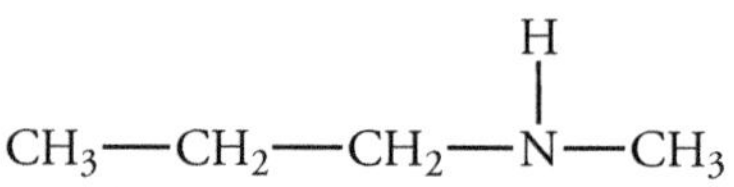

bp 61 °C

2.3 The ethene molecule is planar, and all bond angles are close to 120°. There are three isomeric dichloroethenes. Two isomers have dipole moments and the third does not. Which of the following three is nonpolar? Explain why.

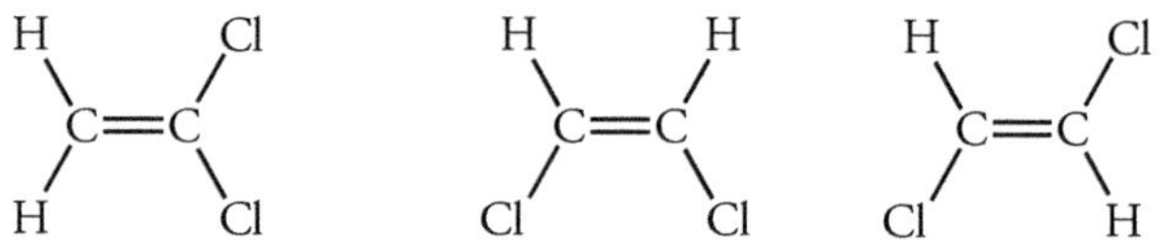

2.4 One of the following compounds has a dipole moment and the other does not. Select the polar compound and explain why the other compound has no dipole moment.

2.5 Propylene glycol is miscible with water. However, the solubility of 1-butanol is only 7.9 g/100 mL of water. Explain why.

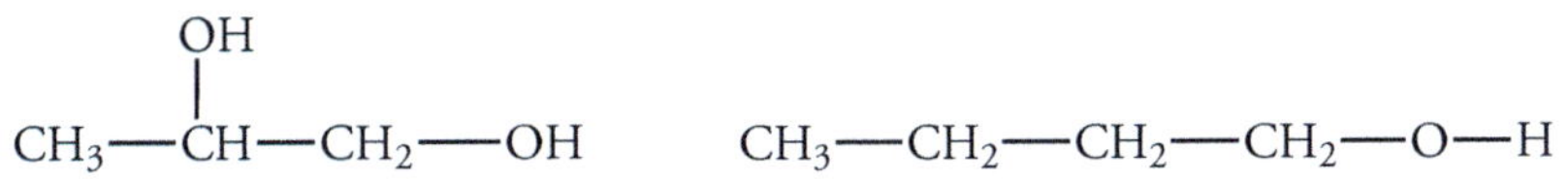

$$CH_3-\underset{\underset{\displaystyle OH}{|}}{CH}-CH_2-OH \qquad\qquad CH_3-CH_2-CH_2-CH_2-O-H$$

propylene glycol   1-butanol

2.6 Explain why butanoic acid is miscible with water, but ethyl ethanoate is not.

$$CH_3-CH_2-CH_2-\overset{\overset{\displaystyle O}{\|}}{C}-OH \qquad\qquad CH_3-\overset{\overset{\displaystyle O}{\|}}{C}-O-CH_2-CH_3$$

butanoic acid   ethyl ethanoate

Acids and Bases

2.7 Write the structure of the conjugate acid of each of the following species:

(a) CH_3-S-CH_3 (b) CH_3-O-CH_3 (c) CH_3-NH_2 (d) CH_3-OH

2.8 Write the structure of the conjugate base of each of the following species:

(a) CH_3-S-H (b) CH_3-NH_2 (c) CH_3-SO_3H (d) $CH\equiv CH$

2.9 Write the structures of the two conjugate acids of hydroxylamine (NH_2-OH). Which is the more acidic?

2.10 Write the structures of the two conjugate bases of hydroxylamine (NH_2-OH). Which is the more basic?

2.11 Identify the Lewis acid and Lewis base in each of the following reactions:

(a) $CH_3-CH_2-Cl + AlCl_3 \longrightarrow CH_3-CH_2^+ + AlCl_4^-$

(b) $CH_3-CH_2-SH + CH_3-O^- \longrightarrow CH_3-CH_2-S^- + CH_3-OH$

(c) $CH_3-CH_2-OH + NH_2^- \longrightarrow CH_3-CH_2-O^- + NH_3$

2.12 Identify the Lewis acid and Lewis base in each of the following reactions:

(a) $CH_3-\ddot{O}-CH_3 + H-\ddot{I}: \longrightarrow CH_3-\underset{\underset{\displaystyle +}{\overset{\overset{\displaystyle H}{|}}{O}}}{}-CH_3 + :\ddot{\underset{..}{I}}:^-$

(b) $CH_3-CH_2^+ + H_2\ddot{O} \longrightarrow CH_3-CH_2-\underset{\underset{\displaystyle ..}{\overset{\overset{\displaystyle H}{|}}{O}}}{\overset{+}{}}-H$

(c) $CH_3-CH=CH_2 + H-\ddot{B}r: \longrightarrow (CH_3)_2CH+ + :\ddot{\underset{..}{B}}r:^-$

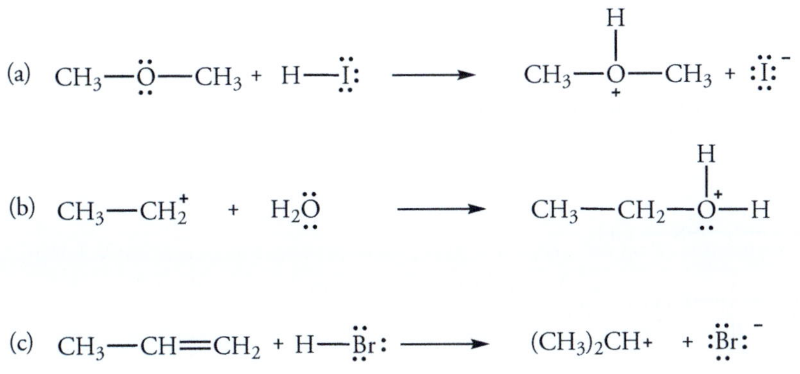

pK Values and Acid Strength

2.13 The approximate pK_a values of CH_4 and CH_3OH are 49 and 16, respectively. Which is the stronger acid? Will the equilibrium position of the following reaction lie to the left or to the right?

$$CH_4 + CH_3{-}O^- \underset{}{\overset{K_{eq}}{\rightleftharpoons}} CH_3^- + CH_3{-}OH$$

2.14 The approximate pK_a values of NH_3 and CH_3OH are 36 and 16, respectively. Which is the stronger acid? Will the equilibrium position of the following reaction lie to the left or to the right?

$$CH_3{-}OH + NH_2^- \underset{}{\overset{K_{eq}}{\rightleftharpoons}} NH_3 + CH_3{-}O^-$$

2.15 The pK_a of acetic acid (CH_3CO_2H) is 4.8. Explain why the carboxylic acid group of Amoxicillin (pK_a = 2.4), a synthetic penicillin, is more acidic than acetic acid, whereas the carboxylic acid group of indomethacin (pK_a = 4.5), an anti-inflammatory analgesic used to treat rheumatoid arthritis, is of comparable acidity.

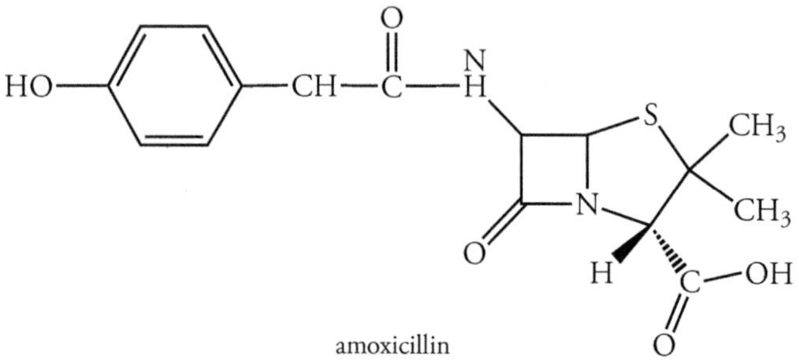

amoxicillin

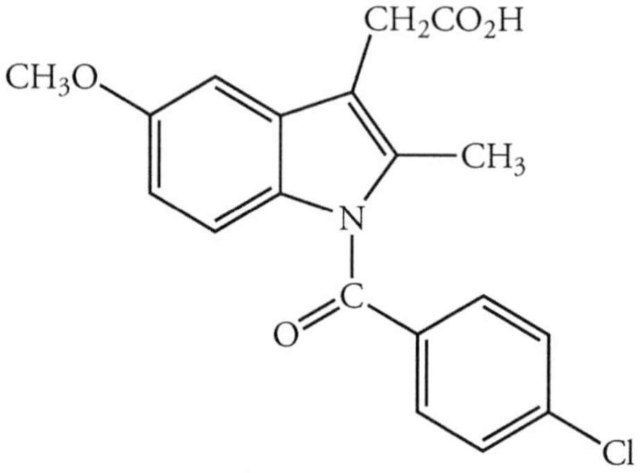

indomethacin

2.16 The pK_a of the OH group of phenobarbital is 7.5, whereas the pK_a of CH_3OH is 16. Explain why phenobarbital is significantly more acidic.

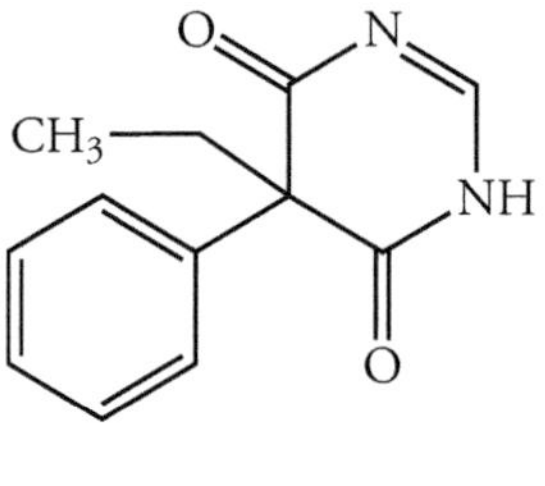

phenobarbital

Oxidation-Reduction Reactions

2.17 Determine whether each of the following reactions is a reduction, an oxidation, or neither.

(a) $CH_3{-}C{\equiv}N \longrightarrow CH_3{-}CH_2{-}NH_2$

(b) $2\ CH_3{-}S{-}H \longrightarrow CH_3{-}S{-}S{-}CH_3$

(c) $CH_3{-}S{-}CH_3 \longrightarrow CH_3{-}\underset{\overset{\|}{O}}{S}{-}CH_3$

2.18   Explain why none of the following reactions is a reduction or an oxidation (although they may appear to be).

(a) 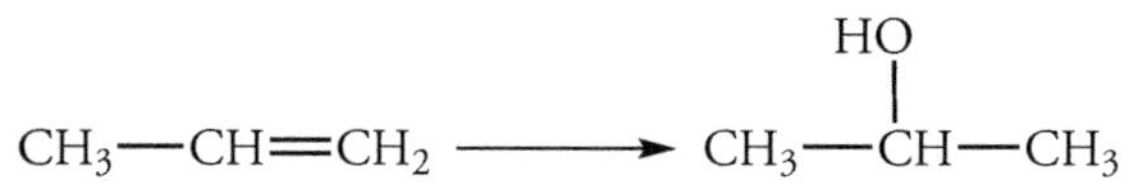 $CH_3-CH=CH_2 \longrightarrow CH_3-\overset{\overset{\displaystyle HO}{|}}{CH}-CH_3$

(b) 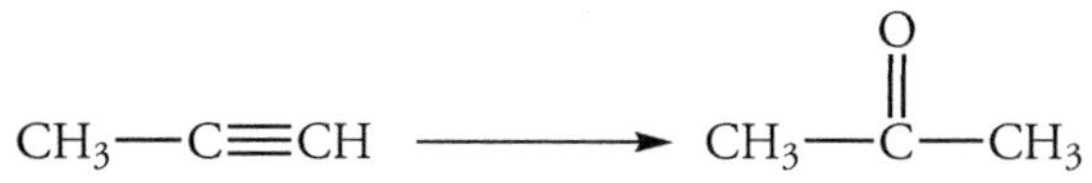 $CH_3-C\equiv CH \longrightarrow CH_3-\overset{\overset{\displaystyle O}{\|}}{C}-CH_3$

(c) $CH_3-\overset{\overset{\displaystyle NH}{\|}}{C}-CH_3 \longrightarrow CH_3-\overset{\overset{\displaystyle O}{\|}}{C}-CH_3$

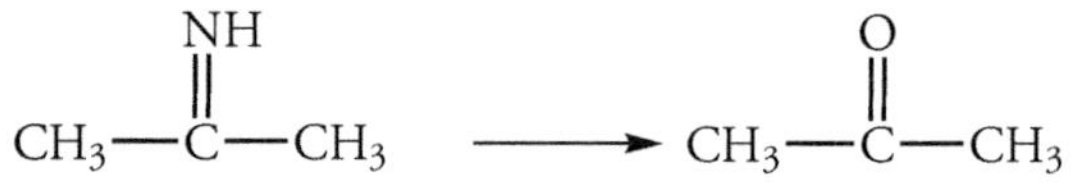

2.19   Consider each of the following reactions for drug metabolism. What type of reaction occurs?

(a)

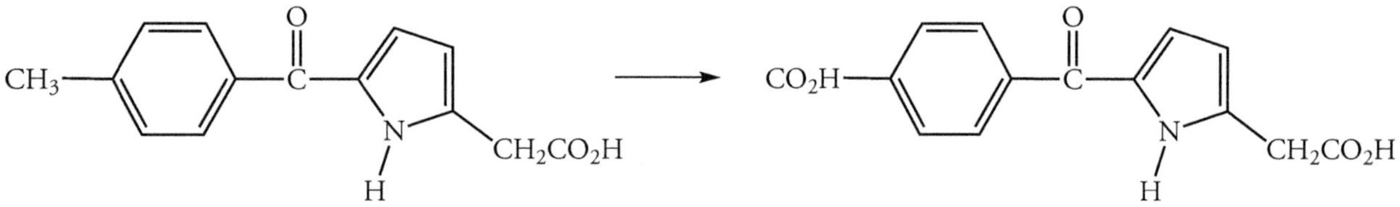

tolmetin, an anti-inflammatory drug

(b)

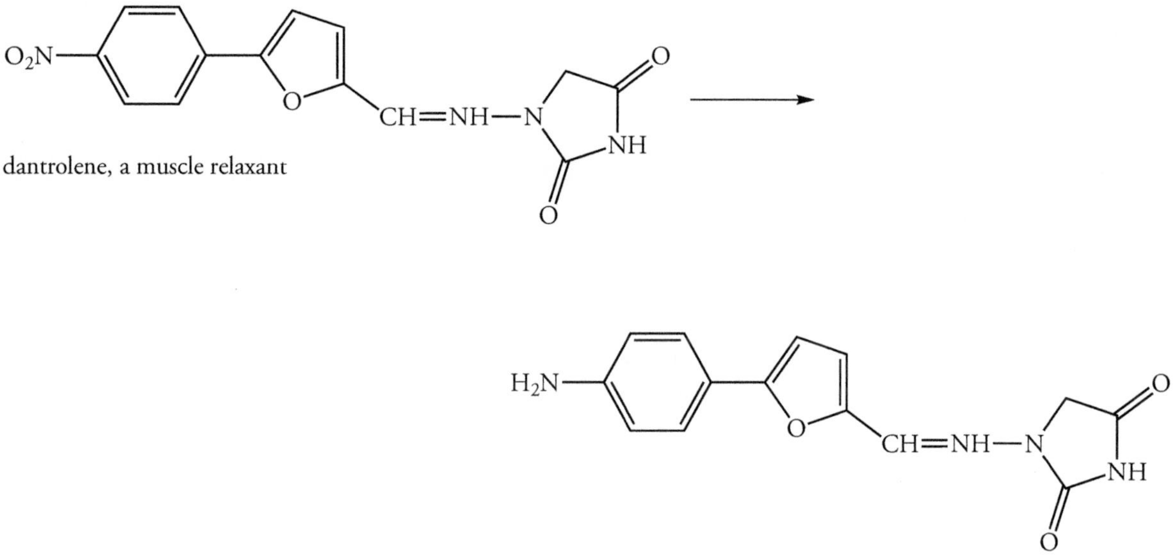

dantrolene, a muscle relaxant

2.20 Consider each of the following reactions for drug metabolism. What type of reaction occurs?

(a)

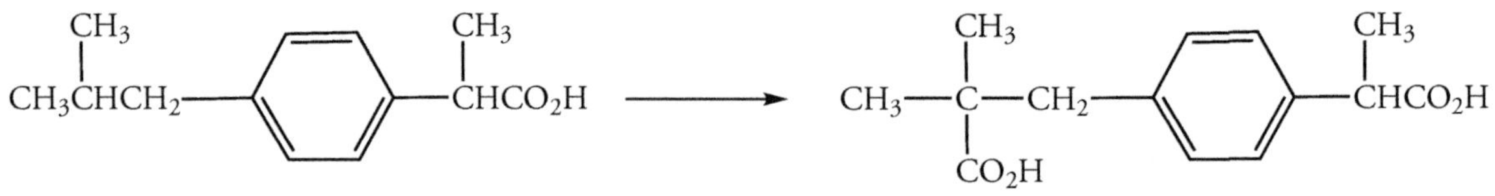

Ibuprofen, an analgesic

(b)

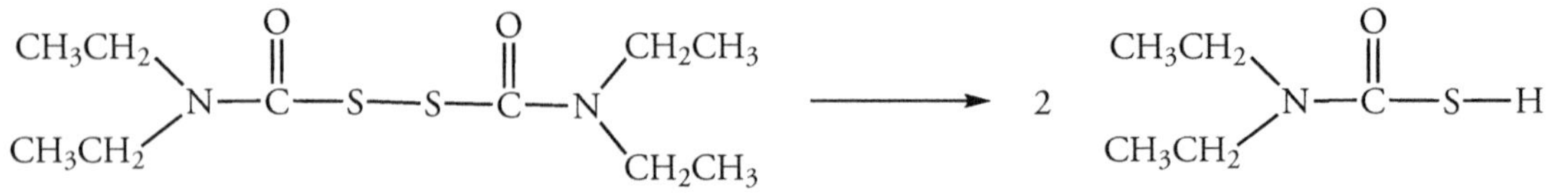

Disulfiram, a drug used to treat alcoholism

Types of Organic Reactions

2.21 Classify the type of reaction by which each of the following unbalanced reactions occurs. Identify the additional types of reactants that would be required to complete each reaction.

(a) $2\ CH_3{-}CH_2{-}OH \longrightarrow CH_3{-}CH_2{-}O{-}CH_2{-}CH_3$

(b) $CH_3{-}C({=}O){-}S{-}CH_3 \longrightarrow CH_3{-}C({=}O){-}OH + CH_3SH$

(c)

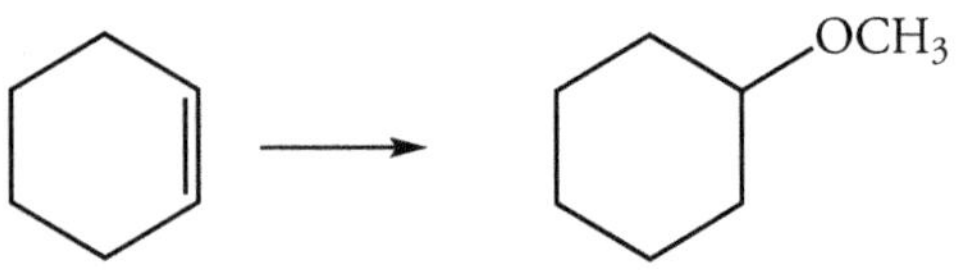

2.22 The metabolism of fatty acids (long-chain carboxylic acids) occurs in a pathway in which the following reactions. Classify the type of reaction in each step. (R is a chain of carbon atoms and CoA is coenzyme A.)

(a) $R{-}CH_2{-}CH_2{-}C({=}O){-}S{-}CoA \longrightarrow R{-}CH{=}CH{-}C({=}O){-}S{-}CoA$

(b) $R{-}CH{=}CH{-}C({=}O){-}S{-}CoA \longrightarrow R{-}CH(OH){-}CH_2{-}C({=}O){-}S{-}CoA$

(c) $R{-}CH(OH){-}CH_2{-}C({=}O){-}S{-}CoA \longrightarrow R{-}CH_2{-}C({=}O){-}CH_2{-}C({=}O){-}S{-}CoA$

2.23 The degradation (catabolism) of glucose occurs in a pathway called glycolysis in which the following reactions occur. Classify the type of reaction in each step.

(a)

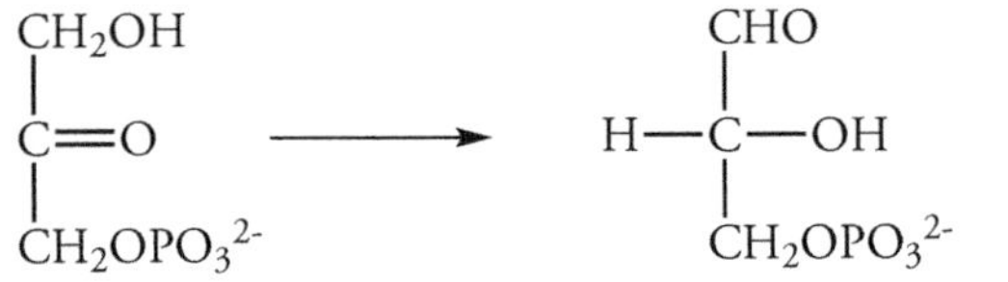

$$CH_2OH - C(=O) - CH_2OPO_3^{2-} \longrightarrow CHO - CH(OH) - CH_2OPO_3^{2-}$$

dihydroxyacetone-phosphate → D-glyceraldehyde-3-phosphate

(b)

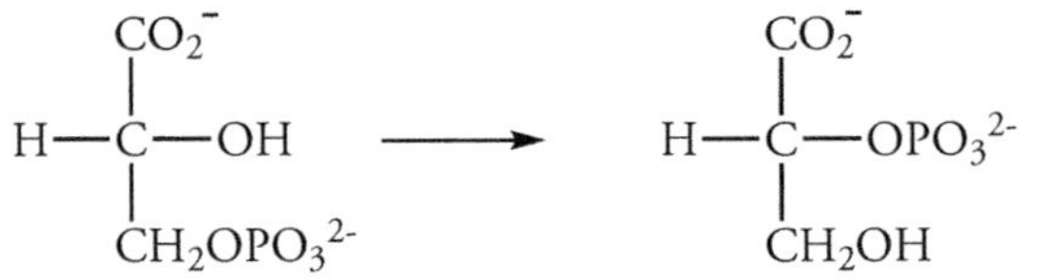

$$CO_2^- - CH(OH) - CH_2OPO_3^{2-} \longrightarrow CO_2^- - CH(OPO_3^{2-}) - CH_2OH$$

D-3-phosphoglycerate → D-2-phosphoglycerate

(c)

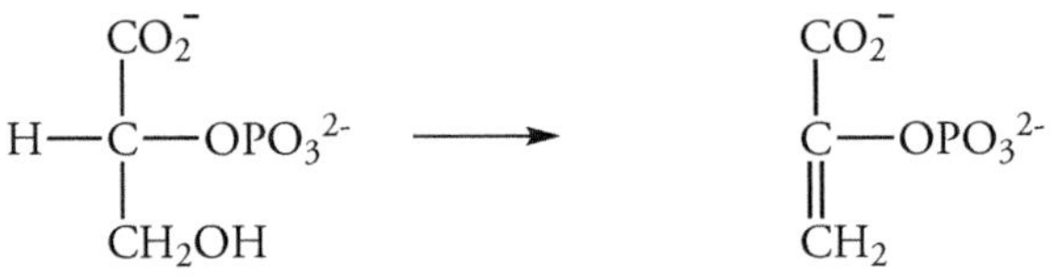

$$CO_2^- - CH(OPO_3^{2-}) - CH_2OH \longrightarrow CO_2^- - C(OPO_3^{2-}) = CH_2$$

phosphoenolpyruvate

2.24 The metabolism of chloroform produces a compound called phosgene, $(COCl_2)$, that causes liver damage. What types of reactions occur in the degradation of phosgene shown below?

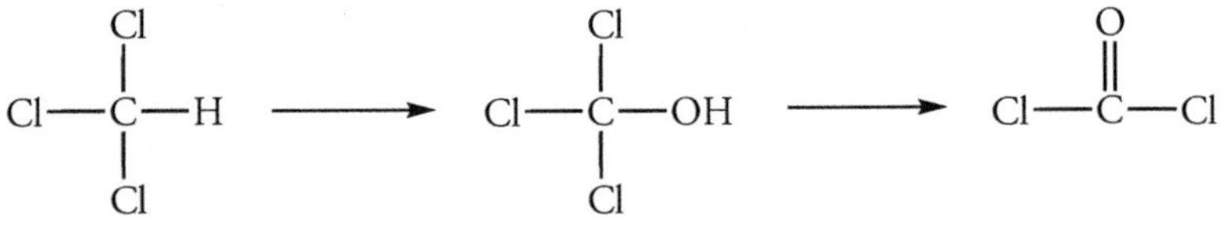

$$Cl-CHCl_2 \longrightarrow Cl-C(Cl)(OH)-Cl \longrightarrow Cl-C(=O)-Cl$$

chloroform → → phosgene

2.25 The sedative-hypnotic chloral hydrate is metabolized as follows. What type of reaction occurs in each step?

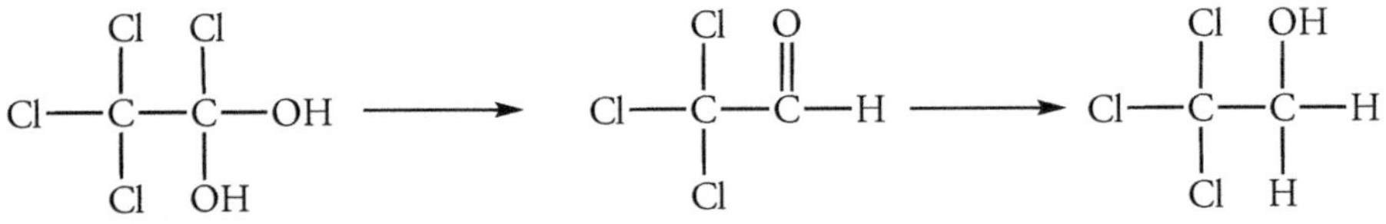

$$Cl-CCl_2-CH(OH)_2 \longrightarrow Cl-CCl_2-CHO \longrightarrow Cl-CCl_2-CH(OH)H$$

chloral hydrate → 2,2,2-trichloroethanal → 2,2,2-trichloroethanol

Equilibria and Rates of Reaction

2.26 A reaction has an equilibrium constant of 10^{-5}. Are the products more or less stable than the reactants?

2.27 Could a reaction have $K = 1$? What relationship would exist between the energies of the reactants and products as shown in the reaction progress diagram?

2.28 Consider the following information about two reactions. Which reaction will occur at the faster rate at a common temperature?

Reaction	$\Delta H°$	E_a
A $\longrightarrow$ X	-125 kJ/mole	+ 104.6 kJ/mole
A $\longrightarrow$ Y	-104.6 kJ/mole	+ 125 kJ/mole

2.29 Given the data in Exercise 2.28, which reaction is more exothermic?

2.30   Identify the processes of bond cleavage and bond formation for each of the following reactions:

(a)

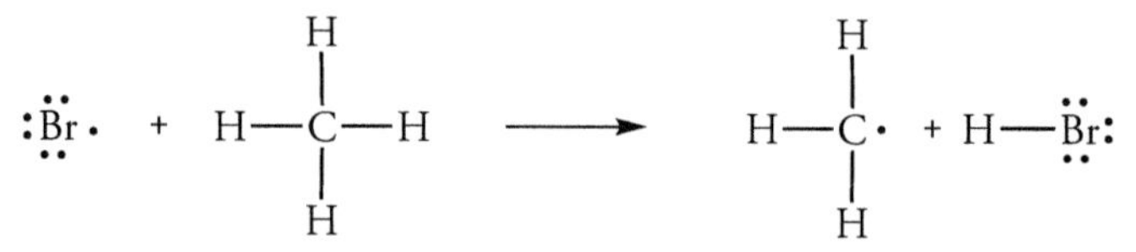

$$:\ddot{B}r\cdot\ +\ H{-}\underset{\underset{H}{|}}{\overset{\overset{H}{|}}{C}}{-}H\ \longrightarrow\ H{-}\underset{\underset{H}{|}}{\overset{\overset{H}{|}}{C}}\cdot\ +\ H{-}\ddot{\underset{\cdot\cdot}{B}}r\!:$$

(b)

$$:\ddot{B}r{-}\ddot{B}r\!:\ +\ H{-}\underset{\underset{H}{|}}{\overset{\overset{H}{|}}{C}}\cdot\ \longrightarrow\ H{-}\underset{\underset{H}{|}}{\overset{\overset{H}{|}}{C}}{-}\ddot{\underset{\cdot\cdot}{B}}r\!:\ +\ :\ddot{B}r\cdot$$

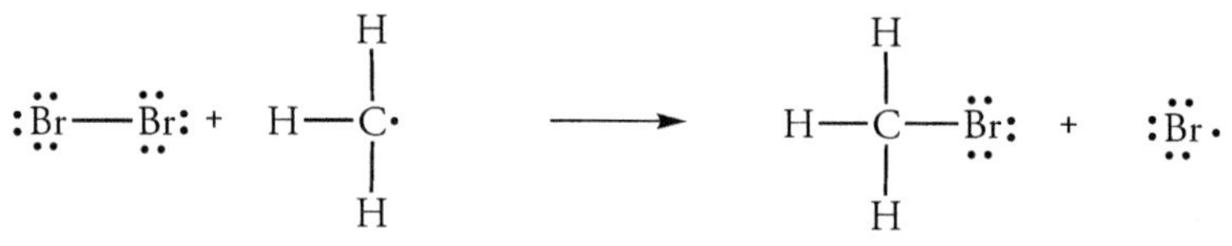

2.31   Identify the processes of bond cleavage and bond formation for each of the following reactions:

(a)

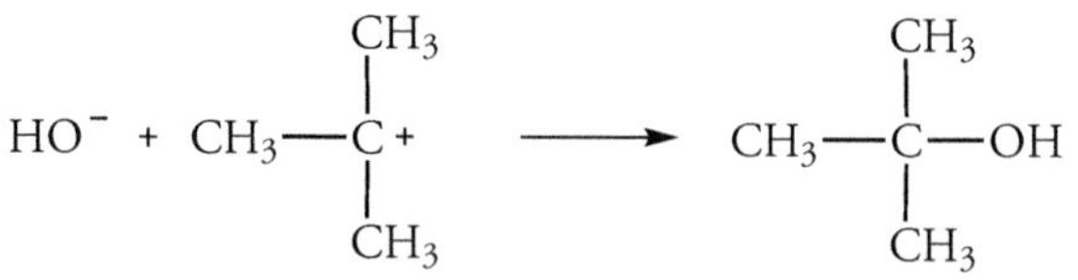

$$HO^-\ +\ CH_3{-}\underset{\underset{CH_3}{|}}{\overset{\overset{CH_3}{|}}{C}}{+}\ \longrightarrow\ CH_3{-}\underset{\underset{CH_3}{|}}{\overset{\overset{CH_3}{|}}{C}}{-}OH$$

(b)

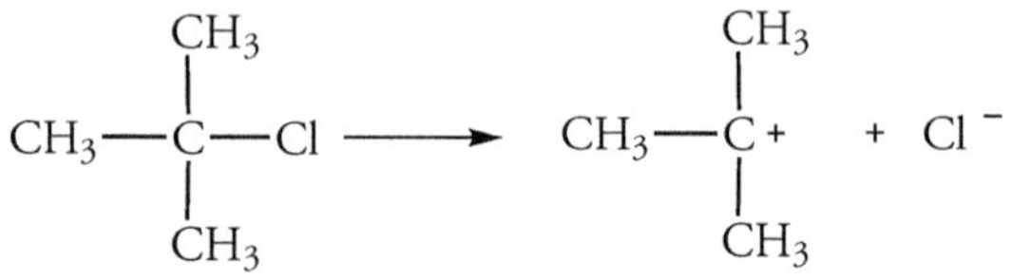

$$CH_3{-}\underset{\underset{CH_3}{|}}{\overset{\overset{CH_3}{|}}{C}}{-}Cl\ \longrightarrow\ CH_3{-}\underset{\underset{CH_3}{|}}{\overset{\overset{CH_3}{|}}{C}}{+}\ +\ Cl^-$$

2.32   Benzoyl peroxide is used in creams to control acne. It is an irritant that causes proliferation of epithelial cells. It undergoes homolytic cleavage of the oxygen-oxygen bond. Write the structure of the product, indicating all of the electrons present on all of the oxygen atoms.

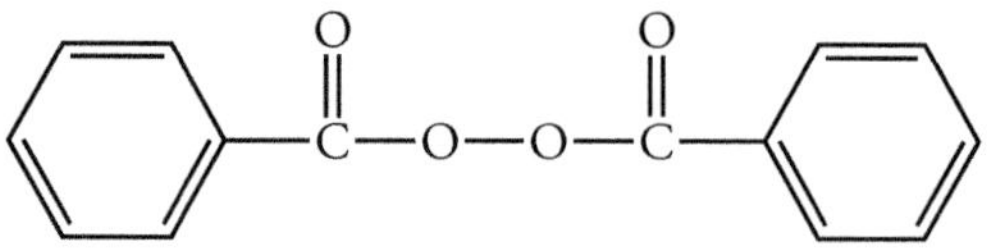

benzoyl peroxide

2.33   The oxygen-chlorine bond of methyl hypochlorite (CH_3–O–Cl) can cleave heterolytically. Based on the electronegativity values of chlorine and oxygen, predict the charges on the cleavage products.

2.34   Hydrogen peroxide (HO—OH) reacts with a proton to give a conjugate acid that undergoes heterolytic cleavage of the oxygen-oxygen bond to give water. What is the second product?

2.35   Chloromethane(CH_3—Cl) reacts with the Lewis acid $AlCl_3$ to give $AlCl_4^-$ and a carbon intermediate. What is the intermediate?

3 ALKANES AND CYCLOALKANES

MOLECULAR MODEL OF THE STEROID RING SYSTEM

Hydrocarbons, as their names indicates, contain only hydrogen and carbon. They occur as mixtures in natural gas, petroleum, and coal, which are collectively known as fossil fuels since they contain the remnants (fossils) of ancient organisms, including plants, animals, and microorganisms that have been buried for millions of years at high temperature and pressure under anaerobic conditions.

Hydrocarbons fall into two broad classes based on the types of bonds between the carbon atoms. A hydrocarbon that has only carbon-carbon single bonds is **saturated**. Hydrocarbons that contain carbon-carbon multiple bonds are **unsaturated**. Alkanes and cycloalkanes are two types of saturated hydrocarbons. **Alkanes** have carbon atoms bonded in chains; **cycloalkanes** have carbon atoms bonded to form a ring.

3.1 CLASSES OF HYDROCARBONS

Compounds that have a chain of carbon atoms, some of which are attached to functional groups, are called **acyclic** compounds, meaning "not cyclic." Compounds that contain rings of carbon atoms, and that may also contain functional groups, are **carbocyclic** compounds, commonly called cycloalkanes. One example is cyclobutane, which is shown below. Some cyclic compounds contain at least one atom in the ring other than carbon; those atoms are called **heteroatoms**. Cyclic compounds containing one or more **heteroatoms** are called **heterocyclic** compounds. The structures of an acyclic compound 2-heptanone, a carbocyclic compound, carvone, and the heterocyclic compound nicotinic acid are shown below.

Aromatic rings, which we briefly introduced in Chapter 2, are also "hydrocarbons," but they are an entirely different class of compounds, which we will discuss in Chapter 4.

$$CH_3—CH_2—CH_2—CH_3$$

butane

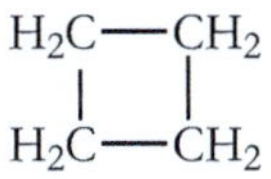

an acyclic compound

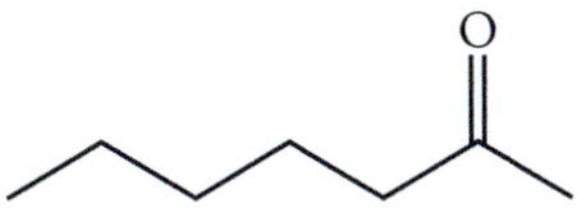

$$\begin{array}{cc} H_2C—CH_2 \\ | \quad\quad | \\ H_2C—CH_2 \end{array}$$

cyclobutane

a cyclic compound

a heterocyclic compound

2-heptanone
in oil of cloves

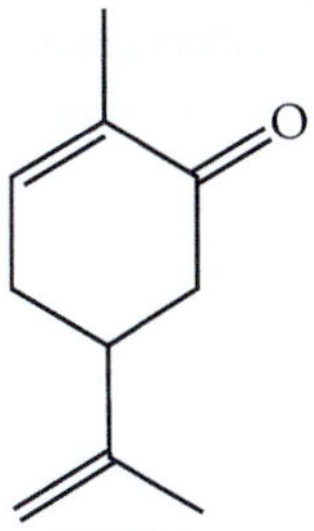

carvone
in spearmint oil

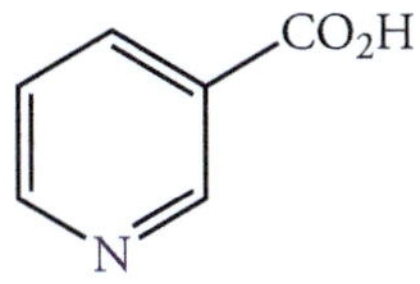

nicotinic acid
a B vitamin

3.2 ALKANES

Hydrocarbons that have no carbon-carbon double or triple bonds are called **saturated hydrocarbons**. They have a continuous chain of carbon atoms, and do not have any "branches." They are called **normal alkanes**. Their structures are often drawn with the carbon chain in a horizontal line.

Principles of Organic Chemistry. http://dx.doi.org/10.1016/B978-0-12-802444-7.00003-3

$$CH_3—CH_2—CH_2—CH_2—CH_2—CH_2—CH_2—CH_3$$
octane (a normal alkane)

The names and condensed structural formulas of 20 normal alkanes are given in Table 3.1. The first four compounds have common names. The names of the higher molecular weight compounds are derived from Greek numbers that indicate the number of carbon atoms in the chain. Each name has the suffix *-ane*, which identifies the compound as an alkane.

Table 3.1
Names of Normal Alkanes

Number of Carbon Atoms	Name	Molecular Formula
1	Methane	CH_4
2	Ethane	C_2H_6
3	Propane	C_3H_8
4	Butane	C_4H_{10}
5	Pentane	C_5H_{12}
6	Hexane	C_6H_{14}
7	Heptane	C_7H_{16}
8	Octane	C_8H_{18}
9	Nonane	C_9H_{20}
10	Decane	$C_{10}H_{22}$
11	Undecane	$C_{11}H_{24}$
12	Dodecane	$C_{12}H_{26}$
13	Tridecane	$C_{13}H_{28}$
14	Tetradecane	$C_{14}H_{30}$
15	Pentadecane	$C_{15}H_{32}$
16	Hexadecane	$C_{16}H_{34}$
17	Heptadecane	$C_{17}H_{36}$
18	Octadecane	$C_{18}H_{38}$
19	Nonadecane	$C_{19}H_{40}$
20	Eicosane	$C_{20}H_{42}$

Saturated hydrocarbons in which one or more groups are bonded to a secondary carbon are called **branched alkanes**. The carbon atom bonded to three or four other carbon atoms is the **branching point**. The carbon atom attached to the chain of carbon atoms at the branching point is part of an **alkyl group**. For example, isobutane is the simplest example of a branched alkane. It has three carbon atoms in the main chain and one branch, a –CH_3 group.

$$CH_3—CH_2—CH_2—CH_3 \qquad CH_3—\underset{\underset{CH_3}{|}}{CH}—CH_3$$

butane   isobutane

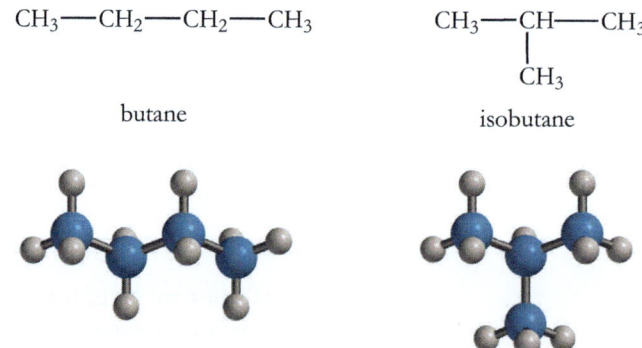

Isopentane and neopentane are isomers of pentane. Isopentane is a branched alkane with four carbon atoms in the main chain and one branching methyl group. Neopentane has three carbon atoms in the main chain and two methyl groups bonded to the central carbon. If an alkane does not have branches, it is said to be a **normal alkane**.

$$CH_3-CH_2-CH_2-CH_2-CH_3 \qquad CH_3-\underset{\underset{CH_3}{|}}{CH}-CH_2-CH_3 \qquad CH_3-\underset{\underset{CH_3}{|}}{\overset{\overset{CH_3}{|}}{C}}-CH_3$$

pentane   isopentane   neopentane

Both normal and branched alkanes have the general molecular formula C_nH_{2n+2}. For example, the molecular formula of hexane is C_6H_{14}.

$$H-\left[\underset{\underset{H}{|}}{\overset{\overset{H}{|}}{C}}-\underset{\underset{H}{|}}{\overset{\overset{H}{|}}{C}}-\underset{\underset{H}{|}}{\overset{\overset{H}{|}}{C}}-\underset{\underset{H}{|}}{\overset{\overset{H}{|}}{C}}-\underset{\underset{H}{|}}{\overset{\overset{H}{|}}{C}}-\underset{\underset{H}{|}}{\overset{\overset{H}{|}}{C}}\right]-H$$

$$H_1 + \qquad\qquad C_2H_{2n} \qquad\qquad + \; H_2$$

Each carbon atom in this normal alkane, where $n = 6$, has at least two hydrogen atoms bonded to it, which accounts for the 2n in the general formula. Each of the two terminal carbon atoms has another hydrogen atom bonded to it, which accounts for the +2 in the subscript on hydrogen in the general formula.

Classification of Carbon Atoms

Hydrocarbon structures are classified according to the number of carbon atoms directly bonded to a specific carbon atom. We will use this classification in later chapters to describe the reactivity of functional groups attached at the various carbon atoms in a structure.

A carbon atom bonded to only one other carbon atom is a **primary carbon atom**, which is designated by the symbol 1°. The carbon atom at each end of a carbon chain is primary. For example, butane has two primary carbon atoms. A carbon atom that is bonded to two other carbon atoms is a **secondary carbon atom**, designated by the symbol 2°. For example, the middle carbon atoms of butane are secondary (Figure 3.1a).

Figure 3.1
Classification of Carbon Atoms

(a) The terminal carbon atoms of butane are primary (1°); they are bonded directly to one other carbon atom. The internal carbon atoms are secondary; they are bonded to two carbon atoms. (b) The terminal carbon atoms of isobutane are primary; they are bonded to one other carbon atom. The internal carbon atom is tertiary (3°); it is bonded to three carbon atoms.

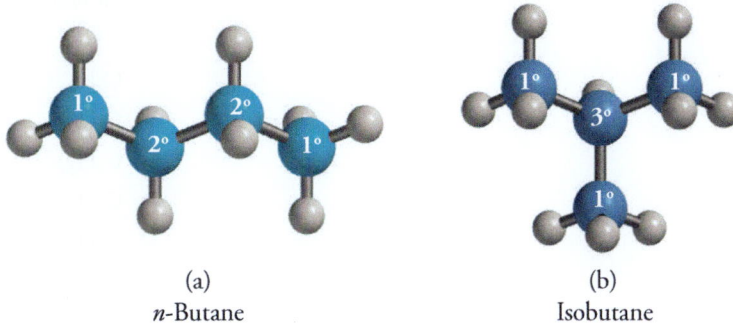

(a)
n-Butane

(b)
Isobutane

A carbon atom bonded to three other carbon atoms is **tertiary**, and is designated by 3°. For example, when we examine the structure of isobutane, we see that one of the four carbon atoms is tertiary; the other three are primary (Figure 3.1b). A **quaternary carbon** atom (4°) is bonded to four other carbon atoms.

Problem 3.1

The following compound is a sex attractant released by the female tiger moth. Classify the carbon atoms in this compound as primary, secondary, or tertiary.

$$\underset{\displaystyle CH_3CHCH_2CH_2CH_2CH_2CH_2CH_2CH_2CH_2CH_2CH_2CH_2CH_2CH_3}{\overset{\displaystyle CH_3}{|}}$$

Solution

Each of the two terminal carbon atoms and the branching $-CH_3$ group are primary carbon atoms, because each is bonded to only one other carbon atom. The second carbon atom from the left is bonded to two atoms in the chain as well as to the branching $-CH_3$ group, so it is tertiary. All 14 remaining carbon atoms are bonded to two carbon atoms, so they are secondary.

Problem 3.2

Pentaerythritol tetranitrate is a drug used to reduce the frequency and severity of angina attacks. Classify the carbon atoms in this compound.

$$NO_2-O-CH_2-\underset{\displaystyle CH_2O-NO_2}{\overset{\displaystyle CH_2O-NO_2}{\underset{|}{\overset{|}{C}}}}-CH_2O-NO_2$$

pentaerythritol tetranitrate

3.3 NOMENCLATURE OF ALKANES

Many alkanes exist as isomers. There are two isomers of C_4H_{10}, called butane and isobutane, and three isomers of C_5H_{12}, called pentane, isopentane, and neopentane. It is easy to learn their names. However, as the number of carbon atoms in an alkane increases, the number of isomers increases geometrically (Table 3.2). Many of these possible isomers have never been found in petroleum or produced in a chemistry laboratory, but each could be made in the laboratory. A system for naming the many isomeric alkanes is clearly necessary.

IUPAC Rules for Naming Alkanes

Alkanes and other organic compounds are named by the rules set forth by the International Union of Pure and Applied Chemistry (IUPAC). When these rules are followed, every chemical compound has a unique name. The IUPAC name consists of three parts: prefix, parent, and suffix.

Table 3.2
Number of Alkane Isomers

Molecular Formula	Number of Isomers
CH_4	1
C_2H_6	1
C_3H_8	1
C_4H_{10}	2
C_5H_{12}	3
C_6H_{14}	5
C_7H_{16}	9
C_8H_{18}	18
C_9H_{20}	35
$C_{10}H_{22}$	75
$C_{20}H_{42}$	336,319
$C_{30}H_{62}$	4,111,846,763
$C_{40}H_{82}$	62,491,178,805,831

The **parent** is the longest continuous carbon chain in a molecule that contains the functional group. The **suffix** identifies the functional group for most classes of organic compounds. A parent alkane has the ending *-ane*. Some functional groups, such as the halogens, are identified in the prefix. For example, the prefixes chloro and bromo identify chlorine and bromine. The prefix also indicates the identity and location of any branching alkyl groups. An alkane that has "lost" one hydrogen atom is called an **alkyl** group. Alkyl groups are named by replacing the *-ane* ending of an alkane with *-yl*. The parent name of CH_4 is methane. Thus, CH_3- is a methyl group. The parent name of C_2H_6 is ethane, so CH_3CH_2- is an ethyl group.

The general shorthand representation of an alkyl group is R–, which stands for the "rest" or "remainder" of the molecule.

The name of an alkane specifies the length of the carbon chain and the location and identity of alkyl groups attached to it.

The IUPAC rules for naming alkanes are as follows:

1.   The longest continuous chain of carbon atoms is the parent.

The parent chain contains five carbons

If two possible parent chains have the same number of carbon atoms, the parent is the one with the larger number of branch points.

The parent chain has six carbons with two branches: a methyl group and an ethyl group

The six carbon parent chain with one branch is not correct

2.   Number the carbon atoms in the longest continuous chain starting from the end of the chain nearer the first branch.

$$CH_3\overset{1}{-}\overset{\underset{\displaystyle CH_3}{\overset{\displaystyle H}{|}}}{\underset{2}{C}}-\overset{3}{C}H_2-\overset{\underset{\displaystyle H}{\overset{\displaystyle CH_2-CH_3}{|}}}{\underset{4}{C}}-\overset{5}{C}H_2-\overset{6}{C}H_3$$

The parent chains contains six carbons, so it is a hexane with two branches: a methyl group at C-2 and an ethyl group at C-4

If the first branch occurs at an equal distance from each end of the chain, number from the end that is nearer the second branch.

$$\underset{8}{C}H_3-\overset{\underset{\displaystyle CH_3}{\overset{\displaystyle H}{|}}}{\underset{7}{C}}-\overset{}{\underset{6}{C}}H_2-\overset{}{\underset{5}{C}}H_2-\overset{\underset{\displaystyle H}{\overset{\displaystyle CH_2CH_3}{|}}}{\underset{4}{C}}-\overset{}{\underset{3}{C}}H_2-\overset{\overset{\displaystyle CH_3}{|}}{\underset{2}{C}}H-\overset{}{\underset{1}{C}}H_3$$

The parent has eight carbons, so it is an octane. It has methyl groups at C-2 and C-7 and an ethyl group at C-4

3. Each branch or substituent has a number that indicates its location on the parent chain. When two substituents are located on the same carbon atom, each must be assigned the same number.

$$\underset{8}{C}H_3-\overset{}{\underset{7}{C}}H_2-\overset{\underset{\displaystyle CH_3}{\overset{\displaystyle H}{|}}}{\underset{6}{C}}-\overset{}{\underset{5}{C}}H_2-\overset{\underset{\displaystyle CH_3}{\overset{\displaystyle CH_2CH_3}{|}}}{\underset{4}{C}}-\overset{}{\underset{3}{C}}H_2-\overset{\overset{\displaystyle CH_3}{|}}{\underset{2}{C}}H-\overset{}{\underset{1}{C}}H_3$$

This octane has methyl groups at C-2, C-4, and C-6, and an ethyl group at C-4

4. The number for the position of each alkyl group is placed immediately before the name of the group and is joined to the name by a hyphen. Alkyl groups and halogen atoms are listed in alphabetical order.

$$CH_3\overset{1}{-}\overset{\underset{\displaystyle CH_3}{\overset{\displaystyle H}{|}}}{\underset{2}{C}}-\overset{3}{C}H_2-\overset{\underset{\displaystyle H}{\overset{\displaystyle CH_2-CH_3}{|}}}{\underset{4}{C}}-\overset{5}{C}H_2-\overset{6}{C}H_3$$

The name of this compound is 4-ethyl-2-methylhexane, *not* 2-methyl-4-ethylhexane

Two or more groups of the same type are indicated by the prefixes di-, tri-, tetra-, and so forth. The numbers that indicate the locations of the branches are separated by commas.

$$CH_3\overset{1}{-}\overset{\underset{\displaystyle CH_3}{\overset{\displaystyle H}{|}}}{\underset{2}{C}}-\overset{3}{C}H_2-\overset{\underset{\displaystyle H}{\overset{\displaystyle CH_3}{|}}}{\underset{4}{C}}-\overset{5}{C}H_2-\overset{6}{C}H_3$$

The name of this compound is 2,4-dimethylhexane

5. The prefixes di-, tri-, tetra-, and so forth do not alter the alphabetical ordering of the alkyl groups.

$$CH_3\overset{1}{-}\overset{\underset{\displaystyle CH_3}{\overset{\displaystyle H}{|}}}{\underset{2}{C}}-\overset{3}{C}H_2-\overset{4}{C}H_2-\overset{\underset{\displaystyle CH_3}{\overset{\displaystyle CH_2CH_3}{|}}}{\underset{5}{C}}-\overset{6}{C}H_2-\overset{7}{C}H_2-\overset{8}{C}H_3$$

The name of this compound is 5-ethyl-2,5-dimethyloctane, *not* 3,5-dimethyl-5-ethyloctane

Names of Alkyl Groups

There is only one alkyl group derived from methane and ethane. However, for a longer chain of carbon atoms, several isomeric alkyl groups are usually possible depending on which carbon atom "loses" a hydrogen atom. Many of these alkyl groups are known by their common names. For example, propane has two primary carbon atoms and a secondary carbon atom. If a primary carbon atom loses a hydrogen atom, a primary alkyl group, propyl, is produced. Propyl and other primary alkyl groups derived from normal alkanes are **normal alkyl groups**. The term "normal" means that the alkane has no branches. If the secondary carbon atom of propane loses a hydrogen atom, a secondary alkyl group known as the isopropyl group is formed. The abbreviations of propyl and isopropyl are *n*-Pr and *i*-Pr.

$$CH_3-CH_2-\overset{1^\circ}{CH_2}- \qquad\qquad CH_3-\overset{\overset{\displaystyle H}{|}}{\underset{|}{\overset{2^\circ}{C}}}-CH_3$$

n-propyl   iso-propyl

Several alkyl groups can be derived from the two isomers of butane, C_4H_{10}. These alkyl groups all have the formula C_4H_9. Two alkyl groups are derived from butane and two from isobutane. If a primary carbon atom of butane loses a hydrogen atom, an *n*-butyl group results; if a secondary carbon atom of butane loses a hydrogen atom, a secondary alkyl group, *sec*-butyl, forms.

$$CH_3-CH_2-CH_2-\overset{1^\circ}{CH_2}- \qquad\qquad CH_3-CH_2-\overset{\overset{\displaystyle CH_3}{|}}{\overset{2^\circ}{CH}}-$$

n-butyl   sec-butyl

Removing a hydrogen atom from a primary carbon atom of isobutane gives a primary alkyl group called the isobutyl group. Removing a hydrogen atom from the tertiary carbon atom of isobutane gives a tertiary alkyl group called the *tert*-butyl (*t*-butyl) group. Thus, there are four isomeric C_4H_9- alkyl groups.

$$CH_3-\overset{\overset{\displaystyle \overset{1^\circ}{CH_2}-}{|}}{\underset{\underset{\displaystyle CH_3}{|}}{C}}-H \qquad\qquad CH_3-\overset{\overset{\displaystyle CH_3}{|}}{\underset{\underset{\displaystyle CH_3}{|}}{\overset{3^\circ}{C}}}-$$

iso-butyl   tert-butyl

Alkyl groups are named by an IUPAC procedure similar to that used to name alkanes with the longest continuous chain beginning at the branch point. For example, the IUPAC name for an isopropyl group is 1-methylethyl, and the IUPAC name for an isobutyl group is 2-methylpropyl. The point of attachment of the alkyl group is numbered carbon 1.

$$CH_3-\overset{\overset{\displaystyle CH_3}{|}}{\underset{\underset{\displaystyle H}{|}}{C}}- \qquad\qquad CH_3-\overset{\overset{\displaystyle CH_2-}{|}}{\underset{\underset{\displaystyle CH_3}{|}}{C}}-H$$

2-methylethyl   2-methylpropyl
(isopropyl)   (sec-butyl)

Complex alkyl groups are enclosed within parentheses when used to name hydrocarbons. Thus, 4-isopropylheptane is also 4-(1-methylethyl)heptane. The methyl within parentheses shows that it modifies ethyl, not heptane. The nonsystematic names for the alkyl groups containing three and four carbon atoms are commonly used, and IUPAC rules allow for their continued use.

Problem 3.3
Name the following compound, which is produced by the alga *Spirogyra*.

$$CH_3CHCH_2CH_2CH_2CHCH_2CH_2CH_2CHCH_2CH_2CH_2CH_3CHCH_2CH_3$$

with CH_3 substituents as drawn.

Solution

The longest continuous chain has 16 carbon atoms, and is named as a substituted hexadecane. The chain is numbered from left to right to locate the four methyl groups at positions 2, 6, 10, and 14. The compound is 2,6,10,14-tetramethylhexadecane.

Problem 3.4

Name the following compound.

$$CH_3-\underset{\underset{CH_3}{|}}{\overset{\overset{CH_3}{|}}{C}}-CH_2-\underset{\underset{H}{|}}{\overset{\overset{CH_2CH_3}{|}}{C}}-CH_3$$

Problem 3.5

Identify the alkyl group on the left of the benzene ring in ibuprofen, an analgesic present in Nuprin®, Advil®, and Motrin®.

$$CH_3-\underset{\underset{H}{|}}{\overset{\overset{CH_3}{|}}{C}}-CH_2-\text{[benzene ring]}-CH-CO_2H \quad (CH_3)$$

Solution

There are four carbon atoms in the alkyl group, which is derived from isobutane—not butane. The benzene ring is bonded to the terminal carbon atom—not the internal carbon atom. This group is the isobutyl group.

$$CH_3-\underset{\underset{H}{|}}{\overset{\overset{CH_3}{|}}{C}}-CH_2-\text{benzene ring}$$

3.4 CONFORMATIONS OF ALKANES

In Chapter 1, we saw that ethane can exist in various spatial arrangements, called **conformations**, which result from rotation of the CH_3 groups around the carbon-carbon σ bond. When the CH_3 groups rotate around the C—C bond, the positions of the hydrogen atoms change with respect to one another, but the connectivities of all the bonds remain the same. Thus, various conformations have different shapes, but are not structural isomers.

The study of the chemical and physical properties of different conformations of organic compounds is called **conformational analysis**. To understand the relationship between structure and physical properties, we need to know how structural differences change the conformations of molecules and which conformations predominate at equilibrium. To understand the relationship between structure and chemical reactivity, we must know the energy difference between the most stable conformation and the conformation required to bring atoms into proximity for reaction. If a substantial conformational change is required to "prepare" a molecule for reaction, then the energy associated with that change affects the rate of the reaction.

In succeeding sections we will study the conformations of small organic molecules, such as ethane, propane, butane, and cyclohexane. At first glance, this might seem to be an unpromising topic. However, once we understand the conformations of small molecules, we will be able to apply conformational concepts to much larger molecules, such as carbohydrates or complex drugs. The conformation of a molecule accounts for its highly specific biological function.

Conformations of Ethane

Rotation around the C-1 to C-2 bond interconverts conformational isomers. Figure 3.2 shows two such arrangements. The hydrogen atoms are in a different spatial relation to one another in the two structures. These two conformational isomers are **conformers**. Since the conformational isomers are interconverted by bond rotation, they are sometimes called **rotamers**.

Ethane can exist in an infinite number of conformations. The conformation in which the hydrogen atoms and the bonding electrons are the farthest away from one another has the lowest energy. This conformation is **staggered**. The conformation in which the hydrogen atoms are closest to one another has the highest energy. This conformation is **eclipsed**. In the eclipsed conformation, each C—H bond on one carbon atom lines up with a C–H bond on another carbon atom, as the moon sometimes eclipses the sun. "Sawhorse" representations of the conformations of ethane are shown in Figure 3.2. These representations are three-dimensional, and show the carbon–carbon bond as well as all of the C—H bonds.

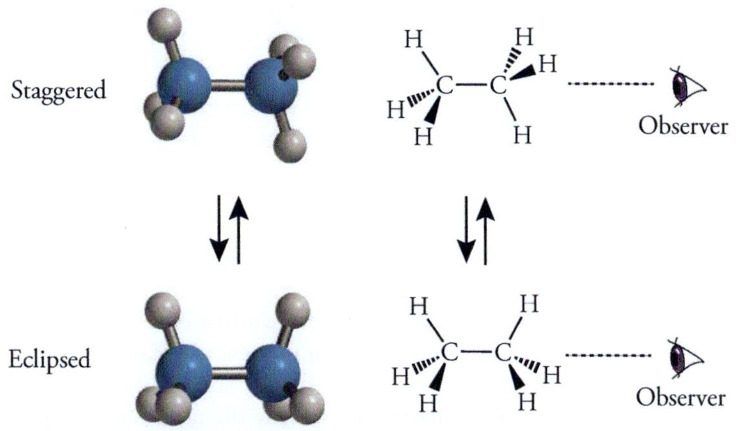

Figure 3.2
Conformations of Ethane
Rotating the methyl group on the right by 60° converts a staggered conformation into an eclipsed conformation. Viewing the carbon-carbon bond end-on in the eclipsed conformation we would see the carbon atom and three hydrogens of the carbon on the right. The left carbon and its three hydrogens would be hidden.

Newman Projection Formulas

Newman projection formulas of structures concentrate on the two carbon atoms around which rotation may occur. The two atoms are viewed end-on. The "front" atom is represented by a point with three bonds. The "back" atom is represented by a circle with three bonds that reach only to the perimeter of the circle. Although there is a bond between the two carbon atoms, it is hidden because it is located along the viewing axis.

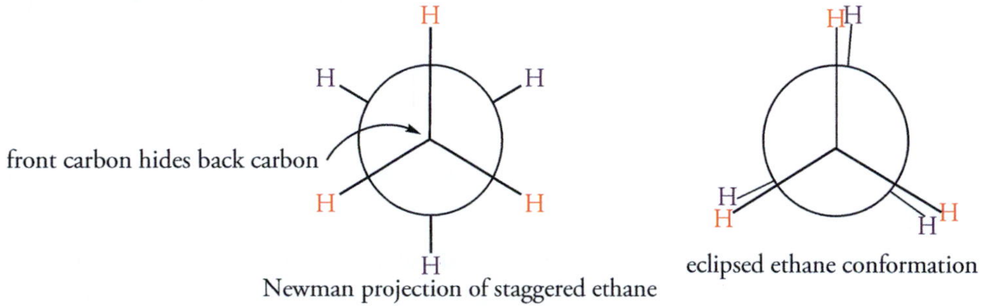

A Newman projection of the eclipsed conformation of ethane shows only the three C—H bonds of the front carbon atom. The bonds and hydrogen atoms at the back are hidden by the front eclipsing bonds and hydrogen atoms. However, the bonded hydrogen atoms of the back carbon atom can be shown by viewing the conformation slightly off the bond axis so that all bonds can be seen.

Why is the eclipsed conformation of higher energy than the staggered conformation? We recall that the VSEPR model predicts that electron pairs and bonded atoms should be separated by the maximum distance. Consider the positions of the bonding pairs of electrons of the C—H bonds on the front and back atoms in the Newman projection formula. The angle between these bonds is called the torsional or dihedral angle. In the eclipsed conformation the torsional angle is 0° whereas in the staggered conformation the torsional angle is 60°. The eclipsed conformation has a higher energy because of torsional strain, which is due to the small repulsion between the bonding electrons in the C—H bonds. Each hydrogen-hydrogen eclipsing interaction amounts to 4 kJ/mol. The total difference in energy between the eclipsed and staggered conformations (12 kJ/mol) is small, and there is rapid interconversion among these conformations. Thus, we say that rotation about the C—C bond is virtually free or unrestricted. Different conformations of ethane are not different compounds but different forms of a single molecule. Ethane is a mixture of conformations, but they cannot be separated from each other. However, at room temperature the staggered conformation is the major form, and any given molecule spends most of its time in a staggered conformation.

eclipsed ethane conformation
(viewed slightly from side)

rotate around C—C bond

staggered ethane conformation

Conformations of Butane

All acyclic alkanes exist as mixtures of conformations that result from rotation about every single bond in their structure. In the case of butane, the staggered conformations are more stable than eclipsed conformations. For butane and higher molecular weight alkanes there is more than one staggered conformation. For butane the most stable staggered conformation has a zigzag arrangement of the carbon-carbon bonds.

A Newman projection viewed along the C(2)-C(3) bond shows that the two methyl groups in this staggered conformation are the maximum distance apart in an anti conformation. A second staggered conformation called the **gauche** conformation is also possible.

anti butane

gauche butane

torsion angle, θ = 60°

There is no torsional strain in either of these two conformations because the torsional angles between the bonding pairs is 60° in both cases. However, the methyl groups are closer to each other in the gauche than in the anti conformation. The interference and repulsion of the electron clouds of the atoms of the two methyl groups with each other is called steric strain. The anti conformation of butane is more stable than the gauche conformation by 3.8 kJ/mole. The anti and gauche conformations interconvert rapidly, but the ratio of anti to gauche conformation is about 2:1. The same types of interactions occur in higher alkanes.

3.5 CYCLOALKANES

Cycloalkanes with one ring have the general formula C_nH_{2n} compared with the general formula $C_nH_{(2n+2)}$ for acyclic alkanes. Cycloalkanes have two fewer hydrogen atoms than alkanes, because another carbon-carbon bond is needed to form the ring. Cycloalkanes are drawn as simple polygons in which the sides represent the carbon-carbon bonds. It is understood that each corner of the polygon is a carbon atom bonded to two hydrogen atoms.

cyclopropane   cyclobutane   cyclopentane   cyclohexane

Multiple rings in a molecule can share one or more common atoms. **Spirocyclic** compounds share one carbon atom between two rings. These compounds are relatively rare in nature. Fused ring compounds share two common atoms and the bond between them. These compounds are prevalent in steroids, which contain four fused rings. **Bridged ring** compounds share two nonadjacent carbon atoms, which are called the bridgehead carbon atoms. These compounds are less prevalent in nature than **fused ring** compounds, but more common than spirocyclic compounds. A few examples are shown below.

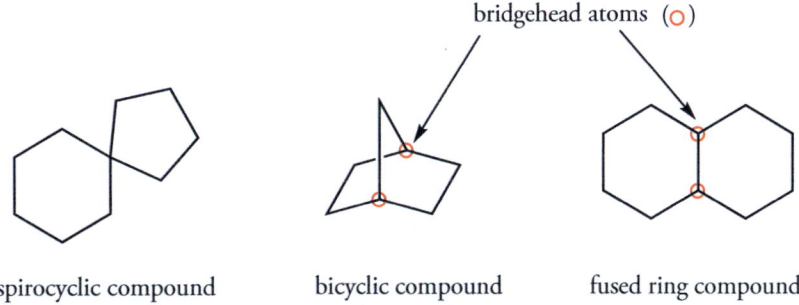

bridgehead atoms (o)

spirocyclic compound   bicyclic compound   fused ring compound

Each ring system shown above has two rings: they are **bicyclic** compounds. This is obvious for the spirocylic compound and the fused ring compound. The bridged ring compound appears at first sight to have three rings, but it has only two. We can determine how many rings are present in a ring system by determining the minimum number of "cuts" that are required to give an acyclic compound. Two such cuts are necessary for the bridged ring system shown above. The first cut gives a single ring, and the second cut gives an acylic compound.

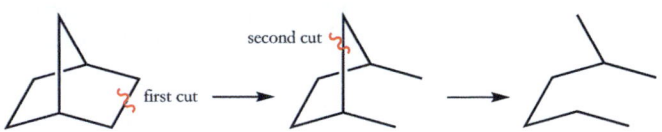

second cut   first cut ⟶   ⟶

Geometric Isomerism

Isomers exist with different carbon skeletons, different functional groups, and different functional group locations. These isomers have different sequential arrangements of atoms. Now let us consider a different type of isomerism. Compounds that have the same sequential arrangement of atoms but different spatial arrangements are **geometric isomers**.

We will begin with cyclopropane, whose three carbon atoms define a plane. Any group attached to the ring may be held "above" or "below" the plane of the ring. If we attach two methyl groups on adjacent carbon atoms on the same side of the plane of the ring, the substance is called a *cis* isomer; it is *cis*-1,2-dimethylcyclopropane. If the two methyl groups are attached on the opposite sides of the plane of the ring, the compound is the ***trans*** isomer. Thus, 1,2-dimethylcyclopropane can exist as both *cis* and *trans* isomers. They are geometric isomers. Note that *cis* and *trans* compounds are not two conformations of the same molecule, but are isomeric substances that have different physical properties. It is impossible to convert one isomer into the other without breaking a bond. In the following structures the cyclopropane ring is viewed as perpendicular to the plane of the page, and the —CH$_2$—group extends "behind the page," away from the viewer (Figure 3.3). Geometric isomers have different physical properties. It is impossible to convert one geometric isomer into the other without breaking a bond.

Figure 3.3
Geometric Isomers of
1,2-dimethylcyclopropane
The ring atoms are in the plane of the page. Heavy wedge-shaped lines indicate groups above the ring plane, and dashed wedges indicate groups below the plane of the ring. The molecular models show the geometries of these isomers more clearly.

cis isomer
The methyl groups are on the same side of the ring plane

trans isomer
The methyl groups are on opposite sides of the ring plane

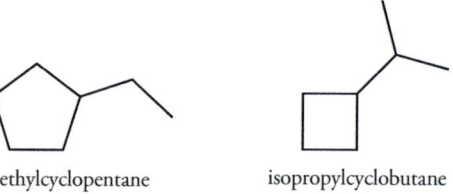

cis-1,2-dimethylcyclopropane   trans-1,2-dimethylcyclopropane

Nomenclature of Cycloalkanes

Cycloalkanes are named according to the IUPAC system by using the prefix *cyclo-*. When only one position contains a functional group or alkyl group, only one compound is possible, as in ethylcyclopentane and isopropylcyclobutane, whose structures are shown below.

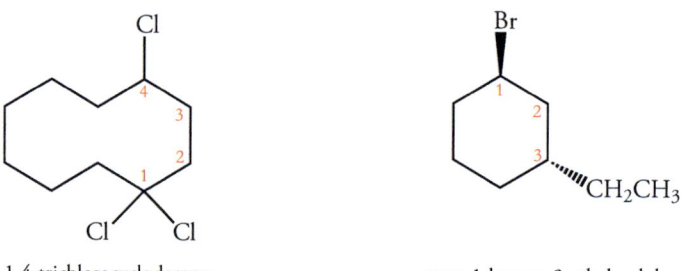

ethylcyclopentane   isopropylcyclobutane

When more than one group is attached to the ring, the ring atoms are numbered. One substituent is at position I, and the ring is numbered in a clockwise or counterclockwise direction to give the lower number to the position with the next substituent attached to the ring, as in 1,1,4-trichlorocyclodecane and *trans*-1-bromo-3-ethylcyclohexane.

1,1,4-trichlorocyclodecane   trans-1-bromo-3-ethylcyclohexane

Problem 3.6
Adamantane has a carbon skeleton also found as part of the structure of diamond. Amantadine, which has an amino group bonded to the adamantane structure , is useful in the prevention of infection by influenza A viruses. What are the molecular formulas of adamantane and amantadine? How many rings are in each structure?

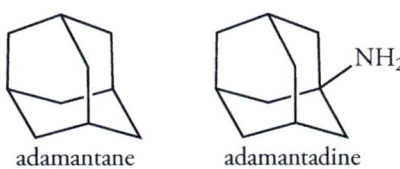

adamantane   adamantadine

Solution

Adamantane has 10 carbon atoms. Four of these carbon atoms are tertiary; they have bonds to three other carbon atoms and one bond to a hydrogen atom. The remaining six carbon atoms are secondary; they have bonds to two other carbon atoms and two bonds to hydrogen atoms. The total number of hydrogen atoms is $4(1) + 6(2) = 16$. The molecular formula of adamantane is $C_{10}H_{16}$. Amantadine has an amino group (—NH2) in place of a hydrogen atom at one of the tertiary carbon atoms. Thus, the molecular formula of amantadine differs from adamantane by one nitrogen atom and one hydrogen atom. The molecular formula is $C_{10}H_{17}N$.

Problem 3.7

Determine the molecular formula of menthol based on the bond-line structure shown below.

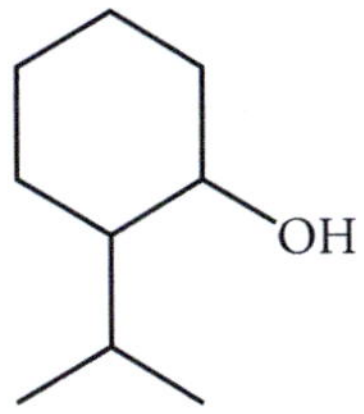

menthol

Problem 3.8

Disparlure, the sex attractant pheromone of the female gypsy moth, has the following general structure. Are geometric isomers possible for this molecule?

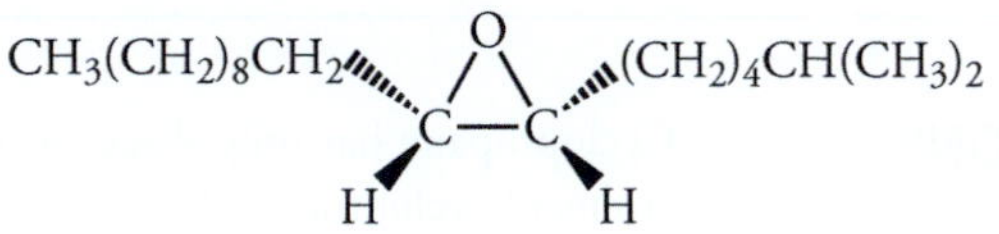

$$CH_3(CH_2)_8CH_2\text{''''} \quad \text{''''}(CH_2)_4CH(CH_3)_2$$

disparlure

Solution

The three-membered heterocyclic ring contains two carbon atoms and one oxygen atom. Each carbon atom of the ring has a hydrogen atom and a large alkyl group bonded to it. These alkyl groups could be located *cis* or *trans* with respect to the plane of the ring. The *cis* isomer is the biologically active compound.

Problem 3.9

Brevicomin, the sex attractant of a species of pine beetle, has the following structure. Write the structure of a geometric isomer of brevicomin.

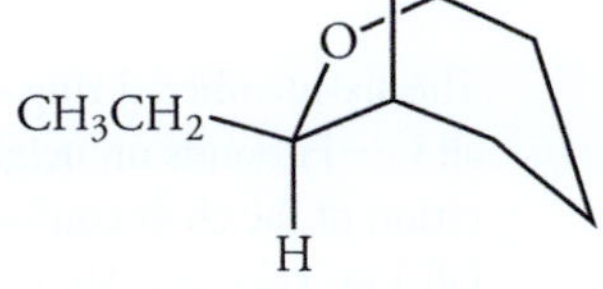

brevicomin

Problem 3.10
What is the name of the following compound?

The ring must be numbered starting from one carbon atom with a chlorine atom and counting toward the other carbon atom with a chlorine atom in the direction that will give the lower position number. Starting with the carbon atom at the "4 o'clock" position and numbering clockwise gives the number 3 to the atom at the "8 o'clock" position.

Solution
The two chlorine atoms are on the same side of the plane of the ring, and the correct name is *cis*-1,3-dichlorocyclohexane.

Problem 3.11
What are the names of the following compounds?

(a)   (b)   (c)

3.6 CONFORMATIONS OF CYCLOALKANES

Cyclopropane has only three carbon atoms, so it is a planar molecule (Figure 3.4). The hydrogen atoms of cyclopropane lie above and below the plane of the ring of carbon atoms, and they eclipse hydrogen atoms on adjacent carbon atoms. All of the C—C—C bond angles are only 60° because the carbon atoms form an equilateral triangle. Cyclobutane and cyclopentane are not planar and exist in slightly "puckered" conformations, which reduce some of the eclipsing of hydrogen atoms on adjacent carbon atoms. The conformations of these compounds will not be considered further. For the purposes of this text, we will treat them as planar rings of carbon atoms.

Figure 3.4
Structure of Cyclopropane
Cyclopropane is planar molecule with essentially no conformational mobility.

Cyclohexane

The six-membered ring of cyclohexane is not planar; it exists in a puckered conformation in which all C—H bonds on neighboring carbon atoms are staggered. Figure 3.5 shows a bond-line representation of the chair conformation of cyclohexane. Note that the hydrogen atoms in this conformation fall into two sets. Six of the hydrogen atoms are axial; they point up or down with respect to the average plane of the ring of carbon atoms.

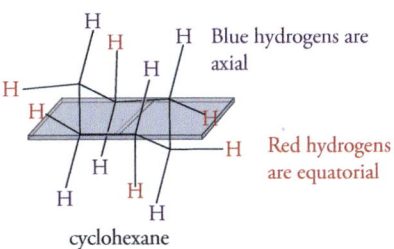

H Blue hydrogens are axial

H Red hydrogens are equatorial

cyclohexane

Three of the axial hydrogen atoms point up and the other three point down. The up-down relationship alternates from one carbon atom to the next. The remaining six hydrogen atoms—called equatorial atoms—lie approximately in the average plane of the ring. Each carbon atom has one equatorial and one axial C—H bond. Cyclohexane is conformationally mobile. Different chair conformations can interconvert and, as a result, the axial hydrogens become equatorial and the equatorial positions become axial (Figure 3.6).

Figure 3.5
Conformation of Cyclohexane
The equatorial C—H bonds lie in a band around the "equator" of the ring. Each carbon atom has one axial hydrogen that is perpendicular to the plane of the ring. The axial hydrogens alternate up and down moving from any axial hydrogen on one carbon to the adjacent carbon.

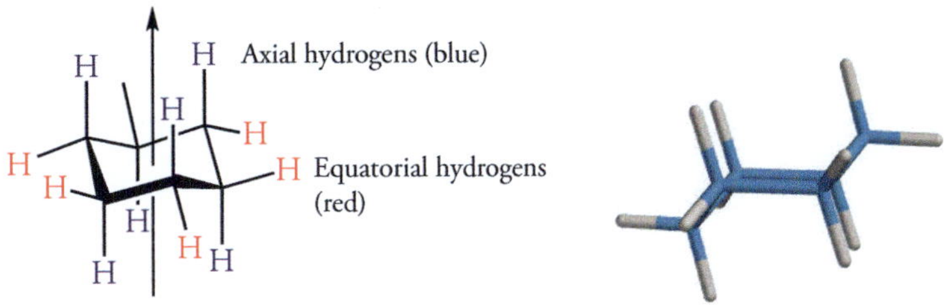

H Axial hydrogens (blue)

H Equatorial hydrogens (red)

axis

We can see this process more clearly by practicing with molecular models. To flip a cyclohexane ring, we hold the four "middle" atoms in place while pushing one "end" carbon downward and the other upward. At each of the two "end" atoms, an equatorial position becomes an axial position and axial one becomes equatorial. And, though it is not easy to see, the hydrogen atoms on every other carbon atom undergo the same transformation.

Figure 3.6
"Ring Flipping" of Cyclohexane
The interconversion of two chair conformations of cyclohexane changes all equatorial hydrogens to axial hydrogens and all axial hydrogens to equatorial hydrogens.

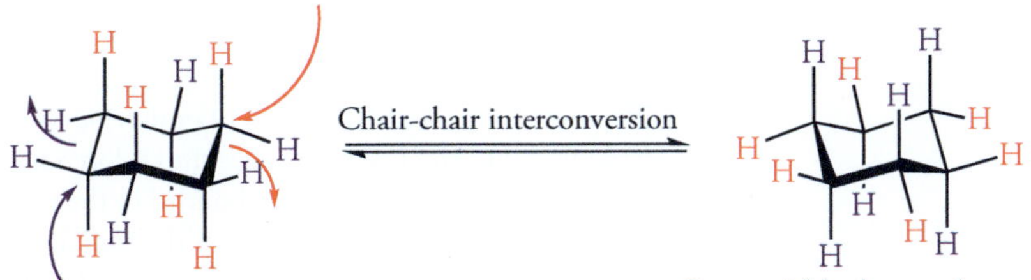

Rotate around bond, top carbon moves down

Chair-chair interconversion

Rotate around bond, bottom carbon moves up

Equatorial hydrogens becomes axial, axial hydrogens become equatorial

Now consider methylcyclohexane in a chair conformation. When the ring flips, the equatorial methyl group moves into an axial position. An equatorial methyl group is more stable than an axial methyl group (Figure 3.7). These two structures are different conformations, *not* isomers. The chair-chair interconversion occurs rapidly. An equatorial methyl group is more stable than an axial methyl group, and 9% of the mixture at equilibrium has an equatorial methyl group. The conformation with an axial methyl group is less stable because there is steric strain between the methyl group and the axial hydrogen atoms at C-3 and C-5. All substituted cyclohexanes behave similarly. Thus, the equatorial conformation is always more stable than the axial one, although the relative amounts of the two conformations vary depending on the substituent.

Figure 3.7
Conformations of Methyl Cyclohexane

Methylcyclohexane rapidly interconverts between two conformations of unequal energy. At room temperature, 95% of the conformations have an equatorial methyl group, and 5% have an axial methyl group. The axial conformation has unfavorable interactions with axial hydrogens at C-3 and C-3'.

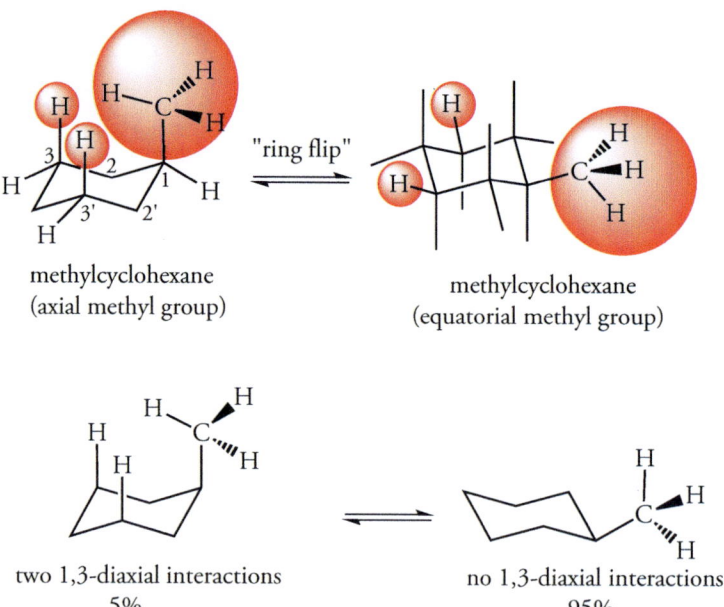

methylcyclohexane (axial methyl group)   methylcyclohexane (equatorial methyl group)

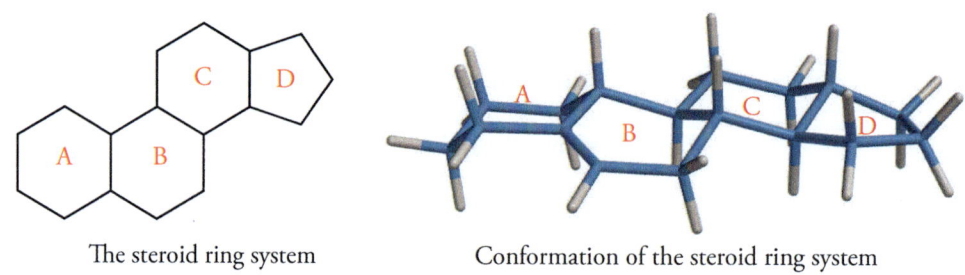

two 1,3-diaxial interactions   no 1,3-diaxial interactions
5%   95%

Steroids

Steroids are tetracyclic compounds containing three six-membered rings and a five-membered ring. Each ring is designated by a letter, and the carbon atoms are numbered by the standard system shown. These compounds contain a variety of functional groups such as hydroxyl, carbonyl, and carbon-carbon double bonds.

The steroid ring system   Conformation of the steroid ring system

The several sites of fusion of two rings may be either *cis* or *trans*, but *trans* fusion is more common. As a result of the *trans* fusion of rings, the cyclohexane rings cannot undergo a ring-flipping process. As a consequence, functional groups bonded to the rings have well-defined positions. The biological activity of steroid hormones depends on the functional groups, their locations on the ring, and whether the groups are axial or equatorial. For example, the hydroxyl group at the α-position in androsterone (a male sex hormone) is located in the axial position. The isomeric compound with an equatorial (β) hydroxyl group, epiandrosterone, has significantly less physiological activity. The activity is related to the spatial arrangement of the hydroxyl group in ring A and the carbonyl group in ring D, which determines whether or not the steroid can bind to specific receptors (Figure 3.8).

Cholesterol is one of the most important steroids best known to the public. Although cholesterol in the diet affects the cholesterol level in blood and presents a health hazard to many people, it is synthesized by most animals and is required in various cell membranes, including those of the brain. A metabolic product of cholesterol called 27-hydroxycholesterol plays a significant part in many cancers. Thus, the metabolism of cholesterol has far-reaching physiological effects.

Cholesterol plays a vital biological role in the formation of other steroid hormones; it is converted to progesterone (a female sex hormone) by several reactions, including the shortening of the chain attached at the C-17 position (ring D). The centrality of progesterone in human physiology is illustrated by its conversion into cortisone as well as testosterone (a male sex hormone).

Figure 3.8
Conformations of Steroid Hormones

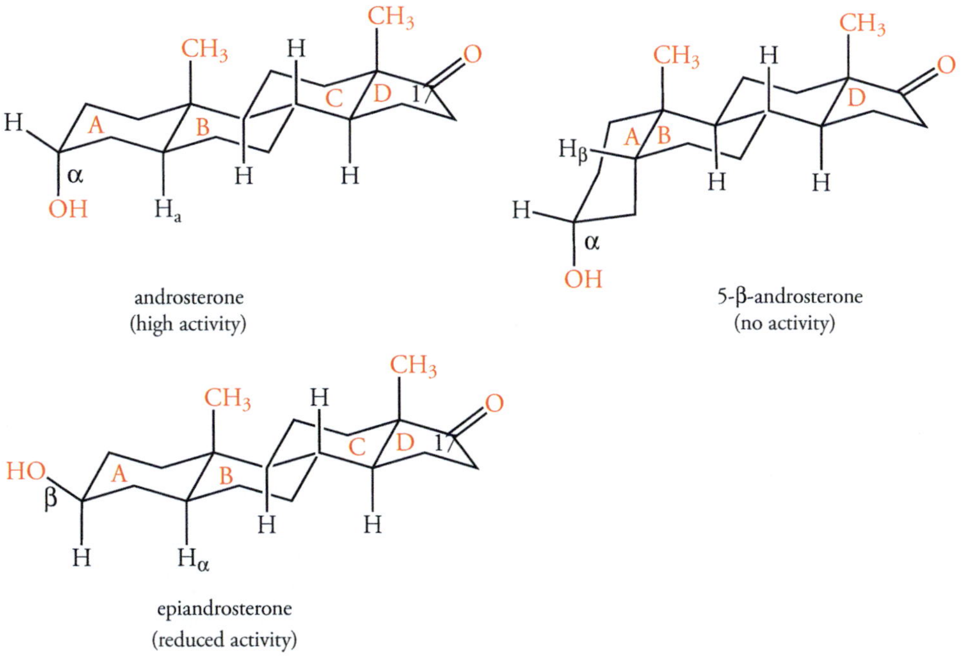

androsterone
(high activity)

5-β-androsterone
(no activity)

epiandrosterone
(reduced activity)

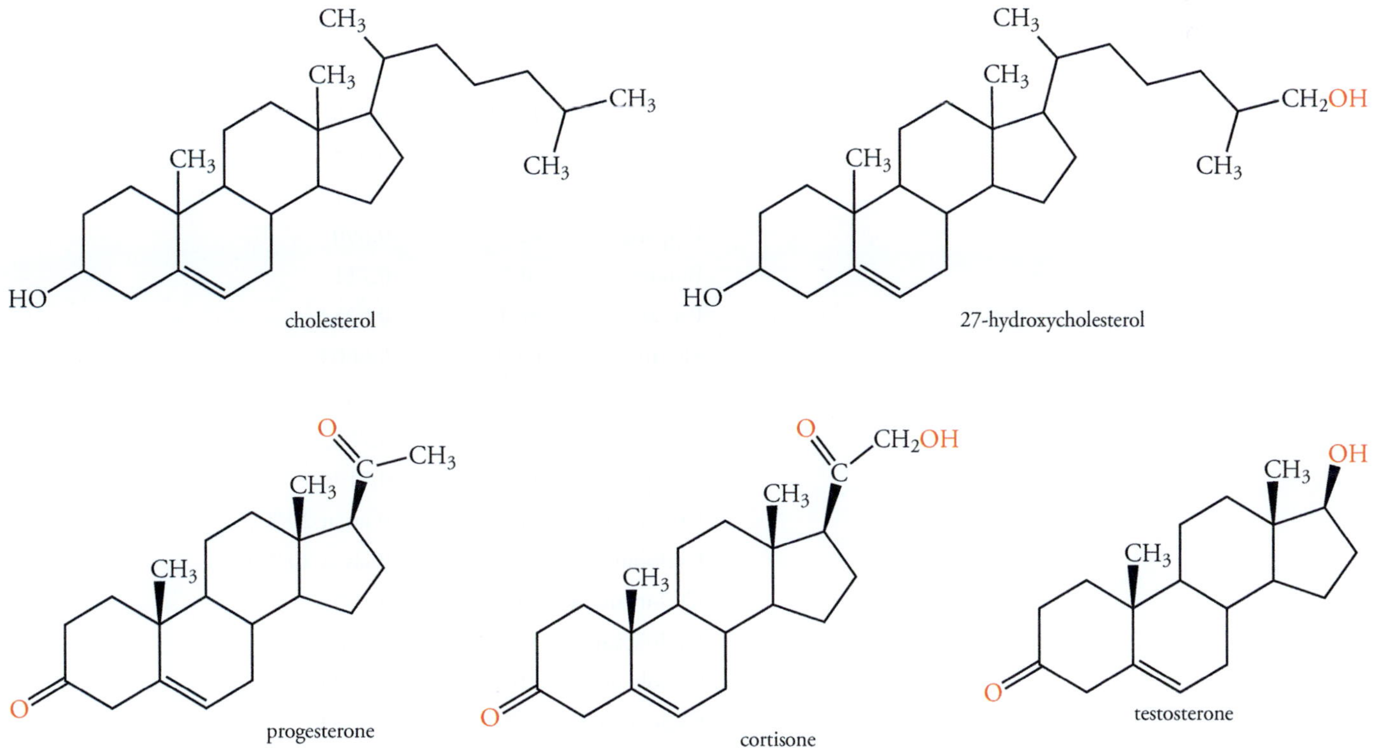

cholesterol

27-hydroxycholesterol

progesterone

cortisone

testosterone

3.7 PHYSICAL PROPERTIES OF ALKANES

Alkanes have densities between 0.6 and 0.8 g/cm³, so they are less dense than water. Thus gasoline, which is largely a mixture of alkanes, is less dense than water, and will float on water. Pure alkanes are colorless, tasteless, and nearly odorless. However, gasoline has an odor and some color because dyes are added to gasoline by refiners to indicate its source and composition. Gasoline also contains aromatic compounds (Chapter 5) that have characteristic odors.

Alkanes contain only carbon-carbon and carbon-hydrogen bonds. Because carbon and hydrogen have similar electronegativity values, the C—H bonds are essentially nonpolar. Thus, alkanes are nonpolar, and they interact only by weak London forces. These forces govern the physical properties of alkanes such as solubility and boiling point.

Alkanes are not soluble in water, a polar substance. The two substances do not meet the usual criterion of solubility: "Like dissolves like." Water molecules are too strongly attracted to each other by hydrogen bonds to allow nonpolar alkanes to slip in between them and dissolve. The boiling points of the normal alkanes increase with increasing molecular weight (Table 3.3). As the molecular weight increases, London forces increase because more atoms are present to increase the surface area or the molecules. Simply put, there are more points of contact between neighboring molecules, and the London forces are stronger.

Normal alkanes have efficient contact between chains, and the molecules can move close together. Branching in alkanes increases the distance between molecules, and the chains of carbon atoms are less able to come close to one another. A branched alkane is more compact and has a smaller surface area than a normal alkane. The order of boiling point of the isomeric C_5H_{12} compounds illustrates this phenomenon. For any group of isomeric alkanes, the most branched isomer has the lowest boiling point. The normal alkane has the highest boiling point.

The physical properties of a series of cycloalkanes of increasing molecular weight are similar to those of a series of alkanes. The densities increase, as do the boiling points (Table 3.3). The boiling points of the cycloalkanes are higher than those of the alkanes containing the same number of carbon atoms.

Table 3.3
Physical Properties of
Alkanes and Cycloalkanes

Hydrocarbon	Boiling Point (°C)	Density (g/mL)
Methane	-164.0	0.678
Ethane	-88.6	0.691
Propane	42.1	0.690
Butane	-0.5	0.711
Pentane	36.1	0.6262
Hexane	68.9	0.6603
Heptane	98.4	0.6837
Octane	125.7	0.7025
Decane	150.8	0.7176
Cyclopropane	-32.7	(Gas at 20 °C)
Cyclobutane	12	(Gas at 20 °C)
Cyclopentane	49.3	0.7457
Cyclohexane	80.7	0.7786
Cycloheptane	110.5	0.8098
Cyclooctane	148.5	0.8349

$$CH_3 - \underset{\underset{CH_3}{|}}{\overset{\overset{CH_3}{|}}{C}} - CH_3$$
neopentane
bp –10 °C

$CH_3CH_2CH_2CH_2CH_3$
pentane
bp 36 °C

$CH_3CH_2CH_2CH_2CH_2CH_3$
hexane
bp 69 °C

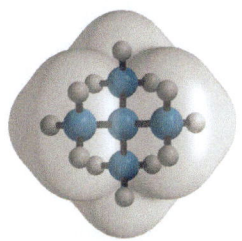
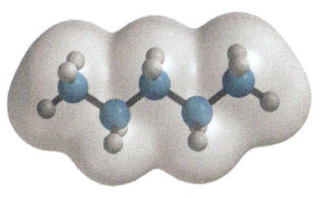

Neopentane   *n*-Pentane

3.8 OXIDATION OF ALKANES AND CYCLOALKANES

The carbon-carbon and carbon-hydrogen bonds of alkanes and cycloalkanes are not very reactive. Alkanes are also called paraffins (from the Latin *parum affinis*, "little activity"). The carbon-carbon bonds are σ bonds, and the bonding electrons are tightly held between the carbon atoms. The carbon-hydrogen bonds are located about the carbon skeleton and are more susceptible to reaction but usually do so only under extreme conditions. One such process is oxidation. Methane, the major component of natural gas, yields 887 kJ/mole when burned. Although the reaction is spontaneous, a small spark or flame is required to provide the activation energy for the reaction. Thus, natural gas can accumulate from a gas leak and not explode. But the methane gas-oxygen mixture is very dangerous.

$$CH^4 + O_2 \longrightarrow CO_2 + H_2O \quad \Delta H° = -887 \text{ kJ/mole}$$

Octane Numbers in Gasoline

In an automobile engine the fuel and air are drawn into the cylinder on its downward stroke, and the piston compresses the mixture on the upward stroke. Ideally, the mixture ignites at the top of the stroke. The resulting explosion drives the piston downward. Normal alkanes are not suitable as fuel in an automobile engine because they tend to ignite prematurely and uncontrollably. Their use results in a knocking or "pinging" sound, which indicates that a force is resisting the upward motion of the piston. Branched hydrocarbons burn more smoothly and are the more efficient fuels.

The burning efficiency of gasoline is rated by an octane number scale (Table 3.4). An octane number of 100 is assigned to 2,2,4-trimethylpentane (also known as isooctane), which is an excellent fuel. Heptane, a poor fuel, has an octane number of zero. Gasoline with the same burning characteristics as a 90% mixture of 2,2,4-trimethylpentane and 10% heptane is rated at 90 octane. Compounds that burn more efficiently than 2,2,4-trimethylpentane have octane numbers greater than 100. Compounds that burn less efficiently than heptane have negative octane numbers. Octane numbers decrease with increasing molecular weight. In isomeric compounds, increased branching increases the octane number.

Table 3.4
Octane Numbers of Alkanes

Formula	Compound	Octane Number
C_4H_{10}	Butane	94
C_5H_{12}	Pentane	62
	2-Methylbutane	94
C_6H_{14}	Hexane	25
	2-Methylpentane	73
	2,2-Dimethylbutane	92
C_7H_{16}	Heptane	0
	2-Methylhexane	42
	2,2-Dimethylpentane	90
C_8H_{18}	Octane	-19
	2-Methylheptane	22
	2,3-Dimethylhexane	100

3.9 HALOGENATION OF SATURATED ALKANES

Alkanes react with halogens at high temperature or in the presence of light to give a substitution product that has a halogen atom in place of a hydrogen atom. For example, methane reacts with chlorine when heated to a high temperature or when exposed to ultraviolet light.

$$CH_3-H \; + \; Cl-Cl \; \longrightarrow \; CH_3-Cl \; + \; H-Cl$$

The reaction is difficult to control because the product also has carbon-hydrogen bonds that can continue to react with additional chlorine to produce several substitution products.

$$CH_3Cl \; + \; Cl_2 \; \longrightarrow \; CH_2Cl_2 \; + \; HCl$$
dichloromethane
(methylene chloride)

$$CH_2Cl_2 \; + \; Cl_2 \; \longrightarrow \; CHCl_3 \; + \; HCl$$
trichloromethane
(chloroform)

$$CH_3Cl \; + \; Cl_2 \; \longrightarrow \; CCl_4 \; + \; HCl$$

tetrachloromethane
(carbon tetrachloride)

Chlorination of higher molecular weight alkanes yields many monosubstituted products. The chlorine atom is so reactive that it is not selective in its substitution for hydrogen atoms. Thus, reaction with 2-methylpropane yields 63% 1-chloro-2-methylpropane and 37% 2-chloro-2-methylpropane. A large number of polysubstituted products is also possible. Bromination of alkane is a very much more selective reaction than the chlorination reaction. The reactivity of C-H bonds decreases in the order 3° > 2° > 1°. Thus, reaction of 2-methyl-propane yields 99% 2-bromo-2-methylpropane.

1-bromo-2-methylpropane
(1%)

2-bromo-2-methylpropane
(99%)

Halogenated hydrocarbons are used for many industrial purposes. Unfortunately, many of these compounds can cause liver damage and cancer. In the past, chloroform was used as an anesthetic, and carbon tetrachloride was used as a dry-cleaning solvent. They are no longer used for these purposes.

Mechanism of Alkane Chlorination

The chlorination of alkanes occurs by a series homolytic bond cleavage and homogenic bond formation steps. In the first step the chlorine molecule absorbs either heat energy or light energy and the bond breaks to give two chlorine atom that are radicals. These species are electron deficient and highly reactive. The chlorination reaction can occur only after this initial homolytic bond cleavage, which has a large activation energy. The reaction that starts the reaction is the **initiation step**.

Step 1. **Initiation**. A chlorine molecule absorbs energy, either from ultraviolet light or high temperatures, and the Cl—Cl bond breaks homolytically to give two chlorine atoms. They are electron-deficient, highly reactive radicals. This step starts the reaction, and is called the initiation step.

$$:\!\ddot{C}l\!-\!\ddot{C}l\!: \; \longrightarrow \; :\!\ddot{C}l\!\cdot \; + \; \cdot\!\ddot{C}l\!:$$

Step 2. **Propagation.** A chlorine atom abstracts a hydrogen atom from methane, breaking a C—H bond and making an H—Cl bond. This step, which continues the reaction by generating a new radical, is called a propagation step.

$$CH_3\text{—}H \ + \ \cdot \ddot{\underset{\cdot\cdot}{Cl}}{:} \ \longrightarrow \ CH_3\cdot \ + \ H\text{—}\ddot{\underset{\cdot\cdot}{Cl}}{:}$$

Step 3. **Propagation.** A Cl—Cl bond breaks and a C—Cl bond forms. A radical reacts and another radical forms. This is also a propagation step.

$$CH_3\cdot \ + \ {:}\ddot{\underset{\cdot\cdot}{Cl}}\text{—}\ddot{\underset{\cdot\cdot}{Cl}}{:} \ \longrightarrow \ CH_3\text{—}Cl \ + \ \cdot\ddot{\underset{\cdot\cdot}{Cl}}{:}$$

The propagation steps, 2 and 3, repeat because one radical generates another in this sequence of reactions. The process continues as long as radicals and a supply of both reactants are present. Therefore, only a few chlorine atoms are required to initiate the reaction.

Any time two radicals recombine, the chain stops. These are termination steps. Since the concentration of radicals is much less than that of either methane or Cl_2, termination steps are relatively rare. One of the termination steps is mechanistically important: the reaction mixture always contains a small amount of ethane, and this only could have occurred if the reaction had proceeded by way of a methyl radical intermediate.

$$CH_3\cdot \ + \ \cdot CH_3 \ \xrightarrow{\ \text{radical chain termination}\ } \ CH_3\text{—}CH_3$$

Physical Properties of Haloalkanes

The physical properties of haloalkanes depend upon the lengths and strengths of the carbon-halogen bonds. The atomic radii of the halogens increases going from top to bottom in the periodic table. This trend is reflected in the bond lengths of the carbon-halogen bond (Table 3.5). As a result of the greater electronegativity of the halogens, the carbon atom of the carbon-halogen bond bears a partial positive charge and the halogen atom has a partial negative charge.

$$\overset{\delta^+}{\underset{\ }{C}}\text{—}X^{\delta^-} \qquad \text{where } X = F, Cl, Br, I$$

The polarizability of the halogen atoms, that is, the ease with which electron density changes because of interactions with other atoms or molecules, increase as we move down the periodic table: F < Cl < Br < I. Because highly polarizable atoms interact more strongly by London forces than less polarizable atoms, the intermolecular forces for haloalkanes increase in the order RF < RCl < RBr < RI. The effect of intermolecular forces is reflected in the boiling points of haloalkanes, which increase in the same order as the polarizability of their halogen components (Table 3.5).

Table 3.5
Physical Properties of Haloalkanes

	CH_3—F	CH_3—Cl	CH_3—Br	CH_3—I
Bond length (pm)	139	178	193	214
Boiling point (°C)	−78.47	−24.2	3.6	42.4

Freons, Free Radicals, and the Ozone Layer

Many polyhalogenated hydrocarbons have been synthesized by the chemical industry, and they have many commercial uses. Based on the individual physical properties, polyhalogenated alkanes have been designed to serve as solvents, dry-cleaning agents, anesthetics, and refrigerants. The extreme stability of these compounds also makes them commercially appealing. For example, halogenated alkanes are less flammable than alkanes. Extensively halogenated compounds such as carbon tetrachloride will not burn at all. At one time, carbon tetrachloride was used in fire extinguishers to provide an inert atmosphere to prevent oxygen from reaching the flames. Today, CF_3Br is one of the gaseous compounds known as halons that are used in environmental fire suppression systems where the use of water is not advised, such as in large computer facilities. They are released in much the same way that water is released in automatic sprinkler systems.

Reduced combustibility makes haloalkanes useful for many purposes. For example, hydrocarbons have been used as refrigerants and as aerosol propellants, but the danger of combustion is a serious drawback. Thus, halogenated alkanes were developed to avoid the danger of explosions. Fluoroalkanes, manufactured under the trade name **Freons**, were found to have a wide range of useful physical properties. These compounds also contain chlorine and are thus also known as chlorofluorocarbons or **CFCs**. In general, Freons are nonflammable, odorless, noncorrosive, and nontoxic. Some Freons have been designed for use as refrigerants in refrigerators and air conditioners. Because the Freons are gases, they have been used as propellants in spray cans. Freons are also used as blowing agents to produce rigid foams, which are used in insulation for ice chests, and flexible foams, which are used in pillows and cushions. The wetting properties of some Freons make them suitable as cleaning fluids for printed circuit boards in computers.

These compounds were used so extensively that billions of tons of CFCs have been released into the atmosphere with huge, deleterious effects on the ozone layer, which protects terrestrial creatures from the sun's ultraviolet radiation. CFC-11 (trichlorofluoromethane) and CFC-12 were two of the most widely used Freons.

trichlorofluoromethane
(CFC-11)

dichlorodifluoromethane
(CFC-12)

CFC-11

CFC-12

The Freons are not biodegraded in the lower atmosphere and were once thought to present no environmental hazard. Unfortunately, although they are inert at Earth's surface, these Freons eventually reach the stratosphere, where they absorb ultraviolet radiation from the sun and decompose to produce radicals resulting from cleavage of the carbon-chlorine bond.

$$CF_2Cl_2 \xrightarrow{\text{UV light}} CF_2\overset{..}{\underset{..}{Cl}}\cdot \quad + \quad :\overset{..}{\underset{..}{Cl}}\cdot$$

This process is partially responsible for the destruction of the ozone layer in the stratosphere, a region from 8 to 30 miles above Earth's surface. Ozone in the stratosphere protects us from solar ultraviolet radiation, which splits ozone molecules into molecular oxygen and atomic oxygen. These products then recombine and release heat energy.

$$:\ddot{O}=\ddot{O}-\ddot{O}: \xrightarrow{\text{UV light}} \cdot\ddot{O}\cdot + :\ddot{O}=\ddot{O}:$$

$$\cdot\ddot{O}\cdot + :\ddot{O}=\ddot{O}: \longrightarrow :\ddot{O}=\ddot{O}-\ddot{O}: + \text{ heat}$$

As a result of the combination of the two reactions, Earth is protected from extensive doses of ultraviolet radiation that is harmful to life and increases the incidence of skin cancer in humans.

Increased ultraviolet radiation also adversely affects plant life and aquatic life at the surface of the world's oceans. The chlorine radical from CFC-12 reacts with ozone in the stratosphere, producing hypochlorite radicals (Cl—O•) that react with atomic oxygen.

$$:\ddot{O}=\ddot{O}-\ddot{O}: + :\ddot{Cl}\cdot \longrightarrow :\ddot{O}=\ddot{O}: + :\ddot{Cl}-\ddot{O}\cdot$$

$$\cdot\ddot{O}\cdot + :\ddot{Cl}-\ddot{O}\cdot \longrightarrow :\ddot{O}=\ddot{O}: + :\ddot{Cl}\cdot$$

Cl—O•

Although international treaties have banned the production of Freons, they are still produced in many countries, and while the depletion of the ozone layer above Antarctica is decreasing, it is estimated that it will be 2050 or later before CFCs in the upper atmosphere will have been eliminated.

Problem 3.12
How many mono-, di-, and trichloro compounds result from the chlorination of ethane?

Solution
The carbon atoms in ethane are equivalent, and only one monochlorinated compound (CH_3CH_2Cl) can result. The two carbon atoms in this product are not equivalent. Substitution by a second chlorine atom results in two isomers.

CH_3CHCl_2	$ClCH_2CH_2Cl$
1,1-dichloroethane	1,2-dichloroethane

In subsequent reactions of these products, three chlorine atoms may be located on a single carbon atom. Another product has two chlorine atoms on the same carbon atom and one on the other.

CH_3CCl_2	$ClCH_2CHCl_2$
1,1,1-trichloroethane	1,1,2-trichloroethane

Problem 3.13
How many different mono- and dichloro compounds can result from chlorination of cyclobutane?

3.10 NOMENCLATURE OF HALOALKANES

Low molecular weight haloalkanes are often named using the name of the alkyl group followed by the name of the halide.

$$CH_3-CH_2-Br$$

ethyl bromide

$$\underset{\displaystyle CH_3-\overset{\displaystyle CH_3}{\overset{|}{CH}}-I}{}$$

isopropyl iodide

$$CH_3-\overset{\displaystyle CH_3}{\underset{\displaystyle CH_3}{\overset{|}{\underset{|}{C}}}}-Cl$$

tert-butyl chloride

Haloalkanes are named in the IUPAC system by an extension of the rules outlined in Section 3.3 for alkanes. Halogen atoms are identified and located using appropriate prefixes. The IUPAC rules are as follows:

1. Identify the longest continuous chain of carbon atoms that includes the hydroxyl group; this is the parent chain.

$$\overset{4}{C}H_3-\overset{3}{\underset{\underset{H}{|}}{\overset{\overset{H}{|}}{C}}}-\overset{2}{\underset{\underset{Cl}{|}}{\overset{\overset{H}{|}}{C}}}-\overset{1}{C}H_3$$

2-chlorobutane

2. If the parent chain has branching alkyl groups, number the chain from the end nearer the first substituent whether it is an alkyl group or a halogen atom.

$$\overset{1}{C}H_3-\overset{2}{\underset{\underset{H}{|}}{\overset{\overset{CH_3}{|}}{C}}}-\overset{3}{\underset{\underset{Cl}{|}}{\overset{\overset{H}{|}}{C}}}-\overset{4}{C}H_2-\overset{5}{C}H_3$$

3-chloro-2-methylpentane

$$\overset{5}{C}H_3-\overset{4}{C}H_2-\overset{3}{\underset{\underset{H}{|}}{\overset{\overset{CH_3}{|}}{C}}}-\overset{2}{\underset{\underset{Br}{|}}{\overset{\overset{H}{|}}{C}}}-\overset{1}{C}H_3$$

2-bromo-3-methylpentane

3. If the compound contains two or more halogen atoms of the same type, indicate them with the prefixes di-, tri-, etc. Give each halogen atom a number that corresponds to its position in the parent chain.

$$\overset{1}{C}H_3-\overset{2}{\underset{\underset{H}{|}}{\overset{\overset{Cl}{|}}{C}}}-\overset{3}{\underset{\underset{Cl}{|}}{\overset{\overset{H}{|}}{C}}}-\overset{4}{C}H_2-\overset{5}{C}H_3$$

2,3-dichloropentane

$$\overset{6}{C}H_3-\overset{5}{C}H_2-\overset{4}{\underset{\underset{H}{|}}{\overset{\overset{Br}{|}}{C}}}-\overset{3}{\underset{\underset{Br}{|}}{\overset{\overset{Br}{|}}{C}}}-\overset{2}{C}H_2-\overset{1}{C}H_3$$

3,3,4-tribromohexane

4. If a compound contains different halogen atoms, number them according to their positions on the chain, and list them in alphabetical order.

$$Cl-\overset{1}{C}H_2-\overset{2}{\underset{\underset{H}{|}}{\overset{\overset{Br}{|}}{C}}}-\overset{3}{C}H_2-\overset{4}{\underset{\underset{H}{|}}{\overset{\overset{CH_3}{|}}{C}}}-\overset{5}{C}H_3$$

2-bromo-1-chloro-4-methylpentane

5. If the chain can be numbered from either end based on the location of the substituents, begin at the end nearer the substituent that has alphabetical precedence, whether it is an alkyl group or a halogen atom.

$$Cl-\overset{1}{C}H_2-\overset{2}{\underset{\underset{}{}}{\overset{\overset{Br}{|}}{C}}H}-\overset{3}{C}H_2-\overset{4}{\underset{\underset{}{}}{\overset{\overset{CH_3}{|}}{C}}H}-\overset{5}{C}H_3$$

2-bromo-1-chloro-4-methylpentane

6. Number halocycloalkanes from the carbon atom bearing the halogen atom unless another functional group, such as a double bond, takes precedence. Number carbon atoms in the ring to give the lower number to the substituent.

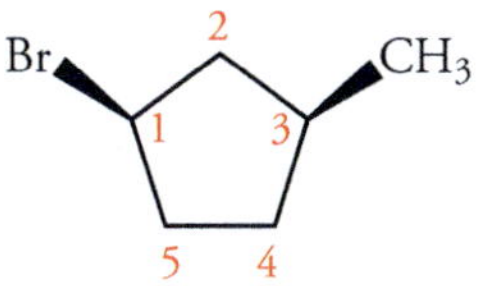

cis-1-bromo-3-methylcyclopentane

SUMMARY OF REACTIONS

1. Free radical chlorination (Section 3.9).

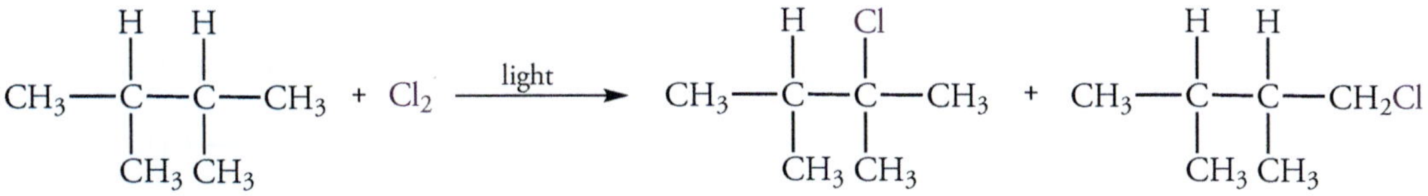

2. Free radical bromination (Section 3.9).

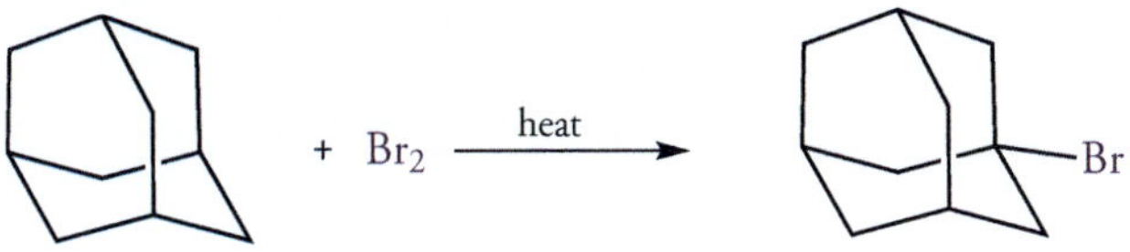

EXERCISES

Molecular Formulas

3.1 Beeswax contains approximately 10% hentriacontane, a normal alkane with 31 carbon atoms. What is the molecular formula of hentriacontane? Write a completely condensed formula of hentriacontane.

3.2 An immensely large normal alkane has 390 carbon atoms. What is the molecular formula for this alkane? Write a completely condensed formula for it.

Structural Formulas

3.3 Redraw each of the following so that the longest continuous chain is written horizontally.

(a) 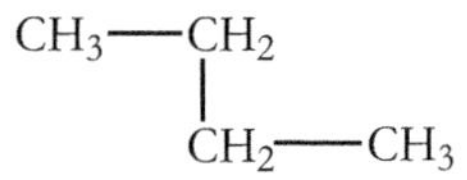
$$CH_3-CH_2$$
$$|$$
$$CH_2-CH_3$$

(b) 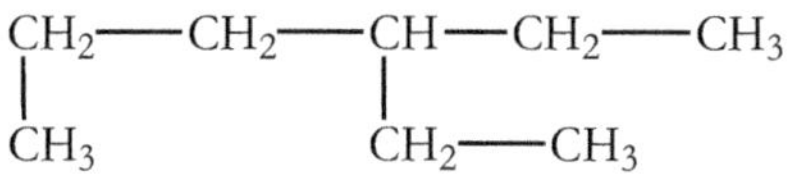
$$CH_2-CH_2-CH-CH_2-CH_3$$
$$\quad\quad\quad\quad | \quad\quad\quad |$$
$$\quad\quad\quad CH_3 \quad CH_2-CH_3$$

(c) 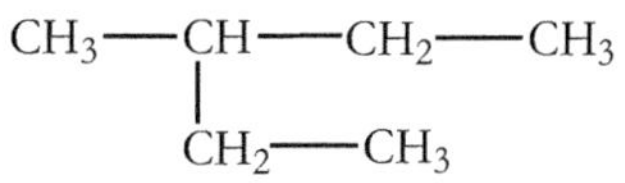
$$CH_3-CH-CH_2-CH_3$$
$$\quad\quad\quad |$$
$$\quad\quad CH_2-CH_3$$

3.4 Redraw each of the following so that the longest continuous chain is written horizontally.

(a) 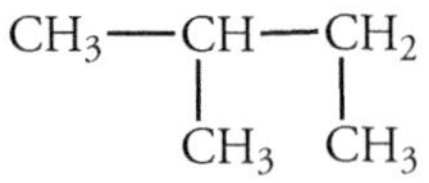
$$CH_3-CH-CH_2$$
$$\quad\quad | \quad\quad |$$
$$\quad CH_3 \quad CH_3$$

(b) 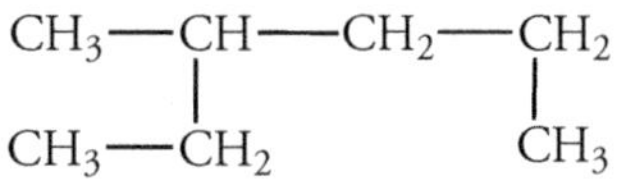
$$CH_3-CH-CH_2-CH_2$$
$$\quad\quad\quad | \quad\quad\quad\quad |$$
$$\quad CH_3-CH_2 \quad\quad CH_3$$

(c) 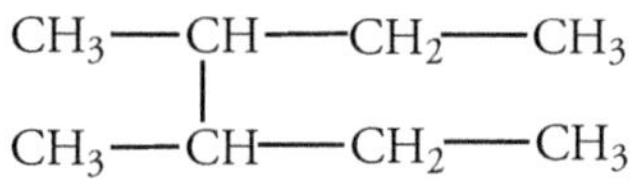
$$CH_3-CH-CH_2-CH_3$$
$$\quad\quad\quad |$$
$$CH_3-CH-CH_2-CH_3$$

Alkyl Groups

3.5 What is the common name for each of the following alkyl groups?

(a) $CH_3-CH_2-CH_2-$

(b) 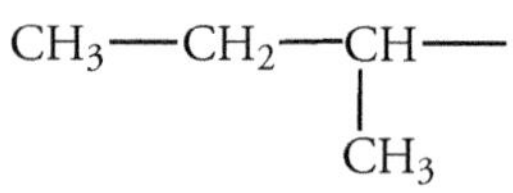
$$CH_3-CH_2-CH-$$
$$\quad\quad\quad\quad |$$
$$\quad\quad\quad\quad CH_3$$

(c) 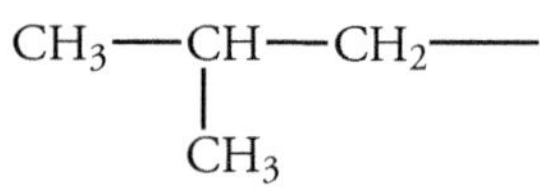
$$CH_3-CH-CH_2-$$
$$\quad\quad | $$
$$\quad CH_3$$

3.6 What is the common name for each of the following alkyl groups?

(a) 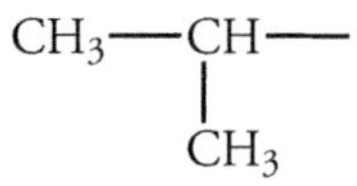
$$CH_3-CH-$$
$$\quad\quad |$$
$$\quad CH_3$$

(b) $CH_3-CH_2-CH_2-CH_2-$

(c) 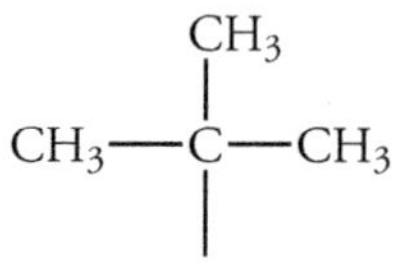
$$\quad\quad\quad CH_3$$
$$\quad\quad\quad |$$
$$CH_3-C-CH_3$$
$$\quad\quad\quad |$$

3.7 What is the common name for each of the following alkyl groups?

(a) 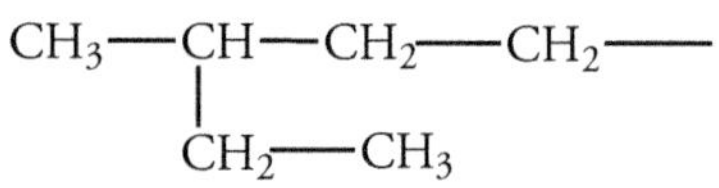
$$CH_3-CH-CH_2-CH_2-$$
$$\quad\quad\quad |$$
$$\quad\quad CH_2-CH_3$$

(b) 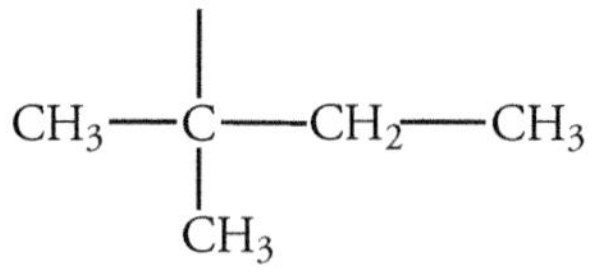
$$\quad\quad\quad |$$
$$CH_3-C-CH_2-CH_3$$
$$\quad\quad\quad |$$
$$\quad\quad\quad CH_3$$

3.8 What is the common name for each of the following alkyl groups?

(a) 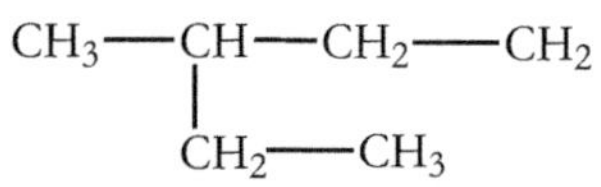
$$CH_3-CH-CH_2-CH_2$$
$$\quad\quad\quad |$$
$$\quad\quad CH_2-CH_3$$

(b) 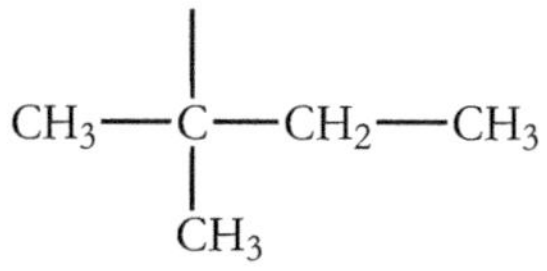
$$\quad\quad\quad |$$
$$CH_3-C-CH_2-CH_3$$
$$\quad\quad\quad |$$
$$\quad\quad\quad CH_3$$

3.10 The spermicide octoxynol-9 is used in diverse contraceptive products. Name the alkyl group to the left of the benzene ring.

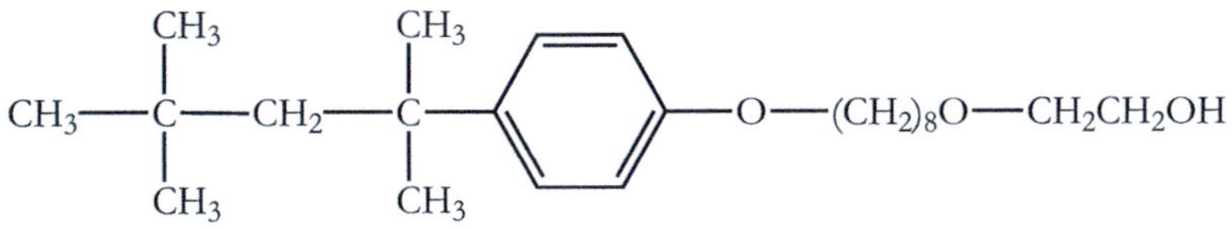

octoxynol-9

3.11 The name vitamin E actually refers to a series of closely related compounds called tocopherols. Name the complex alkyl group present in α-tocopherol.

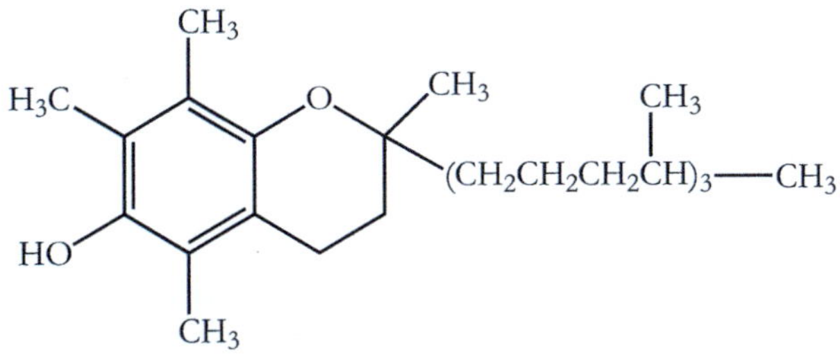

Nomenclature of Alkanes

3.12 Give the IUPAC name for each of the following compounds:

(a) 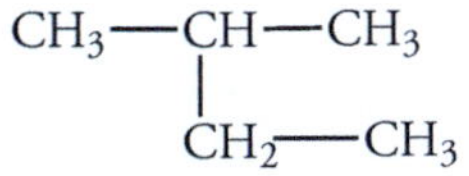 CH$_3$—CH—CH$_3$
       |
       CH$_2$—CH$_3$

(b) 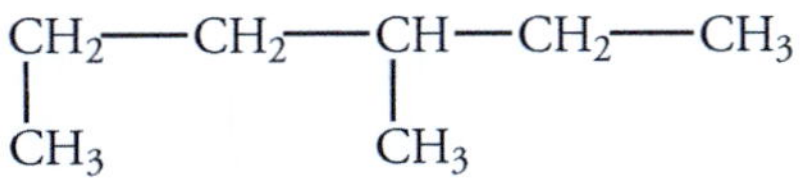 CH$_2$—CH$_2$—CH—CH$_2$—CH$_3$
       | |
       CH$_3$ CH$_3$

(c) 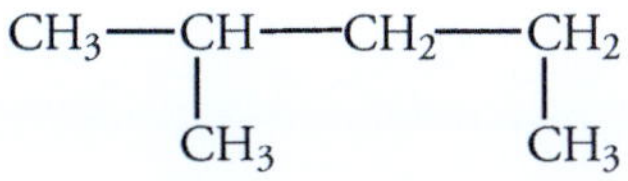 CH$_3$—CH—CH$_2$—CH$_2$
       | |
       CH$_3$ CH$_3$

(d) 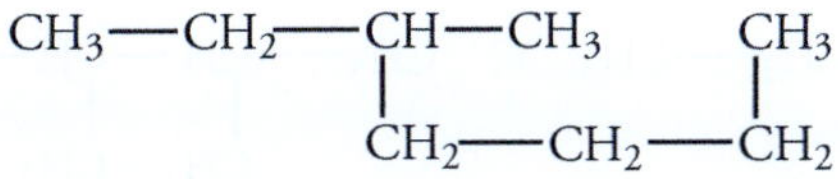 CH$_3$—CH$_2$—CH—CH$_3$ CH$_3$
                        | |
                        CH$_2$—CH$_2$—CH$_2$

3.13 Give the IUPAC name for each of the following compounds:

(a) 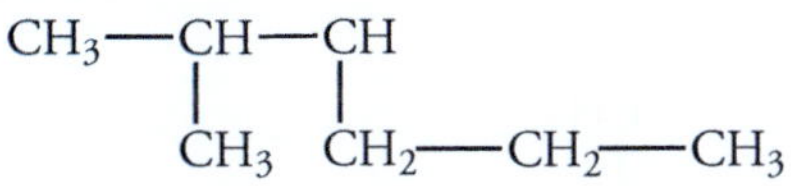 CH$_3$—CH—CH
       | |
       CH$_3$ CH$_2$—CH$_2$—CH$_3$

(b) 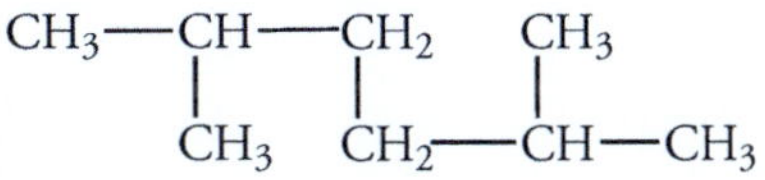 CH$_3$—CH—CH$_2$ CH$_3$
       | | |
       CH$_3$ CH$_2$—CH—CH$_3$

(c) 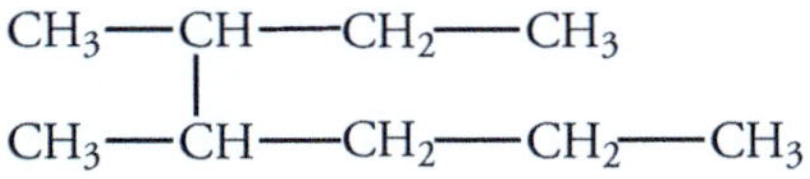 CH$_3$—CH—CH$_2$—CH$_3$
       |
       CH$_3$—CH—CH$_2$—CH$_2$—CH$_3$

(d) 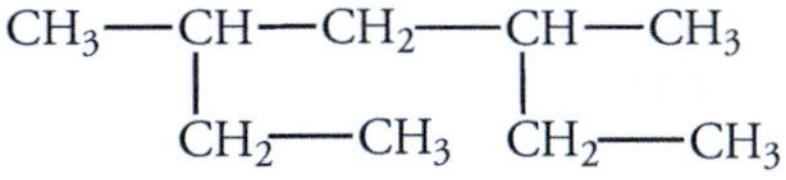 CH$_3$—CH—CH$_2$—CH—CH$_3$
           | |
           CH$_2$—CH$_3$ CH$_2$—CH$_3$

3.14 Give the IUPAC name for the following compound:

CH$_3$—CH$_2$—CH$_2$—CH$_2$—CH—CH$_2$—CH$_2$—CH$_2$—CH$_2$—CH$_3$
                                    |
                     CH$_3$—CH$_2$—CH—CH$_2$—CH$_3$

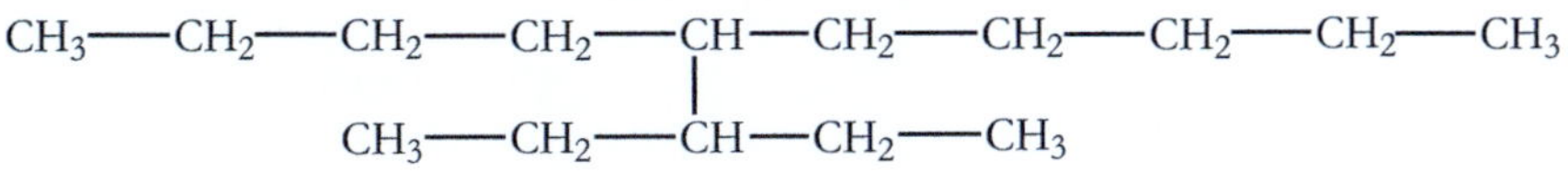

3.15 Give the IUPAC name for the following compound:

CH$_3$—CH$_2$—CH$_2$—CH$_2$—CH—CH$_2$—CH$_2$—CH$_2$—CH$_3$
                                    |
                        CH$_3$—C—CH$_3$
                               |
                        CH$_2$—CH$_3$

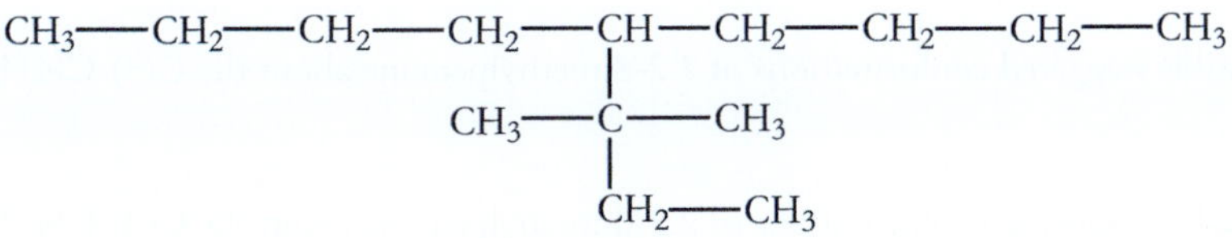

3.16   Write the structural formula for each of the following compounds:
(a) 3,4-dimethylhexane   (b) 2,2,3-trimethylpentane   (c) 2,3,4,5-tetramethylhexane

3.17   Write the structural formula for each of the following compounds:
(a) 3-ethylhexane   (b) 2,2,4-trimethylhexane   (c) 2,2,3,3-tetramethylpentane

3.18   Write the structural formula for each of the following compounds:
(a) 4-(1-methylethyl)heptane   (b) 4-(1,1-dimethylethyl)octane

3.19   Write the structural formula for each of the following compounds:
(a) 4-(2-methylpropyl)octane   (b) 5-(2,2-dimethylpropyl)nonane

Isomers

3.20   There are nine isomeric C_7H_{16} compounds. Name the isomers that have a single methyl group as a branch.

3.21   There are nine isomeric C_7H_{16} compounds. Name the isomers that have two methyl groups as branches and are named as dimethyl-substituted pentanes.

Classification of Carbon Atoms

3.22   Draw the structure of a compound with molecular formula C_5H_{12} that has one quaternary and four primary carbon atoms.

3.23   Draw the structure of a compound with molecular formula C_6H_{14} that has two tertiary and four primary carbon atoms.

3.24   Determine the number of primary, secondary, tertiary, and quaternary carbon atoms in each of the following compounds:

(a) CH_3—$\overset{\overset{\displaystyle CH_3}{|}}{\underset{\underset{\displaystyle CH_3}{|}}{C}}$—$CH_3$

(b) CH_3—$\underset{\underset{\displaystyle CH_3}{|}}{CH}$—$CH_2$—$CH_3$

(c) CH_3—$\underset{\underset{\displaystyle CH_3}{|}}{CH}$—$CH_2$—$CH_2$—$CH_3$

(d) CH_3—$\underset{\underset{\displaystyle CH_3}{|}}{CH}$—$\underset{\underset{\displaystyle CH_3}{|}}{CH}$—$CH_3$

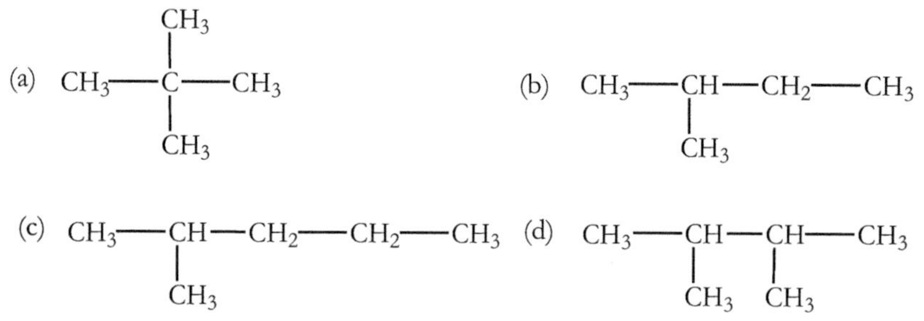

3.25   Determine the number of primary, secondary, tertiary, and quaternary carbon atoms in each of the following compounds:

(a) CH_3—$\overset{\overset{\displaystyle CH_3}{|}}{\underset{\underset{\displaystyle CH_3}{|}}{C}}$—$CH_2$—$\overset{\overset{\displaystyle CH_3}{|}}{\underset{\underset{\displaystyle CH_3}{|}}{C}}$—$CH_3$

(b) CH_3—$\underset{\underset{\displaystyle CH_3}{|}}{CH}$—$CH_2$—$\underset{\underset{\displaystyle CH_3}{|}}{CH}$—$CH_3$

(c) CH_3—CH_2—$\underset{\underset{\displaystyle CH_3}{|}}{CH}$—$CH_2$—$CH_3$

(d) CH_3—$\underset{\underset{\displaystyle CH_3}{|}}{CH}$—$\underset{\underset{\displaystyle CH_3}{|}}{CH}$—$\underset{\underset{\displaystyle CH_3}{|}}{CH}$—$CH_3$

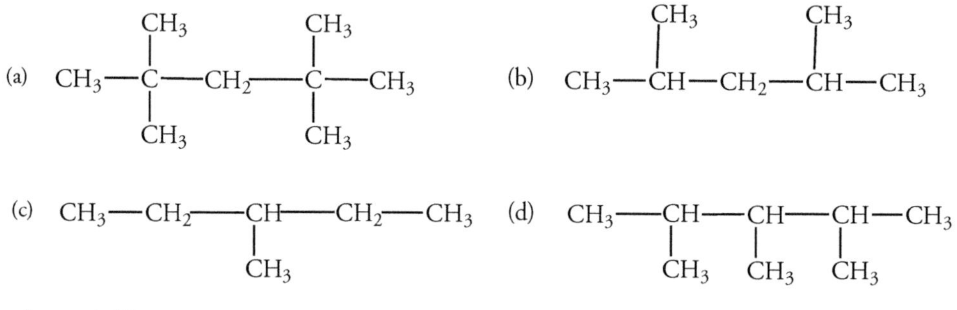

Conformations of Alkanes

3.26   Draw the Newman projection of the staggered conformation of 2,2-dimethylpropane around the C(1)-C(2) bond.

3.27   Draw the Newman projections of the two possible staggered conformations of 2,3-dimethylbutane around the C(2)-C(3) bond.

3.28   Draw the Newman projections of the two possible staggered conformations of 2,2-dimethylpentane about the C(3)-C(4) bond. Which is the more stable?

3.29   Draw the Newman projections of the two possible staggered conformations of 2,2-dimethylpentane about the C(3)-C(4) bond. Which is the more stable?

Cycloalkanes

3.30 Write condensed planar formulas for each of the following compounds.
(a) chlorocyclopropane   (b) 1,1-dichlorocyclobutane   (c) cyclooctane

3.31 Write condensed planar formulas for each of the following compounds.
(a) bromocyclopentane   (b) 1,1-dichlorocyclopropane   (c) cyclopentane

3.32 Name each of the following compounds:

(a) 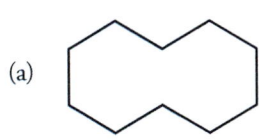   (b) 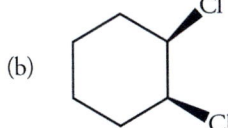   (c)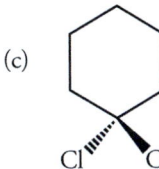

3.33 Name each of the following compounds:

(a) 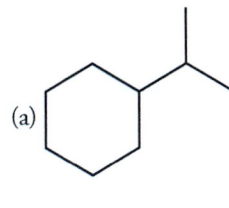   (b) 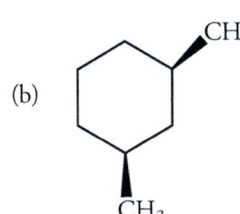   (c)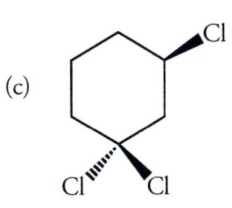

3.34 What is the molecular formula of each of the following compounds?

(a) 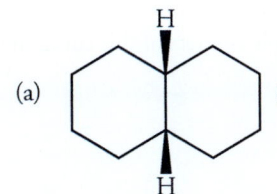   (b) 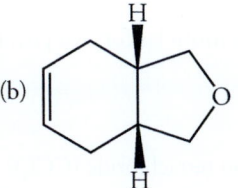   (c)    (d)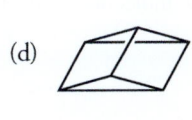

3.35 What is the molecular formula of each of the following compounds?

(a) 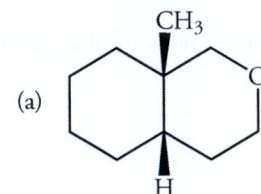   (b) 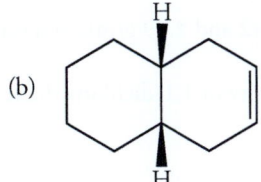   (c) 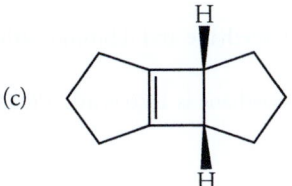   (d)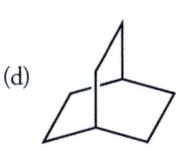

Conformations of Cyclohexanes

3.36 Draw the two chair conformations of fluorocyclohexane. Would you expect the energy difference between these two conformations to be greater or less than the energy difference between the two conformations of methylcyclohexane? Why?

3.37 Draw the two chair conformations of *tert*-butylcyclohexane. Would you expect the energy difference between these two conformations to be greater or less than the energy difference between the two conformations of methylcyclohexane? Why?

3.38 Draw the most stable conformation of each of the following compounds:
(a) *trans*-1,4-dimethylcyclohexane   (b) *cis*-1,3-dimethylcyclohexane

3.39 Draw the most stable conformation of each of the following compounds:
(a) 1,1,4-trimethylcyclohexane   (b) 1,1,3-trimethylcyclohexane

Properties of Hydrocarbons

3.40   Cyclopropane is an anesthetic, but it cannot be used in operations in which electrocauterization of tissue is done. Why?

3.41   Which compound should have the higher octane number, cyclohexane or methylcyclopentane?

3.42   Which of the isomeric C_8H_{18} compounds has the highest boiling point? Which has the lowest boiling point?

3.43   The boiling point of methylcyclopentane is lower than the boiling point of cyclohexane. Suggest a reason why.

Halogenation of Haloalkanes

3.44   How many products can result from the substitution of a chlorine atom for one hydrogen atom in each of the following compounds?

(a) propane   (b) butane   (c) methylpropane   (d) cyclohexane

3.45   How many products can result from the substitution of a chlorine atom for one hydrogen atom in each of the following compounds?

(a) 2-methylbutane   (b) 2,2-dimethylbutane   (c) 2,3-dimethylbutane   (d) cyclopentane

3.46   Write the structure of the radicals that can result from abstraction of a hydrogen atom from pentane.

3.47   Write the structure of the hydrocarbon product that can result from a termination step in the free radical chlorination of ethane.

3.48   Halothane, an anesthetic, has the formula C_2HF_3ClBr. Draw structural formulas for the four possible isomers of this molecular formula.

3.49   A saturated refrigerant has the molecular formula C_4F_8. Draw structural formulas for two possible isomers of this compound.

Properties of Haloalkanes

3.50   Which compound is more polar, methylene chloride (CH_2Cl_2) or carbon tetrachloride (CCl_4)?

3.51   Tribromomethane is more polar than tetrabromomethane, but their boiling points are 150 and 189 °C, respectively. Explain why the more polar compound has the lower boiling point.

3.52   The densities of chloroiodomethane and dibromomethane are 2.42 and 2.49 g/mL, respectively. Why are these values similar?

3.53   The density of 1,2-dichloroethane is 1.26 g/mL. Predict the density of 1,1-dichloroethane.

4

ALKENES AND ALKYNES

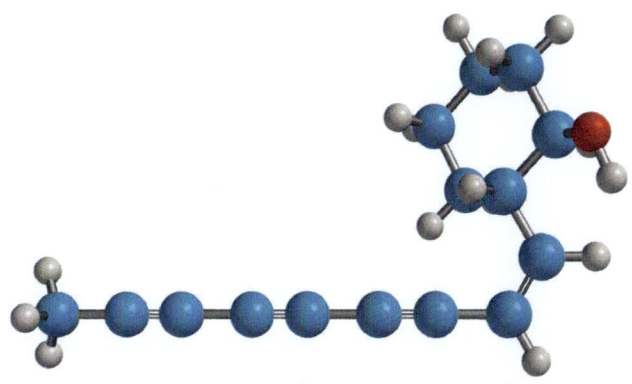

ICHTHYOTHEREOL (AN ANTICONVULSANT)

4.1 UNSATURATED HYDROCARBONS

Organic compounds with carbon-carbon multiple bonds contain fewer hydrogen atoms than structurally related alkanes or cycloalkanes. Thus, these compounds are said to be **unsaturated**. In this chapter we will focus on two classes of unsaturated compounds: **alkenes** and **alkynes**. Alkenes contain a carbon-carbon double bond; alkynes contain a carbon-carbon triple bond. **Aromatic** hydrocarbons, compounds that contain a benzene ring or structural units that resemble a benzene ring, will be discussed in the next chapter.

The π bond of a carbon-carbon double bond and the two π bonds of a carbon-carbon triple bond are the sites of specific reactions. We recall that π bonds involve a side-by-side overlap of 2p orbitals on adjacent carbon atoms. We discussed the hybridization of carbon and the σ and π bonds in ethene (an alkene) and ethyne (an alkyne) in Section 1.8.

In this chapter we will also discuss the chemistry of some **alkadienes**, also called *dienes*. We will discuss compounds in which two double bond units are linked by one single bond. The compounds are called **conjugated dienes**. They undergo reactions that differ from those of individual, or isolated double bonds. Many naturally occurring compounds contain conjugated double bonds. For example, natural rubber is a polymer of isoprene, a conjugated diene. Some synthetic rubbers called neoprenes are produced from chloroprene and are used in products such as industrial hoses.

The compound at the top of this page, ichthyothereol, has three triple bonds in a row that are linked to a double bond that is in turn bonded to a cyclohexane ring with an alcohol group.

$$CH_3-\overset{\displaystyle CH_3}{\underset{\displaystyle \text{isoprene}}{C}}-CH=CH_2 \qquad CH_2=\overset{\displaystyle Cl}{\underset{\displaystyle \text{chloroprene}}{C}}-CH=CH_2$$

Alkenes

The simplest alkene, C_2H_4, is commonly called ethylene. Its IUPAC name is ethene. The IUPAC names of alkenes use the suffix *-ene*. In the structure of ethene shown in Figure 4.1, all six atoms, two carbon atoms and four hydrogen atoms, are located in the same plane. The plane may be written either in the page or perpendicular to it. If the plane is perpendicular to the printed page, the carbon-hydrogen bonds project in front of and in back of the page. As before, wedge-shaped lines represent bonds in front of the page and dashed lines those behind the page.

Principles of Organic Chemistry. http://dx.doi.org/10.1016/B978-0-12-802444-7.00004-5

Figure 4.1
Structures of Ethene and
Ethyne

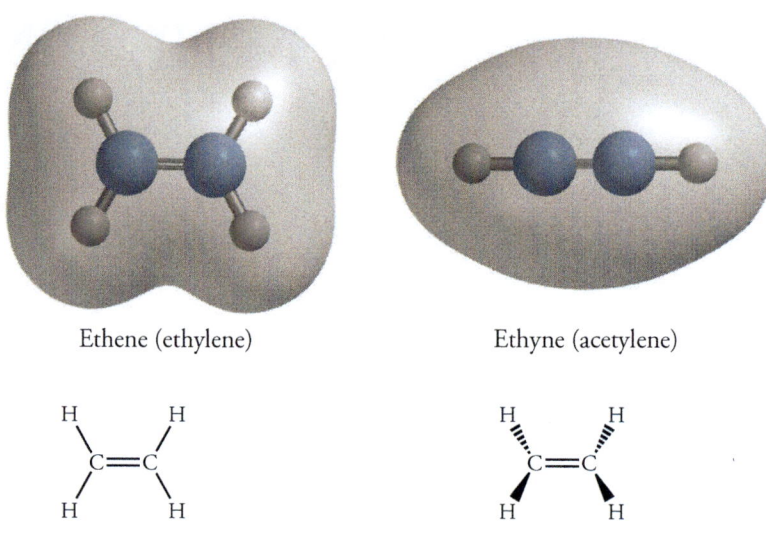

Ethene (ethylene)   Ethyne (acetylene)

Ethene bonds in plane of page   Ethene bonds perpendicular to plane of page

A double bond decreases the number of hydrogen atoms in a molecule by two compared to the number in alkanes, so the general formula for an alkene is C_nH_{2n}. Each additional double bond decreases the number of hydrogen atoms by two.

Alkynes

The simplest alkyne, C_2H_2, has the IUPAC name ethyne. It is also commonly called acetylene. Unfortunately, the common name ends in -ene, which seems to suggest that the compound contains a double bond. Such confusion is one reason why IUPAC names are so important for clear communication in chemistry. The structure of ethyne is shown in Figure 4.1. All four atoms are collinear. Each $H-C\equiv C-$ bond angle is 180°. In other alkynes also, the two triple-bonded carbon atoms and the two atoms directly attached to them are collinear. The triple bond in an alkyne decreases the number of hydrogen atoms in the molecule by four compared to alkanes. As a result, the general molecular formula for alkynes is C_nH_{2n-2}.

Classification of Alkenes and Alkynes

We will often describe alkenes and alkynes in terms of the number of alkyl groups attached to the double or triple bond unit. In the case of double bonds, we speak of the degree of substitution at the site of the double bond. A **monosubstituted** alkene has a single alkyl group attached to one sp^2 hybridized carbon atom of the double bond. An alkene whose double bond is at the end of a chain of carbon atoms is sometimes called a terminal alkene. Alkenes with two, three, and four alkyl groups bonded to the double bond unit are **disubstituted**, trisubstituted, and **tetrasubstituted**, respectively.

Monosubstituted:	$RCH=CH_2$
Disubstituted:	$RCH=CHR$ or $R_2C=CH$
Trisubstituted:	$R_2C=CHR$
Tetrasubstituted:	$R_2C=CR_2$

In general, alkyl groups increase the stability of a double bond in an alkene. Thus, a disubstituted alkene is more stable than a monosubstituted alkene. Therefore, if a chemical reaction can lead to either of these two products, the disubstituted alkene predominates.

The classes of alkynes are more limited. Only one alkyl group can be bonded to each of the two carbon atoms of the triple bond. If one alkyl group is bonded to one sp hybridized carbon atom of the triple bond, the compound is a monosubstituted alkyne ($R-C\equiv C-H$). It is also called a **terminal alkyne**, because the triple bond is on the end of the carbon chain. When alkyl groups are bonded to each carbon atom of the triple bond, the compound is disubstituted or an internal alkyne ($R-C\equiv C-R$).

Physical Properties of Alkenes and Alkynes

The physical properties of the homologous series of alkenes (C_2H_4,) and alkynes (C_nH_{2n-2}) are similar to those of the homologous series of alkanes (C_nH_{2n+2}). The compounds in both classes of unsaturated hydrocarbons are nonpolar. Alkenes and alkynes that contain fewer than five carbon atoms are gases at room temperature. As in the case of alkanes, the boiling points of the alkenes and alkynes increase with an increase in the number of carbon atoms in the molecule because the London forces increase (Table 4.1).

Table 4.1
Boiling Points of Alkanes, Alkenes, and Alkynes

Alkanes	Boiling Point (°C)	Alkenes	Boiling Point (°C)	Alkynes	Boiling Point (°C)
Pentane	36	1-Pentene	30	1-Pentyne	40
Hexane	69	1-Hexene	63	1-Hexyne	71
Heptane	98	1-Heptene	94	1-Heptyne	100
Octane	128	1-Octene	121	1-Octyne	125

Problem 4.1

Caryophyllene, which is responsible for the odor of oil of cloves, contains 15 carbon atoms. The compound has two rings and two double bonds. What is the molecular formula of caryophyllene?

Solution

For $n = 15$, the number of hydrogen atoms for a saturated compound without rings is 32. Each ring and each double bond results in a reduction of two hydrogen atoms. Thus the total number of hydrogen atoms is:

Number of hydrogen atoms = 32 − 2(no. of rings) − 2(no. of double bonds) = 32 − 2(2) − 2(2) = 24

The molecular formula is $C_{15}H_{24}$.

Problem 4.2

β-Carotene, found in carrots, has 40 carbon atoms and contains two rings and 11 double bonds. What is its molecular formula?

Problem 4.3

The urine of the red fox contains a scent marker that is an unsaturated thioether. Classify the degree of substitution of the double bond of the scent marker.

Solution

The compound contains a terminal double bond. The terminal carbon atom has two hydrogen atoms bonded to it. The other carbon atom of the double bond is bonded to a CH_2 group and a CH_2 unit. Thus, the compound contains a disubstituted double bond.

Problem 4.4

Tremorine is used to treat Parkinson's disease. Classify this alkyne.

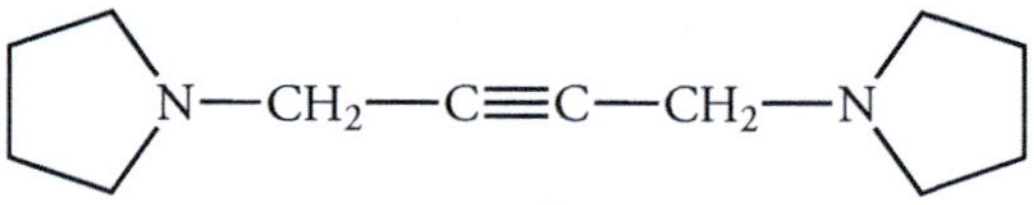

tremorine

Naturally Occurring Alkenes and Alkynes

Alkenes and their chemical cousins, the cycloalkenes, are very common in nature. For example, the common housefly (*Musca domestica*) contains muscalure, an unbranched alkene containing 23 carbon atoms. Muscalure is a pheromone that is released by the female to attract males. Muscalure has been synthesized in the laboratory, and can be used to lure male flies to traps.

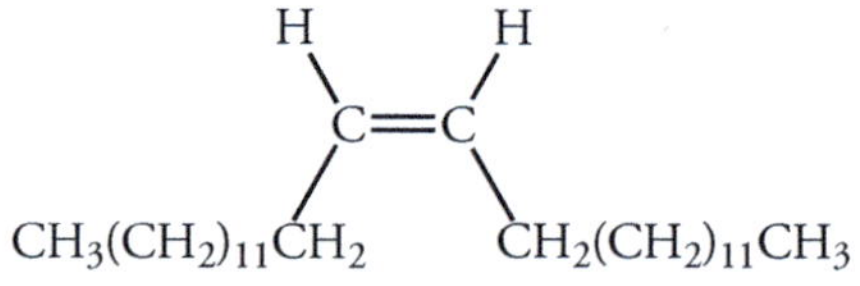

muscalure

Alkynes are not as prevalent in nature, but many of them are physiologically active. For example, the enetriyne ichthyothereol, whose structure we discussed above, is secreted from the skin of a species of frog in the Lower Amazon Basin. This compound is apparently a defensive venom and mucous membrane tissue irritant that wards off mammals and reptiles. The indigenous people of the area use the secretion to coat arrow heads. When the arrow pierces the skin of the target, the compound causes convulsions.

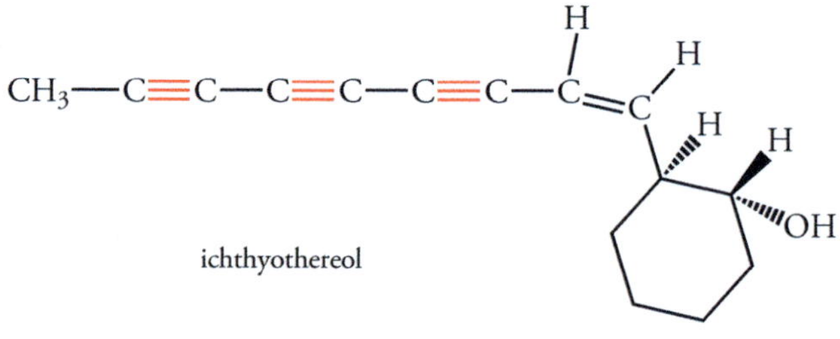

ichthyothereol

An alkene that contains several multiple bonds is said to be **polyunsaturated**. For example, polyunsaturated oils contain several double bonds.

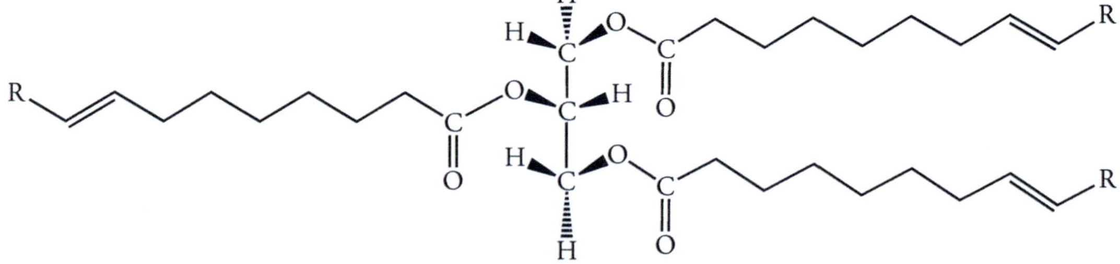

General structure of a polyunsaturated oil

Polyunsaturated compounds are common in nature. For example, vitamin A contains five double bonds.

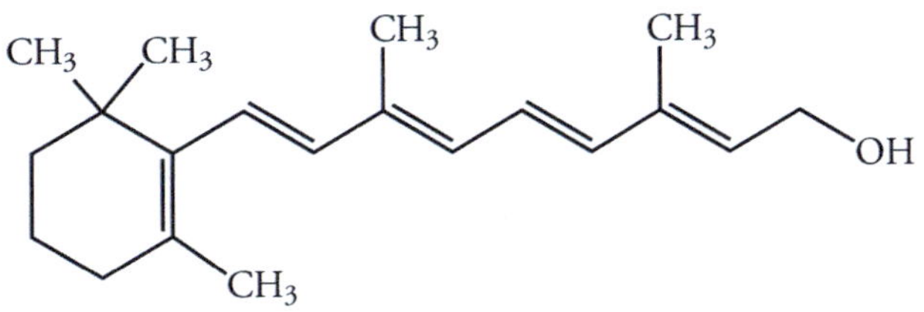

Vitamin A (retinol)

4.2 GEOMETRIC ISOMERISM

We know that there is nearly free rotation about carbon-carbon single bonds—about 12 kJ/mol is sufficient to rotate about the σ bond of ethane. Free rotation cannot occur around the carbon-carbon double bond of an alkene because of its electronic structure. As a result, the groups bonded to the carbon atoms of the double bond can exist in different spatial or geometric arrangements. These isomers have the same connectivity of atoms, but differ from each other in the geometry about the double bond. Hence, these compounds are called **geometric isomers** or *cis-trans* isomers. Consider a general alkene whose formula is CXY=CXY. When we draw a more detailed structural formula, we find that two representations are possible.

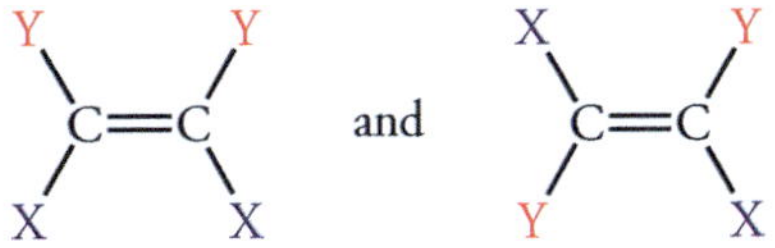

These two structures represent different molecules (Figure 4.2). In the structure on the left, two X groups are on the same "side" of the molecule. This is the *cis* isomer. In the structure on the right, the X groups are on opposite "sides" of the molecule. It is called the *trans* isomer.

Cis and *trans* isomers are possible only if an alkene has two different atoms or groups of atoms attached to each double-bonded carbon atom. For example, in 1,2-dichloroethene, each unsaturated carbon atom has a chlorine atom and a hydrogen atom attached to it. These groups are different, and both *cis* and *trans* isomers are possible.

If one of the unsaturated carbon atoms is attached to two identical groups, *cis-trans* isomerism is not possible. For example, neither chloroethene nor 1,1-dichloroethene can exist as *cis* and *trans* geometric isomers.

Figure 4.2 Geometric Isomers of Alkenes

All six atoms lie in the same plane. In the *cis* isomer two "X" and "Y" groups lie on the same side of the double bond. In the *trans* isomer "X" and "Y" groups lie on opposite sides of the double bond. They do not interconvert because rotation around the π bond does not occur.

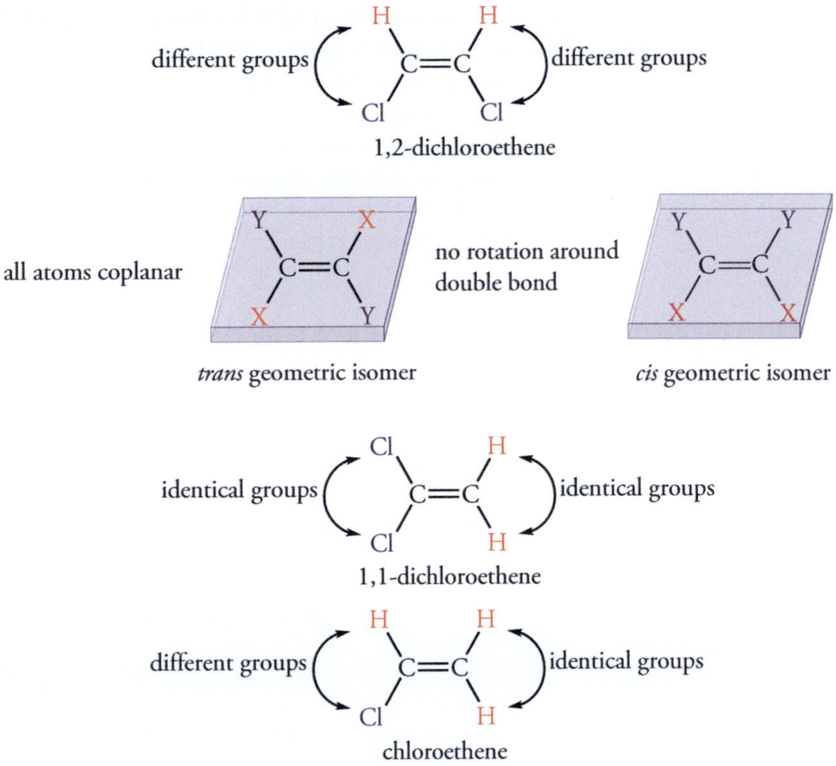

Geometric isomers, like all isomers, have different physical properties. For example, the boiling point of *cis-* and *trans-*1,2-dichloroethene (Figure 4.3) are 60 °C and 47 °C, respectively. The two C—Cl bond moments of the *trans* isomer cancel each other, and the compound has no dipole moment. The *cis* isomer has a dipole moment because the two C—Cl bond moments reinforce each other. As a result, the *cis* isomer is polar and has the higher boiling point.

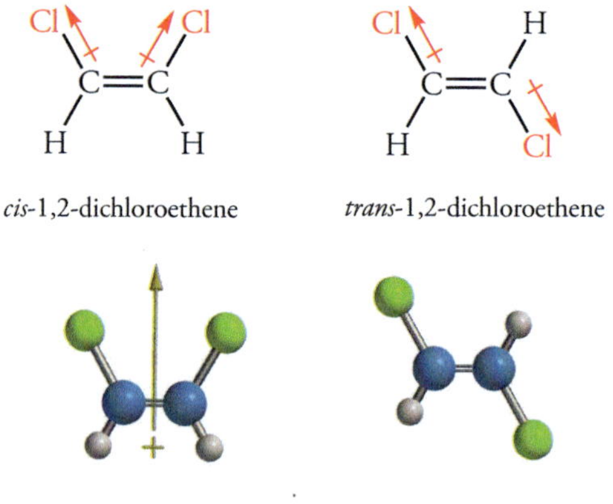

Figure 4.3
cis- and trans-1,1-Dichloroethene

cis- and *trans-*1,2-Dichloroethene have different physical properties. This example shows that the *cis* isomer has a net dipole moment, but the dipole in the *trans* isomer cancels, and it does not have a net dipole moment.

*cis-*1,2-dichloroethene   *trans-*1,2-dichloroethene

Problem 4.5

Is *cis-trans* isomerism possible for either of the double bonds of geraniol, a naturally occurring oil?

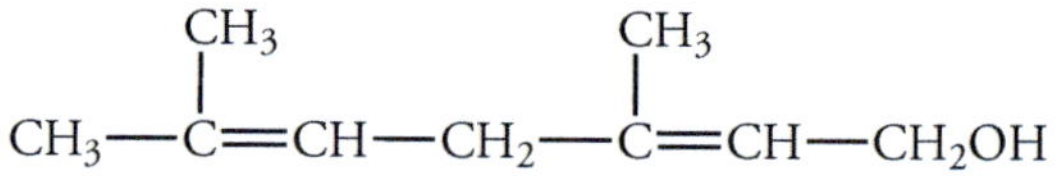

geraniol

Solution

First, consider the double bond near the left end of the molecule. The carbon atom of the double bond on the left has two —CH$_3$ groups bonded to it. On this basis alone geometrical isomerism is not possible at this position, regardless of the groups bonded to the right atom of the double bond. Next, consider the double bond toward the right end of the molecule. The carbon atom of that double bond is bonded to a —CH$_3$ group and a —CH$_2$—group that is part of the parent chain. The groups are different. This carbon atom is bonded to two different groups: a hydrogen atom and a —CH$_2$—group. Thus, geometrical isomerism is possible. The naturally occurring isomer of geraniol has a *trans* arrangement

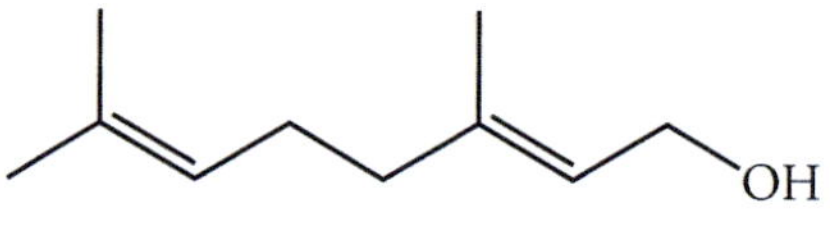

Problem 4.6

Is *cis-trans* isomerism possible about either of the double bonds of bombykol, a pheromone secreted by the female silkworm moth?

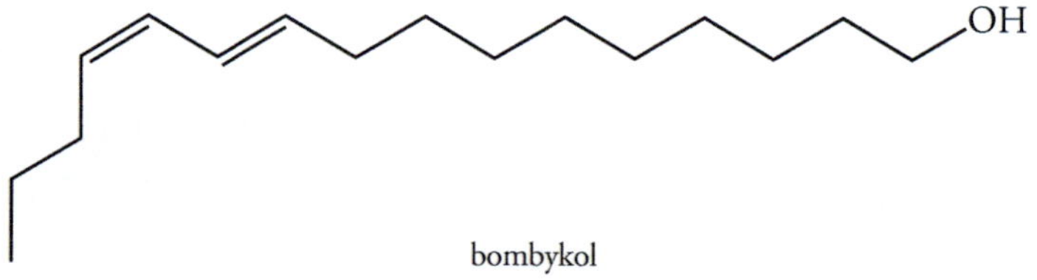

bombykol

4.3 E,Z NOMENCLATURE OF GEOMETRICAL ISOMERS

In the previous section, we used the terms *cis* and *trans* to describe the relationship of two substituents in the disubstituted alkene 1,2-dichloroethene. This type of nomenclature can easily be used for any disubstituted alkenes. Two examples are *cis*-2-butene and *trans*-2-butene.

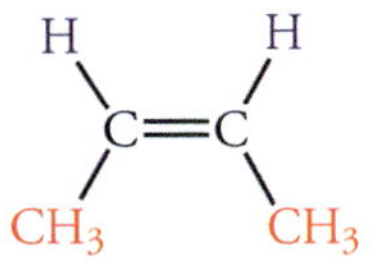

cis-2-butene 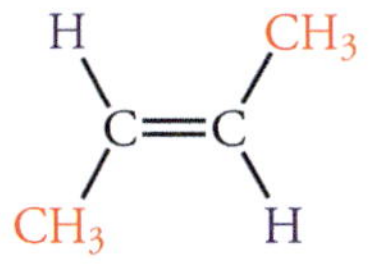   *trans*-2-butene

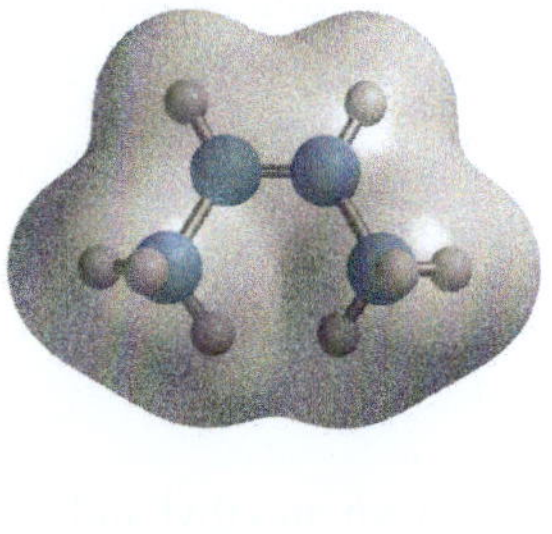

cis-2-butene 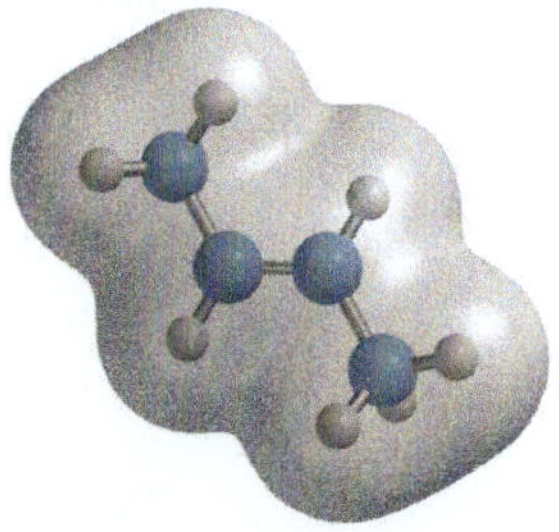   *trans*-2-butene

However, a *cis* and *trans* name does not describe isomeric trisubstituted and tetrasubstituted alkenes because there is no longer a simple reference giving the relationship of groups to one another. For example, even the following relatively simple compounds cannot be named as *cis* and *trans* isomers.

We can distinguish these isomers and all other tri- and tetrasubstituted alkenes by the *E,Z system* of nomenclature. The E,Z system uses **sequence rules** to assign priorities to the groups bonded to the atoms of the double bond of any alkene. The two groups bonded to one carbon atom are ranked by their respective priorities and designated low and high priority. The same consideration is given to the two groups bonded to the other carbon atom. If the higher priority groups on each carbon atom are on the same side of the double bond, the alkene is the Z isomer (German *zusammen*, together). If the higher priority groups on each carbon atom are on opposite sides of the double bond, the alkene is the E isomer (German *entgegen*, opposite).

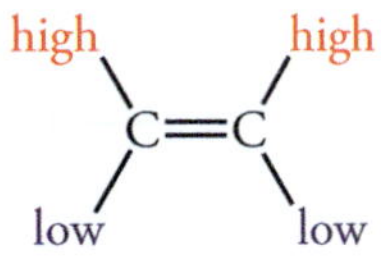

Z isomer 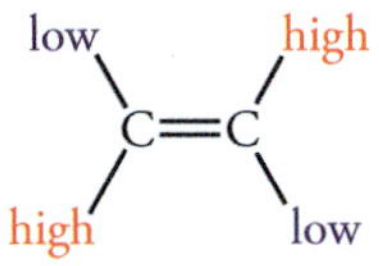   E isomer

Sequence Rules

1. If two atoms with different atomic numbers *are directly bonded* to a double bond, the atom with the higher atomic number receives a higher priority.

The priority order of some common elements is Br > Cl > F > O > N > C > ^{2}H > H. Applying these priorities to the following alkene, which contains several halogen atoms, allows us to make the E,Z assignment. The priority ^{2}H > H tells us that if two isotopes are possible, the one with the higher mass has the higher priority.

high priority Br   CH$_2$CH$_2$Br low priority

low priority Cl   F high priority

C=C

E isomer

The atomic number of bromine is greater than that of chlorine. Therefore, bromine has a higher priority than chlorine. A fluorine atom has a higher priority than a CH$_2$CH$_2$Br group, although the reason for this assignment may not be immediately obvious. The priority of a group depends on the atomic number of the atom *directly bonded* to the carbon atom of the double bond. In this case, we have to compare carbon to fluorine. Because fluorine has a higher atomic number than carbon, it has the higher priority.

2. If the atoms directly attached to the carbon atom of the double bond have the same atomic number, consider the second, third, and farther atoms until a difference is found. Then apply rule 1.

The difference between an ethyl and a methyl group illustrates this rule. They are equivalent at the first directly bonded atom: a carbon atom in each case. The carbon atom of a methyl group is bonded to three hydrogen atoms. The carbon of the ethyl group is bonded to another carbon atom and two hydrogen atoms. Thus, the ethyl group has a higher priority than a methyl group since C > H.

Sometimes the point of first difference is at some distance from the alkene carbon atom. The difference between a —CH$_2$CH$_2$OH group and an *n*-propyl group illustrates rule 2. They are equivalent at the directly bonded atom. They are also identical at the second atom. A difference is not found until the third atom. Because oxygen has a higher priority than carbon, the —CH$_2$CH$_2$OH group has a higher priority than an *n*-propyl group.

If the first point of difference is not the type of atom, but rather the number of those atoms, then the group with the greater number of high-priority atoms is assigned the higher priority. Based on this consideration, the order of alkyl groups is *tert*-butyl > isopropyl > ethyl > methyl.

The third rule of assigning priorities of groups deals with multiple-bonded atoms.

3. A multiple bond is considered equivalent to the same number of single bonds to like atoms. Thus, a double bond is counted as two single bonds for both of the atoms in the double bond. The same principle is used for a triple bond.

C=C   is treated as   —C—C—

—C≡C—   is treated as   —C—C—

Multiple bonds to atoms other than carbon are also "doubled" or "tripled." For example, a carbonyl group is considered a carbon atom with two single bonds to oxygen atoms, but also an oxygen atom bonded to two carbon atoms. A cyano group (—C≡N) is considered a carbon atom with three single bonds to a nitrogen atom and as a nitrogen atom bonded to three carbon atoms.

C=O   is treated as   —C—O   —C≡N   is treated as   H—C—N—

Problem 4.7

Rank the following sets of substituents in order of increasing priority according to the sequence rules.

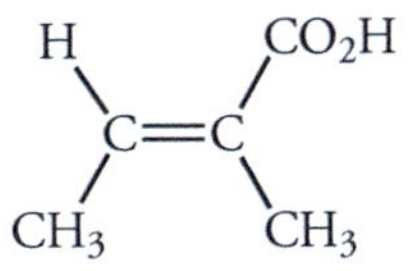

$$\underset{CH_3}{\overset{H}{}}C=C\underset{CH_3}{\overset{CO_2H}{}}$$

tiglic acid

Solution

The double-bonded carbon atom on the left is bonded to a hydrogen atom and a methyl group, whose priorities are low and high, respectively. The right double-bonded carbon atom is bonded to a methyl group and a carboxylic acid (—CO_2H) group. The oxygen atoms of the carboxylic acid group give it a higher priority than the methyl group, which has only hydrogen atoms bonded to the carbon atom. Because the higher priority groups are on opposite sides of the double bond, the compound is E.

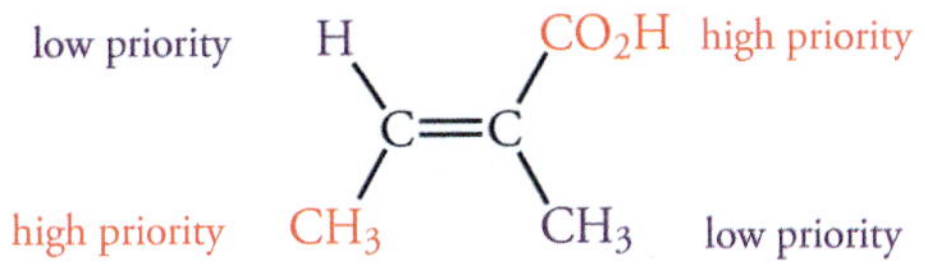

low priority H CO_2H high priority

$$C=C$$

high priority CH_3 CH_3 low priority

tiglic acid

Note that if we consider only the positions of the methyl groups, the compound looks like a *cis* isomer. Thus, one could erroneously classify the compound as the Z isomer. This example illustrates the value of the unambiguous E,Z system.

Problem 4.8

Assign the E or Z configuration to tamoxifen, a drug used in the treatment of breast cancer.

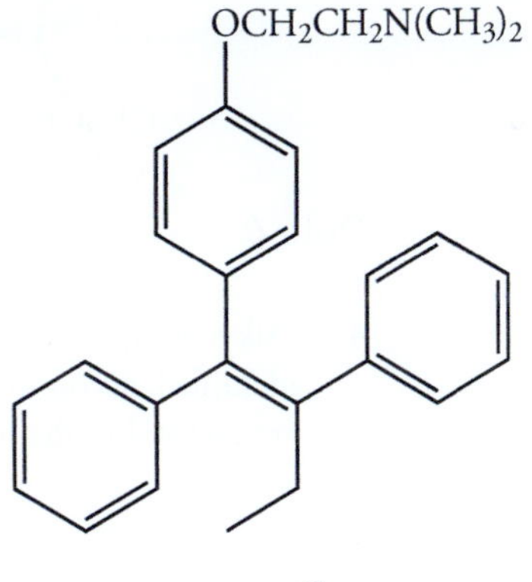

$OCH_2CH_2N(CH_3)_2$

tamoxifen

4.4 NOMENCLATURE OF ALKENES AND ALKYNES

The IUPAC rules for naming alkenes and alkynes are similar to those for alkanes, but the position of the double or triple bond in the chain and the geometric arrangement of substituents around the double bond must be indicated. As in the nomenclature of alkanes, many groups derived from alkenes have common names. Three of the most often encountered are the vinyl, allyl, and isopropenyl groups.

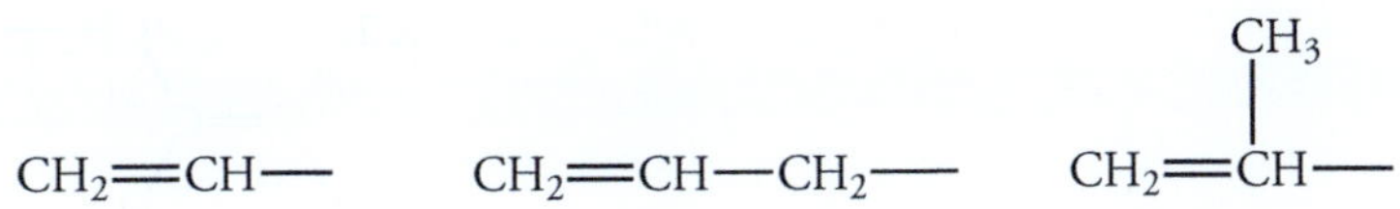

$$CH_2{=}CH{-}$$

vinyl

$$CH_2{=}CH{-}CH_2{-}$$

allyl

$$\underset{}{\overset{CH_3}{\underset{|}{}}}CH_2{=}CH{-}$$

isopropenyl

IUPAC Names of Alkenes

The IUPAC rules for naming alkenes are similar to those for alkanes, but the position of the double bond in the chain and the geometric arrangement of substituents around the double bond must be indicated. As in the case of some simple alkyl groups, a few common names are allowed as part of an IUPAC name, including vinyl, allyl, and isopropenyl.

Rule 1

1. The longest continuous chain of carbon atoms that contains the double bond is the parent alkene.

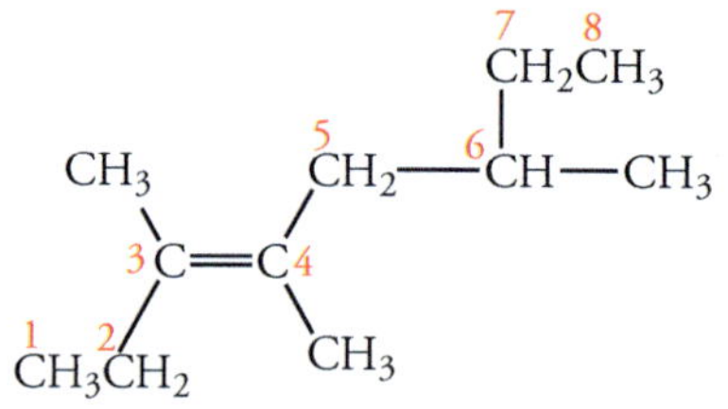

There are eight carbons in this chain, so it is an octene

Rule 2

2. The longest chain is given the same stem name as an alkane, but *-ene* replaces *-ane*. The parent name of the structure shown in rule 1 is octene.

Rule 3

3. Number the carbon atoms in the longest continuous chain starting from the end of the chain nearer the first branch.

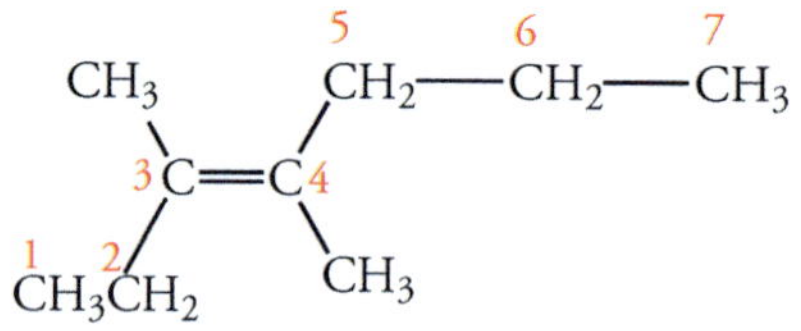

This is a substituted 3-heptene, not a substituted 4-heptene

Rule 4

4. Alkyl groups and other substituents are named, and their positions on the chain are identified, according to the numbering established by rule 3. Names and numbers are prefixed to the parent name.

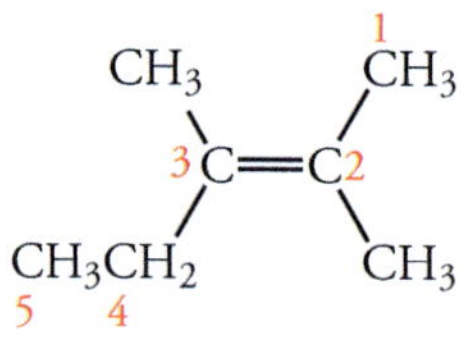

This is 2,3-dimethyl-2-pentene, not 3,4-dimethyl-3-pentene

Rule 5

4. If the compound can exist as an E or Z isomer, the appropriate prefix followed by a hyphen is placed within parentheses in front of the name.

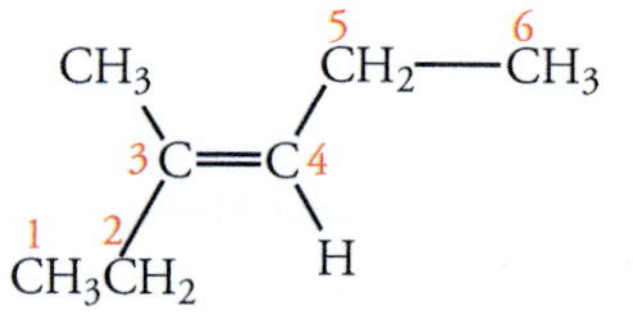

This is (E)-3-methyl-3-hexene

Rule 6

6. If the compound contains more than one double bond, specify the location of each double bond by a number. A prefix to -*ene* indicates the number of double bonds.

$$\overset{1}{CH_2}=\overset{2}{CH}-\overset{3}{CH}=\overset{4}{CH}-\overset{5}{CH}=\overset{6}{CH}-\overset{7}{CH_3}$$

1,3,5-heptatriene

Rule 7

7. Name cycloalkenes by numbering the ring to give the double-bonded carbon atoms the numbers 1 and 2. Choose the direction of numbering so that the first substituent on the ring receives the lower number. The position of the double bond is not given because it is known to be between the C-1 and C-2 atoms.

3-methylcyclopentene   1-methylcyclohexene

IUPAC Names of Alkynes

The IUPAC rules for naming alkynes are similar to those for naming alkenes, using -*yne* to indicate the triple bond, as illustrated by the following examples.

$$\overset{6}{CH_3}-\overset{5}{CH_2}-\overset{4}{CH_2}-\overset{3}{C}\equiv\overset{2}{C}-\overset{1}{CH_3}$$

2-hexyne

$$\overset{6}{CH_3}-\overset{5}{CH_2}-\overset{4}{C}\equiv\overset{3}{C}-\overset{2}{CH}-\overset{1}{CH_3}$$
$$\qquad\qquad\qquad\overset{|}{CH_3}$$

2-methyl-3-hexyne

Compounds with multiple triple bonds are diynes, triynes, and so on. Compounds with both double and triple bonds are enynes, not ynenes. Numbering of compounds with both double and triple bonds starts from the end nearest the first multiple bond regardless of type. When a choice is possible, double bonds are assigned lower numbers than triple bonds.

$$H-\overset{1}{C}\equiv\overset{2}{C}-\overset{3}{CH_2}-\overset{4}{C}\equiv\overset{5}{C}-\overset{6}{CH_3}$$

1,4-hexadiyne

$$\overset{1}{CH_2}=\overset{2}{CH}-\overset{3}{CH_2}-\overset{4}{C}\equiv\overset{5}{C}-\overset{6}{CH_3}$$

1-hexene-4-yne

Problem 4.9
Name the following compound.

Solution

There are six carbon atoms in the longest chain. It is numbered from right to left so that the double bond is at the carbon atom in position 2. The parent is thus 2-hexene. The chlorine and bromine atoms are at 2- and 3-, respectively. The two different groups of atoms on each unsaturated carbon atom make geometric isomers possible. Because both bromine and chlorine have higher priorities the compound is a Z isomer; it's name is (Z)-3-bromo-2-chloro-2-hexene.

high priority Cl   Br   high priority

$$\underset{1}{CH_3} \underset{2}{C} = \underset{3}{C} \underset{4\ 5\ 6}{CH_2CH_2CH_3}$$

low priority CH₃   CH₂CH₂CH₃ low priority

(Z)-3-bromo-2-chloro-2-hexene

Problem 4.10
Draw the structures of the following isomeric compounds.

(a) 5-methyl-1,3-cyclohexadiene   (b) 3-methyl-1,4-cyclohexadiene

Problem 4.11
Why is 2-bromo-4-hexyne an incorrect name for the following compound?

$$CH_3-\underset{\underset{Br}{|}}{CH}-CH_2-C\equiv C-CH_3$$

Solution
There are six carbon atoms in the longest chain. It is numbered from right to left so that the double bond is at the carbon atom in position 2. The parent is then 2-hexene. It's name is (Z)-3-bromo-2-chloro-2-hexene.

$$\underset{6}{CH_3}-\underset{5}{\underset{\underset{Br}{|}}{CH}}-\underset{4}{CH_2}-\underset{3}{C}\equiv\underset{2}{C}-\underset{1}{CH_3}$$

5-bromo-2-hexyne

Problem 4.12
(3E,11E)-1,3,11-Tridecatriene-5,7,9-triyne is a compound found in safflowers that is a chemical defense against nematode infestations. Write the structure of the compound.

4.5 ACIDITY OF ALKENES AND ALKYNES

Although hydrocarbons are extremely weak acids, a very strong base can remove a proton from an alkane, alkene, or alkyne to produce a **carbanion**, an anion with a negative charge on the carbon atom. Hydrocarbons are weaker acids than many acids that you encountered in your first chemistry course. Those acids had the hydrogen atom bonded to an electronegative atom. Because carbon is less electronegative, carbanions are not as stable as the conjugate bases of inorganic acids.

$$CH_3-CH_3 \qquad CH_2=CH_2 \qquad H-C\equiv C-H$$

ethane   ethene   ethyne
$K_a = 10^{-49}$ $K_a = 10^{-44}$ $K_a = 10^{-25}$

The acidity of hydrocarbons is related to the hybridization of the carbon atom. The K_a increases for carbon atoms in the order sp³ < sp² < sp. The order of acidities parallels the contribution of the lower energy of the 2s orbital to the hybrid orbitals in the σ-bond. The energy of a 2s orbital is lower than that of a 2p orbital, and on average a 2s orbital is closer to the nucleus than a 2p orbital. The average distance of hybrid orbitals from the nucleus depends on the percent contribution of the s and p orbitals. For an sp³ hybrid orbital, the contribution of the s orbital is 25%, because one s and three p orbitals contribute to the four hybrid orbitals. Similarly, the contribution of the s orbital is 33% and 50% for the sp² and sp hybrid orbitals, respectively. Because an sp hybrid orbital has more character than an sp³ or sp² orbital, its electrons are located closer to the nucleus. As a result, a proton is more easily removed, and the electron pair remains on the carbon atom.

Hybridization	sp^3	sp^2	sp
% Character	25	33	50
K$_a$	10^{-49}	10^{-44}	10^{-25}

For all practical purposes, only acetylene and terminal alkynes are strong enough acids to be produced using conventional bases. However, even hydroxide ion is not a strong enough base to remove a proton from an alkyne. In fact, the conjugate base of an alkyne is rapidly and quantitatively converted to the alkyne whenever it reacts with compounds containing hydroxyl groups.

$$R—C≡C:^- + H_2O \rightleftharpoons R—C≡C—H + OH^-$$

When we discussed the periodic trends of acidity in Chapter 2, we saw that an N—H bond is a weaker acid than an O—H bond. Therefore, NH_2^-, the conjugate base of ammonia, a very weak acid, is a stronger base than OH^-, the conjugate base of water, a weak acid. The K$_a$ of ammonia is 10^{-36}. The K$_a$ of a terminal alkyne is about 10^{-24}. Thus, an amide ion quantitatively removes a proton from any terminal alkyne.

$$R—C≡C—H + :NH_2^- \longrightarrow R—C≡C:^- + NH_3$$

4.6 HYDROGENATION OF ALKENES AND ALKYNES

We expect alkenes and alkynes to have chemical similarities because both have π bonds. In general, that expectation is correct. Because alkynes have two π bonds, they often react with twice the amount of reagent that reacts with alkenes, which have only one π bond. Both classes of compounds are unsaturated, and they react with hydrogen gas to give more saturated compounds.

Hydrogenation of Alkenes

The reaction of hydrogen gas with an alkene (or cycloalkene) yields a saturated compound. The process is a reduction, but the reaction is also called **hydrogenation**. The hydrogenation of 1-octene yields octane. Hydrogenation requires a catalyst. The catalyst is usually finely divided platinum on finely divided carbon, but nickel and palladium can also be used. The hydrogenation process is heterogeneous; that is, it occurs on the surface of the solid catalyst. The symbol for the catalyst is Pd/C.

The double bond is converted to a single bond

$$CH_2{=}CH(CH_2)_5CH_3 \xrightarrow{\text{H}_2 \text{ / Pd/C}} CH_3{—}CH_2(CH_2)_5CH_3$$

Although functional groups such as ketones or esters also have multiple bonds, they are not normally reduced under the mild conditions used to add hydrogen to a carbon-carbon double bond.

$$CH_3{—}\overset{\overset{\displaystyle O}{\|}}{C}{—}CH_2{—}CH_2{—}CH{=}CH_2 \xrightarrow[\text{1 atm}]{\text{H}_2 \text{ / Pt/C}} CH_3{—}\overset{\overset{\displaystyle O}{\|}}{C}{—}CH_2{—}CH_2{—}CH_2{—}CH_3$$

Catalytic hydrogenation is used commercially to convert liquid vegetable oils into semisolid fats. Fats and oils are structurally related esters and differ in the degree of saturation. The difference between an oil and its companion solid is just the degree of saturation of the carboxylic acids in the esters: The solid is saturated.

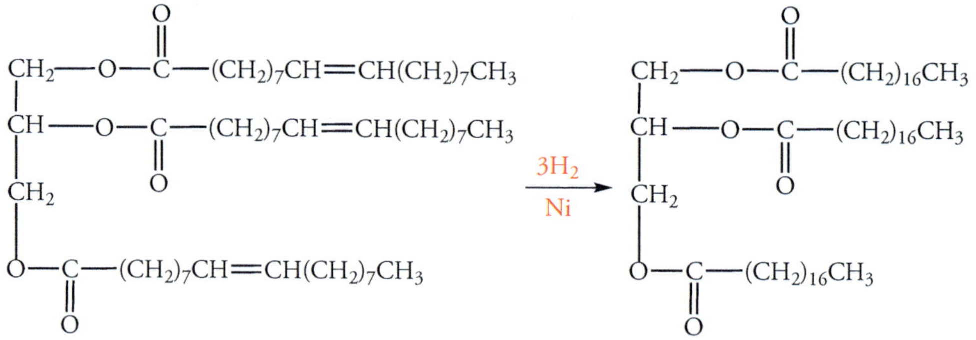

triolein   tristearin

Hydrogenation of an alkene occurs by the addition of hydrogen atoms to the same face of the double bond. For example, the hydrogenation of 1,2-dimethylcyclopentene produces *cis*-1,2-dimethylcyclopentane. The finely divided metal catalyst adsorbs hydrogen gas on the surface, and the hydrogen-hydrogen bond is broken. When the alkene approaches "lands" on the catalyst, the hydrogen atoms must add to the same face of the double bond.

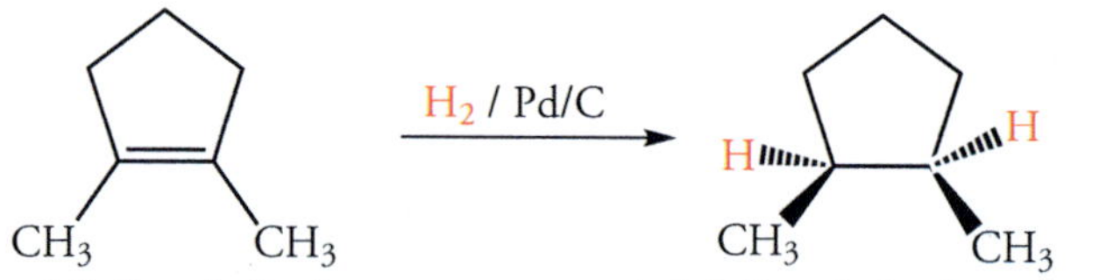

1,2-dimethylcyclopentene   *cis*-1,2-dimethylcyclopentane

Hydrogenation of Alkynes

Alkynes can be completely reduced to alkanes by reaction with two molar equivalents of hydrogen gas in the presence of catalysts such as platinum or palladium.

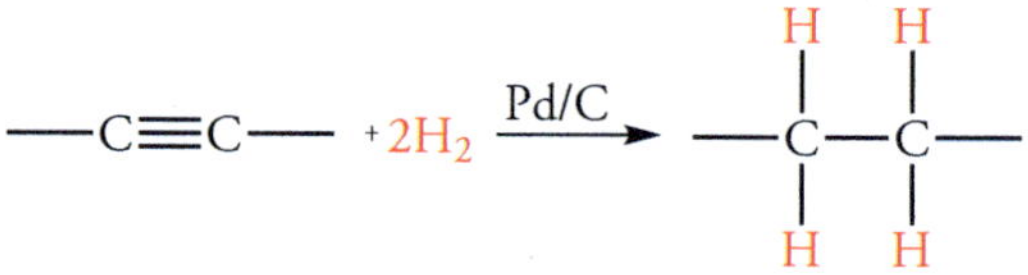

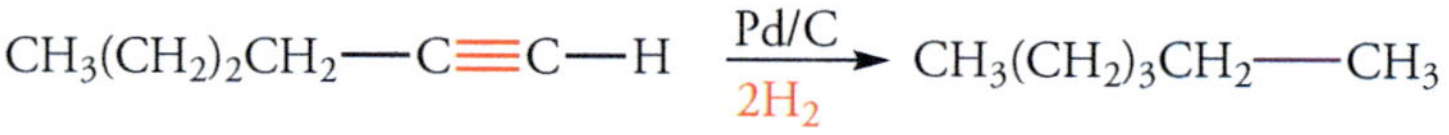

$$CH_3(CH_2)_2CH_2 \!-\! C\!\equiv\!C\!-\!H \xrightarrow[2H_2]{Pd/C} CH_3(CH_2)_3CH_2 \!-\! CH_3$$

1-hexyne   hexane

The hydrogenation of an alkynes can be stopped after adding one molar equivalent of hydrogen gas to form an alkene if the palladium is specially prepared. In the Lindlar catalyst, palladium is coated on calcium carbonate that contains a small amount of lead acetate. Hydrogenation of an alkyne using the Lindlar catalyst gives *cis* alkenes. This form of palladium does not reduce the alkene.

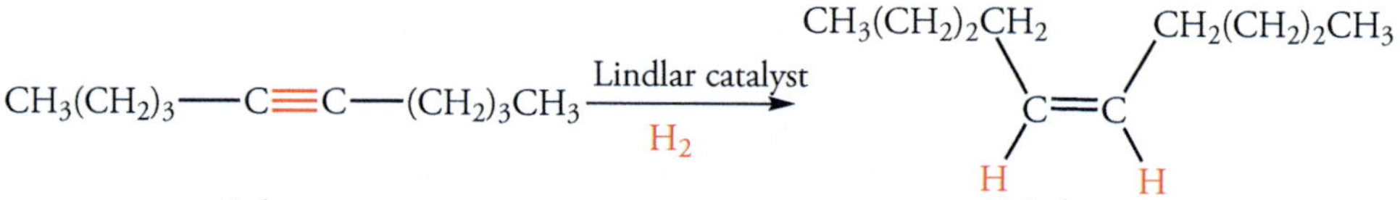

5-decyne   Z-5-decene

In contrast, reduction of an alkyne with lithium or sodium metal as the reducing agent in liquid ammonia as the solvent gives the *trans* isomer. Water is added to the reaction mixture in a second step to consume excess reagent.

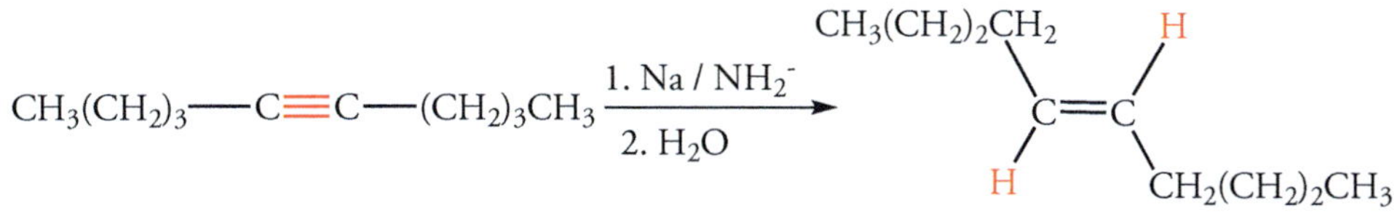

CH3(CH2)3—C≡C—(CH2)3CH3
1. Na / NH2-
2. H2O
CH3(CH2)2CH2
H
C=C
H
CH2(CH2)2CH3
5-decyne
(E)-5-decene

Problem 4.13

How many moles of hydrogen gas will react with cembrene, which is found in pine oil? What is the molecular formula of the product?

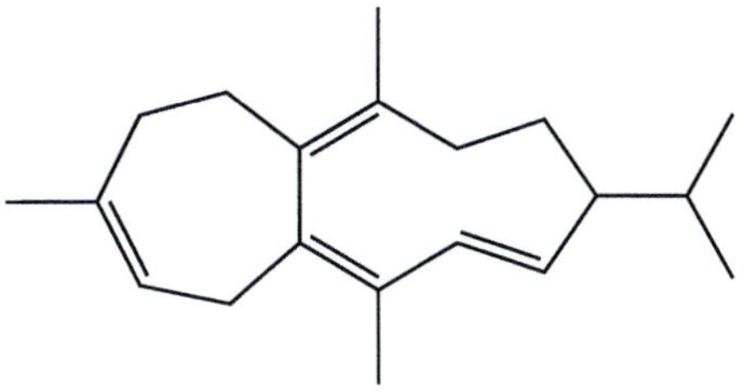

cembrene

Solution

There are four double bonds in the compound. One molar equivalent of cembrene will react with four molar equivalents of hydrogen gas. The product will be a cycloalkane. There are 14 carbon atoms in the ring, three methyl groups, and an isopropyl group for a total of 20 carbon atoms. Because the general molecular formula for a cycloalkane is C_nH_{2n}, the molecular formula of the product is $C_{20}H_{40}$.

Problem 4.14

Write the structure obtained by complete hydrogenation of ipsdienol, a pheromone of the Norwegian spruce beetle.

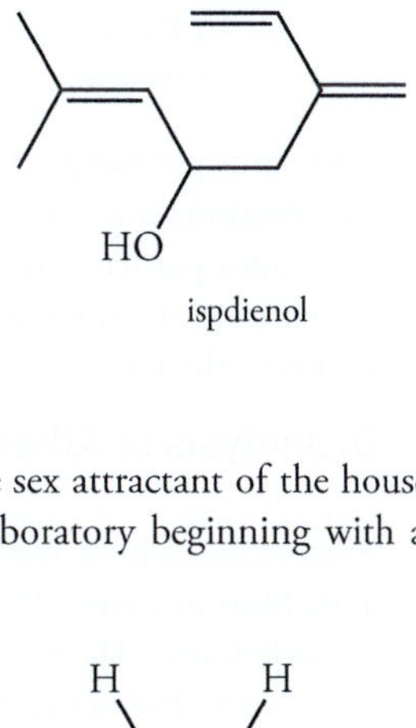

HO

ispdienol

Problem 4.15

The IUPAC name of muscalure, the sex attractant of the housefly, is (Z)-9-tricosene. How can this compound be synthesized in the laboratory beginning with an alkyne. What is the name of the alkyne?

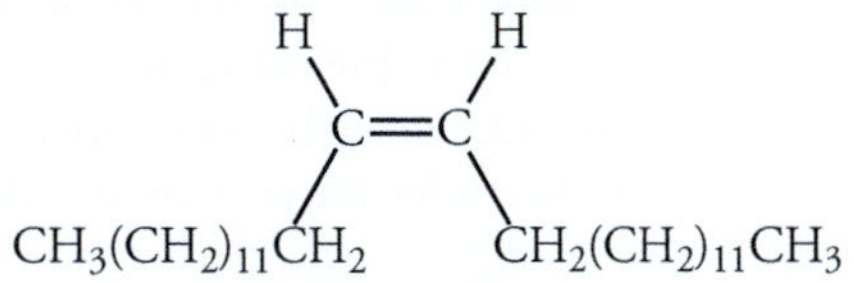

H
H
C=C
CH3(CH2)11CH2
CH2(CH2)11CH3

muscalure

Solution

Alkenes with the Z configuration can be prepared by hydrogenation of an alkyne using the Lindlar catalyst. The required alkyne must have the triple bond at the C-9 position and is named 9-tricosyne.

$$CH_3(CH_2)_{11}CH_2-C \equiv C-CH_2(CH_2)_{11}CH_3 \xrightarrow[\text{H}_2]{\text{Lindlar catalyst}}$$

muscalure

Problem 4.16

(E)-11-Tetradecen-1-ol is an intermediate required to synthesize the sex attractant of the spruce bud-worm moth. How can this compound be prepared from an alkyne? Suggest a name for the alkyne. (The *-ol* ending refers to the hydroxyl group, and the alkyne name must contain the *-ol*.)

(E)-11-tetradecen-1-ol

4.7 OXIDATION OF ALKENES AND ALKYNES

Oxidizing agents react with the π electrons of the double bond of alkenes and the triple bond of alkynes more easily than the electrons in the σ of carbon-carbon single bonds and carbon-hydrogen bonds. Thus, in contrast to alkanes, both alkenes and alkynes can easily be oxidized without destroying the carbon chain.

Hydroxylation of Alkenes

Reaction of an alkene with potassium permanganate ($KMnO_4$) in basic solution yields a product that contains a hydroxyl group on each carbon atom of the original double-bond unit. Thus, the alkene is oxidized. Permanganate is reduced to MnO_2, a brown solid.

| 1-butene (colorless) | (purple solution) | 1,2-butanediol (colorless) | (brown precipitate) |

Potassium permanganate is purple in aqueous solution. Manganese dioxide MnO_2, the product of the reaction, is a brown solid that precipitates from solution. The color change in oxidation with potassium permanganate is a quick chemical test for the presence of a double bond. Alkanes and cycloalkanes are not oxidized by $KMnO_4$, so the purple color remains. When alkenes are oxidized by $KMnO_4$, the purple color fades as a brown precipitate appears.

Ozonolysis of Alkenes and Alkynes

Alkenes and alkyne react rapidly with ozone, O_3. The reaction is carried out in an inert solvent such as dichloromethane. The oxidation occurs in two steps. In the first, an unstable intermediate called an ozonide forms. The intermediate is not isolated but is treated in solution by reacting it with zinc and acetic acid. The net result of the two reactions is the cleavage of the carbon-carbon double bond to produce two carbonyl compounds. The overall process is called **ozonolysis**. Ozonolysis can be used to determine the position of a double bond in an alkene.

carbonyl compounds

If one of the double-bonded carbon atoms has two hydrogen atoms bonded to it, its ozonolysis product is methanal (formaldehyde). If an alkyl group and a hydrogen atom are bonded to the double-bonded carbon atom, the product is an aldehyde; if two alkyl groups are bonded to the double-bonded carbon atom, the product is a ketone.

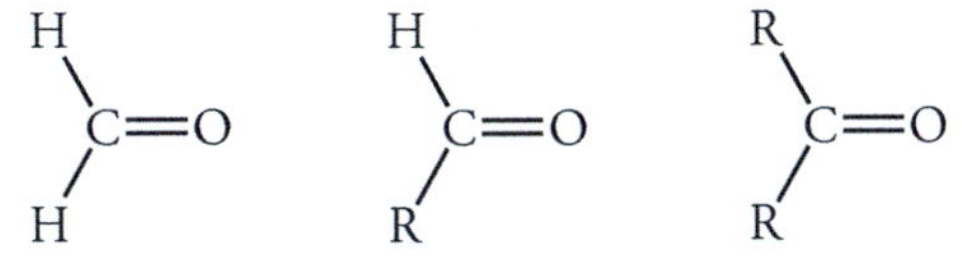

methanal   an aldehyde   a ketone

Although less commonly used, the ozonolysis of alkynes also results in cleavage products. Carboxylic acids are obtained from internal alkynes. A terminal alkyne forms one molar equivalent of CO_2.

$$R-C\equiv C-R' \xrightarrow[\text{2. Zn/H}_3\text{O}^+]{\text{1. O}_3} R-CO_2H \ + \ R'-CO_2H$$

$$R-C\equiv C-H \xrightarrow[\text{2. Zn/H}_3\text{O}^+]{\text{1. O}_3} R-CO_2H \ + \ CO_2$$

4.8 ADDITION REACTIONS OF ALKENES AND ALKYNES

In Section 2.5 we saw that an addition reaction occurs when two reactants combine to form a single product. No atoms are "left over." Examples of addition reactions of ethene (C_2H_4) with some common reagents are shown below.

$$CH_2{=}CH_2 \ + \ Br-Br \longrightarrow Br-CH_2-CH_2-Br$$

$$CH_2{=}CH_2 \ + \ H-Cl \longrightarrow H-CH_2-CH_2-Cl$$

$$CH_2{=}CH_2 \ + \ H-OH \longrightarrow H-CH_2-CH_2-OH$$

Reagents that add to alkenes are classified as symmetrical or unsymmetrical. **Symmetrical reagents** contain two identical groups, as in bromine. **Unsymmetrical reagents** consist of different groups, as in HCl and H_2O.

Addition of Halogens

The reaction of ethene with Br_2 to form 1,2-dibromoethane is an addition reaction. The atoms that add to the double bond are located on adjacent carbon atoms, a common characteristic of addition reactions of alkenes.

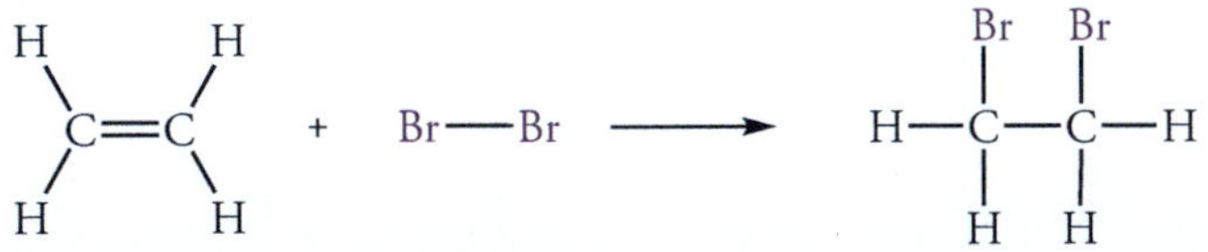

We can easily see the evidence for the addition of bromine to an alkene. Bromine is red-orange. It reacts with alkenes to give a colorless product. If the bromine color disappears when Br_2 is added to a compound, the compound is unsaturated.

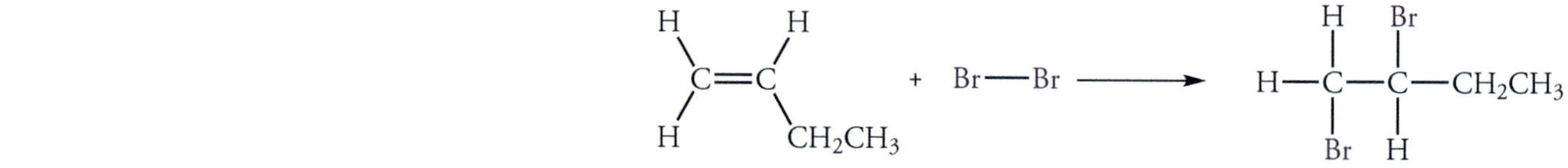

1-butene (colorless) + Br—Br → 1,2-dibromobutane (colorless) (red-orange)

Chlorine also adds to a carbon-carbon double bond, but iodine is not sufficiently reactive to give a good yield of addition product. The reaction of alkenes with fluorine is too reactive to control, and several competing reactions also occur if fluorine is used.

Alkynes react with chlorine or bromine to produce tetrahaloalkanes, which contain two halogen atoms on each of the original carbon atoms of the triple bond. Hence, the reaction consumes two molar equivalents of the halogen. As in alkenes, both chlorine and bromine give halogenated products.

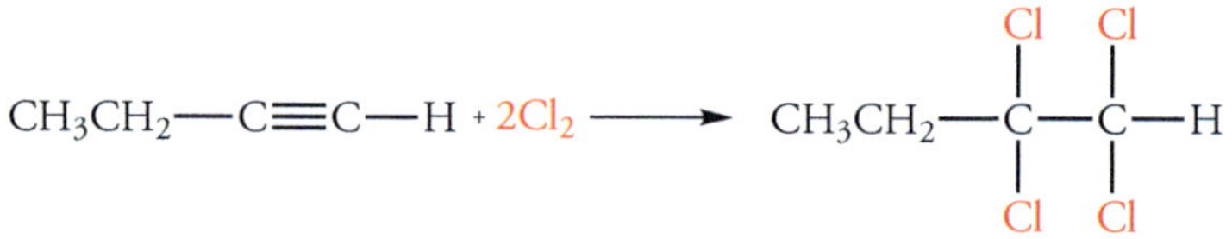

$CH_3CH_2—C\equiv C—H + 2Cl_2 \longrightarrow$

1-butyne   1,1,2,2-tetrachlorobutane

If only one molar equivalent of the halogen is used, the reaction product has the halogen atoms on opposite sides of the double bond.

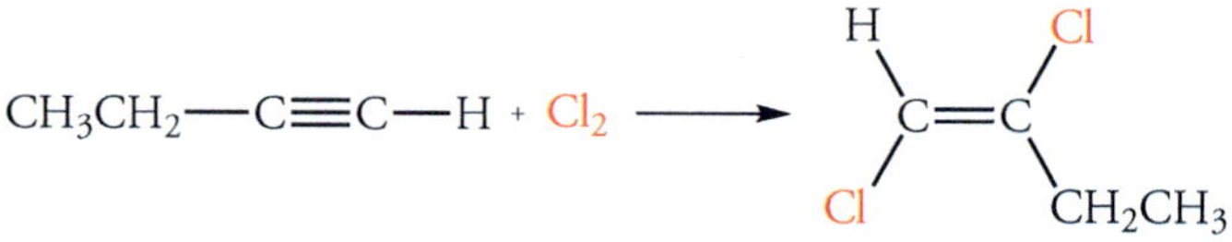

$CH_3CH_2—C\equiv C—H + Cl_2 \longrightarrow$

1-butyne   (E)1,2-diachlorobutane

Addition of Hydrogen Halides

The addition of a symmetrical reagent to an alkene yields only one possible product. It makes no difference which bromine atom bonds to which carbon atom. However, the situation is quite different for an unsymmetrical reagent, such as a hydrogen halide, HX. (Usually only HCl or HBr is used.) With a symmetrical alkene, such as ethylene, only one product is possible because the two carbon atoms are identical.

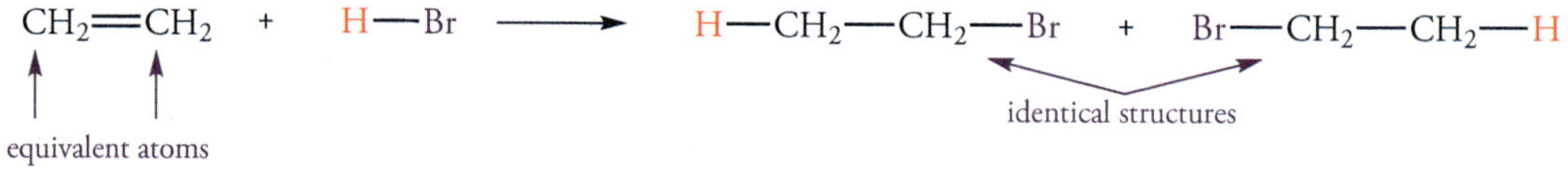

$CH_2=CH_2$ + H—Br → H—CH_2—CH_2—Br + Br—CH_2—CH_2—H

equivalent atoms   identical structures

Although two products could potentially result from the addition of HBr to an unsymmetrical alkene, only one is actually formed. Addition of HBr to propene could yield either 1-bromopropane or 2-bromopropane, but only the latter is formed. The X written through one reaction arrow indicates that the reaction does not occur. Thus, the addition of an unsymmetrical reagent to an alkene is a highly selective process; that is, only one product is formed. A similar selectivity is observed for the reaction of alkynes with hydrogen halides.

propene + H—Br → 2-bromopropane

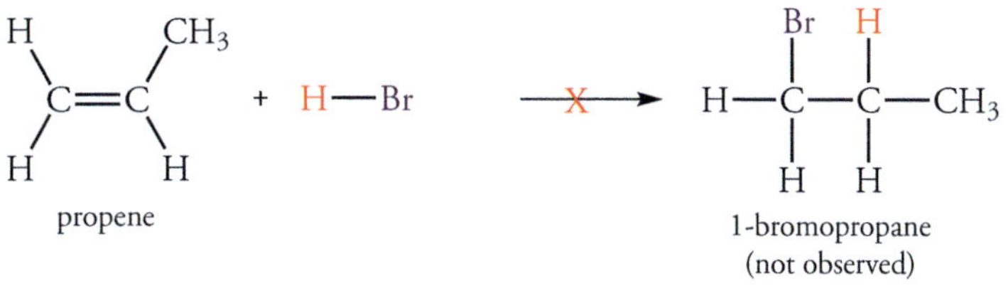

propene   +   H—Br   —X→   1-bromopropane (not observed)

Markovnikov's Rule

In 1870, the Russian chemist Vladimir Markovnikov observed that reagents add to unsymmetrical alkenes in a specific way. **Markovnikov's rule** states that a molecule of the general formula HX adds to a double bond so that the hydrogen atom forms a bond to the unsaturated carbon atom with the largest number of *directly bonded* hydrogen atoms. This is the less substituted double-bonded carbon atom. The addition reactions of HCl with 2-methylpropene and 1-methylcyclohexene with HBr are two examples of Markovnikov's rule.

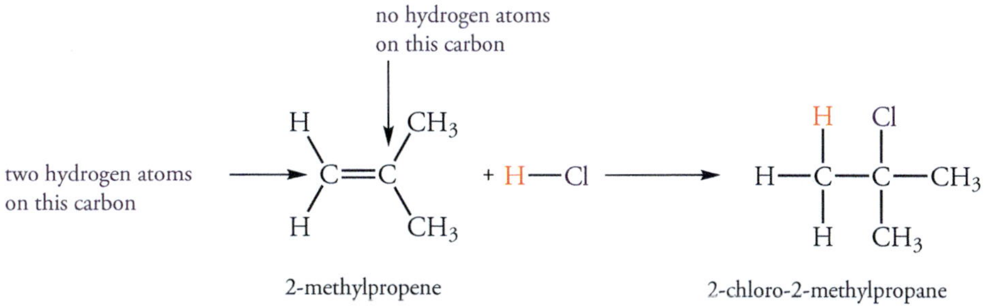

two hydrogen atoms on this carbon   no hydrogen atoms on this carbon

2-methylpropene   + H—Cl   →   2-chloro-2-methylpropane

Problem 4.17
Predict the product that will be formed when HBr is added to 2-methyl-2-butene.

Solution
One of the unsaturated carbon atoms has one attached hydrogen atom, the other no attached hydrogen atoms. The hydrogen atom of HBr bonds to the carbon atom having the greater number of directly bonded hydrogen atoms. The bromine atom bonds to the other carbon atom of the original double bond. The predicted and observed product is 2-bromo-2-methylbutane.

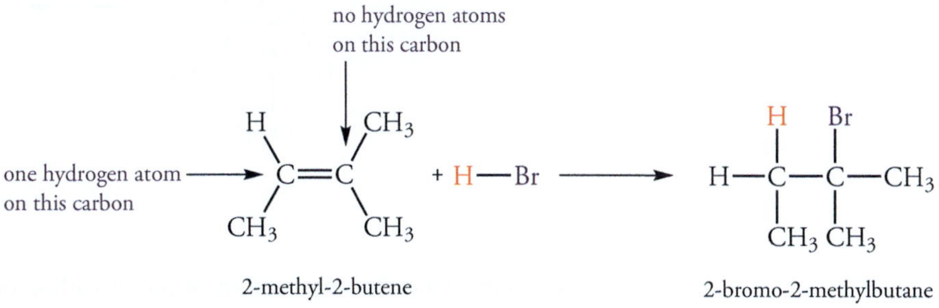

one hydrogen atom on this carbon   no hydrogen atoms on this carbon

2-methyl-2-butene   + H—Br   →   2-bromo-2-methylbutane

Problem 4.18
Predict the product of the addition of HCl to 1-methylcyclohexene.

4.9 MECHANISM OF ADDITION REACTIONS

The specificity of the addition reaction as originally discovered by Markovnikov has now been well established. The π electrons in an alkene act as a **nucleophile**, a species that is attracted to positively charged or partially positively charged nuclei. The nucleophilic π bond reacts with an electron-loving species called an **electrophile**, in this case H⁺.

Consider the reaction of propene with H—Br. We write the first step by using a curved arrow to show the movement of the two electrons in the π bond to form a σ bond to the positively charged hydrogen atom. This step produces a positively charged species called an isopropyl carbocation, which is also an electrophile.

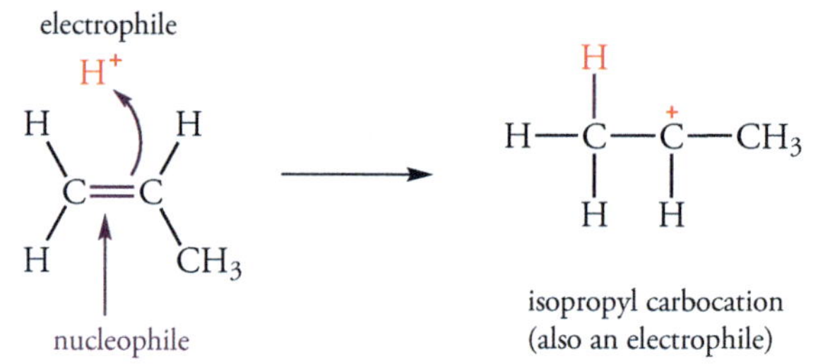

In the second step of the addition reaction, the isopropyl carbocation acts as an electrophile and accepts an electron pair from the bromide ion, which acts as a nucleophile.

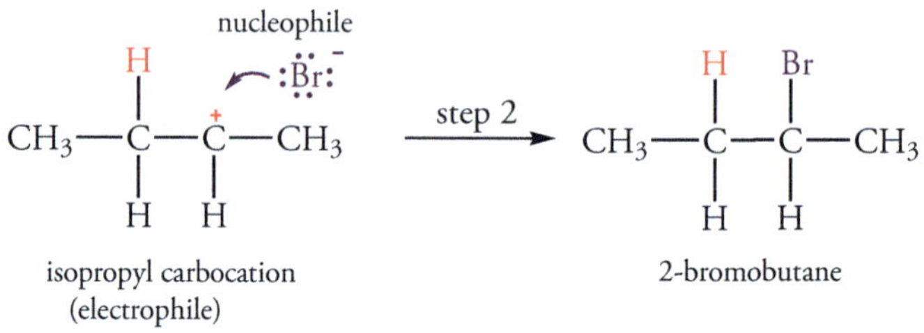

In the first step of this reaction, the hydrogen atom is attached to one of the two possible carbon atoms of the original π bond; this placement accounts for the product predicted by Markovnikov's rule. Why does the isopropyl carbocation form? If the hydrogen atom had bonded to the other side of the double bond, which is bonded to a hydrogen and a methyl group, an *n*-propyl carbocation would have formed.

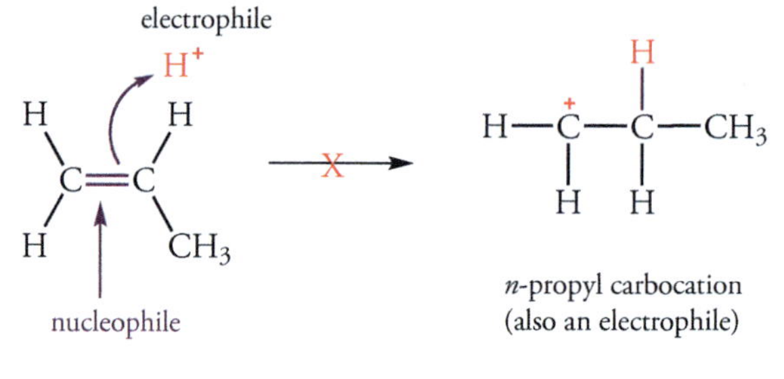

The isopropyl carbocation and the *n*-propyl carbocation are both unstable intermediates. However, the isopropyl carbocation is more stable than the *n*-propyl carbocation, and it forms preferentially. Note the addition of the proton to a carbon-carbon double bond determines the structure of the final product. The second step merely combines the nucleophilic halide ion with the carbocation at the site of the charge generated in the first step.

Why is the isopropyl carbocation formed in preference to the propyl carbocation? Alkyl groups attached to a positively charged carbon atom help stabilize the charge because the electrons in the carbon-carbon bonds are polarized toward the positive center. The isopropyl carbocation has the charge on a secondary carbon atom, that is, one that bonded to two alkyl groups. The propyl carbocation has the charge on a primary carbon atom; it is only bonded to one alkyl group. By the same reasoning, it follows that a tertiary carbocation is more stable than a secondary carbocation, because it has three alkyl groups attached to the positively charged carbon atom.

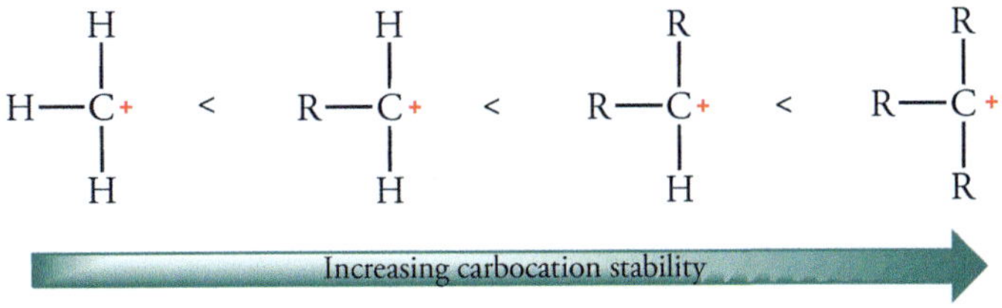

This order of stability accounts for Markovnikov's rule. Addition of the electrophile always occurs to give the most stable carbocation, which determines the identity of the product.

4.10 HYDRATION OF ALKENES AND ALKYNES

In Section 4.8 we saw that water is an unsymmetrical reagent that can add to a π bond. Water fits the general class of H—X compounds (H—OH). Thus, water adds to the double bond of an alkene in accordance with Markovnikov's rule to give 2-propanol, but none of the isomeric 1-propanol. The reaction is **hydration**.

propene 2-propanol

1-propanol
(not observed)

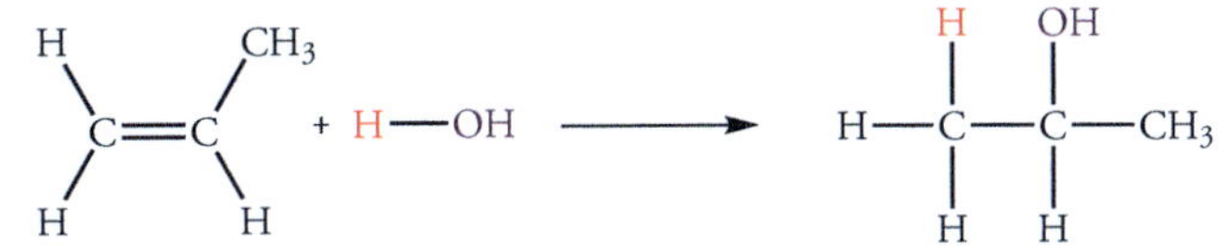

The reverse of the hydration reaction is **dehydration**, which is an example of an elimination reaction as discussed in Section 2.4. The dehydration of an alcohol produces an alkene.

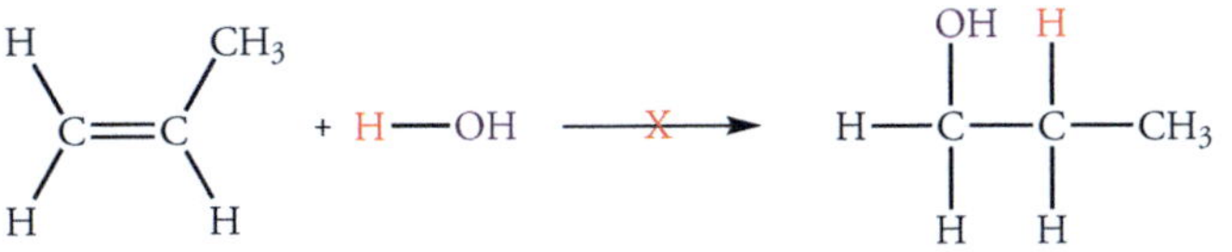

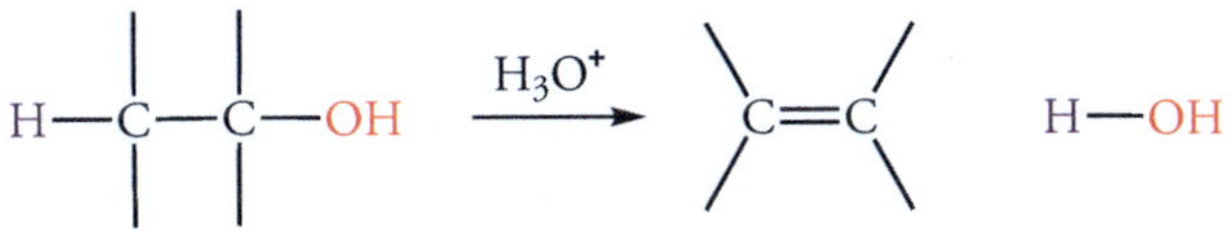

The direction of the reaction—hydration or dehydration—is controlled by conditions governed by Le Châtelier's principle. Conversion of an alkene to an alcohol requires an excess of water. Dehydration occurs if the water concentration is very low, as in concentrated sulfuric acid, which is about 98% H_2SO_4.

Hydration of Alkynes Produces Carbonyl Compounds

Water adds to one of the π bonds of a triple bond in aqueous sulfuric acid in the presence of mercuric sulfate ($HgSO_4$), which acts as a catalyst. However, the alcohol that forms has its —OH group bonded to the double-bonded carbon atom of an alkene. This type of compound is called an **enol**, a name that includes both the *-ene* suffix of a double bond and the alcohol suffix *-ol*.

an enol

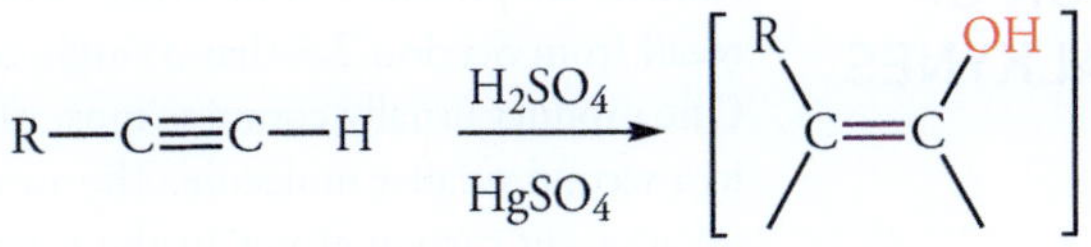

Enols are unstable compounds, and they are rapidly converted to carbonyl compounds in a rearrangement reaction. We will discuss this reaction further in Chapter 10.

an enol → a ketone

The final product of hydration of an alkyne is a ketone. The more substituted carbon atom of the alkyne is converted into a carbonyl carbon atom.

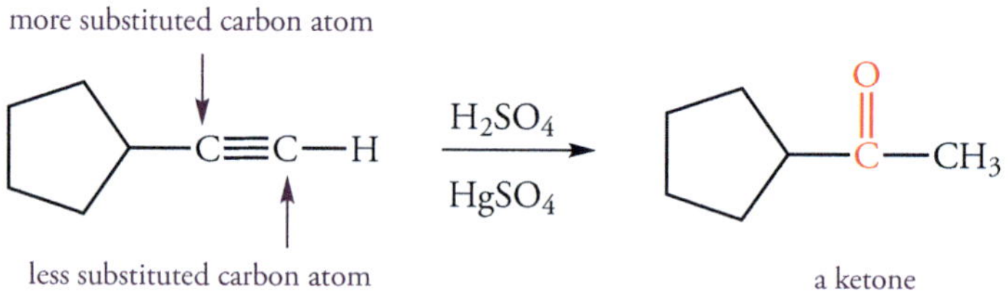

more substituted carbon atom

less substituted carbon atom

a ketone

Problem 4.19

What product(s) will result from the hydration of 2-decyne?

Solution

We have seen that the initial hydration of an alkyne places the hydroxyl group on the more substituted carbon atom. This group is then converted into a carbonyl group of a ketone. In 2-decyne both C-2 and C-3 are substituted to the same degree.

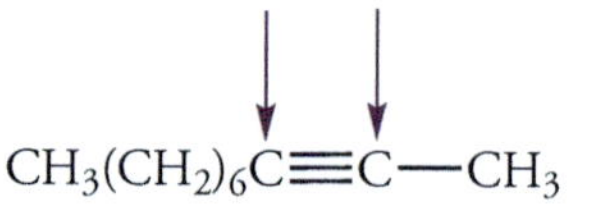

each alkyne carbon is bonded to one alkyl group

$$CH_3(CH_2)_6C \equiv C - CH_3$$

2-decyne

Therefore, hydration can occur either of two ways, and two isomeric products, 3-decanone and 2-decanone, are produced.

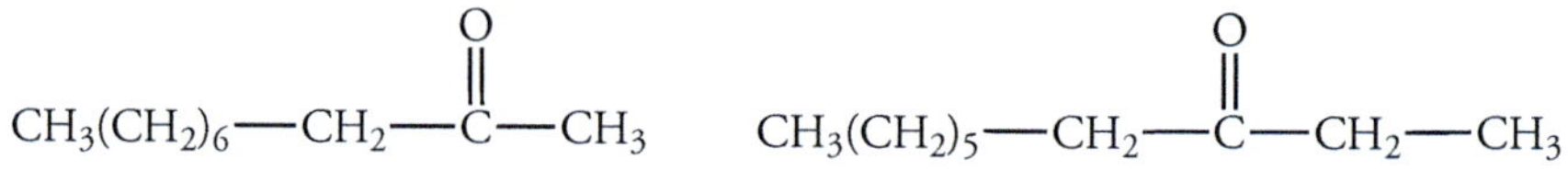

2-decanone

3-decanone

Problem 4.20

What product results from the hydration of 1-methylcyclopentene?

4.11 PREPARATION OF ALKENES AND ALKYNES

Alkenes are prepared from either alcohols or haloalkanes (alkyl halides) by elimination reactions. We recall from Section 2.5 that a single compound splits into two products in an elimination reaction. One product usually contains most of the atoms in the reactant, and the remaining atoms are found in a second smaller molecule. The atoms eliminated to form the smaller molecule are usually located on adjacent carbon atoms in the reactant. We will discuss the mechanisms of elimination reactions in Chapter 7.

Dehydration of Alcohols

The dehydration of 2-propanol produces propene. The reaction requires concentrated acids, such as sulfuric acid, H_2SO_4, or phosphoric acid, H_3PO_4. The reaction is pulled to completion because the water formed in the reaction is solvated with the concentrated acid.

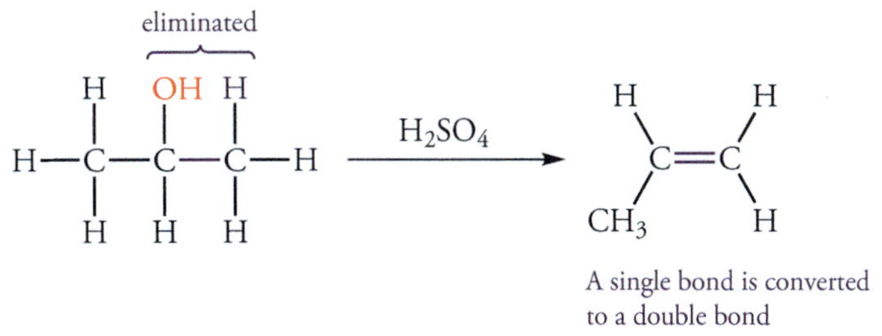

These atoms are eliminated

A single bond is converted to a double bond

The elimination reaction requires breaking the carbon-oxygen bond and a carbon-hydrogen bond on an adjacent carbon atom. For alcohols such as 2-butanol, two different carbon atoms are adjacent to the OH-bearing carbon atom. Each could potentially release a hydrogen atom to form water.

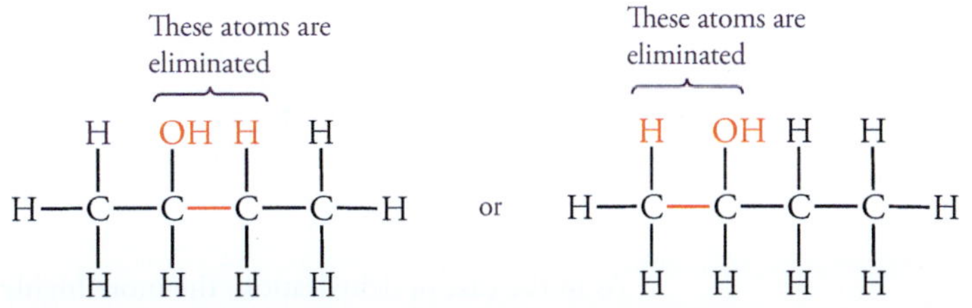

Thus, dehydration produces a mixture of products. The isomer that contains the greater number of alkyl groups attached to the double bond—the more substituted alkene—predominates in the mixture.

Elimination reactions that give the more substituted double bond are said to obey **Zaitsev's rule**. This generalization was discovered by Alexander Zaitsev, a nineteenth-century Russian chemist. We recall from Section 4.1 that increasing the number of alkyl groups bonded to unsaturated carbon atoms of an alkene increases its stability. Thus, Zaitsev observed that the major product of an elimination reaction is the more stable isomer. Zaitsev's rule also applies to mixtures of geometric isomers. For example, 3-pentanol yields a mixture of *cis-* and *trans*-2-pentene. The *trans* isomer is the major product.

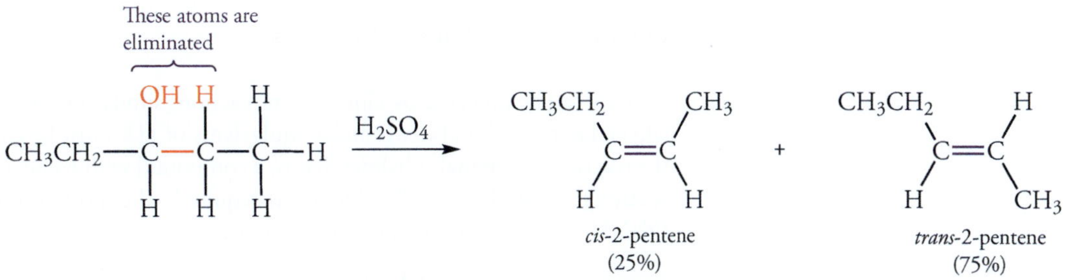

In the *trans* isomer the alkyl groups are well separated, whereas in the *cis* isomer the alkyl groups are near each other. The proximity of the alkyl groups causes a "through-space" interaction called **steric hindrance** (Greek, *steros*, space); that is, the groups repel each other. The *trans* isomer is more stable, so it is the major product (Figure 4.4).

Figure 4.4
Structures of cis-and trans-2-Pentene

In the *cis* isomer the CH_3CH_2— and CH_3— groups repel each other. In the *trans* isomer, the groups lie on opposite sides of the double bond, and therefore do not repel each other. Thus, the *trans* isomer is more stable and is the major product.

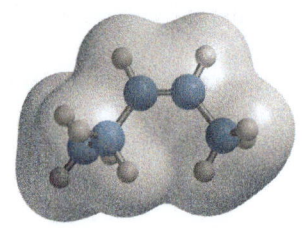

cis-2-pentene

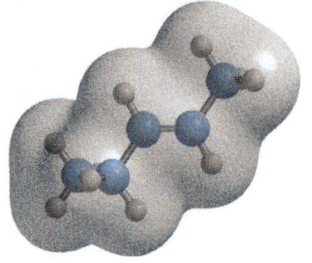

trans-2-pentene

Dehydrohalogenation of Allyl Halides

The elimination of the elements H and X, as in HCl or HBr, from adjacent carbon atoms in an alkyl halide is called **dehydrohalogenation**. The product of the reaction is an alkene. A base is required for the reaction.

Although hydroxide ion is sufficiently basic for this reaction, it is usual to use an alkoxide, the conjugate base of an alcohol, as the base. Sodium ethoxide is commonly used in combination with ethanol as the solvent. We will discuss the mechanism of this reaction in Chapter 7.

As in the case of dehydration, the more highly substituted alkene is the major product when two or more products are possible. When geometric isomers are possible, *trans* isomers are favored over *cis* isomer.

2-methyl-2-butene
(70%)

2-methyl-1-butene
(30%)

Elimination Reactions of Dihalides

Alkynes can be prepared by elimination reactions similar to those used to form alkenes. Because an alkyne has two π bonds, two molar equivalents of HX must be eliminated. The reactant needed for the reaction is a **vicinal** dihalide; that is, a compound with halogen atoms on adjacent carbon atoms. A stronger base than an alkoxide ion is required. The most commonly used base is sodium amide ($NaNH_2$) in liquid ammonia as the solvent.

1,2-dichlorohexane

1-hexyne

4.12 ALKADIENES (DIENES)

Compounds with two double bonds are **alkadienes**, commonly called **dienes**. When one single bond is located between the two double-bonded units, the compounds are chemically different from simple alkenes. These compounds, which are said to be **conjugated**, are the subject of this section. When more than one single bond is located between the two double-bonded units, the compounds are chemically the same as alkenes. The double bonds are said to be **isolated** or **nonconjugated**.

$$CH_2 \overset{1}{=\!=\!=} CH \!-\! \overset{3}{CH} \overset{}{=\!=\!=} CH \!-\! CH_3 \qquad CH_2 \overset{1}{=\!=\!=} CH \!-\! CH_2 \!-\! \overset{4}{CH} \overset{}{=\!=\!=} CH_2$$

1,3-pentadiene, conjugated diene   1,4-pentadiene, nonconjugated diene

Double bonds that share a common atom are said to be cumulated. These compounds are relatively rare, and we will not discuss them further.

$$\overset{1}{CH_2} \overset{}{=\!=\!=} \overset{2}{C} \overset{}{=\!=\!=} C \!-\! CH_2 \!-\! CH_3$$

1,2-pentadiene, cumulated diene

Electrophilic Conjugate Addition Reactions

Addition of electrophilic reagents to nonconjugated alkadienes can occur at one or both double bonds. The products are those predicted by Markovnikov's rule.

$$CH_2 \overset{}{=\!=\!=} CH \!-\! CH_2 \!-\! CH \overset{}{=\!=\!=} CH_2 \xrightarrow{\text{HBr}} CH_3 \!-\! \underset{\underset{Br}{|}}{CH} \!-\! CH_2 \!-\! CH \overset{}{=\!=\!=} CH_2$$

1,4-pentadiene   4-bromo-1-pentene

$$CH_3 \!-\! \underset{\underset{Br}{|}}{CH} \!-\! CH_2 \!-\! CH \overset{}{=\!=\!=} CH_2 \xrightarrow{\text{HBr}} CH_3 \!-\! \underset{\underset{Br}{|}}{CH} \!-\! CH_2 \!-\! \underset{\underset{Br}{|}}{CH} \!-\! CH_3$$

2,4-dibromopentane

Addition of HBr to a conjugated diene is strikingly different. Two products are obtained when one molar equivalent of HBr reacts.

$$CH_2 \overset{}{=\!=\!=} CH \!-\! CH \overset{}{=\!=\!=} CH_2 \xrightarrow{\text{HBr}} \underset{\underset{H}{|}}{CH_2} \!-\! \underset{\underset{Br}{|}}{CH} \!-\! CH \overset{}{=\!=\!=} CH_2 \quad + \quad \underset{\underset{H}{|}}{CH_2} \!-\! CH \overset{}{=\!=\!=} CH \!-\! \underset{\underset{Br}{|}}{CH_2}$$

1,3-butadiene   3-bromo-1-butene   1-bromo-2-butene
                          (1,2-addition, 70%)   (1,4-addition, 30%)

The **3-bromo-1-butene is the product of direct addition to a double bond**, as predicted by Markovnikov's rule. The 1-bromo-2-butene is an unusual product that results from the addition of HBr to C-1 and C-4. Note that the double bond in the product is between C-2 and C-3. This product results from a **1,4-addition reaction**.

To understand the origin of the 1,4-addition product derived from a conjugated diene, let's examine the structure of the intermediate that forms in the first step or the reaction. It is a resonance-stabilized **allylic carbocation** that has two contributing structures.

$$\left[\overset{+}{CH_2} \!-\! CH \overset{}{=\!=\!=} CH_2 \quad \longleftrightarrow \quad CH_2 \overset{}{=\!=\!=} CH \!-\! \overset{+}{CH_2} \right]$$

resonance-stabilized allylic carbocation

In an allylic carbocation, the positive charge is distributed equally between both terminal carbon atoms. Either terminal carbon atom could react with a nucleophile, but the product of the reaction would be the same.

Now let's consider the electrophilic addition of a proton to a conjugated diene to give a carbocation.

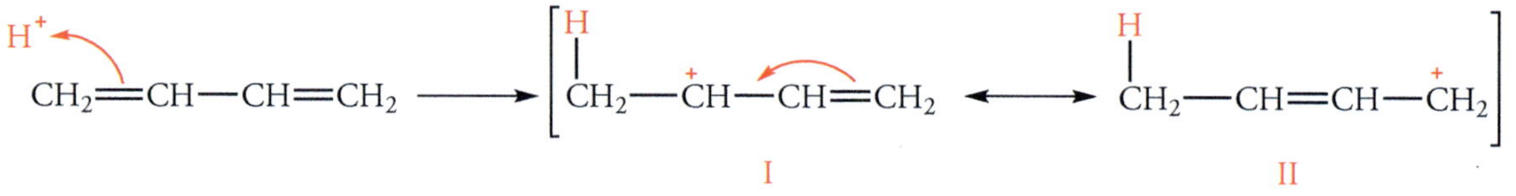

The carbocation intermediate is an allylic carbocation ion that can be represented by two contributing resonance structures, I and II. In the next step in the addition reaction, the nucleophilic bromide ion can form a bond to either of the two carbon atoms bearing a positive charge. Attachment at the secondary carbon atom (resonance structure I) gives the 1,2-addition product. However, if the bromide ion bonds to the primary carbon atom (resonance structure II), the 1,4-addition product forms.

4.13 TERPENES

Terpenes are abundant in the oils of plants and flowers. They have distinctive odors and flavors and are responsible for the odors of pine trees and for the colors of carrots and tomatoes. Terpenes consist of two or more **isoprene** units (2-methyl-1,3-butadiene) that are usually bonded C-1 to C-4, or "head to tail." These compounds may have different degrees of unsaturation and can contain a variety of functional groups. The structures may be acyclic or cyclic. Nevertheless, it is usually easy to identify the isoprene units.

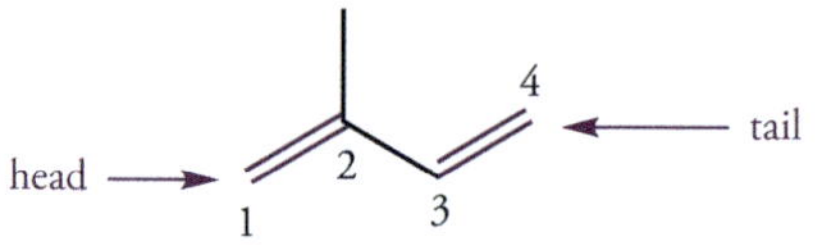

isoprene
(2-methyl-1,3-butadiene)

Farnesol is an acyclic terpene that has three isoprene units joined head to tail. Carvone has two isoprene units but contains a ring. Dashed lines indicate where the isoprene units are joined.

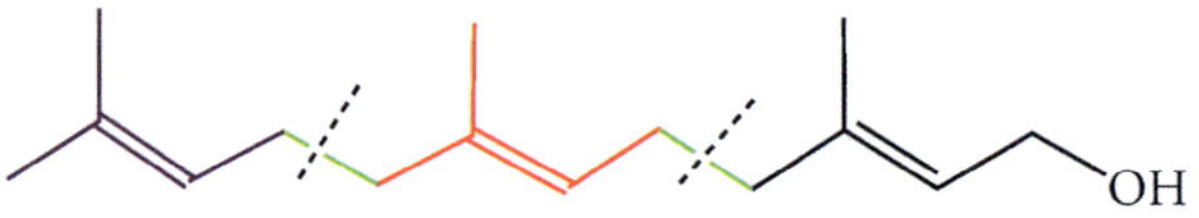

farnesol

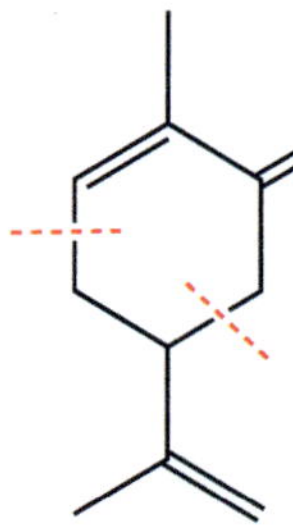

carvone

Terpenes are classified by the number of isoprene units they contain. The **monoterpenes**, the simplest terpene class, contain two isoprene units, and **sesquiterpenes** have three isoprene units. Examples of these structures are shown in Figure 4.5 using bond-line structures. **Diterpenes, triterpenes,** and **tetraterpenes** contain 4, 6, and 8 isoprene units, respectively.

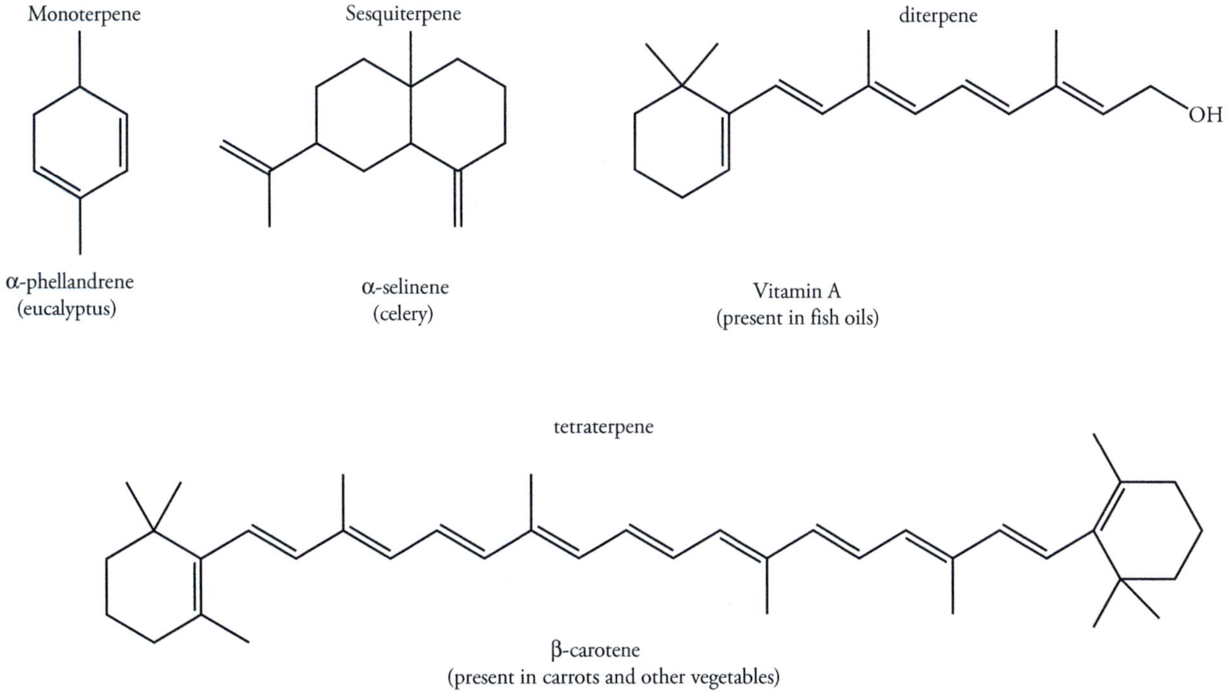

Figure 4.5 Structures and Classification of Terpenes

Allylic Oxidation in Metabolism

The liver enzyme cytochrome P-450 oxidizes many toxic metabolites and drugs at allylic sites. For example, an allylic oxidation is the first step in the degradation of one of the psychoactive ingredients in marijuana. An allyl radical is a likely intermediate in this process.

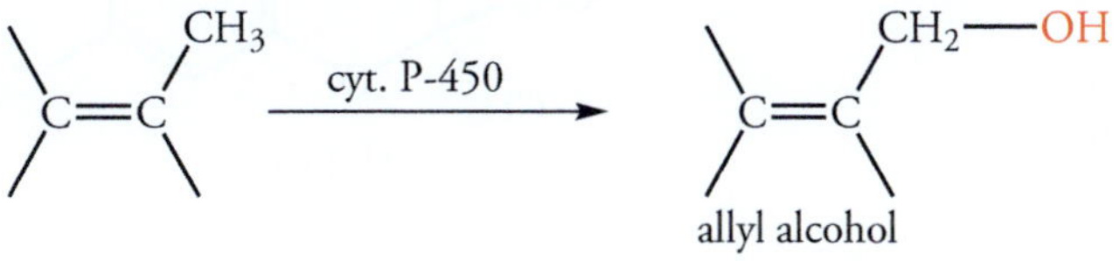

The principle psychoactive component of marijuana contains Δ^1-tetrahydrocannabinol (Δ^1 - THC), which has three allylic centers. The C-3 and C-6 centers are secondary and the C-7 is primary. Allylic oxidation does not occur at C-3 because of steric hindrance caused by the geminal dimethyl groups. Of the other two possible sites, the C-7 product predominates over the C-6 product even though the C-7 atom is primary. However, the difference in the stabilities of radicals is not as large as the difference in the stabilities of carbocations. Thus, other factors such a steric hindrance could play a role in the regioselectivity of this reaction. The C-7 methyl group is sterically more accessible than the secondary C-6 site. Interestingly, the C-7 product is even more psychoactive than Δ^1-THC.

The physiological effects of THF and its metabolic products are mediated by the action of membrane bound proteins called G protein coupled receptors (GPCRs). This regulation occurs via a complex and enormously important cell-signaling mechanism.

The liver ordinarily transforms organic compounds into oxidized products that are more water soluble that can be excreted. However, one of the concerns in the design of drugs is the reactivity of metabolites produced by these metabolic oxidation reactions. Obviously, it is intolerable to have metabolites cause damage to cells and disrupt other life processes. In other words, the cure should not be worse than the disease. The metabolism of hexobarbital (a sedative hypnotic) occurs by allylic oxidation of a methylene group in the cyclohexene ring. Subsequent reactions of this metabolite occur to allow ready elimination of the drug.

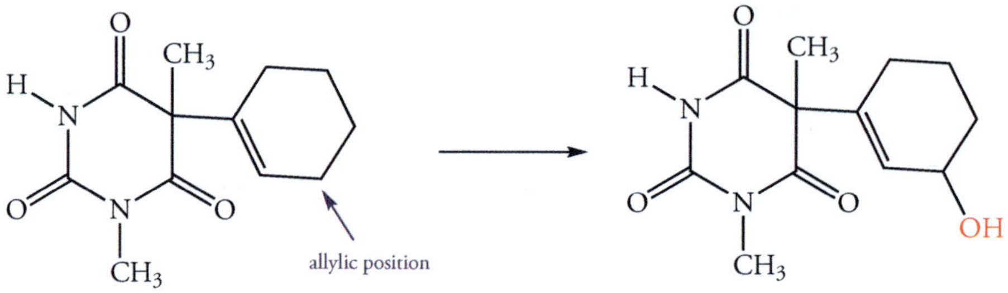

hexobarbital

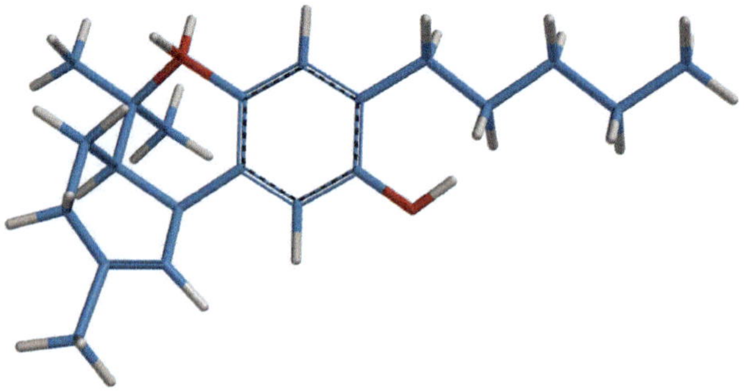

Δ^1-THC

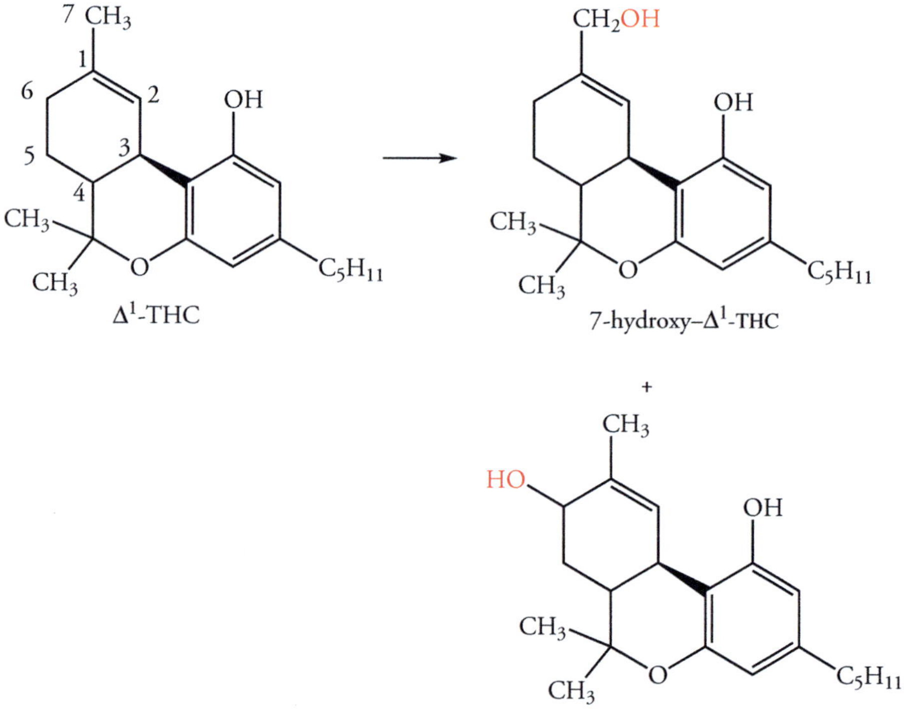

Δ^1-THC

7-hydroxy–Δ^1-THC

+

6-hydroxy–Δ^1-THC

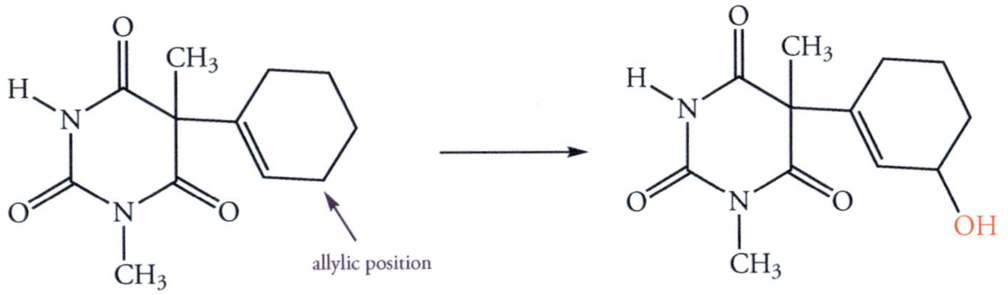

hexobarbital

The anti-arrhythmic drug quinidine is metabolized at the indicated allylic position to give an allylic alcohol. In this case, the metabolite also has anti-arrhythmic activity.

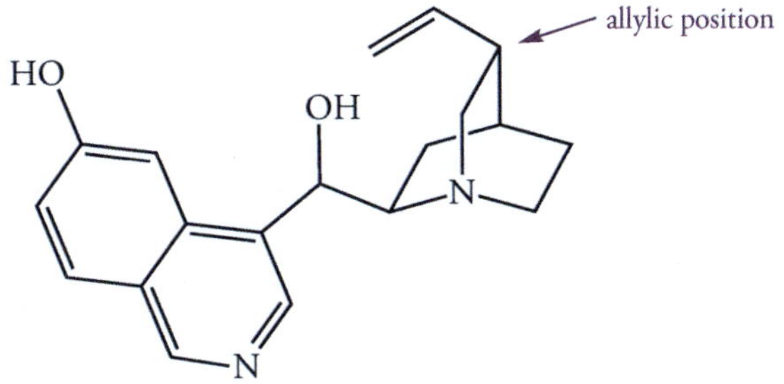

quinidine

SUMMARY OF REACTIONS

1. Hydrogenation of Alkenes (Section 4.6)

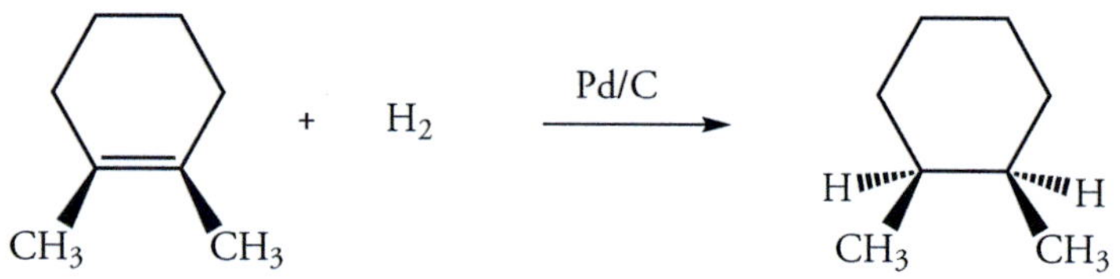

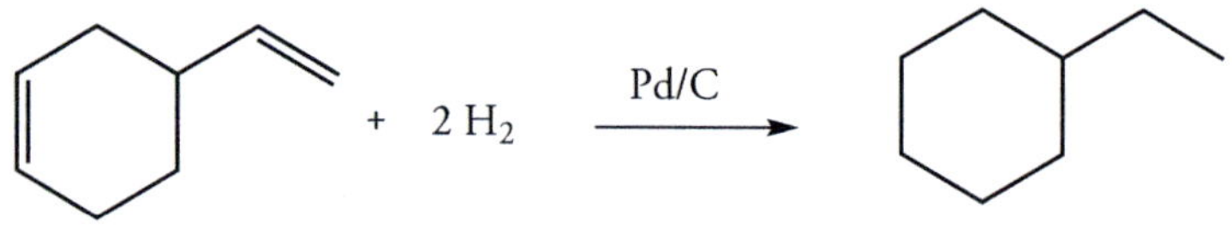

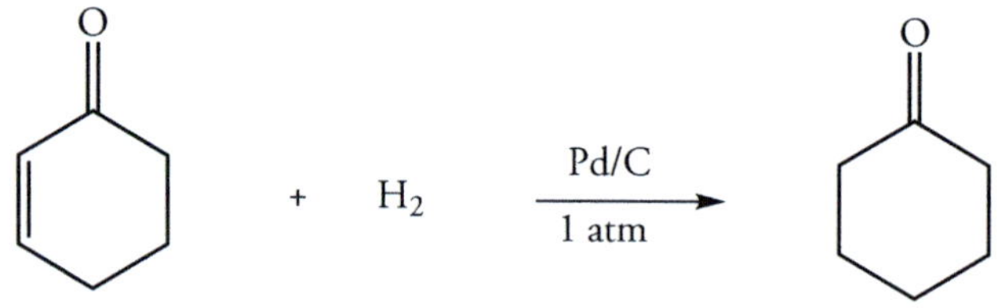

2. Hydrogenation of Alkynes (Section 4.6)

$$CH_3(CH_2)_2CH_2-C\equiv C-H + 2H_2 \xrightarrow{Pd/C} CH_3(CH_2)_3CH_2-CH_3$$

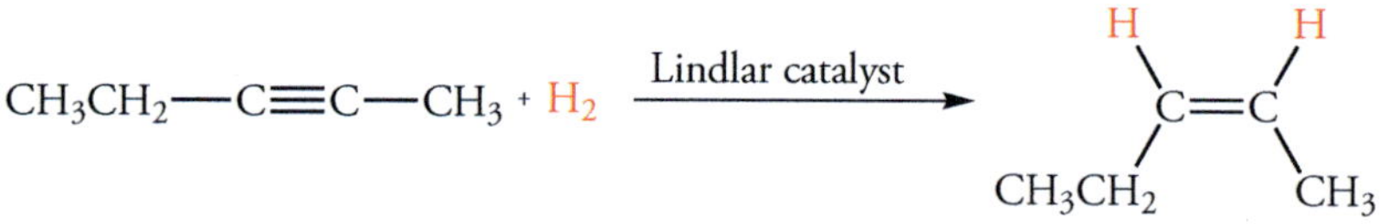

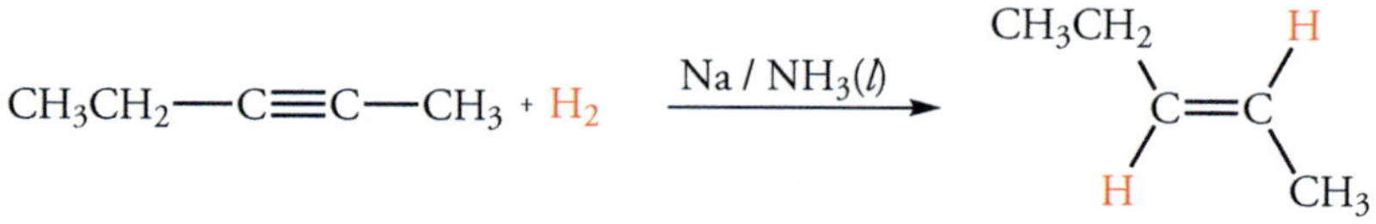

3. Oxidation of Alkenes (Section 4.7)

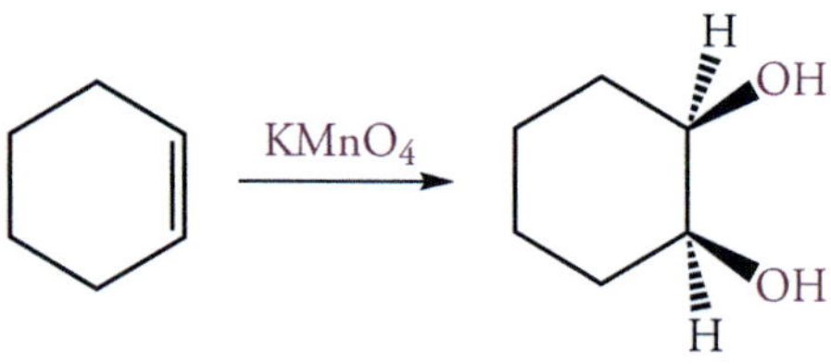

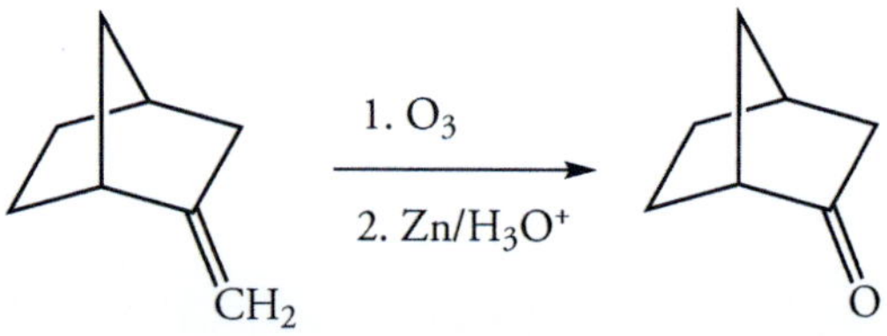

4. Addition of Halogens to Alkenes (Section 4.8)

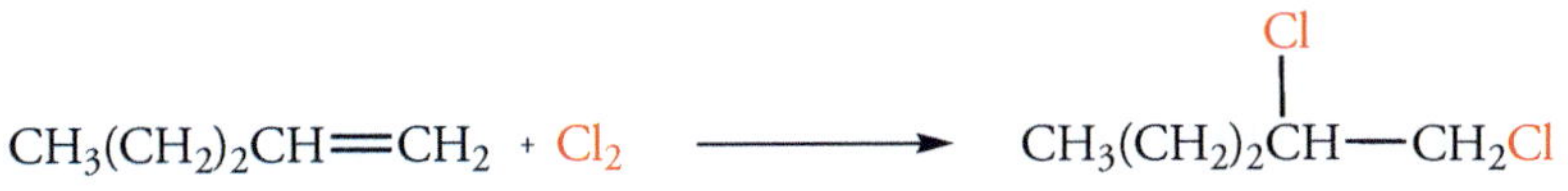

$$CH_3(CH_2)_2CH=CH_2 + Cl_2 \longrightarrow CH_3(CH_2)_2\overset{\overset{\displaystyle Cl}{|}}{CH}-CH_2Cl$$

5. Addition of Hydrogen Halides to Alkenes (Section 4.8)

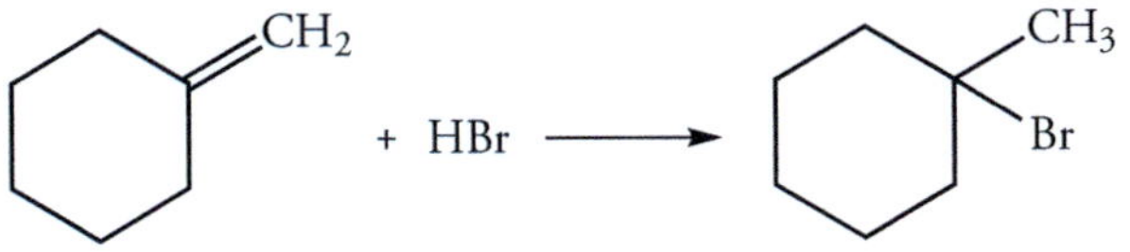

6. Hydration of Alkenes (Section 4.10)

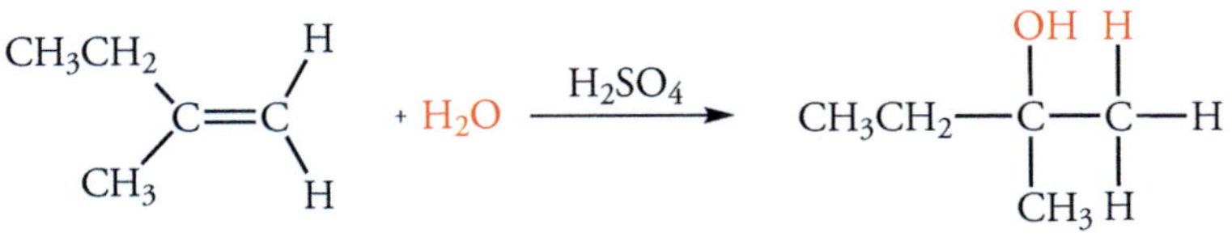

7. Addition of Halogens to Alkynes (Section 4.8)

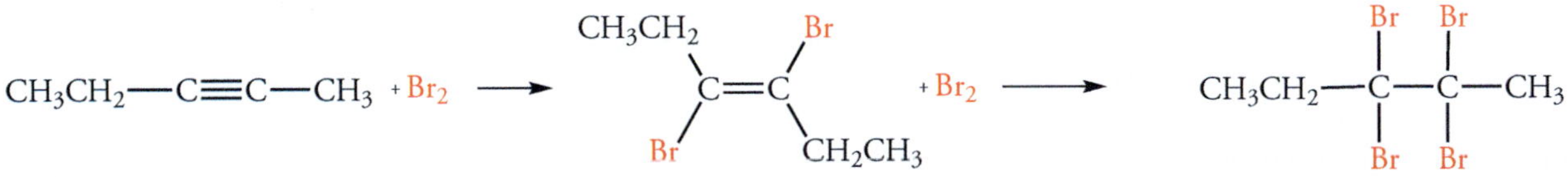

8. Addition of Hydrogen Halides to Alkynes (Section 4.8)

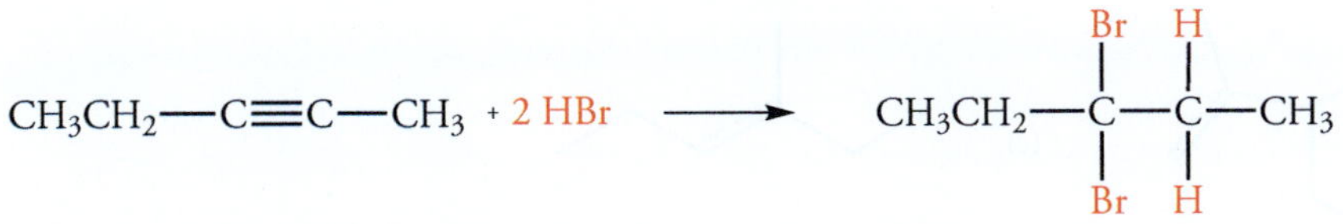

9. Hydration of Alkynes (Section 4.10)

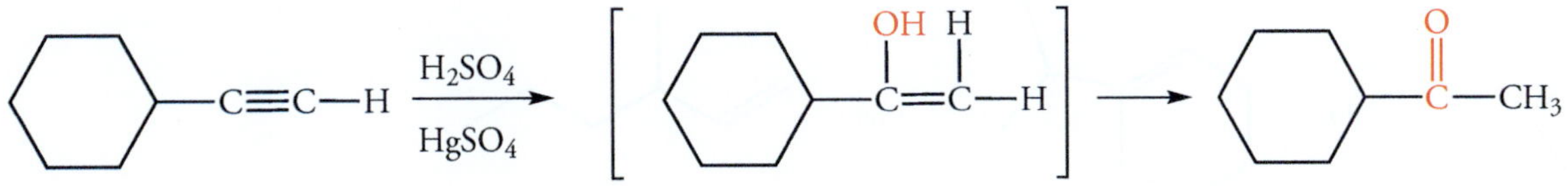

10. Synthesis of Alkenes by Dehydration of Alcohols (Section 4.11)

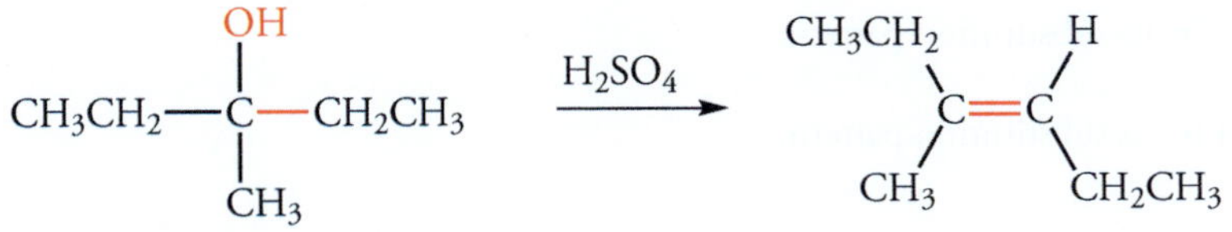

11. Synthesis of Alkenes by Dehydrohalogenation (Section 4.11)

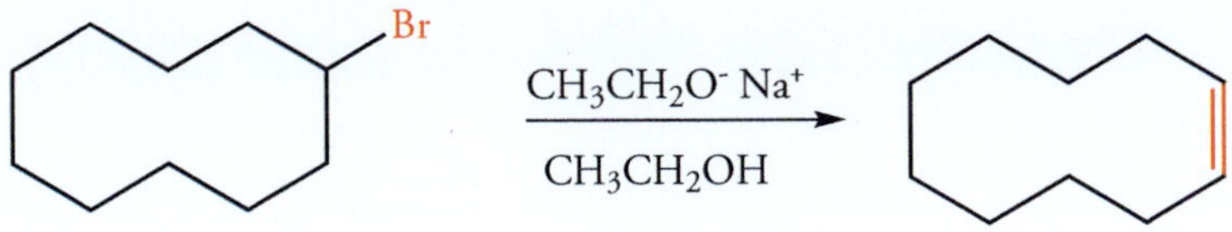

12. Synthesis of Alkynes by Dehydrohalogenation (Section 4.11)

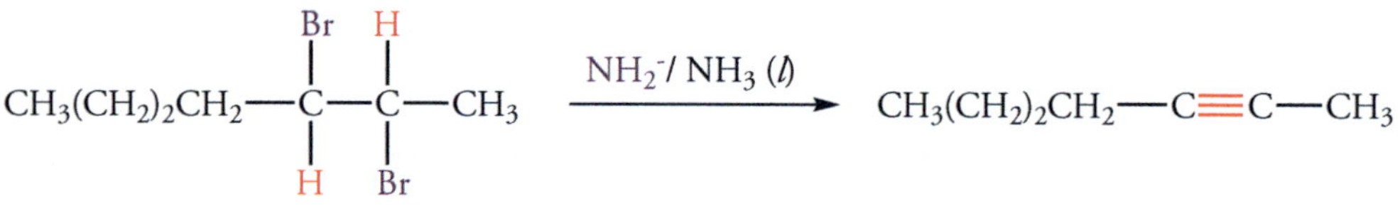

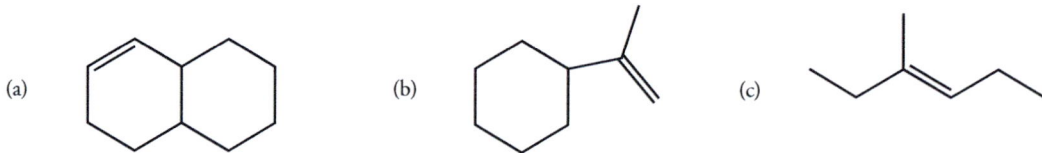

13. Conjugate Addition Reactions of Dienes (Section 4.12)

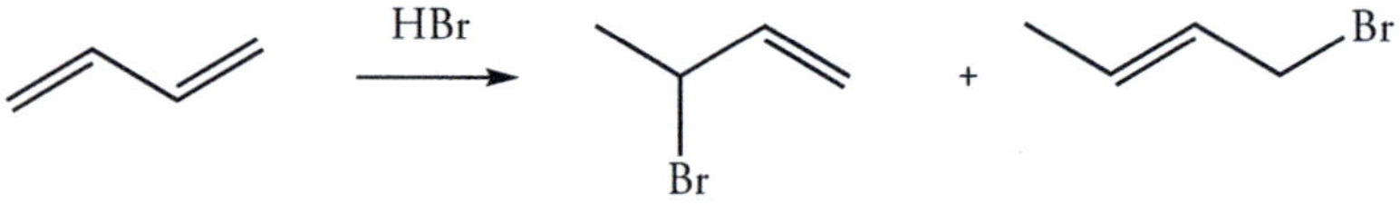

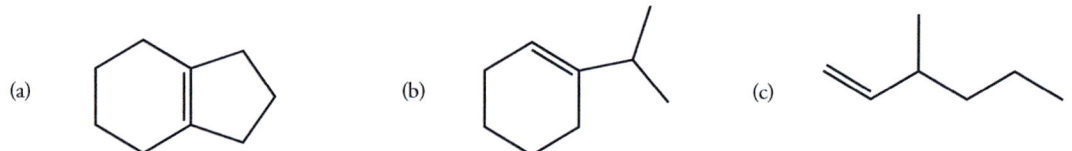

EXERCISES

Molecular Formulas

4.1 What is the molecular formula for the compounds with each of following structural features?
(a) six carbon atoms and one double bond
(b) five carbon atoms and two double bonds
(c) seven carbon atoms, a ring, and one double bond

4.2 What is the molecular formula for a compound with each of the following structural features?
(a) four carbon atoms and two triple bonds
(b) four carbon atoms, a double bond, and a triple bond
(c) ten carbon atoms and two rings

4.3 Write the molecular formula for each of the following compounds:

(a) (b) (c)

4.4 Write the molecular formula for each of the following compounds:

(a) (b) (c)

Classification of Alkenes and Alkynes

4.5 Classify each double bond in the alkenes in Exercise 4.3 by its substitution pattern.

4.6 Classify each double bond in the alkenes in Exercise 4.4 by its substitution pattern.

4.7 Indicate the degree of substitution of the double bond in each of the following compounds:

(a) cholesterol, a steroid required for growth in almost all organisms

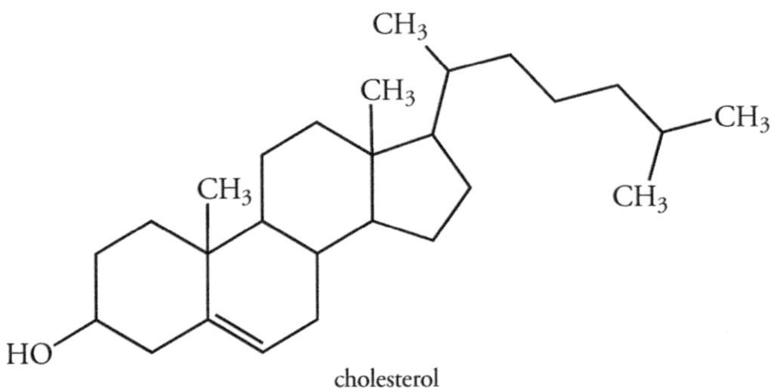

(b) tamoxifen, a drug used in treatment of breast cancer

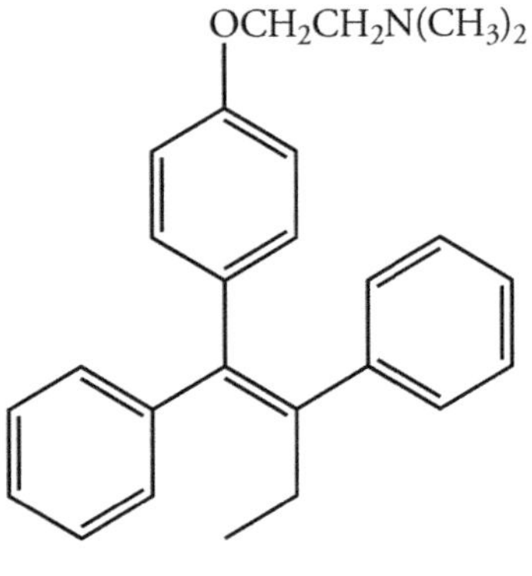

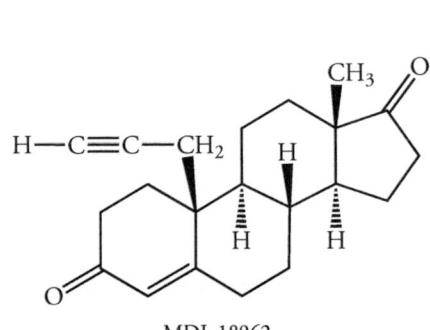

cholesterol

tamoxifen

4.8 Indicate the degree of substitution of the triple bond in each of the following compounds:

(a) MDL, a drug used in breast cancer therapy   (b) MDL, a drug used to induce abortion

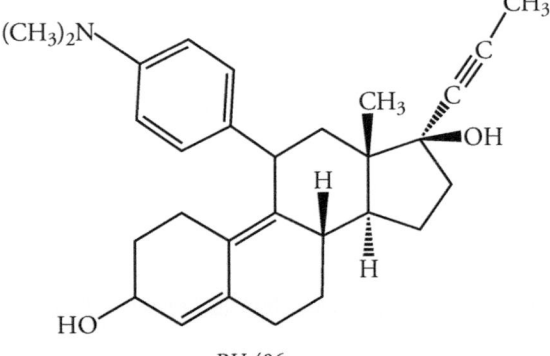

MDL 18962

RU 486

Geometric Isomers

4.9 Which of the following molecules can exist as *cis* and *trans* isomers?
(a) $CH_3CH=CHBr$ (b) $CH_2=CHCH_2Br$ (c) $CH_3CH=CHCH_2Cl$

4.10 Which of the following molecules can exist as *cis* and *trans* isomers?
(a) $CH_3CH=CBr_2$ (b) $CH_2=CHCHBr_2$ (c) $CH_3CH=CHCHCl_2$

4.11 Which of the following molecules can exist as *cis* and *trans* isomers?
(a) 1-hexene   (b) 3-heptene   (c) 4-methyl-2-pentene   (d) 2-methyl-2-butene

4.12 Which of the following molecules can exist as *cis* and *trans* isomers?
(a) 3-methyl-1-hexene   (b) 3-ethyl-3-heptene   (c) 2-methyl-2-pentene   (d) 3-methyl-2-pentene

E,Z System of Nomenclature

4.13 Select the group with the highest priority in each of the following sets:
(a) —$CH(CH_3)_2$,   —$CHClCH_3$,   —CH_2CH_2Br
(b) —$CH_2CH=CH_2$,   —$CH_2CH(CH_3)_2$,   —$CH_2C\equiv CH$
(c) —OCH_3,   —$N(CH_3)_2$,   —$C(CH_3)_3$

4.14   Select the group with the highest priority in each of the following sets:

(a) 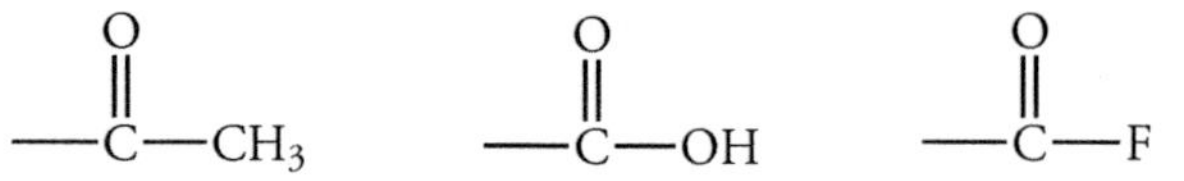 $-\overset{\displaystyle O}{\underset{\displaystyle \|}{C}}-CH_3$ $-\overset{\displaystyle O}{\underset{\displaystyle \|}{C}}-OH$ $-\overset{\displaystyle O}{\underset{\displaystyle \|}{C}}-F$

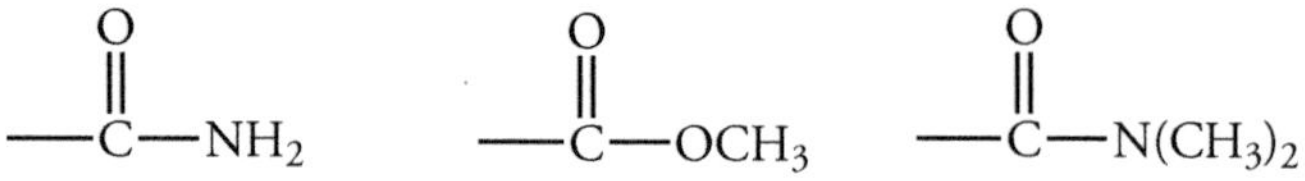

(b) $-\overset{\displaystyle O}{\underset{\displaystyle \|}{C}}-NH_2$ $-\overset{\displaystyle O}{\underset{\displaystyle \|}{C}}-OCH_3$ $-\overset{\displaystyle O}{\underset{\displaystyle \|}{C}}-N(CH_3)_2$

(c) $-\overset{\displaystyle O}{\underset{\displaystyle \|}{C}}-S-CH_3$ $-\overset{\displaystyle O}{\underset{\displaystyle \|}{C}}-O-CH_3$ $-\overset{\displaystyle O}{\underset{\displaystyle \|}{C}}-Cl$

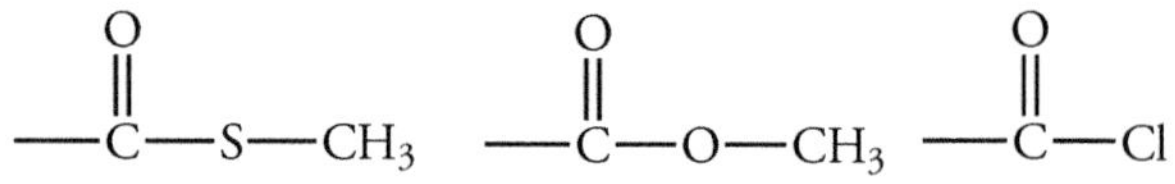

4.15   Assign the E or Z configuration to each of the following antihistamines:

(a)   pyrrobutamine   (b)   triprolidine

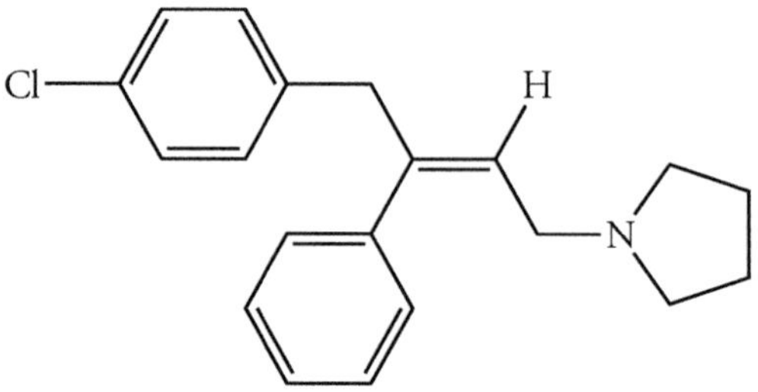

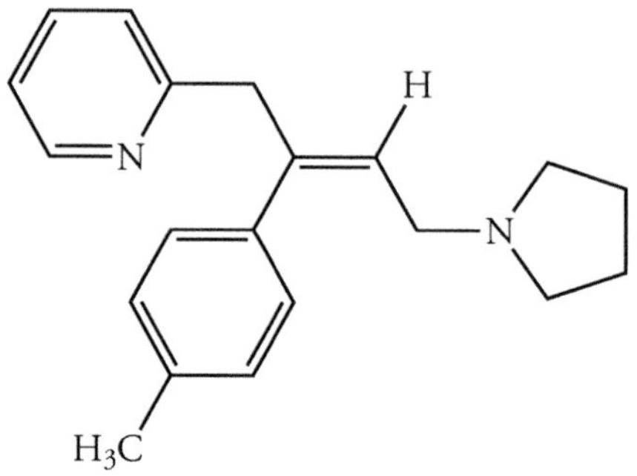

4.16   Assign the E or Z configuration to each of the following hormone antagonists used to control cancer:

(a)   chlomiphene   (b)   nitromifene

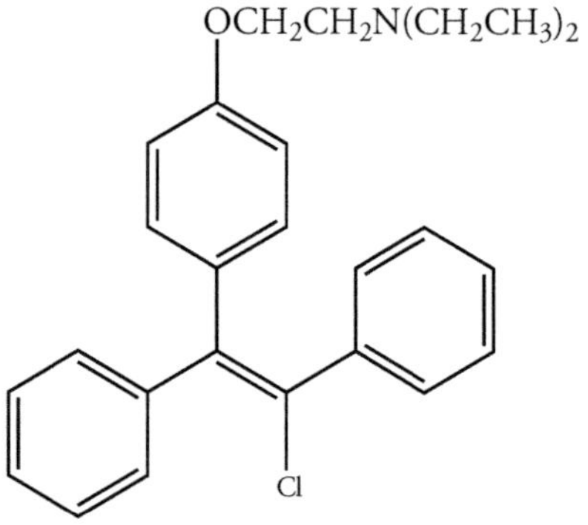

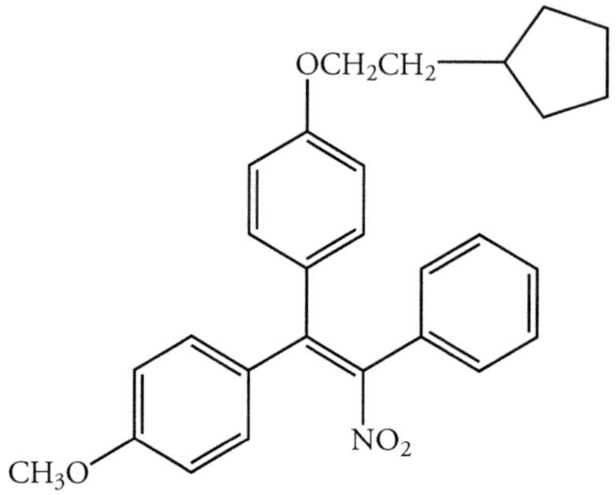

$OCH_2CH_2N(CH_2CH_3)_2$

OCH_2CH_2-

4.17   Draw the structural formula for each of the following pheromones with the indicated configuration.

(a) sex pheromone of Mediterranean fruit fly, E isomer

$CH_3CH_2CH{=}CH(CH_2)_4CH_2OH$

(b) defense pheromone of termite, E isomer

$CH_3(CH_2)_{12}CH{=}CHNO_2$

4.18 Assign the configuration at all double bonds where geometrical isomerism is possible in each of the following sex pheromones:

(a) European vine moth

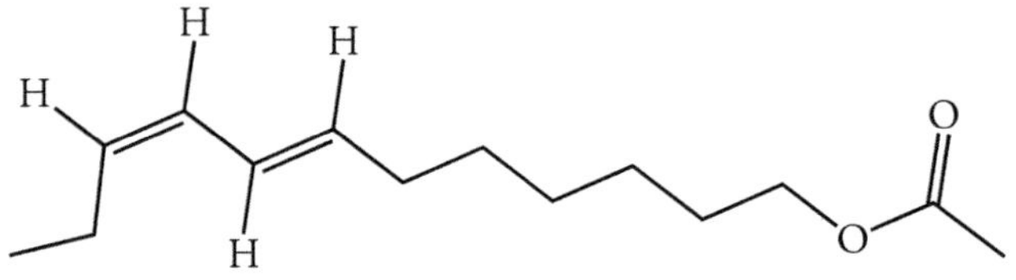

(b) pink bollworm moth

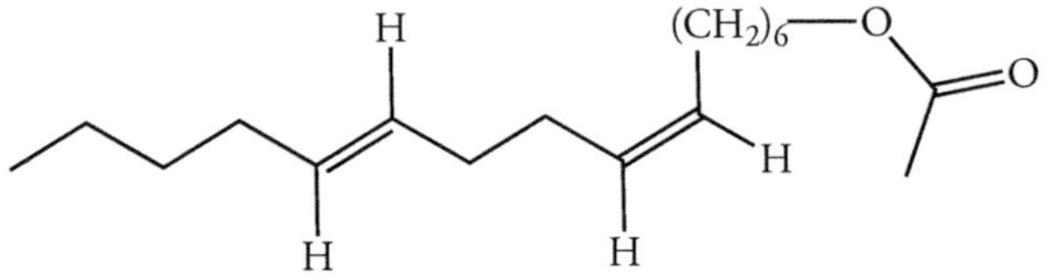

Nomenclature of Alkenes

4.19 Name each of the following compounds:

(a)

(b)

(c)

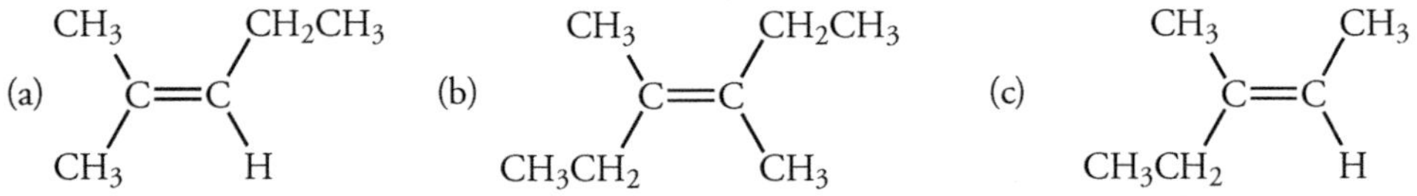

4.20 Name each of the following compounds:

(a)

(b)

(c)

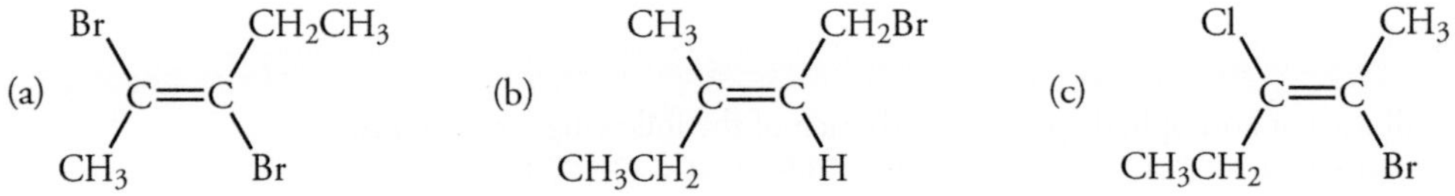

4.21 Name each of the following compounds:

(a)

(b)

(c)

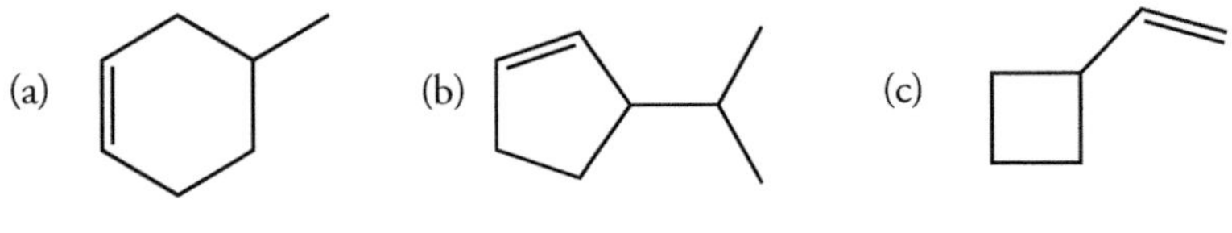

4.22 Name each of the following compounds:

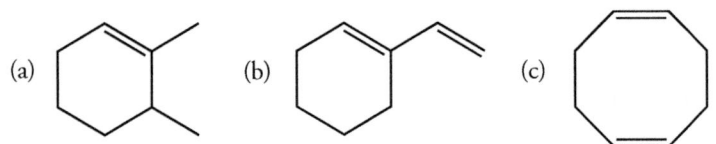

(a)

(b)

(c)

4.23 Draw a structural formula for each of the following compounds:
(a) 2-methyl-2-pentene   (b) *cis*-2-methyl-3-hexene   (c) *trans*-5-methyl-2-hexene

4.24 Draw a structural formula for each of the following compounds:
(a) *trans*-1-chloropropene   (b) *cis*-2,3-dichloro-2-butene (c) 2,4-dimethyl-2-hexene

4.25 Draw a structural formula for each of the following compounds:
(a) 1-methylcyclopentene   (b) 1,2-dibromocyclohexene (c) 4,4-dimethylcyclohexene

4.26 Draw a structural formula for each of the following compounds:
(a) 3-methylcyclohexene   (b) 1,3-dibromocyclopentene (c) 3,3-dichlorocyclopentene

Nomenclature of Alkynes

4.27 Name each of the following compounds:

(a) $CH_3CH_2CH_2C\equiv CH$ (b) $(CH_3)_3CC\equiv CCH_2CH_3$ (c) $CH_3—C\equiv C—CH—CH_3$
$\qquad\qquad\qquad\qquad\qquad\qquad\qquad\qquad\qquad\qquad\qquad\qquad\qquad\qquad\qquad |$
$\qquad\qquad\qquad\qquad\qquad\qquad\qquad\qquad\qquad\qquad\qquad\qquad\qquad\quad CH_2CH_3$

4.28 Name each of the following compounds:

(a) $CH_3CHBrCHBrC\equiv CCH_3$ (b) $Cl(CH_2)_2C\equiv C(CH_2)_3CH_3$ (c) $CH_3—CH—CH_2—C\equiv C—CH—CH_3$
$\qquad\qquad\qquad\qquad\qquad\qquad\qquad\qquad\qquad\qquad\qquad\qquad\qquad\quad\ |\qquad\qquad\qquad\qquad\qquad |$
$\qquad\qquad\qquad\qquad\qquad\qquad\qquad\qquad\qquad\qquad\qquad\qquad\quad CH_2CH_3\qquad\qquad\qquad\quad Cl$

4.29 Draw a structural formula for each of the following compounds:
(a) 2-hexyne   (b) 3-methyl-1-pentyne   (c) 5-ethyl-3-octyne

4.30 Draw a structural formula for each of the following compounds:
(a) 3-heptyne   (b) 4-methyl-1-pentyne   (c) 5-methyl-3-heptyne

Hydrogenation of Alkenes and Alkynes

4.31 How many moles of hydrogen gas will react at atmospheric pressure with each of the following compounds?
(a) $CH_3—CH=CH—C\equiv CH$ (b) $HC\equiv C—C\equiv C—H$ (c) $CH_2=CH—C\equiv C—CH=CH_2$

4.32 How many moles of hydrogen gas will react at atmospheric pressure with each of the following compounds?

(a) ichthyothereol, a convlusant

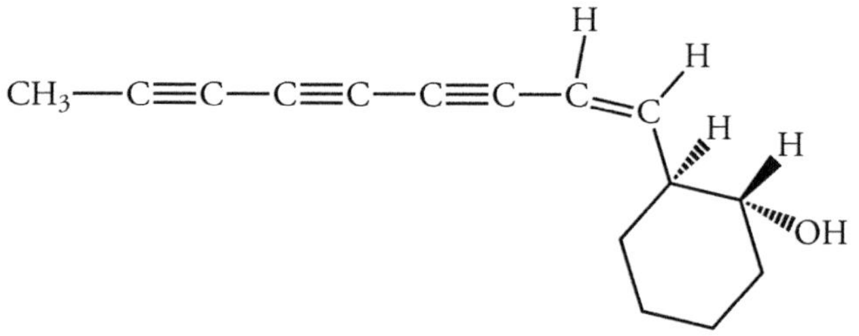

(b) mycomycin, an antibiotic

$H—C\equiv C—C\equiv C—CH=C=CH—CH=CH—CH=CH—CO_2H$

4.33 How could the following unsaturated carboxylic acid be prepared from a structurally related alkyne? What reagent is required?

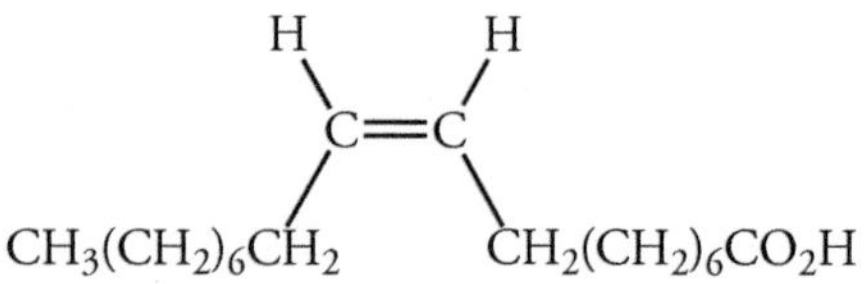

4.34 How could the following compound, which is a constituent of the sex pheromone of the male oriental fruit moth, be prepared from a structurally related alkyne? What reagent is required?

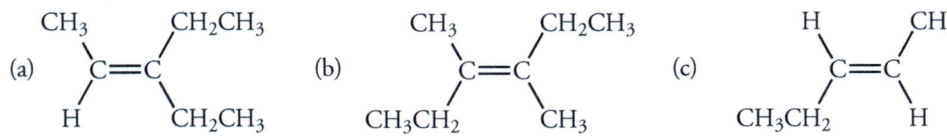

Oxidation of Alkenes

4.35 Describe the observation that is made when *cis*-2-pentene reacts with potassium permanganate. How could this reagent be used to distinguish between *cis*-2-pentene and cyclopentane?

4.36 Write the products of the reaction of vinylcyclohexane as well as allylcyclopentane with potassium permanganate.

4.37 Write the product(s) of the ozonolysis of each of the following compounds:

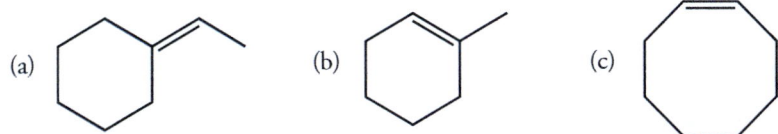

4.38 Write the product(s) of the ozonolysis of each of the following compounds:

(a)　(b)　(c)

Addition Reactions

4.39 Which of the compounds in Exercise 4.37 would give a single product when reacted with HBr? Why?

4.40 Write the product of the reaction of HBr with each of the compounds in Exercise 4.38.

4.41 The addition of HBr to an alkyne results in the trans addition of hydrogen and bromine. Write the product(s) of the reaction of one mole of HBr with each of the following alkynes.

(a) $CH_3CH_2C\equiv CH$ (b) $CH_3CH_2C\equiv CCH_3$ (c) $CH_3CH_2C\equiv CCH_2CH_3$

4.42 Write the product(s) of the reaction of two moles of HBr with each of the compounds in Exercise 4.41.

4.43 Write the product of hydration of each of the compounds in Exercise 4.38.

4.44 Hydration of one of the following two compounds yields a single ketone product. The other compound yields a mixture of ketones. Which one yields only one ketone product? Why?

Preparation of Alkenes and Alkynes

4.45 How many alkenes would be formed by dehydrohalogenation of each of the following alkyl bromides? Which compound should be the major isomer?

(a) $CH_3CH_2CHBrCH_2CH_3$　　(b) $(CH_3)_3CCHBrCH_3$　(c) $CH_3(CH_2)_3CHBrCH_3$

4.46   Write the structure of a bromo compound that exclusively gives each of the following alkenes by dehydrohalogenation:

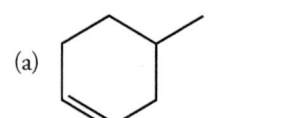 (a)

(b)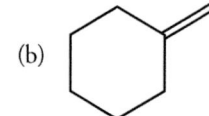

(c)

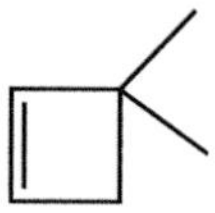

4.47   Write the structure of a compound that would yield the following alkyne upon dehydrohalogenation:

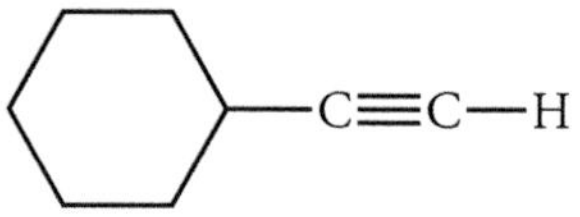

—C≡C—H

4.48   Alkynes can be prepared by elimination reactions of geminal dihalides, compounds with two halogen atoms bonded to the same carbon atom. Would the following reaction provide a good yield of the indicated product? Explain.

$$CH_3CH_2CH_2CBr_2CH_3 \xrightarrow[NH_3]{NaNH_2} CH_3CH_2C{\equiv}CCH_3$$

Polyunsaturated Compounds

4.49   Which of the following compounds has conjugated double bonds?

(a)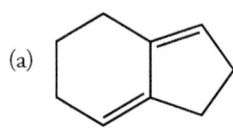

(b)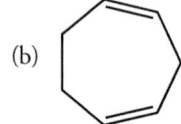

(c)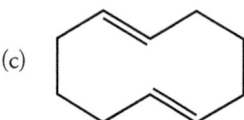

4.50   Which of the following compounds has conjugated double bonds?

(a)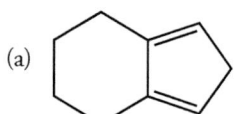

(b)

(c)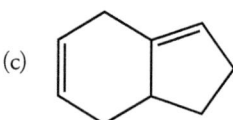

Terpenes

4.51   Classify each of the following terpenes and divide it into isoprene units.

(a)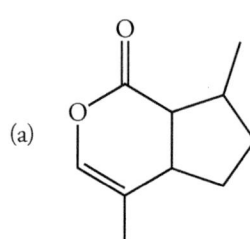

(b)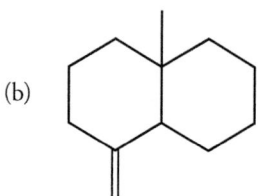

4.52   Classify each of the following terpenes and divide it into isoprene units.

(a)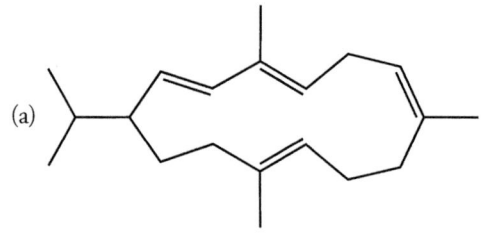

(b)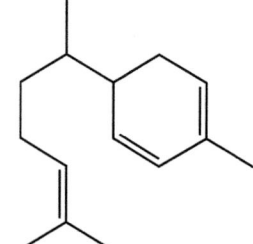

5

AROMATIC COMPOUNDS

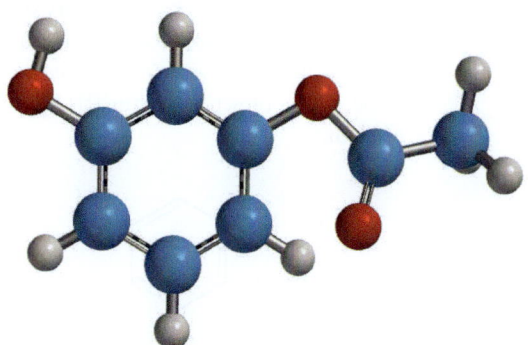

RESORCINOL MONOACETATE (A GERMICIDE)

5.1 AROMATIC COMPOUNDS

The term aromatic means "fragrant" (Ancient Greek, aroma). For this reason many fragrant substances were called "aromatic compounds." Many of these compounds contain a benzene ring (Section 1.6) that is bonded to one or more substituents. Oil of sassafras, oil of wintergreen, and vanillin are well-known examples of fragrant, aromatic compounds.

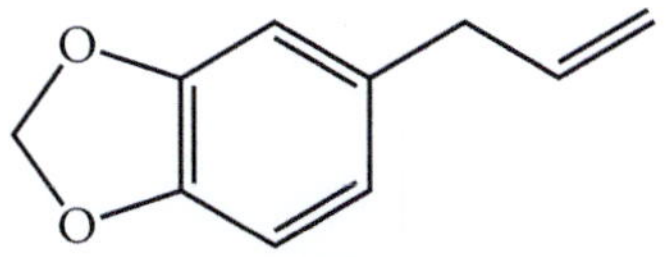

safrole
(oil of sassafras)

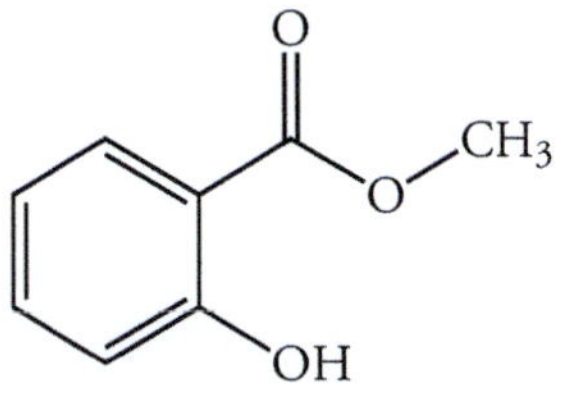

methyl salicylate
(oil of wintergreen)

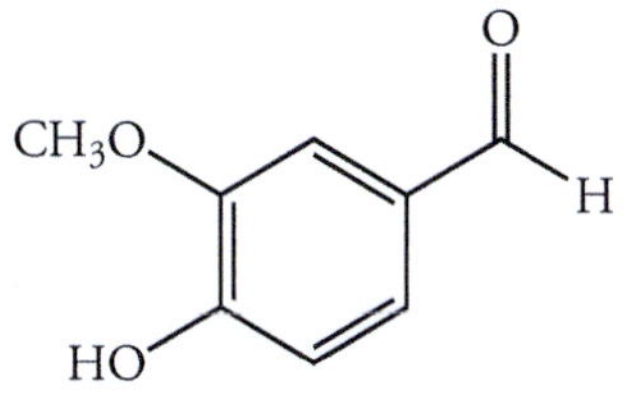

vanillin
(vanilla)

Today, the classification of aromatic compounds is no longer based on odor because many compounds containing a benzene ring are not fragrant. Many aromatic compounds are solids that have little or no odor. Solid aromatic compounds include the pain relievers, or analgesics, aspirin, ibuprofen, and acetaminophen, and the antibiotic chloramphenicol.

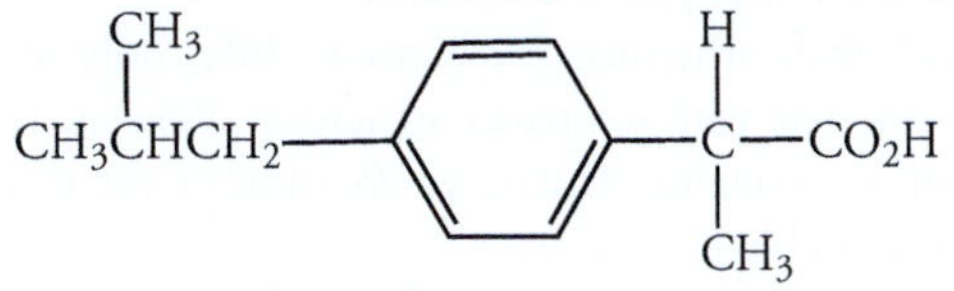

ibuprofen

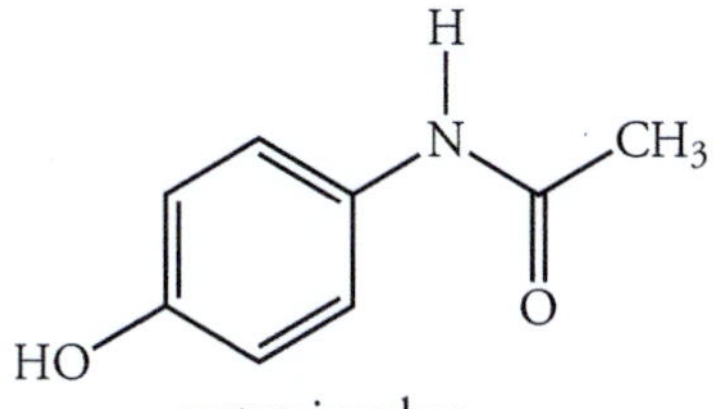

acetaminophen

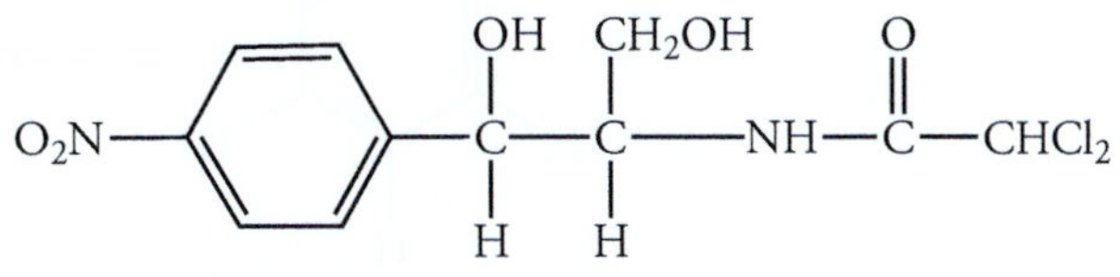

chloramphenicol

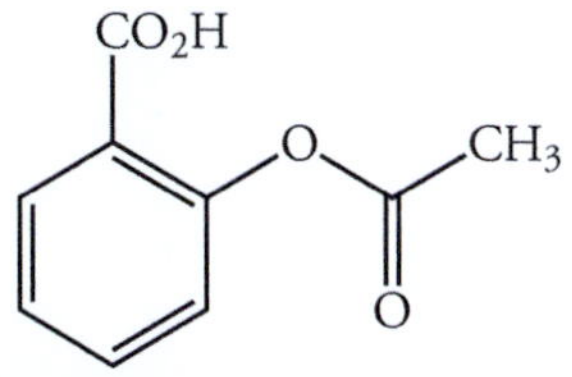

aspirin

The common feature of aromatic compounds is not their odor, but the benzene ring. This six-carbon unit is usually not affected by reactants that alter the rest of the structure. The distinguishing characteristic of aromatic compounds is the very low reactivity of the benzene ring.

Principles of Organic Chemistry. http://dx.doi.org/10.1016/B978-0-12-802444-7.00005-7

5.2 AROMATICITY

Benzene, C_6H_6, is highly unsaturated—it has six fewer hydrogen atoms than cyclohexane, C_6H_{12}—its cyclic saturated counterpart. Although benzene is represented by a hexagon that contains three double bonds, unlike alkenes it does not undergo addition reactions with reagents such as bromine, HBr, or water. The lack of reactivity of benzene contradicts what we know about unsaturated compounds. That is, benzene does not behave like the "triene" depicted by its Lewis structure. Benzene typically undergoes *substitution* reactions, a reaction not typical of alkenes. Benzene reacts with bromine, in the presence of iron(III) bromide as a catalyst, to give a single monosubstituted product, C_6H_5Br.

This result indicates that all six hydrogen atoms of benzene are chemically equivalent. There are three possible isomeric dibromobenzenes, $C_6H_4Br_2$. The electronic structure of benzene explains these facts.

Kekulé's Concept of Benzene

In 1865, a German chemist, August Kekulé, suggested that benzene is a ring of six carbon atoms linked by alternating single and double bonds. He proposed that benzene actually exists as two structures differing only in the arrangement of the single and double bonds that oscillate around the ring.

Kekulé proposed that the rapid oscillation of single and double bonds somehow made benzene resist addition reactions. This concept "explained" why only one bromobenzene (C_6H_5Br) forms in the substitution reaction of benzene with Br_2 in the presence of an iron(III) catalyst. He reasoned that the rapid oscillation of single and double bonds around the ring makes all six carbon atoms, and therefore all six hydrogen atoms, equivalent. However, benzene does *not* exist as a mixture of equilibrating structures. However, benzene exists as a single structure.

Resonance Theory and Benzene

The two Kekulé structures for benzene differ only in the arrangement of electron pairs, a feature that we associate with resonance structures. Benzene is a *resonance hybrid* that is represented by two contributing resonance structures. We indicate the relationship between the contributing structures by a single double-headed arrow.

equivalent contributing structures for the resonance hybrid of benzene

Benzene is a planar molecule in which all carbon-carbon bonds are the same; the bond angles of the ring are all 120°. Thus, the σ bonds in benzene are made with sp²-hybridized carbon atoms. Each carbon atom shares one electron in each of its three σ bonds: two σ bonds are to adjacent carbon atoms: the third σ bond is to a hydrogen atom. The fourth electron is in a 2p orbital perpendicular to the plane of the benzene ring (Figure 5.1). A set of six 2p orbitals (one from each carbon atom) overlap to share their six electrons in a π system that extends over the entire ring of carbon atoms. The π electrons are located both above and below the plane of the ring. The sharing of electrons over many atoms is called **delocalization**. This delocalization of electrons accounts for the unique chemical stability of benzene.

The structure of benzene is usually represented in chemical equations as one of the two possible Kekulé structures. Each corner of the hexagon represents a carbon atom with one attached hydrogen atom, which is often not shown.

The Hückle Rule

Most aromatic compounds contain a benzene ring or a related structure. What is responsible for the characteristic stability of benzene and its unique reactivity? Several general criteria must be met if a molecule is to be aromatic.

1. An aromatic molecule must be *cyclic*.
2. An aromatic molecule must be *planar*.
3. An aromatic ring must contain *only* sp²-hybridized atoms that can form a delocalized system of π molecular orbitals.
4. The number of π electrons in the delocalized π system must equal 4n + 2, where n is an integer.

The "4n + 2 rule" was proposed by E. Hückel, and is known as the **Hückel rule**. The theoretical basis for this rule is beyond the scope of this text. However, based on the Hückel rule, cyclic π systems with 6 (*n* = 1), 10 (*n* = 2), and 14 (*n* = 3) electrons are aromatic. Benzene meets the criteria for aromaticity for *n* = 1. We will discuss examples in the following section for various 6, 10, and 14 π electron systems. We will also see that aromatic compounds can contain atoms other than carbon (Section 5.3).

Figure 5.1
Structure of Benzene
Benzene is a planar molecule. (a) The six π electrons of benzene are distributed around the ring and lie above and below the plane of the ring. (b) The electrons in the σ bonds lie in the plane of the ring.

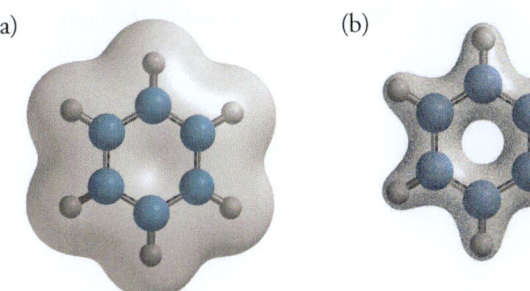

Some cyclic polyenes with alternating single and double bonds are not aromatic. These compounds do not obey the Hückel rule. Two examples are cyclobutadiene (four π electrons) and cyclooctatetraene (eight π electrons). Both compounds undergo addition reactions of the kind we have discussed, and have none of the characteristics of benzene.

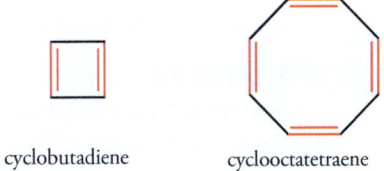

cyclobutadiene   cyclooctatetraene

Problem 5.1

What are the structural and electronic similarities of anthracene and phenanthrene?

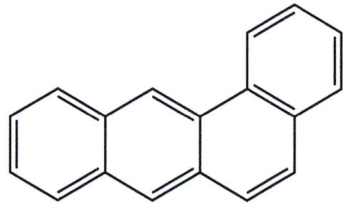

anthracene   phenanthrene

Solution

Both compounds have 14 carbon atoms and 10 hydrogen atoms; they are isomers. Both molecules consist of three fused rings that are shown with a series of alternating single and double bonds; each has 14 π electrons—a number that fits the Hückel rule for $n = 3$. Thus, both of these compounds are aromatic.

Problem 5.2

1,2-Benzanthracene is a carcinogen present as a combustion product in tobacco smoke. What is its molecular formula? Is the compound aromatic based on the Hückel rule?

benzanthracene

Problem 5.3

Histamine is released in the body in persons with allergic hypersensitivities such as hay fever. Describe the distribution of the electrons in orbitals on the two nitrogen atoms in the heterocyclic ring.

histamine

Solution

The ring nitrogen atom bonded to a hydrogen atom resembles the nitrogen atom of pyrrole. Three σ bonds are shown. Each has one valence electron contributed from the nitrogen atom. Thus, the remaining two valence electrons of nitrogen are located in a 2p orbital. These two electrons, along with the four electrons of the two π bonds shown in the structure, account for six electrons of an aromatic system. The other nitrogen atom in the ring has one single and one double bond as shown. This nitrogen atom resembles the nitrogen atom of pyridine. Two of its electrons participate in the two π bonds to carbon atoms; one electron is contributed to the π bond with a carbon atom. The remaining two electrons are in an sp² hybrid orbital projecting out from the plane of the ring.

Problem 5.4

Adenine is present in nucleic acids (DNA and RNA). Which nitrogen atoms in the rings have nonbonding electrons in sp² hybrid orbitals? How many electrons does each nitrogen atom contribute to the aromatic π system?

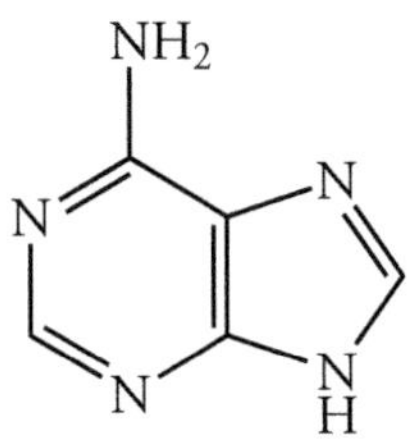

NH_2 structure

adenine
(6-aminopurine)

5.3 NOMENCLATURE OF AROMATIC COMPOUNDS

Benzene is the "parent" of many aromatic compounds, which have both common and IUPAC names. The common names of substituted benzenes often came from their sources. One example is toluene, which used to be obtained from the South American gum tree, *Toluifera balsamum*. A few benzene compounds are shown below. Their common names are shown in parentheses below their IUPAC names. The common names have been used for so long that they have become accepted by IUPAC.

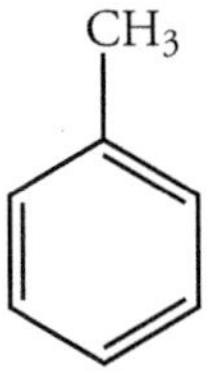

methylbenzene
(toluene)

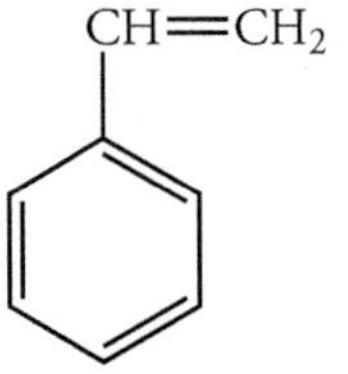

vinylbenzene
(styrene)

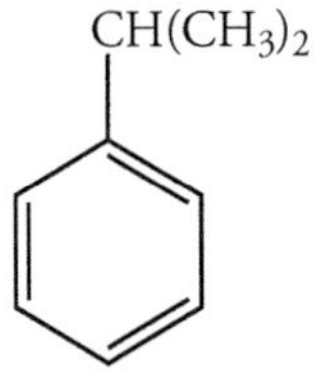

isopropylbenzene
(cumene)

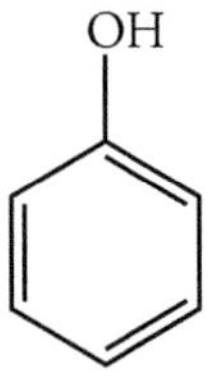

benzenol
(phenol)

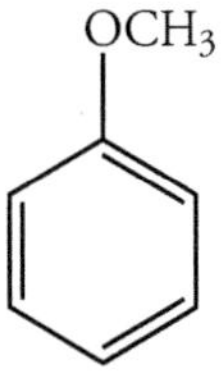

methoxybenzene
(anisole)

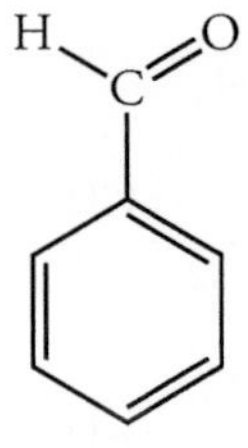

benzenecarboxaldehyde
(benzaldehyde)

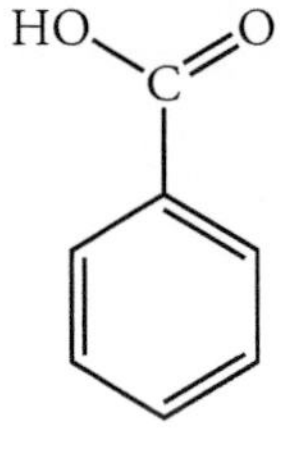

benzenecarboxylic acid
(benzoic acid)

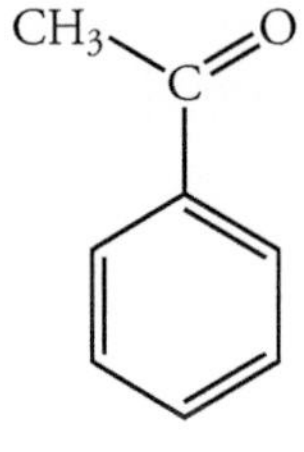

1-phenylethanone
(acetophenone)

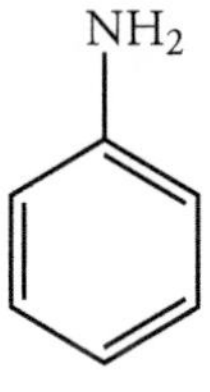

benzenamine
(aniline)

The IUPAC system of naming substituted aromatic hydrocarbons uses the names of the substituents as prefixes to benzene. A few examples are shown below.

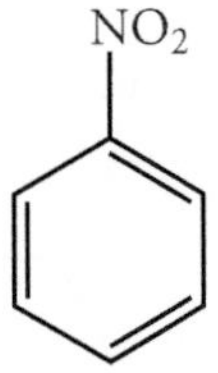

nitrobenzene

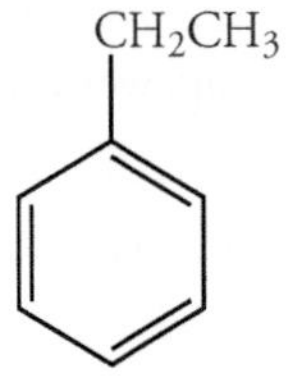

ethylbenzene

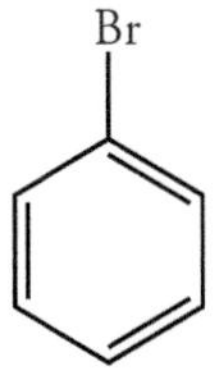

bromobenzene

The two substituents of disubstituted benzene compounds can be arranged in a benzene ring in three different ways to give three different isomers. These isomers are designated *ortho*, *meta*, and *para*. As prefixes these terms are abbreviated as *o*-, *m*-, and *p*-, respectively.

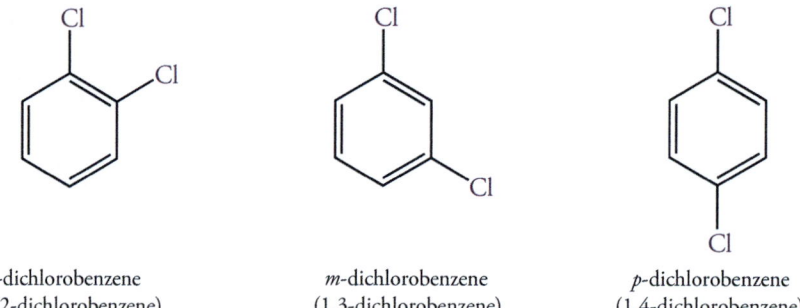

o-dichlorobenzene
(1,2-dichlorobenzene)

m-dichlorobenzene
(1,3-dichlorobenzene)

p-dichlorobenzene
(1,4-dichlorobenzene)

The *ortho* isomer has two groups on adjacent carbon atoms—that is, in a 1,2 relationship. In the *meta* and *para* isomers, the two groups are in a 1,3 and a 1,4 relationship, respectively. The IUPAC names of disubstituted aromatic compounds are obtained by numbering the benzene ring to give the lowest possible numbers to the carbon atoms bearing the substituents. When three or more substituents are present, the ring carbon atoms must be numbered.

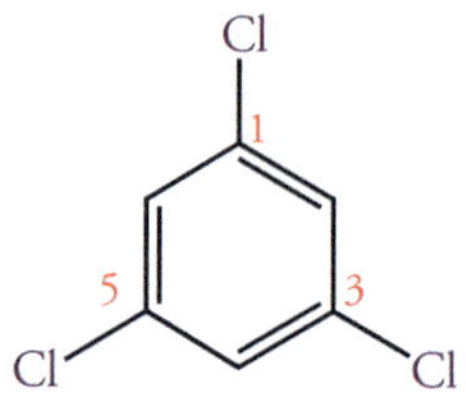

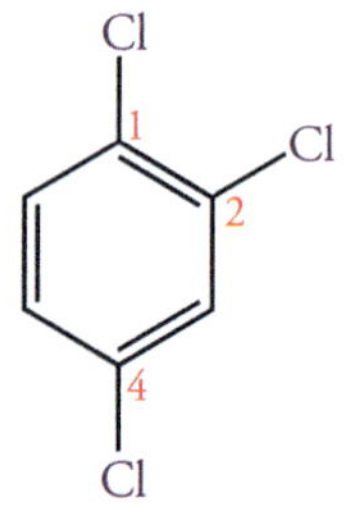

1,3,5-trichlorobenzene

1,2,4-trichlorobenzene

Many derivatives of benzene are named with the common name of a monosubstituted aromatic compound as the parent. The position of the substituent of the "parent" is automatically designated 1, but the number is not used in the name. The remaining substituents are prefixed in alphabetical order to the parent name along with numbers indicating their locations.

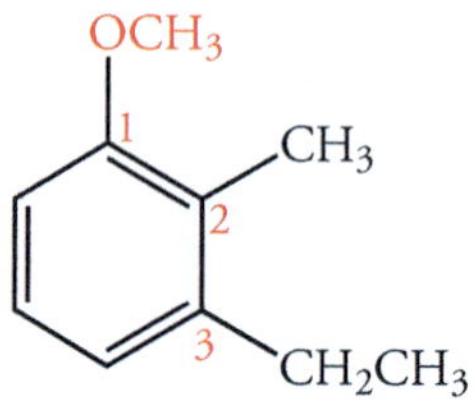

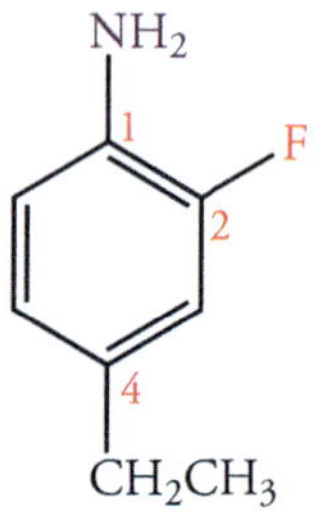

3-ethyl-2-methylanisole

4-ethyl-2-fluoroanline

Aromatic hydrocarbons belong to a general class called **arenes**. An aromatic ring residue attached to a larger parent structure is an aryl group; it is symbolized as Ar (not to be confused with argon), just as the symbol R is used for an alkyl group. Two groups whose names unfortunately do not make much sense are the phenyl (fen'-nil) and benzyl groups. We might reasonably expect that the aryl group derived from benzene (C_6H_5—) would be named benzyl. Alas, it is a phenyl group. A benzyl group, which is derived from toluene, has the formula $C_6H_5CH_2$—.

C_6CH_5——   or

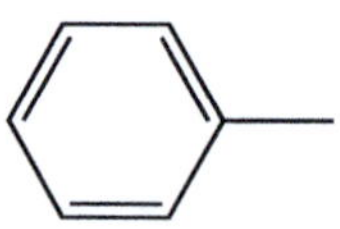

phenyl group

$C_6CH_5CH_2$——   or   —CH_2—

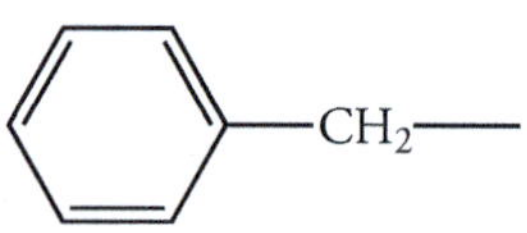

benzyl group

Problem 5.5

Show how a name could be derived for the following trisubstituted compound, known as BHA, which is used as an antioxidant in some food products.

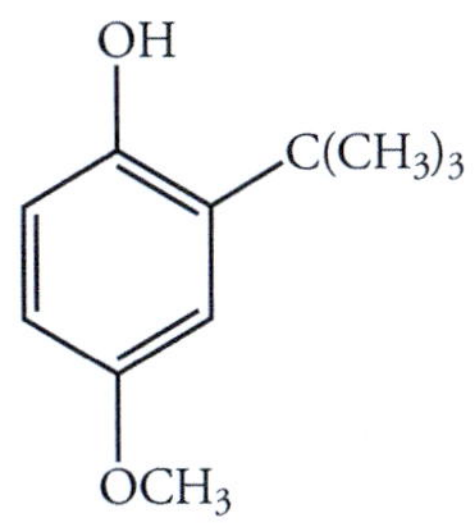

$$\text{OH}$$
$$\text{C(CH}_3)_3$$
$$\text{OCH}_3$$

Solution

Either the —OH group or the —OCH$_3$ group could provide the parent name, which would be phenol or anisole, respectively. Let's assume that the compound is a substituted anisole. When we assign the number 1 to the carbon atom bearing the —OCH$_3$ group at the "six o'clock" position, we must number the ring in a counterclockwise direction. A *tert*-butyl group is located at the 3 position and a hydroxyl group at the 4 position, so the compound is 3-*tert*-butyl-4-hydroxyanisole.

Problem 5.6

Name the following compound, which is used to make a local anesthetic.

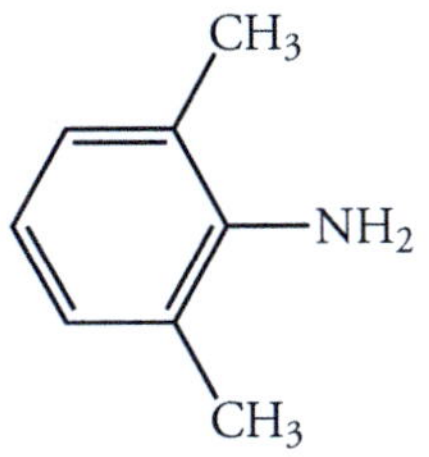

$$\text{CH}_3$$
$$\text{—NH}_2$$
$$\text{CH}_3$$

Problem 5.7

What is the name of the following compound?

$$\text{CH}_3\text{CH}_2\text{CH}_2 \quad \text{H}$$
$$\text{C}=\text{C}$$
$$\text{CH}_2\text{CH}_3$$

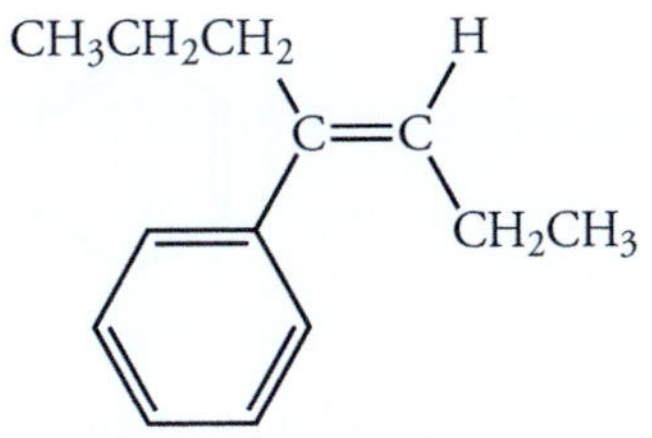

Solution

The compound is an alkene with an aromatic ring as a substituent. First, we determine that the chain has seven carbon atoms. Next, we number the chain from right to left so that the double bond is assigned to C-3. The phenyl group is then located on C-4. The phenyl group has a higher priority than the propyl group. At C-3, the ethyl group has higher priority than the hydrogen atom. The compound is the Z isomer because the higher priority phenyl group is on the same side of the double bond as the ethyl group. The complete name is (Z)-4-phenyl-3-heptene.

Problem 5.8

Name the following compound as a substituted pyridine.

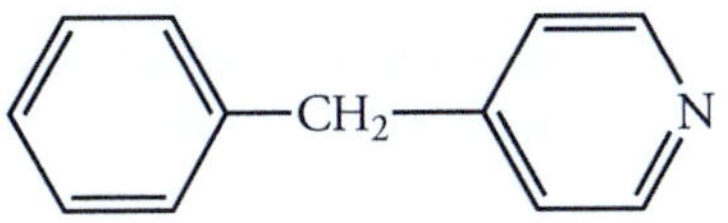

$$\text{—CH}_2\text{—} \quad \text{N}$$

5.4 ELECTROPHILIC AROMATIC SUBSTITUTION

Aromatic rings do not undergo the electrophilic addition reactions we discussed for alkenes. Instead, they react with electrophiles—and even then only in the presence of a catalyst—to give a substitution product. In these reactions an electrophile (E$^+$) substitutes for H$^+$. The general process is shown below.

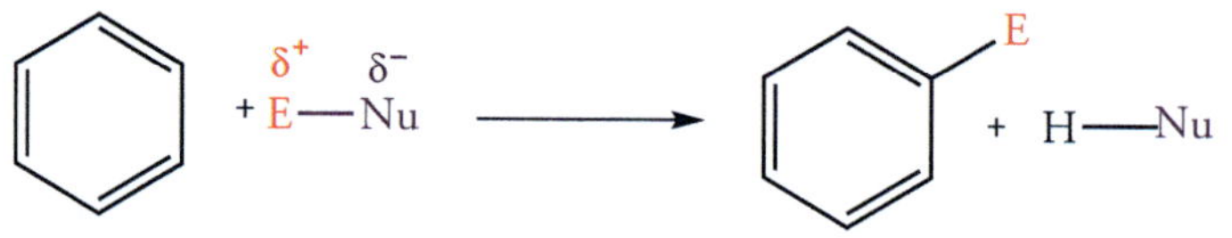

Mechanism of Electrophilic Aromatic Substitution

The first step of electrophilic aromatic substitution is similar to the first step in the addition of electrophiles to alkenes. The electrophile accepts an electron pair from the aromatic ring.

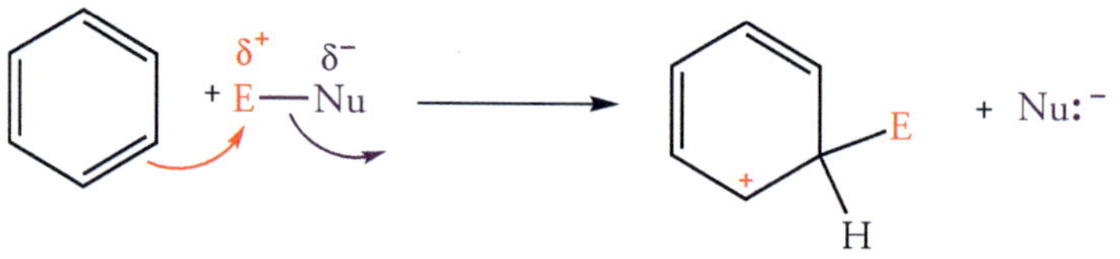

The carbocation is resonance stabilized. However, the carbocation is not aromatic—it has only four π electrons, and it has an sp^3-hybridized carbon atom. Therefore, it is less stable than the original aromatic ring, which had a six-electron aromatic π system. The intermediate is a cyclohexadienyl carbocation.

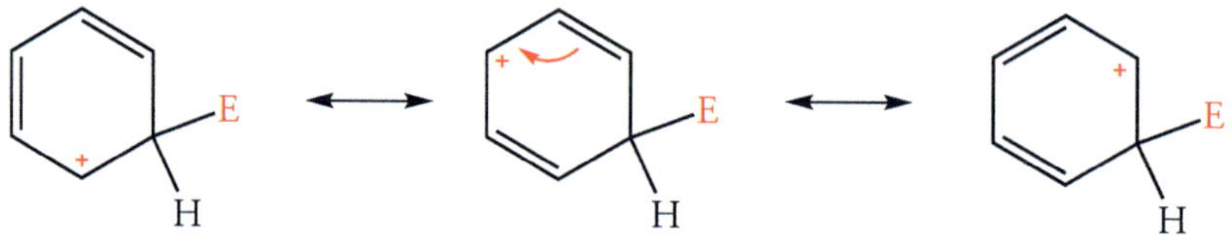

resonance forms of a cyclohexadienyl carbocation

In the second step of the reaction, the proton bonded to the same carbon atom as the newly arrived electrophile departs, which restores the six-electron aromatic π system. The nucleophile acts as a base to extract the proton.

Typical Electrophilic Aromatic Substitution Reactions

In the preceding discussion we used a generic electrophile, E^+. In this section we will consider some specific examples of electrophiles that react with aromatic rings. Our first examples are **bromination** and **chlorination**. Bromination requires both Br_2 and a Lewis acid catalyst, $FeBr_3$. Chlorination proceeds in the same manner as bromination, but uses the Lewis acid catalyst, $FeCl_3$. In bromination the catalyst forms an intermediate that has a polarized Br—Br bond.

Lewis-acid-Lewis-base
complex

Although this intermediate is the electrophile, it is convenient to use just the Br^+ formed by heterolytic cleavage of the Lewis-acid-Lewis-base complex to write the mechanism of the substitution reaction.

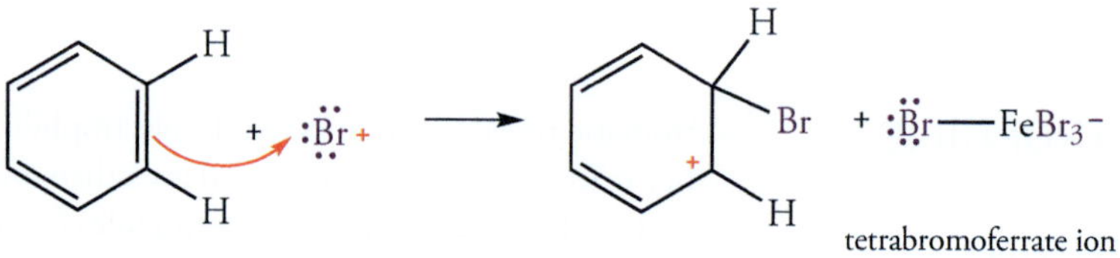

tetrabromoferrate ion

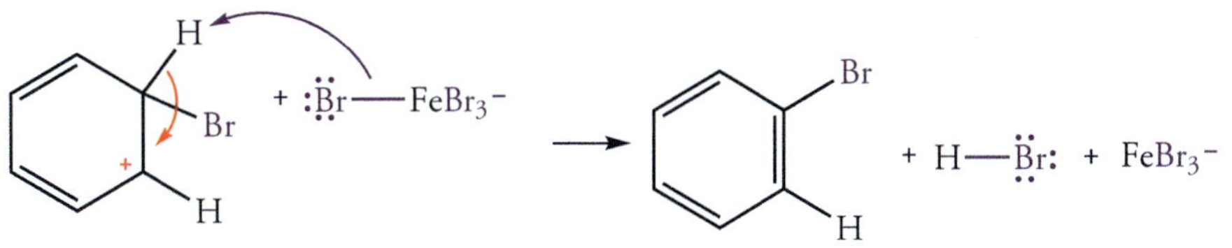

In **nitration**, a nitro group, $-NO_2$, is introduced onto an aromatic ring using nitric acid, HNO_3, with sulfuric acid as the catalyst. The electrophile is the nitronium ion, $-NO_2^+$, which is produced by the reaction of nitric acid with sulfuric acid.

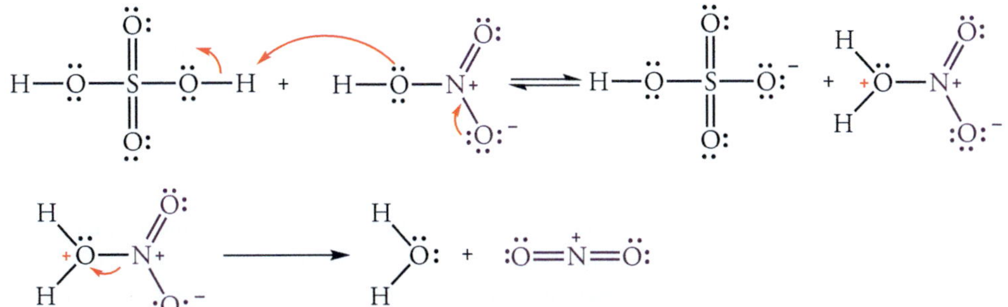

The net reaction for nitration of benzene is the result of electrophilic attack by the nitronium ion on the aromatic ring followed by loss of a proton.

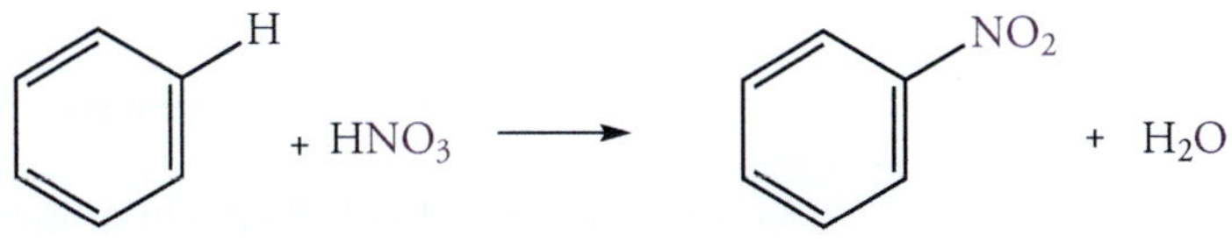

A sulfonic acid group, $-SO_3H$, can also be introduced onto an aromatic ring by electrophilic aromatic substitution. The process, called **sulfonation**, requires a mixture of SO_3 and sulfuric acid, called fuming sulfuric acid, to form $^+SO_3H$.

$$SO_3 + H_2SO_4 \rightleftharpoons {}^+SO_3H + HSO_4^-$$

The net reaction for sulfonation of benzene is the result of electrophilic attack by $^+SO_3H$ followed by loss of a proton.

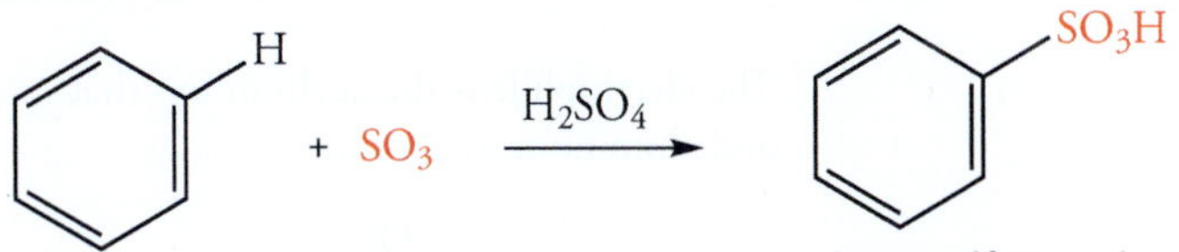

benzensulfonic acid

An alkyl group can be substituted for a hydrogen atom of an aromatic ring by a reaction called the **Friedel-Crafts alkylation**. This reaction requires an alkyl halide with the corresponding aluminum trihalide as the catalyst. The catalyst produces an alkyl carbocation, which is the electrophile.

$$(CH_3)_2CHCl + AlCl_3 \longrightarrow (CH_3)_2CH^+ + AlCl_4^-$$

The reaction is commonly carried out only with alkyl bromides or alkyl chlorides. The net reaction for alkylation of benzene by 2-chloropropane is shown below.

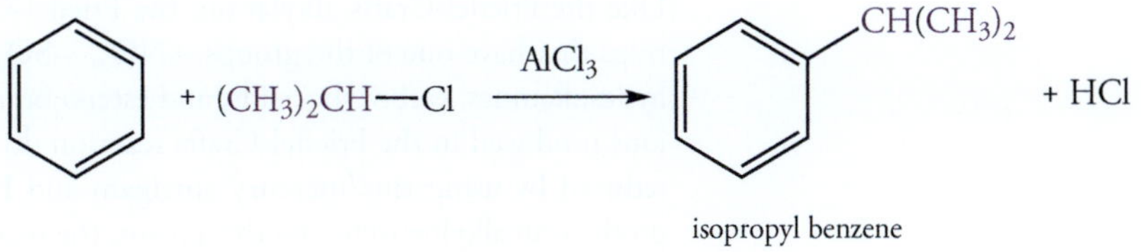

isopropyl benzene

The Friedel-Crafts alkylation does not occur on aromatic rings that have one of the groups $-NO_2$, $-SO_3H$, $-C\equiv N$, or any carbonyl-containing group (aldehydes, ketones, carboxylic acids, and esters) bonded directly to the aromatic ring. All of these substituents make the benzene ring less reactive, a subject we will discuss in the next section.

The Friedel-Crafts alkylation is also limited by the possible structural rearrangement of the alkyl carbocation generated from the alkyl halide. For example, the reaction with 1-chloropropane in the presence of $AlCl_3$ yields a small amount of propylbenzene but a larger amount of the isomer, isopropyl benzene.

isopropyl benzene
(major product)

n-propyl benzene
(minor product)

Isomerization of carbocations occurs in the Friedel-Crafts reaction. Isomerization often occurs by a hydride (H:⁻) shift that converts a less stable carbocation into a more stable one. We recall from Section 4.9 that the order of carbocation stability is tertiary > secondary > primary. Thus, if the Friedel-Crafts reaction is carried out with 1-chloropropane, a relatively unstable propyl carbocation forms initially. But this intermediate rearranges by shifting a hydrogen atom along with the bonding electron pair, H:⁻, from C-2 to C-1. This rearrangement accounts for the formation of isopropyl benzene.

An acyl group can replace hydrogen in an aromatic ring by a reaction called the **Friedel-Crafts acylation**. The reaction requires an acyl halide and the corresponding aluminum trihalide. It is commonly performed only with acyl chlorides. The reaction yields a ketone.

propanoyl chloride

propiophenone

The electrophile is the acylium ion that results from an acid-base reaction between the acyl chloride and aluminum trichloride.

(an acyl chloride)

(an acylium ion)

Like the Friedel-Crafts alkylation, the Friedel-Crafts acylation reaction does not occur on aromatic rings that have one of the groups $-NO_2$, $-SO_3H$, $-C\equiv N$, or any carbonyl-containing group (aldehydes, ketones, carboxylic acids, and esters) bonded directly to the aromatic ring. However, acylium ions produced in the Friedel-Crafts reaction do not rearrange. The acyl group in the product can be reduced by using zinc/mercury amalgam and HCl, a reaction called a **Clemmensen reduction**, to produce an alkylbenzene. By this means, the rearrangement of primary alkyl groups that occurs in the

Friedel-Crafts alkylation reaction is circumvented. For example, propylbenzene can be synthesized by reduction of propiophenone, the product of the acylation of benzene with propanoyl chloride.

propiophenone   Zn(Hg) / HCl   *n*-propyl benzene

5.5 STRUCTURAL EFFECTS IN ELECTROPHILIC AROMATIC SUBSTITUTION

To this point, we have discussed only electrophilic substitution reactions of benzene itself. Now let's examine the effect that a substituent already bonded to the aromatic ring has on the introduction of a second substituent. The substituent could affect the rate of the reaction compared to benzene. Also, any of three possible products—the *ortho*, *meta*, and *para* isomer—could be produced.

Effects of Ring Substituents on Reaction Rates

The relative rates of nitration of benzene and several substituted benzenes are given in the following list. The difference in rate between substituting a nitro group onto phenol and substituting a nitro group onto nitrobenzene is a phenomenal factor of 10^{10}. (For comparison, this rate difference is like the difference between the speed of light and the speed of walking!)

Table 5.1
Effect of Substituents on Aromatic Substitution

Strongly activating
—NH_2, —NHR, —NR_2
— OH, —OCH_3

Weakly activating
—CH_3, —CH_2CH_3, —R

Weakly deactivating
—F, —Cl, —Br

Strongly deactivating
—CO—R, —CO_2H
—CN
—NO_2, —CF_3, —CCl_3

Relative rates of nitration of benzene and its derivatives

	phenol	toluene	benzene	chorobenzene	nitrobenzene
rel. rate	10^3	25	1	3×10^{-2}	10^{-7}

The hydroxyl group and the methyl group, which make the aromatic ring more reactive compared to benzene, are called **activating groups.** The chloro and nitro groups make the aromatic ring less reactive and are called **deactivating groups.** Table 5.1 lists some common substituents and divides them into activating and deactivating groups toward electrophilic aromatic substitution.

Orientation Effects of Ring Substituents

Now let's consider the distribution of products formed in the nitration of toluene. The *ortho* and *para* isomers predominate, and very little of the *meta* isomer forms. The methyl group is said to be an *ortho,para* director. That is, the methyl group directs or orients the incoming substituent into positions *ortho* and *para* to itself. It turns out that all activating groups are *ortho,para* directors. Halogens, which are weakly deactivating, are also *ortho,para* directors.

minor product

A second class of ring substituents directs incoming substituents into the *meta* position. These groups, known as **meta directors**, include nitro, trifluoromethyl, cyano, sulfonic acid, and all carbonyl-containing groups. For a nitration reaction, the trifluoromethyl group orients the incoming nitro group into a position *meta* position to itself. Very small amounts of the *ortho* and *para* isomers are formed. All deactivating groups (except halogens) are *meta* directing groups.

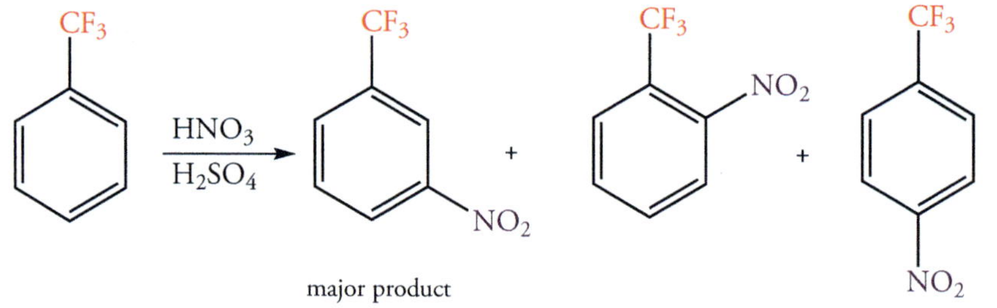

major product

Problem 5.9

Predict whether the following compounds react faster or slower than benzene in a reaction with bromine and FeBr₃.

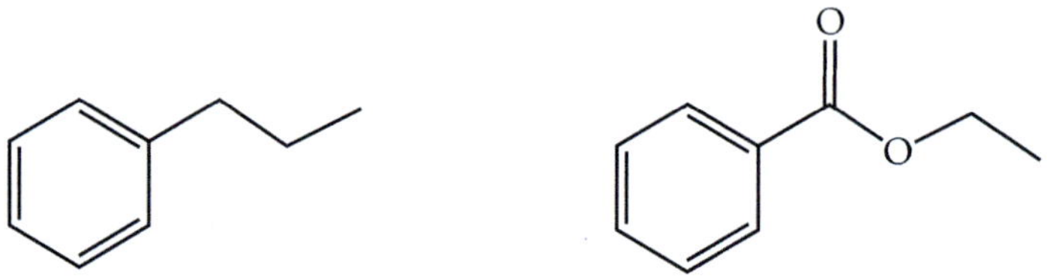

n-propylbenzene   ethyl benzoate

Solution

Propylbenzene contains an alkyl substituent. As in the case of a methyl group, the propyl group is slightly activating. Thus, propylbenzene is more reactive than benzene. Ethyl benzoate has a carbonyl carbon atom bonded directly to the aromatic ring. As a result, its rate of bromination will be significantly slower than that of benzene.

Problem 5.10

Predict whether the following compounds react faster or slower than benzene in a reaction with bromine and FeBr₃.

(a)   (b)   (c)

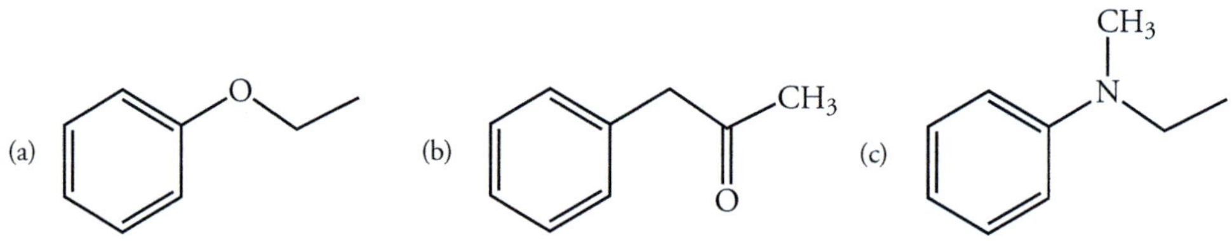

Problem 5.11

Predict the structure of the product(s) formed in the bromination of each of the following compounds.

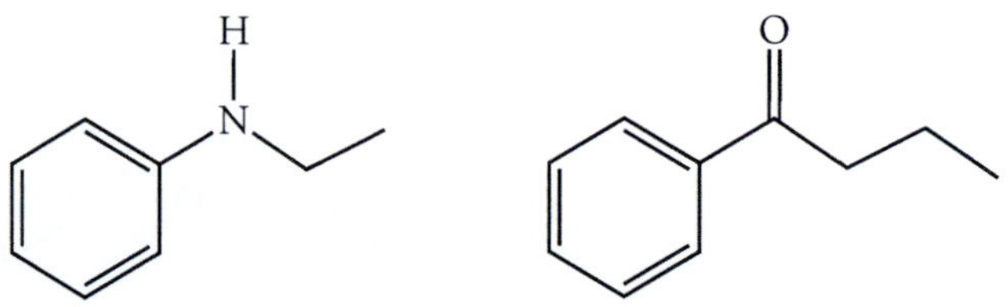

N-ethylaniline   butyrophenone

Solution

N-Ethylaniline has a nitrogen atom with a nonbonded electron pair. The compound resembles aniline, and its substituent should direct the bromine to the *ortho* or *para* positions. Two isomeric compounds should result. Butyrophenone has a carbonyl group bonded to the benzene ring. Therefore, the substituent should direct the bromine to the *meta* position.

Problem 5.12

Predict the structure of the product(s) formed in the bromination of each of the following compounds.

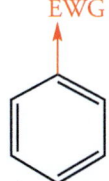

5.6 INTERPRETATION OF RATE EFFECTS

The influence of a substituent on both the rate and distribution of products in electrophilic aromatic substitution reactions can be understood with one model based on the ability of the substituents to either donate or withdraw electrons from the aromatic ring. Let's consider the effect of two types of groups on the electron density of the benzene ring.

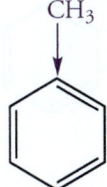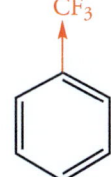

An electron donating group (EDG), increases electron density in the ring

An electron donating group (EWG), decreases electron density in the ring

All activating groups listed in Table 5.1 are electron-donating groups; deactivating groups are electron-withdrawing. Substituents can donate or withdraw electron density by inductive or resonance effects or a combination of these effects.

Inductive Effects of Ring Substituents

Inductive effects are perhaps more easily visualized than resonance effects because they are related to the concept of electronegativity. As discussed in Chapter 4, alkyl groups are electron donating relative to hydrogen, and tend to stabilize double bonds; they also stabilize carbocations. Alkyl groups transfer electron density through the σ bonds to sp²-hybridized carbon atoms. Thus, they also donate electron density to the benzene ring by an inductive effect. The trifluoromethyl group, whose fluorine atoms pull electron density away from the carbon atom to which they are bonded, withdraws electron density from the ring by an inductive effect.

CH_3

A methyl group activates the aromatic ring

CF_3

A trifluoromethyl group deactivates the aromatic ring

The halogens directly bonded to an aromatic ring withdraw electron density and deactivate the compound toward electrophilic aromatic substitution. Any functional group that has a formal positive charge on the atom bonded to the aromatic ring, such as the nitro group, withdraws electron density from the ring. Groups such as the carbonyl group or the nitrile group have a partial positive charge on the atom bonded to the aromatic ring, so they also withdraw electron density from the ring. All such groups deactivate the aromatic ring.

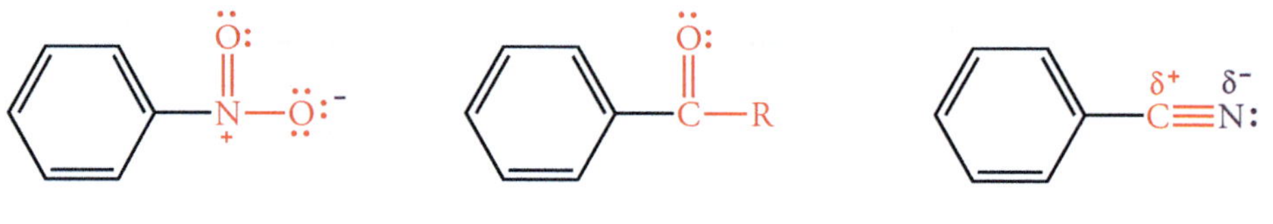

Resonance Effects of Ring Substituents

Next, let's consider how electron density is shifted into or out of a benzene ring by resonance effects. We do this by "moving" electrons and drawing alternate resonance forms. Consider the nitro group or any carbonyl-containing groups. They have sp^2-hybridized atoms that are conjugated with the ring. The oxygen atoms of the nitro group are more electronegative than nitrogen. Hence, we can write resonance forms by "shifting" an electron pair in a nitrogen-oxygen double bond onto the oxygen atom followed by "shifting" an electron pair out of the ring to make a carbon-nitrogen double bond, leaving a positive charge on the aromatic ring. Because a positive charge develops in the ring, the ring is less reactive toward electrophiles. A similar effect explains why the acyl group also makes the ring less reactive toward electrophiles.

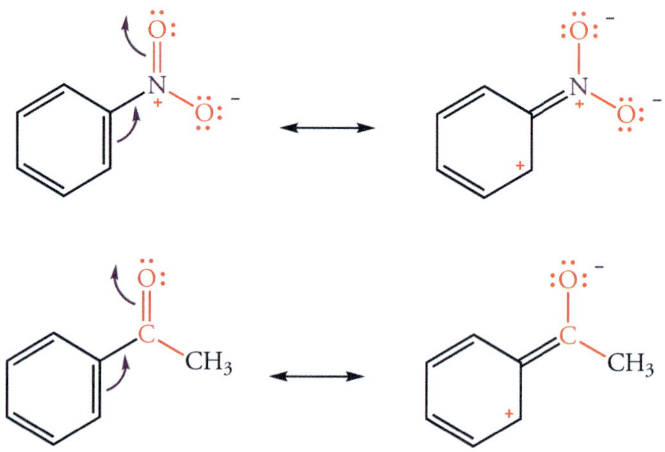

Now let's consider ring substituents that have atoms with lone pair electrons that can be donated to the ring by resonance. When electrons are donated to the ring, the ring develops a negative charge and is therefore more reactive toward electrophiles. Groups that have an unshared electron pair on the atom attached to the ring include the hydroxyl (−OH); alkoxy, such as methoxy (−OCH$_3$); and amino (−NH$_2$) or any substituted amino group (−NHR, −NR$_2$). All of these groups donate electrons to the aromatic ring by resonance.

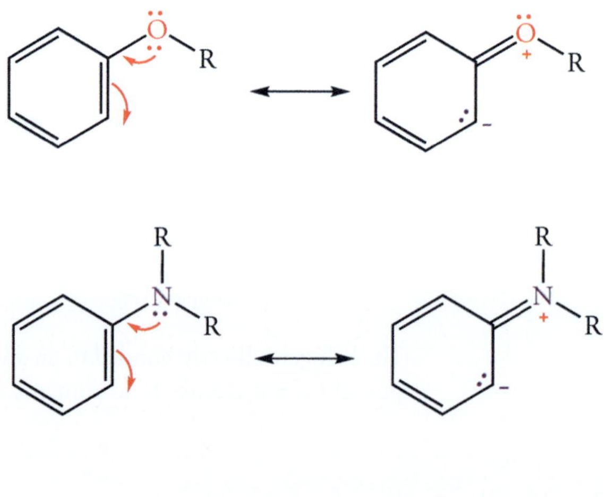

These electron-releasing groups are also electronegative. Therefore, they also withdraw electron density from the ring by an inductive effect. Thus, these substituents inductively take electron density from the ring while giving electron density back by resonance. A group that donates electrons by resonance, such as an amino or hydroxyl group, interacts with the ring through its 2p orbital, which very effectively overlaps with the 2p orbital of a ring carbon atom. Thus, donation of electrons by resonance is very effective and is more important than inductive electron withdrawal. As a result, these groups activate the aromatic ring. This situation, however, does not hold for chlorine or bromine. These electronegative atoms pull electron density out of the aromatic ring by an inductive effect. However, neither the 3p orbital of chlorine nor the 4p orbital of bromine overlaps effectively with the 2p orbital of carbon, so electron donation by resonance is less effective. As a result, the halogens have the net effect of deactivating the aromatic ring.

Substituent Effects on the Metabolism of Benzene Derivatives

We have seen that benzene is remarkably unreactive even under very strong reaction conditions. We might therefore expect benzene to be inert in living cells at pH 7 and 37 °C, and it is. Benzene itself is not metabolized in most human cells; rather it accumulates in the liver, where it does great harm. Benzene is carcinogenic and extremely toxic. Oxidation of benzene and aromatic compounds by the enzyme cytochrome P-450 often yields phenols. Although the process appears to be aromatic hydroxylation, the reaction actually occurs via a three-membered heterocyclic ring called an epoxide (Chapter 9). The epoxide intermediates, called arene oxides, rearrange to phenols.

(an arene oxide)   (a phenol)

Arene oxide intermediates are very reactive, and undergo several types of reactions besides the rearrangement reaction to form phenols. Arene oxides react with proteins, RNA, and DNA. As a result, serious cellular disruptions can occur. These processes will be discussed in Chapter 9, when we will discuss the chemistry of epoxide. Some drugs containing aromatic rings are hydroxylated at the *para* position when metabolized. Phenytoin, an anti-convulsive drug, is an example. The phenolic compounds react further to form water-soluble derivatives.

phenytoin   *p*-hydroxyphenytoin

When some drugs are hydroxylated in the liver, they are also pharmacologically active. For example, the site of hydroxylation of phenylbutazone, an anti-inflammatory drug, is at the *para* position. This hydroxylated product has been produced in the laboratory, and is now marketed under the trade names Tandearil® and Oxalid®.

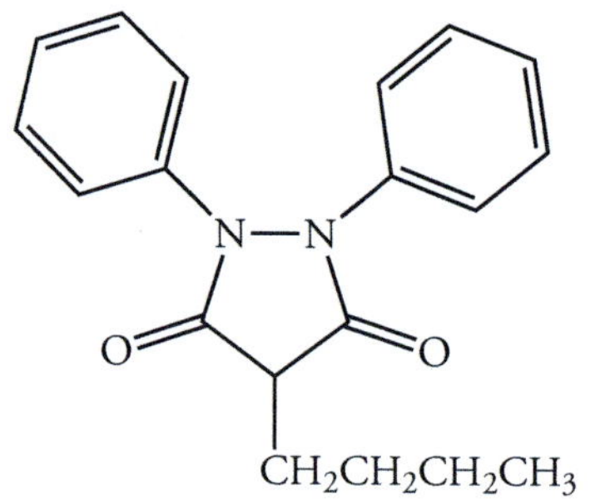

phenylbutazone, used to treat gout

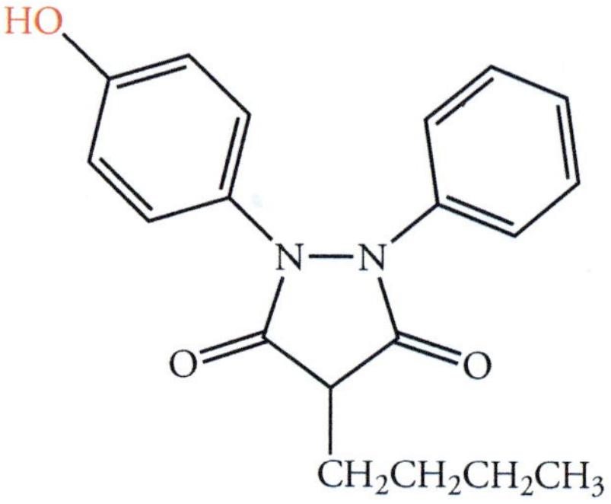

Tandearil

Groups that deactivate aromatic rings toward aromatic substitution affect their metabolic reactions. Dioxin (2,3,7,8-tetrachlorodibenzo-p-dioxin) and PCBs (polychlorinated biphenyls) have multiple deactivating groups. Dioxin is an impurity formed in the production of the herbicide 2,4,5-trichlorophenoxyacetic acid. Millions of tons of PCBs have produced. They were used as heat exchange agents in products such as electrical transformers.

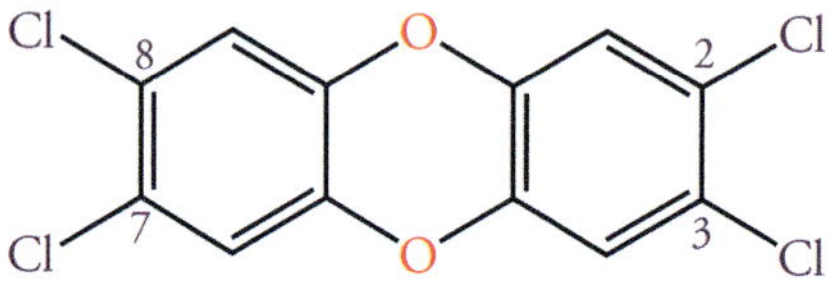

2,3,7,8-tetrachlorodibenzo-*p*-dioxin

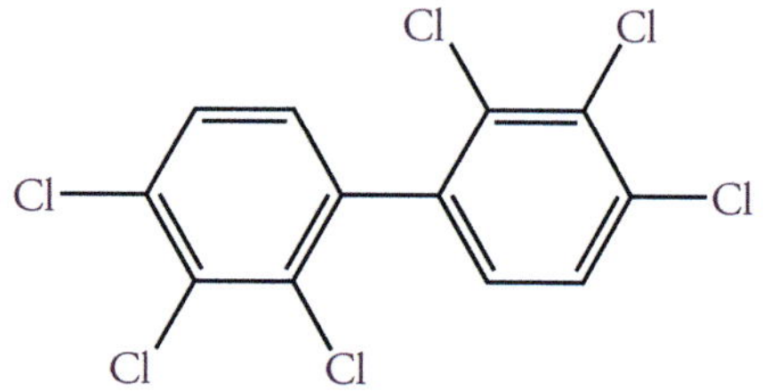

(a polychlorinated biphenyl compound)

Because the aromatic ring loses electrons in the oxidation reaction, the electron-withdrawing groups slow the rate of biological oxidation. These halogenated compounds are nonpolar and quite soluble in fatty tissue. Thus, they tend to persist in the bodies of organisms that inadvertently ingest them. As a consequence they accumulate in the food chain. For example, fish concentrate PCBs in their tissues from contaminated waters. The amount of PCBs increases in the tissues of birds or humans who then eat the fish with highly deleterious consequences.

5.7 INTERPRETATION OF DIRECTING EFFECTS

We noted earlier that, with the exception of the halogens, *ortho,para* directors activate the ring toward electrophilic substitution because they supply electrons to the ring. But why are the *ortho* and *para* positions especially susceptible to attack? To answer this question, let's examine the resonance forms of the intermediate carbocations resulting from reaction at the *ortho* and *para* positions and compare them with those resulting when an electrophile reacts at the *meta* position. First, consider the resonance forms for the intermediate formed by nitration of toluene at the *ortho* and *para* positions.

least stable contributing structure

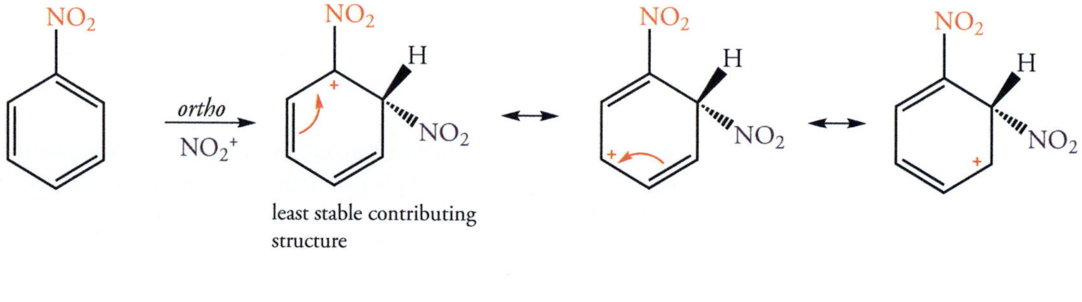

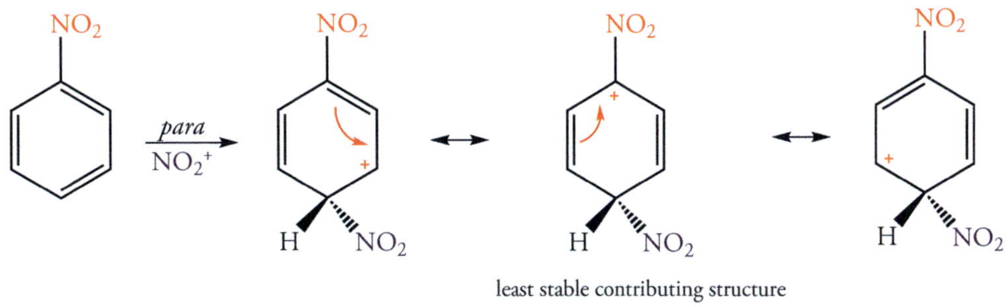

least stable contributing structure

Attack at either the *ortho* or the *para* position results in one resonance structure with a positive charge on the ring carbon atom bonded to the methyl group. This tertiary carbocation makes a major contribution to the stability of the resonance hybrid.

Now consider nitration at the *meta* position. None of the resonance structures have the positive charge on the carbon atom attached to the methyl group.

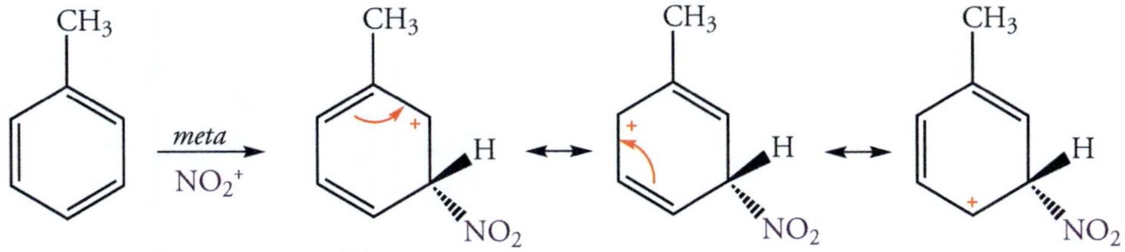

All resonance forms are secondary carbocations. As a result, this intermediate is less stable than the intermediates formed from attack at either the *ortho* or *para* position. As a consequence, reaction at the *ortho* or *para* position is favored over a reaction at the *meta* position.

Next, let's consider the effect of an *ortho,para*-directing hydroxyl group that can donate an unshared pair of electrons to the ring by resonance. An attack either *ortho* or *para* to the hydroxyl group by an electrophile, such as the bromonium ion, leads to an intermediate that is resonance stabilized by the lone pair electrons donated by the oxygen atom. The most stable resonance form has a positive charge on the oxygen atom, but all atoms have an octet of electrons.

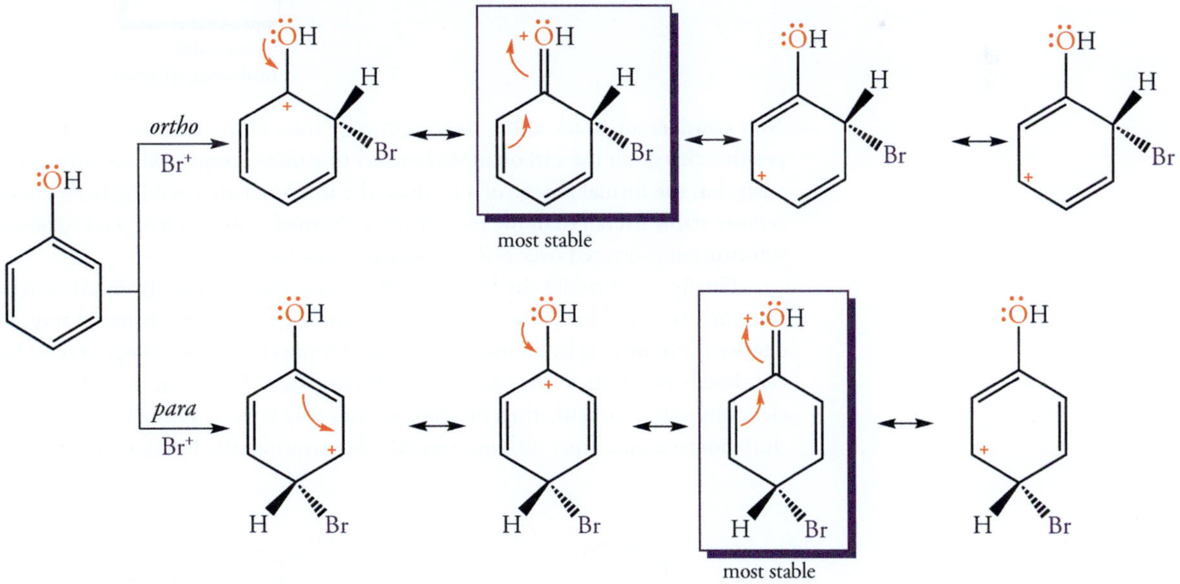

No such stabilization is possible for a group that attacks *meta* to the hydroxyl substituent. Hence, *ortho,para* substitution is preferred over *meta* substitution.

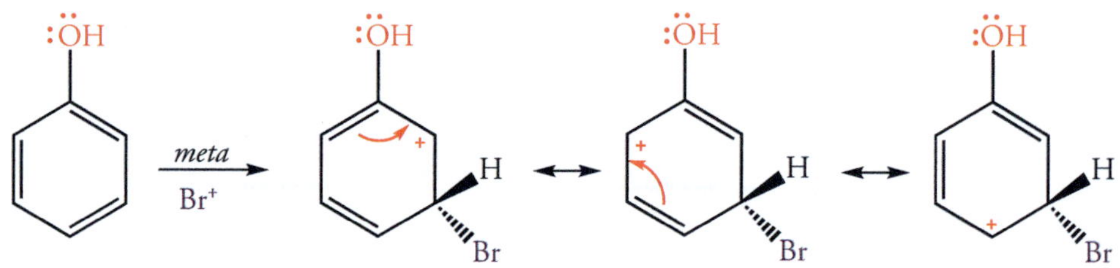

Substituents that strongly deactivate the ring toward electrophilic aromatic substitution are *meta* directors. To explain this relationship, let's consider the possible nitration of nitrobenzene at the *ortho* and *para* positions. In one of the resonance forms for both *ortho-* and *para*-substituted intermediates, a positive charge is located at a carbon atom containing the original nitro group. The nitrogen atom of the nitro group also has a formal positive charge, and such resonance forms containing positive charges on adjacent atoms are not favorable. As a result, these intermediates, formed by attack at either the *ortho* or *para* position, are less stable than those formed in the attack of benzene itself.

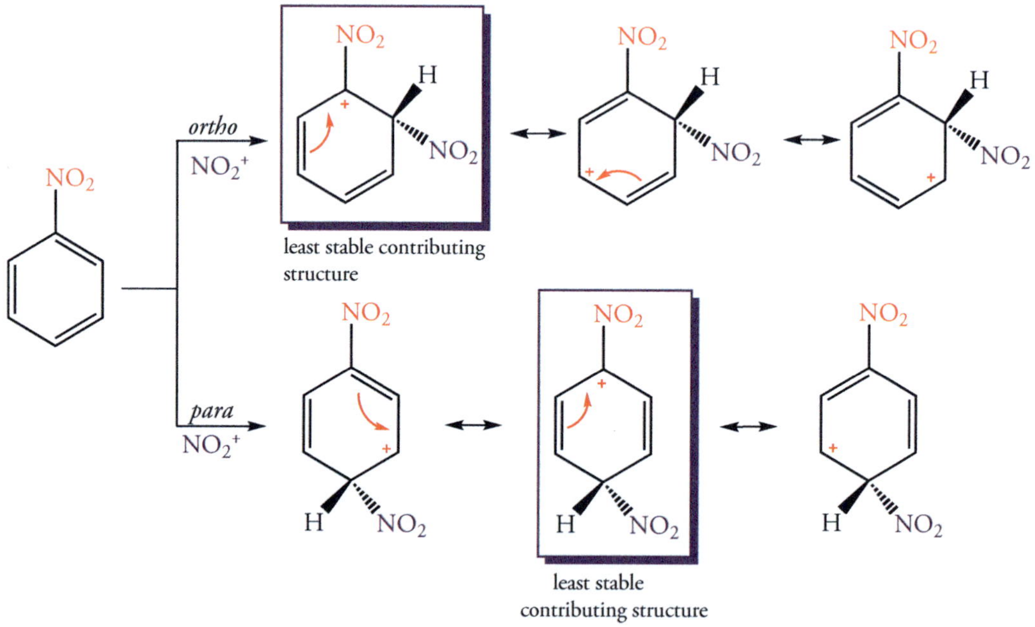

Now consider an attack at the *meta* position. None of the resonance forms of the intermediate has a positive charge on the carbon atom bonded to a nitro group—whose nitrogen atom, we noted previously, has the formal charge of +1. Thus, the intermediate resulting from attack at the *meta* position is more stable overall than the intermediates formed from *ortho* or *para* substitution. As a result, *meta* substitution is favored over *ortho* or *para* substitution.

Finally, we consider the halogen substituents that are weakly deactivating, but yet are *ortho,para* directors. Because the halogens are more electronegative than a benzene ring, they withdraw electron density from the ring by an inductive effect. However, because halogens have lone pair electrons, they can donate electrons to the carbocation intermediate. However, this resonance effect, which supplies electrons to the aromatic ring, only comes into play if the entering electrophile attacks *ortho* or *para* to the halogen atom. Thus, although weakly deactivating, the halogens are *ortho,para* directors.

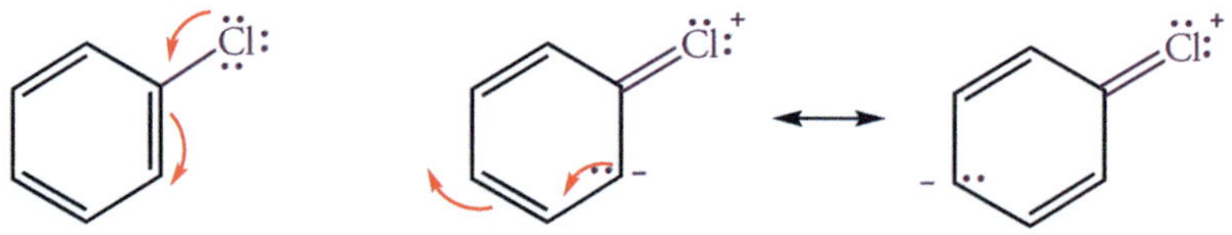

5.8 REACTIONS OF SIDE CHAINS

A group of carbon atoms bonded to an aromatic ring is called a *side chain*. Side-chain carbon atoms that are separated from the aromatic ring by two or more σ bonds behave independently of the aromatic ring. However, carbon atoms directly bonded to the aromatic ring are influenced by the ring.

For example, a benzyl carbocation has a positively charged carbon atom directly attached to a benzene ring and is more stable than primary and secondary carbocations, and of comparable stability to a tertiary carbocation. A benzyl carbocation is resonance stabilized in the same manner as the allylic carbocation we discussed in the previous chapter. That is, the positive charge at the benzylic carbon atom is delocalized among the carbon atoms of the benzene ring.

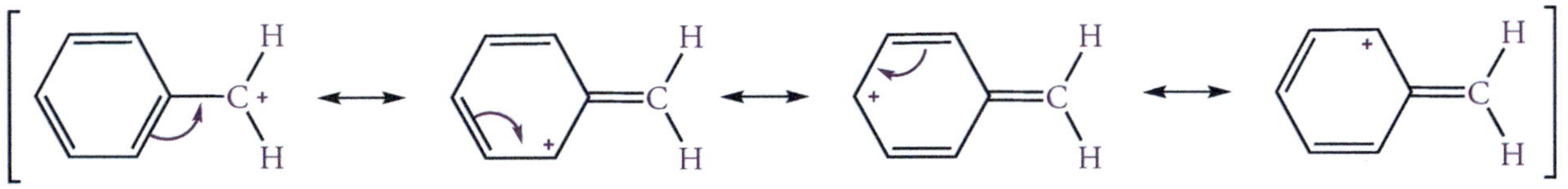

The effect of resonance stabilization of a benzyl carbocation is illustrated in the addition reaction of HBr to indene. Electrophilic attack of H^+ at one of the carbon atoms of the carbon-carbon double bond gives a secondary benzyl carbocation. Attack at the other carbon atom directly would produce a much less stable secondary carbocation.

less stable secondary carbocation

more stable benzyl carbocation

indene

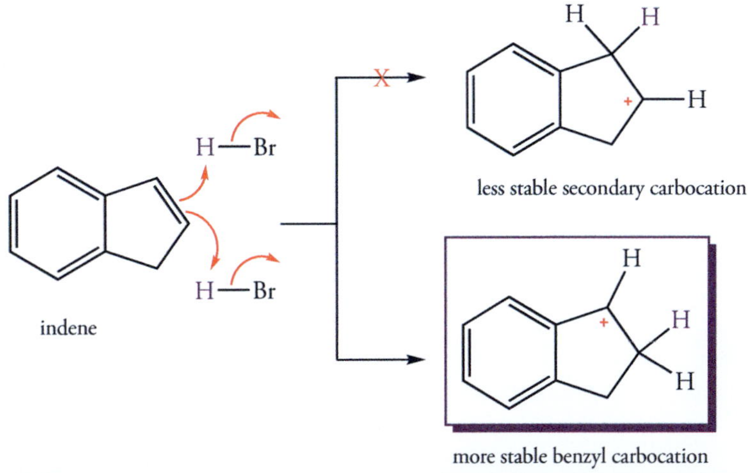

The only product obtained in the reaction is derived from the more stable carbocation, which then reacts with the bromide ion.

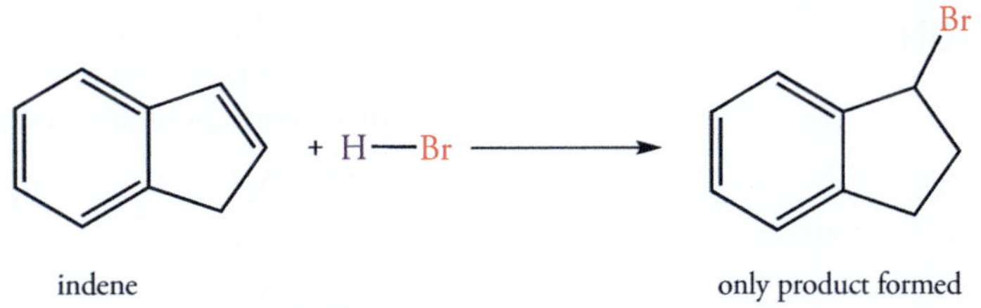

indene

only product formed

Although the aromatic ring causes special reactivity on the side chain, the ring itself does not react with many reagents, and remains intact. We recall that potassium permanganate (Section 4.7) reacts with the π bonds of an alkene under conditions where the σ bonds of the saturated part of the molecule do not react. The benzene ring, in spite of being "unsaturated," is not oxidized by potassium permanganate even under vigorous conditions that totally oxidize side chains to produce a carboxylic acid at the site of the alkyl group.

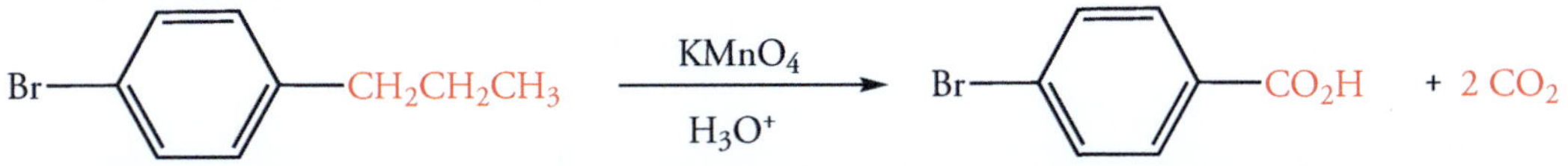

$$Br-\text{\Large\bigcirc}-CH_2CH_2CH_3 \xrightarrow[H_3O^+]{KMnO_4} Br-\text{\Large\bigcirc}-CO_2H \ + \ 2\ CO_2$$

Problem 5.13
Predict whether the following compounds react faster or slower than benzene in a reaction with bromine and $FeBr_3$.

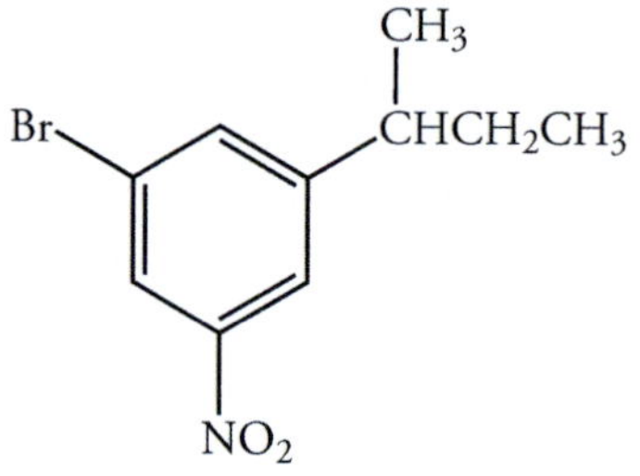

Solution

Potassium permanganate oxidizes the side chain of a substituted aromatic compound completely and forms a carboxylic acid. The sec-butyl group is oxidized, and a —CO_2H group results. The product of the reaction is 3-bromo-5-nitrobenzoic acid.

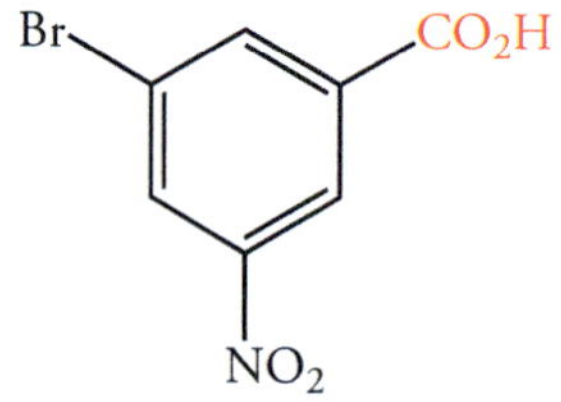

3-bromo-5-nitrobenozic acid

Problem 5.14

A compound with molecular formula $C_{10}H_{14}$ is oxidized to give the following dicarboxylic acid. What are the structures of two possible compounds that would give this result?

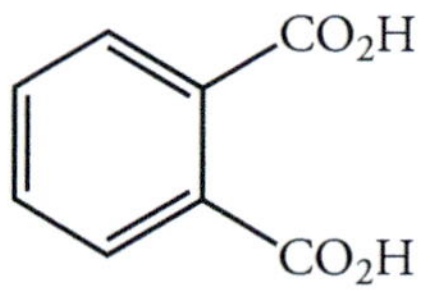

phthalic acid

5.9 FUNCTIONAL GROUP MODIFICATION

Functional group modifications are important because only a few functional groups can be placed directly on an aromatic ring by electrophilic aromatic substitution. The remaining groups are obtained by modifying a group already bonded to the aromatic ring. One example, which we discussed earlier, is the conversion of a methyl group into a carboxylic acid group.

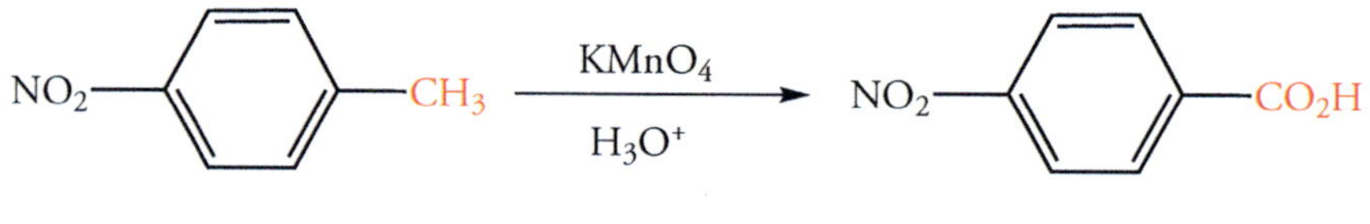

p-nitrotoluene   p-nitrobenzoic acid

Conversion of an Acyl Group to an Alkyl Group

An acyl group bonded to a benzene ring can be converted into an alkyl group by reduction with a zinc-mercury amalgam in HCl. Since an acyl group has a carbonyl carbon atom directly attached to the ring, it is a deactivating, *meta*-directing substituent. However, an alkyl group is an activating, *ortho,para*-directing group.

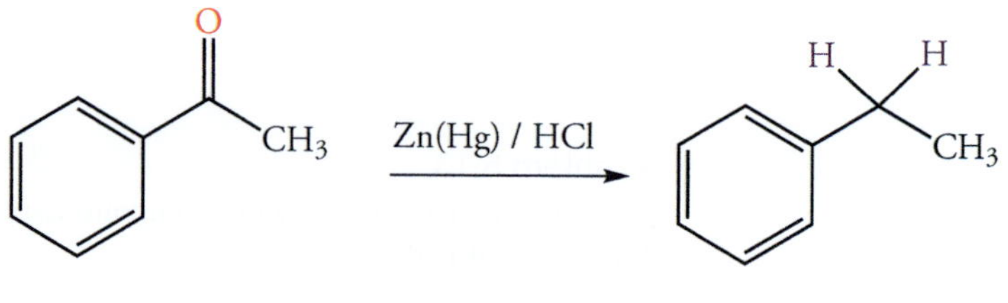

Reduction of a Nitro Group to an Amino Group

Electrophilic aromatic substitution can attach a nitro group directly to a benzene ring, but cannot attach an amino group in one step. However, after a nitro group is introduced, it can easily be reduced to an amino group, producing an aniline. This reaction transforms a strongly deactivating *meta*-directing nitro group into a strongly activating *ortho,para*-directing amino group.

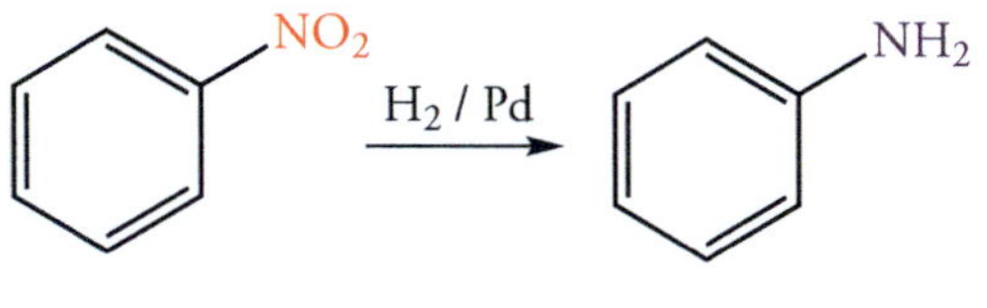

We can also convert a nitro group to an amino group by treating it with tin in the presence of hydrochloric acid.

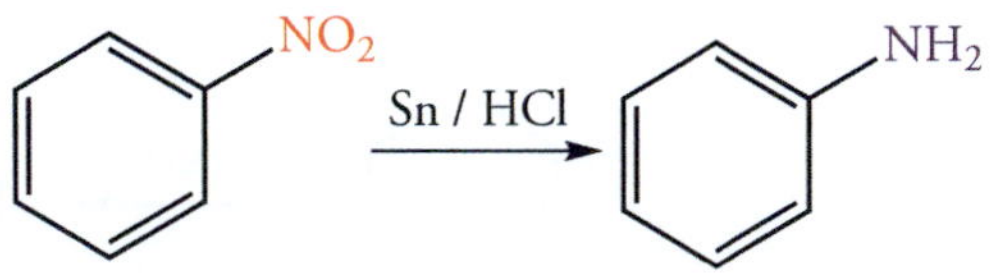

Converting an Amino Group to a Diazonium Ion: The Sandmeyer Reaction

If an aromatic ring has an amino group, the possibilities for further functional group modifications increase dramatically. The amino groups of anilines can be converted into many other groups. The door to other functional groups is opened by converting the amino group into an aryl diazonium ion, $Ar-N_2^+$. The diazonium ion results from the reaction of an aniline with nitrous acid (HNO_2), prepared by treating sodium nitrite with sulfuric acid. This step is called **diazotization**.

$$Ar-NH_2 \xrightarrow{HNO_2} Ar-\overset{+}{N}\equiv N:$$
(a diazonium ion)

In 1884, the German chemist Traugott Sandmeyer found that diazonium ions react with nucleophiles supplied in the form of a Cu(I) salt. These nucleophiles replace the diazonium group and release nitrogen gas. These reactions are known collectively as the **Sandmeyer reaction**.

$$Ar-\overset{+}{N}\equiv N: + Nu:^- \longrightarrow Ar-Nu + N_2$$
(a diazonium ion)

For example, an aromatic diazonium ion can be treated with Cu_2Cl_2 to yield chlorobenzene or with Cu_2Br_2 to give bromobenzene.

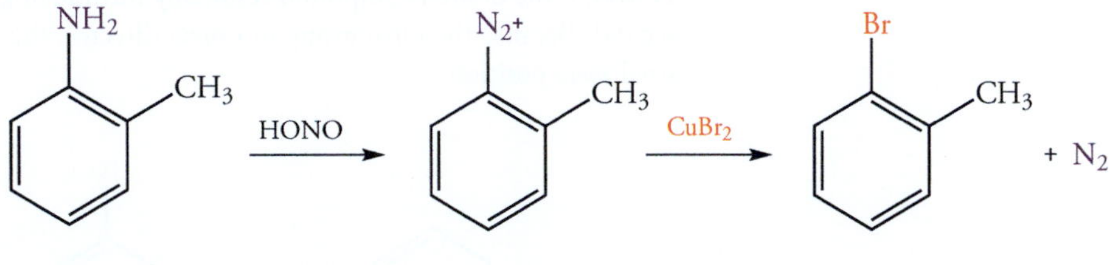

Copper(I) cyanide reacts with aryl diazonium ions to give aryl nitriles.

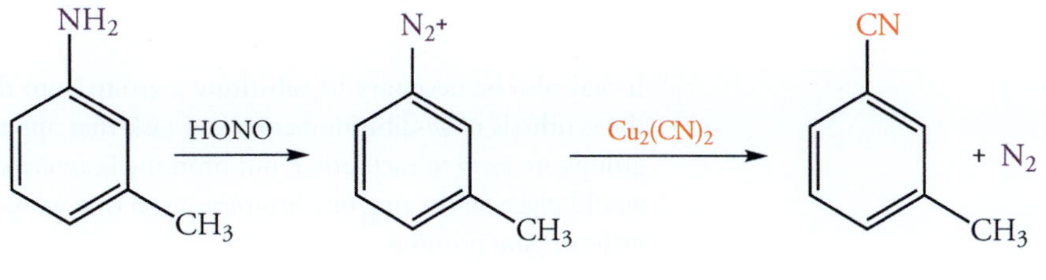

Aryl diazonium compounds react with hot aqueous acid to give phenols. This is the best way to attach an −OH group to an aromatic ring.

$$\text{(m-toluidine)} \xrightarrow{\text{HONO}} \text{(m-methyl benzenediazonium)} \xrightarrow{\text{H}_3\text{O}^+} \text{(m-cresol)} + \text{N}_2$$

Treating aryl diazonium compounds with hypophosphorous acid (H_3PO_2) replaces the diazonium group with a proton. This process can be used to remove the amino substituent from the aromatic ring after its role as a directing group in a synthesis concludes.

$$\text{(m-toluidine)} \xrightarrow{\text{HONO}} \text{(diazonium)} \xrightarrow{\text{H}_3\text{PO}_2} \text{(toluene)} + \text{N}_2$$

5.10 SYNTHESIS OF SUBSTITUTED AROMATIC COMPOUNDS

Chemists often design benzene derivatives with two or more substituents strategically placed around the ring. A project of this type begins with an analysis of the *ortho,para-* or *meta*-directing characteristics of the substituents. For example, consider the problem of synthesis of *m*-chloronitrobenzene. A nitro group is *meta*-directing; a chloro group is *ortho,para*-directing. The order in which we add these groups is clearly important. If chlorination precedes nitration, the entering nitro group will be largely directed to form *o*-chloronitrobenzene and *p*-chloronitrobenzene. Very little of the desired *meta* isomer will form.

$$\text{benzene} \xrightarrow[\text{FeCl}_3]{\text{Cl}_2} \text{chlorobenzene} \xrightarrow[\text{H}_2\text{SO}_4]{\text{HNO}_3} \text{p-chloronitrobenzene} + \text{o-chloronitrobenzene}$$

However, the desired compound results by introducing the nitro group first and the chlorine group second. Because the nitro group is a *meta* director, the entering chlorine atom is directed to the desired *meta* position.

$$\text{benzene} \xrightarrow[\text{H}_2\text{SO}_4]{\text{HNO}_3} \text{nitrobenzene} \xrightarrow[\text{FeCl}_3]{\text{Cl}_2} \text{m-chloronitrobenzene}$$

It may also be necessary to substitute a group onto the aromatic ring and then modify it. Consider the synthesis of *m*-dibromobenzene, a task that appears at first glance to be impossible. The bromo groups are *meta* to each other, but bromine is an *ortho, para* director! Direct bromination of benzene would place on the ring one bromine atom that would then direct the second bromine atom into the *ortho* or *para* position.

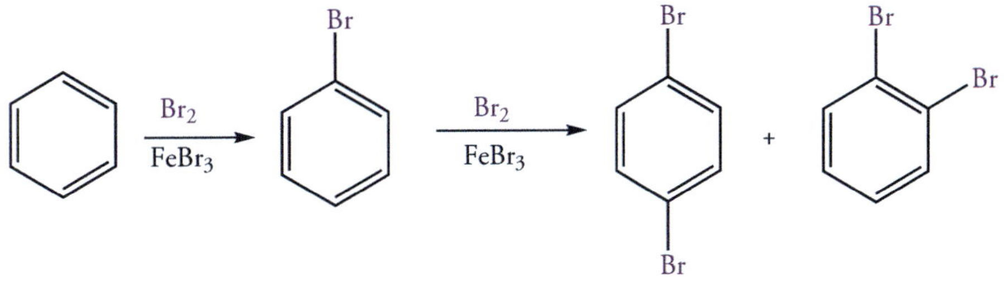

We know that a nitro group is a *meta* director. So, we first make nitrobenzene, then brominate it to obtain *m*-bromonitrobenzene.

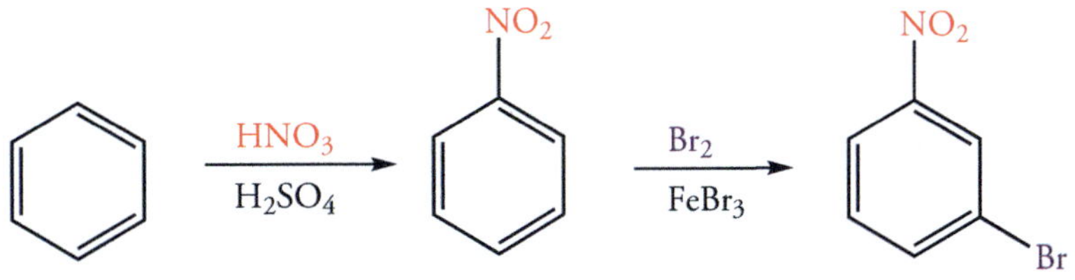

We have seen that a nitro group can be converted to a bromo group by (1) reducing the nitro group to an amino group, (2) converting the amino group to a diazonium group, and (3) treating the diazonium compound with copper(I) bromide. The procedure requires several steps, but it accomplishes the apparently impossible task of preparing *m*-dibromobenzene.

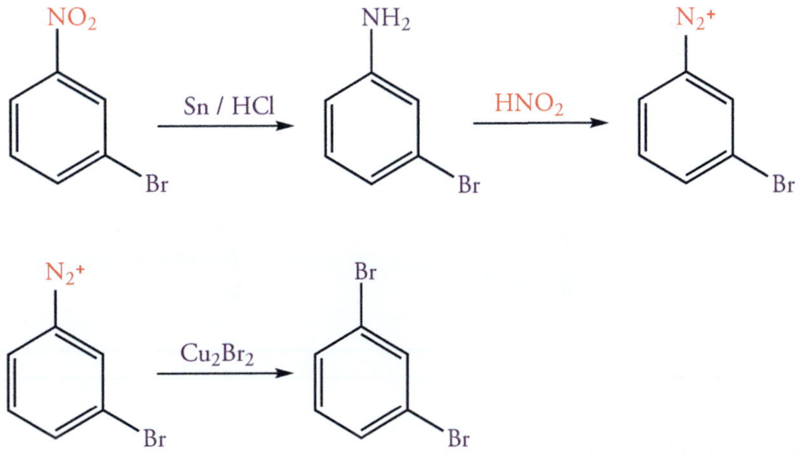

Problem 5.15

Devise a synthesis of *m*-bromoaniline starting from benzene.

Solution

Bromine, which is an *ortho,para*-director, can be introduced directly onto the benzene ring by reaction with bromine and FeBr$_3$. The amino group, −NH$_2$, of aniline is also an *ortho,para*-director. It can be introduced indirectly by first nitrating benzene and then reducing the nitro compound. Recall that the nitro group is a *meta* director. Bromination of benzene followed by nitration gives a mixture of *ortho*- and *para*-bromonitrobenzene. The desired *meta* isomer is not formed. Nitration of benzene gives nitrobenzene—a compound that now directs subsequent electrophilic substitution reactions to produce the *meta* isomer. Thus, bromination of nitrobenzene followed by reduction of the product gives the desired *m*-bromoaniline.

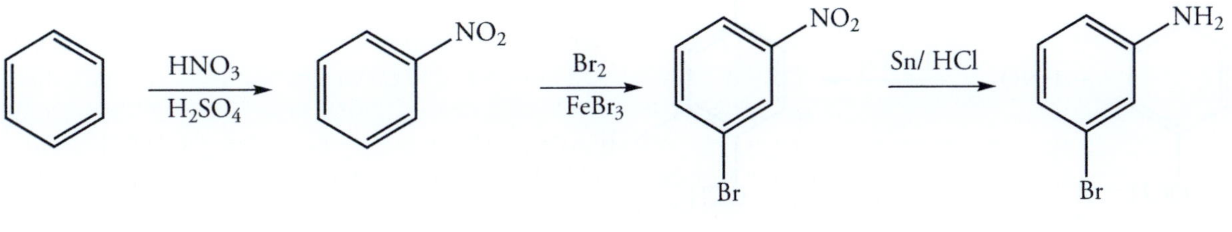

Problem 5.16

Propose a method to synthesize *p*-nitrobenzoic acid starting from benzene.

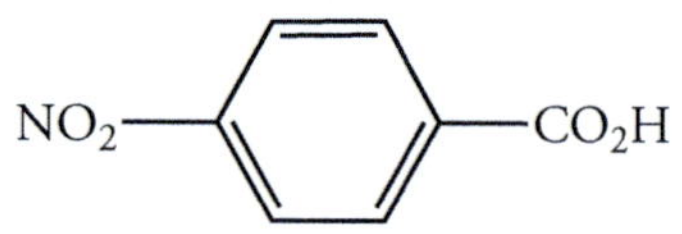

$$NO_2 - C_6H_4 - CO_2H$$

p-nitrobenzoic acid

Problem 5.17

Propose a synthesis of *p*-bromophenol starting from benzene.

Solution

The hydroxyl and bromo groups are both *ortho,para* directors. Thus, the synthesis must proceed via an intermediate compound that has a *meta* director such as the nitro group. Nitration of benzene followed by bromination gives *m*-bromonitrobenzene.

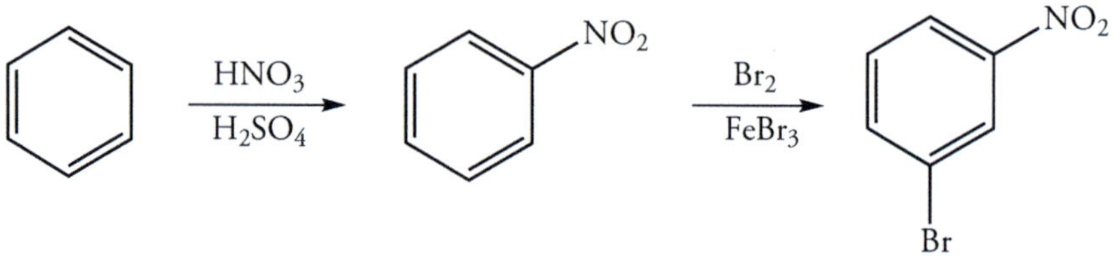

Reduction of the nitro group followed by conversion of the amino group into a diazonium ion gives an intermediate that can be transformed into a phenol.

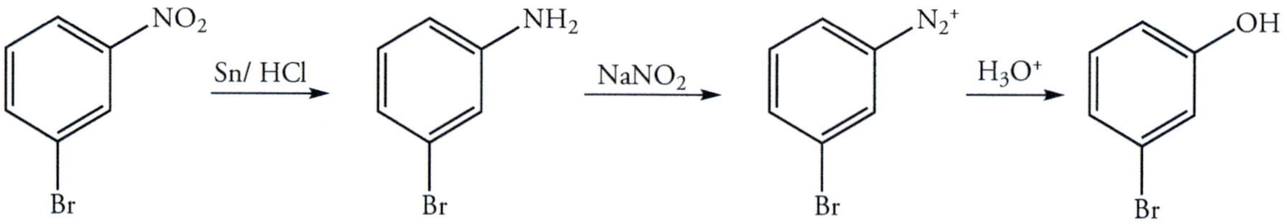

SUMMARY OF REACTIONS

1. Halogenation (Section 5.5)

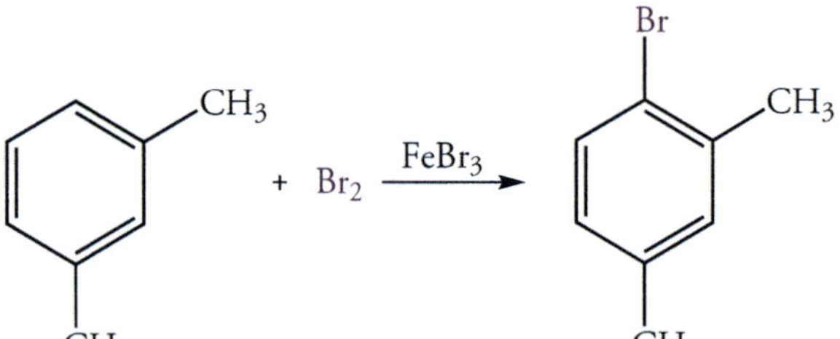

2. Nitration (Section 5.5)

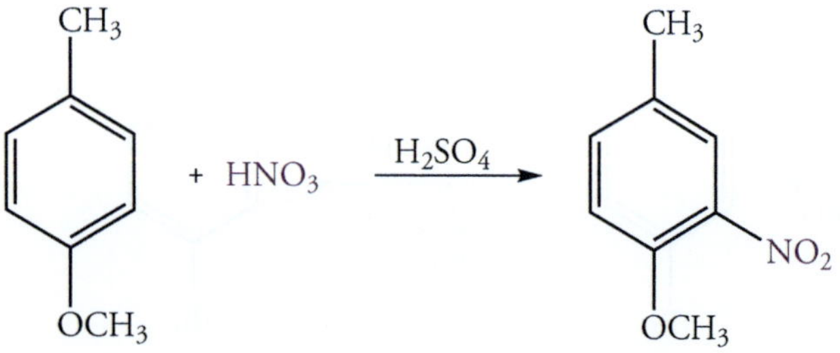

3. Sulfonation (Section 5.5)

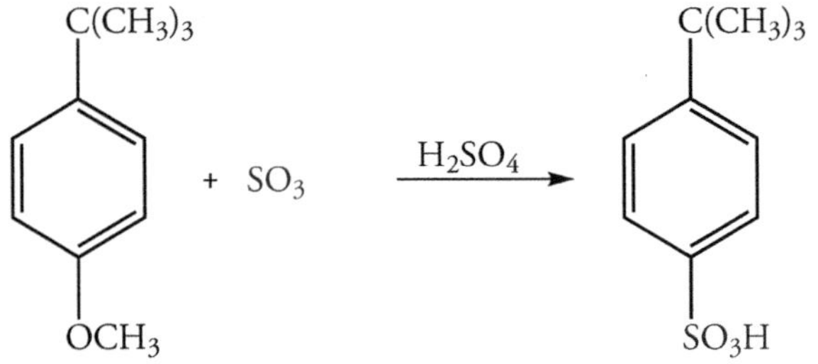

4-tert-butylphenol methyl ether + SO$_3$ $\xrightarrow{\text{H}_2\text{SO}_4}$ sulfonated product with SO$_3$H

4. Alkylation (Friedel-Crafts) (Section 5.5)

benzene + Cl—cyclohexane $\xrightarrow{\text{AlCl}_3}$ phenylcyclohexane

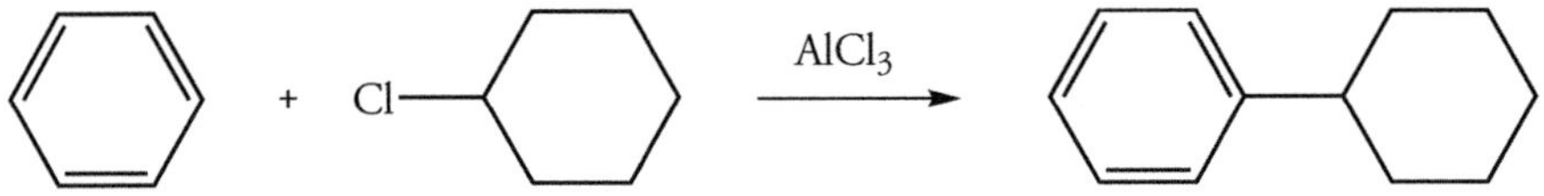

5. Acylation (Friedel-Crafts) (Section 5.5)

benzene + benzoyl chloride $\xrightarrow{\text{AlCl}_3}$ benzophenone

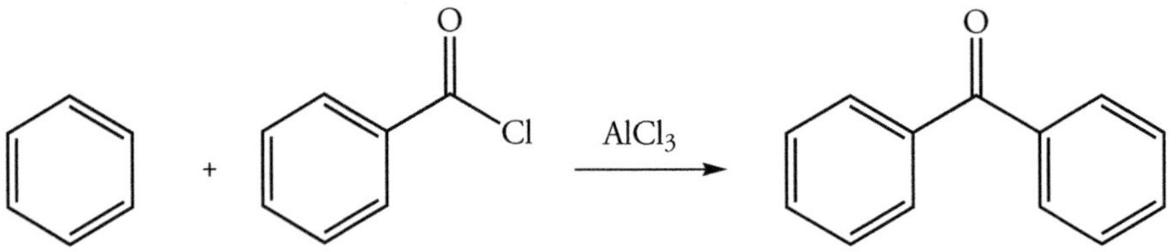

6. Side-Chain Oxidation (Section 5.8)

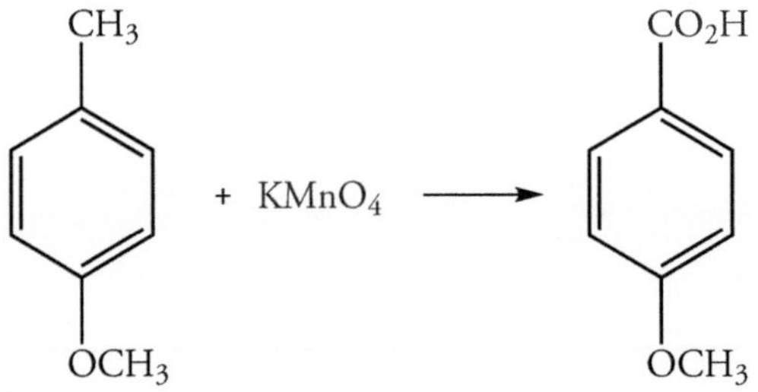

4-methylanisole + KMnO$_4$ $\longrightarrow$ 4-methoxybenzoic acid

7. Reduction of Acyl Side Chains (Section 5.5)

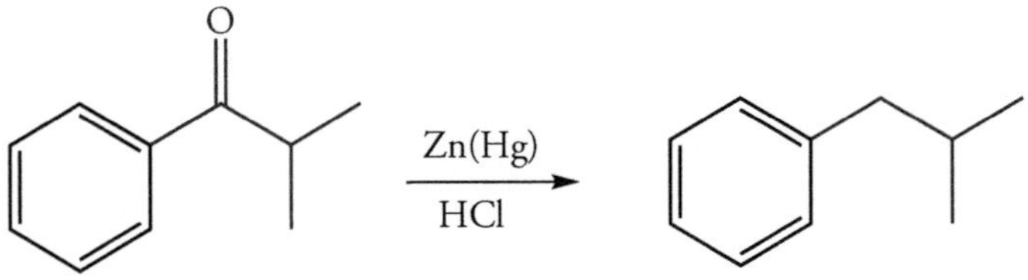

$\xrightarrow[\text{HCl}]{\text{Zn(Hg)}}$

8. Reduction of Nitro Groups (Section 5.10)

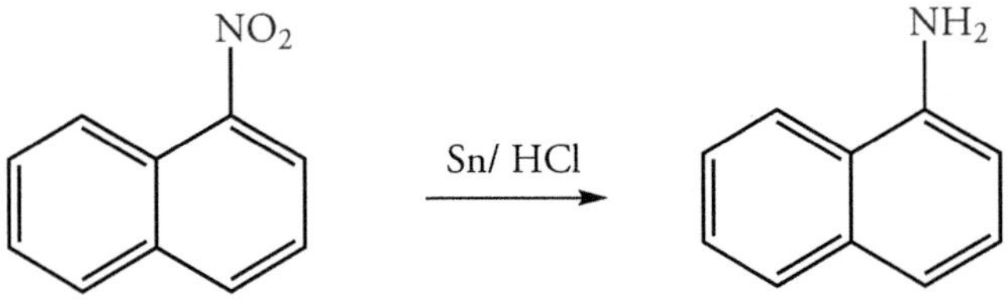

1-nitronaphthalene $\xrightarrow{\text{Sn/ HCl}}$ 1-aminonaphthalene

9. Reactions of Amino Groups via Diazonium Ions (Section 5.10)

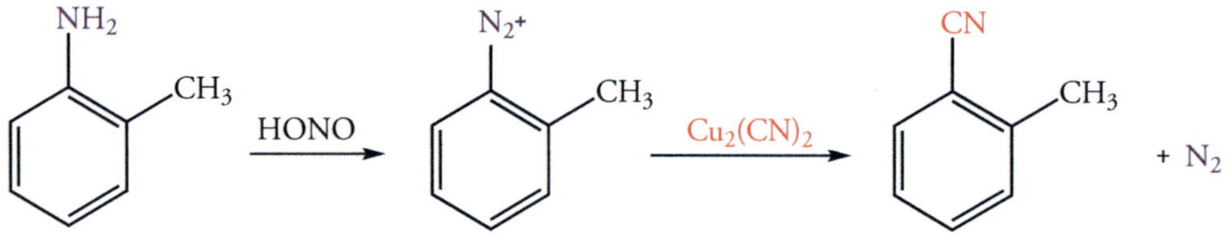

NH₂ → [CH₃, HONO] → N₂⁺ → [CH₃, Cu₂(CN)₂] → CN [CH₃] + N₂

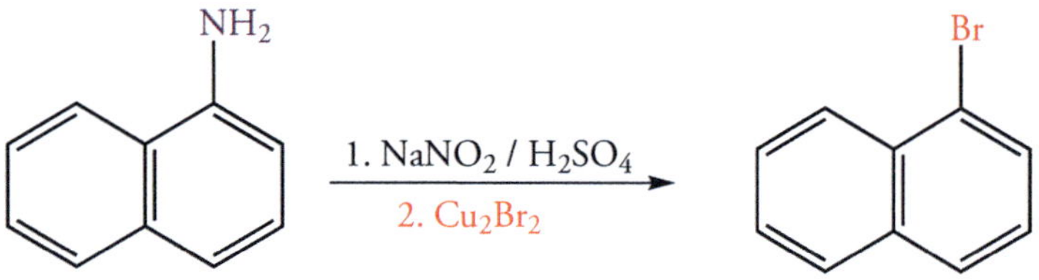

NH₂ → [1. NaNO₂ / H₂SO₄, 2. Cu₂Br₂] → Br

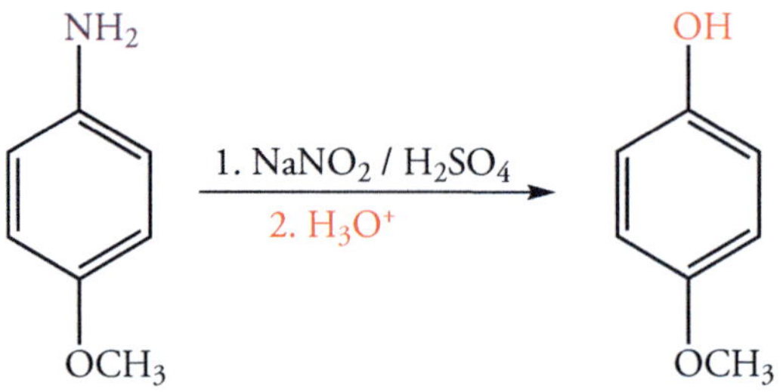

NH₂ [OCH₃] → [1. NaNO₂ / H₂SO₄, 2. H₃O⁺] → OH [OCH₃]

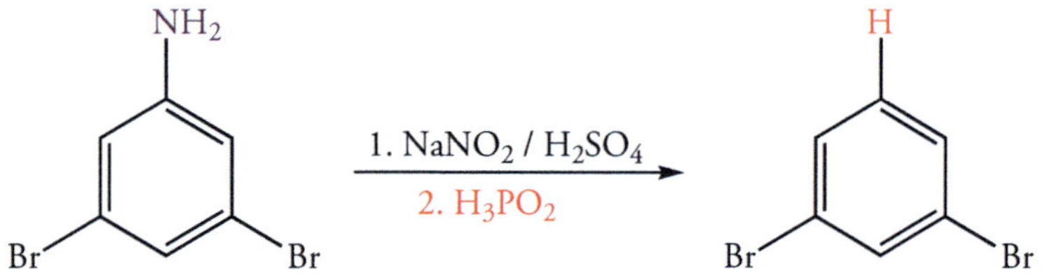

NH₂ [Br, Br] → [1. NaNO₂ / H₂SO₄, 2. H₃PO₂] → H [Br, Br]

EXERCISES

Aromaticity

5.1 Determine whether each of the following is an aromatic compound:

(a)   (b)   (c)

5.2 Determine whether each of the following is an aromatic compound:

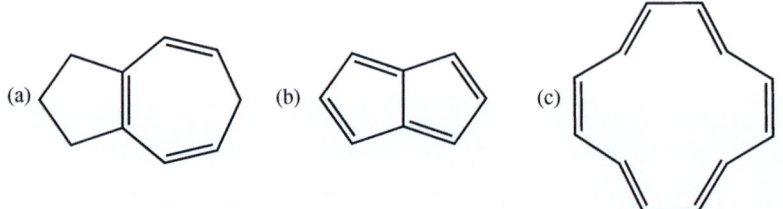

(a)   (b)   (c)

Polycyclic and Heterocyclic Aromatic Compounds

5.3   There are two isomeric bromonaphthalenes. Draw their structures.

5.4   There are three isomeric bromoanthracenes. Draw their structures.

5.5   There are three isomeric diazines, $C_4N_2H_4$, that resemble benzene, but have two nitrogen atoms in place of carbon in the ring. Draw their structures. Which of the isomers should have no dipole moment?

5.6   There are three isomeric triazines, $C_3N_3H_3$, that resemble benzene but have three nitrogen atoms in place of carbon in the ring. Draw their structures. Which of the isomers should have no dipole moment?

5.7   **How many electrons does each heteroatom contribute to the π system in each of the following compounds?**

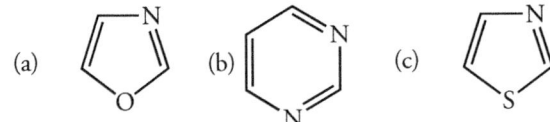

5.8   How many electrons does each heteroatom contribute to the π system in each of the following compounds?

(a)   (b)   (c)   (d) ethionamide, an anti-tubercular agent

5.9   Identify the heterocyclic ring structure in each of the following compounds that have been investigated as possible male contraceptives.

(a)   (b)

5.10   Identify the aromatic heterocyclic ring structure contained in each of the following compounds:

(a) tolmetin, a drug used to lower blood sugar levels

Isomerism in Aromatic Compounds

5.11   There are three isomeric dichlorobenzenes. One compound is nonpolar. Which one?

5.12   The boiling points of benzyl alcohol ($C_6H_5CH_2OH$) and anisole ($C_6H_5COCH_3$)) are 205 and 154°C, respectively. Explain this difference.

5.13   There are four isomeric substituted benzene compounds with the molecular formula C_8H_{10}. Name each compound.

5.14   There are three isomeric substituted benzene compounds with the molecular formula $C_6H_3Br_3$. Name each compound.

Nomenclature of Aromatic Compounds

5.15 Identify each of the following as an *ortho-*, *meta-*, or *para*-substituted compound:

(a) methylparaben, a food preservative

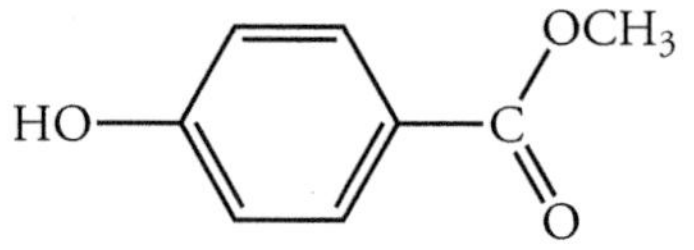

(b) DEET, an insect repellent

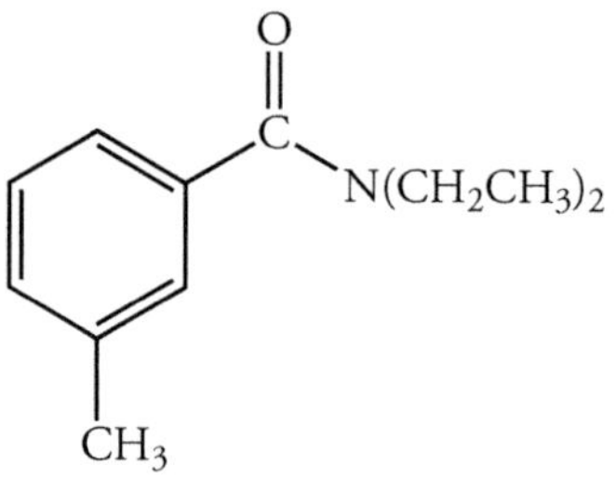

5.16 Identify each of the following as an *ortho-*, *meta-*, or *para*-substituted compound:

(a) resorcinol monoacetate, a germicide used to treat skin conditions

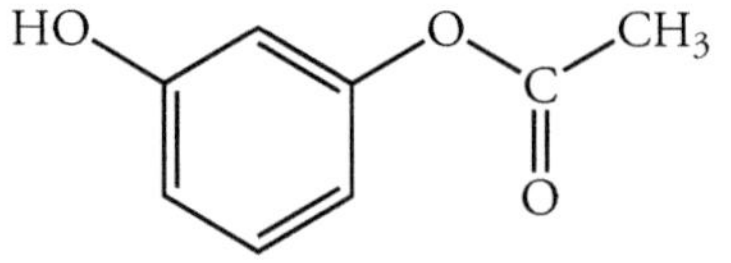

(b) salicylamide, an analgesic

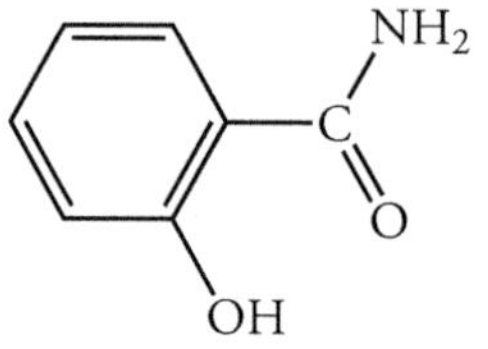

5.17 Name each of the following compounds:

(a)   (b) 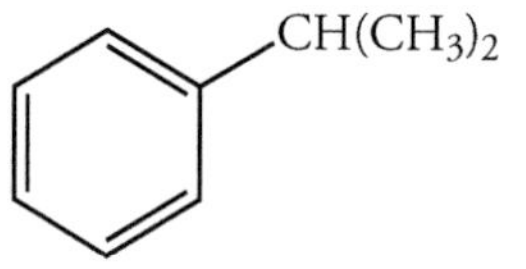   (c)

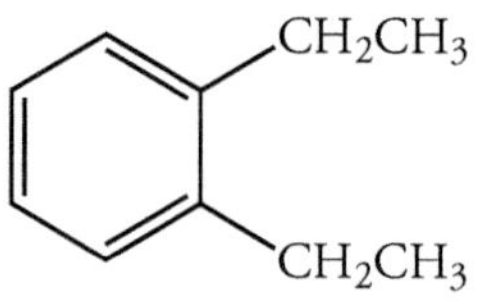

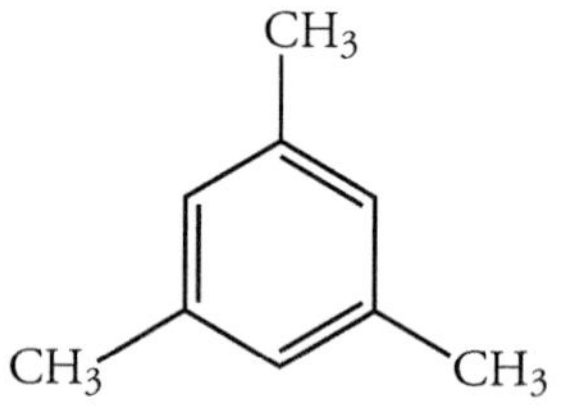

5.18 Name each of the following compounds:

(a)   (b) 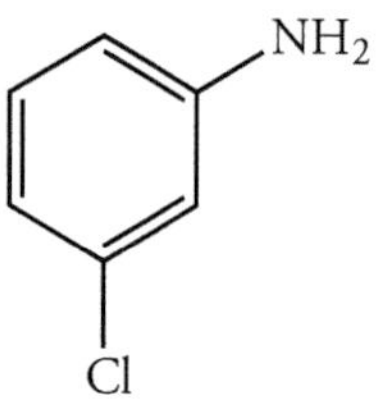   (c)

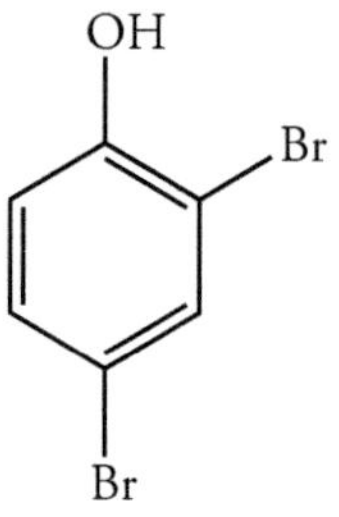

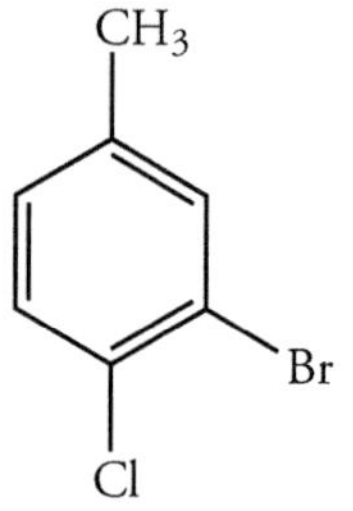

5.19 Name each of the following compounds:

(a) an antiseptic agent used to treat athlete's foot

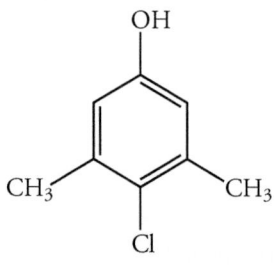

(b) a disinfectant

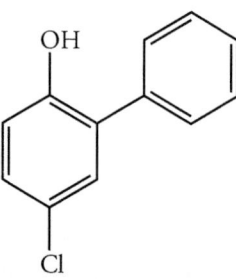

5.20 Draw the structure of 3,4,6-trichloro-2-nitrophenol, a lampricide used to control sea lampreys in the Great Lakes.

5.21 Draw the structure of each of the following compounds:
(a) 5-isopropyl-2-methylphenol, found in oil of marjoram
(b) 2-isopropyl-5-methylphenol, found in oil of thyme
(c) 2-hydroxybenzyl alcohol, found in the bark of the willow tree

5.22 N,N-Dipropyl-2,6-dinitro-4-trifluoromethylaniline is the IUPAC name for Treflan®, a herbicide. Draw its structure. (The prefix N signifies the location of substituents on a nitrogen atom.)

Electrophiles and Ring Substituents

5.23 Predict whether –SCH₃ is an activating or deactivating group. Will it be an *ortho,para*-directing group or a *meta*-directing group?

5.24 The sulfonamide group is found in sulfa drugs. Is it an activating or deactivating group. Will it be *ortho,para*-directing or *meta*-directing?

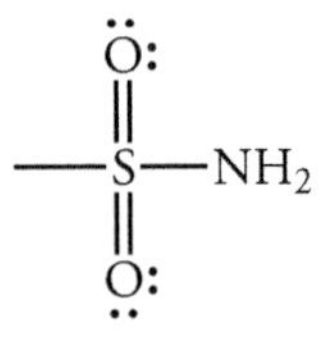

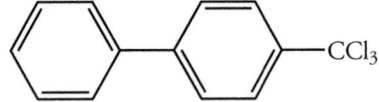

5.25 What product results from the Friedel-Crafts alkylation of benzene using 1-chloro-2-methylpropane and aluminum trichloride?

5.26 Alkylation of benzene can be accomplished using an alkene such as propene and an acid catalyst. Identify the electrophile and the product.

5.27 Indicate on which ring and at what position bromination of each compound will occur.

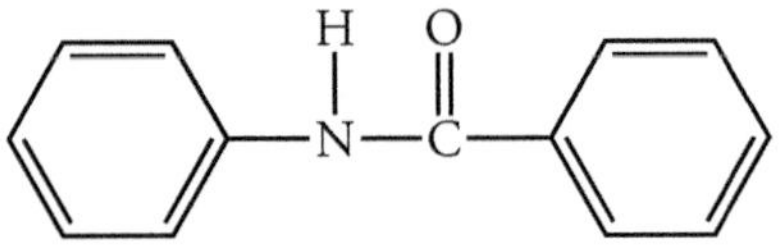

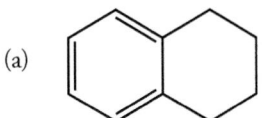

5.28 Indicate on which ring and at what position nitration of each compound will occur.

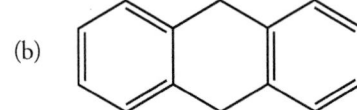

Isomerism in Aromatic Compounds

5.29 Draw the oxidation product of each compound in Exercise 5.17 when reacted with potassium permanganate.

5.30 Predict the oxidation product of each of the following compounds when reacted with potassium permanganate.

(a) (b) (c)

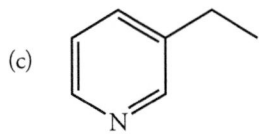

5.31 Free radical bromination of propylbenzene yields essentially one product. What is its structure? Suggest a reason why other isomers are not formed.

5.32 Treatment of allyl benzene with dilute acid causes isomerization to an isomeric compound, Suggest its structure and propose a mechanism for its formation.

Synthesis of Aromatic Compounds

5.33 What reagent is required for each of the following reactions? Will an *ortho,para* mixture of products or the *meta* isomer predominate?
(a) nitration of bromobenzene   (b) sulfonation of nitrobenzene
(c) bromination of ethylbenzene (d) methylation of anisole

5.34 What reagent is required for each of the following reactions? Will an *ortho,para* mixture of products or the *meta* isomer predominate?
(a) bromination of benzoic acid (b) acetylation of isopropylbenzene
(c) nitration of acetophenone (d) nitration of phenol

5.35 Starting with benzene, describe the series of reagents and reactions required to produce each of the following compounds.
(a) *p*-bromonitrobenzene (b) *m*-bromonitrobenzene (c) *p*-bromoethylbenzene (d) *m*-bromoethylbenzene

5.36 Starting with benzene, describe the series of reagents and reactions required to produce each of the following compounds.
(a) *m*-bromobenzenesulfonic acid (b) *p*-bromobenzenesulfonic acid (c) *p*-nitrotoluene (d) *p*-nitrobenzoic acid

5.37 Starting with either benzene or toluene, describe the series of reagents and reactions required to produce each of the following compounds.
(a) 3,5-dinitro-l-chlorobenzene (b) 2,4,6-trinitrotoluene (c) 2,6-dibromo-4-nitrotoluene

5.38 Starting with either benzene or toluene, describe the series of reagents and reactions required to produce each of the following compounds.
(a) 2,4,6-tribromobenzoic acid (b) 2-bromo-4-nitrotoluene (c) 1-bromo-3,5-dinitrobenzene

5.39 Starting with either benzene or toluene, describe the series of reagents and reactions required to produce each of the following compounds.
(a) *m*-bromophenol (b) *m*-bromoaniline (c) *p*-methylphenol

5.40 Starting with either benzene or toluene, describe the series of reagents and reactions required to produce each of the following compounds.
(a) *m*-bromophenol (b) *m*-bromoaniline (c) *p*-methylphenol

5.41 Starting with either benzene or toluene, describe the series of reagents and reactions required to produce each of the following compounds.

(a)
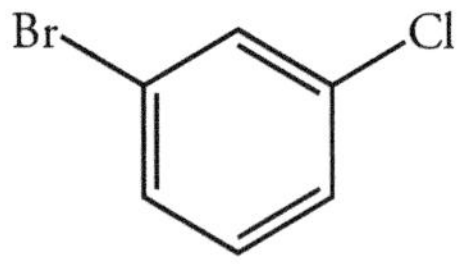

Br
Cl

(b)
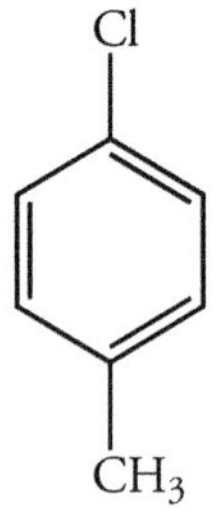

Cl
CH_3

(c)
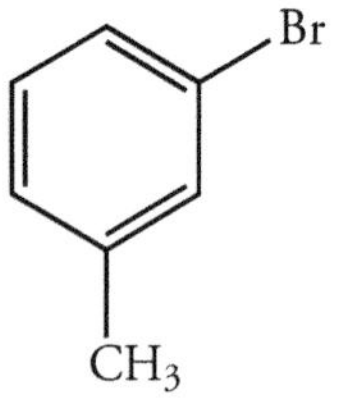

Br
CH_3

Metabolic Oxidation of Aromatic Compounds

5.42 Explain why aromatic hydroxylation of chlorpromazine, an anti-psychotic drug, occurs at the indicated position and in that ring.

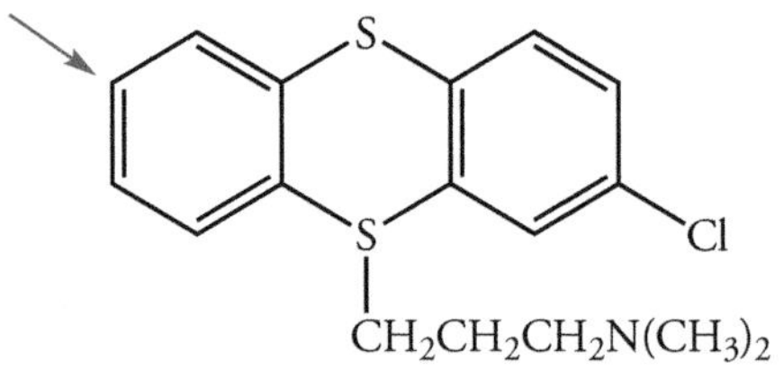

S
S
Cl
$CH_2CH_2CH_2N(CH_3)_2$

chlorprozamine

5.43 Why doesn't aromatic hydroxylation of probenecid, a drug used to treat chronic gout, occur?

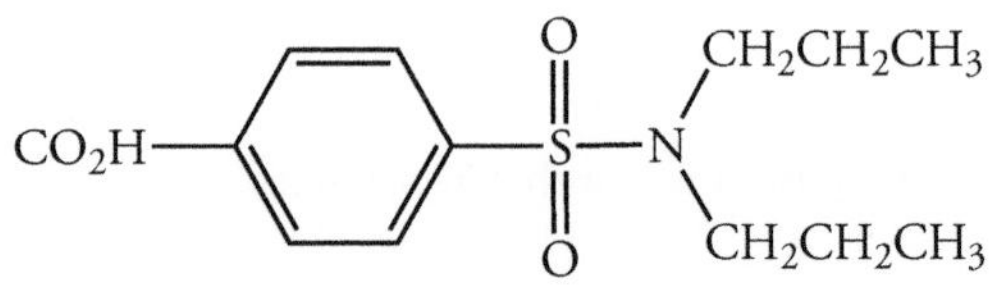

CO_2H
O
S
O
N
$CH_2CH_2CH_3$
$CH_2CH_2CH_3$

probenecid

6 STEREOCHEMISTRY

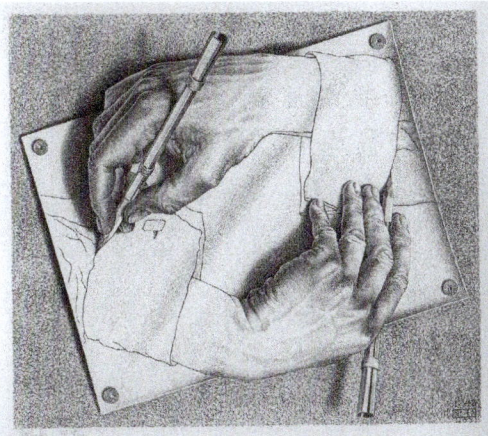

M. C. ESCHER, DRAWING HANDS, 1948

6.1 CONFIGURATION OF MOLECULES

In Chapters 3 and 4 we considered the structures of geometric isomers, which are one of a general class of stereoisomers. **Stereoisomers** have the same connectivity—the same sequence of bonded atoms—but different arrangements of the atoms in space. The different three-dimensional arrangements of atoms in space determine their configurations. Geometric isomers have different configurations. The **configuration** of a molecule plays a major role in its biological properties. Stereoisomers often have entirely different biological properties. Geometric isomers invariably elicit different responses in organisms. For example, bombykol, the sex attractant of the male silk-worm moth, has a (Z)/(E) arrangement about the double bonds at C-10 and C-12. It is 10^9 to 10^{13} times more potent than the other three possible geometric isomers. Disparlure, the sex attractant of the female gypsy moth, is biologically active only if the alkyl groups bonded to the three-membered ring are in a *cis* configuration.

bombykol

disparlure

Geometric isomerism is only one type of stereoisomerism. Another type of stereoisomerism is the result of the minor image relationships between molecules, the subject of this chapter. These molecules differ in configuration about an sp^3-hybridized, "tetrahedral carbon" atom bearing four different groups of atoms, which is called a **stereogenic center**. This phenomenon is not as easily visualized as geometric isomers, but its consequences are even more vital to life processes.

6.2 MIRROR IMAGES AND CHIRALITY

The fact that we live in a three-dimensional world has important personal consequences. In the simple act of looking into a mirror, you see someone who does not actually exist—namely, your mirror image. Every object has a mirror image, but this reflected image need not be identical to the actual object. Let's consider a few common three-dimensional objects. A simple wooden chair looks exactly like its mirror image (Figure 6.1). When an object and its mirror image exactly match, we say that they are **superimposable**. Superimposable objects can be "placed" on each other so that each three-dimensional feature of one object coincides with an equivalent three-dimensional feature in the mirror image.

Now let's consider some objects and their mirror images that are not identical. These are said to be **nonsuperimposable**. One example is the side-arm chair found in many classrooms. When a chair with a "right-handed arm" is reflected in a mirror, it becomes a chair with a "left-handed arm" (Figure 6.1). We can convince ourselves of this by imagining sitting in the chair or its mirror image, or we could stop by a classroom and do the experiment ourselves.

Principles of Organic Chemistry. http://dx.doi.org/10.1016/B978-0-12-802444-7.00006-9

Figure 6.1 Objects and Their Mirror Images

In (a), the chair and its mirror image are identical. They can be superimposed. In (b), the mirror image, side-arm chairs cannot be superimposed. One chair has a "right-handed" arm, the other has a "left-handed" arm. (These chairs were designed by George Nakashima.)

Our hands are related as nonsuperimposable mirror images. We know that we cannot superimpose our hands, as is deftly shown by the M.C. Escher in the lithograph found at the beginning of this chapter and in Figure 6.2. An object that is not superimposable on its mirror image is **chiral** (Greek *chiron*, hand). Objects such as gloves and shoes also have a "handedness," and they are also chiral. We can determine whether or not an object is chiral without trying to superimpose it on its mirror image. If an object has a *plane of symmetry*, it is not chiral. A plane of symmetry bisects an object so that one half is the mirror image of the other half. For example, a cup has a plane of symmetry that divides it so that one half is the mirror image of the other half. The chair in part (a) of Figure 6.1 it is achiral because it has a plane of symmetry. *The presence or absence of a plane of symmetry tells us whether an object is chiral or achiral.*

Figure 6.2 Chiral Hands

M.C. Escher's "Drawing Hands" show the relationship between a pair of hands. If you lay your hands on a surface such as a piece of paper you will find the same arrangement. (M.C. Escher's "Drawing Hands" © 2014 The M.C. Escher Company-The Netherlands. All rights reserved.)

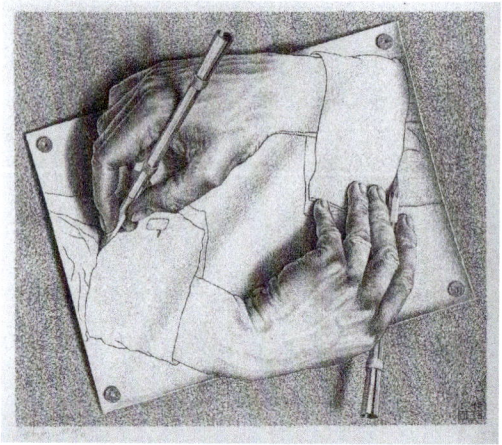

M.C. ESCHER, DRAWING HANDS, 1948

Chiral Molecules

We can extend the concept of chirality from macroscopic objects to molecules. *A molecule is chiral if it contains at least one carbon atom attached to four different atoms or groups.* Such a carbon atom is a **stereogenic** center. A stereogenic center is sometimes called a **chiral center,** and the carbon atom is sometimes called a chiral carbon atom, although it is the molecule that is chiral, not a single carbon atom within it. Most molecules produced in living organisms are chiral, nearly all drugs are chiral, and the synthesis of chiral molecules in the laboratory is a significant part of organic synthesis.

The four atoms or groups at a stereogenic center can be arranged in two ways to give two stereoisomers. The stereoisomers of bromochlorofluoromethane provide an example. Bromochlorofluoromethane does not have a plane of symmetry. Figure 6.3 shows that it can exist as a pair of nonsuperimposable mirror image isomers. Therefore, bromochlorofluoromethane is chiral. In contrast, neither dichloromethane (Figure 6.4) nor bromochloromethane (Figure 6.5) have a plane of symmetry. Neither is chiral.

Figure 6.3
Nonsuperimposable Mirror Image Molecules

Bromochlorofluoromethane does not have a plane of symmetry. Therefore, it is chiral, and it exists as a pair of nonsuperimposable mirror image isomers. (a) Schematic diagram; (b) Ball-and-stick molecular models.

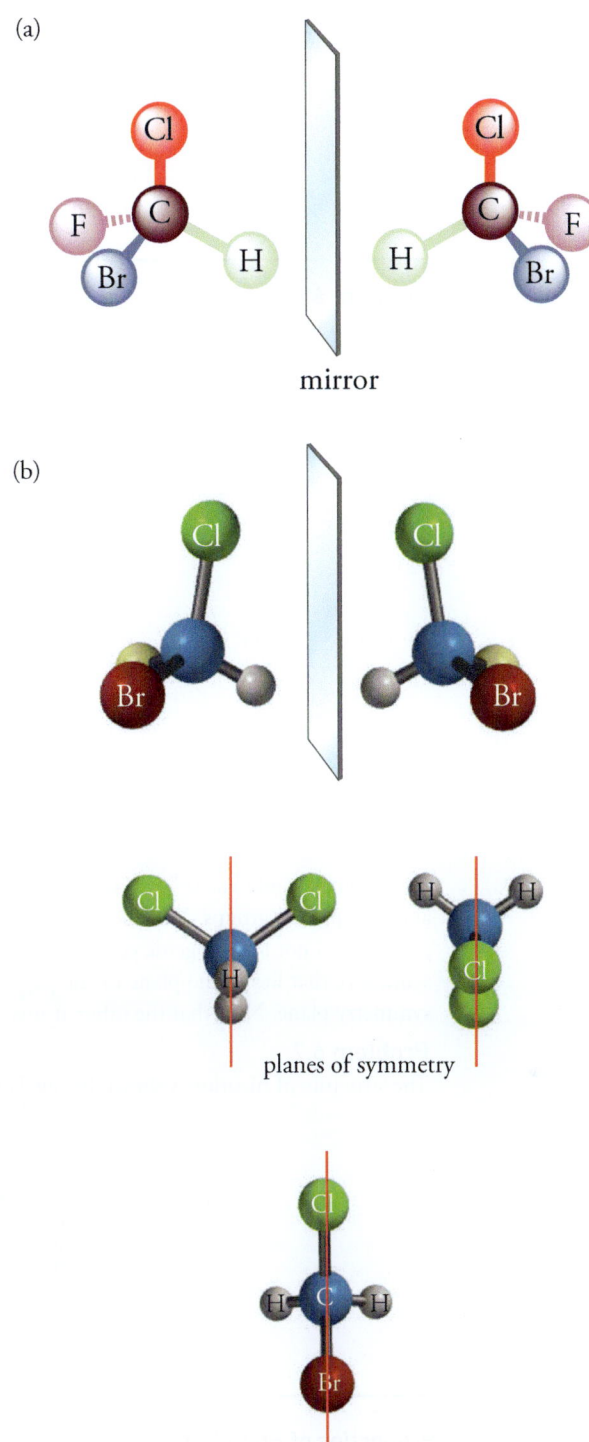

(a)

mirror

(b)

Figure 6.4
Planes of Symmetry in Dichloromethane

Dichloromethane, which has not one, but two planes of symmetry can be superimposed on its mirror image. It is achiral.

planes of symmetry

Figure 6.5
Plane of Symmetry in Bromochloromethane

Bromochloromethane has a plane of symmetry, and therefore it can be superimposed on its mirror image. It is achiral.

plane of symmetry

Mirror Image Isomers Are Enantiomers

Two stereoisomers related as nonsuperimposable mirror images are called **enantiomers** (Greek *enantios*, opposite + *meros*, part). We can tell that a substance is chiral and predict that two enantiomers exist by identifying the substituents on each carbon atom. A carbon atom with four different substituents is a stereogenic center, and a molecule with a stereogenic center is chiral. It can exist as either of a pair of enantiomers. For example, 2-bromobutane is chiral because C-2 is attached to four different groups (CH₃—, CH₃CH₂—, Br—, and H—). In contrast, no carbon in 2-bromopropane is bonded to four different groups; C-2 is bonded to two methyl groups. Thus, 2-bromopropane is not chiral.

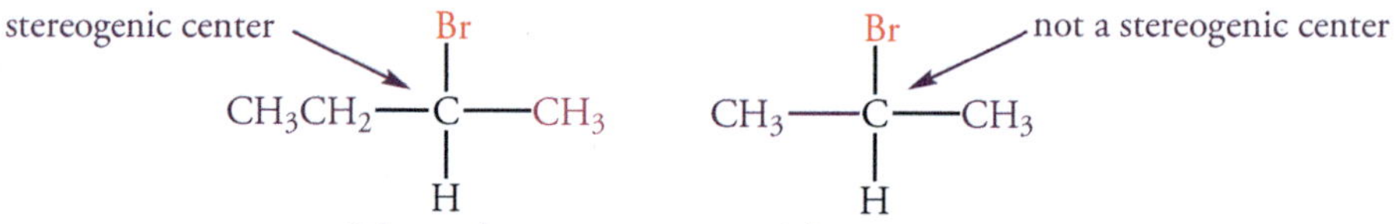

stereogenic center → Br / not a stereogenic center

CH₃CH₂—C—CH₃ ... CH₃—C—CH₃

2-bromobutane
(a chiral molecule)

2-bromopropane
(an achiral molecule)

Problem 6.1

Phenytoin has anticonvulsant activity. Is phenytoin chiral or achiral? Determine your answer by identifying the number of different groups bonded to its tetrahedral carbon atoms; then determine whether or not it has a plane of symmetry.

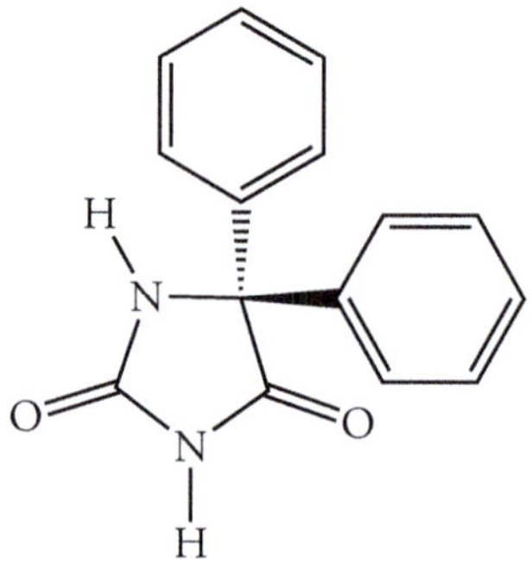

phenytoin

Solution

Phenytoin has only one sp^3-hybridized carbon atom. It is bonded to a nitrogen atom, a carbonyl group, and two phenyl groups. Because the sp^3-hybridized carbon atom is attached to two identical phenyl groups, it is not a stereogenic center, and as a result the molecule is achiral. Phenytoin has a plane of symmetry that lies in the plane of the page. The phenyl groups of phenytoin lie above and below the symmetry plane. Note that the other atoms of phenytoin are bisected by this plane.

Problem 6.2

The structure of nicotine is shown below. Is nicotine chiral?

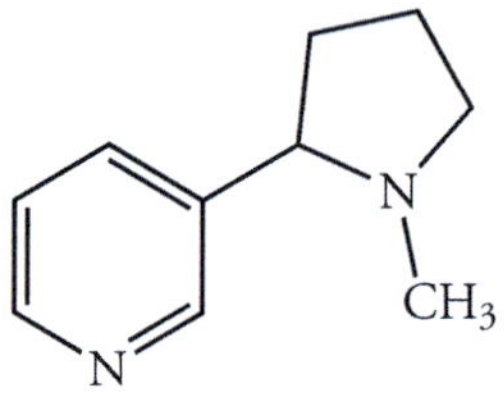

nicotine

Properties of Enantiomers

We can regard hands as analogous to the enantiomers of a chiral molecule. Let's consider the interaction of hands with a symmetrical object such as a pair of tweezers. The tweezers are symmetrical. They can be used equally well with either hand because there is no preferred way to pick up or manipulate a pair of tweezers. However, even if blindfolded, we could easily use our hands to distinguish right- and left-handed gloves. Our hands are "a chiral environment," and in this environment, mirror image gloves do not interact with hands in the same way. The right glove will fit only the right hand. *We can distinguish chiral objects only because we are chiral.*

Pairs of enantiomers have the same physical and chemical properties: they have that same heats of formation, density, melting point, and boiling point. They also have the same chemical properties, and undergo the same reactions in an achiral environment. However, enantiomers can be distinguished in a chiral environment. This difference is important in many processes in living cells. Only one of a pair of enantiomers fits into a specific site in a biological molecule such as an enzyme catalyst because the site on the enzyme that binds the enantiomer is chiral. The binding of this enantiomer is **stereospecific**.

6.3 OPTICAL ACTIVITY

Although enantiomers have identical chemical properties in achiral environments, they differ in one important physical property: Enantiomers behave differently toward plane-polarized light. This difference allows us to distinguish a chiral molecule from its enantiomer in the laboratory.

Plane-Polarized Light

A beam of light consists of electromagnetic waves oscillating in an infinite number of planes at right angles to the direction of propagation of the light. When a light beam passes through a polarizing filter, it is converted to *plane-polarized light* whose electromagnetic waves oscillate in a single plane. We are familiar with this phenomenon in everyday life: Plane-polarized light can be produced by certain sunglasses, which reduce glare by acting as a polarizing filter. They partly block horizontally oscillating light reflecting off the surfaces of various objects. Some camera lenses also have polarizing filters to reduce glare in brightly lit photographs.

Plane-polarized light interacts with chiral molecules. This interaction can be measured by an instrument called a **polarimeter** (Figure 6.6). In a polarimeter, light with a single wavelength—that is, *monochromatic light*—passes through a polarizing filter. The polarized light then traverses a tube containing a solution of the compound to be examined. Plane-polarized light is not affected by achiral molecules. However, the plane of polarized light rotates when it is absorbed by chiral molecules. When the plane-polarized light leaves the sample tube, it passes through a second polarizing filter called an analyzer. The analyzer is rotated in either clockwise or counterclockwise direction to match the rotated polarization plane, so that it passes through the filter with maximum intensity. An angle, α, is read off the analyzer. This angle is called the *observed optical rotation*, α_{obs}. It equals the angle by which the light has been rotated by the chiral compound. Because chiral molecules rotate plane-polarized light, they are **optically active**. Achiral molecules do not rotate plane-polarized light, so they are **optically inactive**.

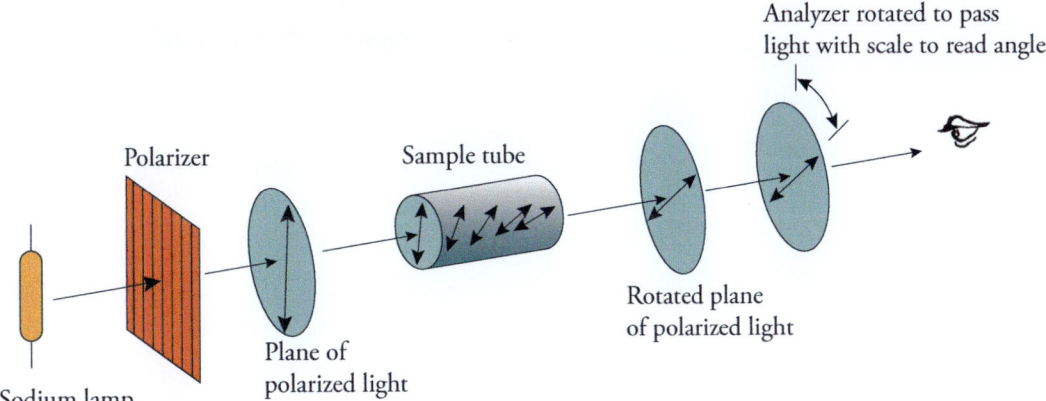

Figure 6.6
Schematic Diagram of a Polarimeter
Plane-polarized light is obtained by passing light through a polarizing filter. Any chiral compound in the sample tube rotates the plane-polarized light. The direction and magnitude of the rotation are determined by rotating the analyzer to allow the light to pass through with maximum brightness. In a modern instrument this is all done electronically, but the basic principle is the same.

Specific Rotation

The amount of rotation observed in a polarimeter depends on the structure of the substance and on its concentration. The optical activity of a pure chiral substance is reported as its *specific rotation*, symbolized by $[\alpha]_D$. It is the number of degrees of rotation of a solution at a concentration measured in g mL^{-1} in a tube 1 dm (10 cm) long. The standard conditions selected for polarimetry measurements are 25 °C, and a wavelength of 589 nm. This yellow light is the D line of a sodium vapor lamp.

$$[\alpha]_D = \frac{\alpha_{obs}}{l \times c}$$

If a chiral substance rotates plane-polarized light to the right—that is, in a positive (+) or clockwise direction—the substance is *dextrorotatory* (Latin *dextra*, right). If a chiral substance rotates plane-polarized light to the left—in a negative (–) or counterclockwise direction—the substance is *levorotatory* (Latin *laevus*, left). The enantiomers of a chiral substance—called dextrorotatory and levorotatory isomers—rotate polarized light the same number of degrees, but in opposite directions. Therefore, they are sometimes called **optical isomers**.

We often refer to an enantiomer by prefixing the sign of the optical rotation at 589 nm to the name of the compound. For example, one of the enantiomers of 2-iodobutane has $[\alpha]_D = -15.15$. It is called (–)-2-iodobutane. The other enantiomer is (+)-2-iodobutane, $[\alpha]_D = +15.15$.

(+)-2-iodobutane   (–)-2-iodobutane

The (+) isomer is sometimes called the *d* form because it is dextrorotatory; the (–) isomer is sometimes called the *l* form because it is levorotatory. Earlier, we encountered levodopa, so named because it is levorotatory. It is also called L-dopa and (–)-dopa. The specific rotation of L-dopa is –13.1°. Table 6.1 lists the specific rotations of some common substances.

Table 6.1
Specific Rotations of
Common Compounds

Compound	$[\alpha]_D$
Azidothymidine (AZT)	+99°
Cefotaxin (a cephalosporin)	+55°
Cholesterol	–31.5°
Cocaine	–16°
Codeine	–136°
Epinephrine (adrenaline)	–5.0°
Levodopa	–13.1°
Monosodium glutamate (MSG)	+25.5°
Morphine	–132°
Oxacillin (a penicillin)	+201°
Progesterone	+172°
Sucrose	+66°
Testosterone	+109°

6.4 FISCHER PROJECTION FORMULAS

Drawing molecules in three dimensions is time consuming. Furthermore, it is not easy to "read" the resulting perspective structural formulas, especially for compounds that contain several chiral centers (Section 6.6). However, the structural formula of a chiral substance can be conveniently drawn as a **Fischer projection**, which was introduced by the German chemist Emil Fischer more than a century ago. The configuration of a chiral substance in a Fischer projection formula is obtained by comparing it to the configuration of a *reference compound* whose common name is glyceraldehyde.

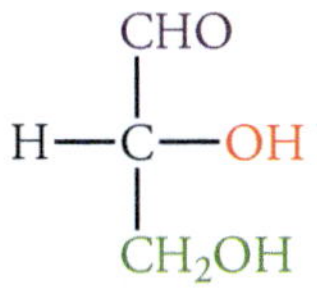

$$\underset{\text{glyceraldehyde}}{\overset{\displaystyle CHO}{\underset{\displaystyle CH_2OH}{H-C-OH}}}$$

Glyceraldehyde contains a carbon atom bonded to four different groups, so it can exist as either of two enantiomers (Figure 6.7). The enantiomers of glyceraldehyde in a Fischer projection are drawn according to the following conventions:

1. Arrange the carbon chain vertically with the most oxidized group (CHO in glyceraldehyde) at the "top."
2. Place the carbon atom at the chiral center in the plane of the paper. It is C-2 in glyceraldehyde.
3. Because C-2 is bonded to four groups, the CHO group and the CH_2OH group extend behind the plane of the page, and the hydrogen atom and the hydroxyl group extend up and out of the plane.
4. Project these four groups onto a plane. The carbon atom at the chiral center is usually not shown in this convention. It is located at the point where the bond lines cross. The vertical lines project away from the viewer. The horizontal lines project toward the viewer.

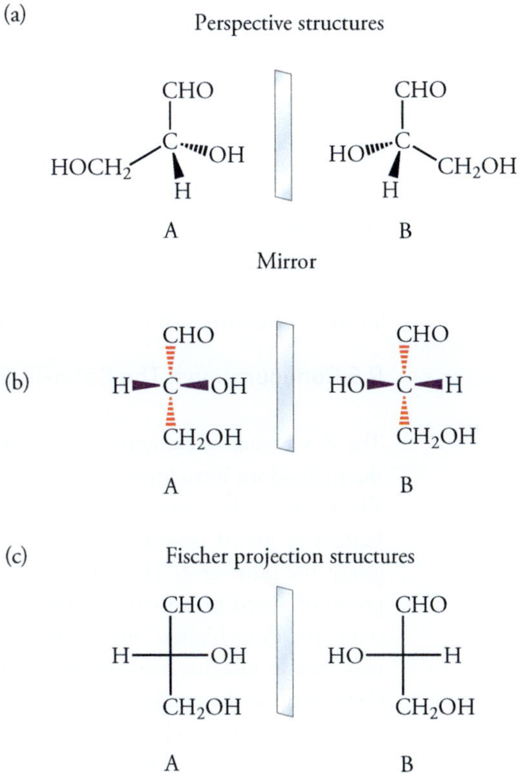

Figure 6.7
Fischer Projection Structures of Glyceraldehyde

(a) Perspective structures of glyceraldehyde. (b) Projection structures. (c) Fisher projection structures of the enantiomers glyceraldehyde. The chiral center is located at the point where the bond lines intersect. The carbon atom is not usually shown. The vertical lines extend away from the viewer, behind the plane of the page; horizontal lines extend toward the viewer, out of the plane of the page, as shown in part (b).

A Fischer projection formula is a two-dimensional representation. It might appear that if we lifted one formula out of the plane and rotated it 180° around the carbon backbone, we would obtain the structure of the enantiomer. However, if this were done for molecule A in Figure 6.7, the carbonyl group and the hydroxymethyl group, originally behind the plane, would be in front of the plane. These groups would not occupy identical positions with respect to the carbonyl group and hydroxymethyl group of molecule B, which are behind the plane. Therefore, to avoid the error of apparently achieving a two-dimensional equivalence of nonequivalent three-dimensional molecules, we *cannot* lift the two-dimensional representations out of the plane of the paper.

Fischer projection formulas can be drawn for any pair of enantiomers. These formulas imply that we know the configuration at the chiral carbon atom. However, the true configuration could not be determined by early chemists because there was no way to determine the arrangement of the atoms in space. Therefore, Fischer arbitrarily assigned a configuration to one member of the enantiomeric pair of glyceraldehydes. The dextrorotatory enantiomer of glyceraldehyde, which rotates plane-polarized light in a clockwise direction (+13.5°), was assigned to the Fischer projection with the hydroxyl group on the right side. Fischer called the compound **D-glyceraldehyde**. The mirror image compound, (–)-glyceraldehyde, corresponds to the structure in which the hydroxyl group is on the left. It rotates plane-polarized light in a counterclockwise direction (–13.5°). Fischer called the compound **L-glyceraldehyde.**

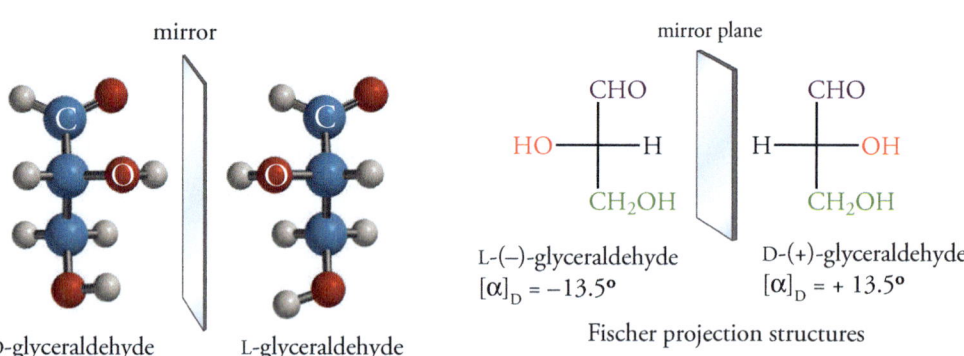

6.5 ABSOLUTE CONFIGURATION

The configuration of the stereogenic center of some molecules, such as amino acids and carbohydrates, can easily be compared to that of D(+)-glyceraldehyde. But this procedure is not easily applied to molecules whose structures differ considerably from the reference compound. To circumvent this difficulty. R.S. Cahn, K.C. Ingold, and V. Prelog established a set of rules that define the **absolute configuration** of any stereogenic center. The configuration is designated by placing the symbol R or S within parentheses in front of the name of the compound.

R,S Configurations: The Cahn-Ingold-Prelog System of Configurational Nomenclature

The R,S system of configurational nomenclature for describing absolute configurations is related to the method we introduced in Section 4.3 to describe the positions of groups in geometric isomers of alkenes. In the R,S system, the four groups bonded to each stereogenic carbon atom are arranged from highest to lowest priority. The highest priority group is assigned the number 1, the lowest priority group the number 4. Then the molecule is oriented so that the bond from the carbon atom to the group of lowest priority is arranged directly along our line of sight (Figure 6.8). When this has been done, the three highest priority groups point toward us and lie on the circumference of a circle. (It may help to imagine holding the lowest priority group in your hand like the stem of a flower as you then examine the petals.) Consider the path taken as we trace the groups ranked 1 to 3. In Figure 6.8 this direction is clockwise. Therefore, the configuration is designated R (Latin, *rectus*, right). If we trace a counterclockwise path from groups ranked 1 to 3, the configuration is designated S (Latin, *sinister*, left).

**Figure 6.8
Cahn-Ingold-Prelog System of Configurational Nomenclature**
Place the lowest priority atom or group away from your eye, and view the chiral site along the axis of the carbon-bond to the lowest priority group. (The diagram of the eye in this figure is from a drawing in the notebooks of Leonardo da Vinci.)

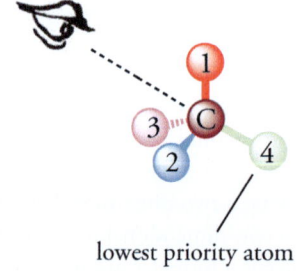

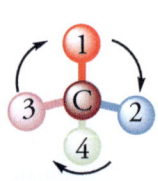

lowest priority atom

Clockwise rotation from 1, to 2, to 3 gives R configuration

Priority Rules

The priority rules we defined in Chapter 4 for describing the configuration of geometric isomers also apply to R,S configurational nomenclature for chiral compounds.

1. *Atoms:* Rank the four atoms bonded to a chiral carbon atom in order of decreasing atomic number; the lower the atomic number, the lower the priority. Isotopes are ranked in order of decreasing mass. For example, ^2H (deuterium) > ^1H.

$$I > Br > Cl > F > O > N > C > {}^2H > {}^1H$$

Highest priority   Lowest priority

2. *Groups of atoms:* If a chiral atom is attached to two or more identical atoms, move down the chain until a difference is encountered. Then apply rule 1. Using this rule, we find that the priority of alkyl groups is

$$(CH_3)_3C\text{—} > (CH_3)_2CH\text{—} > CH_3CH_2\text{—} > CH_3\text{—}$$

3. *Multiple bonds:* If a group contains a double bond, both atoms are doubled. That is, a double bond is counted as two single bonds to each of the atoms of the double bond. The same principle is used for a triple bond. Thus, the order is

$$HC\equiv C\text{—} > CH_2{=}CH\text{—} > CH_3CH_2\text{—}$$

The priority order for common functional groups containing oxygen is

$$\text{—}CO_2H \text{ (carboxylic acid)} > \text{—}CHO \text{ (aldehyde)} > \text{—}CH_2OH \text{ (alcohol)}$$

We can use the R,S system to describe the configuration of the enantiomers of alanine, which has a chiral center bonded to a hydrogen atom, a methyl group, a carboxylic acid group, and an amino group (NH_2). A perspective drawing of the enantiomer of alanine isolated from proteins is shown below. It has an S configuration.

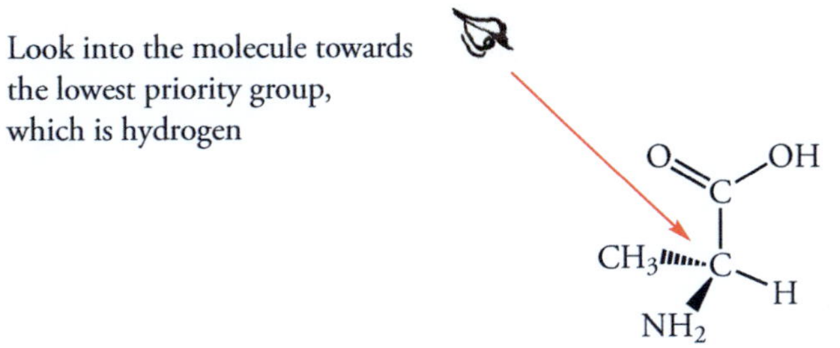

Look into the molecule towards
the lowest priority group,
which is hydrogen

We recall that the direction or magnitude of the optical rotation of a stereoisomer does not determine its absolute configuration. That is, a (+) optical rotation does *not* mean that a molecule has an R configuration. For example, the optical rotation of (S)-2-butanol is clockwise, (+). This isomer is S-(+)-2-butanol.

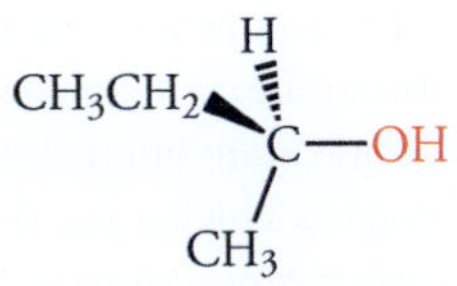

(S)-(+)2-butanol

Problem 6.3

Arrange the groups at the stereogenic center of ethchlorvynol, a sedative-hypnotic, in order from low to high priority.

$$Cl \overset{}{\underset{H}{C}} = CH - \overset{OH}{\underset{CH_2CH_3}{C}} - C \equiv CH$$

Solution

The stereogenic center has three bonds to carbon atoms and one to an oxygen atom. Because oxygen has a higher atomic number than carbon, the —OH group has the highest priority. The priorities of the other three groups are determined by going to the next atom in the chain. In each case the next atom is carbon. However, there are different numbers of carbon-carbon bonds. The ethyl group has the lowest priority, followed by the —CH=CHCl group and the —C≡CH group in order of increasing priority. Thus, the groups around the stereogenic carbon atom increase in priority in the order —CH_2CH_3 < —CH=CHCl < —C≡CH < —OH.

Problem 6.4

Arrange the groups at the stereogenic center of baclofen, an antispastic drug, in order from low to high priority.

baclofen

Problem 6.5

Warfarin is an anticoagulant drug. Warfarin is used both to treat thromboembolic disease and, in larger doses, as a rat poison. Assign its configuration.

Solution

Warfarin contains only one carbon atom that is attached to four different groups. That stereogenic carbon atom is attached to a hydrogen atom, a —C_6H_5 group (a phenyl group), and two other more complex groups. The lowest priority group is the hydrogen atom. The remaining three groups are linked through carbon atoms. One of them, the methylene group at the 12 o'clock position, has the next lowest priority (3) because it is attached to two hydrogen atoms. Next, we assign the priorities of the phenyl group and the ring system to the left. Both groups are attached to the stereogenic carbon atom by a carbon atom with a single and a double bond. Therefore, we must move to the next atom. When we do this in the complex ring, we find a carbon atom bonded to oxygen, which has a higher priority than the carbon atom bonded to hydrogen at the equivalent position in the phenyl ring. Therefore, the complex ring has a higher priority (1) than phenyl (2). Looking into the carbon-hydrogen bond at the stereogenic carbon atom, so that the hydrogen atom points away from us, we trace a counterclockwise path from group 1 to group 2 to group 3. This enantiomer of warfarin has the S configuration.

Problem 6.6

What is the configuration of epinephrine, commonly known as adrenaline?

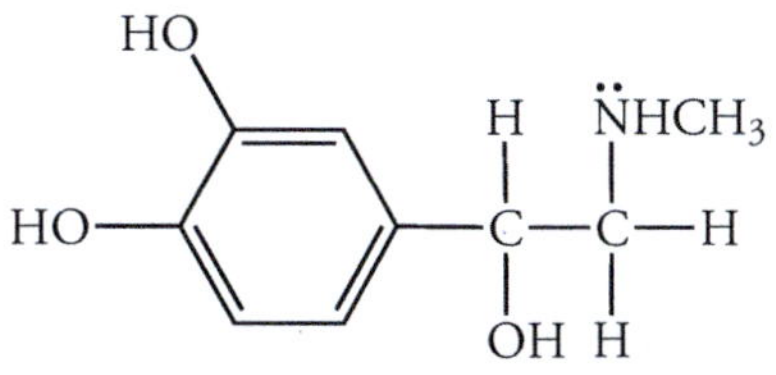

epinephrine

6.6 MOLECULES WITH MULTIPLE STEREOGENIC CENTERS

Many compounds contain several stereogenic centers. For example, the antibiotic erythromycin contains 18 stereogenic centers! (Figure 6.9) How is the number of stereoisomers in a molecule with two or more stereogenic centers related to the number of stereogenic carbon atoms? The chirality of a molecule with two or more stereogenic centers depends on whether the centers are equivalent or *nonequivalent*. The term nonequivalent means that the stereogenic carbon atoms are not bonded to identical sets of substituents.

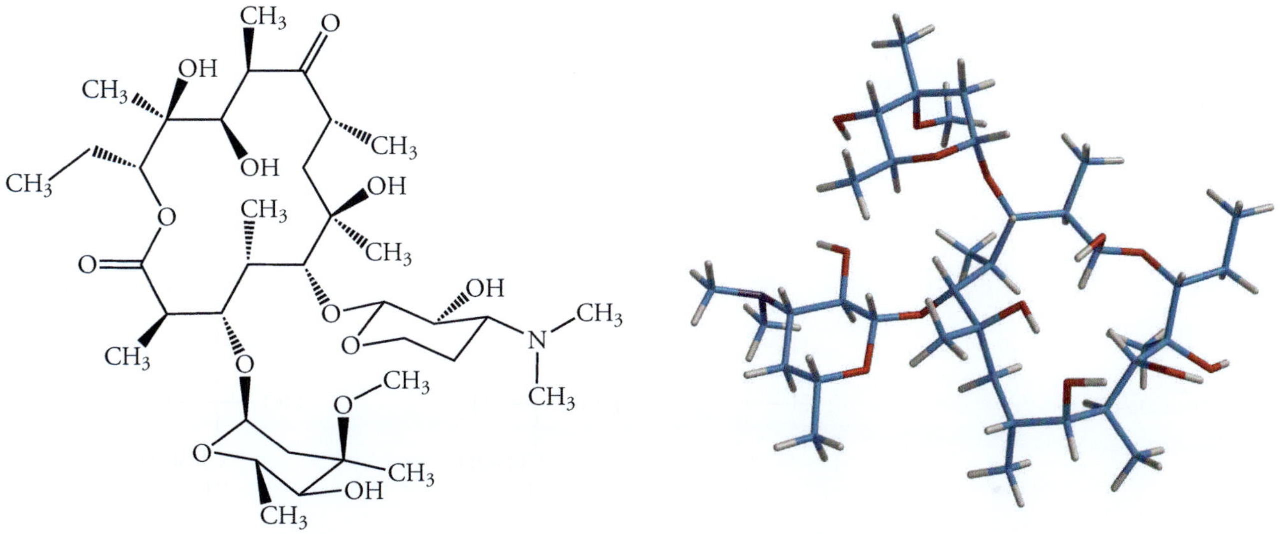

Figure 6.9
Erythromycin—A Chiral Antibiotic
Erythromycin A has 18 chiral centers. Each one is designated with dashed or solid wedge-shaped lines. The hydrogen atoms at the stereogenic centers have been omitted for clarity.

Nonequivalent Stereogenic Centers

If a molecule contains two or more stereogenic centers, and if they are not bonded to identical groups, the stereogenic centers are **nonequivalent**. For n nonequivalent centers, the number of stereoisomers equals 2^n. The following example, 2,3,4-trihydroxybutanal, illustrates the general principle.

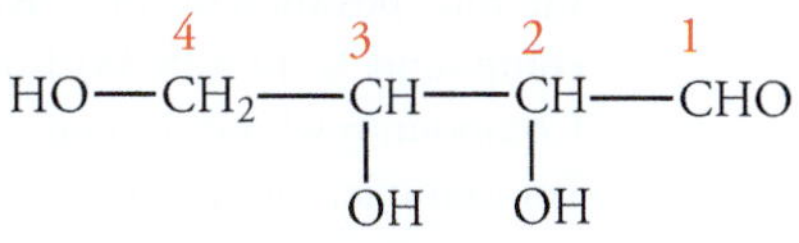

2,3,4-trihydroxybutanal

C-2 and C-3 are chiral. They are nonequivalent because they are not bonded to identical groups. Therefore, the configurations at C-2 and at C-3 can be R or S. Without even drawing the structures, we predict that the four stereoisomers calculated from the 2^n rule can be identified as (2R,3R), (2S,3S), (2R,3S), and (2S,3R). Figure 6.10 shows these configurations in Fischer projection formulas.

Figure 6.10
Enantiomers and Diastereomers

A molecule that contains two nonequivalent chiral centers, such as 2,3-4-trihydroxybutanal, can exist as four stereoisomers. They exist as two pairs of enantiomers. Stereoisomers that are not enantiomers are diastereomers.

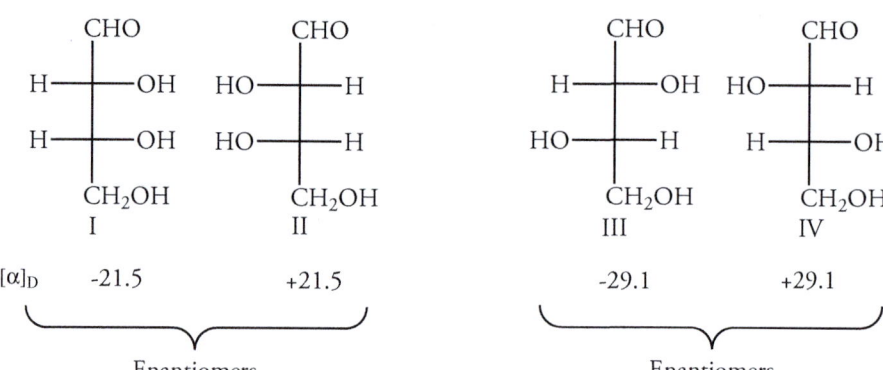

The relationships between the stereoisomeric 2,3,4-trihydroxybutanals are established with mirror planes. Imagine a mirror placed between I and II. Structures I and II are nonsuperimposable mirror images; they are enantiomers. Structures III and IV are also nonsuperimposable mirror images. Like all enantiomers, they rotate plane-polarized light in equal and opposite directions.

Structures I and III are stereoisomers, but they are not enantiomers. *Stereoisomers that are not enantiomers are called* **diastereomers**. The pairs II and III, I and IV, and II and IV are diastereomers. In contrast to enantiomers, which have the same chemical and physical properties, diastereomers have different chemical and physical properties. For example, the enantiomers I and II both are liquids at room temperature and are very soluble in ethanol. The enantiomers III and IV both melt at 130°C and are only slightly soluble in ethanol.

Nomenclature of Diastereomers

The name of a compound with two or more stereogenic centers must indicate the configuration of every center. The configuration of each stereogenic carbon atom is indicated by a number, which corresponds to its position in the carbon chain, and the letter R or S, separated by commas. Figure 6.10 shows the structures of the four stereoisomers of 2,3,4-trihydroxybutanol. Each structure has two stereogenic carbon atoms: C-2 and C-3. Each of these stereogenic carbon atoms can be R or S. Thus, the four possibilities are (2R,3R), (2S.3S), (2S,3R), and (2R,3S), The enantiomer of the (2R,3R) compound is the (2S,3S) isomer, which has the opposite configuration at each stereogenic center Compounds whose configurations differ at only one of the two stereogenic centers are diastereomers. For example, the (2R.3R) compound is diastereomer of the (2S,3R) isomer.

Achiral Diastereomers

Compounds that have two or more equivalent stereogenic centers, but are nevertheless achiral, are called **meso compounds** (Greek, *meso*, middle). Meso compounds are not optically active. Let's consider compounds with two equivalently substituted stereogenic centers, as in the tartaric shown in Figure 6.11. In each structure, the C-2 and C-3 atoms are connected to four different groups. But instead of the four diastereomers that would exist if the stereogenic centers were nonequivalent, only three stereoisomers exist, and one is optically inactive. The compounds labeled (2S,3S) and (2R,3R) are enantiomers. Therefore, they are optically active. But look at the structures labeled (2R,3S) and (2S,3R). Although the structures are drawn as "mirror images," these mirror images are, in fact, superimposable and are identical. To show that this is so, rotate one structure 180° in the plane of the paper: the resulting structure is superimposable on the original structure. Thus the two structures represent the same molecule, which is not optically active. The structures labeled (2R,3S) and (2S,3R) have two equivalent stereogenic carbon atoms. Each of these structures has a plane of symmetry. We recall from Section 6.2 that a structure with a plane of symmetry is achiral and that it is superimposable on it mirror image. In the case of the tartaric acid, the plane of symmetry between the C-2 and C-3 atoms on the top half of the molecule is the mirror image of the bottom half.

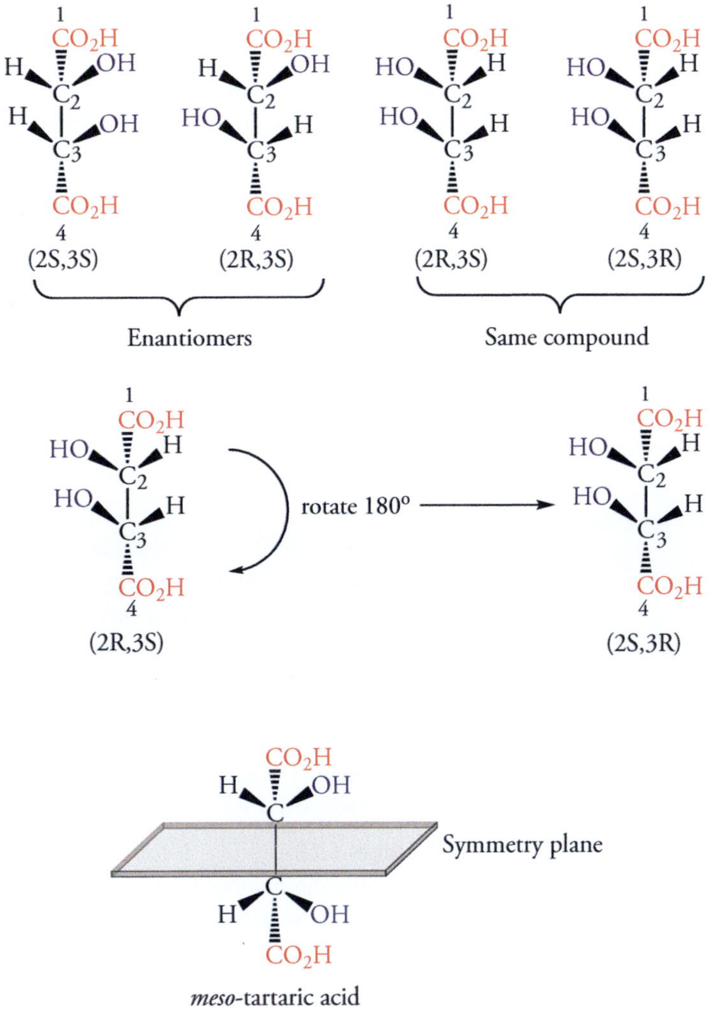

Figure 6.11
Configurations of Optically Active Tartaric Acids and Meso Compounds

Only three stereoisomers exist for tartaric acid because it has two equivalent chiral centers. Two of the stereoisomers are enantiomers. The third has a plane of symmetry, is optically inactive, and is called a *meso* compound; that is, *meso*-tartaric acid.

Problem 6.7
Threonine, an amino acid isolated from proteins, has the following condensed molecular formula. Write the Fischer projections of the possible stereoisomers. What is the configuration at each stereogenic center in each stereoisomer?

$$\underset{4}{CH_3}\underset{3}{CH(OH)}\underset{2}{CH(NH_2)}\underset{1}{CO_2H}$$

Solution

C-2 and C-3 are each bonded to four different substituents. Therefore, threonine has two chiral centers. Because the chiral centers are nonequivalent, four diastereomers are possible. The Fischer projections are written by placing the carboxyl group at the top of the vertical chain. The amino and hydroxyl groups can be on the right or left sides of the projection formula. The structure of threonine isolated from proteins is given by the Fischer projection at the right. Its configuration is 2S,3R.

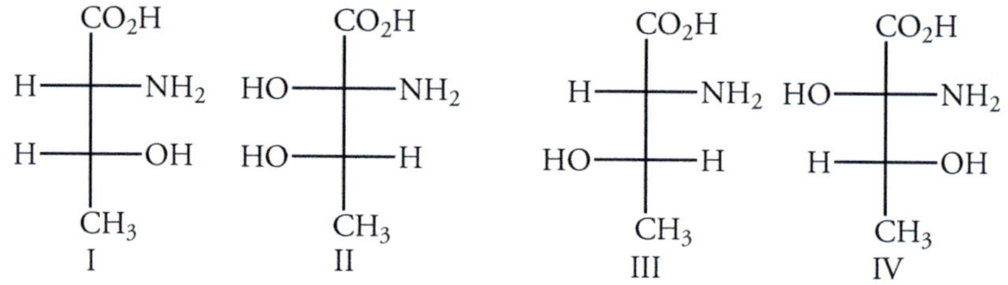

Problem 6.8

Write the Fischer projection formulas of the stereoisomeric 2,3-dibromobutanes. What relationships should exist between the optical activities of these isomers?

Problem 6.9

Determine the number of chiral centers in vitamin K_1. How many stereoisomers are possible?

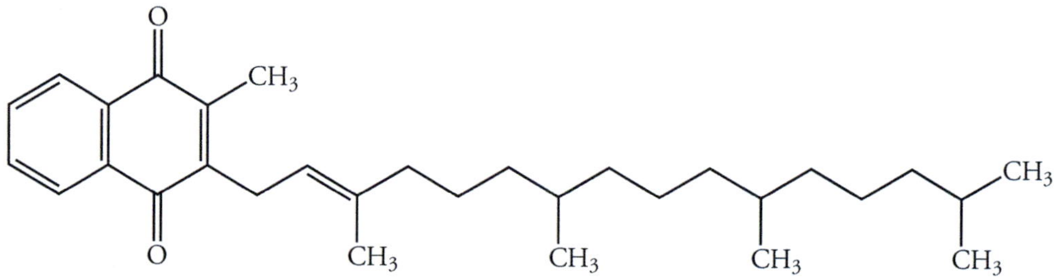

Solution

The carbon atoms in the two rings are not chiral because neither one has a tetrahedral carbon atom. The long alkyl chain contains eight methylene units, none are chiral centers because a carbon atom in a methylene group is bonded to two hydrogen atoms. The tertiary carbon atom near the end of the alkyl chain, which has two methyl groups, is not chiral either.

Next, consider the positions in the middle of the alkyl chain that have methyl group branches. The methyl group on the left is bonded to a double-bonded carbon atom, which does not have four groups bonded to it; therefore it is not chiral. The next two methyl groups are located on chiral centers. Because there are two chiral carbon atoms, $2^2 = 4$ stereoisomers are possible.

Problem 6.10

Determine the number of stereogenic centers in nootkatone, found in grapefruit oil. How many stereoisomers are possible?

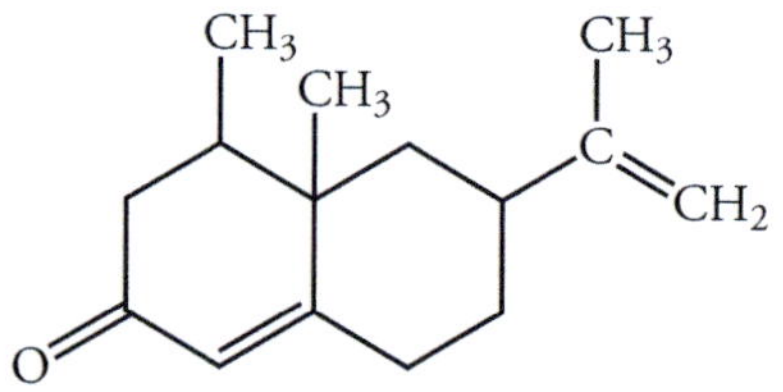

Metabolic Variations Within and Among Species

The metabolism of drugs is often species dependent—a fact that must be considered because drugs are usually tested on animals prior to human trials. Even within the same species there are often strain differences, which are common among inbred test animals such as mice and rabbits. Genetic differences in drug metabolism have been clearly established in humans. The differences between African Americans, Northern Europeans, Eskimos, and Asians must be considered in prescribing certain drugs. For example, the antituberculosis agent isoniazid is metabolized at different rates by individuals with different genetic backgrounds. Eskimos metabolize the drug far faster than Egyptians. Drug metabolism also varies by sex. Some oxidative processes are controlled by sex hormones, particularly the androgens. This factor is important and drugs are tested in men and women. Metabolism in men is more easily studied because of smaller hormonal changes day to day. Also of concern is the possible effect of drugs on a very early fetus before a woman knows she's pregnant.

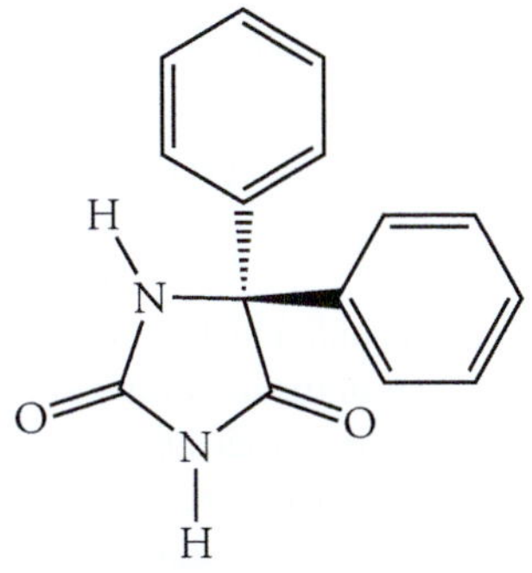

phenytoin

The anticonvulsant phenytoin shows a dramatic difference in metabolism depending on species. This achiral compound is oxidized to a chiral phenol. In humans, the hydroxylation occurs at the para position of one ring, and the compound has the S configuration. In dogs, the hydroxylation occurs at the *meta* position of the other ring, and the compound has the R configuration.

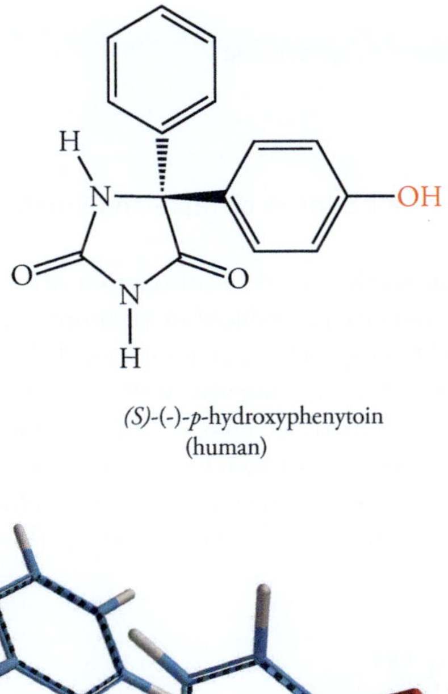

(S)-(-)-*p*-hydroxyphenytoin
(human)

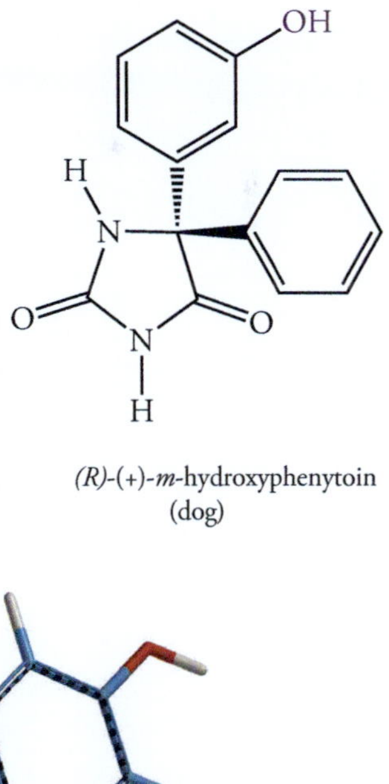

(R)-(+)-*m*-hydroxyphenytoin
(dog)

6.7 SYNTHESIS OF STEREOISOMERS

Reaction of an achiral reactant with an achiral reagent to produce a compound with a stereogenic center gives a 50:50 mixture of enantiomers called a **racemic mixture**. Consider the reduction of pyruvic acid with $NaBH_4$. C-2 atom is a carbonyl carbon atom. The atoms directly bonded to the carbonyl carbon atom are arranged in a trigonal plane. Addition of hydrogen to the carbon atom can occur from either face of the molecular plane. Thus the tetrahedral carbon atom of the lactic acid formed can have two possible configurations.

pyruvic acid   (R)-(+)-lactic acid   (S)-(−)-lactic acid

The individual lactic acid molecules produced are optically active. But a solution containing the products of the reaction is not optically active because the rotation of plane-polarized light by the (R)-lactic acid is canceled by the opposite optical rotation of (S)-lactic acid, which is formed in equal amounts. Note that there is a difference between a racemic mixture and a *meso* compound. A racemic mixture contains optically active components; the *meso* compound is a single achiral substance. If we wish to synthesize a chiral product from an achiral reactant, the reaction must occur in a chiral environment. Protein catalysts called enzymes are examples of chiral reagents. Reduction of pyruvic acid using the liver enzyme lactate dehydrogenase yields exclusively (S)-lactic acid. The reducing agent for the reaction is nicotinamide adenine dinucleotide, NADH.

pyruvic acid   (S)-(−)-lactic acid

Chirality and Our Senses of Taste and Smell

Our senses are sensitive to the configuration of molecules. Both the sense of taste and the sense of smell result from changes induced in a sensory receptor when it binds a specific small molecule (ligand). Ligand binding causes a conformational change that triggers a sequence of events culminating in transmission of a nerve impulse to the brain by sensory neurons. The brain interprets the input from sensory neurons as the "odor" of, say, spearmint.

Diastereomers interact with highly specific sensory receptors. For example, D-mannose, a carbohydrate, exists in two diastereomeric forms that differ in the configuration of a hydroxyl group at one center. The two isomers are designated α and β. The α form tastes sweet, but the β form tastes bitter.

α-D-mannose
(The hydroxy group is axial in the α isomer)

β-D-mannose
(The hydroxy group is equatorial in the β isomer)

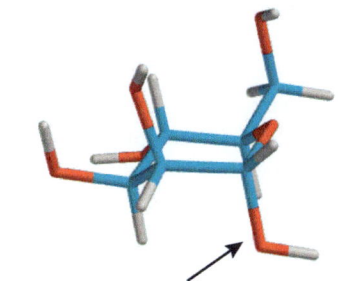

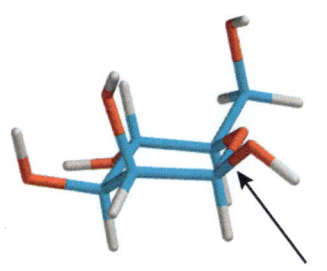

α-D-mannose (axial hydroxyl group)   β-D-mannose (equatorial hydroxyl group)

Sensory receptors also readily distinguish enantiomers. The specificity of response is similar to the relationship between our hands and how they fit into gloves. Because sensory receptors are chiral, they interact stereospecifically with only one of a pair of enantiomers. The two enantiomeric forms of carvone have very different odors. (+)-Carvone is present in spearmint oil, imparting its odor. In contrast, its enantiomer, (–)-carvone, is present in caraway seed. It has the familiar odor associated with rye bread.

R-(–)-carvone   S-(+)-carvone

Problem 6.11
Based on the data for the conversion of (R)-2-bromooctane into (S)-2-octanol using NaOH, predict the product of the reaction of (S)-2-bromooctane with NaOH.

Solution
Nucleophilic attack at the side opposite the bond of the displaced leaving group from (S)-2-brmooc-tane gives a product with inversion of configuration. Thus, the enantiomeric R compound should react likewise, and give an inverted product, (S)-2-octanol.

Problem 6.12
Free radical chlorination of (S)-2-bromobutane yields a mixture of compounds with chlorine substituted at any of the four carbon atoms. Write the structure of the 2-bromo-1-chlorobutane formed. Determine the configuration(s) of the stereogenic center(s). Is the product optically active?

6.8 REACTIONS THAT PRODUCE STEREOGENIC CENTERS

We have studied several reactions that yield products with stereogenic centers from compounds with no stereogenic centers. What prediction can we make about the configuration of the product? The reaction of an achiral radical described previously shows that chiral products cannot form from the reaction of achiral reactants. Molecules with stereogenic centers can form, however, the enantiomers form in equal amounts.

Stereochemistry of Markovnikov Addition to Alkenes

Let's examine the stereochemistry of the addition of HBr to 1-butene to give 2-bromobutane. We know that this is Markovnikov addition. A proton adds to 1-butene at C-1 to give a secondary carbocation. It is achiral because it has a plane of symmetry (Figure 6.12). The carbocation is attacked by the nucleophilic bromide ion with equal probability from the top or bottom side of the planar intermediate. Attack at the top gives the S enantiomer, attack at the bottom gives the R enantiomer, and a racemic mixture results.

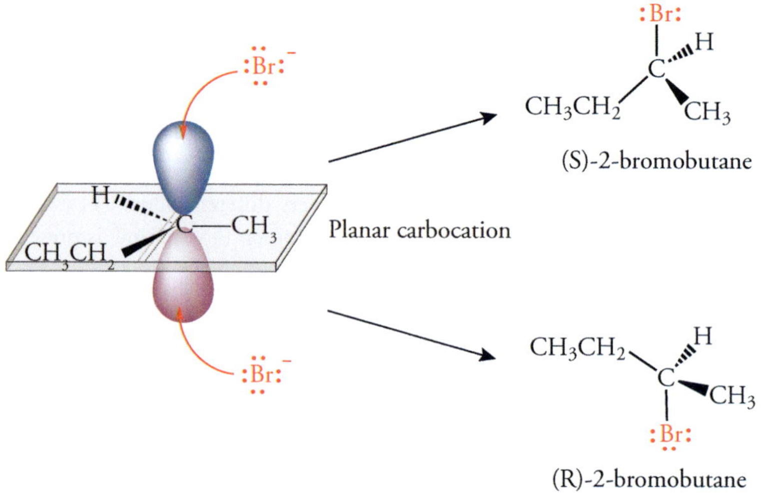

Figure 6.12
Stereochemistry of Markovnikov Addition of HBr to 1-Butene

A proton adds to the double bond of 1-butene to give an intermediate secondary carbocation. It is achiral because it has a plane of symmetry. Bromide ion can attack with equal probability from the top or the bottom to give a racemic mixture.

(S)-2-bromobutane

Planar carbocation

(R)-2-bromobutane

Biochemical processes are catalyzed by enzymes that have multiple stereogenic centers, and are therefore chiral. Enzymes provide a chiral environment in which to form stereogenic centers. Thus, only one enantiomer forms from an enzyme-catalyzed reaction, even if the reactant is achiral. For example, fumaric acid reacts with water in an addition reaction catalyzed by the enzyme fumarase in the citric acid cycle to give only (S)-malic acid. We show the carboxylic acids as their conjugate bases because they are ionized at pH 7. These ionic compounds are called "fumarate" and "malate." This reaction converts fumarate to (S)-malate.

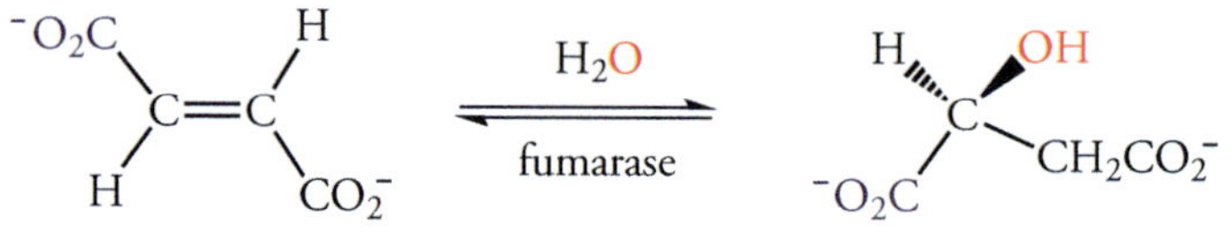

fumarate
(the conjugate base of fumaric acid)

(S)-malate
(the conjugate base of (S)-malic acid)

Only one enantiomer forms in the reaction, and only the *trans* geometric isomer reacts in the presence of fumarase. The *cis* unsaturated isomer is not converted to a hydrated product by fumarase. In fact, it does not bind to the enzyme at all.

Stereochemistry of Alkene Bromination

We recall that the reaction of bromine with an alkene gives a product with bromine atoms on adjacent carbon atoms (Section 6.6). For example, 2-butene reacts with bromine to give 2,3-dibromobutane. Two equivalently substituted stereogenic centers form in this reaction. There are three stereoisomers for such compounds, a pair of enantiomers and a *meso* compound. Which products would we predict based on the reaction mechanism we discussed in Section 4.6? Put another way, how do the observed products support the proposed mechanism of the reaction?

$$CH_3CH{=}CCH_3 \xrightarrow{Br_2} CH_3CHBr{-}BrCCH_3$$

The configuration of the addition product depends on the configuration of the 2-butene, which can be *cis*- or *trans*-, and on the stereochemistry of the *anti* addition reaction that occurs in the second step. Bromine adds to *cis*-2-butene to give a mixture of the enantiomeric (2R,3R)- and (2S,3S)-dibromobutanes (Figure 6.13a). Although the bromonium ion could form by attack equally well on the top or bottom, let's examine the intermediate obtained from attack on the top. (The intermediate obtained from attack on the bottom is the same because it is achiral.) Subsequent attack of bromide ion can occur at either the right or left carbon atom. Attack at the right carbon atom gives the 2R,3R isomer. Attack at the left carbon atom gives the 2S,3S isomer. Both paths of attack are equally probable, and a racemic mixture results.

Now let's consider the consequences of formation of the cyclic bromonium ion derived from *trans*-2-butene followed by nucleophilic attack by bromide ion (Figure 6.13b). The bromonium ion results from attack on the top. Bromide ion attacks equally well at the right and left carbon atoms, giving the 2S,3R and 2R,3S structures, respectively. This pattern corresponds to two equivalently substituted chiral carbon atoms in a molecule with a plane of symmetry; thus, this isomer corresponds to a single *meso* compound.

These results support the mechanism for addition of bromine to alkenes because it agrees with the experimental facts. We have again found that achiral reactants—in this case either *cis*- or *trans*-2-butene and bromine—always form optically inactive products. Remember: the products have two stereogenic centers; the reaction produces either a racemic mixture or a *meso* compound.

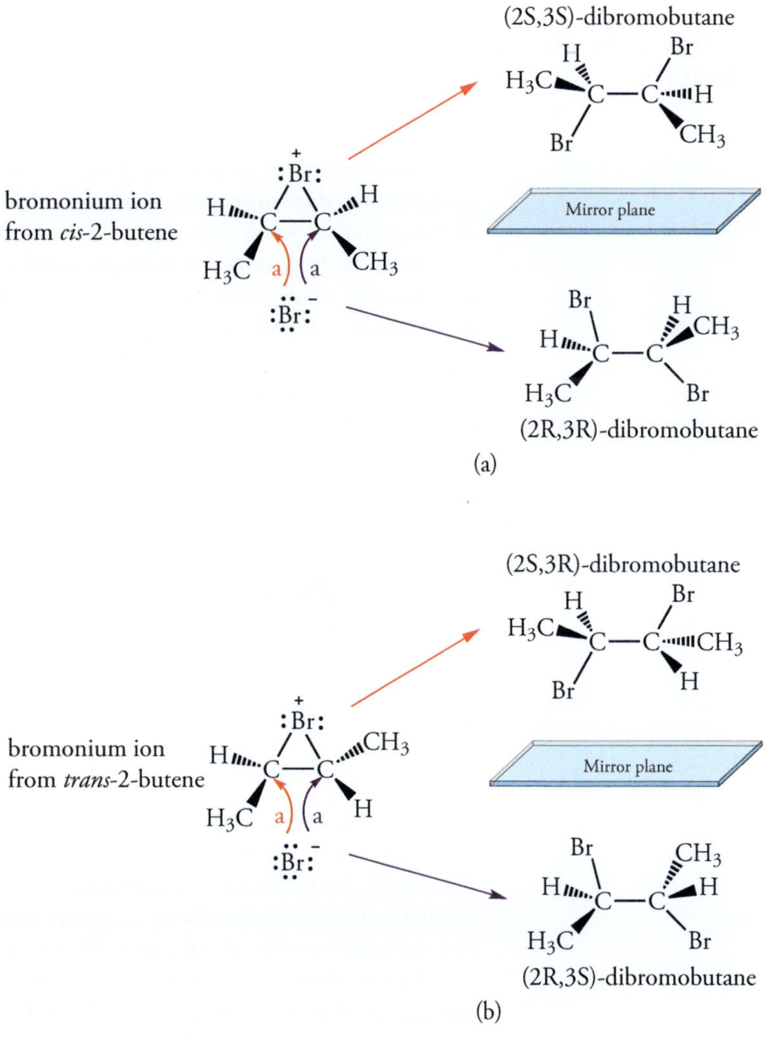

Figure 6.13
Stereochemistry of Bromine Addition to Alkenes

The reaction of bromine with an alkene produces a bromonium ion intermediate. This intermediate reacts with bromide ion in a process that results in net anti addition of bromine. The stereochemical consequences for adding bromine to *cis*-2-butene and *trans*-2-butene are different. *cis*-2-Butene yields a pair of enantiomers; *trans*-2-butene yields a *meso* compound.

Problem 6.13

Sodium borohydride (NaBH$_4$) reacts with the C-2 carbonyl carbon atom of pyruvic acid to give lactic acid. What is the optical rotation of the product(s)?

pyruvic acid → lactic acid

Problem 6.14

Reduction of pyruvic acid by NADH using the liver enzyme lactate dehydrogenase yields exclusively (S)-lactic acid. Write the Fischer projection of this product. Why does only a single product form?

pyruvic acid → (S)-(-)-lactic acid

6.9 REACTIONS THAT FORM DIASTEREOMERS

In the previous section, we discussed the formation of compounds with one or two stereogenic centers from achiral reactants. Now, we'll see what happens when a second stereogenic center forms in a chiral molecule. Diastereomers could result. A molecule with one stereogenic site, designated A$_R$, that forms a second stereogenic site at B within the molecule could give A$_R$B$_R$ and A$_R$B$_S$. We recall that a single enantiomer results when a stereogenic center forms in a molecule in a chiral environment, such as that provided by an enzyme. Similarly, a chiral site in a molecule should affect the stereochemistry of the second site when diastereomers form.

In the hydrogenation of an alkene using a transition metal catalyst, the planar molecule binds to the surface of the metal. If the alkene is achiral, the "side" presented to the surface of the metal is not important. The alkene can be hydrogenated from the "top" or "bottom" to give the hydrogenated product. If the alkene contains a chiral carbon atom near the double bond, however, two products are possible. Consider the catalytic hydrogenation of (R)-2-methylmethylenecyclohexane. Two stereoisomers, 1S,2R and 1R,2R, form, but in unequal amounts. Approximately 70% of the product is the *cis* isomer (1S,2R).

Because the alkene is chiral, there is a difference between the steric environment of the two faces of the double bond. The methyl group above the plane decreases the probability of hydrogenation from that face of the double bond. Hydrogenation from the less hindered bottom side "pushes" the newly formed methyl group up, and the *cis* isomer results. The two stereoisomers form in unequal amounts as a consequence of the chiral center. The reaction is **stereoselective**.

Similar observations show that one enantiomer reacts with an achiral reagent to give unequal amounts of diastereomeric products. The relative yields of the diastereomers often depend on the structure of the existing stereogenic center and its proximity to the newly formed stereogenic center. Many stereogenic centers are present in an enzyme catalyst. They create a chiral environment, which leads to high stereoselectivity. Usually only one diastereomer forms in enzyme-catalyzed reactions.

Problem 6.15

Based on the percent composition of the products for the hydrogenation of 2-methyl-methylene-cyclohexane, predict the product(s) of the hydrogenation of 2-*tert*-butylmethylenecyclohexane.

Solution

The 2-*tert*-butyl group on the "top" of the molecule decreases the probability of hydrogenation from that face. Hydrogenation tends to occur from the less hindered side and "pushes" the newly formed methyl group up. The methyl and *tert*-butyl groups are *cis*. The *cis/trans* ratio is larger than the 70:30 obtained from 2-methylmethylenecyclohexane because the larger *tert*-butyl group hinders attack by hydrogen more than the smaller methyl group.

Problem 6.16

Write the structure of the oxirane (epoxide) that forms when (Z)-2-butene reacts with *m*-chloroperbenzoic acid (mCPBA). Assign the configurations of the stereogenic centers.

Exercises

Chirality

6.1 Which of the following isomeric methylheptanes has a chiral center?

(a) 2-methylheptane   (b) 3-methylheptane   (c) 4-methylheptane

6.2 Which of the following isomeric bromohexanes has a chiral center?

(a) l-bromohexane   (b) 2-bromohexane   (c) 3-bromohexane

6.3 Which of the compounds with molecular formula $C_5H_{11}Cl$ has a chiral center?

6.4 Which of the compounds with molecular formula $C_3H_5Cl_2$ has a chiral center?

6.5 Which of the following isomeric methylheptanes has a chiral center?

(a) 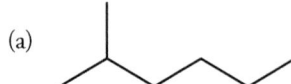   (b) 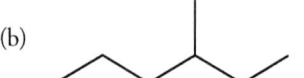   (c)

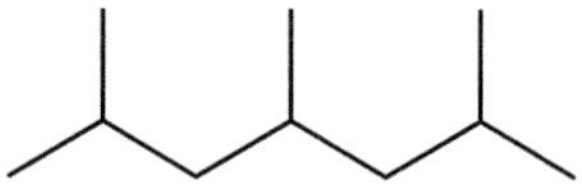

6.6 How many chiral centers does each of the following cyclic compounds have?

(a) 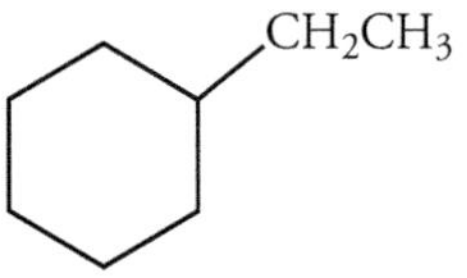   CH₂CH₃   (b)   CH₂CH₃   (c) 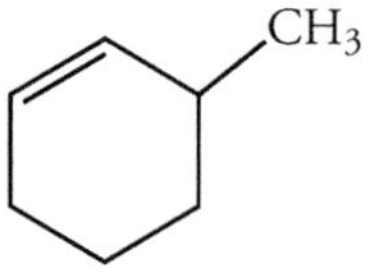   CH₃

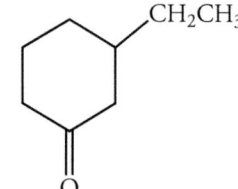

6.7 How many chiral centers does each of the following barbiturates have?

(a)   phenobarbital   (b)   secobarbital 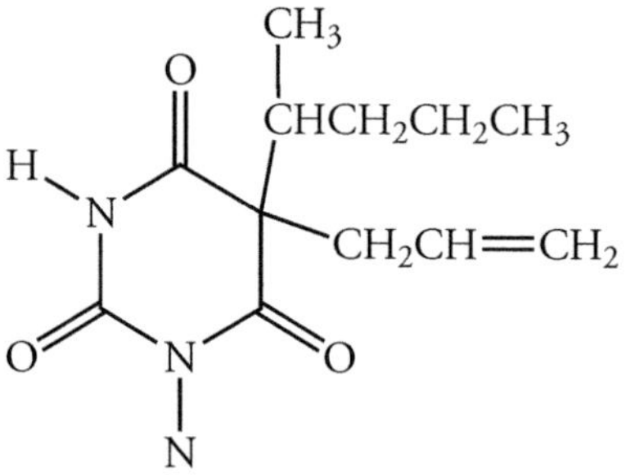   (c)   hexobarbital

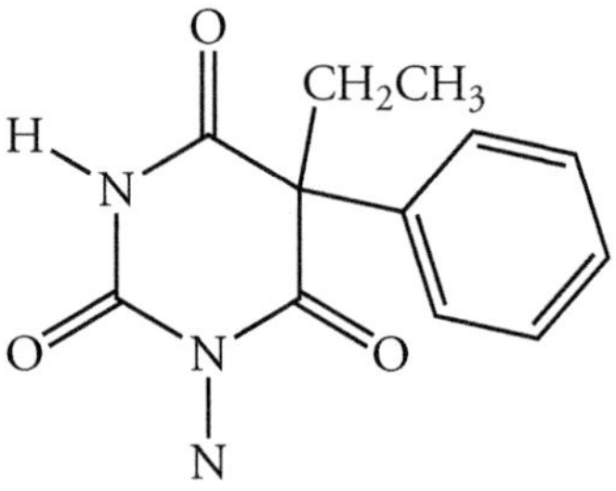

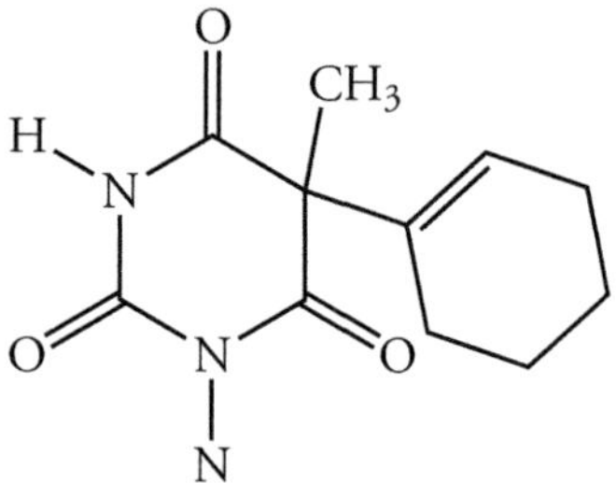

6.8 How many chiral centers does each of the following drugs have?

(a)   ibuprofen, an analgesic   (b)   chloramphenicol, an antibiotic

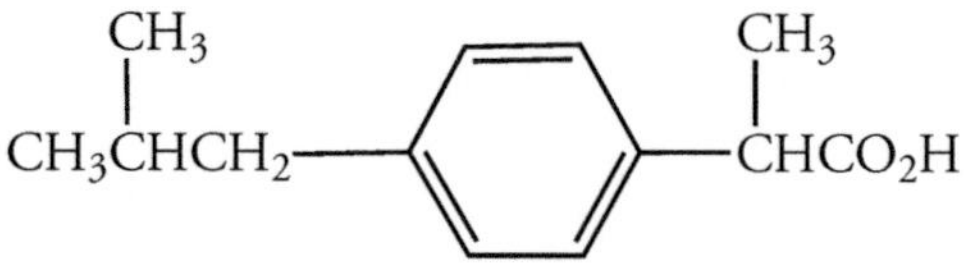

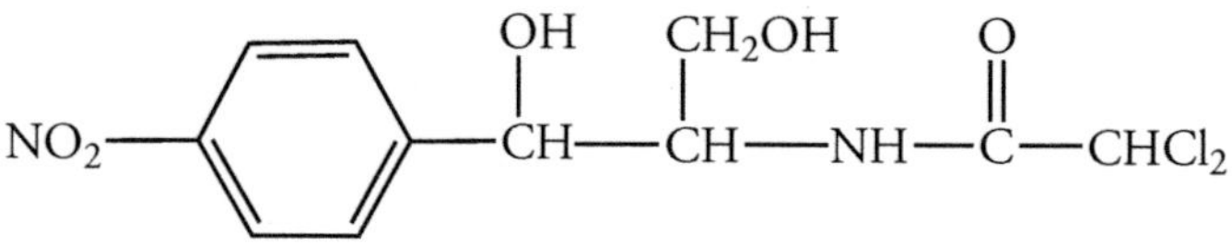

6.9 Arrange the groups in each of the following sets in order of increasing priority:

(a) —OH, —SH, —SCH₃, —OCH₃   (b) —CH₂Br, —CH₂Cl, —Cl, —Br

(c) —CH₂—CH=CH₂, —CH₂—O—CH₃, — CH₂—C≡CH, —C≡C—CH₃

(d) —CH₂CH₃, —CH₂OH, —CH₂CH₂Cl, —OCH₃

6.10 Arrange the groups in each of the following sets in order of increasing priority:

(a) 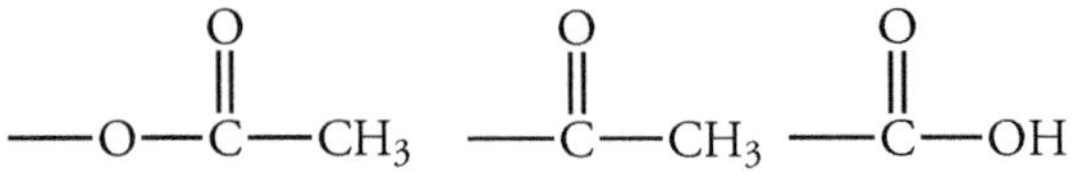 $-O-\overset{\overset{\displaystyle O}{\|}}{C}-CH_3 \qquad -\overset{\overset{\displaystyle O}{\|}}{C}-CH_3 \qquad -\overset{\overset{\displaystyle O}{\|}}{C}-OH$

(b) $-O-\overset{\overset{\displaystyle O}{\|}}{C}-CH_3 \qquad -NH-\overset{\overset{\displaystyle O}{\|}}{C}-CH_3 \qquad -\overset{\overset{\displaystyle O}{\|}}{C}-NH_2$

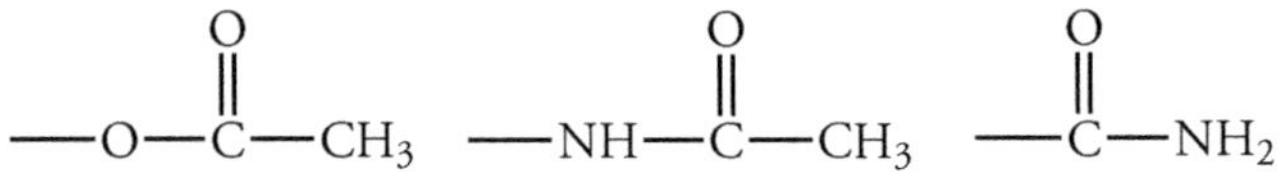

(c) $-S-\overset{\overset{\displaystyle O}{\|}}{C}-CH_3 \qquad -O-\overset{\overset{\displaystyle O}{\|}}{C}-CH_2Br \qquad -\overset{\overset{\displaystyle O}{\|}}{C}-Cl$

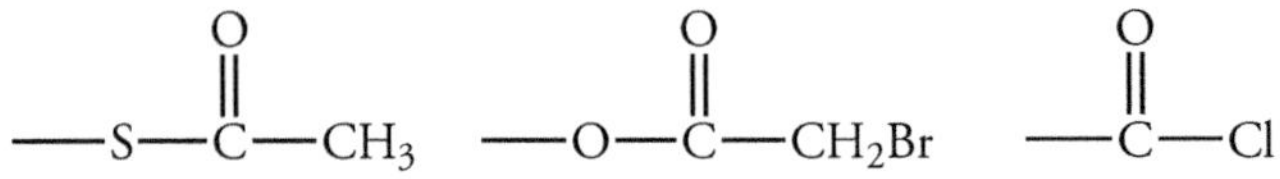

(d)   —C≡H   —C≡N   —N≡C

6.11   Examine the chiral carbon atom in each of the following drugs and arrange the groups from low to high priority:

(a) chlorphenesin carbamate, a muscle relaxant

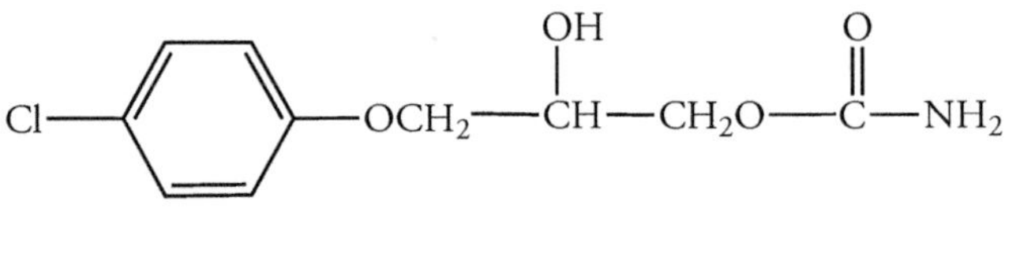

$Cl-\overset{}{\underset{}{\bigcirc}}-OCH_2-\overset{\overset{\displaystyle OH}{|}}{CH}-CH_2O-\overset{\overset{\displaystyle O}{\|}}{C}-NH_2$

(b) mexiletine, an antiarrhythmic

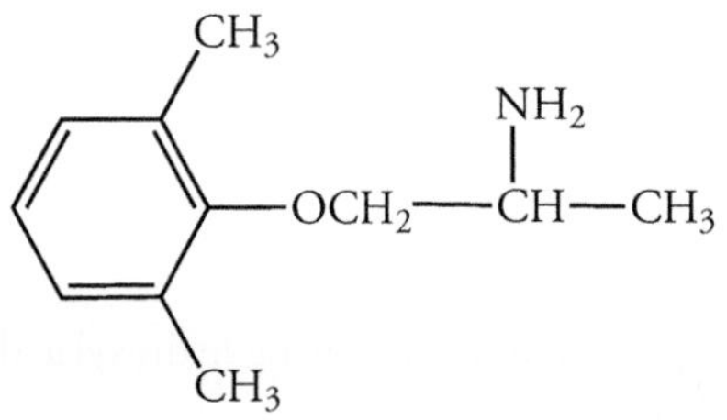

$-OCH_2-\overset{\overset{\displaystyle NH_2}{|}}{CH}-CH_3$

6.12   Examine the chiral carbon atom in each of the following drugs and arrange the groups from low to high priority:

(a) brompheniramine, an antihistamine

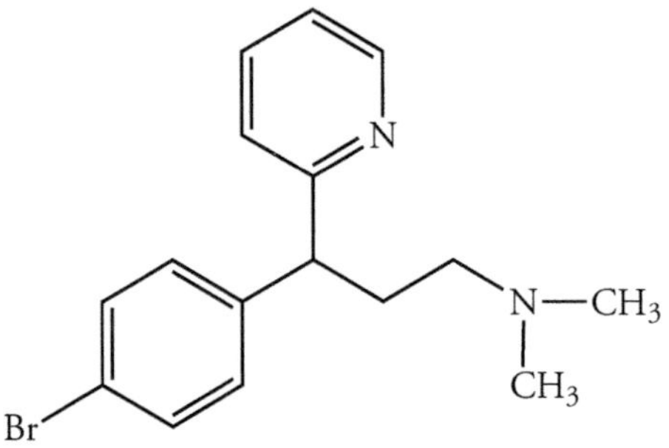

(b) fluoxetine, an antidepressant

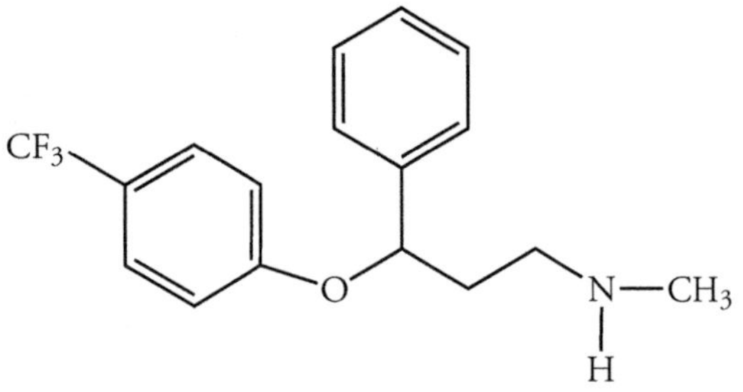

R,S Configuration

6.13 Draw the structure of each of the following compounds:

(a) *(R)-2-chloropentane*   (b) *(R)-3-chloro-1-pentene* (c) *(S)-3-chloro-2-methylpentane*

6.14 Draw the structure of each of the following compounds:

(a) *(S)-2-bromo-2-phenylbutane*   (b) *(S)-3-bromo-1-hexyne* (c) *(R)-2-bromo-2-chlorobutane*

6.15 Assign the configuration of each of the following compounds:

(a)

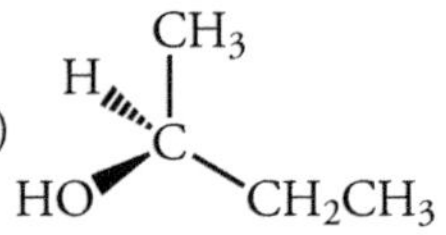

(b)

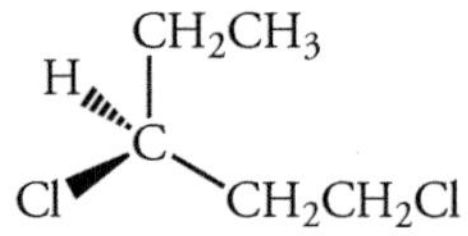

(c)

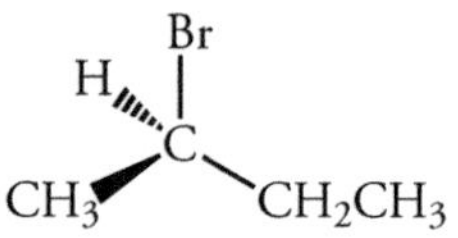

6.16 Assign the configuration of each of the following compounds:

(a)

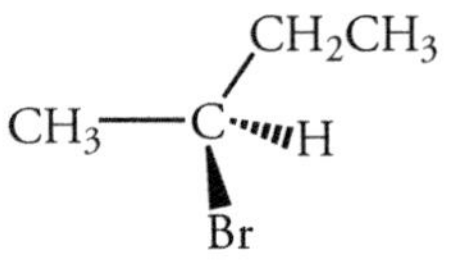

(b)

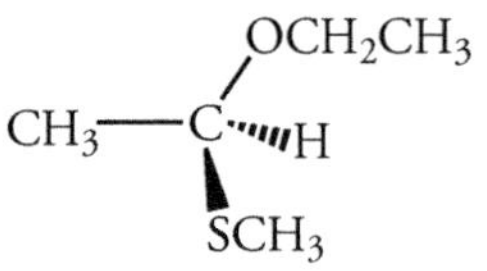

(c)

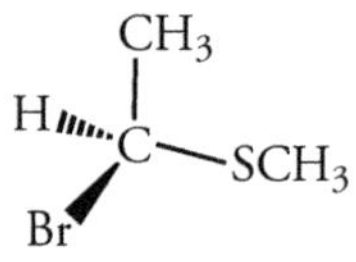

6.17 Assign the configuration of terbutaline, a drug used to treat bronchial asthma.

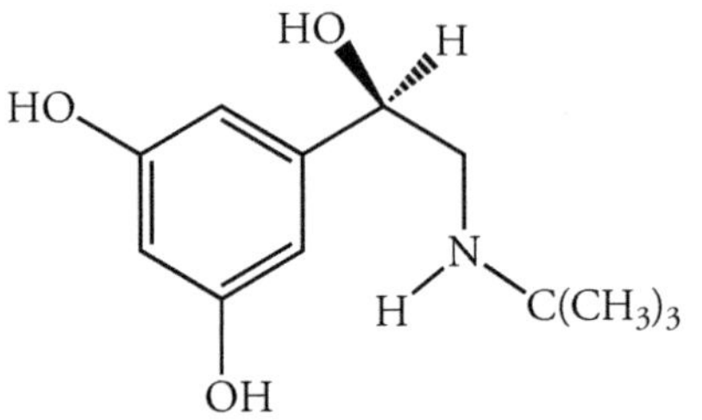

terbutaline

6.18 Assign the configuration of the following hydroxylated metabolite of diazepam, a sedative:

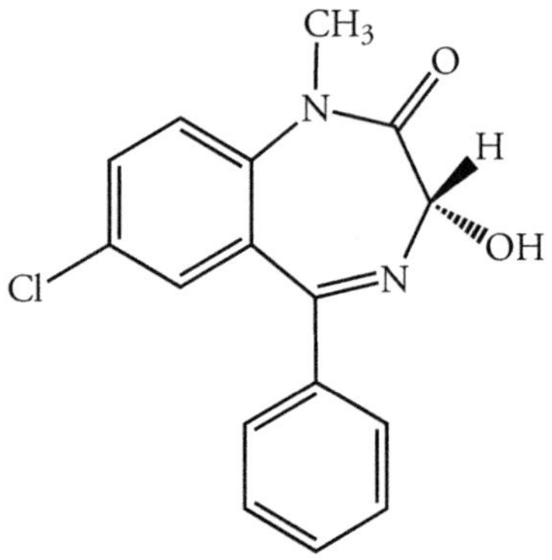

Optical Activity

6.19 The naturally occurring form of glucose has a specific rotation of +53. What is the specific rotation of its enantiomer?

6.20 The naturally occurring form of the amino acid threonine has a specific rotation of +26.3. What is the specific rotation of its enantiomer?

6.21 What do the various prefixes in *(R)*-(+)-glyceraldehyde and *(S)*-(–)-lactic acid mean?

6.22 *(R)*-(–)-lactic acid is converted into a methyl ester when it reacts with methanol. What is the configuration of the ester? Can you predict its sign of rotation?

6.23   Consider the following four projection formulas. Determine the two missing rotations.

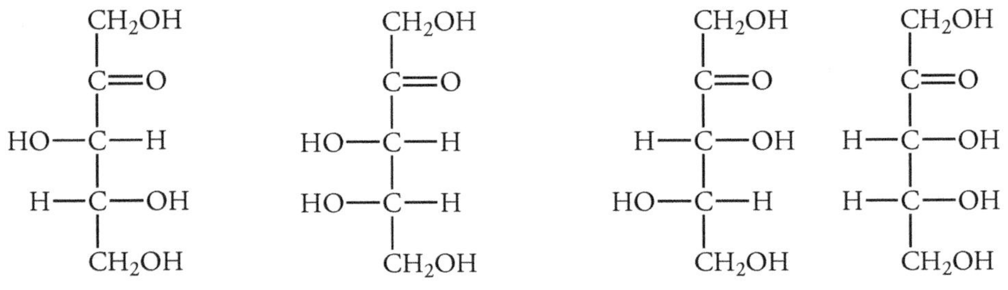

$$
\begin{array}{cccc}
\text{CH}_2\text{OH} & \text{CH}_2\text{OH} & \text{CH}_2\text{OH} & \text{CH}_2\text{OH} \\
| & | & | & | \\
\text{C}{=}\text{O} & \text{C}{=}\text{O} & \text{C}{=}\text{O} & \text{C}{=}\text{O} \\
| & | & | & | \\
\text{HO}{-}\text{C}{-}\text{H} & \text{HO}{-}\text{C}{-}\text{H} & \text{H}{-}\text{C}{-}\text{OH} & \text{H}{-}\text{C}{-}\text{OH} \\
| & | & | & | \\
\text{H}{-}\text{C}{-}\text{OH} & \text{HO}{-}\text{C}{-}\text{H} & \text{HO}{-}\text{C}{-}\text{H} & \text{H}{-}\text{C}{-}\text{OH} \\
| & | & | & | \\
\text{CH}_2\text{OH} & \text{CH}_2\text{OH} & \text{CH}_2\text{OH} & \text{CH}_2\text{OH}
\end{array}
$$

$[\alpha] = +19.6°$ $[\alpha] = -14.8°$

6.24   What stereochemical relationship exists between any and all pairs of the following structures of carbohydrates?

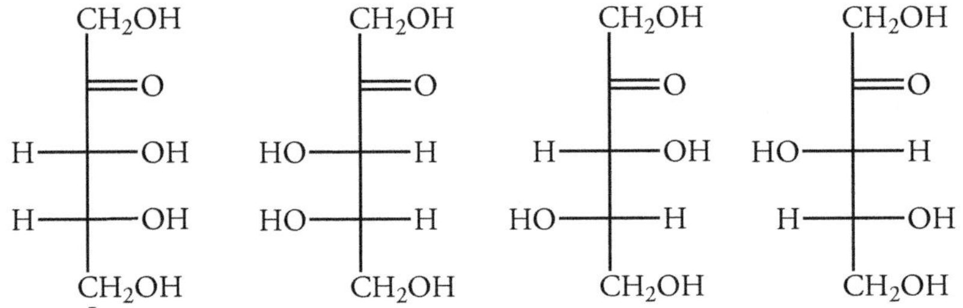

$$
\begin{array}{cccc}
\text{CH}_2\text{OH} & \text{CH}_2\text{OH} & \text{CH}_2\text{OH} & \text{CH}_2\text{OH} \\
\|{=}\text{O} & \|{=}\text{O} & \|{=}\text{O} & \|{=}\text{O} \\
\text{H}{-}\text{OH} & \text{HO}{-}\text{H} & \text{H}{-}\text{OH} & \text{HO}{-}\text{H} \\
\text{H}{-}\text{OH} & \text{HO}{-}\text{H} & \text{HO}{-}\text{H} & \text{H}{-}\text{OH} \\
\text{CH}_2\text{OH} & \text{CH}_2\text{OH} & \text{CH}_2\text{OH} & \text{CH}_2\text{OH} \\
\text{I} & \text{II} & \text{III} & \text{IV}
\end{array}
$$

Diastereomers

6.25 5-Hydroxylysine is an amino acid isolated from collagen. Determine the number of possible stereoisomers.

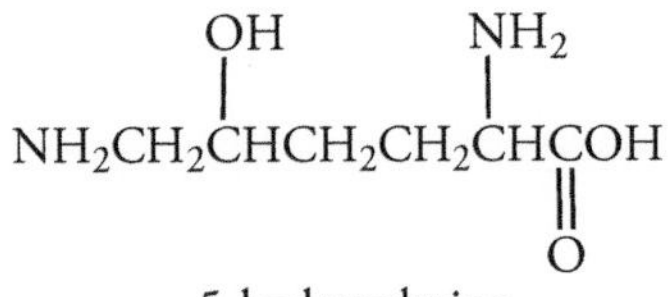

$$
\begin{array}{cc}
\text{OH} & \text{NH}_2 \\
| & | \\
\text{NH}_2\text{CH}_2\text{CHCH}_2\text{CH}_2\text{CHCOH} \\
& \| \\
& \text{O}
\end{array}
$$
5-hydroxylysine

6.26   Consider the structure of the following tripeptide, and determine the number of stereoisomers possible.

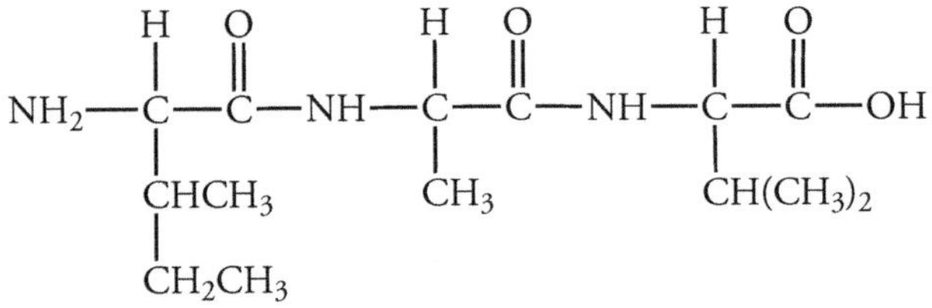

$$
\text{NH}_2{-}\underset{\underset{\underset{\text{CH}_2\text{CH}_3}{|}}{\underset{\text{CHCH}_3}{|}}}{\overset{\overset{\text{H}}{|}}{\text{C}}}{-}\overset{\overset{\text{O}}{\|}}{\text{C}}{-}\text{NH}{-}\underset{\underset{\text{CH}_3}{|}}{\overset{\overset{\text{H}}{|}}{\text{C}}}{-}\overset{\overset{\text{O}}{\|}}{\text{C}}{-}\text{NH}{-}\underset{\underset{\text{CH(CH}_3)_2}{|}}{\overset{\overset{\text{H}}{|}}{\text{C}}}{-}\overset{\overset{\text{O}}{\|}}{\text{C}}{-}\text{OH}
$$

6.27 Ribose is optically active, but ribitol, its reduction product, is optically inactive. Why?

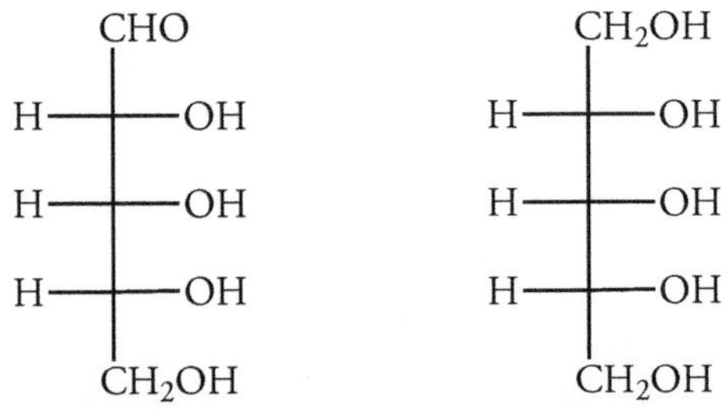

$$
\begin{array}{cc}
\text{CHO} & \text{CH}_2\text{OH} \\
\text{H}{-}\text{OH} & \text{H}{-}\text{OH} \\
\text{H}{-}\text{OH} & \text{H}{-}\text{OH} \\
\text{H}{-}\text{OH} & \text{H}{-}\text{OH} \\
\text{CH}_2\text{OH} & \text{CH}_2\text{OH} \\
\text{ribose} & \text{ribitol}
\end{array}
$$

6.28 There are four isomeric 2,3-dichloropentanes but only three isomeric 2,4-dichloropentanes. Explain why.

Chemical Reactions

6.29   Addition of HBr to 1-butene yields a racemic mixture of 2-bromobutanes. Explain why.

6.30   Reduction of acetophenone with $NaBH_4$ produces a racemic mixture of 1-phenyl-1-ethanols. Explain why.

6.31   How many products are possible when HBr adds to the double bond of *(R)*-3-bromo-1-butene? Which are optically active?

6.32   How many products are possible when HBr adds to the double bond of 4-methylcyclohexene? Which are optically active?

Stereoisomers in Biochemistry

6.33   D-Glucose is a sugar that the body can metabolize. Suggest what would happen if one were to eat its enantiomer.

6.34   The mold *Penicillium glaucum* can metabolize one enantiomer of optically active tartaric acid. Explain what would happen if a racemic mixture of tartaric acid were added to the mold.

6.35   Natural adrenaline is levorotatory. The enantiomer has about 5% of the biological activity of the natural compound. Explain why.

6.36   The following isomer of hydroxycitronellal has the odor of lily of the valley. Its mirror image has a minty odor. Explain why.

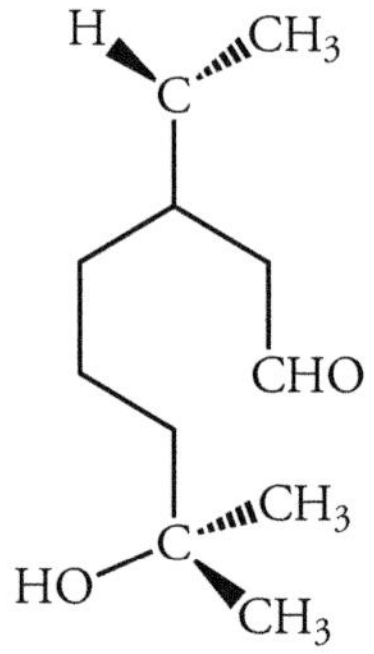

7 NUCLEOPHILIC SUBSTITUTION AND ELIMINATION REACTIONS

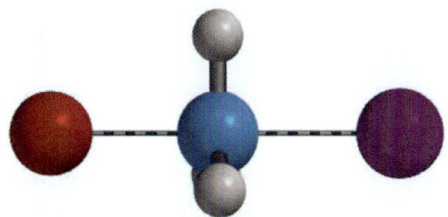

TRANSITION STATE FOR AN S_N2 REACTION

7.1 REACTION MECHANISMS AND HALOALKANES

We introduced the concept of functional groups and their role in the organization of the structures of organic molecules in Section 1.9. We described the importance of reaction mechanisms as an organizational device to classify chemical reactions in Section 2.9. The details of the electrophilic addition reactions of alkenes (Section 4.9) and electrophilic substitution reactions of aromatic compounds (Section 5.5) are examples of two important reaction mechanisms. In this chapter we examine two more types of reactions mechanisms—nucleophilic substitution and elimination reactions. These mechanisms often occur in competition with one another and describe the reactions of several classes of compounds, such as haloalkanes (also called alkyl halides) and alcohols. In this chapter we focus on the substitution and elimination reactions of haloalkanes. These reactions illustrate the role of structure in determining the degree to which a given reaction mechanisms occurs.

Reactivity of Haloalkanes

Haloalkanes have a halogen atom bonded to an sp^3-hybridized carbon atom. As a result of the greater electronegativity of the halogens, the carbon atom of the carbon-halogen bond bears a partial positive charge and the halogen atom has a partial negative charge.

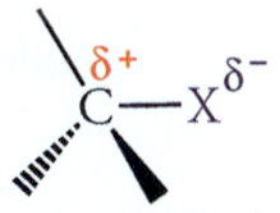

where X = F, Cl, Br, I

Since a carbon-halogen bond is polar, a haloalkane has two sites of reactivity. One is at the carbon atom bonded to the halogen atom. This carbon atom is electropositive and reacts with nucleophiles. The second site of reactivity in a haloalkane is the hydrogen atom bonded to the carbon atom adjacent to the carbon atom bonded to the halogen atom. This hydrogen atom is more acidic than the hydrogen atoms in alkanes because the halogen atom on the adjacent carbon atom withdraws electron density by an inductive effect.

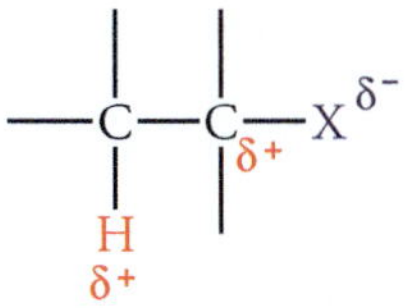

First, let's consider the reaction of the hydroxide ion with a carbon atom bonded to a halogen atom. Hydroxide ion can displace the halide ion in a substitution reaction to produce an alcohol. However, the hydroxide ion is not only a nucleophile, but also a strong base that can remove a proton from the carbon atom adjacent to the one bonded to the halogen atom. Abstraction of a proton and the departure of a halide ion in an elimination reaction give an alkene.

Principles of Organic Chemistry. http://dx.doi.org/10.1016/B978-0-12-802444-7.00007-0

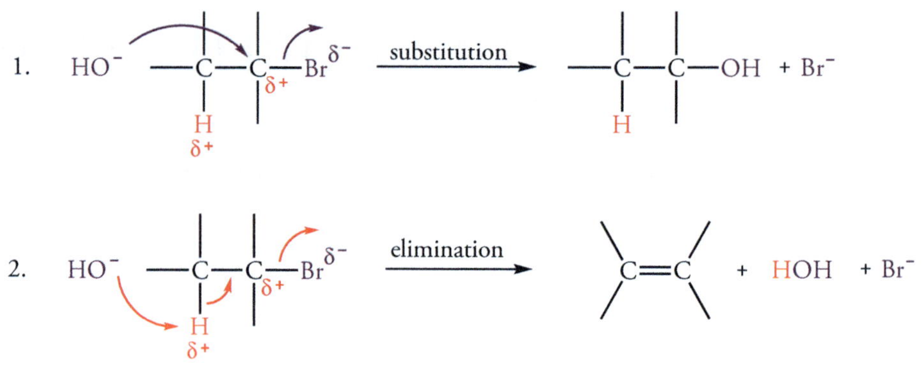

1.
$$HO^- \quad -\overset{|}{\underset{\underset{\delta^+}{H}}{C}}-\overset{|}{\underset{\delta^+}{C}}\!-Br^{\delta^-} \quad \xrightarrow{\text{substitution}} \quad -\overset{|}{\underset{H}{C}}-\overset{|}{C}\!-OH \;+\; Br^-$$

2.
$$HO^- \quad -\overset{|}{C}-\overset{|}{\underset{\delta^+}{C}}\!-Br^{\delta^-} \quad \xrightarrow{\text{elimination}} \quad \overset{}{\underset{}{C}}\!=\!\overset{}{\underset{}{C} \quad +\; HOH \;+\; Br^-}$$

The substitution and elimination reactions usually occur concurrently, and mixtures of products result. In the following sections we will see how one reaction can be favored over the other.

Nomenclature of Haloalkanes

Low molecular weight haloalkanes are often named using the name of the alkyl group followed by the name of the halide.

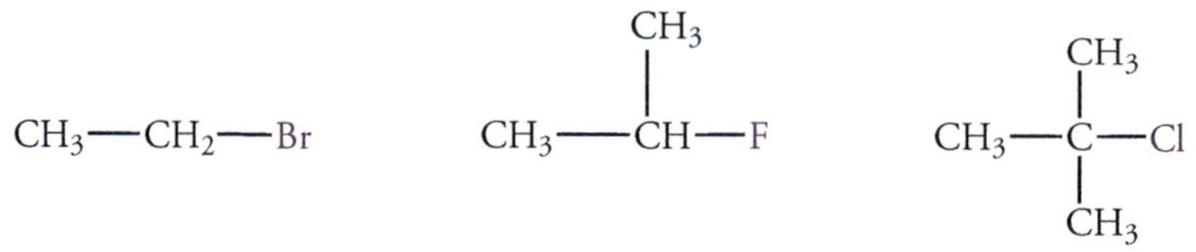

$$CH_3-CH_2-Br \qquad\qquad CH_3-\overset{\overset{\displaystyle CH_3}{|}}{CH}-F \qquad\qquad CH_3-\overset{\overset{\displaystyle CH_3}{|}}{\underset{\underset{\displaystyle CH_3}{|}}{C}}-Cl$$

ethyl bromide   isoproyl fluoride   tert-butyl chloride

Allyl and benzyl halides also have a halogen atom bonded to an sp³-hybridized carbon atom and have reactivities similar to those of haloalkanes.

an allyl halide   a benzyl halide

We described the IUPAC rules for naming haloalkanes in Section 3.10. The halogen atoms have a lower priority than all other functional groups, such as double and triple bonds. Thus, the double bond in 4-chlorocyclohexene is used to number the cyclohexene ring, which then determines the number of the carbon atom bonded to the chlorine atom.

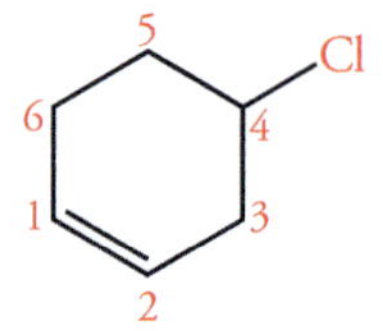

Problem 7.1
(E)-8-Bromo-3,7-dichloro-2,6-dimethyl-1,5-octadiene is produced by a species of red algae. Draw its structure.

Solution

Draw the eight-carbon-atom parent chain, and select a direction for numbering it. Place double bonds between C-1 and C-2 and between the C-5 and C-6.

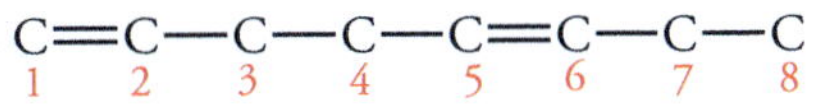

$$\underset{1}{C}=\underset{2}{C}-\underset{3}{C}-\underset{4}{C}-\underset{5}{C}=\underset{6}{C}-\underset{7}{C}-\underset{8}{C}$$

Next, place a bromine atom at C-8, chlorine atoms at C-3 and C-7, and methyl groups at C-2 and C-6.

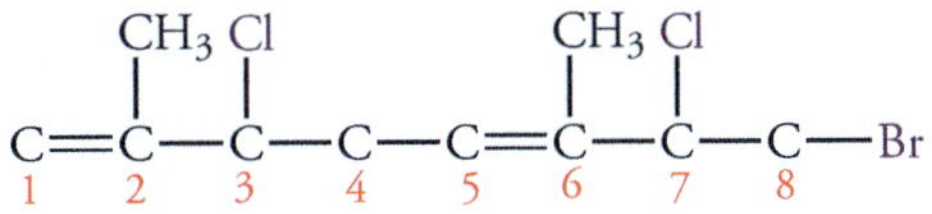

Next, arrange the groups about the double bond between C-5 and C-6 to give the E configuration. The higher priority groups at both C-5 and C-6 are the alkyl groups that are part of the parent carbon chain. Indicate the (E) configuration by placing the atoms of the carbon chain on the opposite sides of the double bond. Then, fill in the requisite hydrogen atoms.

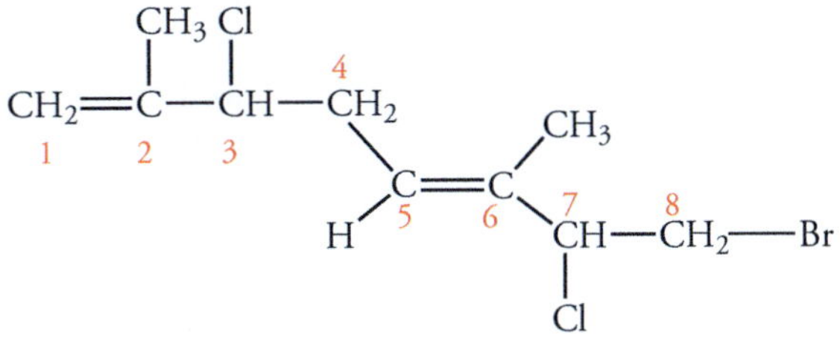

(E)-8-bromo-3,7-dichloro-2,6-dimethyl-1,5-octadiene

Problem 7.2
Name the following compound.

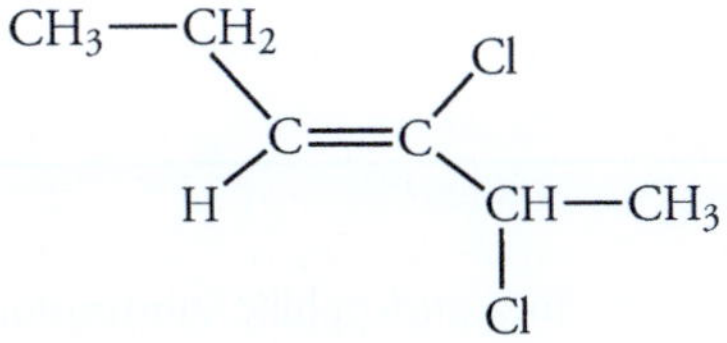

Halogen Compounds in Ocean Organisms

Many pharmaceutical compounds were developed initially from naturally occurring compounds found in terrestrial plants. The chemistry of the organisms living in the ocean has also been examined in the search for new compounds that might be useful in drug design. Given the diversity of species in the ocean, it seems likely that a wealth of potentially useful compounds may be discovered. As in terrestrial plants and animals, most of the heterocyclic compounds in the ocean have five- and six-membered rings. However, in contrast to compounds obtained from terrestrial sources, a high percentage of the heterocyclic compounds isolated from ocean organisms are halogenated. For example, a Bahamian sponge produces several chlorinated indoles or indole-derived compounds, such as batzelline.

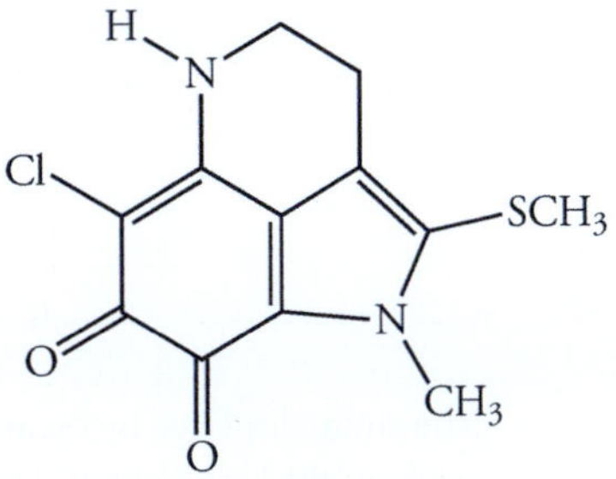

batzelline

Halogen-containing compounds occur widely in other marine organisms, such as algae and mollusks. The compounds made by these species have unusual structures, and some of them are clinically useful as antimicrobial, antifungal, and antitumor agents. Why do various marine species produce these halogen-containing compounds, many of which are cytotoxic?

The answer may be that they are used for defense in an environment where every organism must find a food source. On the coral reef, many invertebrates are limited to a confined space. Organisms that are essentially stationary, or even those that move slowly, apparently use halogen-containing compounds as part of a chemical defense mechanism to avoid predators. Thus, they can better survive in an environment where virtually every organism is simultaneously predator and prey. For example, red algae produce halogen compounds as a chemical shield that repels most herbivores. However, the sea hare, a soft-bodied, shell-less mollusk, is not repelled by compounds from the red algae, which are a source of food for the sea hare.

(from red algae)   (from sea hare)

Most mollusks are protected from predators by a hard shell, so the shell-less sea hare might seem to have little prospect of survival in the face of large carnivores. However, the sea hare converts the halogen-containing compounds in red algae into closely related substances and secretes them in a mucous coating. This coating protects its soft body against carnivorous fish, which are repelled by the compounds.

7.2 NUCLEOPHILIC SUBSTITUTION REACTIONS

In a nucleophilic substitution reaction, the nucleophile donates an electron pair to the electrophilic carbon atom to form a carbon-nucleophile bond. The nucleophile reacts with a haloalkane, which is called the substrate; that is, the compound upon which the reaction occurs. The nucleophile may be either negatively charged, as in the case of OH^-, or neutral, as in the case of NH_3. These two types of nucleophiles are commonly represented as $Nu:^-$ and $Nu:$, respectively. If the nucleophile is negatively charged, the product has no net charge. If the nucleophile is neutral, the product is positively charged.

The group displaced by the nucleophile is called the **leaving group**. It has an electron pair that was originally in the C—X bond. Haloalkanes can react with a halide anion, as in the case of the nucleophilic substitution of bromide for iodide in the substrate iodomethane.

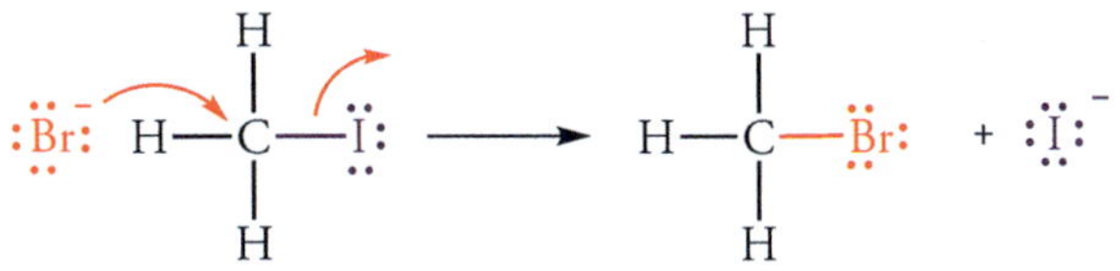

$$\ddot{:}\!\overset{..}{Br}\!:^- \quad H-\overset{\displaystyle H}{\underset{\displaystyle H}{\overset{|}{\underset{|}{C}}}}-\overset{..}{\underset{..}{I}}: \longrightarrow H-\overset{\displaystyle H}{\underset{\displaystyle H}{\overset{|}{\underset{|}{C}}}}-\overset{..}{\underset{..}{Br}}: \quad + \quad :\overset{..}{\underset{..}{I}}:^-$$

A similar reaction occurs when the hydroxide ion replaces a halide ion to produce an alcohol. When the oxygen-containing nucleophile is an alkoxide ion (RO⁻), the product is an ether.

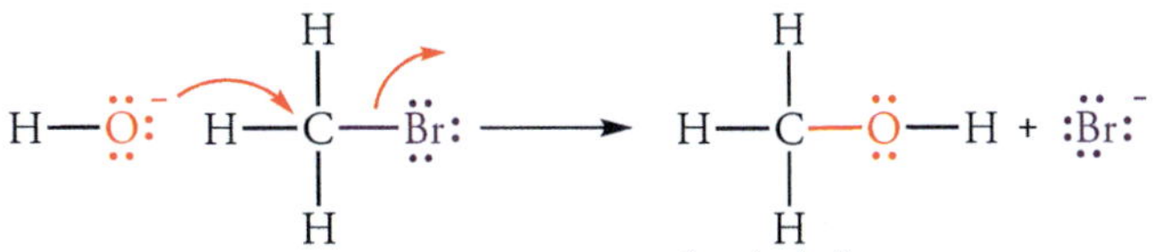

$$H-\overset{..}{\underset{..}{O}}:^- \quad H-\overset{\displaystyle H}{\underset{\displaystyle H}{\overset{|}{\underset{|}{C}}}}-\overset{..}{\underset{..}{Br}}: \longrightarrow H-\overset{\displaystyle H}{\underset{\displaystyle H}{\overset{|}{\underset{|}{C}}}}-\overset{..}{\underset{..}{O}}-H \; + \; :\overset{..}{\underset{..}{Br}}:^-$$

(an alcohol)

$$CH_3CH_2-\overset{..}{\underset{..}{O}}: \quad H-\overset{\displaystyle H}{\underset{\displaystyle H}{\overset{|}{\underset{|}{C}}}}-\overset{..}{\underset{..}{Br}}: \longrightarrow H-\overset{\displaystyle H}{\underset{\displaystyle H}{\overset{|}{\underset{|}{C}}}}-\overset{..}{\underset{..}{O}}-CH_2CH_3 \; + \; :\overset{..}{\underset{..}{Br}}:^-$$

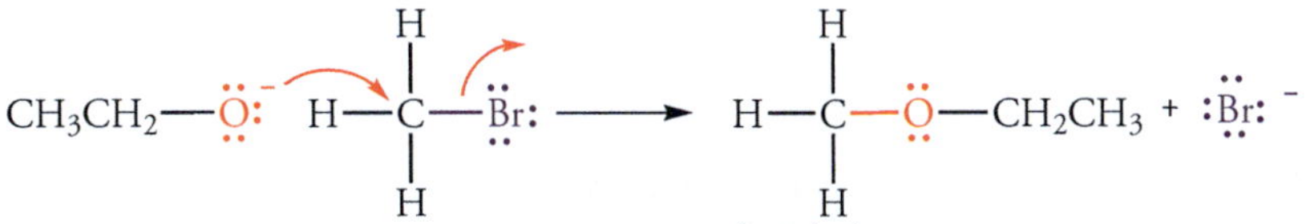

(an ether)

Nucleophilic substitution reactions by sulfur-containing nucleophiles, such as hydrogen sulfide ion, HS⁻, and thiolate ions, RS⁻, also occur. These reactions yield sulfur analogs of alcohols and ethers—namely, thiols and thioethers (Chapter 8).

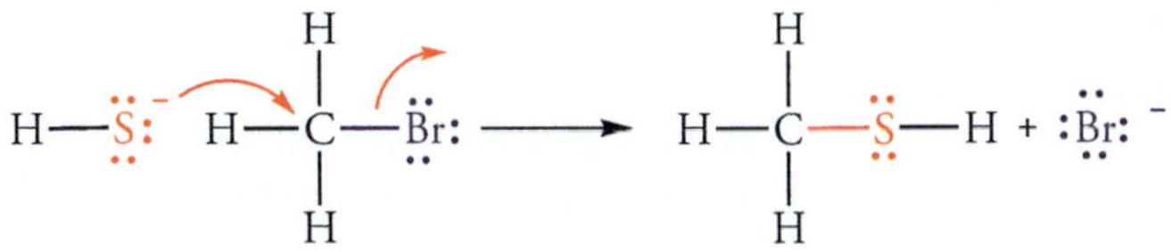

$$H-\overset{..}{\underset{..}{S}}:^- \quad H-\overset{\displaystyle H}{\underset{\displaystyle H}{\overset{|}{\underset{|}{C}}}}-\overset{..}{\underset{..}{Br}}: \longrightarrow H-\overset{\displaystyle H}{\underset{\displaystyle H}{\overset{|}{\underset{|}{C}}}}-\overset{..}{\underset{..}{S}}-H \; + \; :\overset{..}{\underset{..}{Br}}:^-$$

(a thiol)

$$CH_3CH_2-\overset{..}{\underset{..}{S}}: \quad H-\overset{\displaystyle H}{\underset{\displaystyle H}{\overset{|}{\underset{|}{C}}}}-\overset{..}{\underset{..}{Br}}: \longrightarrow H-\overset{\displaystyle H}{\underset{\displaystyle H}{\overset{|}{\underset{|}{C}}}}-\overset{..}{\underset{..}{S}}-CH_2CH_3 \; + \; :\overset{..}{\underset{..}{Br}}:^-$$

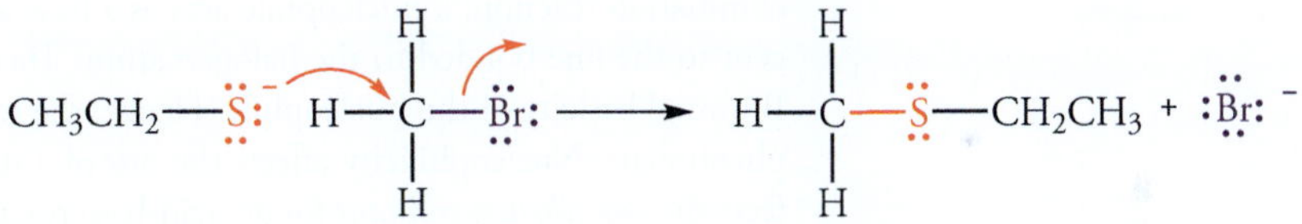

(a thioether)

Haloalkanes also react with carbon-containing nucleophiles to form carbon-carbon bonds, which increase the length of the carbon chain. The cyanide ion, C≡N⁻, is a carbon-containing nucleophile. In the reaction shown below it produces a nitrile with the formula RCN, which extends the carbon chain by one carbon atom. Nitriles can be transformed into carboxylic acids (Chapter 12) and amines (Chapter 14). Carbon-containing nucleophiles derived from alkynes are called **alkynide** ions. These nucleophiles, the conjugate bases of alkynes (Chapter 4), react to form alkynes containing the carbon atoms of both the haloalkane and the alkynide.

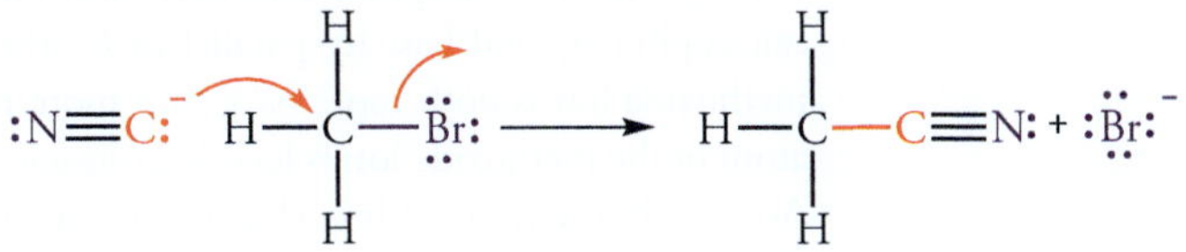

$$:N\!\equiv\!C:^- \quad H-\overset{\displaystyle H}{\underset{\displaystyle H}{\overset{|}{\underset{|}{C}}}}-\overset{..}{\underset{..}{Br}}: \longrightarrow H-\overset{\displaystyle H}{\underset{\displaystyle H}{\overset{|}{\underset{|}{C}}}}-C\!\equiv\!N: \; + \; :\overset{..}{\underset{..}{Br}}:^-$$

(a nitrile)

$$R-C\!\equiv\!C:^- \quad H-\overset{\displaystyle H}{\underset{\displaystyle H}{\overset{|}{\underset{|}{C}}}}-\overset{..}{\underset{..}{Br}}: \longrightarrow H-\overset{\displaystyle H}{\underset{\displaystyle H}{\overset{|}{\underset{|}{C}}}}-C\!\equiv\!C-R \; + \; :\overset{..}{\underset{..}{Br}}:^-$$

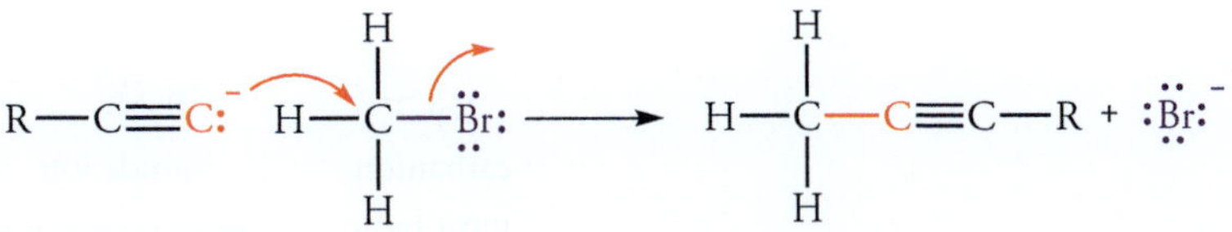

(an alkyne)

Problem 7.3

Using compounds containing no more than three carbon atoms, propose two ways to prepare CH_3CH_2—S—$CH_2CH_2CH_3$

Solution

This compound is a thioether. It can be prepared by reaction of a thiolate with a primary haloalkane. It has two different alkyl groups bonded to sulfur. One alkyl group can be bonded to the sulfur atom in the thiolate, and the other can be in the haloalkane. Thus, two possible combinations of reactants can yield the product.

$$CH_3CH_2—\overset{..}{\underset{..}{S}}:^- \quad + \quad :\overset{..}{\underset{..}{Br}}—CH_2CH_2CH_3 \longrightarrow CH_3CH_2—\overset{..}{\underset{..}{S}}—CH_2CH_2CH_3$$

ethylthiolate 　　　　1-bromopropane

$$CH_3CH_2—\overset{..}{\underset{..}{Br}}: \quad + \quad ^-:\overset{..}{\underset{..}{S}}—CH_2CH_2CH_3 \longrightarrow CH_3CH_2—\overset{..}{\underset{..}{S}}—CH_2CH_2CH_3$$

bromoethane 　　　　n-propylthiolate

Problem 7.4

Propose two ways to prepare 2-pentyne using nucleophilic substitution reactions.

7.3 NUCLEOPHILICITY VERSUS BASICITY

Haloalkanes undergo substitution reactions, in which a halide ion is displaced by a nucleophile, or elimination reactions, in which a halide ion and a hydrogen ion are removed from adjacent carbon atoms to give an alkene. The type of reaction that occurs depends on two properties of the nucleophile, termed nucleophilicity and basicity.

The property of the nucleophile that is a measure of its ability to displace the leaving group is called **nucleophilicity**. The nucleophiles that supply an electron pair to carbon are also bases. In an elimination reaction, a nucleophile acts as a base and abstracts a proton from the carbon atom adjacent to the one bonded to the halogen atom. Therefore, the elimination reaction depends upon the Brønsted basicity of the nucleophile. Hence, the terms nucleophilicity and basicity describe different phenomena. Nucleophilicity affects the *rate* of a substitution reaction at a carbon center. Basicity affects the *equilibrium constant* for an acid-base reaction between the hydrogen atom of the haloalkane and the nucleophile in an elimination reaction.

The nucleophilicities and basicities of a series of structurally or chemically related nucleophiles such as halide ions, oxygen-containing anions, and sulfur-containing anions are not always related in a simple way. However, trends are evident based on periodic properties of the elements. Table 7.1 gives the relative rates of the reaction of various nucleophiles with iodomethane.

Trends in Nucleophilicity Within a Period

For a group of nucleophilic ions derived from elements in the same period of the periodic table, the nucleophilicity and basicity parallel each other and decrease from left to right in the period. Thus, methoxide ion is both more basic and more nucleophilic than fluoride ion (Table 7.1). The oxygen atom of the methoxide ion is less electronegative, and holds its electrons less tightly than fluoride ion. As a result, the nonbonding electrons of the oxygen atom can be more easily donated to carbon in a nucleophilic substitution reaction. The periodic trend of nucleophilicities parallels the basicities for organic anions.

$R_3C:^-$	$R_2\overset{..}{N}:^-$	$R\overset{..}{\underset{..}{O}}:^-$	$:\overset{..}{\underset{..}{F}}:^-$
carbanion	amide ion	alkoxide ion	fluoride ion
most basic	←		least basic
most nucleophilic		→	least nucleophilic

Table 7.1

Relative Rates of Reaction of
Nucleophiles with Iodomethane

Nucleophile	Relative Rate
CH_3OH	1
NO_3^-	30
F^-	5×10^2
SO_4^{-2}	3×10^3
$CH_3CO_2^-$	2×10^4
Cl^-	2.5×10^4
NH_3	3.2×10^5
N_3^-	6×10^5
Br^-	6×10^5
CH_3O^-	2×10^6
I^-	2.5×10^7
CH_3S^-	1×10^9

Trends in Nucleophilicity Within a Group

For a group of nucleophiles containing atoms in the same group of the periodic table, the order of nucleophilicity runs opposite to the order of basicity. First, consider the nucleophilicities of thiolate and alkoxide ions. Thiolate is more nucleophilic but less basic than alkoxide ion (Table 7.1).

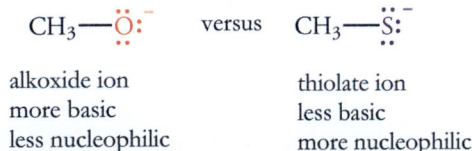

alkoxide ion   thiolate ion
more basic   less basic
less nucleophilic   more nucleophilic

Similarly, when we compare the nucleophilicities and basicities of the halides, we find that the least basic one, iodide, is the most nucleophilic, whereas the most basic one, fluoride, is the least nucleophilic.

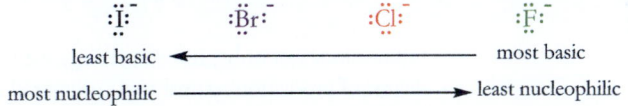

least basic ⟵   most basic
most nucleophilic ⟶   least nucleophilic

The order of nucleophilicities reflects the polarizability of the atoms. We recall that the atomic radii of elements increase going down a group in the periodic table. As a result, the electrons are more polarizable going down a group. Thus, iodide ion is more polarizable and therefore more nucleophilic than bromide ion. The polarizability of a nucleophile is important in a nucleophilic substitution reaction because an electron pair of the nucleophile forms a bond to the electrophilic carbon atom during the reaction. Basicity, which is a measure of the ability to bond to a proton, a center of highly concentrated positive charge, is not as sensitive to polarizability.

Effect of Charge on Nucleophilicity

When a nucleophile can exist either as an anion or as its uncharged conjugate acid, the anion is more nucleophilic (Table 7.1). Thus, alkoxide ions (RO^-) are better nucleophiles than alcohols (ROH), and hydroxide ion is a better nucleophile than water. Similarly, thiolate ions (RS^-) are better nucleophiles than thiols (RSH).

R—Ö:⁻   >   R—ÖH   R—S̈:⁻   >   R—S̈H

alkoxide ion more basic   alcohol less basic   thiolate ion more basic   thiol less basic
more nucleophilic   less nucleophilic   more nucleophilic   less nucleophilic

Biological Substitution Reactions of Sulfur Compounds

Many cellular molecules possess a nucleophilic sulfur atom. Of these, one of the most important is glutathione, which contains a sulfhydryl group (−SH). Glutathione participates in several enzyme-catalyzed reactions in which the sulfhydryl group reacts as a nucleophile with various toxic intermediates that are produced when drugs are metabolized in liver cells.

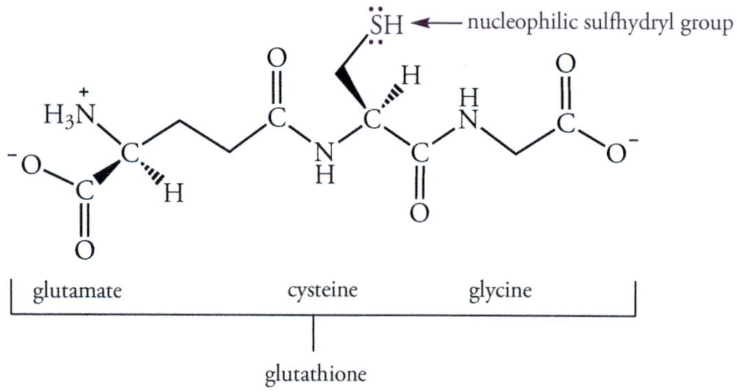

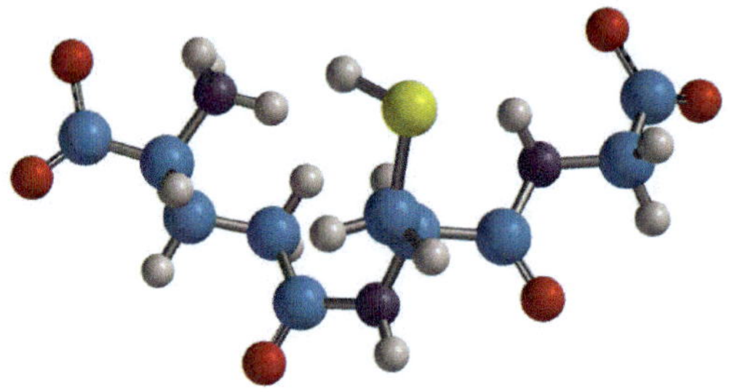

glutathione (the sulfur atom is shown in yellow)

The sulfhydryl group of glutathione, often represented as GSH, displaces substituents bonded to carbon. The various leaving groups of reactive metabolites are each represented as L. They are all strongly electron-withdrawing groups, and they make the carbon atom to which they are bonded partially positive and susceptible to nucleophilic attack by an essential macromolecule with a nucleophilic center (M−Nuc) or by glutathione (GSH). Glutathione protects cells by reacting with toxic metabolites, represented here by R−L, before they react with cellular macromolecules (M−Nu:).

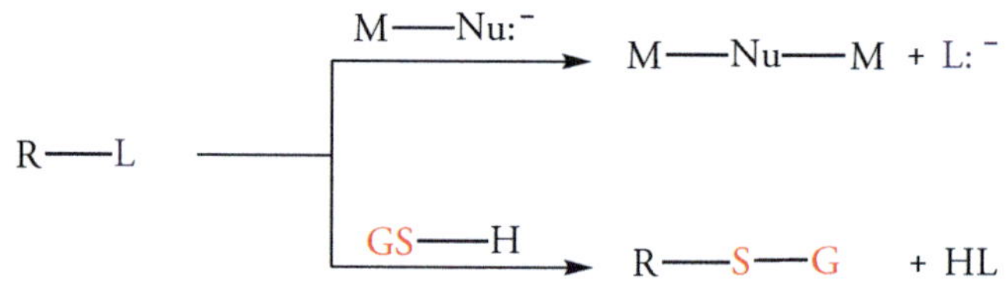

Glutathione not only protects cells against toxic by-products of drug metabolism, but also provides some degree of protection against toxic industrial chemicals, such as benzyl, allyl, and methyl halides. However, long-term exposure to these chemicals eventually overwhelms the protection provided by glutathione and damages the organism.

Sulfur also plays an important role in substitution reactions in all living cells requiring transfer of a methyl group. The sulfur atom in S-adenosylmethionine (SAM) has three carbon atoms bonded to it, forming a positively charged sulfur atom, known as a sulfonium ion.

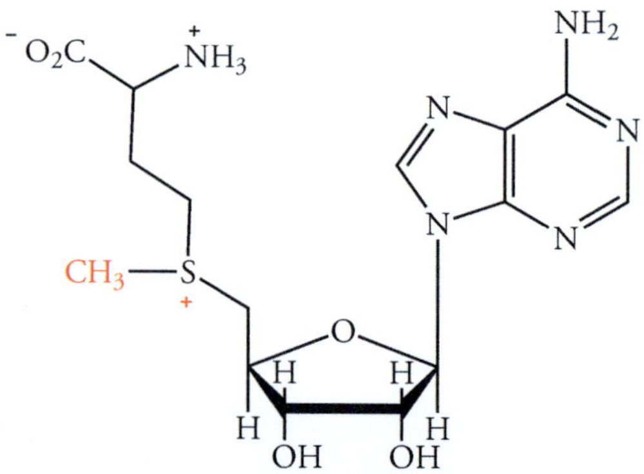

S-adenosylmethionine

The sulfur atom is part of a large leaving group called S-adenosylhomocysteine, which results from transfer of the methyl group to a substrate. This nucleophilic substitution reaction is shown here with a generic nucleophile (Nuc⁻) and an abbreviated representation of SAM.

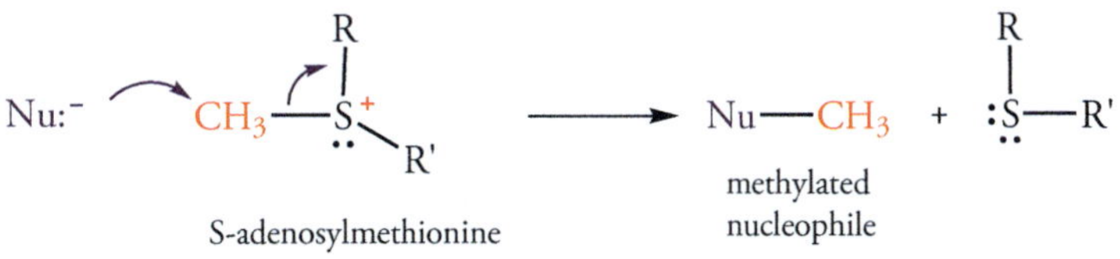

S-adenosylmethionine

methylated
nucleophile

An important example of methyl group transfer from SAM to a nucleophile occurs in the biosynthesis of the neurotransmitter epinephrine. In this reaction, an amino group of norepinephrine attacks the methyl group carbon atom of S-adenosylmethionine in a nucleophilic substitution reaction to produce epinephrine.

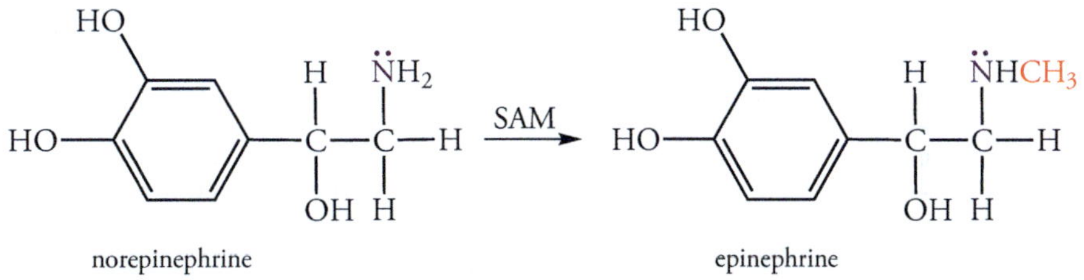

norepinephrine

epinephrine

7.4 MECHANISMS OF SUBSTITUTION REACTIONS

Nucleophilic substitution reactions can occur by either of two mechanisms. These mechanisms are described by the symbols S_N2 and S_N1, where the term S_N means *substitution, nucleophilic*. The numbers 2 and 1 refer to the number of reactants that are present in the transition state for the rate-determining step of the reaction.

The S_N2 Mechanism

In the S_N2 mechanism, a nucleophile attacks the substrate and the leaving group, L, departs simultaneously. The reaction occurs in one step and is therefore **concerted**. The substrate and the nucleophile are both present in the transition state for this step. Because two molecules are present in the transition state, the reaction is **bimolecular**, as indicated by the number 2 in the S_N2 symbol. As a result, the reaction rate depends on the concentrations of both the nucleophile and the substrate. If the substrate concentration is doubled, the reaction rate is doubled. Similarly, if the concentration of the nucleophile is doubled, the rate again doubles. This relationship between the rate and the concentration of the reactants exists because the reactants must collide in the rate-determining step for the reaction to occur. The probability that the nucleophile will collide with the substrate increases if the concentration of either or both species is increased.

Let's consider the S_N2 reaction of the hydroxide ion with chloromethane to give methanol and chloride ion. This reaction is shown with the energy diagram in Figure 7.1. In this plot, the transition state, which occurs at the point of highest energy, contains the hydroxide ion and the substrate. As the reaction proceeds through the transition state, a bond between carbon and the hydroxide ion forms and the bond between carbon and chlorine breaks. The rates of reaction for haloalkanes via the S_N2 mechanism decrease in the order primary > secondary >> tertiary. This trend is observed because alkyl groups block the approach of the nucleophile and slow the rate of the reaction (Figure 7.2). This crowding of alkyl groups around the reactive carbon atom is called **steric hindrance**.

Figure 7.1
Activation Energy and the S$_N$2 Reaction

The reaction of hydroxide ion with chloromethane occurs in a single step.

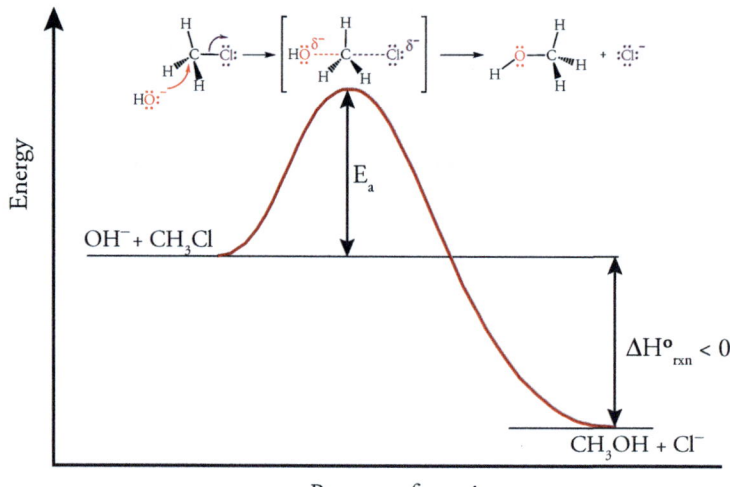

Figure 7.2
Steric Effects in S$_N$2 Reactions

Alkyl substituents decrease the rates of S$_N$2 reactions by interfering with the approach of the nucleophile. In 1-bromo-2,2-dimethylpropane, steric hindrance at the tertiary carbon prevents an S$_N$2 reaction.

(R)-2-bromobutane $\xrightarrow{\text{S}_N2}$ *(S)*-2-butanol

When *(R)*-2-bromobutane reacts with sodium hydroxide, the substitution product is *(S)*-2-butanol. The reaction occurs with **inversion of configuration**. Thus, the nucleophile approaches the electrophilic carbon atom from the back and the leaving group simultaneously departs from the front of the substrate in the S$_N$2 mechanism.

1-bromopropane

1-bromo-2-methylpropane

Steric hindrance by methyl group

1-bromo-2,2-dimethylpropane

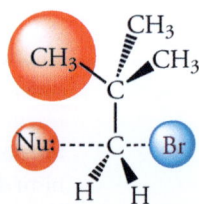

The S$_N$1 Mechanism

A nucleophilic substitution reaction that occurs by an S$_N$1 mechanism proceeds in two steps. In the first step, the bond between the carbon atom and the leaving group breaks to produce a carbocation and, most commonly, an anionic leaving group. In the second step, the carbocation reacts with the nucleophile to form the substitution product.

Step 1.

substrate → a carbocation + L:$^-$ Slow

Step 2.

Nu:$^-$ → product   Fast

The formation of a carbocation is the slow, or *rate-determining*, step. The subsequent step, formation of a bond between the nucleophile and the carbocation, occurs very rapidly. Because the slow step of the reaction involves only the substrate, the reaction is **unimolecular.** Because only the substrate is present in the transition state, the rate of the reaction depends only on its concentration, and not on the concentration of the nucleophile.

Figure 7.3 shows an energy diagram tracing the progress of a reaction that occurs by an S$_N$1 mechanism. The rate of the reaction reflects the activation energy required to form the carbocation intermediate. The activation energy required for step 2, addition of the nucleophile to the carbocation, is much smaller, so step 2 is very fast. The rate of step 2 has no effect on the overall rate of the reaction.

The rates of S$_N$1 reactions decrease in the order 3° > 2° > 1°, which is the reverse of the order observed in S$_N$2 reactions. The relative reactivity of haloalkanes in S$_N$1 reactions corresponds to the relative stability of carbocation intermediates that form during the reaction. We recall from Chapter 4 that the order of stability of carbocations is tertiary > secondary > primary. A tertiary carbocation forms faster than a less stable secondary carbocation, which in turn forms very much faster than a highly unstable primary carbocation. However, S$_N$1 mechanisms are also favored by resonance-stabilized primary carbocations such as benzyl and allyl.

In contrast to S$_N$2 reactions at stereogenic centers, which occur with inversion of configuration, an S$_N$1 reaction gives a racemic mixture of enantiomers that has no optical rotation. For example, (S)-3-bromo-3-methylhexane reacts with water to give a racemic mixture of 3-methyl-3-hexanols. The reaction occurs via an achiral carbocation intermediate with a plane of symmetry (Figure 7.4). The carbocation intermediate's plane of symmetry allows the nucleophile to attack equally well from either side. The product is then a racemic mixture of enantiomers. Thus, a chiral substrate loses chirality in a reaction that occurs by an S$_N$1 mechanism.

Figure 7.3
Activation Energy and the S$_N$1 Reaction

The reaction of 2-bromo-2-methylpropane occurs in two steps with formation of an intermediate carbocation. It forms in the rate-determining step, which does not involve the nucleophile. In the second, fast step, the carbocation reacts with a nucleophile such as water to form the product.

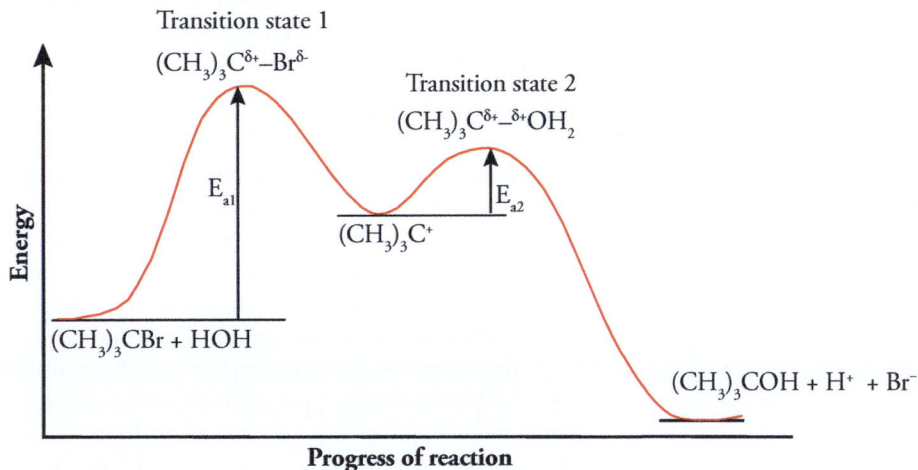

Figure 7.4 Stereochemical Effects in S$_N$1 Reactions

A chiral starting material, (S)-3-methyl-3-bromohexane, reacts with water to give a tertiary carbocation. This intermediate is planar and can be attacked by water either from the top or bottom side to give a racemic mixture of products. The reaction proceeds by an S$_N$1 mechanism.

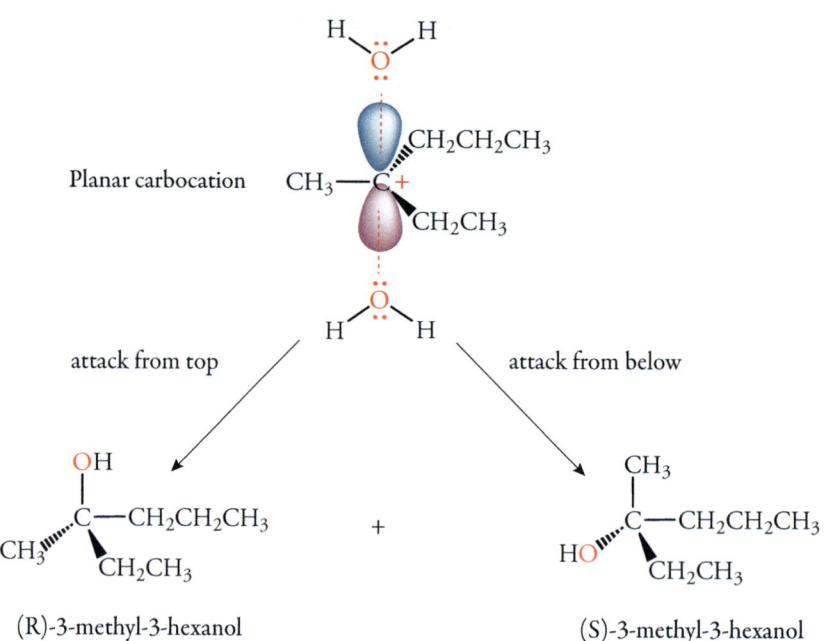

(R)-3-methyl-3-hexanol   (S)-3-methyl-3-hexanol

7.5 S$_N$2 VERSUS S$_N$1 REACTIONS

Now that we have outlined the general properties of S$_N$2 and S$_N$1 reactions, let's see if we can predict which one is likely to occur. We will consider (1) the structure of the substrate, (2) the nucleophile, and (3) the nature of the solvent.

Structure of the Substrate

Primary haloalkanes almost always react in nucleophilic substitution reactions by the S$_N$2 mechanism, whereas tertiary haloalkanes react by the S$_N$1 mechanism. Secondary haloalkanes may react by either mechanism depending on the nature of the nucleophile and the solvent.

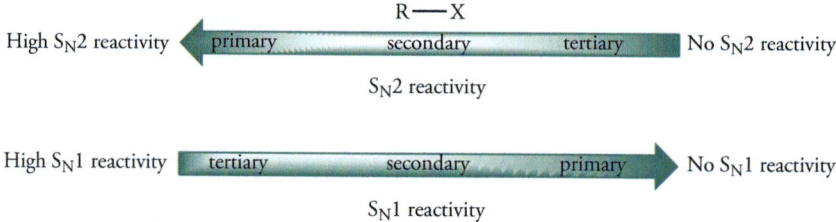

Effect of the Nucleophile

The nature of the nucleophile sometimes determines the mechanism of a nucleophilic substitution reaction. If the nucleophile is a highly polarizable species such as thiolate ion, RS⁻, it tends to react by an S$_N$2 mechanism. If the nucleophile is an uncharged species such as H$_2$O or CH$_3$OH, an S$_N$1 mechanism is more likely.

Effect of the Solvent

Until now, we have neglected the role of solvent in nucleophilic substitution reactions, but the choice of solvent can tip the balance in favor of a particular substitution mechanism. We have seen that secondary haloalkanes can react by either an S$_N$1 or an S$_N$2 mechanism. In these cases, the polarity of the solvent plays an important role. The S$_N$1 process forms a carbocation intermediate. A polar solvent stabilizes charged species better than a nonpolar solvent, therefore it increases the rate of S$_N$1 reactions much more than it affects the rate of S$_N$2 reactions.

The solvent also affects nucleophilicity. Solvents that have proton-donating ability, such as alcohols, are called **protic solvents**. A protic solvent interacts strongly with nucleophilic anions by forming hydrogen bonds with the unshared pairs of electrons on the nucleophiles. When the nucleophile is hydrogen-bonded to the solvent, its nucleophilicity decreases, which decreases the likelihood of an S$_N$2 reaction.

Solvents that do not have protons available for hydrogen bonding to nucleophiles are called **aprotic solvents**. Examples of polar aprotic solvents include dimethylformamide (DMF) and dimethyl sulfoxide (DMSO).

dimethylformamide (DMF)   dimethyl sulfoxide

The electron pairs of the oxygen atoms of these aprotic solvents enable them to solvate cations but not anions. For example, these solvents tie up the cation of KCN, but leave the C≡N⁻ ion free. Therefore, the nucleophilicity of C≡N⁻ is greater in dimethyl sulfoxide than in ethanol (CH_3CH_2OH). An aprotic solvent such as dimethyl sulfoxide favors an S_N2 reaction.

Problem 7.5

Explain why the reaction of 3-bromocyclohexene with methanol (CH_3OH) is faster than the reaction of bromocyclohexane with methanol.

bromocyclohexane   3-bromocyclohexene

Solution

Both substrates are secondary bromides. The reaction with methanol, a neutral nucleophile, will tend to occur by an S_N1 process. Although both carbocations are secondary, the one derived from 3-bromocyclohexene is also a resonance-stabilized allylic carbocation. Its resonance stabilization enhances its rate of formation. No such stabilization occurs in the reaction of bromocyclohexane.

resonance-stabilized allylic carbocation

Problem 7.6

The relative rates of reaction of 1-iodobutane with chloride ion in methanol, formamide, and dimethylformamide are 1, 12, and 1.2×10^6, respectively. Explain the relatively small difference in rate between methanol and formamide and the large difference in rate between formamide and dimethylformamide.

methanol   formamide   dimethylformamide

7.6 MECHANISMS OF ELIMINATION REACTIONS

We saw in Section 7.1 that when a haloalkane undergoes a nucleophilic substitution reaction, a competing elimination reaction may also occur because a nucleophile is also a base. In a substitution reaction, the nucleophile attacks a carbon atom in the substrate. When a haloalkane undergoes an elimination reaction, the nucleophile acts as a base and removes a proton from a carbon atom adjacent to the carbon atom bearing the halogen. Both the proton and the leaving group are eliminated and a carbon-carbon

π bond forms. When a proton and a halogen are lost by a substrate, the elimination reaction is called **dehydrohalogenation**.

An elimination reaction is denoted by the symbol E. Elimination reactions can occur by two mechanisms, designated E2 and E1. An E1 or an E2 process may compete with S_N1 and S_N2 reactions.

The E2 Mechanism

Like the S_N2 reaction mechanism, the E2 mechanism is a concerted process. In an E2 dehydrohalogenation reaction, the base (nucleophile) removes a proton on the carbon atom adjacent to the carbon atom containing the leaving group. As the proton is removed, the leaving group departs and a double bond forms.

Transition state
for E2 elimination

Like the S_N2 reaction, an E2 reaction requires a precise molecular arrangement. The anticonformation of the hydrogen and halogen atoms is preferred for the reaction because it aligns the orbitals properly for the formation of the π bond. We can visualize the process as the removal of the proton to provide an electron pair that attacks the neighboring carbon atom from the back to displace the leaving group. An E2 reaction occurs at a rate that depends on the concentrations of both the substrate and the base. If the substrate concentration is doubled, the reaction rate also doubles, as in S_N2 processes. Thus, both E2 and S_N2 mechanisms are affected in the same way, and the two mechanisms compete with each other.

The E1 Mechanism

We recall that an S_N1 reaction proceeds in two steps, and that the rate-determining step is formation of a carbocation intermediate. Similarly, an E1 mechanism occurs in two steps, and the rate-determining step is the formation of a carbocation. Thus, just as E2 and S_N2 mechanisms compete with each other, an E1 mechanism competes with an S_N1 mechanism. Because the rate-determining step of an E1 reaction involves only the substrate, the formation of the carbocation is a unimolecular reaction. If the carbocation reacts with a nucleophile at the positively charged carbon atom, the result is substitution. But if the nucleophile acts as a base and removes a proton from the carbon atom adjacent to the cationic center, the net result in the loss of HL and the formation of a π bond—that is, an elimination reaction.

7.7 EFFECT OF STRUCTURE ON COMPETING REACTIONS

Let's now examine the variety of product mixtures that result from competing substitution and elimination processes. We will divide our discussion according to the type of haloalkane. These results are summarized in Table 7.2.

Table 7.2
Summary of Substitution and Elimination Reactions

Structure	S_N1	S_N2	E1	E2
RCH_2X	Does not occur	Very favored	Does not occur	Occurs readily with strong bases
R_2CHX	Occurs for allylic and benzylic compounds	E2 reactions occur in competition with substitution	Occurs for allylic and benzylic compounds	Occurs readily with strong bases
R_3CX	Occurs readily in polar solvents	Does not occur	Occurs in competition with S_N1 reaction	Occurs readily with bases

Competing Substitution and Elimination Reactions of Tertiary Haloalkanes

Tertiary haloalkanes can undergo substitution reactions only by an S_N1 mechanism because there is too much steric hindrance for an S_N2 reaction to occur. A tertiary haloalkane can undergo an elimination reaction by either an E2 or E1 mechanism. The mechanism depends on the basicity of the nucleophile and the polarity of the solvent. If the nucleophile is a weak base, S_N1 and E1 processes compete, and the amount of the two types of products depends only on the carbocation formed. For example, 2-bromo-2-methylbutane reacts in ethanol to give about 64% of the product from an S_N1 process.

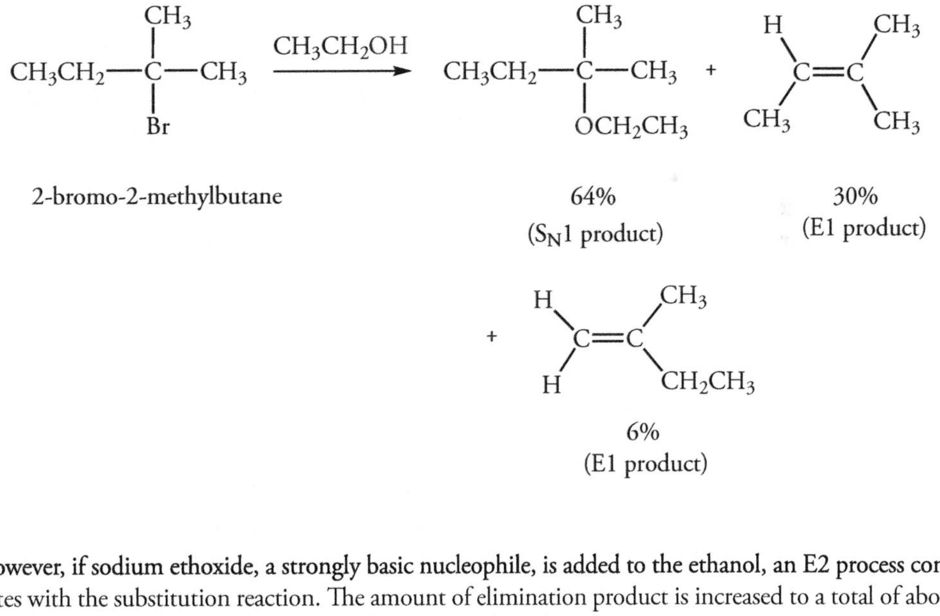

2-bromo-2-methylbutane

64%
(S_N1 product)

30%
(E1 product)

6%
(E1 product)

However, if sodium ethoxide, a strongly basic nucleophile, is added to the ethanol, an E2 process competes with the substitution reaction. The amount of elimination product is increased to a total of about 93% of the product mixture; only 7% of the S_N1 product forms.

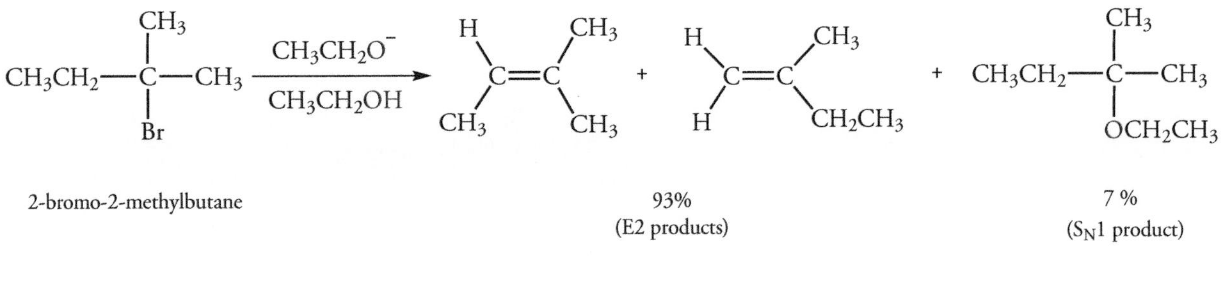

2-bromo-2-methylbutane

93%
(E2 products)

7 %
(S_N1 product)

Competing Substitution and Elimination Reactions of Primary Haloalkanes

Primary haloalkanes can undergo either S_N2 or E2 reactions. They do not undergo S_N1 or E1 reactions because a primary carbocation is very unstable. Primary haloalkanes react with strongly nucleophilic, weakly basic reactants, such as ethyl thiolate ($CH_3CH_2S^-$), exclusively by an S_N2 process. However, a primary haloalkane reacts with ethoxide ion, a weaker nucleophile, but a stronger base than ethyl thiolate, to give some elimination product, but mostly S_N2 product.

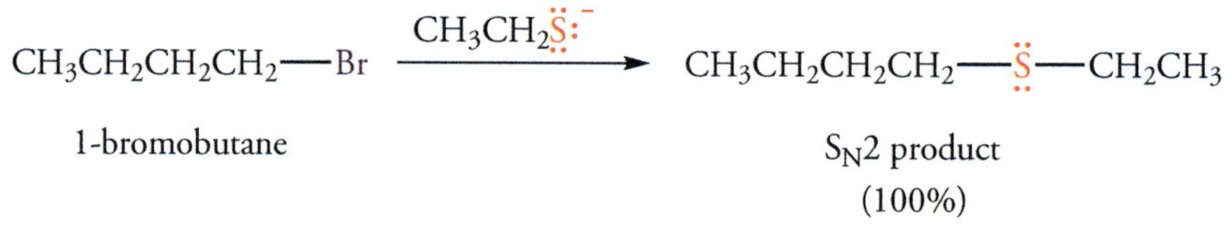

$$CH_3CH_2CH_2CH_2 - Br \xrightarrow{CH_3CH_2\ddot{S}:^-} CH_3CH_2CH_2CH_2 - \ddot{S} - CH_2CH_3$$

1-bromobutane
S_N2 product
(100%)

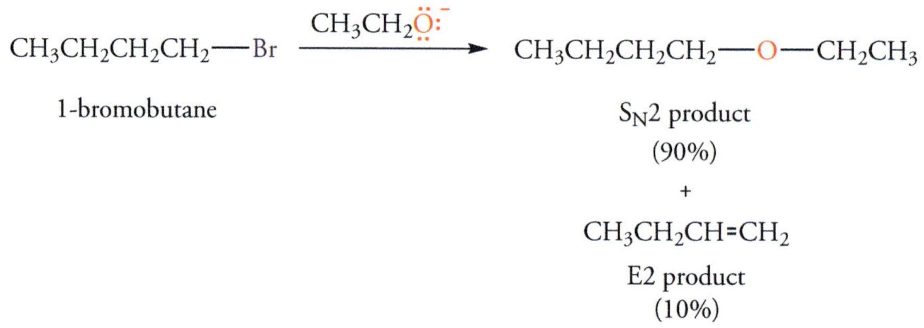

$$CH_3CH_2CH_2CH_2 - Br \xrightarrow{CH_3CH_2\ddot{O}:^-} CH_3CH_2CH_2CH_2 - O - CH_2CH_3$$

1-bromobutane
S_N2 product
(90%)

+

$$CH_3CH_2CH=CH_2$$

E2 product
(10%)

If a primary haloalkane is treated with *tert*-butoxide ion instead of ethoxide, the amount of elimination product increases significantly. The *tert*-butoxide ion is not only more basic than the ethoxide ion, it is also much more sterically hindered. The combination of these two factors favors elimination by an E2 process over substitution by an S_N2 process.

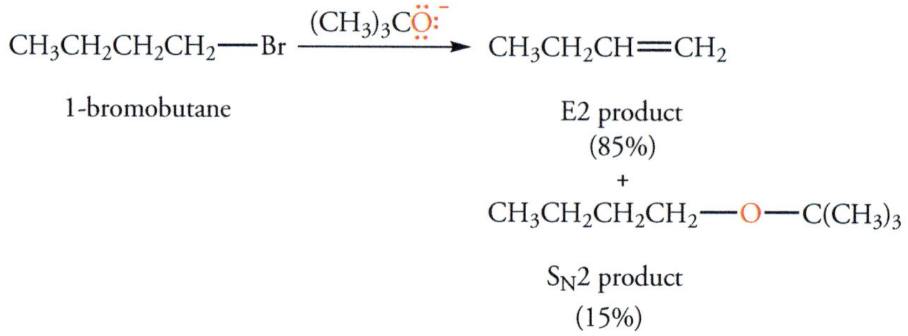

$$CH_3CH_2CH_2CH_2 - Br \xrightarrow{(CH_3)_3C\ddot{O}:^-} CH_3CH_2CH=CH_2$$

1-bromobutane
E2 product
(85%)

+

$$CH_3CH_2CH_2CH_2 - O - C(CH_3)_3$$

S_N2 product
(15%)

Competing Substitution and Elimination Reactions of Secondary Haloalkanes

Secondary haloalkanes can react by S_N2, E2, S_N1, and E1 mechanisms, and it is sometimes difficult to predict which of these processes will occur in a given reaction. However, secondary haloalkanes tend to react with strong nucleophiles that are weak bases, such as thiolates or cyanide ion, by an S_N2 process.

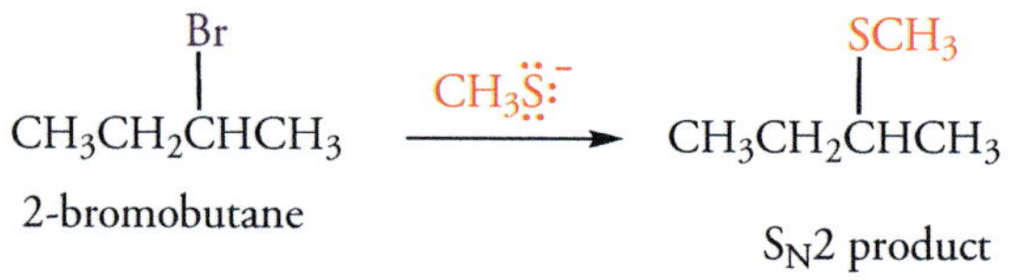

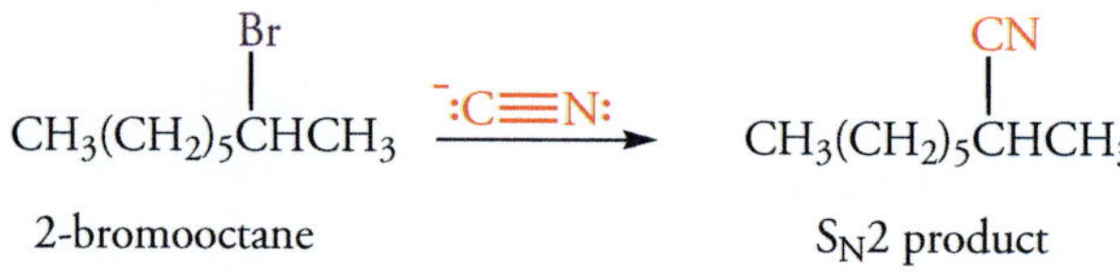

$$\underset{\text{2-bromobutane}}{CH_3CH_2\overset{\overset{\displaystyle Br}{|}}{C}HCH_3} \xrightarrow{CH_3\ddot{S}:^-} \underset{S_N2 \text{ product}}{CH_3CH_2\overset{\overset{\displaystyle SCH_3}{|}}{C}HCH_3}$$

$$\underset{\text{2-bromooctane}}{CH_3(CH_2)_5\overset{\overset{\displaystyle Br}{|}}{C}HCH_3} \xrightarrow{^-:C\equiv N:} \underset{S_N2 \text{ product}}{CH_3(CH_2)_5\overset{\overset{\displaystyle CN}{|}}{C}HCH_3}$$

On the other hand, a secondary haloalkane tends to react with weak nucleophiles that are also weak bases, like ethanol, by an S_N1 mechanism with some accompanying E1 product.

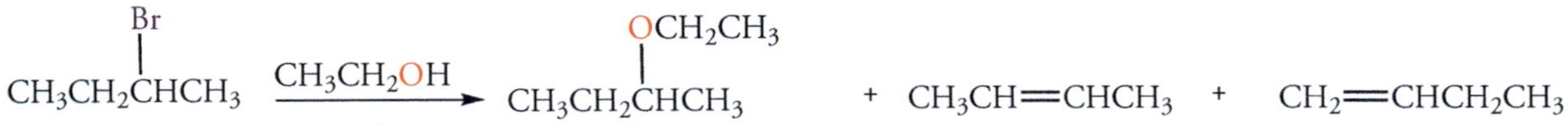

2-bromobutane $\xrightarrow{CH_3CH_2OH}$ S_N1 product (95%) + E1 Product (4%) + E1 Product (1%)

We can tip the scales in the other direction by adding sodium ethoxide to ethanol. By adding this strong base, we find that the product of the S_N1 reaction drops to 18% of the total, and E2 products account for the rest.

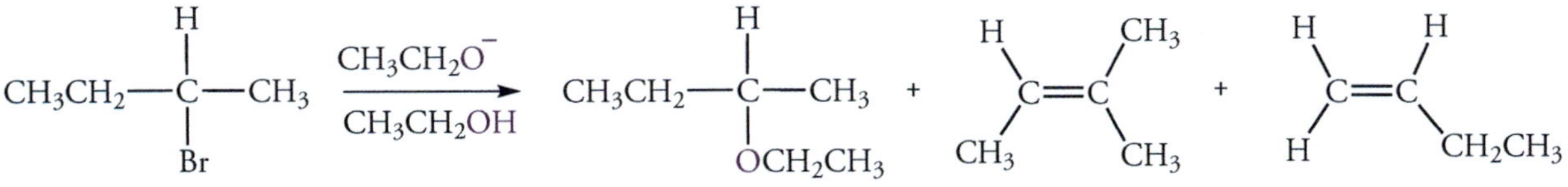

2-bromobutane $\xrightarrow[CH_3CH_2OH]{CH_3CH_2O^-}$ 18% (S_N1 product) + 66% (E2 product) + 16% (E2 product)

Problem 7.7
Which of the following two methods of preparing $CH_3CH_2OCH(CH_3)_2$ will give the better yield?

$$CH_3CH_2{-}O^- + Br{-}CH(CH_3)_2 \longrightarrow CH_3CH_2{-}O{-}CH(CH_3)_2$$

ethoxide   2-bromopropane

$$CH_3CH_2{-}Br + {^-}O{-}CH(CH_3)_2 \longrightarrow CH_3CH_2{-}O{-}CH(CH_3)_2$$

bromoethane   isopropoxide

Solution
An ether can be prepared by treating a haloalkane with an alkoxide in an S_N2 reaction. This ether has two different alkyl groups bonded to oxygen: one from the alkoxide and the other from the haloalkane.

The first reaction is a nucleophilic displacement at a secondary center by a nucleophile that is also a strong base. A competing elimination reaction to yield propene will also occur. The second reaction occurs by an S_N2 mechanism at a primary center, which tends to occur with little competition from an elimination reaction. Therefore, the better way to make the desired product is by the second reaction.

Problem 7.8
The amount of elimination product for the reaction of 1-bromodecane with an alkoxide in the corresponding alcohol solvent is about 1% for methoxide ion and 85% for *tert*-butoxide ion. Explain these data.

SUMMARY OF REACTIONS

1. Nucleophilic Substitution of Haloalkanes (Section 7.2)

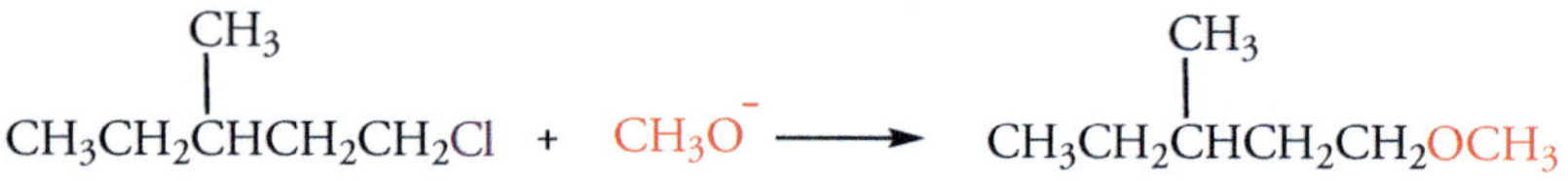

$$\underset{\underset{CH_3}{|}}{CH_3CH_2CHCH_2CH_2Cl} + CH_3O^- \longrightarrow \underset{\underset{CH_3}{|}}{CH_3CH_2CHCH_2CH_2OCH_3}$$

$$\text{(benzyl ring)}-CH_2CH_2Br + CH_3CH_2S^- \longrightarrow \text{(benzyl ring)}-CH_2CH_2SCH_2CH_3$$

$$\text{(cyclopentyl)}-CH_2CH_2Br + CN^- \longrightarrow \text{(cyclopentyl)}-CH_2CH_2CN$$

$$\text{(cyclopentyl)}-CH_2CH_2Br + CH_3CH_2-C\equiv C^- \longrightarrow \text{(cyclopentyl)}-CH_2CH_2-C\equiv C-CH_2CH_3$$

2. Dehydrohalogenation of Haloalkanes (Section 7.6)

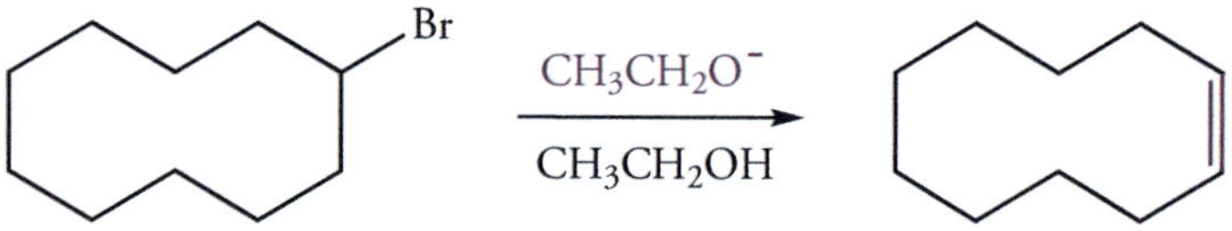

$$\xrightarrow[CH_3CH_2OH]{CH_3CH_2O^-}$$

EXERCISES

Nomenclature of Haloalkanes

7.1   What is the IUPAC name for each of the following compounds?

   (a) vinyl fluoride   (b) allyl chloride   (c) propargyl bromide

7.2   What is the IUPAC name for each of the following compounds?

   (a) $(CH_3)_3CCH_2Cl$ (neopentyl chloride)   (b) $(CH_3)_2CHCH_2CH_2Br$ (isoamyl bromide)   (c) $C_6H_5CH_2CH_2F$ (phenethyl fluoride)

7.3   Draw the structure of each of the following compounds:

   (a) *cis*-l-bromo-2-methylcyclopentane   (b) 3-chlorocyclobutene   (c) (E)-1-fluoro-2-butene   (d) (Z)-l-bromo-1-propene

7.4   What is the IUPAC name for each of the following compounds?

(a) 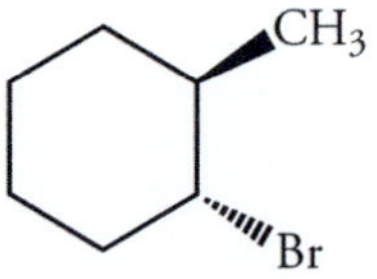   (b)    (c)

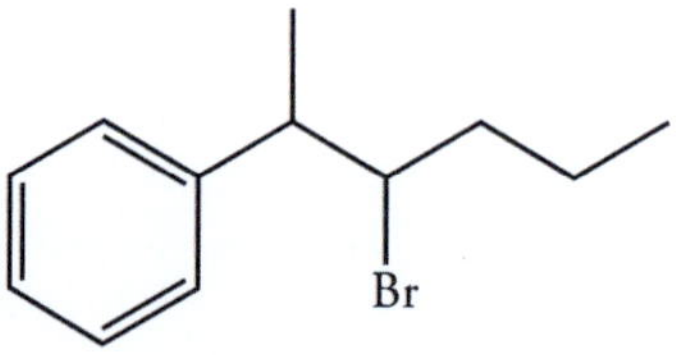

Nucleophilic Substitution Reactions

7.5 Write the structure of the product obtained for each of the following combinations of reactants:

(a) 1-chloropentane and sodium iodide (b) 1,3-dibromopropane and excess sodium cyanide

(c) p-methylbenzylchloride and sodium acetylide (d) 2-bromobutane and sodium hydrosulfide (NaSH)

7.6 What haloalkane and nucleophile are required to produce each of the following compounds?

(a) $CH_3CH_2CH_2C{\equiv}CH$ (b) $(CH_3)_2CHCH_2CN$ (c) $CH_3CH_2SCH_2H_3$

7.7 Alcohols (ROH) are converted into alkoxides (RO⁻) by reaction with NaH. Treatment of the following compound with sodium hydride yields C_4H_8O. What is the structure of the product? How is it formed?

$$HOCH_2CH_2CH_2CH_2{-}Br + NaH \longrightarrow C_4H_8O + H_2 + NaBr$$

7.8 Treatment of the following compound with sodium sulfide yields C_4H_8S. What is the structure of the product? How is it formed?

$$Cl{-}CH_2CH_2CH_2CH_2{-}Cl + Na_2S \longrightarrow C_4H_8S + 2\,NaCl$$

Structure and Rates of Nucleophilic Substitution Reactions

7.9 Which compound in each of the following pairs reacts with sodium iodide at the faster rate in an S_N2 reaction?

(a) 1-bromobutane or 2-bromobutane   (b) 1-chloropentane or chlorocyclopentane

(c) 2-bromo-2-methylpentane or 2-bromo-4-methylpentane

7.10 The rate of reaction of *cis*-1-bromo-4-*tert*-butylcyclohexane (I) with methyl thiolate(CH_3S^-) is faster than for the *trans* isomer (II). Suggest a reason for this difference.

7.11 Predict which compound in each of the following pairs reacts with methanol (CH_3OH) at the faster rate in an S_N1 reaction. Explain why.

(a) bromocyclohexane or 1-bromo-1-methylcyclohexane

(b) isopropyl iodide or isobutyl iodide

(c) 3-bromo-1-pentene or 4-bromo-1-pentene

7.12 Predict which compound in each of the following pairs reacts with ethanol (CH_3CH_2OH) at the faster rate in an S_N1 reaction. Explain why.

(a) l-bromo-1-phenylpropane or 2-bromo-1-phenylpropane

(b) 3-chlorocyclopentene or 4-chlorocyclopentene

(c) benzyl bromide or p-methylbenzyl bromide

7.13 *p*-Methylbenzyl bromide reacts faster than p-nitrobenzyl bromide with ethanol to form an ether product. Suggest a reason for this observation.

7.14 Although 1-bromo-2,2-dimethylpropane is a primary halide, S_N2 reactions of this compound are about 10,000 times slower than those of 1-bromopropane. Suggest a reason.

Stereochemistry of Nucleophilic Substitution Reactions

7.15   Write the structure of the product of the reaction of (R)-2-bromobutane with sodium cyanide.

7.16   Write the structure of the product of the reaction of *cis*-1-bromo-2-methylcyclopentane with methylthiolate.

7.17   Write the structure of the product of the reaction of *(S)*-3-bromo-3-methylhexane with ethanol (CH₃CH₂OH).

7.18   Optically active 2-iodooctane slowly becomes racemic when treated with sodium iodide in an inert polar solvent. Explain why.

7.19   Optically active 2-butanol, (CH₃CH₂CHOHCH₃), slowly becomes racemic when treated with dilute acid. Explain why.

Elimination Reactions

7.20   What alkene will be formed in the E2 reaction of the following compound with sodium methoxide (CH₃O⁻Na⁺)?

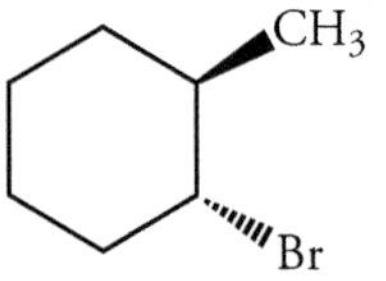

7.21   How many isomeric alkenes can be formed by the elimination of HBr from the following compound by an E1 process? Which one will predominate?

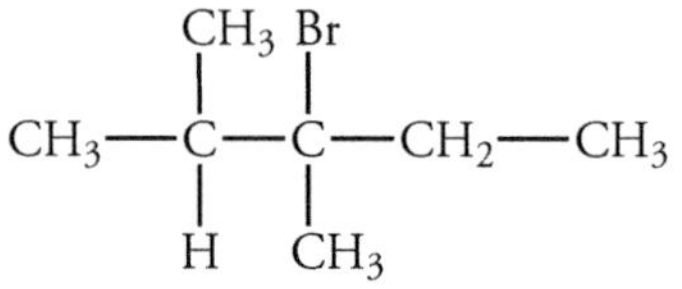

7.22   The rate of elimination of hydrogen bromide in the reaction of *cis*-1-bromo-4-*tert*-butylcyclohexane with hydroxide ion is faster than for the *trans* isomer. Suggest a reason for this difference.

7.23   Eight diastereomers of 1,2,3,4,5,6-hexachlorocyclohexane are possible. The following isomer undergoes an E2 reaction about 1000 times slower than any of the others. Why?

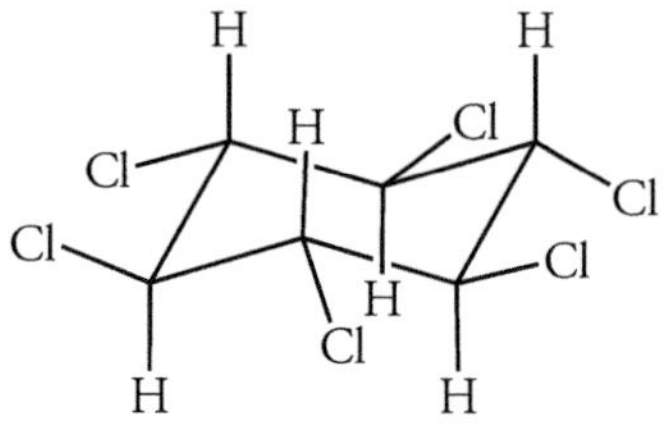

7.24   What is the configuration of the alkene formed by the elimination of one molar equivalent of HBr from the following compound?

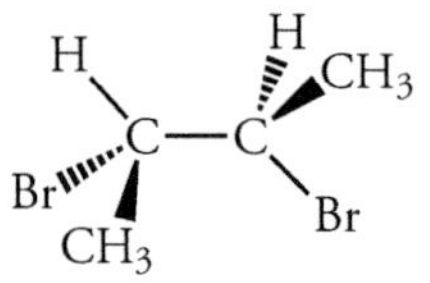

7.25   How many isomers can result from the elimination of HBr from the following compound? Which one will predominate?

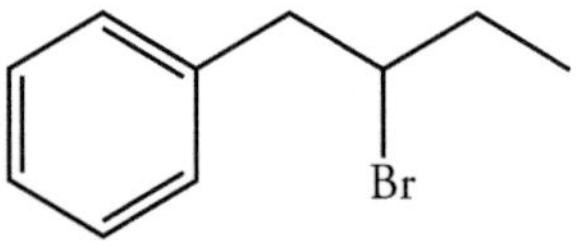

8 ALCOHOLS AND PHENOLS

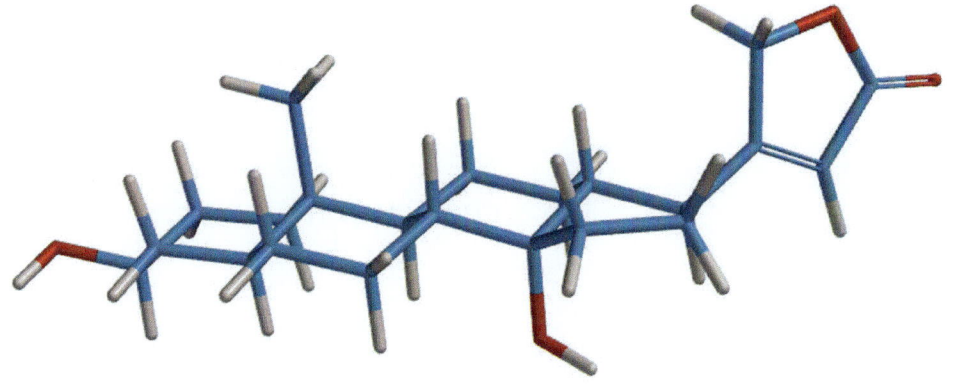

DIGITOXIGENIN

8.1 THE HYDROXYL GROUP

Families of organic compounds that have functional groups containing oxygen include alcohols, phenols, ethers, aldehydes, ketones, acids, esters, and amides. Alcohols and phenols both contain a hydroxyl group (−OH). A hydroxyl group is also present in carboxylic acids, but it is bonded to a carbonyl carbon atom. As a result, the chemistry of carboxylic acids, the subject of Chapter 12, is substantially different from the chemistry of alcohols and phenols. Alcohols and phenols can be viewed as organic "relatives" of water in which one hydrogen atom is replaced by an alkyl group or an aryl group. Alcohols contain a hydroxyl group bonded to an sp^3-hybridized carbon atom. Phenols have a hydroxyl group bonded to an sp^2-hybridized carbon atom of an aromatic ring.

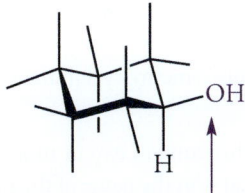

This hydroxyl group is bonded to an sp^3-hybridized carbon atom. The compound is an alcohol

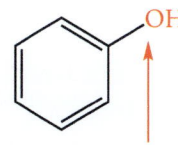

This hydroxyl group is bonded to an sp^2-hybridized carbon atom. The compound is a phenol

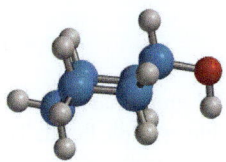

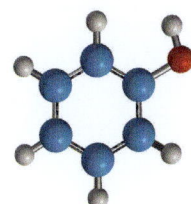

Common Names of Alcohols

The common names of alcohols consist of the name of the alkyl group (Section 3.3) followed by the term *alcohol*. For example, CH_3CH_2OH is ethyl alcohol and $CH_3CH(OH)CH_3$ is isopropyl alcohol. Other common names are allyl alcohol and benzyl alcohol, whose structures are shown below.

$$CH_2{=\!=}CH-CH_2-OH$$

allyl alcohol

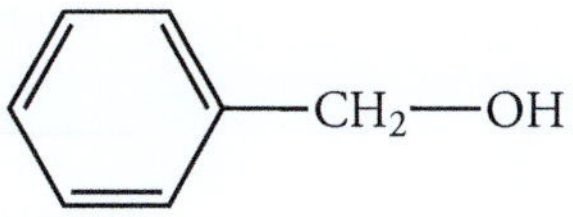

benzyl alcohol

Principles of Organic Chemistry. http://dx.doi.org/10.1016/B978-0-12-802444-7.00008-2

The IUPAC system of naming alcohols is based on the longest chain of carbon atoms that includes the hydroxyl group as the parent chain. The parent name is obtained by substituting the suffix -ol for the final -e of the corresponding alkane. The IUPAC rules are as follows:

1. The position of the hydroxyl group is indicated by the number of the carbon atom to which it is attached. The chain is numbered so that the carbon atom bearing the hydroxyl group has the lower number.

$$\overset{4}{CH_3}-\overset{3}{CH}-\overset{2}{CH}-\overset{1}{CH_3}$$

CH$_3$ OH

3-methyl-2-butanol

The longest chain that contains the hydroxyl group has 4 carbon atoms. An OH group is at C-2 and a methyl group is at C-3. So, the name is 3-methyl-2-butanol

2. When the hydroxyl group is attached to a ring, the ring is numbered starting with the carbon atom bearing the hydroxyl group. Numbering continues in the direction that gives the lowest numbers to carbon atoms with substituents such as alkyl groups. The number 1 is not used in the name to indicate the position of the hydroxyl group.

trans-2-methylcyclobutanol

3,3-dimethylcyclohexanol

3. Alcohols that contain two or more hydroxyl groups are called diols, triols, and so on. Retain the terminal -*e* in the name of the parent alkane, and add the suffix -diol or -triol. Indicate the positions of the hydroxyl groups in the parent chain by numbers as a prefix.

$$HO-\overset{1}{CH_2}-\overset{2}{CH_2}-\overset{3}{CH_2}-\overset{4}{CH}-\overset{5}{CH_3}$$

OH

1,4-pentanediol

4. The hydroxyl group has precedence in numbering the carbon chain when an alcohol contains a double or triple bond. The number that indicates the position of the multiple bond is located in the prefix of the name of the alkene (or alkyne). Drop the final -*e*. The number that indicates the position of the hydroxyl group hydroxyl group is appended to the name of the alkene (or alkyne) along with the suffix -*ol*.

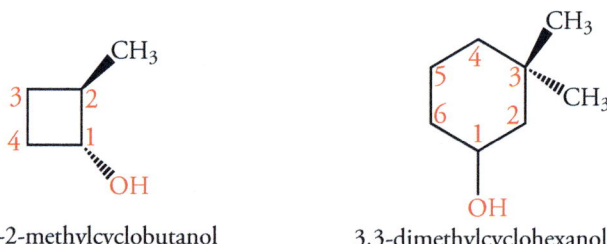

$$\overset{5}{CH_2}=\overset{4}{CH}-\overset{3}{CH_2}-\overset{2}{CH_2}-\overset{1}{CH_2}-OH$$

4-pentene-1-ol

$$\overset{6}{CH_3}-\overset{5}{C}\equiv\overset{4}{C}-\overset{3}{CH_2}-\overset{2}{CH}-\overset{1}{CH_2}-OH$$

OH

4-hexyne-1,2-diol

Problem 8.1
Classify the alcohol functional groups in the broad-spectrum antibiotic chloramphenicol.

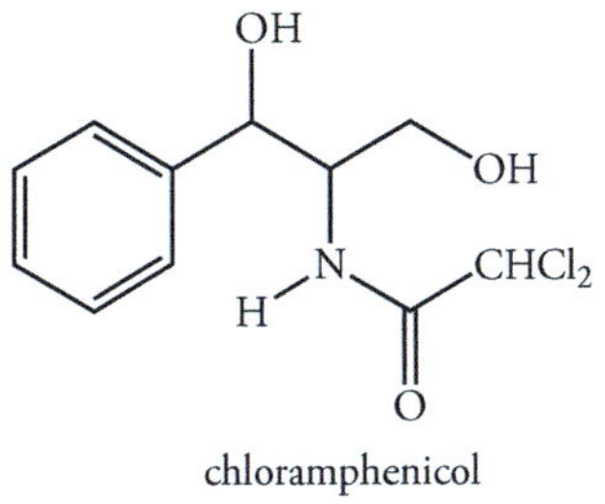

chloramphenicol

Solution

First, locate the oxygen atoms in the structure. The five oxygen atoms appear at four sites. Two oxygen atoms are bonded to a nitrogen atom in the nitro group; the double-bonded oxygen atom is part of a carbonyl group of an amide. The hydroxyl group on the right is bonded to a carbon atom that has two hydrogen atoms and a carbon atom bonded to it; this is a primary alcohol. The hydroxyl group in the middle of the molecule is bonded to a carbon atom with two other carbon atoms bonded to it; one carbon atom is part of a substituted alkyl group, the other carbon atom is part of an aryl group. This alcohol is secondary.

Problem 8.2

Classify the alcohol functional groups in riboflavin (vitamin B_2).

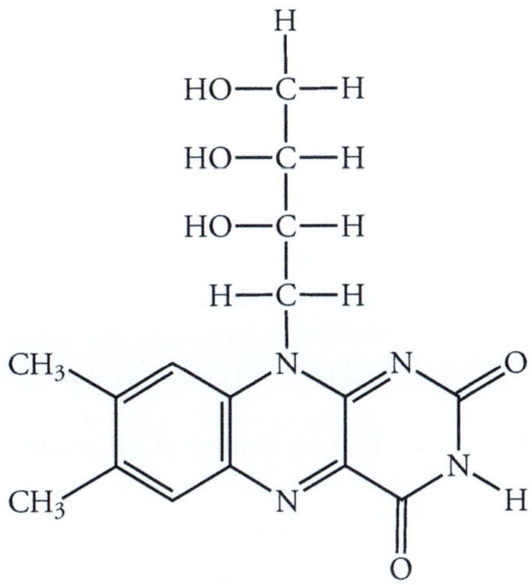

riboflavin (vitamin B_2)

Problem 8.3

(E)-1-Chloro-3-ethyl-1-penten-4-yn-3-ol is a sedative-hypnotic drug known as ethchlorvynol. Draw its structure.

Solution

The term *1-penten* tells us that the parent chain contains five carbon atoms and has a double bond between C-1 and C-2. The term *4-yn* informs us that there is a triple bond between the C-4 and C-5. Using this information, we draw the five-carbon-atom chain and select a direction for numbering it. Then we place the multiple bonds in the proper places.

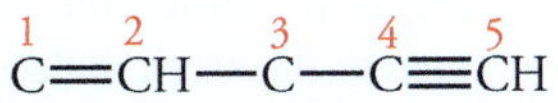

The IUPAC name also tells us that a chlorine atom is located at C-1, an ethyl group is at C-3, and a hydroxyl group at C-3.

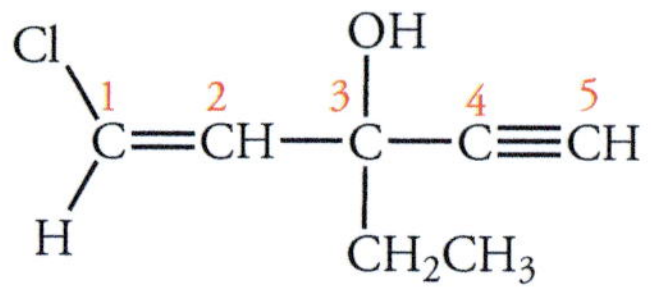

Next, we draw the *(E)* configuration by placing the chlorine atom and the C-3 atom on opposite sides of the double bond with the necessary hydrogen atoms attached to the compound.

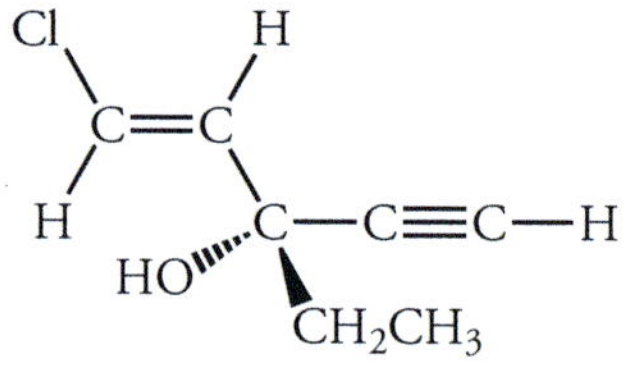

Problem 8.4
Assign the IUPAC name for citronellol, a compound found in geranium oil that is used in perfumes.

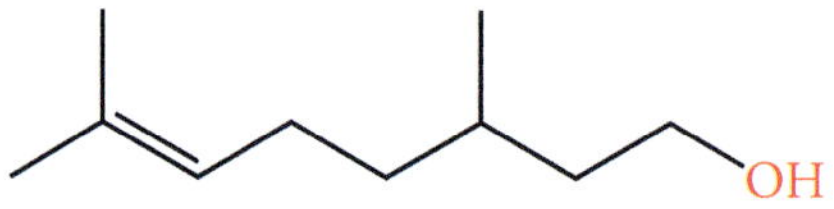

citronellol

8.2 PHYSICAL PROPERTIES OF ALCOHOLS

The dipole moments of ethanol and propane are 1.69 and 0.08 D, respectively. Alcohols are much more polar than alkanes because they have both a polar C—O bond and a polar O—H bond. However, the strong intermolecular forces in alcohols are the result of hydrogen bonds, which have an enormous influence on the physical properties of alcohols.

Boiling Points of Alcohols

Alcohols boil at dramatically higher temperatures than alkanes of comparable molecular weight. For example, propane boils at –42°C, whereas ethanol boils at 78°C. These two compounds have approximately the same London forces, but ethanol also has dipole-dipole forces of attraction as well as large forces of attraction due to hydrogen bonding between hydroxyl groups. The hydroxyl group of an alcohol can serve as both a hydrogen bond donor and a hydrogen bond acceptor. As a result, much more energy is needed to separate hydrogen-bonded alcohol molecules than is required to disrupt the relatively weak London forces in alkanes.

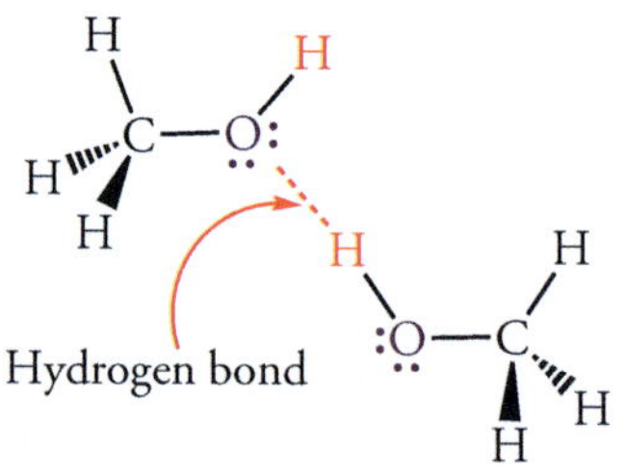

Hydrogen bond

Figure 8.1 shows the differences in the boiling points of primary alcohols and alkanes of approximately the same molecular weight. As the molecular weights of alkanes and alcohols increase, the two curves approach each other. In alcohols with high molecular weights, hydrogen bonding is still possible, but interactions due to London forces increase due to the longer carbon chain. As a result, the difference in boiling point between an alcohol and an alkane of comparable molecular weight decreases.

Table 8.1
Boiling Points and Solubilities of Alcohols

Compound	Boiling Point (°C)	Solubility (g/100 mL water)
Methanol	65	Miscible
Ethanol	78	Miscible
1-Propanol	97	Miscible
1-Butanol	117	7.9
1-Pentanol	137	2.7
1-Hexanol	158	0.59
1-Heptanol	176	0.09
1-Octanol	194	Insoluble
1-Nonanol	213	Insoluble
1-Decanol	229	Insoluble

Figure 8.1
Comparison of the Boiling Points of Alcohols and Alkanes

The boiling points of alkanes and alcohols increase with increasing chain length. Alcohols have higher boiling points than alkanes of comparable molecular eight. This difference decreases as the length of the carbon chain increases.

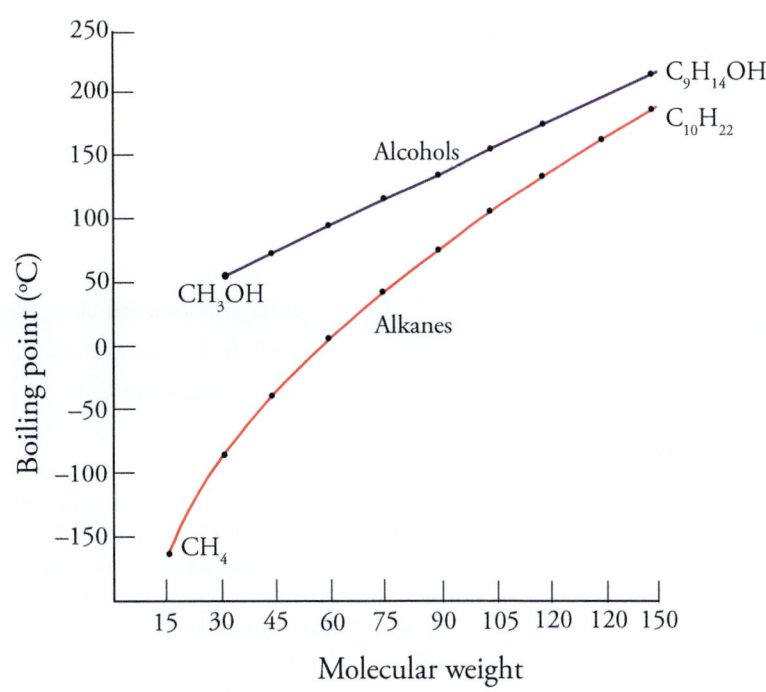

Solubility of Alcohols in Water

The lower molecular weight alcohols are miscible with water (Table 8.1). These molecules, like water, are highly polar, and we know that "like dissolves like." Furthermore, the hydroxyl group can form three hydrogen bonds to water. Two can form between the two sets of lone pair electrons of the alcohol oxygen atom, which are hydrogen bond acceptors, and the hydrogen atoms of water. The third can form between the hydrogen atom of the hydroxyl group, which is a hydrogen bond donor, and the lone pair electrons on the oxygen atom in water. As the size of the alkyl group increases, alcohols more closely resemble alkanes, and the hydroxyl group has a smaller effect on their physical properties. Water can still can form hydrogen bonds to the hydroxyl group. However, the long chain interferes with other water molecules and prevents them from hydrogen bonding to each other. As a result, the solubility of alcohols decreases with increasing size of the alkyl group.

Alcohols as Solvents

Ethanol is an excellent solvent for many organic compounds, especially those with lone pair electrons that are hydrogen bond acceptors. Polar compounds dissolve readily in the "like" polar solvent. Nonpolar compounds dissolve in alcohols to some extent, but the solubility is often limited because the extensive hydrogen-bonding network of the alcohol must be broken to accommodate the solute.

8.3 ACID-BASE REACTIONS OF ALCOHOLS

We know that water can act as a proton donor (an acid) in some reactions and as a proton acceptor (a base) in other reactions depending on conditions. Molecules that can act in this way are *amphoteric*. Alcohols can also act as acids or bases. Thus, alcohols are also amphoteric.

Alcohols are slightly weaker acids than water; the K_a of ethanol is 1.3×10^{-16} (pK_a = 16) and the K_a of water is 1.8×10^{-16} (pK_a = 8.7). The pK_a values of some common alcohols are listed in Table 8.2. We recall that a strong acid has a large K_a and a small pK_a.

$$CH_3CH_2OH + H_2O \rightleftharpoons CH_3CH_2O^- + H_3O^+ \qquad K_a = 1.3 \times 10^{-16}$$

$$H_2O + H_2O \rightleftharpoons HO^- + H_3O^+ \qquad K_a = 1.8 \times 10^{-16}$$

Table 8.2
Effect of Structure on Acidity of Alcohols

Alcohol	Formula	K_a	pK_a
Methanol	CH_3OH	3.2×10^{-16}	8.5
Ethanol	CH_3CH_2OH	1.3×10^{-16}	8.9
Isopropyl alcohol	$(CH_3)_2CHOH$	1×10^{-18}	18.0
tert-Butyl alcohol	$(CH_3)_3COH$	1×10^{-19}	19.0
2-Chloropropanol	$ClCH_2CH_2OH$	5×10^{-15}	14.3
2,2,2-Trifluoroethanol	CF_3CH_2OH	4×10^{-13}	12.4
3,3,3-Trifluoropropanol	$CF_3CH_2CH_2OH$	2.5×10^{-15}	14.6
4,4,4-Trifluorobutanol	$CF_3CH_2CH_2CH_2OH$	4×10^{-16}	8.4

The acidity of alcohols increases when electronegative substituents are added to a carbon atom near the hydroxyl group. Such substituents withdraw electron density from the oxygen atom by an inductive effect that weakens the O—H bond, which destabilizes the alcohol. The substituents also stabilize the negative charge of the conjugate base. Replacing a hydrogen atom at C-2 of ethanol with a chlorine atom decreases the pK_a from 8.9 to 14.3, which means that K_a increases by a factor of 50. Replacing all three hydrogen atoms at C-2 of ethanol with the more electronegative fluorine atoms decreases the pK_a to 12.4, which corresponds to an increase in acidity of more than 1000 fold. The effect of the electron-withdrawing CF_3 group decreases with distance from the oxygen atom. The pK_a of 4,4,4-trifluorobutanol, for example, is similar to the pK_a of a primary alcohol such as ethanol.

When an alcohol loses a proton, an alkoxide ion forms. Because alcohols are weaker acids than water, and alkoxides are stronger bases than hydroxide ion. Alkoxides are used as bases in organic solvents because they are more soluble than hydroxide salts. Alkoxide ions can be easily prepared by adding an alkali metal to an alcohol.

$$CH_3OH + Na \rightleftharpoons \underset{\text{methoxide ion}}{CH_3O^-} + 1/2\ H_2 + Na^+$$

Alcohols can act as bases because they have two sets of lone pair electrons on the oxygen atom. But alcohols are very weak bases and can only be protonated by strong acids to form the conjugate acid, an oxonium ion. The formation of an **oxonium ion** is analogous to the reaction of water with a strong acid to give the hydronium ion. Oxonium ions are intermediates in many reactions catalyzed by strong acids.

$$CH_3CH_2OH + HA \rightleftharpoons CH_3CH_2OH_2^+ + A^-$$
alkoxonium ion

$$H_2O + HA \rightleftharpoons H_3O^+ + A^-$$
hydronium ion

8.4 SUBSTITUTION REACTIONS OF ALCOHOLS

The hydroxyl group of an alcohol can be replaced by a halogen atom in either an S_N2 or S_N1 reaction (Chapter 7). For example, treating a primary alcohol with hydrogen bromide, HBr, produces an alkyl bromide. Similarly, treating a primary alcohol with HCl in the presence of $ZnCl_2$, which is required as a catalyst, produces an alkyl chloride.

$$CH_3CH_2CH_2OH + HBr \longrightarrow CH_3CH_2CH_2Br + H_2O$$

$$CH_3CH_2CH_2CH_2OH + HCl \xrightarrow{ZnCl_2} CH_3CH_2CH_2CH_2Cl + H_2O$$

These reactions also occur when secondary and tertiary alcohols are the substrates. Their relative rates depend on the type of alcohol. The rate of reaction decreases in the order tertiary > secondary > primary alcohols.

Reaction Mechanisms

The reaction mechanism depends on the structure of the alkyl group. Primary alcohols react by an S_N2 mechanism, whereas secondary and tertiary alcohols react by an S_N1 mechanism. However, in each case the leaving group is not hydroxide ion, but a water molecule. The acid catalyst is required to form the conjugate acid of the alcohol. The departure of water, a neutral leaving group, from a developing carbocation center requires less energy than the departure of hydroxide ion, a negatively charged leaving group.

Water is a better leaving group than hydroxide ion because water is the weaker base. There is a general correlation between Lewis basicity and leaving-group ability. A weak base is a better leaving group than a stronger base in both S_N1 and S_N2 reactions.

Primary and secondary alcohols, which react only slowly with HBr and HCl, react readily with thionyl chloride and phosphorus trihalides, such as phosphorus tribromide, to give the corresponding alkyl halides. The products of these reactions are easily separated from the inorganic by-products. Thionyl chloride produces hydrogen chloride and sulfur dioxide, which are released from the reaction as gases; the chloroalkane remains in solution. The reaction with phosphorus tribromide produces phosphorous acid, which has a high boiling point and is water-soluble. Thus, the bromoalkane can be separated from the reaction mixture by distillation or by adding water and extracting the haloalkane in a solvent such as diethyl ether.

$$R\!-\!OH + SOCl_2 \longrightarrow R\!-\!Cl + SO_2 (g) + HCl (g)$$

$$R\!-\!OH + PBr_3 \longrightarrow R\!-\!Br + H_3PO_3$$

These reactions are commonly written as unbalanced equations that don't show the inorganic products and contain the reagent placed over the reaction arrow.

$$\underset{\text{2-butanol}}{\underset{|}{CH_3CH_2\overset{OH}{C}HCH_3}} \xrightarrow{SOCl_2} \underset{\text{2-chlorobutane}}{\underset{|}{CH_3CH_2\overset{Cl}{C}HCH_3}}$$

$$\underset{\text{2-octanol}}{\underset{|}{CH_3(CH_2)_5\overset{OH}{C}HCH_3}} \xrightarrow{PBr_3} \underset{\text{2-chlorooctane}}{\underset{|}{CH_3(CH_2)_5\overset{Br}{C}HCH_3}}$$

8.5 DEHYDRATION OF ALCOHOLS

The removal of a water molecule from an alcohol is a **dehydration** reaction, which is an example of an elimination reaction. Dehydration of an alcohol requires an acid catalyst, such as sulfuric acid or phosphoric acid, and is illustrated by the formation of ethylene from ethanol.

These groups are eliminated

$$H-\underset{\underset{H}{|}}{\overset{\overset{H}{|}}{C}}-\underset{\underset{H}{|}}{\overset{\overset{OH}{|}}{C}}-H \xrightarrow{H_2SO_4} \overset{H}{\underset{H}{>}}C=C\overset{H}{\underset{H}{<}} + H-OH$$

In a dehydration reaction, the carbon-oxygen bond of one carbon atom and a carbon-hydrogen bond of an adjacent carbon atom both break. Because there is often more than one adjacent carbon atom, such as in 2-butanol, mixtures of alkenes can result.

These groups are eliminated

$$CH_3-\underset{\underset{H}{|}}{\overset{\overset{H}{|}}{C}}-\underset{\underset{H}{|}}{\overset{\overset{OH}{|}}{C}}-\underset{\underset{H}{|}}{\overset{\overset{H}{|}}{C}}-H \quad \text{or} \quad CH_3-\underset{\underset{H}{|}}{\overset{\overset{H}{|}}{C}}-\underset{\underset{H}{|}}{\overset{\overset{OH}{|}}{C}}-\underset{\underset{H}{|}}{\overset{\overset{H}{|}}{C}}-H$$

In general, when two or more products are possible in a dehydration reaction, the product formed in higher yield is the isomer that contains the greater number of alkyl groups attached to the double bond (the more substituted alkene). This more stable alkene is often called the **Zaitsev product** (Section 4.12). If the products are a mixture of geometric isomers, the more stable *trans* isomer predominates.

$$\underset{\underset{|}{CH_3CH_2\overset{OH}{C}HCH_2CH_3}}{} \xrightarrow{H_2SO_4} \underset{\underset{\text{25\%}}{cis\text{-2-pentene}}}{\overset{CH_3}{\underset{H}{>}}C=C\overset{CH_2CH_3}{\underset{H}{<}}} + \underset{\underset{\text{75\%}}{trans\text{-2-pentene}}}{\overset{H}{\underset{CH_3}{>}}C=C\overset{CH_2CH_3}{\underset{H}{<}}}$$

Dehydration reactions (and the reverse hydration reactions) occur in metabolism. These reactions are catalyzed by enzymes and are rapid, even though high concentrations of acids are not present and the reaction temperature is only 37°C. One example is the dehydration of citric acid, catalyzed by the enzyme aconitase, to give *cis*-aconitic acid.

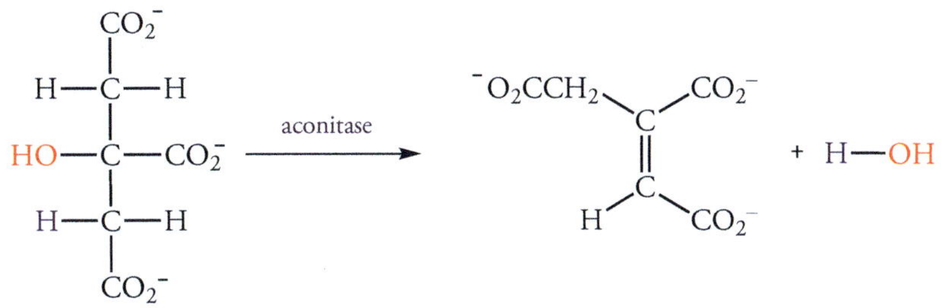

citric acid
(At pH 7 citric acid exists as the
anion, called citrate, as shown above)

cis-aconitate
(The carboxylic acid groups
exist as anions at pH 7)

Reaction Mechanisms

The dehydration of alcohols occurs by mechanisms that depend on the structure of the alcohol. Tertiary alcohols undergo acid-catalyzed dehydration by an E1 mechanism; primary alcohols are dehydrated by an E2 mechanism. In either mechanism, the first step is the protonation of the oxygen atom (an acid-base reaction) to produce an oxonium ion. The acid is represented as HA.

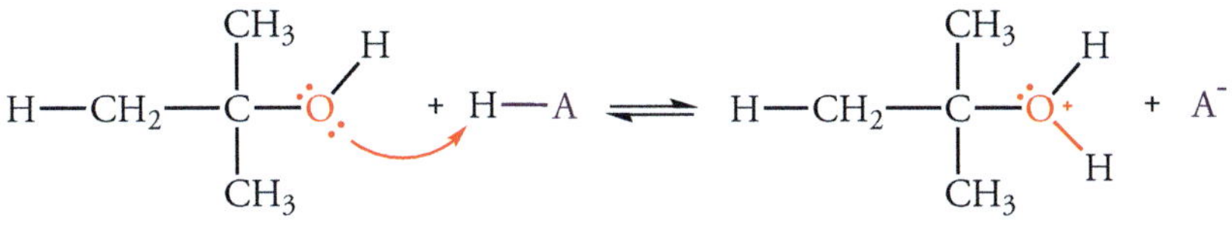

an oxonium ion

A tertiary alcohol loses water by an S_N1 process to produce a tertiary carbocation. The tertiary carbocation then loses a proton from the carbon atom adjacent to the carbon atom bearing the positive charge, producing an alkene.

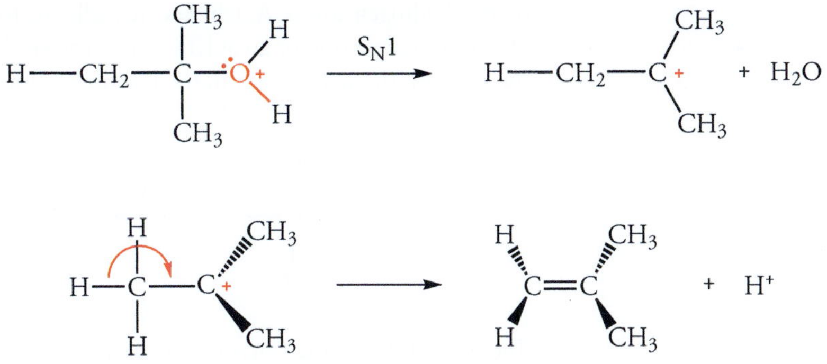

Primary alcohols are dehydrated by an E2 mechanism. The first step again is protonation of the −OH group to give an oxonium ion. However, in the second step, a base removes a proton from the carbon atom adjacent to the carbon atom bearing the oxygen atom. An electron pair in the C—H bond "moves" to form a carbon-carbon double bond, and the electron pair of the C—O bond is retained by the oxygen atom. Formation of the double bond and loss of water occur in a single concerted step.

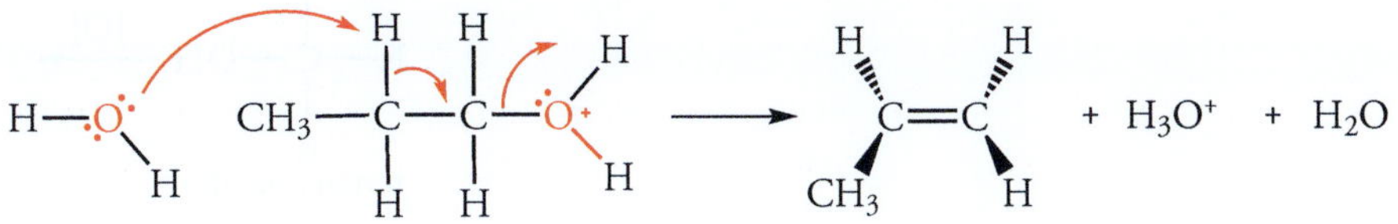

Problem 8.5
Predict the product(s) of the dehydration of 1-methylcyclohexanol.

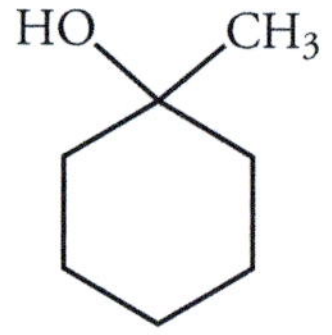

Solution
This tertiary alcohol has three carbon atoms adjacent to the carbon atom bearing the hydroxyl group. Each carbon atom can lose a hydrogen atom in the dehydration reaction. However, loss of a hydrogen atom from either C-2 or C-6 results in the same product. Thus, only two isomers form.

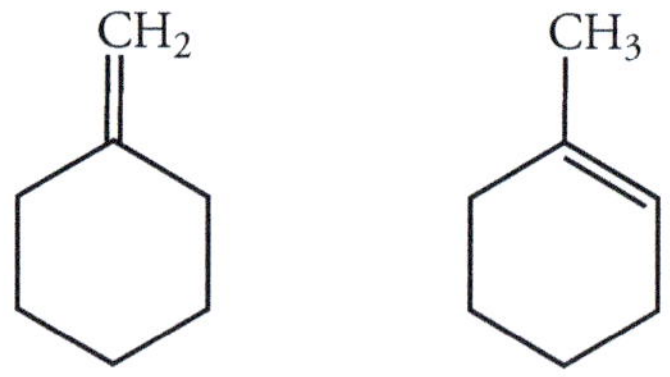

Problem 8.6
Write the structures of the products of dehydration of 4-methyl-2-pentanol. Which of the isomeric alkenes should be the major product?

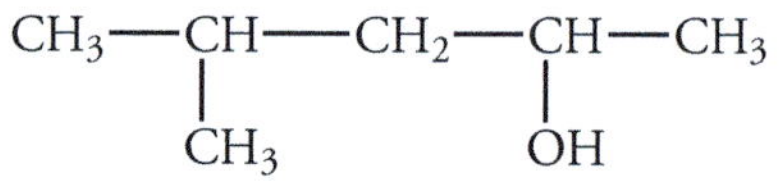

$$CH_3-CH-CH_2-CH-CH_3$$
with CH_3 below the first CH and OH below the second CH

8.6 OXIDATION OF ALCOHOLS

Many oxidizing agents react with primary and secondary alcohols. Primary alcohols, which have the general formula RCH_2OH, can be oxidized to aldehydes. The oxidation reaction occurs with the loss of two hydrogen atoms. Aldehydes are easily oxidized and may react further to produce carboxylic acids. In the conversion of an aldehyde to a carboxylic acid ($RCOOH$), the second oxygen comes from the solvent. The symbol [O] in the reactions shown below denotes an oxidation reaction.

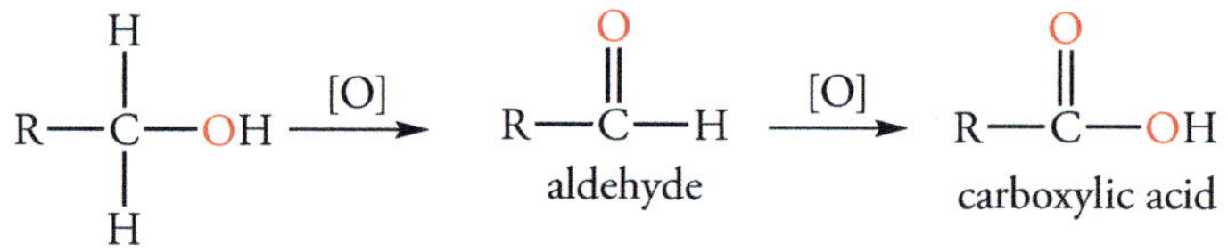

aldehyde → carboxylic acid

The symbol [O] in the reactions shown below denotes an oxidation reaction.

secondary alcohol → ketone

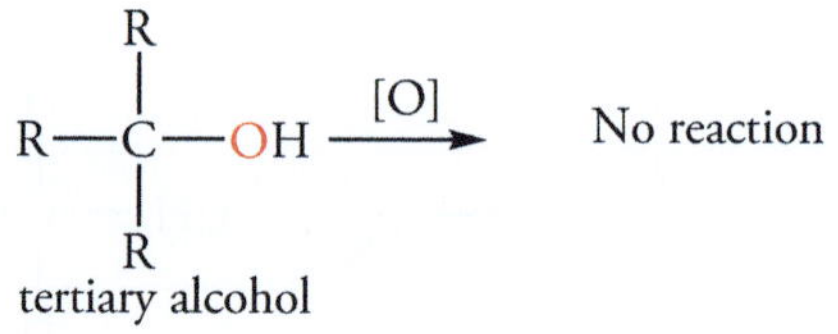

tertiary alcohol → No reaction

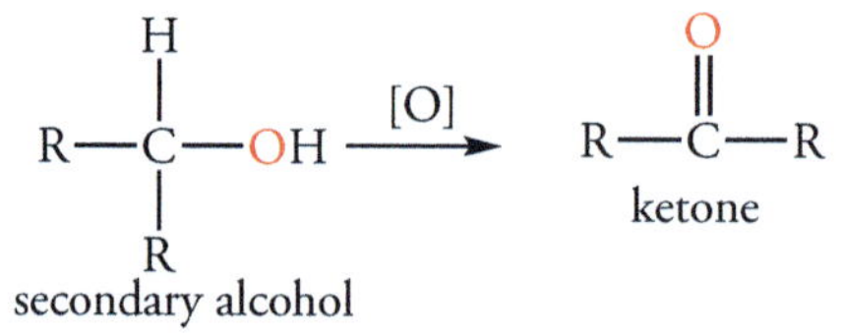

Alcohols are oxidized by the Jones reagent, which consists of chromium trioxide in aqueous sulfuric acid and acetone. The Jones reagent oxidizes primary alcohols to carboxylic acids. This reagent also converts secondary alcohols to ketones.

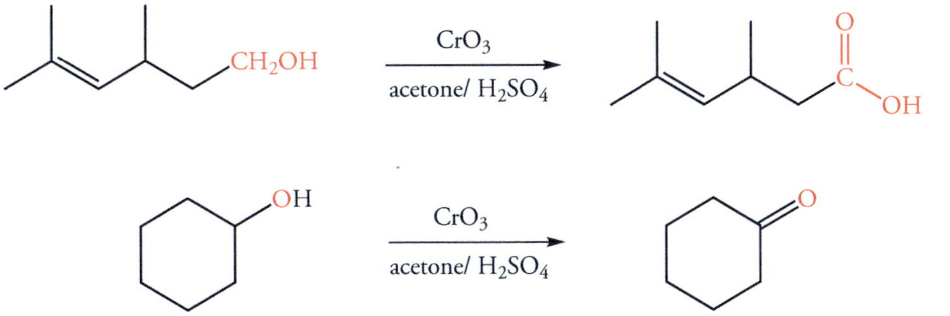

Alcohols are also oxidized by pyridinium chlorochromate (PCC) in methylene chloride (CH_2Cl_2) as solvent. Secondary alcohols are oxidized to ketones. However, in contrast to the Jones reagent, oxidation of primary alcohols by PCC yields aldehydes without continued oxidation to carboxylic acids.

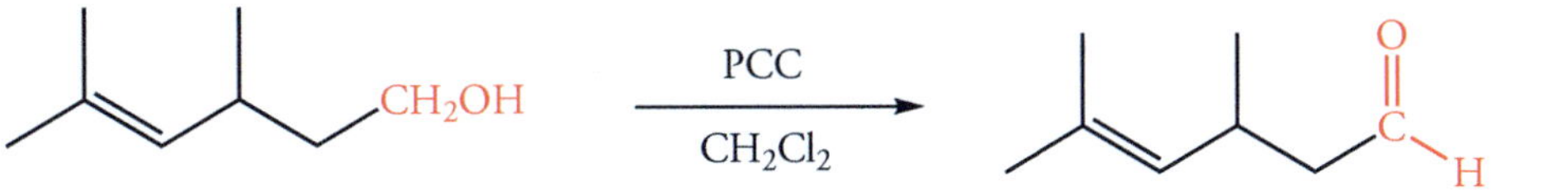

PCC is prepared by dissolving CrO_3 in hydrochloric acid and then adding pyridine. The isolated PCC is used in CH_2Cl_2 as solvent.

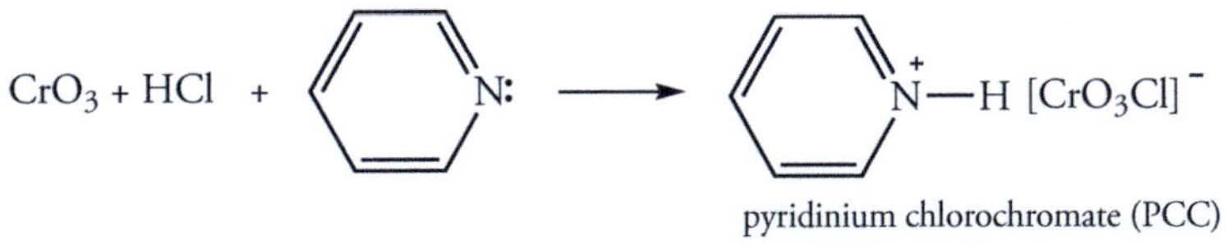

pyridinium chlorochromate (PCC)

Problem 8.7
Which of the isomeric $C_4H_{10}O$ alcohols react with Jones reagent to produce a ketone, C_4H_8O?

Solution
There are four isomeric alcohols because there are four C_4H_9- alkyl groups (Section 3.3). The *n*-butyl and isobutyl groups are primary: the *tert*-butyl group is tertiary. Only the *sec*-butyl group provides a secondary alcohol, and only secondary alcohols yield ketones when oxidized. Thus, only 2-butanol (*sec*-butyl alcohol) yields a ketone.

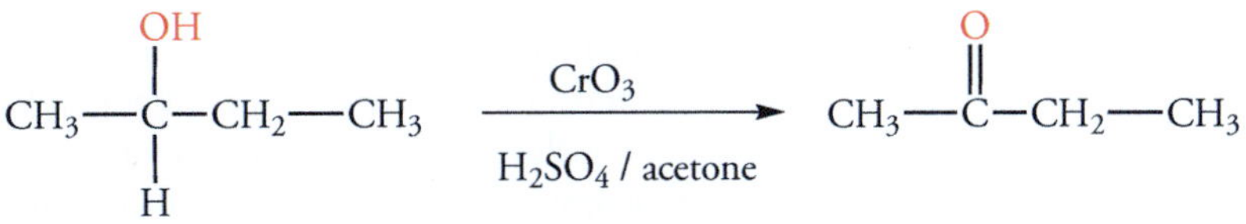

Problem 8.8
Which of the isomeric $C_5H_{12}O$ alcohols react with PCC to produce a ketone, $C_5H_{10}O$?

Toxicity of Alcohols

Methanol is highly toxic. Drinking as little as 15 mL of pure methanol can cause blindness; 30 mL will cause death. Prolonged breathing of methanol vapor is also a serious health hazard. Although ethanol is the least toxic of the simple alcohols, it is still a poisonous substance, and must be oxidized in the liver to prevent high blood alcohol levels, which can poison the brain. The liver enzyme alcohol dehydrogenase (LADH) oxidizes methanol and ethanol. LADH requires a coenzyme, nicotinamide adenine dinucleotide (NAD⁺), as an oxidizing agent. The coenzyme can exist in an oxidized form, NAD⁺, and a reduced form, NADH. NAD⁺-dependent LADH oxidizes ethanol to ethanal (acetaldehyde). Subsequent oxidation of ethanal yields ethanoic acid (acetic acid), which is nontoxic in small concentrations.

$$CH_3-\underset{\underset{H}{|}}{\overset{\overset{H}{|}}{C}}-OH \ + \ NAD^+ \ \xrightarrow{\text{LADH}} \ \underset{CH_3}{}\overset{O}{\underset{}{\overset{||}{C}}}{}_H \ + \ NADH + H^+$$

ethanal
(acetaldehyde)

$$\underset{CH_3}{}\overset{O}{\underset{}{\overset{||}{C}}}{}_H \ + \ NAD^+ \ \xrightarrow{\text{LADH}} \ \underset{CH_3}{}\overset{O}{\underset{}{\overset{||}{C}}}{}_{OH} \ + \ NADH + H^+$$

ethanoic acid
(acetic acid)

The oxidation products of some other alcohols are toxic. In the case of methanol, oxidation catalyzed by LADH gives methanal (formaldehyde), and then methanoic acid (formic acid).

$$H-\underset{\underset{H}{|}}{\overset{\overset{H}{|}}{C}}-OH \ + \ NAD^+ \ \xrightarrow{\text{LADH}} \ \underset{H}{}\overset{O}{\underset{}{\overset{||}{C}}}{}_H \ + \ NADH + H^+$$

methanal
(formaldehyde)

$$\underset{H}{}\overset{O}{\underset{}{\overset{||}{C}}}{}_H \ + \ NAD^+ \ \xrightarrow{\text{LADH}} \ \underset{H}{}\overset{O}{\underset{}{\overset{||}{C}}}{}_{OH} \ + \ NADH + H^+$$

methanoic acid
(formic acid)

Formaldehyde travels in the blood throughout the body and reacts with proteins, destroying their biological function. Methanol causes blindness because formaldehyde destroys an important visual protein. Formaldehyde reacts with an amine functional group of the amino acid lysine in a protein, called rhodopsin. Formaldehyde also reacts with amino groups in other proteins, including many enzymes, and the loss of the function of these biological catalysts causes death.

Ethylene glycol is also toxic. If this sweet-tasting substance is left in open containers, oxidation occurs to give oxalic acid, a toxic substance that causes kidney failure.

$$H-\underset{\underset{H}{|}}{\overset{\overset{OH}{|}}{C}}-\underset{\underset{H}{|}}{\overset{\overset{OH}{|}}{C}}-H \ \longrightarrow \ H-\overset{\overset{O}{||}}{C}-\overset{\overset{O}{||}}{C}-H \ \longrightarrow \ HO-\overset{\overset{O}{||}}{C}-\overset{\overset{O}{||}}{C}-OH$$

ethylene glycol   glyoxal   oxalic acid
1,2-ethane diol

Physicians treat methanol or ethylene glycol poisoning with intravenous injections of ethanol before substantial oxidation has occurred. LADH binds more tightly to ethanol than to methanol or ethylene glycol, and the rate of oxidation of ethanol is about six times faster than that of ethanol. The ethanol concentration can be kept higher because it is directly injected. As a result, neither methanol nor ethylene glycol is competitively oxidized to toxic products, and the kidneys can slowly excrete them.

8.7 SYNTHESIS OF ALCOHOLS

We have already discussed two methods of preparing alcohols: (1) substitution of a halide by hydroxide and (2) addition of water to an alkene. These reactions often have low yields. The yield in the substitution reaction is diminished by the competing elimination reaction (Chapter 7). The yield in the hydration of an alkene is somewhat limited because the reaction is reversible (Chapter 4).

In this section we will discuss two types of reactions that give excellent yields of alcohols. The first type is the reduction of carbonyl compounds; the second is an "indirect" hydration of alkenes. In both reactions, the functional group converted to the hydroxyl group is located on the proper hydrocarbon skeleton. In Chapter 10 we will examine reactions to simultaneously form alcohols and build new hydrocarbon skeletons.

Reduction of Carbonyl Compounds

Alcohols can be produced by reducing the carbonyl group of aldehydes and ketones with hydrogen gas at approximately 100 atm pressure in the presence of a metal catalyst such as palladium, platinum, or a special reactive form of nickel called Raney nickel. Aldehydes yield primary alcohols; ketones yield secondary alcohols.

an aldehyde → a primary alcohol

a ketone → a secondary alcohol

The reduction reaction occurs by the transfer of hydrogen atoms bound to the surface of the metal catalyst to the carbonyl oxygen and carbon atoms. Note that the same catalysts are used for the hydrogenation of alkenes, which is a much faster reaction that occurs at 1 atm pressure. Thus, any carbon-carbon double bonds in a molecule are reduced under the conditions required to reduce a carbonyl group.

A carbonyl group can be reduced to an alcohol selectively by reagents attracted to the highly polarized carbonyl group. The carbonyl carbon atom has a partial positive charge and tends to react with nucleophiles.

The carbon-carbon double bond of alkenes is not polar, so it does not react with nucleophiles. This difference in reactivity is the basis for the reduction of carbonyl compounds by metal hydrides, such as sodium borohydride, $NaBH_4$, and lithium aluminum hydride, $LiAlH_4$, neither of which reacts with alkenes. Both reagents are a source of a nucleophilic hydride ion. Sodium borohydride is less reactive than lithium aluminum hydride, but both easily reduce both aldehydes and ketones. Sodium borohydride can be used in ethanol as the solvent.

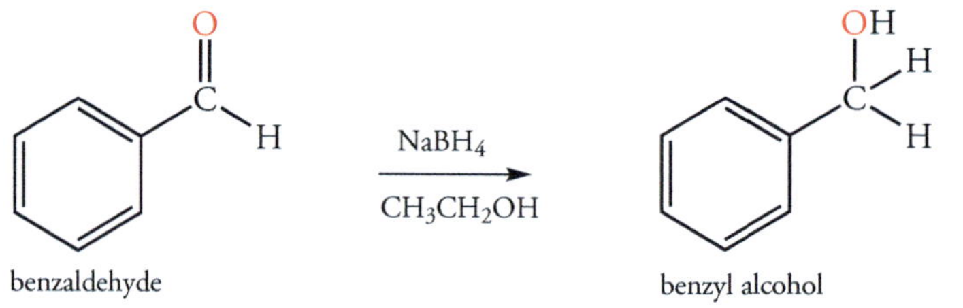

benzaldehyde → benzyl alcohol

In reduction by sodium borohydride, a hydride ion of the borohydride ion, BH_4^-, is transferred to the carbonyl carbon atom from boron, and the carbonyl oxygen atom is protonated by the ethanol.

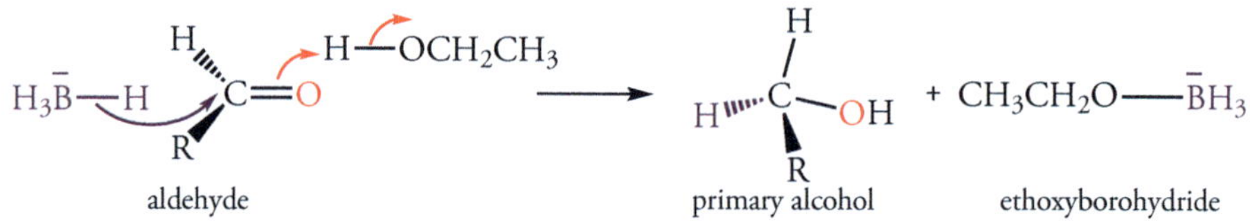

aldehyde → primary alcohol + ethoxyborohydride

The ethoxyborohydride product in the above reaction has three remaining hydride ions available for further reduction reactions, and the ultimate boron product is tetraethoxyborohydride, $(RO)_4B^-$. Thus, one mole of $NaBH_4$ reduces four moles of a carbonyl compound. A dilute solution of acid is used to destroy any excess reagent as part of the standard work-up procedure.

When lithium aluminum hydride is used to reduce carbonyl compounds, an ether, such as diethyl ether, $(CH_3CH_2)_2O$, is used as the solvent. The reduction of a carbonyl group by lithium aluminum hydride occurs by transfer of a hydride anion from AlH_4^- to the carbonyl carbon atom. The carbonyl oxygen atom forms an alkoxyaluminate salt.

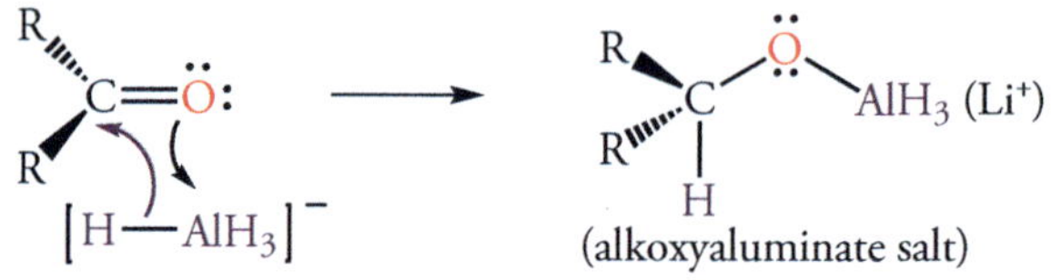

(alkoxyaluminate salt)

The initial alkoxyaluminate has three remaining hydride ions available for further reduction reactions, and the ultimate aluminum product is tetraalkoxyaluminate, $(RO)_4Al^-$. Thus, one mole of $LiAlH_4$ reduces four moles of a carbonyl compound. The tetraalkoxyaluminate is hydrolyzed with aqueous acid in a separate, second step.

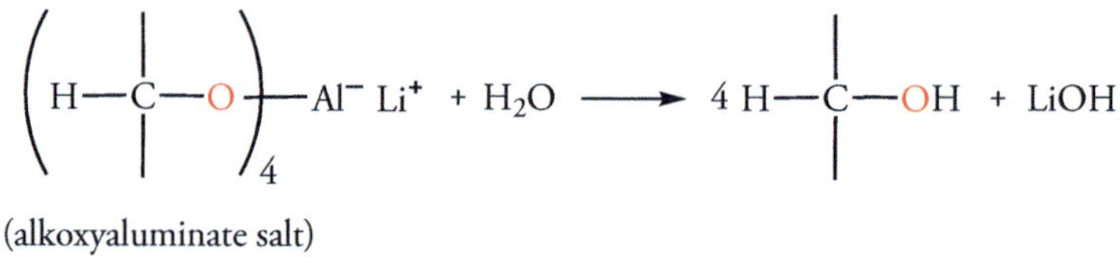

(alkoxyaluminate salt)

Reduction of carbonyl compounds in metabolism occurs via NADH, the reduced form of the coenzyme NAD$^+$, nicotinamide adenine dinucleotide. The structures of the two forms of the coenzyme are shown with an R group representing a portion of the molecule that is not directly involved in its action.

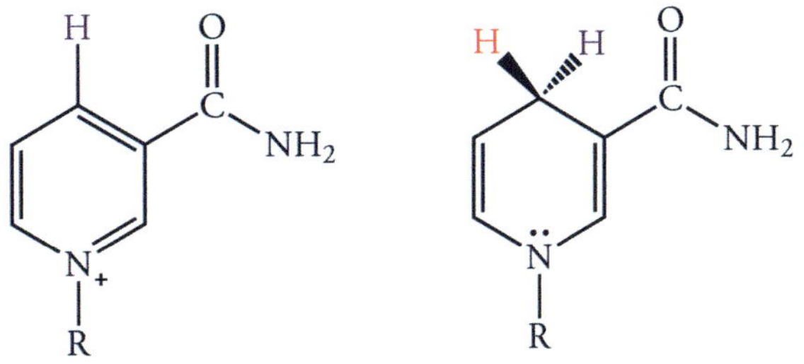

NAD
(oxidized form)

NADH
(reduced form)

The reduced form formally behaves as a hydride source in which an enzyme called a dehydrogenase transfers a hydrogen atom and a bonding pair of electrons from NADH to a carbonyl carbon atom.

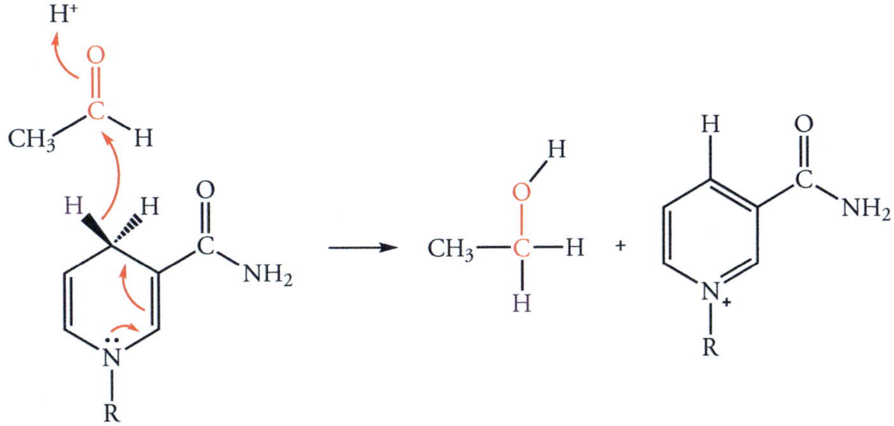

NADH

NAD$^+$

Indirect Hydration of Alkenes

In Chapter 4 we discussed the "direct" electrophilic addition of water to alkenes to give alcohols. In this section we consider two "indirect" ways to add the elements of water to a double bond. These methods are indirect because the hydroxyl group, the hydrogen atom, or both originate in reagents other than water. One such reaction, oxymercuration-demercuration of an alkene, gives a product that corresponds to Markovnikov addition of water. Hydroboration-oxidation, another indirect reaction, gives the equivalent of an anti-Markovnikov addition of water to the alkene.

In an oxymercuration-demercuration reaction, an alkene is treated with mercuric acetate, $Hg(OAc)_2$, and the product is treated with sodium borohydride. The net result is a **Markovnikov addition product** in which the $-OH$ group is bonded to the more substituted carbon atom of the alkene.

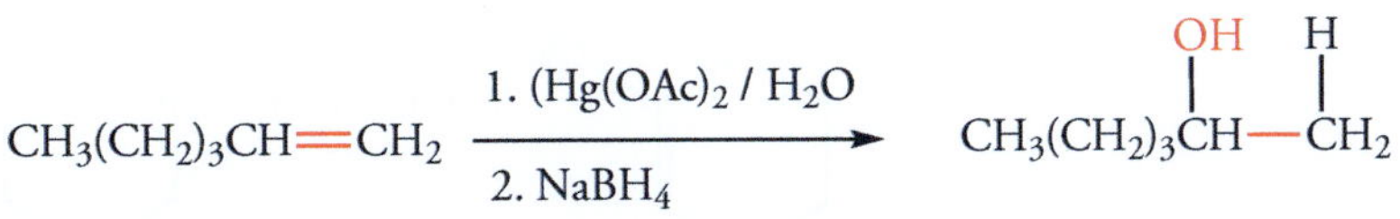

$$CH_3(CH_2)_3CH\!=\!CH_2 \xrightarrow[\text{2. NaBH}_4]{\text{1. (Hg(OAc)}_2 \text{ / H}_2\text{O}} CH_3(CH_2)_3\overset{\text{OH}}{CH}\!-\!\overset{\text{H}}{CH_2}$$

1-hexene   2-hexanol

In the first step, an electrophilic HgOAc$^+$ ion adds to the double bond followed by attack by a nucleophilic water molecule. The net result is the bonding of $-HgOAc$ and a hydroxyl group on adjacent carbon atoms. The product corresponds to a Markovnikov addition because the electrophile attacks the less substituted carbon atom.

$$CH_3(CH_2)_3CH\!=\!CH_2 + {}^+Hg(OAc) \xrightarrow{\text{H}_2\text{O}} CH_3(CH_2)_3\overset{\text{OH}}{CH}\!-\!\overset{\text{HgOAc}}{CH_2} + H^+$$

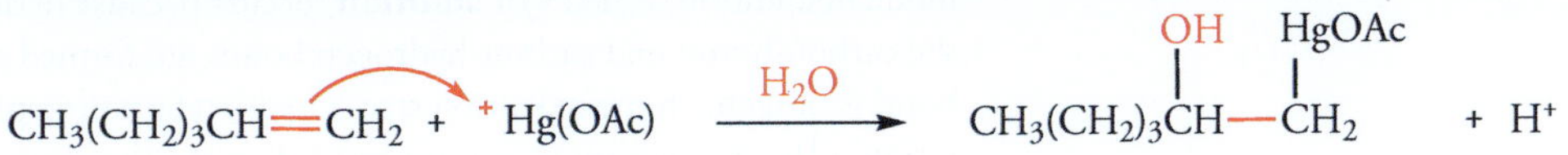

The organomercury compound is then reduced with sodium borohydride, and the —HgOAc group is replaced by a hydrogen atom. Thus, oxymercuration-demercuration gives the product that would result from direct hydration of an alkene.

$$\underset{\text{OH}}{\underset{|}{\text{CH}_3(\text{CH}_2)_3\text{CH}}}\text{—}\underset{\underset{|}{\text{HgOAc}}}{\text{CH}_2} \quad \xrightarrow{\text{NaBH}_4} \quad \underset{\text{OH}}{\underset{|}{\text{CH}_3(\text{CH}_2)_3\text{CH}}}\text{—}\underset{\underset{|}{\text{H}}}{\text{CH}_2} \ + \ \text{Hg} \ + \ \text{OAc}^-$$

Hydroboration-oxidation of alkenes, which was developed by the American chemist H.C. Brown, also requires two steps. The sequence of reactions adds the elements of water to a double bond to give a product that corresponds to **anti-Markovnikov addition**.

$$\text{CH}_3(\text{CH}_2)_3\text{CH}{=}\text{CH}_2 \quad \xrightarrow[\text{2. H}_2\text{O}_2 \ / \ \text{OH}^-]{\text{1. B}_2\text{H}_6} \quad \underset{\underset{|}{\text{H}}}{\text{CH}_3(\text{CH}_2)_3\text{CH}}\text{—}\underset{\underset{|}{\text{OH}}}{\text{CH}_2}$$

anti-Markovnikov addition product

In the hydroboration step, an alkene is treated with diborane, B_2H_6. Diborane acts as if it were the monomeric species called borane, BH_3. The compound is usually prepared in an ether solvent such as diethyl ether or tetrahydrofuran (Chapter 9). It adds to the carbon-carbon double bond of one alkene and then adds successively to two more alkenes to produce a trialkylborane, R_3B. These steps are hydroboration reactions.

In the oxidation step, the trialkylborane is treated with hydrogen peroxide and base to oxidize the organoborane to form an alcohol.

trialkylborane

The hydroboration-oxidation of 1-methylcyclohexene gives the anti-Markovnikov product; the hydrogen atom is added to the more substituted carbon atom, and the hydroxyl group is on the less substituted carbon atom. (Only a single addition of BH_3 is shown for the sake of simplicity.)

trialkylborane   syn addition

Note that the hydrogen atom and BH_2 unit are introduced from the same side of the double bond. This mode of addition, called **syn addition**, occurs because hydroboration is a concerted process. That is, the carbon-boron and carbon-hydrogen bonds are formed at the same time that the boron-hydrogen bond is broken. In the oxidation step a hydroxyl group replaces the boron with retention of configuration. (The mechanism of this process is beyond the scope of this text.)

Borane reacts with alkenes for two reasons. First, the boron atom in borane is an electron-deficient species—it has only six electrons. Thus, the boron atom has a vacant 2p orbital and is an electrophile. It bonds to the least substituted carbon atom—much like a proton. Second, boron is more electropositive than hydrogen. Therefore, the hydrogen atom of the boron-hydrogen bond has a partial negative charge. This hydrogen atom behaves like a hydride ion, not like a proton. In summation, two properties of BH_3—the electrophilic character of the boron atom and the hydride character of the hydrogen atom—account for anti-Markovnikov addition of BH_3 to alkenes.

Problem 8.9

What product forms from the following alkene by oxymercuration-demercuration? What product is formed by hydroboration-oxidation?

Solution

The alkene is disubstituted, and both alkyl groups are bonded to the same carbon atom. The double-bonded CH_2 is the less substituted carbon atom; the ring carbon atom is the more substituted carbon atom. An oxymercuration-demercuration reaction places a hydrogen atom at the CH_2 site and a hydroxyl group on the ring carbon atom. This is equivalent to Markovnikov addition of water to the alkene.

The hydroboration-oxidation product has a hydroxyl group at the CH_2 site and a hydrogen atom at the ring carbon atom. This process is equivalent to anti-Markovnikov addition of water to the double bond.

Problem 8.10

Write the product of the oxymercuration-demercuration and for hydroboration-oxidation of 3,3-dimethyl-1-butene.

Industrial Synthesis of Alcohols

About 10 billion pounds of methanol are produced annually in the United States by reaction of carbon monoxide and hydrogen in a 1:2 mixture known as *synthesis gas*.

$$C\equiv O \ + \ 2\,H_2 \ \xrightarrow[\substack{400\,°C \\ 600\,atm}]{Cr/Zn} \ CH_3OH$$

Synthesis gas is obtained by adjusting the ratio of carbon monoxide to hydrogen of the water gas mixture obtained by the reaction of water with methane.

$$CH_4 \ + \ H_2O \ \xrightarrow[700\,°C]{NO} \ CO \ + \ 3\,H_2O$$

Methanol is used as a solvent and as an antifreeze in windshield wiper fluid. It is also used as a fuel in racing cars because it is a clean-burning fuel and has an octane number of 116. However, most of the methanol produced is converted to other compounds. About 50% is catalytically oxidized by oxygen in air to methanal (formaldehyde) for use in the synthesis of resins and plastics such as Bakelite®.

$$2\ CH_3OH\ +\ O_2\ \xrightarrow[700\ °C]{ZnO/Cr_2O_3}\ 2\ \underset{H}{\overset{H}{C}}{=}O\ +\ 2\ H_2O$$

The gasoline additive MTBE (*tert*-butyl methyl ether) is produced by adding methanol to 2-methyl-propene. This mixture is heated at high temperature over an acidic catalyst to give MTBE, the addition product predicted by Markovnikov's rule. MTBE increases the octane number of gasoline by about 5 units.

Ethanol has been prepared by distillation of the mixture obtained by fermentation of grains and sugar. However, currently only 5% of ethanol is produced by this industrial method. Most ethanol is obtained by acid-catalyzed hydration of ethylene. Ethylene is a chemical intermediate for the production of many industrial products.

$$:\!\ddot{O}\!-\!H\ +\ CH_2{=}CH_2\ \xrightarrow[300\ °C]{H_3PO_4}\ H{-}\underset{H}{\overset{CH_3}{C}}{-}\ddot{O}H$$

Isopropyl alcohol is also produced by an acid-catalyzed hydration reaction. It is used as a solvent and as an intermediate for other industrial syntheses, such as the production of propanone (acetone), which is an important solvent.

$$:\!\ddot{O}\!-\!H\ +\ CH_2{=}CH{-}CH_3\ \xrightarrow[300\ °C]{H_3PO_4}\ H{-}\underset{CH_3}{\overset{CH_3}{C}}{-}\ddot{O}H$$

$$H{-}\underset{CH_3}{\overset{CH_3}{C}}{-}\ddot{O}H\ \xrightarrow[400\ °C]{ZnO}\ CH_3{-}\overset{:O:}{C}{-}CH_3$$

Million of tons of ethylene oxide are produced annually from ethylene by direct air oxidation over a silver oxide catalyst. Most of this product is used to produce ethylene glycol in an acid-catalyzed hydrolysis reaction.

$$CH_2{=}CH_2\ \xrightarrow{O_2/AgO}\ \triangle\!\!O\ \xrightarrow{H_3O^+}\ HOCH_2{-}CH_2OH$$

ethylene oxide   ethylene glycol

8.8 PHENOLS

The chemistry of phenols is quite different from the chemistry of alcohols because it is very difficult to break the C—O bond in phenols. The carbon atom is sp^2-hybridized and the C—O bond in phenols is shorter and stronger than the C—O bond of alcohols, where the carbon atom is sp^3-hybridized. As a result, neither S_N2 nor S_N1 processes occur.

Acidity of Phenols

Phenols are stronger acids than alcohols, but they are still quite weak acids. A typical alcohol has a pK_a of 16-17. In contrast, phenol is 10 million times more acidic: its pK_a is 10.

$$R{-}O{-}H\ +\ H_2O\ \rightleftharpoons\ R{-}O^-\ +\ H_3O^+\qquad pK_a\ 17$$

$$Ar{-}O{-}H\ +\ H_2O\ \rightleftharpoons\ Ar{-}O^-\ +\ H_3O^+\qquad pK_a\ 10$$

Phenol is more acidic than cyclohexanol and acyclic alcohols because the phenoxide ion is more stable than the alkoxide ion. In an alkoxide ion, such as the one derived from cyclohexanol, the negative charge is localized at the oxygen atom. However, in a phenoxide ion, the negative charge is delocalized over the benzene ring; that is, it is resonance-stabilized.

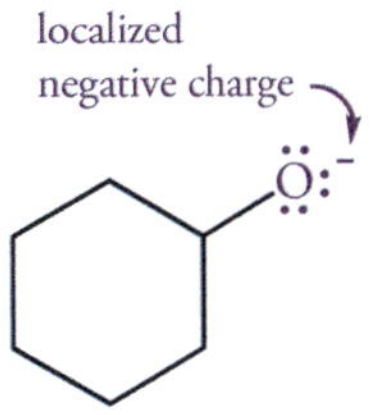

localized
negative charge

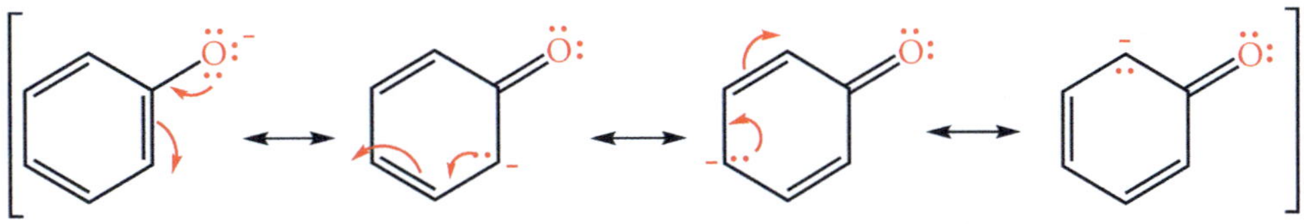

delocalized resonance contributing structures

Phenols are more acidic when the ring is substituted with electron-withdrawing groups. These substituents stabilize the phenoxide ion by further delocalizing the negative charge. Phenols substituted with electron-donating groups are less acidic than phenol.

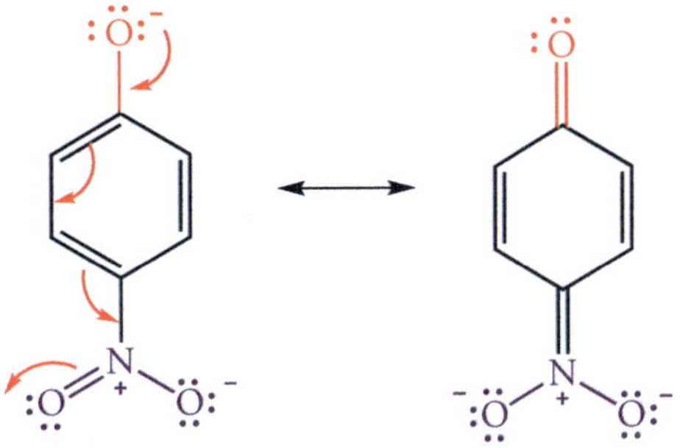

Phenols are much more acidic than alcohols, and as a result they dissolved in basic solutions. However, phenols are not sufficiently acidic to react with aqueous sodium bicarbonate. In contrast to organic acids such as carboxylic acids ($pK_a = 5$) phenols react with aqueous sodium bicarbonate to form carbon dioxide gas and a carboxylate anion.

Oxidation of Phenols

Phenols are oxidized to give conjugated 1,4-diketones called quinones. Hydroquinone, a phenol with two hydroxyl groups, is very easily oxidized. It will reduce silver bromide that has been activated by exposure to light in photographic film emulsions.

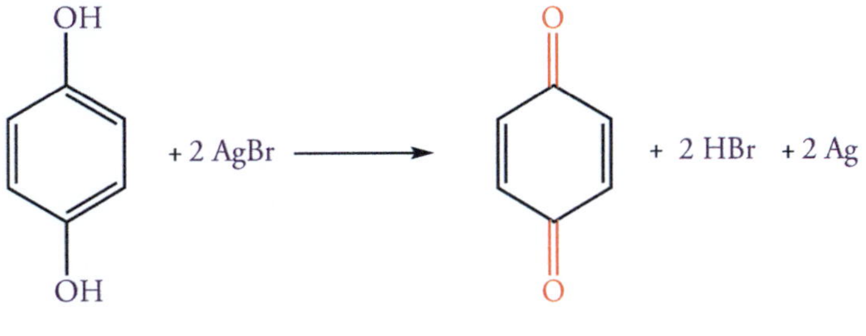

OH

+ 2 AgBr $\longrightarrow$

O

+ 2 HBr + 2 Ag

OH

hydroquinone

O

p-benzoquinone
(quinone)

In classical, film-based photography, hydroquinone is used as a developer. It reacts faster with light-activated silver bromide than with unexposed grains of the salt. As a result, silver precipitates in the film at points where exposure to light occurred. The result is a "negative" image. Areas in the film negative that are clear correspond to unlit areas in the scene that was photographed. Areas that are very dark in the negative correspond to well-lit areas in the scene photographed.

Quinones are widely distributed in nature, where along with the reduced hydroquinone form, they serve as reducing and oxidizing agents. Coenzyme Q, also called ubiquinone, is a quinone found within the mitochondria of cells. Coenzyme Q, in its oxidized form, oxidizes NADH to regenerate

NAD, a common oxidizing agent in biological reactions. The side chain represented by R in the following structure is a polyunsaturated unit that consists of isoprene units. Although the isoprene units are required for biological activity, they do not participate in the redox reaction.

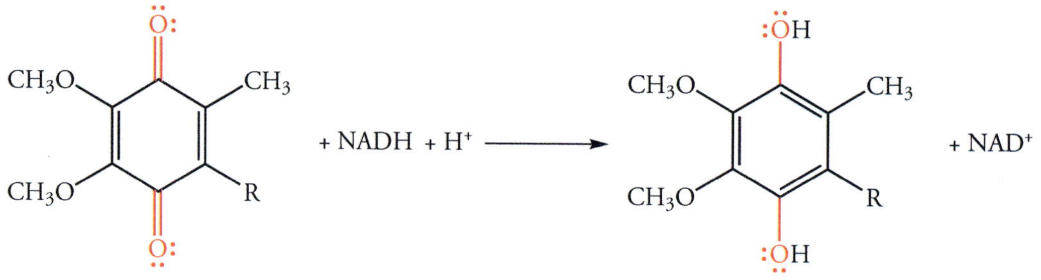

coenzyme Q (oxidized form)   + NADH + H⁺ ⟶   coenzyme Q (reduced form)   + NAD⁺

Phenols are more acidic when the ring is substituted with electron-withdrawing groups. These substituents stabilize the phenoxide ion by further delocalizing the negative charge. Phenols substituted with electron-donating groups are less acidic than phenol.

Phenols As Germicides

Germicides are compounds that are classified as disinfectants and antiseptics. Disinfectants are compounds that decrease the bacterial count on objects such as medical instruments. Antiseptics also inhibit bacterial growth but are used on living tissue. The English surgeon Joseph Lister used phenol itself as a hospital disinfectant in the late nineteenth century. However, phenol is no longer used as an antiseptic, because it causes severe burns. A 2% solution was formerly used to decontaminate medical instruments. However, this use has also been largely discontinued, because substituted phenols and other compounds are more effective.

The efficiency of a germicide is measured in terms of its **phenol coefficient (PC)**. Phenol itself has a PC value of 1. If a 1% solution of a germicide is as effective as a 10% solution of phenol, it has a PC value of 10. If a phenol is substituted with alkyl groups, its germicidal action increases. For example, the methyl-substituted phenols called *ortho-*, *meta-*, and *para*-cresol are used in the commercial disinfectant Lysol®. Another phenol, called thymol, is sometimes used by dentists to sterilize a tooth before filling it.

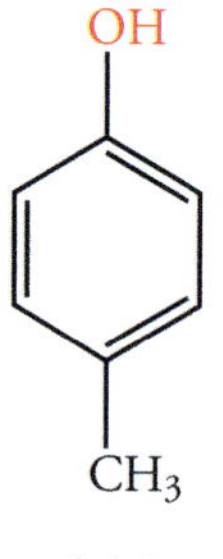

p-methylphenol
(p-cresol)

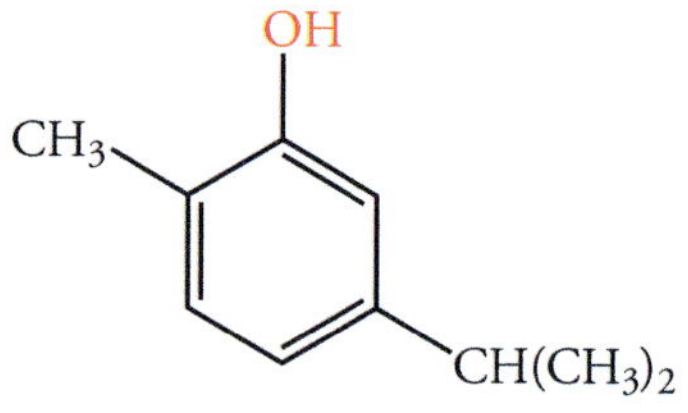

5-isopropyl-2-methyl-phenol
(thymol)

The phenol coefficient increases when the phenol is halogenated, particularly if the halogen is *para* to the hydroxyl group. The structures of a few halogenated phenols are shown in the following structures. Chlorophene is a more effective germicide than *o*-phenylphenol. 4-Chloro-3,5-dimethylphenol is more effective than the cresols; it is used in topical preparations for athlete's foot and jock itch. Hexachlorophene (PC = 120) has been used in some toothpastes, deodorants, and soaps. However, it is toxic to infants and is no longer used in commercial products, although it is still used as a surgical scrub.

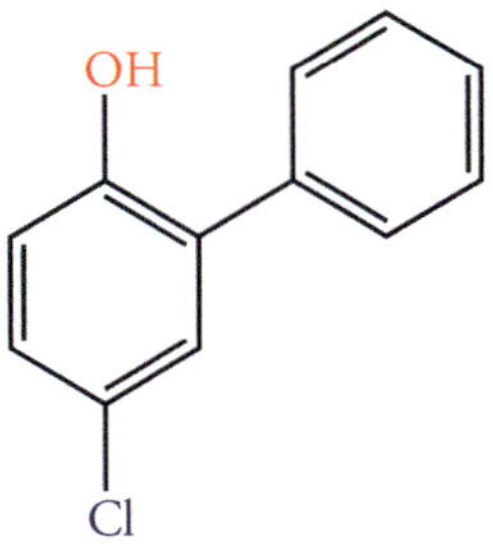

chlorophene

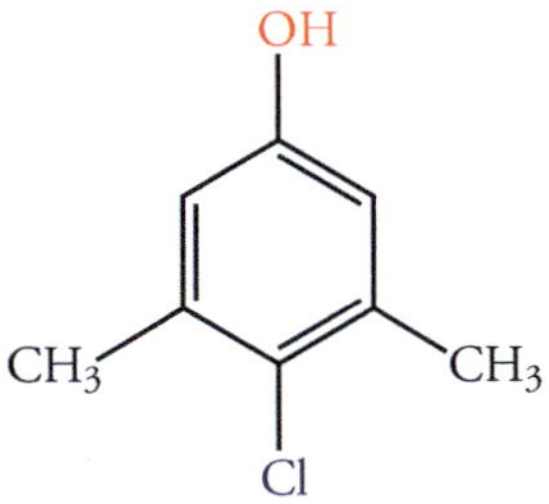

4-chloro-3,5-dimethylphenol

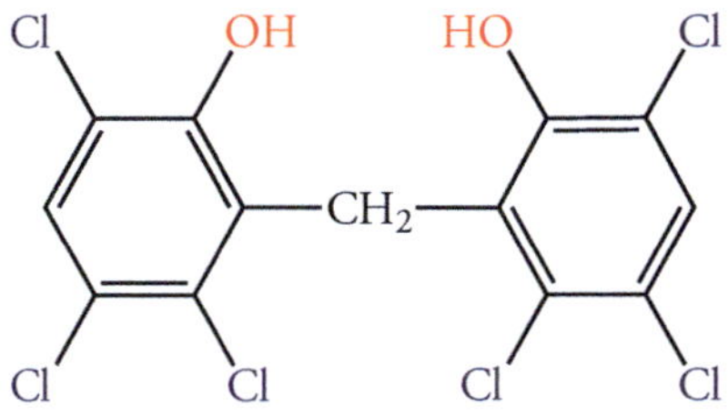

hexachlorophene

Phenols that contain two hydroxyl groups are also germicides. When the two hydroxyl groups are *meta*, the phenols are called resorcinols. Resorcinol and hexylresorcinol are effective germicides. Resorcinol has a PC of only 0.4, but it is useful in the treatment of psoriasis and seborrhea. As in the case of phenols, alkyl substitution increases the PC. Hexylresorcinol has a PC of 98. It is used in throat lozenges.

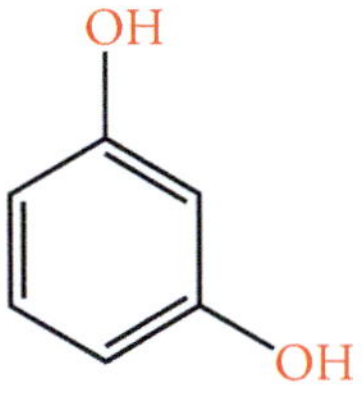

resorcinol

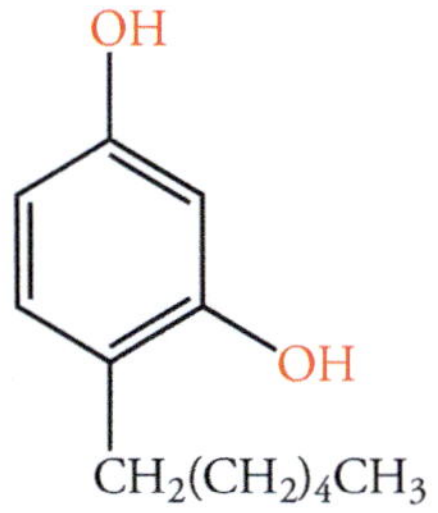

hexylresorcinol

The sensitivity of biological function to differences in chemical structure is illustrated by the properties of phenols having hydroxyl groups in a 1,2 arrangement (pyrocatechols). A series of compounds where the R group at the 3-position is a polyunsaturated chain of 15 carbon atom are urushiols. They are the skin irritant found in ivy and poison oak!

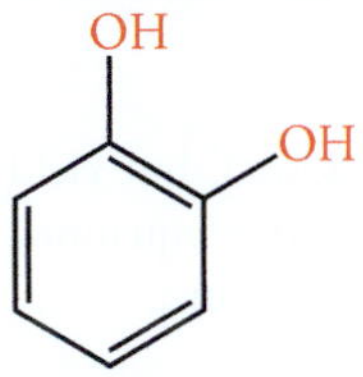

pryocatechol

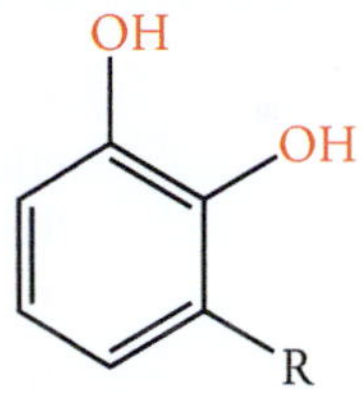

various urushiols

8.9 SULFUR COMPOUNDS: THIOLS AND THIOETHERS

Sulfur is in the same group of the periodic table as oxygen and forms compounds structurally similar to alcohols. Compounds containing an −SH group, called a **sulfhydryl group**, are named as **mercaptans** or **thiols**. The nomenclature of these compounds resembles that of alcohols, except that the suffix *-thiol* replaces the suffix *-ol* and the *-e* of the alkane name is retained.

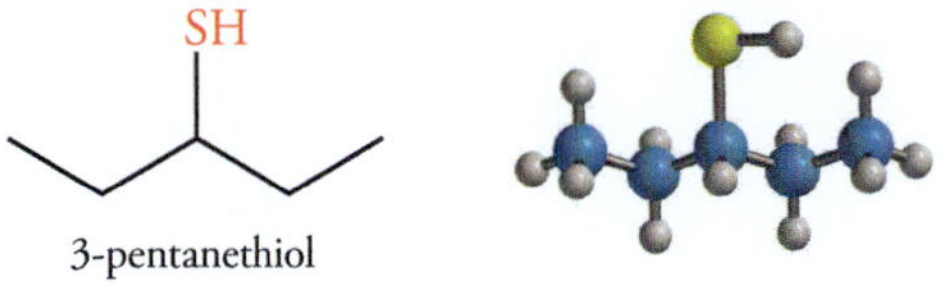

3-pentanethiol

Properties of Thiols

Alcohols and thiols resemble each other in many ways, but they also differ in some significant respects. For example, thiols have lower boiling points than the corresponding alcohols because sulfur does not form intermolecular hydrogen bonds. We recall from Chapter 2 that only nitrogen, oxygen, and fluorine form hydrogen bonds.

1-butanethiol
bp = 98 °C

1-butanol
bp = 117 °C

Some alcohols have rather sweet odors, but one of the distinguishing properties of thiols is their strong, disagreeable odor. The odor of the striped skunk (*Memphitis mephitis*) is due to 3-methyl-1-butanethiol. The human nose can detect thiols at a few parts per billion in air. There is a positive side to the obnoxious odors of thiols: Small amounts of thiols are added to natural gas to aid in easy detection of leaks. A skunk takes more drastic measures to deter predators. Although thiols are weak acids, they are far stronger than alcohols.

$$CH_3CH_2SH + H_2O \rightleftharpoons CH_3CH_2S^- + H_3O^+ \quad pK_a = 8$$

$$CH_3CH_2OH + H_2O \rightleftharpoons CH_3CH_2O^- + H_3O^+ \quad pK_a = 17$$

The sulfhydryl group is acidic enough to react with hydroxide ions to form thiolate salts. Thiolate anions are excellent nucleophiles.

$$R—\ddot{S}—H + NaOH \longrightarrow R—\ddot{S}:^- + Na^+ + H_2O$$
a thiolate

Reactions of Thiols

Thiols can be obtained from haloalkanes by nucleophilic displacement of halide ion with the sulfhydryl ion (HS⁻). Thioethers, the sulfur analogs of ethers, are obtained from haloalkanes by nucleophilic displacement with thiolates.

$$R—Br + HS^- \longrightarrow R—S—H + Br^-$$

$$R—Br + R'—S^- \longrightarrow R—S—R' + Br^-$$
a thioether

Thiols are easily oxidized, but yield disulfides rather than the structural analogs of aldehydes and ketones. In the following equation, the symbol [O] represents an unspecified oxidizing agent that removes the hydrogen atoms.

$$2 R—S—H \xrightarrow{[O]} R—S—S—R$$
a disulfide

This reaction is of great biochemical importance because many proteins contain the amino acid cysteine. Oxidation of the sulfhydryl group gives a disulfide bond in an amino acid called cystine.

2 H₃N⁺—C—H with CO₂⁻ and CH₂—S—H (cysteine) → [O] → H₃N⁺—C—H with CO₂⁻ and CH₂—S—S—CH₂ and H₃N⁺—C—H with CO₂⁻ (cystine)

In some cases, the sulfhydryl groups in an enzyme must be maintained in the reduced state for proper biological function. If an essential cysteine sulfhydryl group is oxidized, the enzyme becomes inactive. Sulfhydryl groups in enzymes (E—S—H) also react with salts of lead and mercury. Because these reactions inactivate enzymes, lead and mercury salts are highly toxic.

$$2 E—S—H + Hg^{+2} \longrightarrow E—S—Hg—S—E + 2 H^+$$

SUMMARY OF REACTIONS

1. Synthesis of Haloalkanes from Alcohols (Section 8.5)

$$CH_3CH_2\overset{\displaystyle CH_3}{\underset{|}{C}}HCH_2CH_2OH + HCl \xrightarrow{Zn} CH_3CH_2\overset{\displaystyle CH_3}{\underset{|}{C}}HCH_2CH_2Cl$$

(tetralin with OH) + HBr ⟶ (tetralin with Br)

(cyclooctane with OH) + SOCl$_2$ ⟶ (cyclooctane with Cl)

$$3 \text{ (cyclopentyl)}-CH_2CH_2OH + PBr_3 \longrightarrow \text{(cyclopentyl)}-CH_2CH_2-Br + H_3PO_3$$

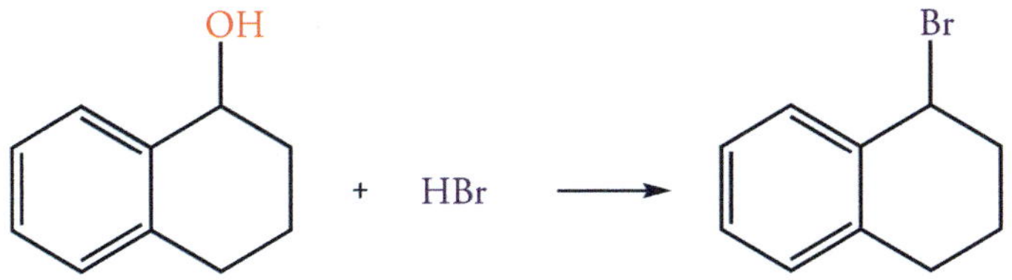

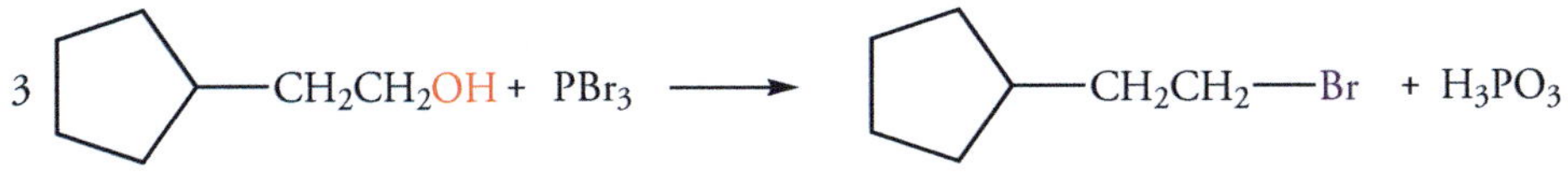

2. Dehydration of Alcohols (Section 8.6)

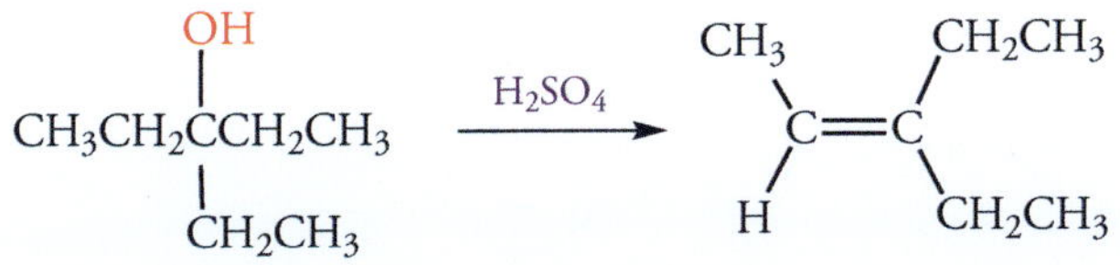

$$CH_3CH_2\overset{\displaystyle OH}{\underset{\underset{\displaystyle CH_2CH_3}{|}}{\overset{|}{C}}}CH_2CH_3 \xrightarrow{H_2SO_4} \underset{H}{\overset{CH_3}{>}}C=C\underset{CH_2CH_3}{\overset{CH_2CH_3}{<}}$$

3. Oxidation of Alcohols (Section 8.7)

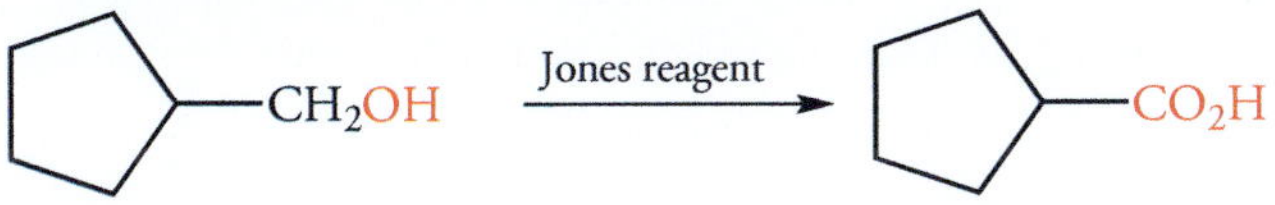

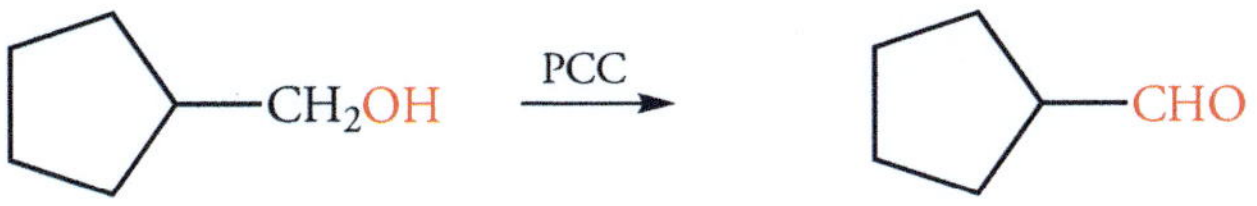

(cyclopentyl)−CH$_2$OH $\xrightarrow{\text{Jones reagent}}$ (cyclopentyl)−CO$_2$H

(cyclooctane with OH) $\xrightarrow{PCC}$ (cyclooctanone, =O)

(cyclopentyl)−CH$_2$OH $\xrightarrow{PCC}$ (cyclopentyl)−CHO

4. Synthesis of Alcohols by Reduction of Carbonyl Compounds (Section 8.8)

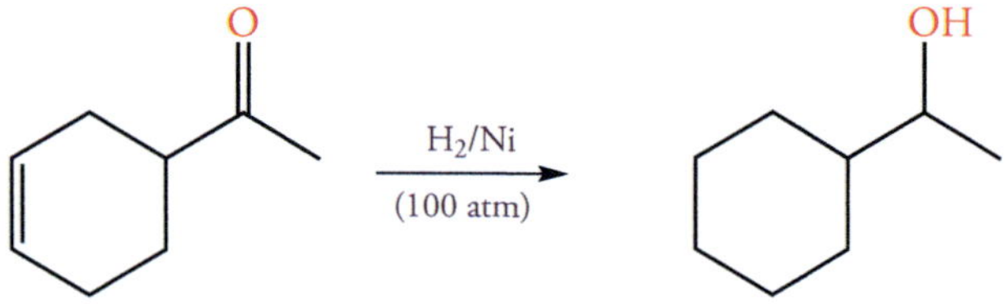

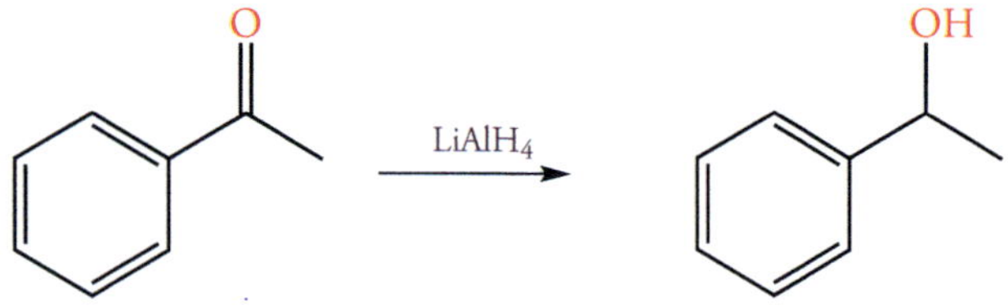

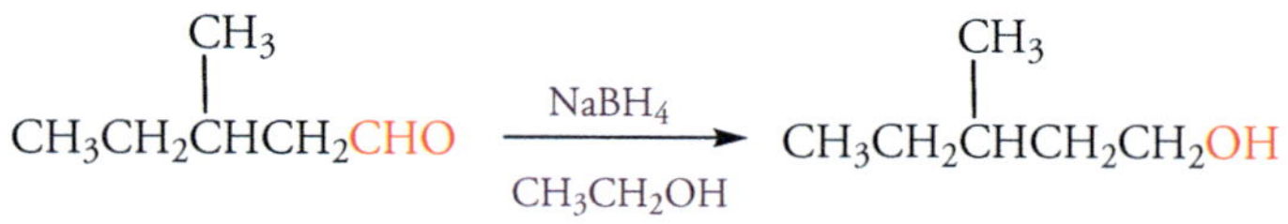

$$\text{H}_2/\text{Ni} \quad (100 \text{ atm})$$

$$\text{LiAlH}_4$$

$$\underset{\text{CH}_3}{\text{CH}_3\text{CH}_2\overset{|}{\text{CH}}\text{CH}_2\text{CHO}} \quad \xrightarrow[\text{CH}_3\text{CH}_2\text{OH}]{\text{NaBH}_4} \quad \underset{\text{CH}_3}{\text{CH}_3\text{CH}_2\overset{|}{\text{CH}}\text{CH}_2\text{CH}_2\text{OH}}$$

5. Synthesis of Alcohols from Alkenes Compounds (Section 8.8)

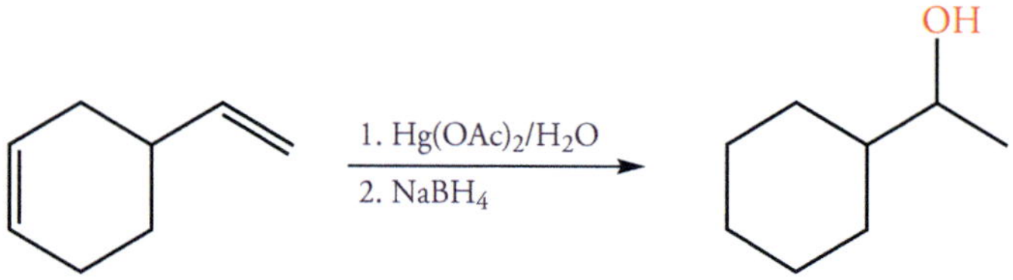

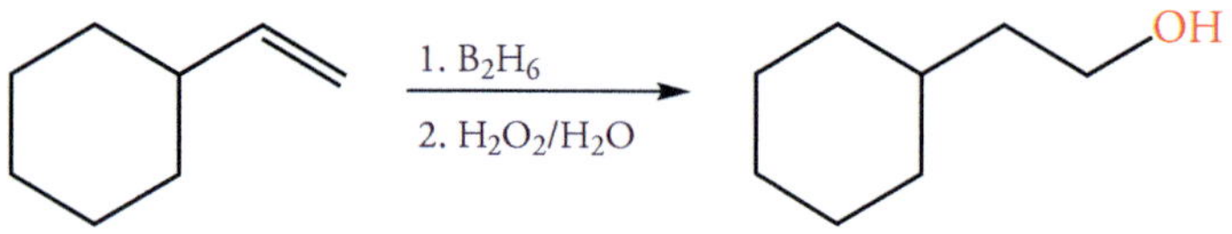

$$1.\ \text{Hg(OAc)}_2/\text{H}_2\text{O}$$
$$2.\ \text{NaBH}_4$$

$$1.\ \text{B}_2\text{H}_6$$
$$2.\ \text{H}_2\text{O}_2/\text{H}_2\text{O}$$

6. Synthesis of Sulfides (Section 8.10)

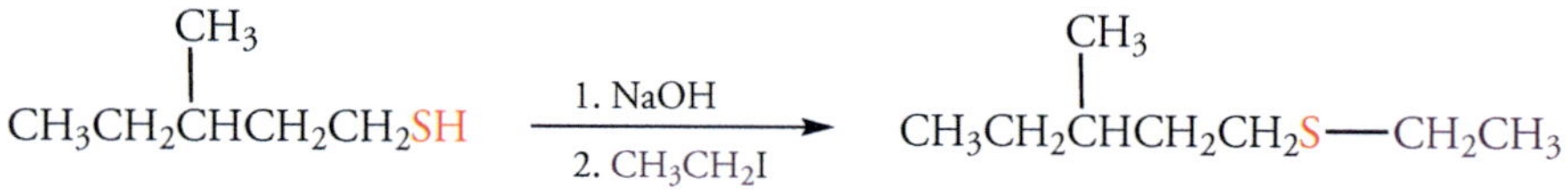

$$\underset{\text{CH}_3}{\text{CH}_3\text{CH}_2\overset{|}{\text{CH}}\text{CH}_2\text{CH}_2\text{SH}} \quad \xrightarrow[\text{2. CH}_3\text{CH}_2\text{I}]{\text{1. NaOH}} \quad \underset{\text{CH}_3}{\text{CH}_3\text{CH}_2\overset{|}{\text{CH}}\text{CH}_2\text{CH}_2\text{S}-\text{CH}_2\text{CH}_3}$$

EXERCISES

Nomenclature of Alcohols

8.1 Write the structural formula of each of the following compounds:

(a) 2-methyl-2-pentanol   (b) 2-methyl-1-butanol   (c) 2,3-dimethyl-1-butanol

8.2 Write the structural formula of each of the following compounds:

(a) 2-methyl-3-pentanol   (b) 3-ethyl-3-pentanol   (c) 4-methyl-2-pentanol

8.3 Write the structural formula of each of the following compounds:

(a) 1-methylcyclohexanol   (b) *trans*-2-methylcyclohexanol   (c) *cis*-3-ethylcyclopentanol

8.4 Write the structural formula each of the following compounds?

(a) 1.2-hexanediol   (b) 1.3-propanediol   (c) 1,2,4-butanetriol

8.5 Name each of the following compounds:

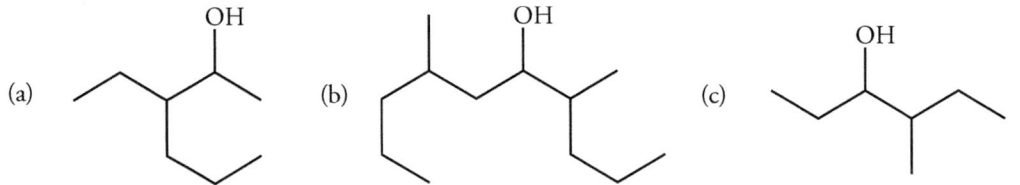

8.6 Name each of the following compounds:

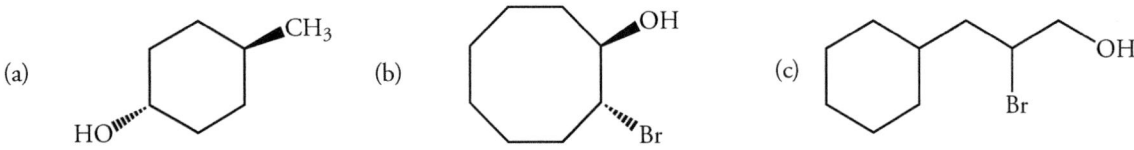

8.7 Name the following compound, which is the sex attractant of the Mediterranean fruit fly:

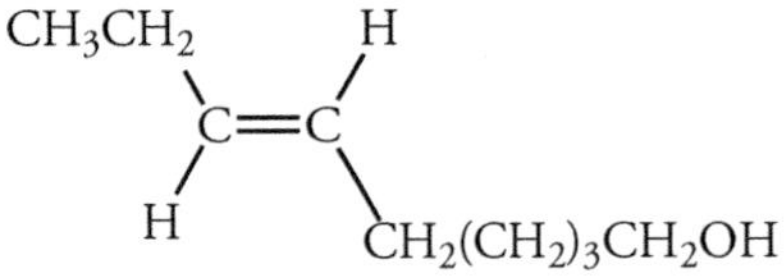

CH₃CH₂ $\quad$ H

$$\overset{\displaystyle CH_3CH_2}{\underset{\displaystyle H}{}}C=C\overset{\displaystyle H}{\underset{\displaystyle CH_2(CH_2)_3CH_2OH}{}}$$

8.8 Name the following compound, which is used as a mosquito repellent:

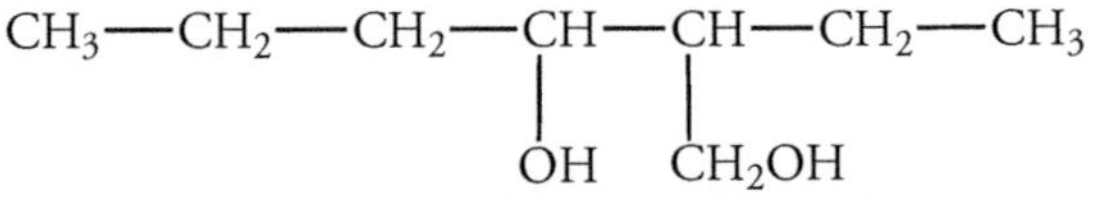

CH₃—CH₂—CH₂—CH—CH—CH₂—CH₃
$\qquad\qquad\qquad\quad$ OH $\quad$ CH₂OH

Classification of Alcohols

8.9 Classify each of the following alcohols.

(a) 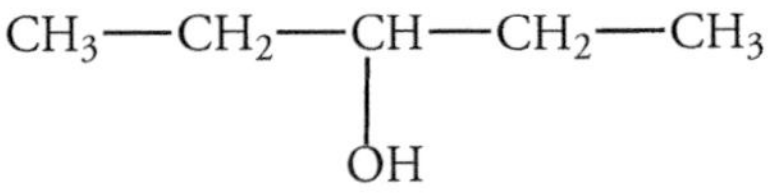 CH₃—CH₂—CH—CH₂—CH₃
$\qquad\qquad\quad$ OH

(b) 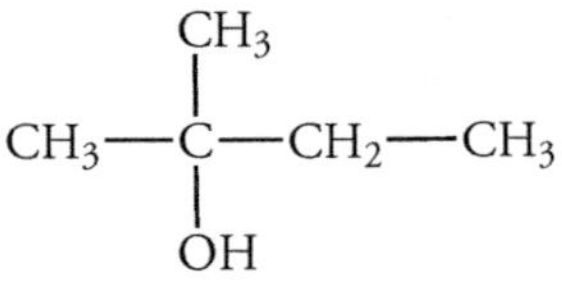 CH₃—C(CH₃)(OH)—CH₂—CH₃

(c) CH₃—C(CH₃)(OH)—CH₃

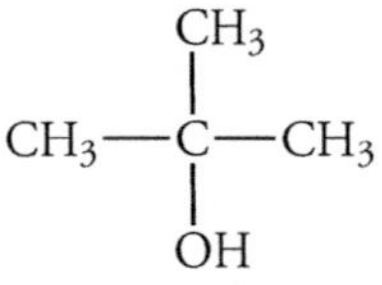

8.10 Classify each of the following alcohols:

(a) 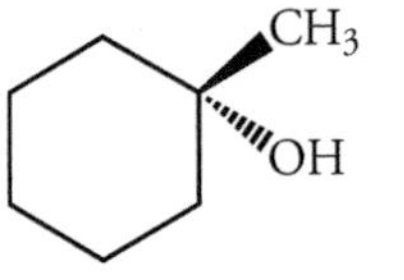 cyclohexane with CH₃ and OH

(b) 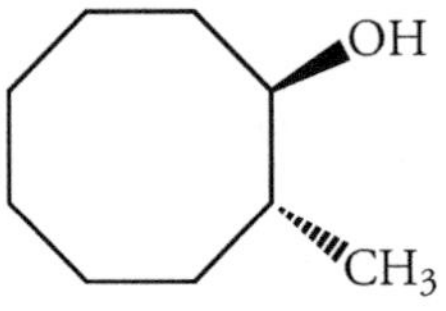 cyclooctane with OH and CH₃

(c) 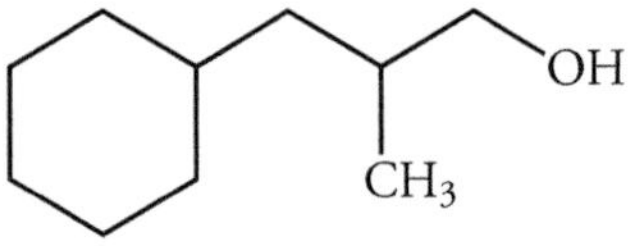 cyclohexyl-CH(CH₃)-CH₂-OH

8.11 Classify each of the hydroxyl groups in the following vitamins:

(a) pyridoxal (vitamin B₆)

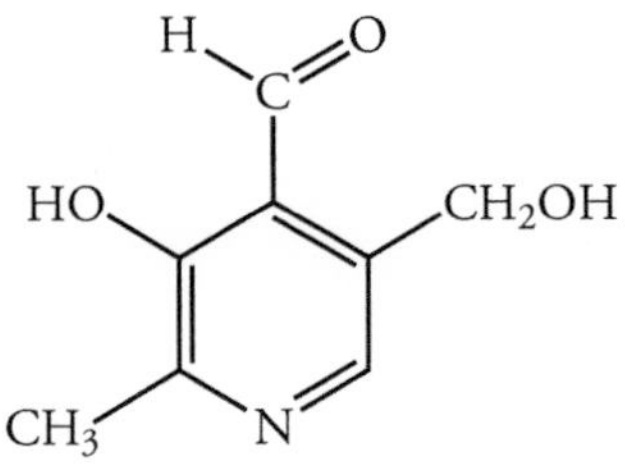

(b) thiamine (vitamin B₁)

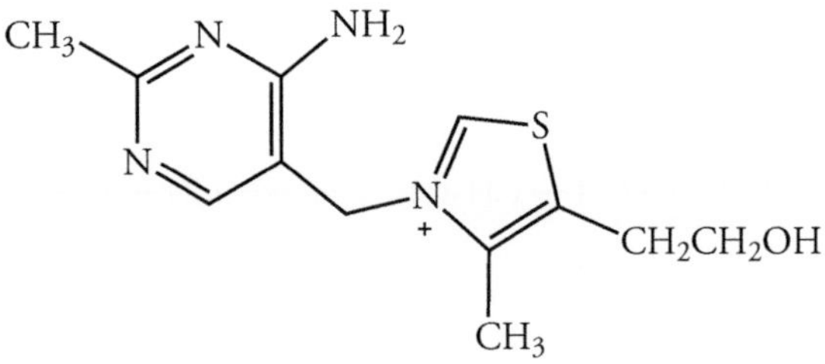

8.12 Classify each of the hydroxyl groups in the following steroids:

(a) digitoxigenin, a cardiac glycoside

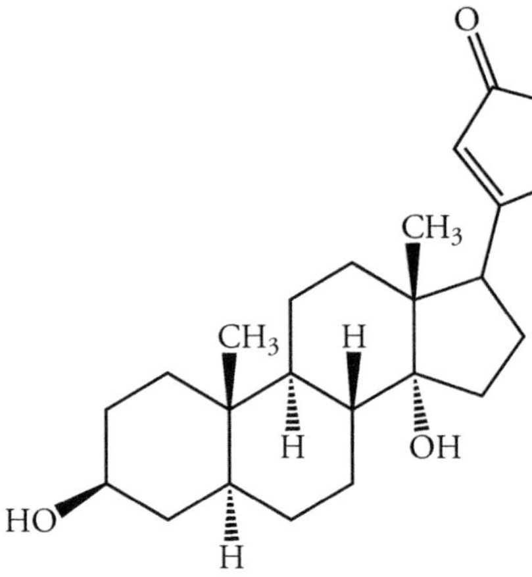

(b) hydrocortisone, an anti-inflammatory drug

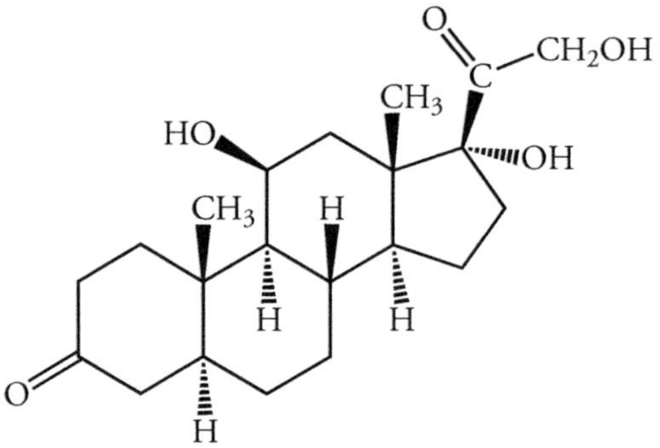

Physical Properties of Alcohols

8.13 1,2-Hexanediol is very soluble in water, but 1-heptanol is not. Explain why these two compounds with similar molecular weights have different solubilities.

8.14 Ethylene glycol and 1-propanol boil at 198 and 97 °C, respectively. Explain why these two compounds with similar molecular weights have different boiling points.

8.15 Explain why 1-butanol is less soluble in water than 1-propanol.

8.16 Suggest a reason why 2-methyl-1-propanol is much more soluble in water than 1-butanol.

Acid-Base Properties of Alcohols

8.17 1,1,1-Trichloro-2-methyl-2-propanol is used as a bacteriostatic agent. Compare its pK_a to that of 2-methylpropanol.

8.18 Based on the date in Table 8.2, estimate the K_a of 2-bromoethanol.

8.19 Based on the date in Table 8.2, estimate the K_a of cyclohexanol.

8.20 Which base is the stronger, methoxide ion or *tert*-butoxide ion? Explain your reasoning.

Formation of Alkyl Halides from Alcohols

8.21 Rank the following compounds according to their rates of reaction with HBr:

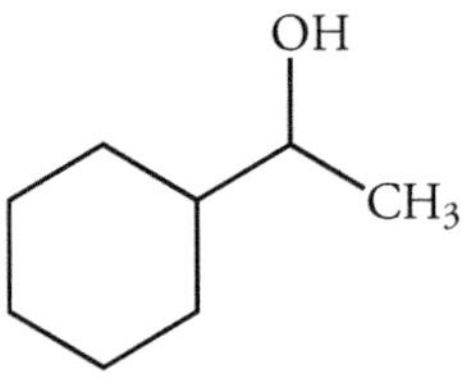

I

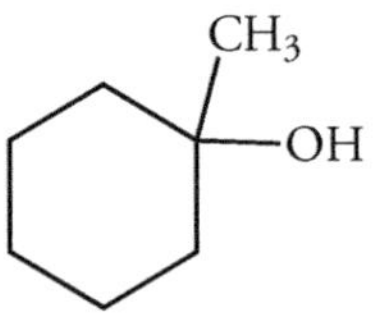

II

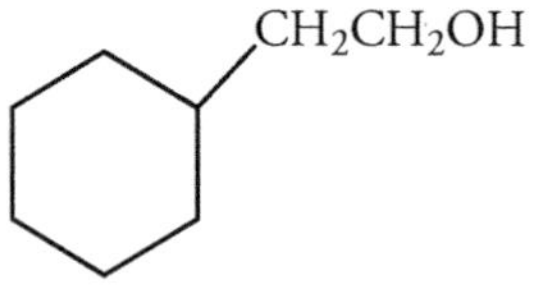

III

8.22 Rank the following compounds according to their rates of reaction with HCl and ZnCl$_2$:

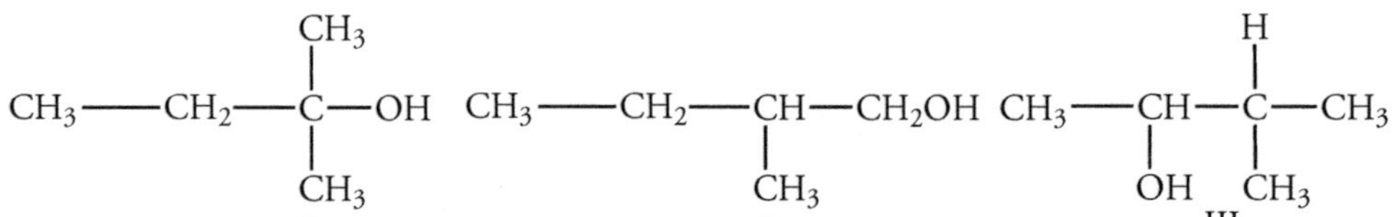

I

II

III

8.23 Write the structure of the product of reaction for each of the following compounds with PBr_3:

$$CH_3-\overset{\overset{\displaystyle CH_3}{|}}{\underset{\underset{\displaystyle CH_3}{|}}{C}}-CH_2-CH_2-OH \qquad CH_3-\overset{\overset{\displaystyle OH}{|}}{\underset{\underset{\displaystyle CH_2CH_3}{|}}{C}}-CH_2-CH_3$$

$$\text{I} \qquad\qquad\qquad \text{II}$$

$$CH_3-\overset{\overset{\displaystyle H}{|}}{\underset{\underset{\displaystyle CH_3}{|}}{C}}-CH_2-\overset{}{\underset{\underset{\displaystyle OH}{|}}{C}H}-CH_3$$

$$\text{III}$$

8.24 Write the structure of the product of reaction for each of the following compounds with $SOCl_2$:

(a) [structure: phenyl-CH(OH)-CH_3] (b) [structure: cyclopentyl-OH] (c) [structure: phenyl-CH_2CH_2OH]

8.25 Reaction of 3-buten-1-ol with HBr yields a mixture of two products: 3-bromo-l-butene and 1-bromo-2-butene. Explain why. (*Hint:* The reaction of this allyl alcohol occurs via an S_N1 process.)

8.26 The rate of reaction of the following unsaturated alcohol (I) with HBr is faster than the rate of reaction of the saturated alcohol (II). Explain why.

(I) [structure: cyclohexene ring with OH and CH_3]   (II) [structure: cyclohexane ring with OH and CH_3]

Dehydration of Alcohols

8.27 Draw the structure of the dehydration product(s) when each of the following compounds reacts with sulfuric acid. If more than one product is formed, predict the major isomer.

$$CH_3-\overset{\overset{\displaystyle OH}{|}}{\underset{\underset{\displaystyle CH_3}{|}}{C}}-CH_2-CH_3 \qquad CH_3CH_2-\overset{\overset{\displaystyle H}{|}}{\underset{\underset{\displaystyle CH_3}{|}}{C}}-CH_2CH_2OH \qquad CH_3CH_2-\overset{\overset{\displaystyle H}{|}}{\underset{\underset{\displaystyle OH}{|}}{C}}-CH_2CH_3$$

$$\text{I} \qquad\qquad\qquad \text{II} \qquad\qquad\qquad \text{III}$$

8.28 Draw the structure of the dehydration product(s) when each of the following compounds reacts with sulfuric acid. If more than one product is formed, predict the major isomer.

(a) [structure: cyclohexane with OH and CH_3]   (b) [structure: cyclopentane with HO and CH_3]   (c) [structure: bicyclic with CH_3, OH, H]

8.29 Write the expected product of the acid-catalyzed dehydration of 1-phenyl-2-propanol. The reaction is more rapid than the rate of dehydration of 2-propanol. Explain why.

8.30 The dehydration of *trans*-2-methylcyclopentanol occurs via an E2 process to give predominantly 3-methylcyclopentene rather than 1-methylcyclopentene. What does this information indicate about the stereochemistry of the elimination reaction?

Oxidation of Alcohols

8.31 Write the product formed from the oxidation of each of the compounds in Exercise 8.21 with the Jones reagent.

8.32 Write the product formed from the oxidation of each of the compounds in Exercise 8.22 with the Jones reagent.

8.33 Write the product formed from the oxidation of the sex attractant of the Mediterranean fruit fly in Exercise 8.7 with PCC.

8.34 Write the product formed by oxidation with PCC of the mosquito repellent in Exercise 8.8.

8.35 Which of the compounds in Exercises 8.23 and 8.24 will give a ketone when oxidized by Jones reagent?

8.36 Which of the compounds in Exercises 8.23 and 8.24 will give an acid product when oxidized by Jones reagent?

Preparation of Alcohols

8.37 Name the final product of oxymercuration-demercuration of each of the following compounds:

(a)
H, CH$_3$ / C=C / H, CH$_3$

(b)
CH$_3$, CH$_3$ / C=C / H, CH$_3$

(c)
H, CH$_2$CH$_3$ / C=C / H, H

8.38 Name the final product of hydroboration-oxidation of each of the compounds in Exercise 8.37.

8.39 Draw the structure of the hydroboration-oxidation product of each of the following compounds:

(a) [bicyclic (decalin-type) ring with CH$_3$ substituent on double bond]

(b) [cyclobutane ring with CH=CH$_2$ substituent]

(c) [cyclohexane ring with =C(H)CH$_3$ substituent]

8.40 Draw the structure of the oxymercuration-demercuration product of each compound in Exercise 8.39.

Synthetic Sequences

8.41 Write the structure of the final product of each of the following sequences of reactions:

(a) [cyclopentane ring]—CH$_2$CH=CH$_2$ $\xrightarrow[\text{2. NaBH}_4]{\text{1.Hg(OAc)}_2\,/\,H_2O}$ $\xrightarrow[\text{H}_2\text{SO}_4]{\text{CrO}_3}$

(b) [naphthalene ring]—CH=CH$_2$ $\xrightarrow[\text{2. H}_2\text{O / OH}^-]{\text{1. B}_2\text{H}_6}$ $\xrightarrow{\text{PCC}}$

(c) CH$_3$CH$_2$CH$_2$CH=CH$_2$ $\xrightarrow[\text{2. H}_2\text{O / OH}^-]{\text{1. B}_2\text{H}_6}$ $\xrightarrow[\text{H}_2\text{SO}_4]{\text{CrO}_3}$

8.42 Write the structure of the product of reaction for each of the following compounds with $SOCl_2$:

(a) 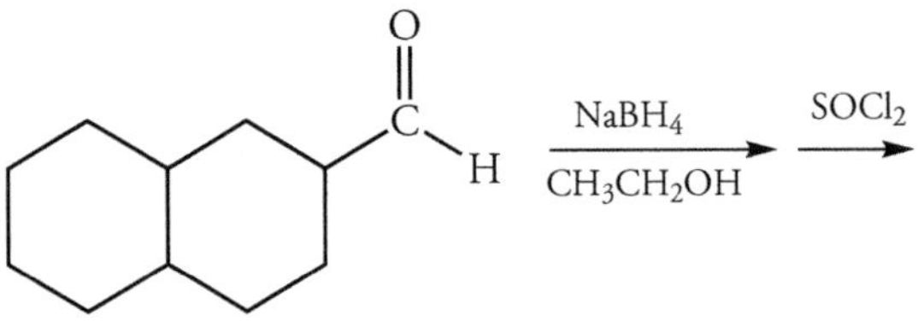
$$\xrightarrow[\text{CH}_3\text{CH}_2\text{OH}]{\text{NaBH}_4} \xrightarrow{\text{SOCl}_2}$$

(b) 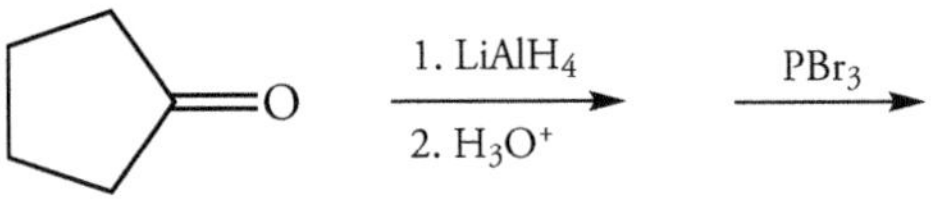
$$\xrightarrow[\text{2. H}_3\text{O}^+]{\text{1. LiAlH}_4} \xrightarrow{\text{PBr}_3}$$

(c) 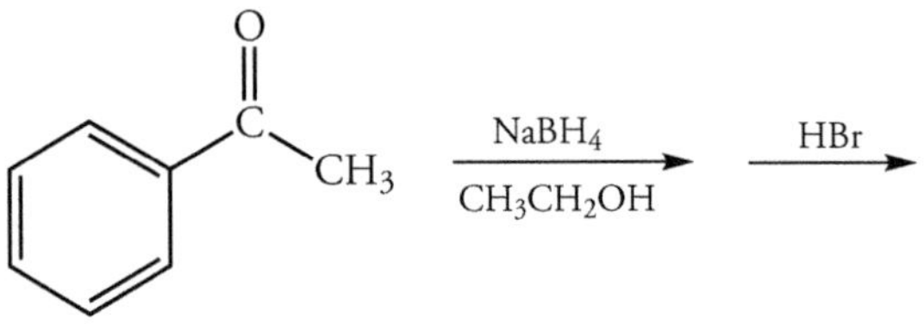
$$\xrightarrow[\text{CH}_3\text{CH}_2\text{OH}]{\text{NaBH}_4} \xrightarrow{\text{HBr}}$$

Phenols

8.43 *p*-Nitrophenol is a much stronger acid than phenol. Explain why.

8.44 Which phenoxide is the stronger base, *p*-ethylphenoxide or *p*-chlorophenoxide? Explain your choice.

8.45 Draw the structure of the quinone obtained from the oxidation of the following substituted naphthalene:

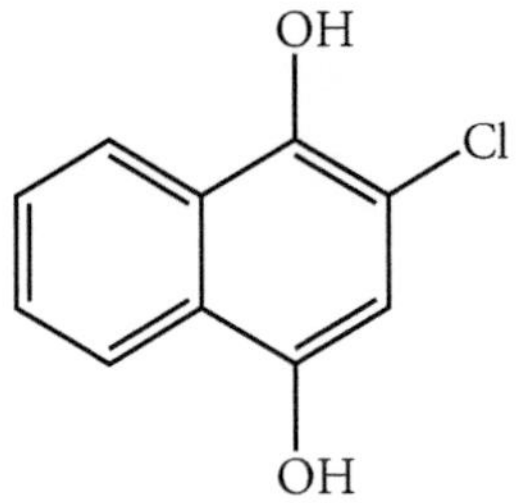

8.46 2-Methylhydroquinone is more easily oxidized to a quinone than 2-cblorohydroquinone. Explain why.

Sulfur Compounds

8.47 There are four isomeric compounds $C_4H_{10}S$ with an —SH group. Draw the structures of the compounds.

8.48 There are three isomeric compounds C_3H_8S. Draw their structures.

8.49 Draw the structure of each of the following compounds:
(a) l-propanethiol   (b) 2-methyl-3-pentanethiol   (c) cyclopentanethiol

8.50 Draw the structure of each of the following compounds:
(a) 2-propanethiol   (b) 2-methyl-1-propanethiol   (c) cyclobutanethiol

8.51 Addition of sodium hydroxide to an aqueous solution of $CH_3CH_2CH_2SH$ eliminates the odor. Explain why.

8.52 The boiling points of ethanethiol and dimethyl sulfide are 35 °C and 37 °C, respectively. Why are the boiling points similar? What types of intermolecular forces are responsible for this similarity?

CH_3—CH_2—SH
ethanethiol

CH_3—S—CH_3
dimethyl sulfide

8.53 Indicate two methods to produce the scent marker of the red fox using a thiol as one of the reactants.

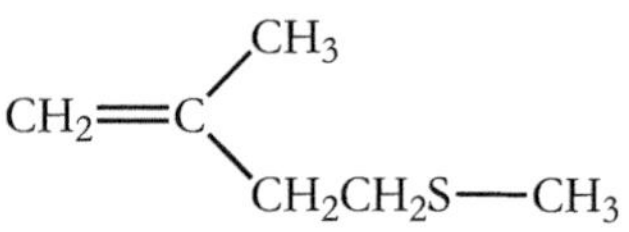

$$CH_2\!\!=\!\!C\!\!\begin{array}{c} \diagup CH_3 \\ \diagdown CH_2CH_2S\!-\!CH_3 \end{array}$$

8.54 Outline a series of reactions to produce the compound used for defense by the skunk, starting with 3-methyl-1-butene.

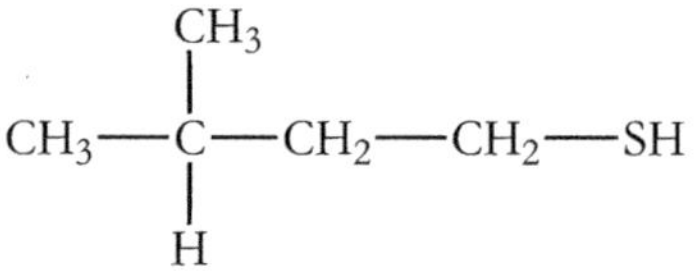

$$CH_3\!-\!\underset{\underset{H}{|}}{\overset{\overset{CH_3}{|}}{C}}\!-\!CH_2\!-\!CH_2\!-\!SH$$

8.55 Dimethyl disulfide is emitted by female hamsters as an attractant for male hamsters. How might this compound be produced?

8.56 Minks secrete the following disulfide. How could this compound be prepared in a laboratory synthesis starting from 1-bromo-3-methylbutane?

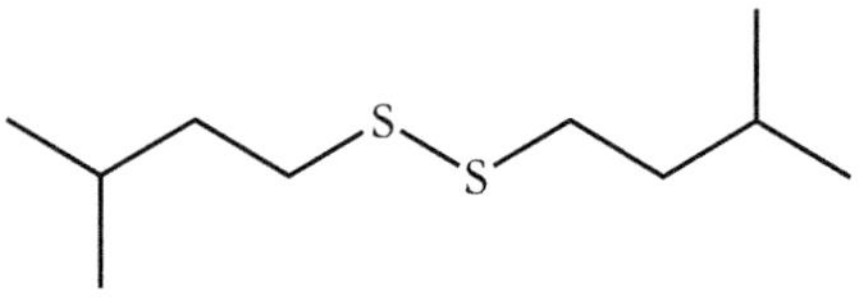

9 ETHERS AND EPOXIDES

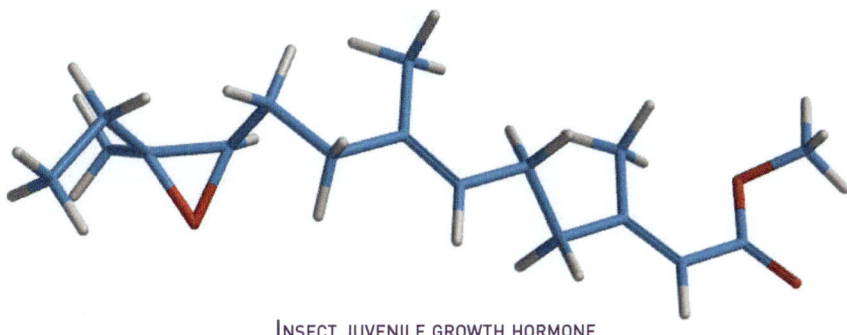

INSECT JUVENILE GROWTH HORMONE

9.1 STRUCTURE OF ETHERS

We can view ethers, like alcohols, as organic "cousins" of water. Ethers contain two groups, which may be alkyl or aryl groups, bonded to an oxygen atom. The two alkyl or aryl groups are identical in a **symmetrical ether** and different in an **unsymmetrical ether**.

CH₃CH₂—O—CH₂CH₃

diethyl ether
(a symmetrical ether)

〈phenyl〉—O—CH₂CH₂CH₃

phenyl propyl ether
(an unsymmetrical ether)

〈phenyl〉—O—〈phenyl〉

diphenyl ether
(a symmetrical ether)

The oxygen atom of an ether is sp³-hybridized, and the C—O—C bond angle is approximately the tetrahedral bond angle. The bond angle in dimethyl ether is 112° (Figure 9.1). The two O—C bonds are directed to two of the corners of a tetrahedron. Based on the hybridization of oxygen, the lone pair electrons in the remaining two sp³ hybrid orbitals are directed to the remaining "corners" of the tetrahedron.

We can predict the most stable conformations of ethers by comparing them to the structures of similar hydrocarbons. Imagine creating an ether by replacing a —CH₂— group of an alkane with an oxygen atom. For example, replacing the C-3 methylene group of pentane with an oxygen atom would give diethyl ether. Diethyl ether has an antiarrangement of all carbon and oxygen atoms in its most stable conformation.

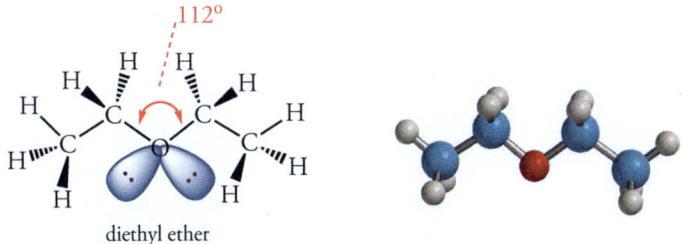

diethyl ether

Figure 9.1
Structure of Dimethyl Ether

The oxygen atom of methanol is sp³-hybridized. The C—O—C bond angle, 112°, is close to the tetrahedral bond angle (109.5°). The two sets of lone pair electrons are in sp³ hybrid orbitals that are directed to two of the corners of a tetrahedron.

A similar situation prevails when we compare conformations of cyclic ethers and cycloalkanes. For example, tetrahydropyran, the ether analog of cyclohexane, also exists in a chair conformation. The tetrahedral oxygen atom has two lone pairs that occupy positions in space corresponding to the axial and equatorial C—H bonds of cyclohexane. The conformation of tetrahydropyran is particularly important because many carbohydrates contain tetrahydropyran rings (Chapter 13).

Principles of Organic Chemistry. http://dx.doi.org/10.1016/B978-0-12-802444-7.00009-4

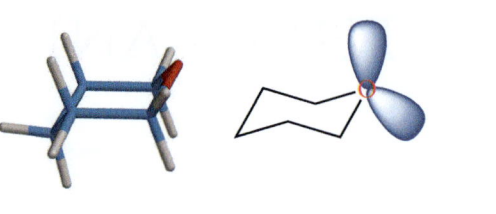
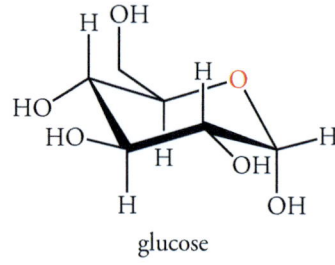

tetrahydropyran

glucose

9.2 NOMENCLATURE OF ETHERS

Common Names

Simple ethers are named as alkyl alkyl ethers. To obtain the name we list the alkyl (or aryl) groups in alphabetical order and appending the name ether. For example, an unsymmetrical ether that has an *n*-butyl group and a methyl group bonded to an oxygen atom is called *n*-butylmethyl ether. Symmetrical ethers are named by using the prefix *di-* with the name of the alkyl group to indicate that the alkyl groups are the same. For example, the name of an ether with two isopropyl groups bonded to an oxygen atom is diisopropyl ether.

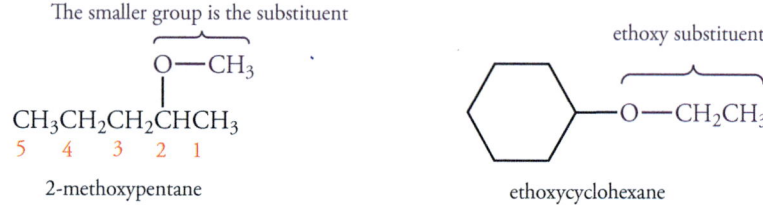

$CH_3-O-CH_2CH_2CH_2CH_3$

n-butylmethyl ether

diisopropyl ether

IUPAC Names

Ethers are named in IUPAC nomenclature as **alkoxyalkanes**, where the smaller alkyl group and the oxygen atom constitute an **alkoxy** group. An alkoxy group is treated as a substituent on the larger parent alkane chain or cycloalkane ring. For example, a five-carbon chain with an CH_3O- group at C-2 is named 2-methoxypentane and a cyclohexane ring with an $-OCH_2CH_3$ group is named ethoxycyclohexane.

The smaller group is the substituent

ethoxy substituent

$O-CH_3$

$CH_3CH_2CH_2CHCH_3$

5 4 3 2 1

2-methoxypentane

$O-CH_2CH_3$

ethoxycyclohexane

Cyclic Ethers

The three- through six-membered cyclic ethers have common names. In all ring systems, the oxygen atom is assigned the number 1, and the rings are numbered in the direction that gives the first substituent the lower number. Cyclic ethers with three-atom rings are called **epoxides**. Because these compounds are formed from the oxidation of an alkene, the common name of an epoxide is obtained by adding *oxide* to the name of the alkene.

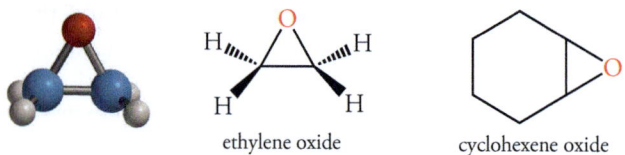

ethylene oxide

cyclohexene oxide

The four-membered ether ring compounds, called trimethylene oxides, are not common substances. The name of the five-membered ring ether is tetrahydrofuran (THF), based on its relationship to the aromatic compound furan. Similarly, tetrahydropyran (THP), a six-membered ring ether, is related to pyran, an unsaturated ether.

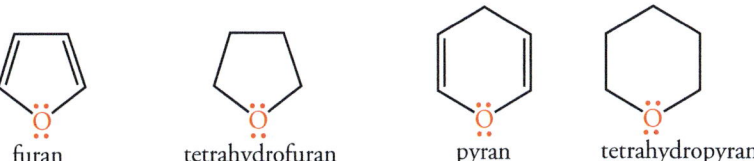

furan   tetrahydrofuran   pyran   tetrahydropyran

The IUPAC names for cyclic ethers having three-, four-, five-, and six-membered rings are *oxirane*, *oxetane*, *oxolane*, and *oxane*, respectively. The oxygen atom in each of these rings receives the number 1. As we noted above, the ring is numbered in the direction that gives the lowest numbers to substituents.

2,2-dimethyloxirane   2-ethyloxetane   3-ethoxyoxolane   4-chlorooxane

Problem 9.1

What are the common and IUPAC names of the following compounds?

Sample Solution

The common name of an unsymmetrical ether is based on the names of the two alkyl (or aryl) groups. In this case they are phenyl and ethyl. The names of the two groups are arranged alphabetically, and the word ether is added to give the name ethylphenyl ether. The IUPAC name is obtained by first identifying the smaller of the two groups bonded to the oxygen atom and then selecting it as the alkoxy group on the larger parent alkane chain. In this case, the benzene ring is selected as the parent and the ethoxy group is the substituent. The IUPAC name is ethoxybenzene.

Problem 9.2

What are the IUPAC names of the following compounds?

(a)   (b)   (c)

9.3 PHYSICAL PROPERTIES OF ETHERS

Diethyl ether—often called ethyl ether or just "ether"—was used as a general anesthetic as early as 1842. It is administered as a vapor, and acts as a depressant on the central nervous system, causing unconsciousness. However, its high flammability and volatility present hazards in the operating room. Ethers such as ethyl vinyl ether, divinyl ether, and methyl propyl ether have also been used as anesthetics. All low molecular weight ethers are potentially explosive when mixed with oxygen.

Dipole Moments and Boiling Points

Ethers have two polar C—O bonds at a tetrahedral angle. Therefore, ethers have substantial dipole moments. Ethers are more polar than alkanes and somewhat less polar than alcohols (Table 9.1).

Ethers do not have an O—H bond, and therefore cannot serve as hydrogen bond donors. As a consequence, ether molecules do not hydrogen bond to each other. Thus, ethers have boiling points substantially lower than those of alcohols of comparable molecular weight (Table 9.1). The boiling points of ethers are very close to the boiling points of alkanes of similar molecular weight.

Because ethers are polar, they are more soluble in water than alkanes of similar molecular weight. The solubility of ethers in water results from hydrogen bonds between the water molecules and the nonbonding electron pairs of ether molecules. The solubility of diethyl ether is about 7.5 g per 100 mL of water, much higher than that of pentane, which dissolves only slightly in water (Table 9.1). The solubility of diethyl ether in water is similar to that of 1-butanol, because water can form hydrogen bonds to each compound.

Table 9.1
Physical Properties of Ethers

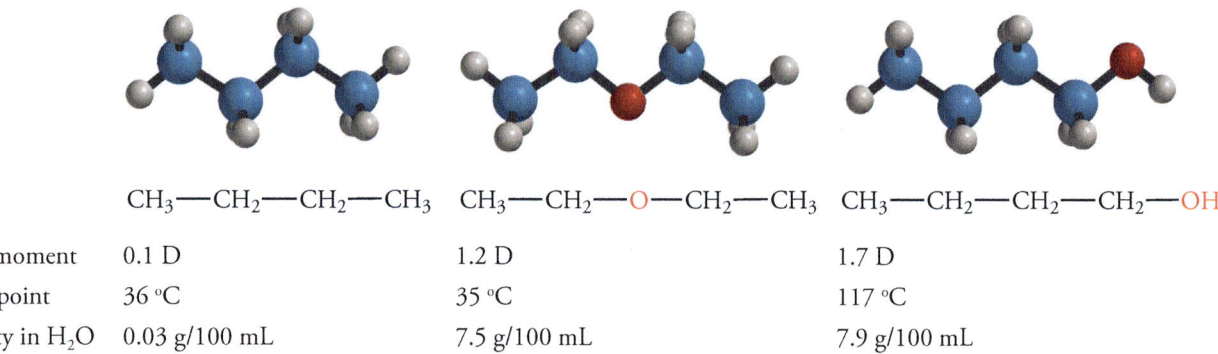

	$CH_3—CH_2—CH_2—CH_3$	$CH_3—CH_2—O—CH_2—CH_3$	$CH_3—CH_2—CH_2—CH_2—OH$
Dipole moment	0.1 D	1.2 D	1.7 D
Boiling point	36 °C	35 °C	117 °C
Solubility in H_2O	0.03 g/100 mL	7.5 g/100 mL	7.9 g/100 mL

Ethers as Solvents

Ethers such as diethyl ether are good solvents for a wide range of polar and nonpolar organic compounds. Nonpolar compounds are generally more soluble in diethyl ether than in alcohols such as ethanol because ethers do not have a hydrogen bonding network that would have to be broken up to dissolve the solute.

Because diethyl ether has a dipole moment, polar substances readily dissolve in it. Polar compounds that can serve as hydrogen bond donors dissolve in diethyl ether because they can form hydrogen bonds to the nonbonding electron pairs of the ether oxygen atoms.

Ethers are aprotic. Thus, basic substances, such as Grignard reagents, can be prepared in ether and tetrahydrofuran. These ethers solvate the magnesium ion, which is coordinated to the lone pair electrons of the ether.

The nonbonding electron pairs of ethers can stabilize electron-deficient species such as BF_3 and borane, BH_3. For example, the borane-THF complex is used in the hydroboration of alkenes (Section 8.8).

BF₃-THF complex   BH₃-THF complex

9.4 THE GRIGNARD REAGENT AND ETHERS

Ethers such as diethyl ether or tetrahydrofuran are excellent solvents for certain reagents that would otherwise react with protons supplied by protic solvents. One such example is the **Grignard reagent**, represented as R—Mg—X, which can be prepared from haloalkanes as well as from aryl halides.

$$R—X \xrightarrow[\text{ether}]{Mg(s)} R—Mg—X$$

The oxygen atom of diethyl ether (or THF) forms a complex with the magnesium atom of the Grignard reagent. These reagents in ether solution are very useful in organic synthesis.

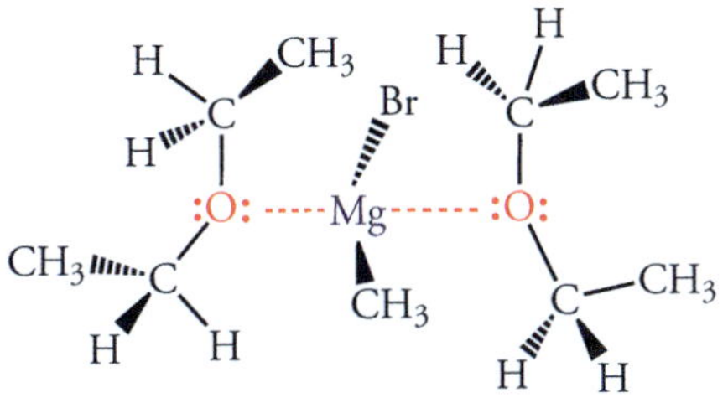

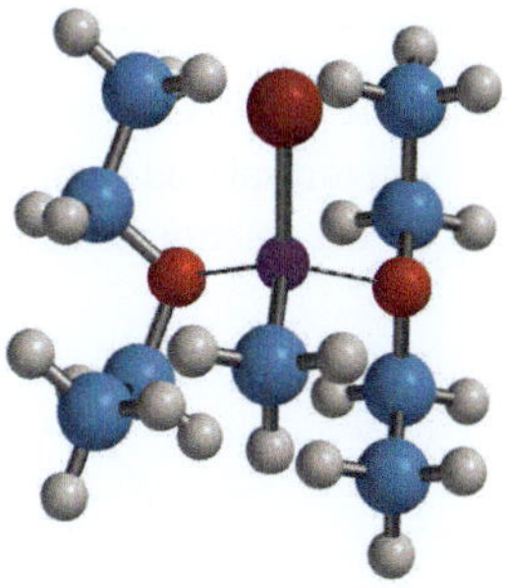

The French chemist Victor Grignard received the Nobel Prize in 1912 for developing the methods to prepare these organomagnesium compounds. In a Grignard reagent, the R group may be a 1°, 2°, or 3° alkyl group as well as a vinyl or aryl group. The halogen may be Cl, Br, or I. Fluorine compounds do not form Grignard reagents.

A Grignard reagent has a very polar carbon-magnesium bond in which the carbon atom has a partial negative charge and the metal a partial positive charge.

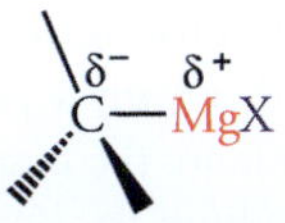

This bond polarity is opposite that of the carbon-halogen bond of haloalkanes. Because the carbon atom in a Grignard reagent has a partial negative charge, it resembles a carbanion, and it reacts with electrophiles. Grignard reagents are very reactive reactants that are used synthetically to form new carbon-carbon bonds. We will discuss these reactions in Section 10.6.

Grignard reagents react rapidly with acidic hydrogen atoms in molecules such as alcohols and water to produce alkanes. Thus, formation of the Grignard reagent followed by reaction with water provides a way to convert a haloalkane to an alkane in two steps.

$$R{-}X \xrightarrow[\text{ether}]{Mg(s)} R{-}MgX \xrightarrow{H_2O} R{-}H$$

Problem 9.3

Devise a synthesis of $CH_3CH_2CHDCH_3$ starting from 1-butene and "heavy water" (D_2O).

Solution

Reaction of a Grignard reagent R—MgBr with D_2O yields R—D. The necessary Grignard reagent is obtained from the corresponding bromoalkane R—Br.

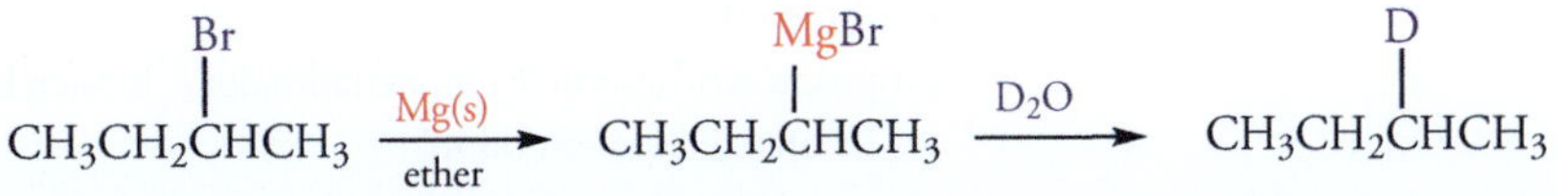

$$\underset{CH_3CH_2CHCH_3}{\overset{Br}{|}} \xrightarrow[\text{ether}]{Mg(s)} \underset{CH_3CH_2CHCH_3}{\overset{MgBr}{|}} \xrightarrow{D_2O} \underset{CH_3CH_2CHCH_3}{\overset{D}{|}}$$

The required 2-bromobutane can be prepared from 1-butene by adding HBr. This reaction occurs according to Markovnikov's rule; that is, a hydrogen atom adds to the less substituted carbon atom of the double bond.

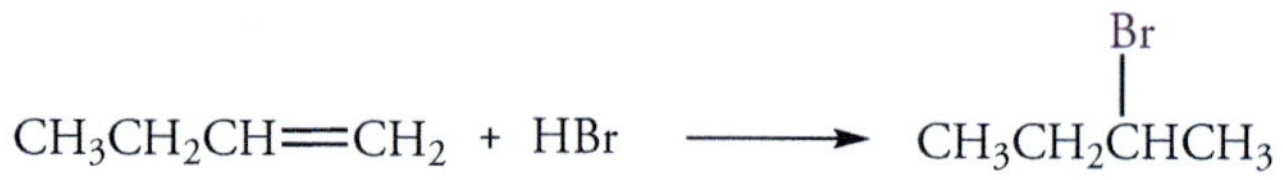

$$CH_3CH_2CH\!=\!CH_2 + HBr \longrightarrow CH_3CH_2\overset{\displaystyle Br}{\underset{\displaystyle |}{C}}HCH_3$$

Problem 9.4

Devise a synthesis of the following compound starting from benzene.

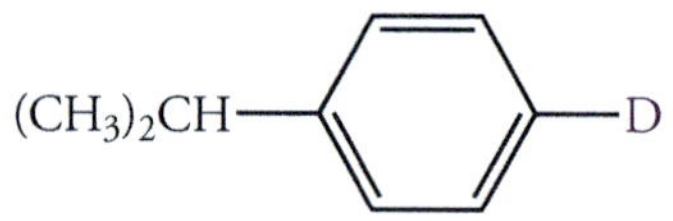

$$(CH_3)_2CH\!-\!\!\!\bigcirc\!\!\!-D$$

9.5 SYNTHESIS OF ETHERS

Ethers can be prepared by a method called the **Williamson ether synthesis**. In this reaction a halide ion is displaced from an alkyl halide by an alkoxide ion in an S_N2 reaction. The alkoxide ion is prepared by the reaction of an alcohol with a strong base such as sodium hydride.

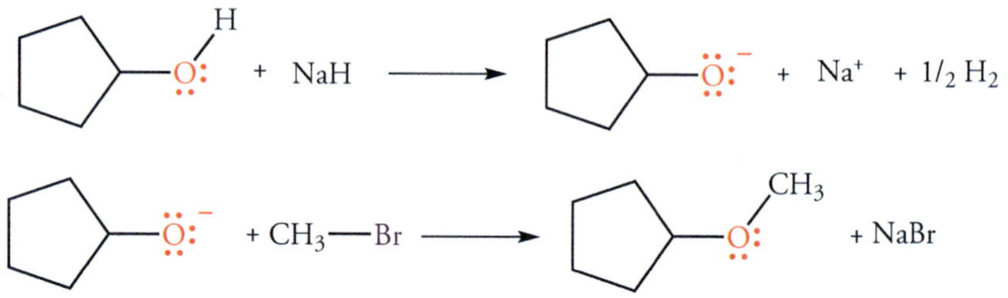

The Williamson synthesis gives the best yields with methyl or primary halides because the reaction occurs by an S_N2 displacement in which a halide ion is the leaving group. The yield is lower for secondary alkyl halides because they also react with the alkoxide ion in a competing elimination reaction. The Williamson synthesis cannot be used with tertiary alkyl halides because they undergo elimination reactions instead of participating in S_N2 reactions. Thus, to make an unsymmetrical ether with a primary and a tertiary alkyl group, a primary alkyl halide and a tertiary alkoxide ion are the best reagents. For example, *tert*-butyl methyl ether can be prepared by the reaction of sodium *tert*-butoxide with methyl iodide, but not by the reaction of sodium methoxide with 2-iodo-2-methylpropane.

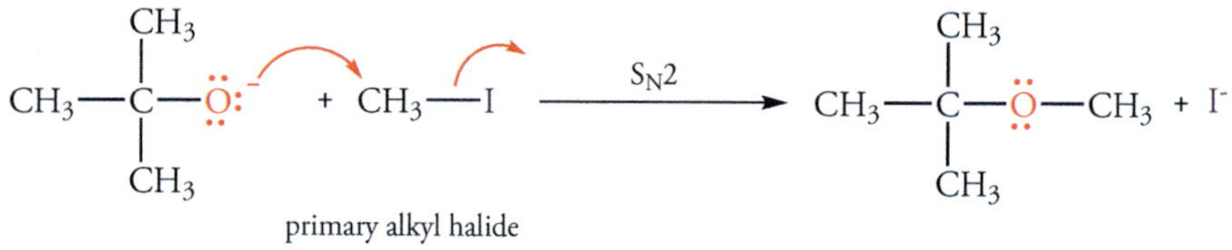

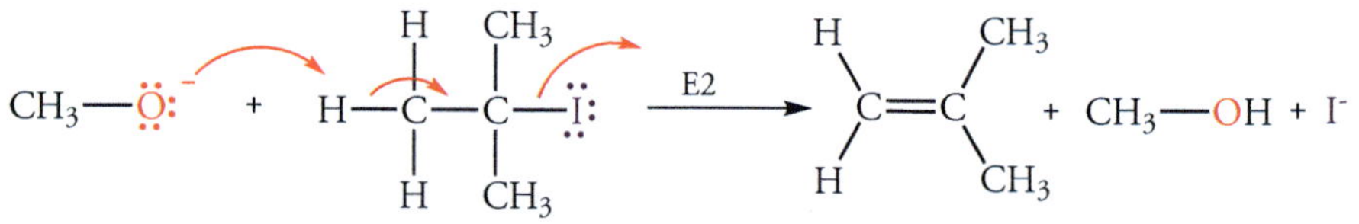

In a subsequent reaction, the alcohol product reacts with a second mole of an alkyl halide. Both alkyl groups of the ether are eventually converted into alkyl halides if sufficient hydrogen halide is used.

Problem 9.5

Propose a synthesis of 2-ethoxynapthalene, known by its trade name Nerolin II. It is used in perfumery for its orange blossom odor.

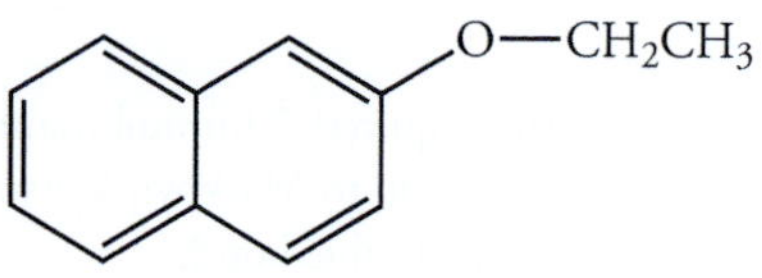

Consider reactions of the following two possible combinations of reagents.

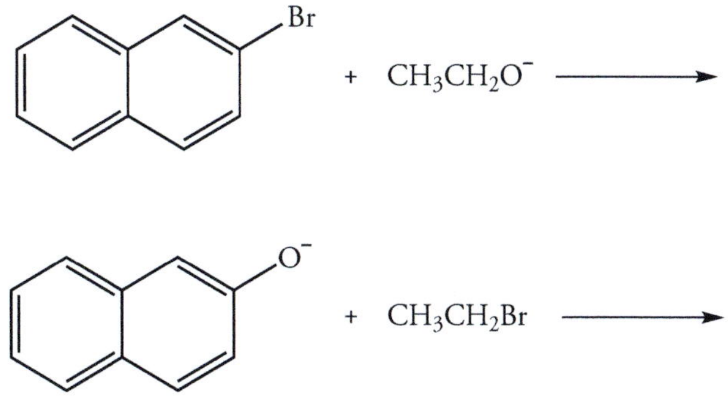

$+ \quad CH_3CH_2O^-$ ⟶

$+ \quad CH_3CH_2Br$ ⟶

The first combination will not give the ether product because S_N2 reactions cannot occur at the sp^2-hybridized carbon atom of this aromatic compound. However, the second reaction of a phenoxide type ion with bromoethane occurs readily because bromoethane is a primary alkyl halide. The required nucleophile is formed by reacting the following phenol with sodium hydride.

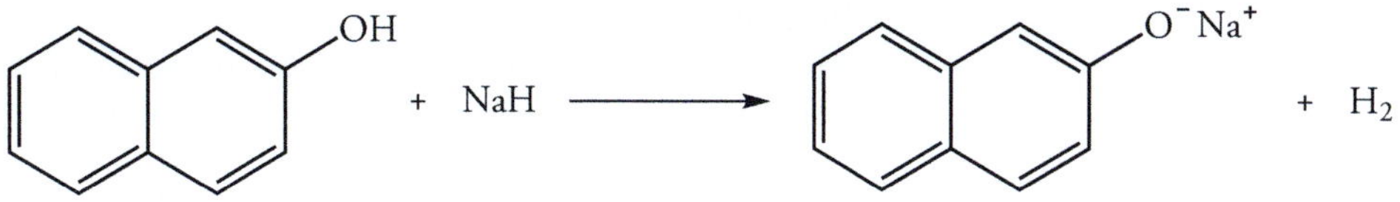

$+ \quad NaH$ ⟶ $+ \quad H_2$

Problem 9.6
Propose a synthesis of benzyl *tert*-butyl ether using the Williamson synthesis.

9.6 REACTIONS OF ETHERS

Ethers are very stable compounds that react with few common reagents. They do not react with bases, but do react with strong acids whose conjugate bases are good nucleophiles. For example, ethers react with HI (or with HBr) with cleavage of the carbon-oxygen bond to produce alkyl iodides (or bromides).

$$R{-}O{-}R' \xrightarrow{HX} R{-}O{-}H + R'{-}X \xrightarrow{HX} R{-}X + R'{-}X$$

In general, the less substituted halide is formed by an S_N2 reaction. The halide ion attacks the less hindered carbon atom, and the oxygen atom of the displaced alkoxy group remains bonded to the more substituted carbon atom. The cleavage reaction does not occur with a halide salt. A proton from the halogen acid must protonate the oxygen atom, providing an alcohol as the leaving group.

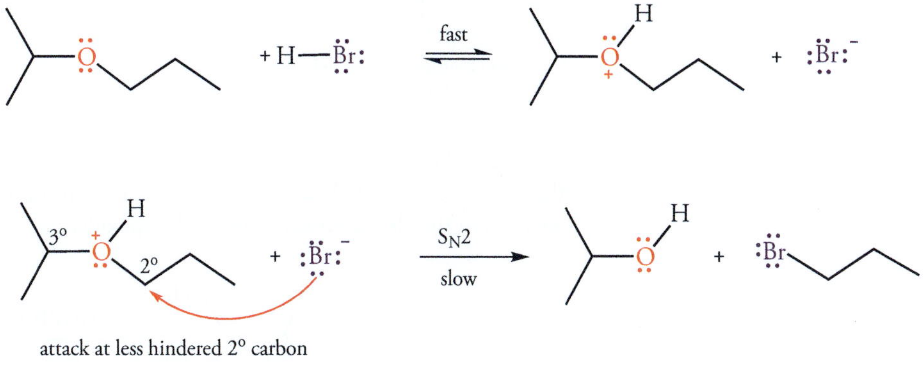

attack at less hindered 2° carbon

Problem 9.7
Based on the mechanism of ether cleavage, write the products of the reaction of HI with phenyl propyl ether.

Solution
First, the strong acid protonates the ether oxygen atom to give an oxonium ion.

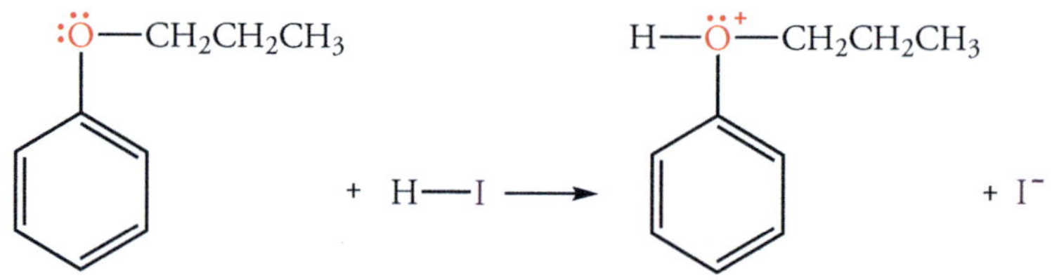

Subsequent nucleophilic attack by the iodide ion can occur only at the methylene carbon atom of the propyl group bearing the oxygen atom. An S_N2 reaction at the carbon atom of the benzene ring that bears the oxygen atom is not possible. The phenol produced also will not react further with HI for the same reason.

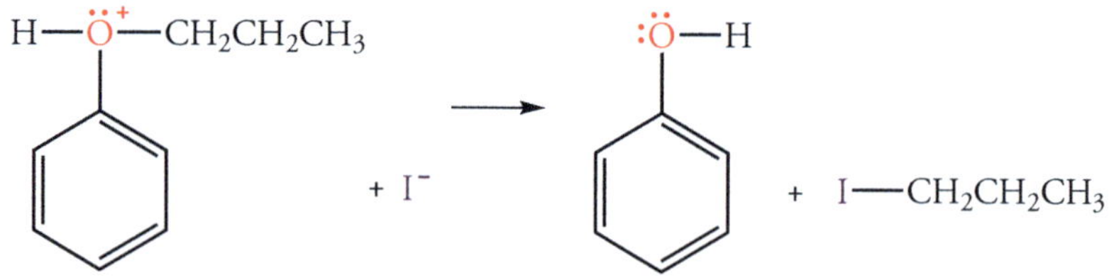

Problem 9.8
Based on the mechanism of ether cleavage, predict the structure of the bromo alcohol that forms in the first step of the reaction of the following ether with HBr.

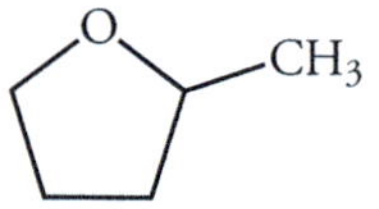

9.7 SYNTHESIS OF EPOXIDES

The synthesis and reactions of cyclic ethers containing four or more atoms are similar to those of acyclic ethers. The three-membered cyclic ethers, which are important intermediates in synthesis (Section 9.8), can be synthesized by oxidizing an alkene with a peroxyacid (RCO_3H). Peroxyacetic acid, CH_3CO_3H, is used in industry, but *m*-chloroperoxybenzoic acid (MCPBA) is used to prepare smaller amounts of epoxides in the laboratory.

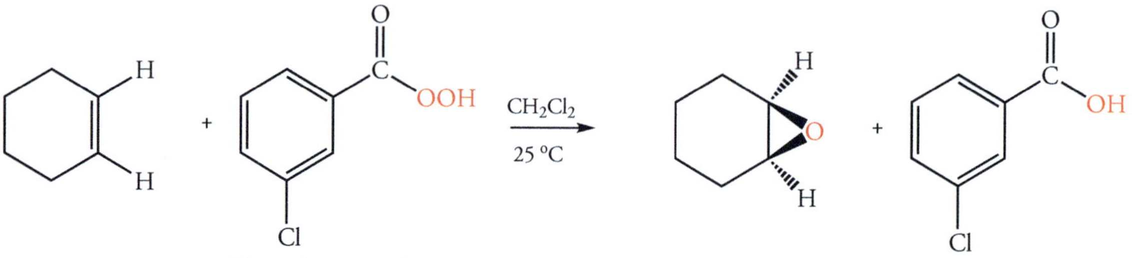

m-chloroperbenzoic acid
(MCPBA)

In the epoxidation of alkenes with *m*-chloroperoxybenzoic acid, the stereochemistry of the groups in the alkene is retained. That is, groups that are *cis* in the alkene are *cis* in the epoxide, and groups that are *trans* in the alkene remain *trans* in the epoxide.

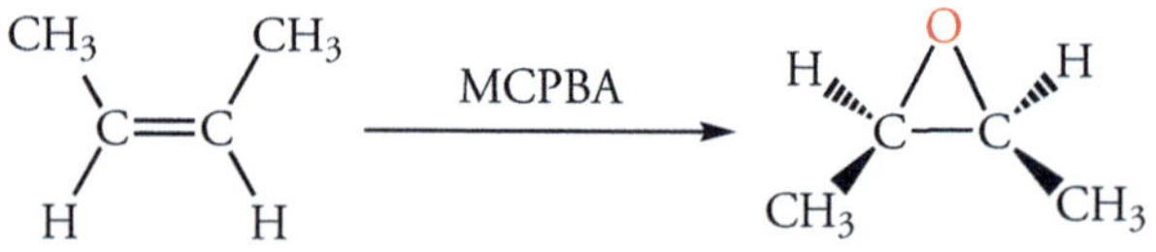

9.8 REACTIONS OF EPOXIDES

The cyclic ethers tetrahydrofuran and tetrahydropyran are as unreactive as acyclic ethers and are often used as solvents. In contrast, epoxides are highly reactive because the three-membered ring has considerable bond angle strain. The products of the ring-opening reactions have normal tetrahedral bond angles and are not strained.

Acid-Catalyzed Ring Opening

Consider the reaction of water with ethylene oxide to form ethylene glycol.

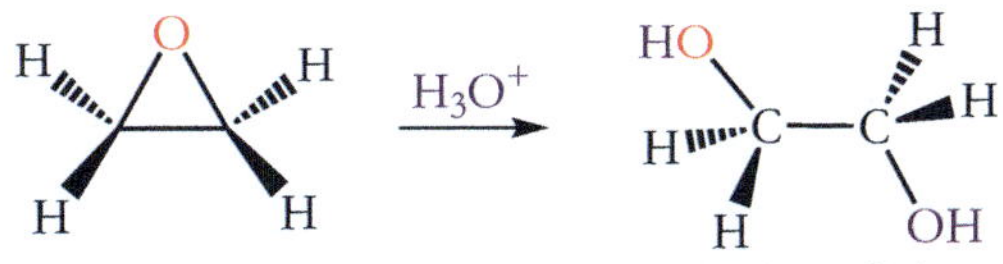

1,2-ethanediol
(ethylene glycol)

The acid-catalyzed ring opening of epoxides occurs by an S_N2 mechanism in which water is the nucleophile and the "leaving group" is the protonated oxygen atom of the epoxide.

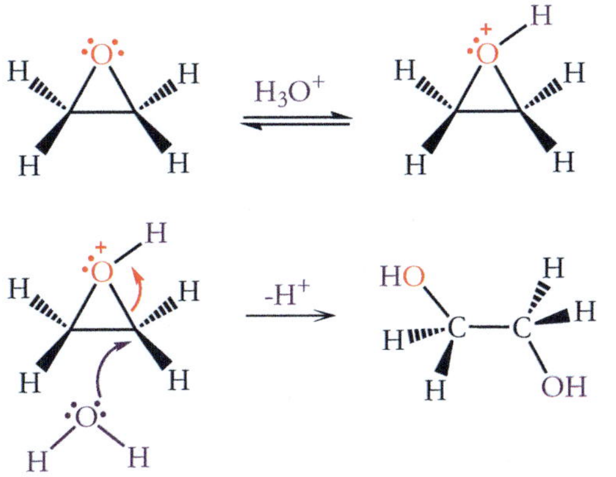

Ring Opening by Nucleophiles

Ethers do not generally react with nucleophiles under neutral or basic conditions. However, epoxides are so strained that the C—O bond of the ring is cleaved even by nucleophiles such as OH^-, SH^-, or NH_3 or the related organic species RO^-, RS^-, and RNH_2. For example, the reaction of ethylene oxide with ammonia gives 2-amino-ethanol, a compound used commercially as a corrosion inhibitor.

Similar reactions occur in the sterilization of temperature-sensitive equipment when it is exposed to ethylene oxide gas. The epoxide ring reacts with a variety of nucleophilic functional groups in bacterial macromolecules, and the bacteria die as a result.

Epoxides react with Grignard reagents to produce alcohols with two more carbon atoms than the starting alkyl halide. The sequence of reactions is shown below.

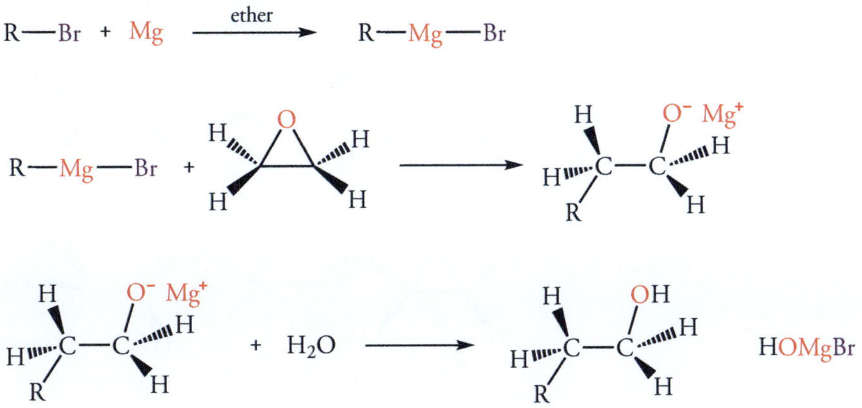

Direction of Ring Opening

Unsymmetrical epoxides give different products under acid- and base-catalyzed conditions.

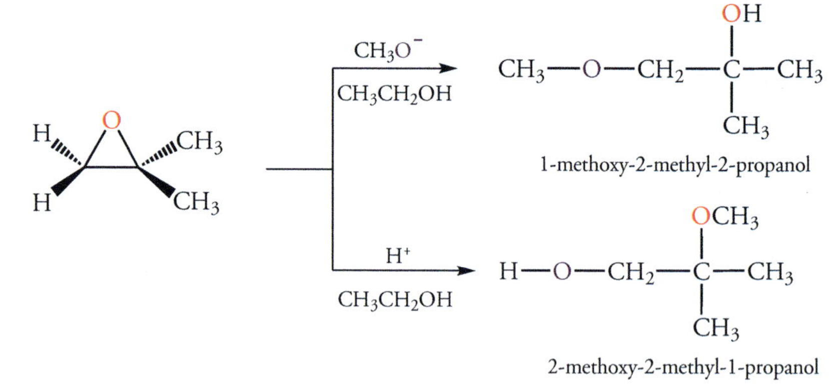

CH3O−
CH3CH2OH
H+
CH3CH2OH
1-methoxy-2-methyl-2-propanol
2-methoxy-2-methyl-1-propanol

In the case of the ring opening by a nucleophile under basic conditions, the reaction is controlled by features of the S_N2 displacement reactions we considered in Chapter 7. The nucleophile attacks the less hindered primary carbon atom instead of the tertiary carbon atom. The resulting alkoxide ion then exchanges a proton with the solvent, and the methoxide base is regenerated.

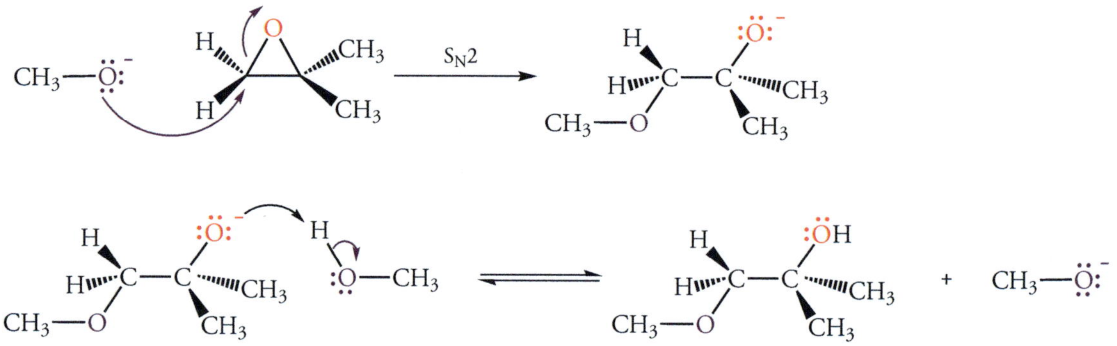

SN2

Under acid-catalyzed conditions, the epoxide oxygen is protonated, and the ring opens in the opposite direction when the epoxide reacts with methanol.

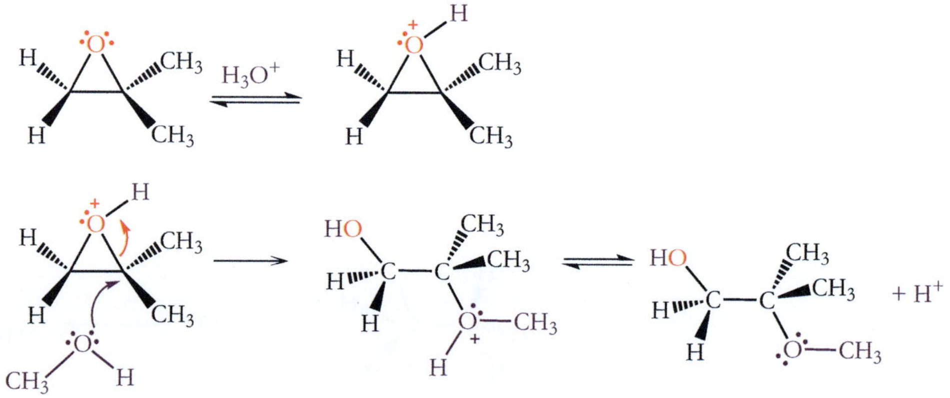

H3O+

We can understand the acid-catalyzed reaction by considering the resonance forms shown below. The positive charge is more stable on the tertiary carbon atom than on the primary carbon atom. The tertiary carbocation resonance form is a more important contributor than the primary one. Thus, the tertiary carbocation center combines with the nucleophile to give the conjugate acid of the observed product of the acid-catalyzed ring opening. The conjugate acid then exchanges a proton with the solvent.

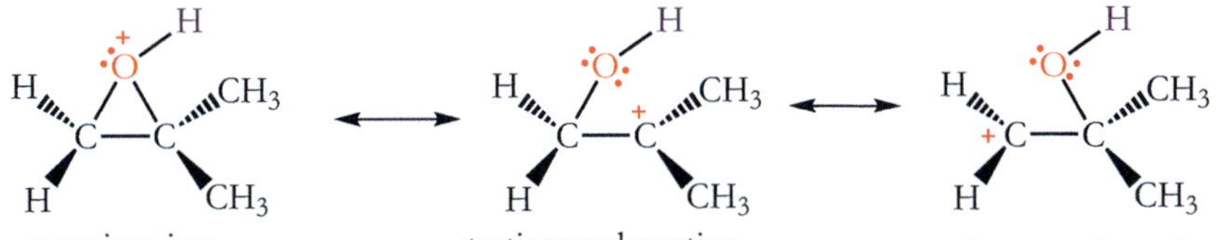

oxonium ion   tertiary carbocation   primary carbocation

Problem 9.9

Predict the product of the reaction of 2-methyloxirane with the Grignard reagent prepared from bromoethane.

Solution

The Grignard reagent reacts as a nucleophile. Thus, the Grignard reagent of bromoethane behaves as an ethyl carbanion. Nucleophilic attack of the ethyl carbanion on 2-methyloxirane occurs at the primary rather than the secondary carbon atom. Subsequent hydrolysis of the magnesium alkoxide yields 2-pentanol.

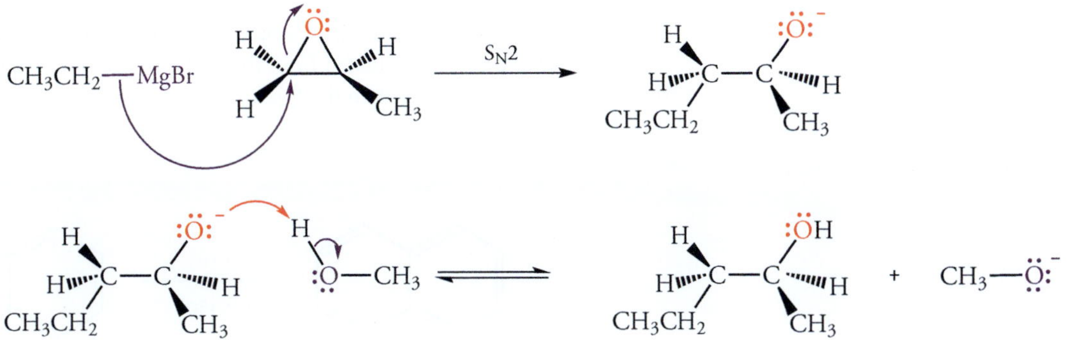

Problem 9.10

Predict the product of the reaction of styrene oxide (phenyloxirane) in an acid-catalyzed reaction with methanol.

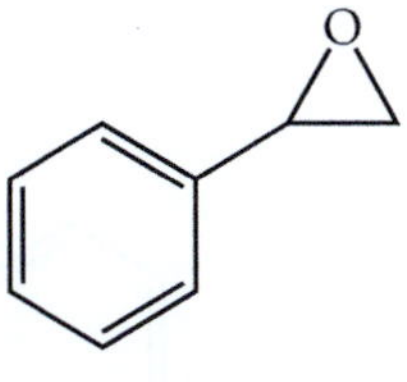

styrene oxide

Biochemical Reactions of Epoxides

Epoxides are produced biologically as oxidation products of alkenes and aromatic compounds. Epoxides that are formed in the liver by cytochrome P-450 undergo ring-opening reactions. The oxygen of the epoxide comes from molecular oxygen. If the epoxide reacts with a biological macromolecule, the result is potentially devastating. The epoxides made from aromatic compounds are called arene oxides. These molecules can undergo four kinds of reactions. With one exception, the reactions give products that do not harm the organism.

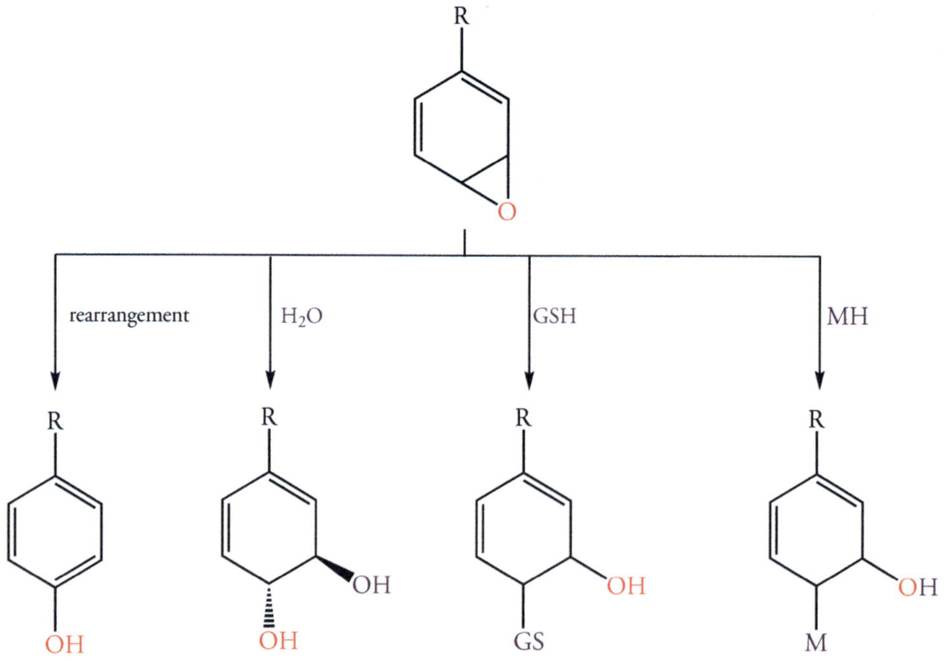

The rearrangement of an arene oxide gives a water-soluble phenol that is easily eliminated from the body as polar derivatives. Both sulfate esters or acetals of glucuronic acid (Chapter 13) are typical derivatives. Hence, this pathway does not lead to the accumulation of toxic by-products. Ring opening of the arene oxide by water gives a *trans* 1,2-diol by an S_N2 mechanism. These diols tend to be water-soluble and easily eliminated from the body. In the case of naphthalene, the epoxide forms at the C-1 to C-2 bond rather than the C-2 to C-3 bond.

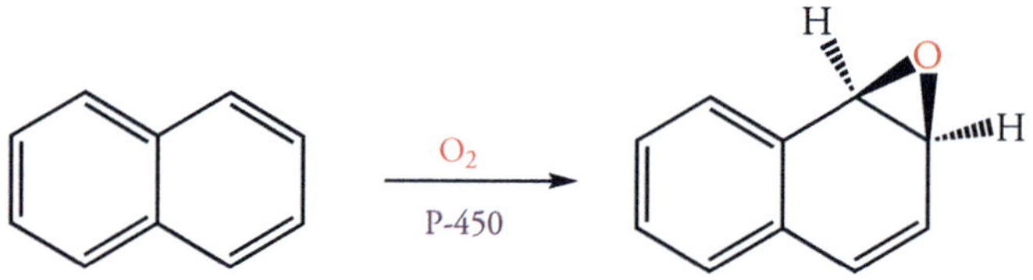

Isotopic labeling studies have shown that the oxygen atom of the benzylic hydroxyl group in the diol formed by ring opening of the epoxide is the original oxygen atom of the epoxide, and the other hydroxyl group comes from water.

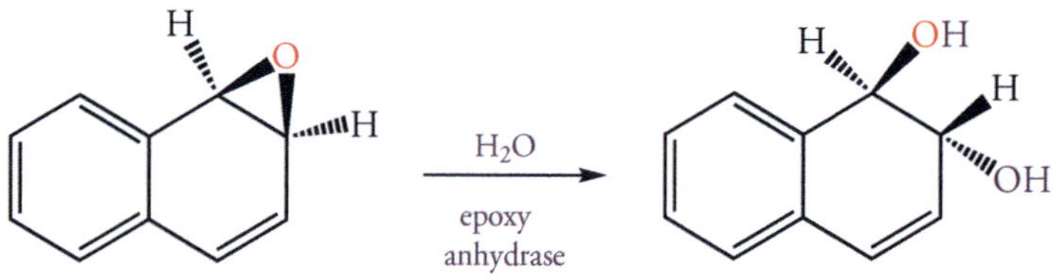

We recall that glutathione contains a nucleophilic sulfhydryl group. It acts as a scavenger that reacts with toxic metabolites. Glutathione (GSH) reacts with arene oxides in a ring-opening reaction. The product contains many polar functional groups, so it is water soluble and easily eliminated from the body.

Synthetic Sequences

9.26 Write the structure of the final product of each of the following sequences of reactions:

(a)
$$\text{CH}_2\text{CH}=\text{CH}_2$$

$$\xrightarrow[\text{2. HNaBH}_4]{\text{1. Hg(OAc)}_2/\text{H}_2\text{O}} \xrightarrow[\text{2. CH}_3\text{Br}]{\text{1. NaH}}$$

(b)
$$\text{—CH}=\text{CH}_2 \xrightarrow[\text{2. H}_2\text{O}_2/\text{OH}^-]{\text{1. B}_2\text{H}_6} \xrightarrow[\text{2. CH}_3\text{CH}_2\text{Br}]{\text{1. NaH}}$$

(c)
$$\text{CH}_2\text{CH}=\text{CH}_2$$

$$\xrightarrow[\text{2. H}_2\text{O}_2/\text{OH}^-]{\text{1. B}_2\text{H}_6} \xrightarrow[\substack{\text{2. ethylene oxide} \\ \text{3. H}_3\text{O}^+}]{\text{1. NaH}}$$

9.27 Write the structure of the final product of each of the following sequences of reactions:

(a)
$$\xrightarrow[\text{CH}_3\text{CH}_2\text{OH}]{\text{NaBH}_4} \xrightarrow[\text{2. CH}_3\text{CH}_2\text{Br}]{\text{1. NaH}}$$

(b)
$$=\text{O} \xrightarrow[\text{2. H}_3\text{O}^+]{\text{1. LiAlH}_4} \xrightarrow[\text{2. CH}_3\text{Br}]{\text{1. NaH}}$$

(c)
$$\xrightarrow[\text{CH}_3\text{CH}_2\text{OH}]{\text{NaBH}_4} \xrightarrow[\substack{\text{2. methyloxirane} \\ \text{3. H}_3\text{O}^+}]{\text{1. NaH}}$$

Reactions of Ethers

9.28 A compound of formula $C_5H_{12}O_2$ reacts with HI to give a mixture of iodomethane, iodoethane, and 1,2-diiodoethane. What is the structure of the compound?

9.29 A compound of formula $C_5H_{12}O_2$ reacts with HI to give a mixture of iodomethane and 1,3-diiodopropane. What is the structure of the compound?

9.30 A compound of formula $C_5H_{10}O$ reacts with HI to give only 1,5-diiodopentane. What is the structure of the compound?

9.31 A compound of formula $C_4H_8O_2$ reacts with HI to give only 1 ,2-diiodoethane. What is the structure of the compound?

Synthesis of Epoxides

9.32 Two products can be obtained by the epoxidation of the following bicycloalkene with MCPBA. Draw their structures.

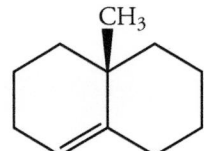

9.33   What alkene is required to synthesize the following compound, which is the sex attractant of the Gypsy moth, using MCPBA?

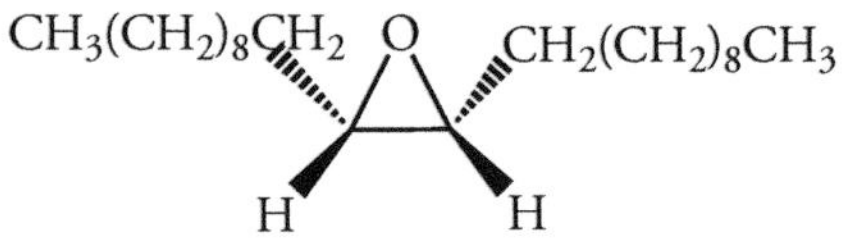

disparlure

Reactions of Epoxides

9.34   Ethyl cellosolve, $CH_3CH_2—O—CH_2CH_2—OH$, is an industrial solvent. Suggest a synthesis of this compound using ethylene oxide as one of the reactants.

9.35   2-Phenylethanol is used in some perfumes. How can this compound be prepared starting from bromobenzene?

9.36   What is the product of the reaction of 2-methyloxirane with methanol in the presence of an acid catalyst?

9.37   Epoxide rings can be cleaved by metal hydrides. Write the product of the reaction of cyclohexene oxide and $LiAlD_4$.

9.38   A mixture of 2,2-dimethyloxirane and ethanethiol is treated with sodium hydroxide. Write the structure of the expected product.

9.39   Epoxide rings can be cleaved by phenoxides. Propose a synthesis of the muscle relaxant methocarbamol using this fact.

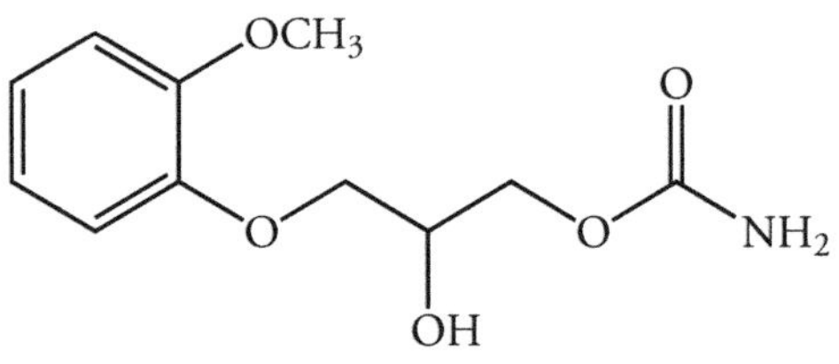

methocarbamol

10 ALDEHYDES AND KETONES

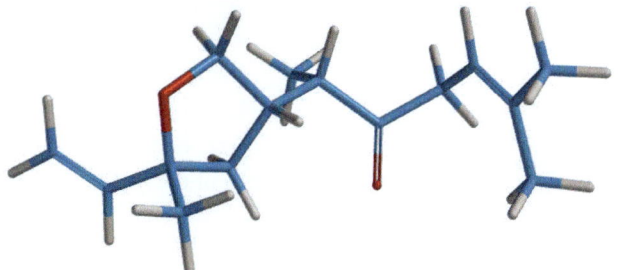

DAVANONE (A COMPONENT OF SOME PERFUMES)

10.1 THE CARBONYL GROUP

A carbonyl group consists of a carbonyl carbon atom and a carbonyl oxygen atom linked by a double bond. The simplest compound with a carbonyl group is formaldehyde, CH_2O. Although the lone pairs on the oxygen atom of the carbonyl group are shown in the structure below, they are not always shown in drawings of molecular structures.

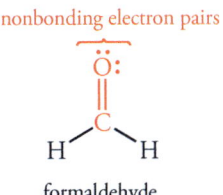

formaldehyde

The carbonyl carbon atom is sp^2-hybridized. It contributes one electron to each of the three hybrid orbitals, forming three σ bonds. Formaldehyde has two σ bonds to hydrogen atoms and one σ bond to the carbonyl oxygen atom. These coplanar bonds lie at approximately 120° to one another. The fourth electron of the carbonyl carbon atom occupies a 2p orbital perpendicular to the plane of the three sp^2 hybrid orbitals. The carbonyl oxygen atom, also sp^2-hybridized, contributes one of its six valence electrons to the sp^2 hybrid orbital that forms a π bond with the carbonyl carbon atom. Four valence electrons remain as two sets of nonbonding electron pairs in the other two sp^2 hybrid orbitals. They lie at approximately 120° to each other and to the carbon-oxygen bond (Figure 10.1). The last valence electron occupies a 2p orbital perpendicular to the plane of the sp^2 hybrid orbitals. The 2p orbitals of the carbon and oxygen atoms overlap side-by-side to form a π bond.

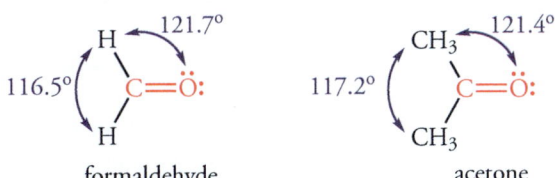

formaldehyde   acetone

An aldehyde can be represented as a resonance hybrid that has two nonequivalent resonance structures. Contributing structure (1) is more important because each atom has a Lewis octet. However, the dipolar structure (2) is often used to account for the chemical reactions of aldehydes and ketones. We will see that the carbonyl carbon atom reacts with nucleophiles, and the carbonyl oxygen atom reacts with electrophiles.

Principles of Organic Chemistry. http://dx.doi.org/10.1016/B978-0-12-802444-7.00010-0

Figure 10.1
Structure of Formaldehyde

The carbonyl carbon and oxygen atoms of formaldehyde are sp²-hybridized. The H—C—H bond angle is close to 120º.

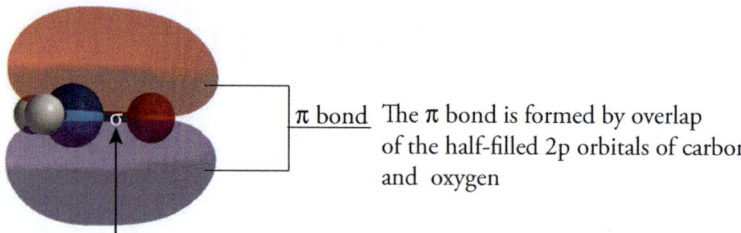

π bond The π bond is formed by overlap of the half-filled 2p orbitals of carbon and oxygen

σ bond
An electron of the sp² hybrid orbital of the carbonyl carbon atom and an electron of the sp² hybrid orbital of the oxygen atom form a σ bond

Carbonyl Compounds

When a carbonyl carbon atom is bonded to at least one hydrogen atom, the resulting compound is an **aldehyde**. Formaldehyde, the simplest aldehyde, has two hydrogen atoms bonded to the carbonyl carbon atom. The carbonyl group in other aldehydes is bonded to one hydrogen atom and either an alkyl group (R) or an aromatic, aryl group (Ar). Although the bond angles around the carbonyl carbon atom are approximately 120°, structures are often written with the carbonyl oxygen atom at 90° to a linear arrangement of atoms.

general formulas for an aldehyde

A carbonyl carbon atom is bonded to two other carbon atoms of either alkyl or aryl group in a ketone. As with aldehydes, a ketone has 120° bond angles at the carbonyl carbon atom, but structures are often written with the carbonyl oxygen atom at 90° to a linear arrangement of atoms.

general formulas for a ketone

An aldehyde can be written with the condensed formula RCHO or ArCHO, where the symbol CHO indicates that both the hydrogen and oxygen atoms are bonded to the carbonyl carbon atom. A ketone has the condensed formula RCOR'. In this condensed formula the symbol CO represents the carbonyl group, and the two R groups flanking the CO group are bonded to the carbonyl carbon atom.

Naturally Occurring Aldehydes and Ketones

The carbonyl group is prevalent in compounds isolated from biological sources. Aldehydes and ketones often have pleasant odors. For that reason, they are used in commercial products such as air fresheners, soaps, and perfumes. For example, α-ionone and jasmone are fragrant ketones responsible for the scent of irises and jasmine, respectively (Figure 10.2). The relationship between structure and the physiological response of odor depends upon their interactions with membrane proteins in neurons called guanine nucleoside coupled protein receptors (GCPRs).

At one time the extraction of fragrant compounds from flowers and other plants was the sole source of materials for products such as perfumes. However, it is now more economical to synthesize these compounds in the laboratory. Chemical synthesis also allows for the production of compounds with new odors.

Figure 10.2
Structures of Naturally Occurring Aldehydes and Ketones

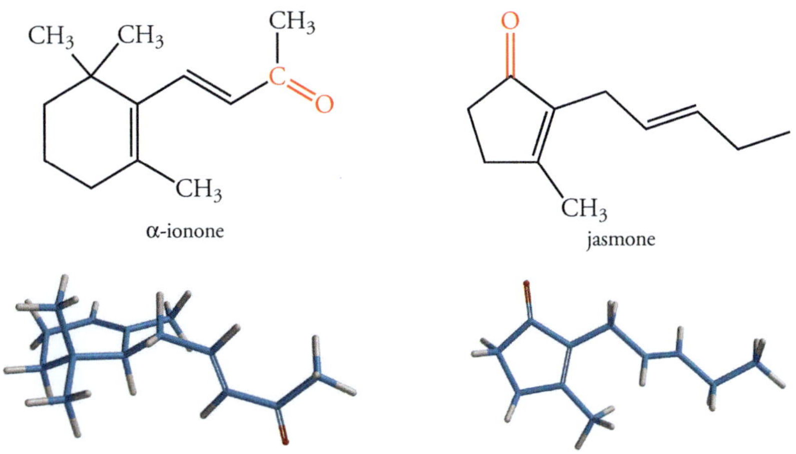

α-ionone

jasmone

10.2 NOMENCLATURE OF ALDEHYDES AND KETONES

Common Names of Aldehydes and Ketones

Aldehydes and ketones with low molecular weights are often referred to by their common names. The names of aldehydes are derived from the common names of related acids (Chapter 12).

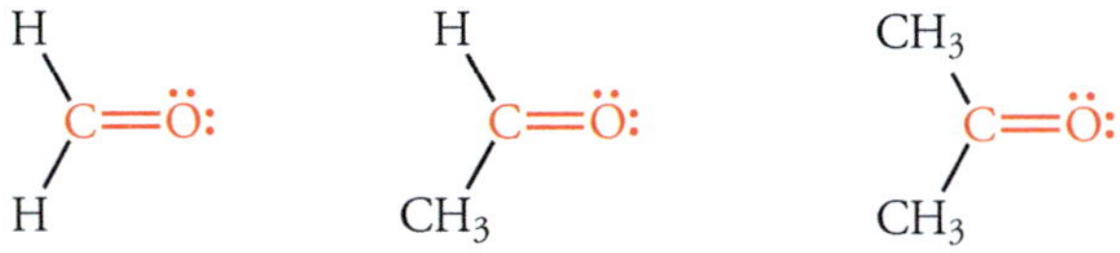

formaldehyde   acetaldehyde   acetone

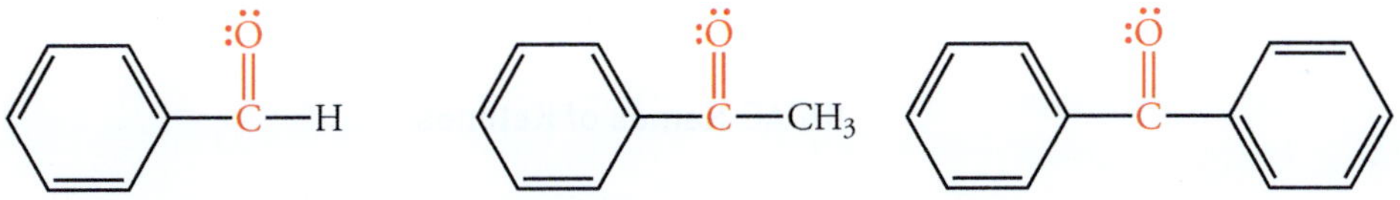

benzaldehyde   acetophenone   benzophenone

IUPAC Names of Aldehydes

1. Aldehydes are named by IUPAC rules similar to those outlined for alcohols. The final -*e* of the parent hydrocarbon corresponding to the aldehyde is replaced by the ending -*al*.

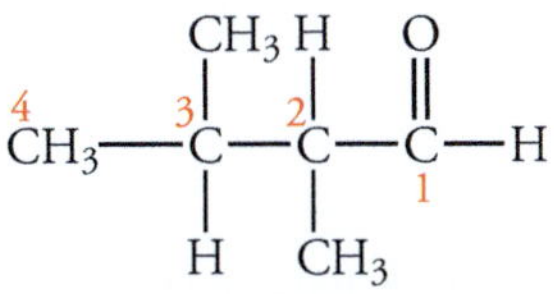

This is 2,3-dimethylbutanal, *not* 2,3-dimethyl-1-butanal

2. The aldehyde functional group has a higher priority than alkyl, halogen, hydroxyl, and alkoxy groups. The names and positions of these groups are indicated as prefixes to the name of the parent aldehyde.

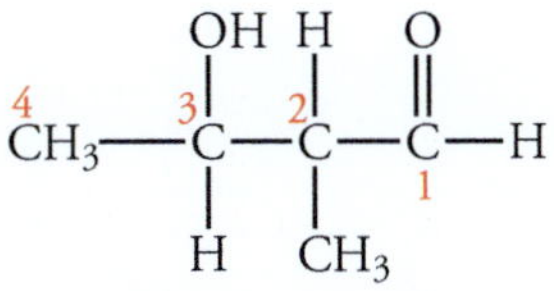

3-hydroxy-2-methylbutanal

3. The aldehyde functional group has a higher priority than double or triple bonds. When the parent chain contains a double or triple bond, replace the final *-e* of the name of the parent alkene or alkyne with the suffix *-al*. Indicate the position of the multiple bond with a prefix.

$$CH_3-\underset{4}{\overset{\overset{\displaystyle CH_3}{|}}{CH}}-\underset{3}{C}\equiv\underset{2}{C}-\underset{1}{\overset{\overset{\displaystyle O}{\|}}{C}}-H$$
$$\underset{5}{CH_3}$$

4-methyl-2-pentynal

4. If an aldehyde or ketone contains other groups with a higher priority, such as carboxylic acids, give the carbonyl group the prefix *-oxo*. Use a number to indicate the position of the *-oxo* group. The priority order is carboxylic acid > aldehyde > ketone.

$$CH_3-\underset{4}{\overset{\overset{\displaystyle O}{\|}}{C}}-\underset{3}{\overset{\overset{\displaystyle H}{|}}{\underset{\underset{\displaystyle CH_3}{|}}{C}}}-\underset{1}{\overset{\overset{\displaystyle O}{\|}}{C}}-H$$

2-methyl-3-oxobutanal

5. If an aldehyde group is attached to a ring, use the suffix *-carbaldehyde*.

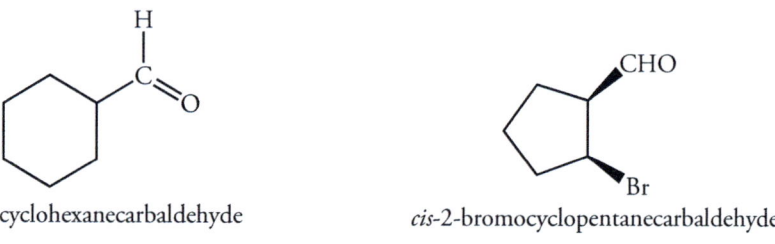

cyclohexanecarbaldehyde   *cis*-2-bromocyclopentanecarbaldehyde

IUPAC Names of Ketones

The IUPAC rules for naming ketones are similar to those used for aldehydes. The final *-e* of the parent hydrocarbon is replaced with the ending *-one*. However, the carbonyl group in a ketone is not on a terminal carbon atom, and we must indicate its position.

1. Number the carbon chain so that the carbonyl carbon atom has the lower number. This number appears as a prefix to the parent name. The identity and location of substituents are indicated with a prefix to the parent name.

$$CH_3-\underset{5}{\overset{\overset{\displaystyle CH_3}{|}}{\underset{\underset{\displaystyle H}{|}}{C}}}-\underset{4}{\overset{\overset{\displaystyle H}{|}}{\underset{\underset{\displaystyle H}{|}}{C}}}-\underset{3}{\overset{\overset{\displaystyle O}{\|}}{C}}-\underset{1}{CH_3}$$

This is 4-methyl-2-pentanone,
not 2-methyl-4-pentanone

2. Name cyclic ketones as cycloalkanones. The carbonyl carbon is C-1. Number the ring in the direction that gives the lower number to the first substituent encountered.

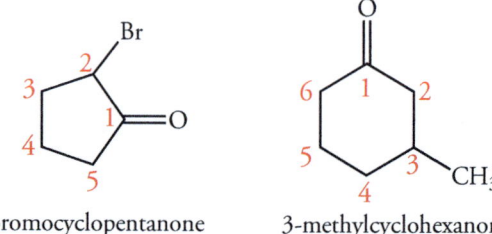

2-bromocyclopentanone   3-methylcyclohexanone

3. Halogen, hydroxyl, alkoxy groups, and multiple bonds have lower priorities than the ketone group. These substituted ketones are named using the same method described for aldehydes.

Problem 10.1

The IUPAC name for capillin, a drug used against skin fungi, is 1-phenyl-2,4-hexadiyn-1-one. Draw its structure.

Solution

When we dissect the name, we see that it has the suffix -1-one and the stem name *hexa*, indicating that the parent chain is a ketone containing six carbon atoms. We write the carbon skeleton and number the chain. Place the carbonyl oxygen atom on C-1.

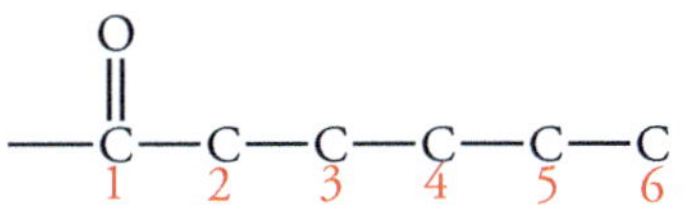

The name has the prefix 1-phenyl. Therefore, we add a phenyl group at C-1. Note that the presence of the phenyl group makes the compound a ketone. A carbonyl carbon atom at the end of a chain would otherwise be an aldehyde.

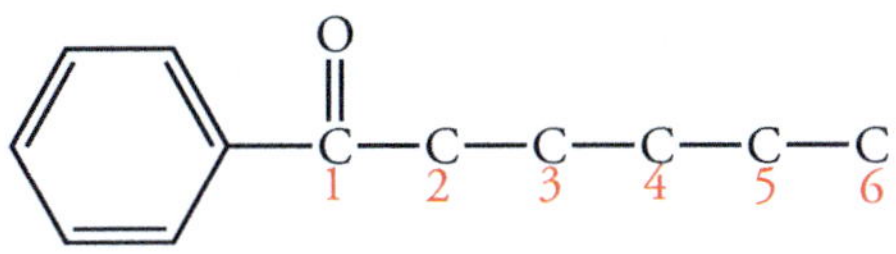

The name "*diyn*" tells us that there are two triple bonds; they are located at C-2 and C-4. Fill in the requisite hydrogen atoms.

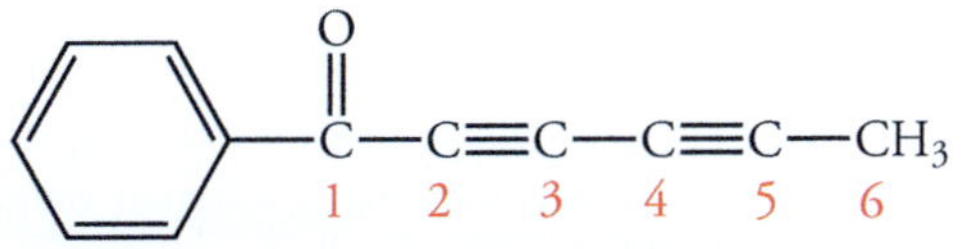

Problem 10.2

What is the IUPAC name for the following compound, which is an alarm pheromone in some species of ants.

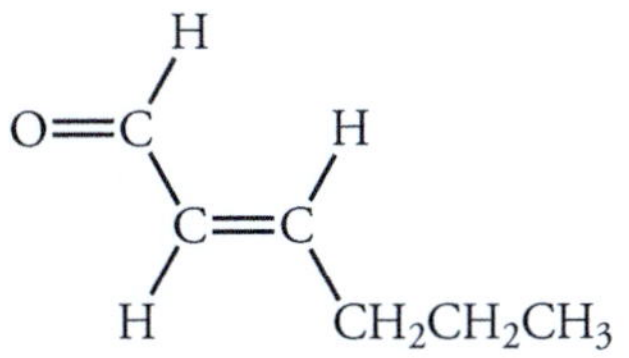

10.3 PHYSICAL PROPERTIES OF ALDEHYDES AND KETONES

Because oxygen is more electronegative than carbon, the electrons in the carbonyl bond are pulled toward the oxygen atom, and the carbonyl group is polar. This polarity is also expected based on the contributing charged structure of the resonance hybrid.

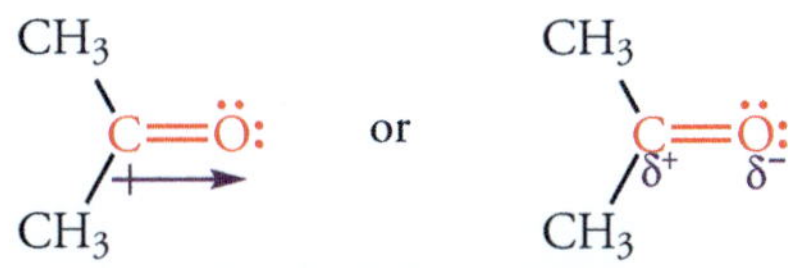

2-propanone (2.9 D)
(acetone)

As a result, the dipole moment for propanal, a typical aldehyde, is larger than the dipole moment of butane and 1-propanol (Table 10.1). The dipole moment reflects the polarity of the carbonyl group.

Physical Properties of Aldehydes and Ketones

Aldehydes and ketones have higher boiling points than the alkanes of similar molecular weight (Table 10.1) because of dipole-dipole intermolecular forces due to the carbonyl group. However, alcohols have higher boiling points than aldehydes and ketones even though alcohols have smaller dipole moments than carbonyl compounds. This order of boiling points is the result of hydrogen bonding in alcohols that is not possible in carbonyl compounds. As the molecular weights of the carbonyl compounds increase, their dipole-dipole attractive forces become less important than the London forces of the hydrocarbon skeleton. As a result, the physical properties of aldehydes and ketones become more like those of hydrocarbons as chain length increases. The boiling point differences become smaller, although the order of boiling points is still alcohol > carbonyl compound > alkane.

Table 10.1
Comparisons of Physical Properties

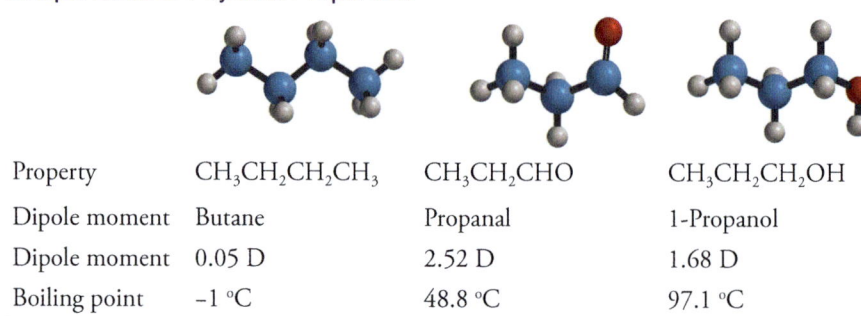

Property	$CH_3CH_2CH_2CH_3$	CH_3CH_2CHO	$CH_3CH_2CH_2OH$
Dipole moment	Butane	Propanal	1-Propanol
Dipole moment	0.05 D	2.52 D	1.68 D
Boiling point	–1 °C	48.8 °C	97.1 °C

Solubility of Aldehydes and Ketones in Water

Aldehydes and ketones cannot form hydrogen bonds with one another because they cannot function as hydrogen bond donors. However, the carbonyl oxygen atom has lone pair electrons that can serve as hydrogen bond acceptors. Thus, carbonyl groups can form hydrogen bonds with water. Hence, the lower molecular weight compounds formaldehyde, acetaldehyde, and acetone are soluble in water in all proportions.

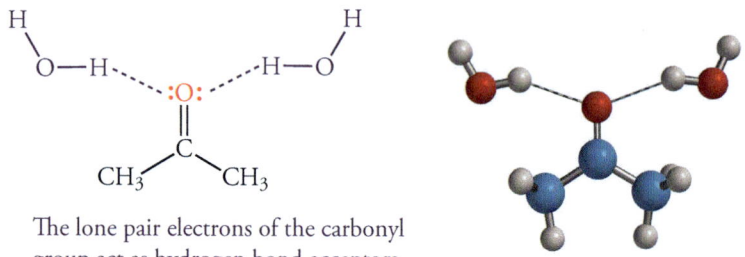

The lone pair electrons of the carbonyl group act as hydrogen bond acceptors

However, the solubility of carbonyl compounds in water decreases as the chain length increases, and their solubilities become more like those of hydrocarbons. Both acetone and 2-butanone (known in industry as methyl ethyl ketone, or MEK) are excellent solvents for many organic compounds. These polar solvents dissolve polar solutes because "like dissolves like." These solvents also readily dissolve protic solutes such as alcohols and carboxylic acids because the carbonyl group acts as a hydrogen bond acceptor for these compounds, as shown in the molecular model of acetone hydrogen bonded to methanol, below.

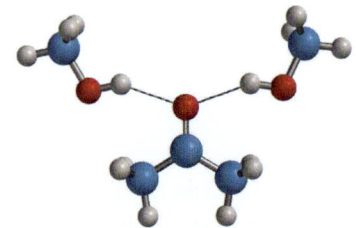

10.4 OXIDATION-REDUCTION REACTIONS OF CARBONYL COMPOUNDS

The carbonyl group is in an oxidation state between that of an alcohol and a carboxylic acid. Thus, a carbonyl group can be reduced to an alcohol or oxidized to a carboxylic acid.

Oxidation of Aldehydes

In Chapter 8 we saw that primary alcohols are oxidized to aldehydes, which are subsequently easily oxidized to acids. Under the same conditions, secondary alcohols are oxidized to ketones, which are not oxidized further. This difference in reactivity distinguishes these classes of compounds. For example, Tollens reagent and Fehling's solution are mild oxidizing reagents. Each of these converts aldehydes to carboxylic acids; neither of them oxidizes ketones. These reagents therefore provide a simple qualitative way of distinguishing aldehydes from ketones.

Tollens reagent is a basic solution of a silver ammonia complex ion. When Tollens reagent is added to a test tube that contains an aldehyde, the aldehyde is oxidized and metallic silver is deposited as a mirror on the wall of the test tube.

Fehling's solution contains cupric ion, Cu^{2+}, as a complex ion in a basic solution. It oxidizes aldehydes to carboxylic acids as the Cu^2 is reduced to Cu^+, which forms a brick-red precipitate, Cu_2O. Fehling's solution has the characteristic blue color of Cu^{2+}, which fades as the red precipitate of Cu_2O forms. Because Fehling's solution is basic, the carboxylic acid product is formed as its conjugate base.

Reduction of Aldehydes and Ketones to Alcohols

In Chapter 4 we saw that carbon-carbon double bonds can be reduced by hydrogen gas with nickel, palladium, or platinum as catalysts. In Chapter 8 we found that aldehydes and ketones can be catalytically reduced to alcohols by hydrogen gas and catalysts such as palladium and Raney nickel. However, the reduction of aldehydes and ketones with hydrogen gas requires more severe conditions than are required to reduce alkenes. We also saw that both lithium aluminum hydride, $LiAlH_4$, and sodium borohydride, $NaBH_4$, reduce carbonyl groups, but neither reagent reduces carbon-carbon double or triple bonds (Section 8.8). Thus, either of these reagents may be used to reduce a carbonyl group selectively in compounds with carbon-carbon multiple bonds.

Reduction of a Carbonyl Group to a Methylene Group

A carbonyl group can be reduced directly to a methylene group by using either the Clemmensen reduction or the Wolff-Kishner reduction. The former uses a zinc amalgam (Zn/Hg) and HCl, and the latter uses hydrazine (NH_2NH_2) and base.

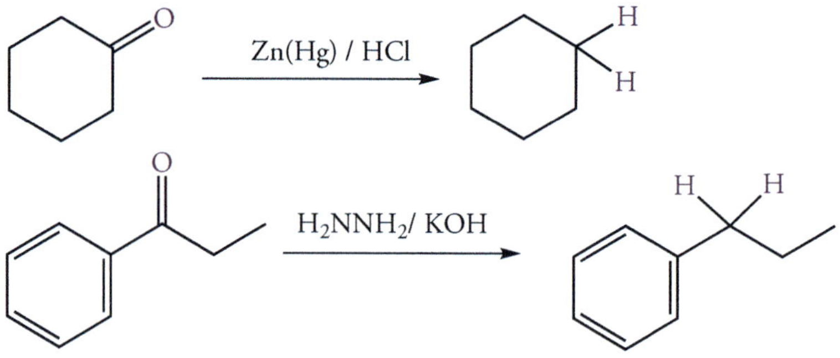

We introduced the reduction of a carbonyl group to a methylene group in Chapter 5 as a method of converting the product of a Friedel-Crafts acylation to an alkyl group that could not be produced by direct Friedel-Crafts alkylation.

Problem 10.3
Can the isomeric carbonyl-containing compounds of molecular formula C_4H_8O be distinguished by Tollens reagent?

Solution
There are three isomeric carbonyl-containing compounds—two aldehydes and one ketone.

$$\underset{\underset{CH_3}{|}}{CH_3CHCHO} \qquad CH_3CH_2CH_2CHO \qquad CH_3CH_2CH_2\overset{\overset{O}{||}}{C}CH_3$$

Both aldehydes react with Tollens reagent to produce a silver mirror. Therefore, these two compounds cannot be distinguished from each other with this reagent. However, the ketone does not react with Tollens reagent. Therefore, the compound that does not yield a silver mirror is 2-butanone.

Problem 10.4
Explain how the following two isomeric compounds could be distinguished using Fehling's solution.

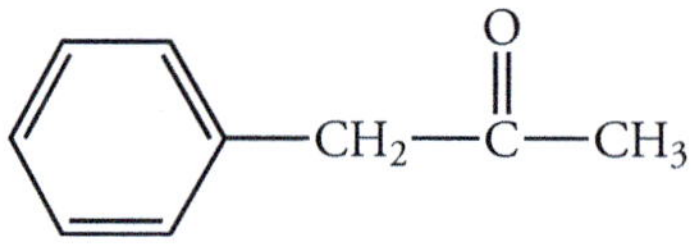

(1) 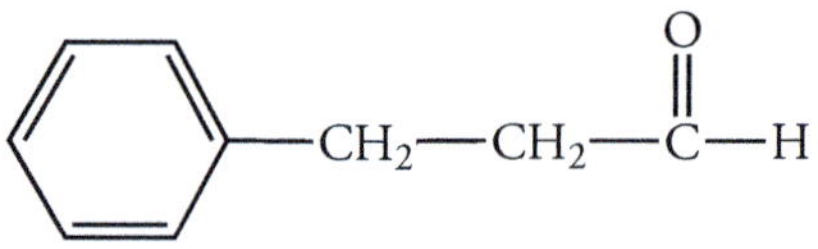   (2)

Problem 10.5
Reduction of the following ketone using sodium borohydride gives a mixture of two isomeric compounds. Draw their structures.

Solution
The hydride ion can attack from either the right or left of the planar carbonyl group. Attack from the left places the hydroxyl group on the same side of the ring as the carbon-carbon double bond. Attack from the right gives an isomer with the hydroxyl group on the side opposite that of the carbon-carbon double bond.

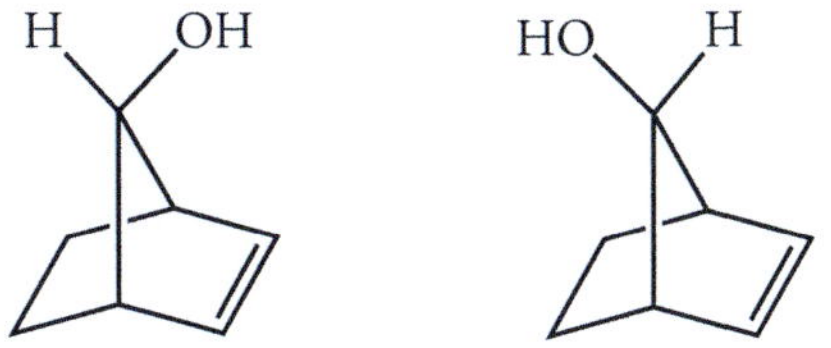

Problem 10.6
Draw the structure of the product of the reaction of the following compound with each of the following reducing agents: (a) hydrazine and base, (b) palladium and hydrogen at 1 atm, and (c) Raney nickel and hydrogen at 100 atm.

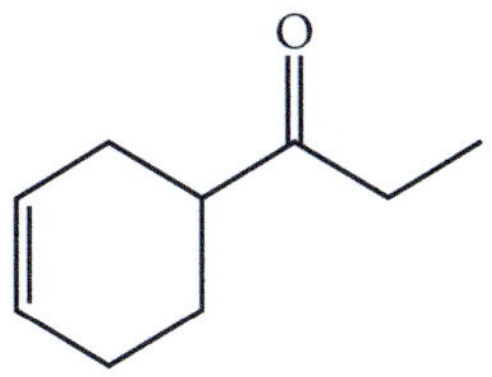

10.5 ADDITION REACTIONS OF CARBONYL COMPOUNDS

Aldehydes and ketones contain a π bond that undergoes addition reactions. Because the carbonyl bond is polar, an unsymmetrical reagent reacts with it so that the electrophilic part bonds to the carbonyl oxygen atom and the nucleophilic part to the carbonyl carbon atom.

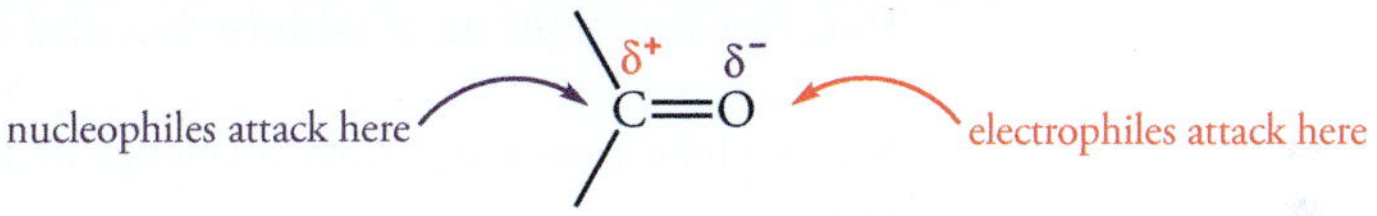

Many reagents that add to carbonyl compounds can be represented as H—Nu. The electrophilic part of the reagent is H⁺; the nucleophilic part is Nu:⁻. The addition reaction occurs in several steps. The order of the steps depends on whether the reaction is acid- or base-catalyzed.

For the base-catalyzed reaction, the nucleophile attacks the carbonyl group, giving a tetrahedral intermediate that is subsequently protonated.

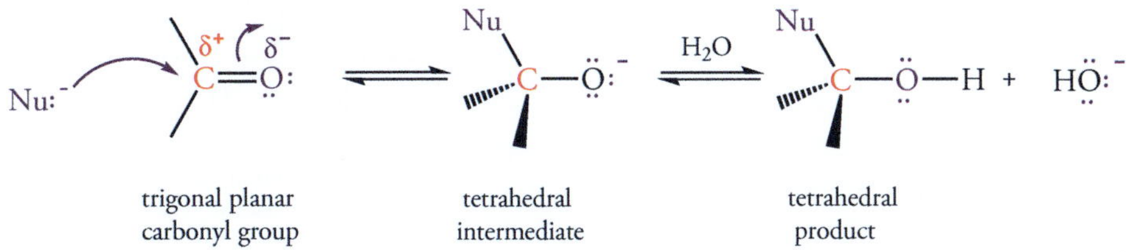

trigonal planar carbonyl group   tetrahedral intermediate   tetrahedral product

The net reaction is the addition of a nucleophile and a proton across the π bond of the carbonyl group. The negatively charged nucleophile, Nu:⁻, is formed by an acid-base reaction of H—Nu with OH⁻. Note that OH⁻ is regenerated as a result of protonation of the oxygen atom of the tetrahedral intermediate.

The acid-catalyzed reaction occurs by initial protonation of the carbonyl oxygen atom to produce a resonance-stabilized intermediate.

trigonal planar
carbonyl group

resonance-stabilized oxocarbocation

As a result, the carbonyl carbon atom is more susceptible to attack by even weak nucleophiles. This reaction can be shown using either of the two contributing resonance forms. Subsequent loss of the proton bonded to the original nucleophile in an acid-base reaction with solvent regenerates the proton required in the first step.

tetrahedral
intermediate

Equilibria in Nucleophilic Addition Reactions

Nucleophiles that add to carbonyl compounds are classed according to the reversibility of the reaction. Nucleophiles that add irreversibly include hydride ion derived from lithium aluminum hydride or sodium borohydride (Section 10.4) and the carbanion derived from a Grignard reagent (Section 10.6). These nucleophiles are conjugate bases of weak acids and are thus poor leaving groups. As a result, once they add to the carbonyl carbon atom the reverse reaction is unfavorable. Nucleophiles that add reversibly are conjugate bases of strong acids and are good leaving groups. We will consider examples of these reactions in Sections 10.7–10.10. These reversible reactions are usually driven to completion by adjusting reaction conditions based on Le Châtelier's principle.

Relative Reactivities of Aldehydes and Ketones

Nucleophiles react faster with aldehydes than with ketones because of both electronic and steric effects.

First, we will consider electronic effects. A ketone has two alkyl groups attached to the carbonyl carbon atom. These groups donate electron density to the carbonyl carbon atom and stabilize its partial positive charge. (We recall that alkyl groups stabilize carbocations in the same way.) The carbonyl carbon atom of an aldehyde is attached to only one alkyl group. Therefore, the carbonyl carbon atom of an aldehyde has a larger partial positive charge than that of a ketone. As a result, nucleophiles react faster with aldehydes than with ketones.

less stabilization
more reactive

more stabilization
less reactive

The sizes of groups also affect the reactivity of aldehydes and ketones. Because a ketone has two alkyl groups attached to the carbonyl carbon atom, it is sterically hindered relative to the carbonyl carbon atom of an aldehyde, which has only a hydrogen atom and one alkyl group bonded to it. Thus, a nucleophile can approach the carbonyl group of an aldehyde more readily, so the reaction occurs faster with an aldehyde (Figure 10.3). Therefore, both electronic and steric effects reinforce each other to make aldehydes more reactive than ketones.

Figure 10.3
Steric Effects on Addition
Reactions

The carbonyl group of an aldehyde (a) is less sterically hindered than the carbonyl group of a ketone (b). Therefore, ketones react more slowly than aldehydes in nucleophilic addition reactions. The nucleophile here is water.

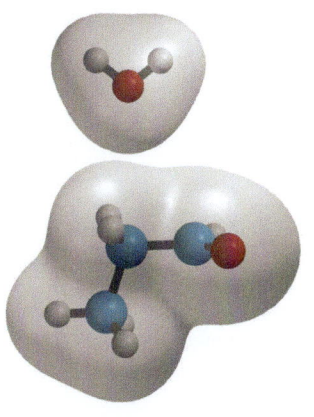

(a) Hydration of an aldehyde

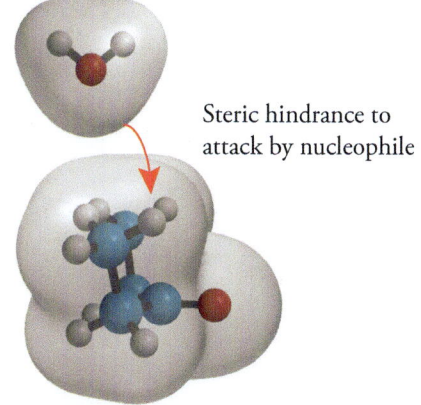

Steric hindrance to attack by nucleophile

(b) Hydration of a ketone

10.6 SYNTHESIS OF ALCOHOLS FROM CARBONYL COMPOUNDS

Reactions of Aldehydes and Ketones with Grignard Reagents

In Chapter 9 we discussed the reaction of haloalkanes with magnesium to produce Grignard reagents.

$$R-X \xrightarrow[\text{ether}]{\text{Mg(s)}} R-Mg-X$$

These reagents are used to form carbon-carbon bonds in the synthesis of complex molecules from simpler molecules. We recall that Grignard reagents contain a very strongly polarized carbon-magnesium bond in which the carbon atom has a partial negative charge.

$$\overset{\delta^-}{C}-\overset{\delta^+}{MgX}$$

The carbon atom of the Grignard reagent resembles a carbanion. It reacts as a nucleophile and adds to the electrophilic carbon atom of a carbonyl group in an aldehyde or ketone. The magnesium ion forms a salt with the negatively charged oxygen atom. The resulting product is a magnesium alkoxide, which is hydrolyzed to obtain an alcohol.

$$R-MgX \quad \overset{\delta^+}{C}=\overset{\delta^-}{O} \longrightarrow \underset{\text{a magnesium alkoxide}}{R-C-\ddot{O}-MgX} \xrightarrow{H_2O} \quad R-C-\ddot{O}-H \quad + \quad HOMgX$$

A Grignard reagent adds to various types of carbonyl compounds to give primary, secondary, and tertiary alcohols. Primary alcohols are synthesized by reacting the Grignard reagent, R—MgX, with formaldehyde.

Secondary alcohols are obtained by reacting the Grignard reagent with an aldehyde, RCHO. Note that the carbon atom bearing the hydroxyl group is bonded to the alkyl groups from both the Grignard reagent and the aldehyde.

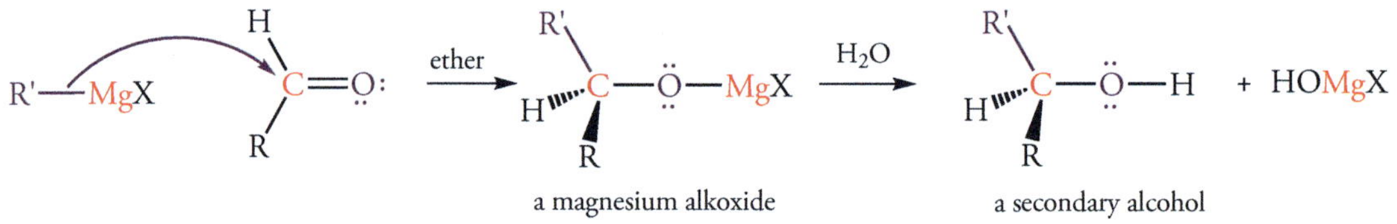

a magnesium alkoxide   a secondary alcohol

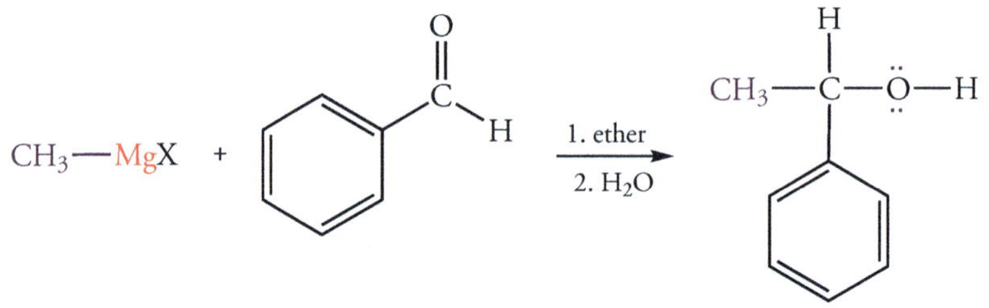

Tertiary alcohols are made by reacting the Grignard reagent with a ketone. Two of the alkyl groups bonded to the carbon atom bearing the hydroxyl group were part of the ketone; one alkyl group is provided from the Grignard reagent.

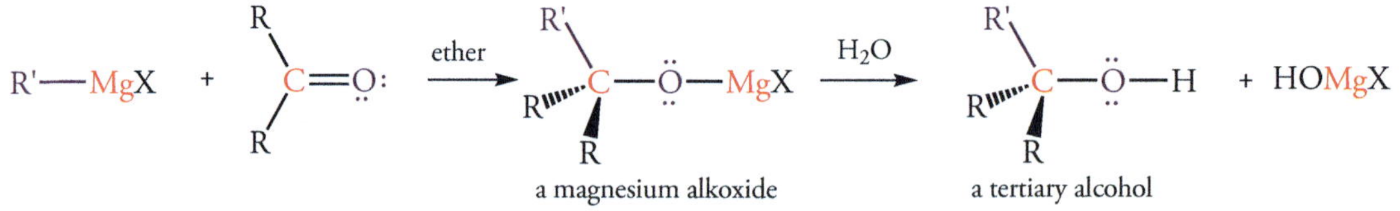

a magnesium alkoxide   a tertiary alcohol

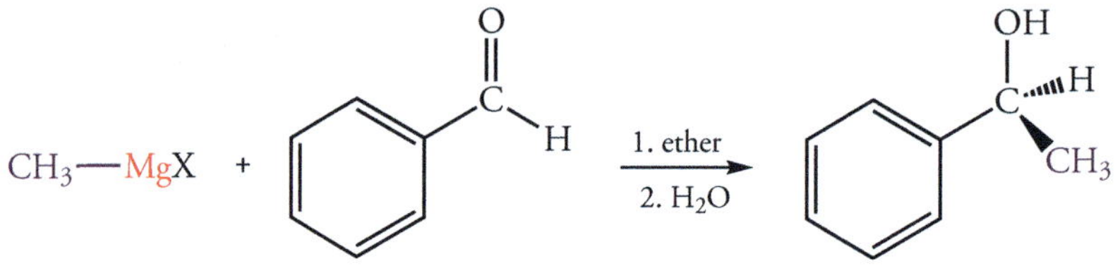

Acetylenic Alcohols

Alkynide ions, the conjugate bases of alkynes, react with carbonyl groups in much the same way as Grignard reagents do. The alkynides are prepared in an acid-base reaction with acetylene or a terminal alkyne using sodium amide in ammonia. If a carbonyl compound is then added to the reaction mixture, an alcohol forms after acid workup. If the alkynide is derived from acetylene, an acetylenic alcohol forms.

$$H-C\equiv C-H + NaNH_2 \longrightarrow H-C\equiv C-Na + NH_3$$

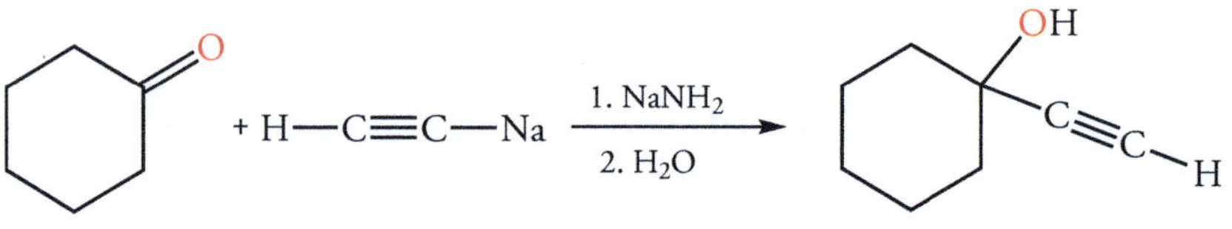

1-ethynylcyclohexanol

Problem 10.7

The European bark beetle produces a pheromone that causes beetles to aggregate. Describe two ways that the compound could be synthesized in the laboratory by a Grignard reagent.

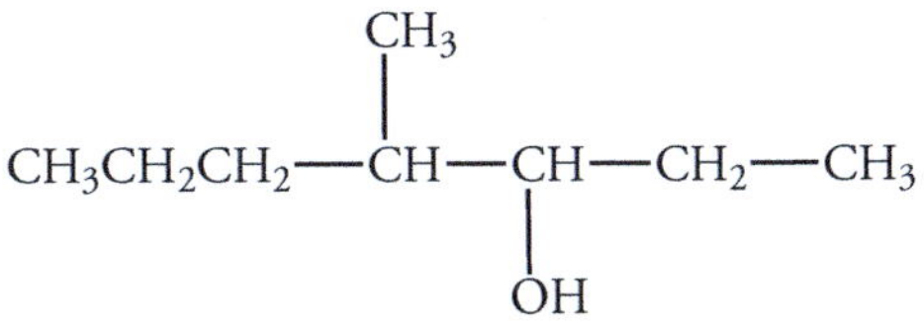

$$CH_3CH_2CH_2-\overset{\overset{\displaystyle CH_3}{|}}{CH}-\overset{\overset{}{}}{CH}-CH_2-CH_3$$
$$\underset{OH}{|}$$

Solution

The compound is a secondary alcohol that can be made from an aldehyde and a Grignard reagent. The carbon atom bonded to the hydroxyl group must be chosen as the carbonyl carbon atom. One of the two alkyl groups bonded to this carbon atom must be part of the aldehyde. The other must be the alkyl group of the Grignard reagent. The two possible components that could be introduced by a Grignard reagent are an ethyl group or a 1-methylbutyl group.

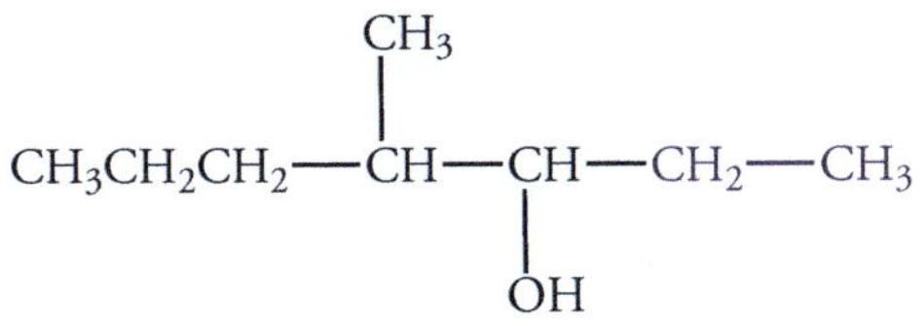

$$CH_3CH_2CH_2-\overset{\overset{\displaystyle CH_3}{|}}{CH}-CH-CH_2-CH_3$$
$$\underset{OH}{|}$$

ethyl

$$CH_3CH_2CH_2-\overset{\overset{\displaystyle CH_3}{|}}{CH}-CH-CH_2-CH_3$$
$$\underset{OH}{|}$$

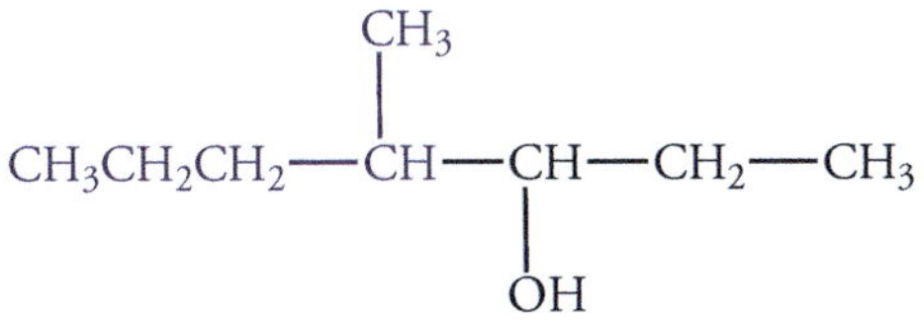

1-methylbutyl

If ethyl magnesium bromide is the Grignard reagent, the required aldehyde for the reaction is 2-methylpentanal. If the Grignard reagent is 1-methylbutyl magnesium bromide, propanal is required as the aldehyde.

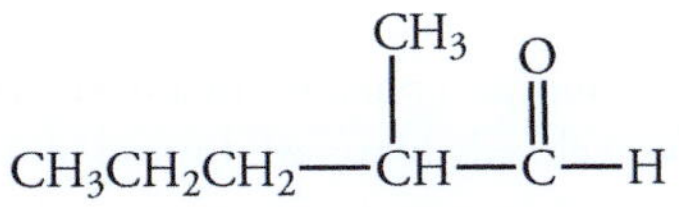

$$CH_3CH_2CH_2-\overset{\overset{\displaystyle CH_3}{|}}{CH}-\overset{\overset{\displaystyle O}{\|}}{C}-H$$

2-methylpentanal

$$H-\overset{\overset{\displaystyle O}{\|}}{C}-CH_2-CH_3$$

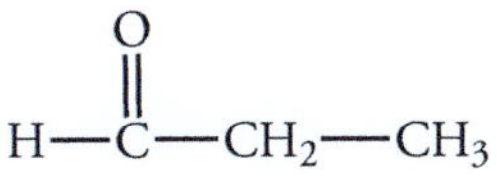

propanal

Problem 10.8

The methyl Grignard reagent reacts with 4-*tert*-butylcyclohexanone to give a mixture of two isomeric products. Draw their structures.

Oral Contraceptives

Alkynide ions, the conjugate bases of alkynes, react with carbonyl groups in much the same way as the female sex hormones that are collectively called estrogens. Estrogens, such as estradiol, are released during pregnancy and inhibit further ovulation. Oral contraceptives are designed to mimic this effect of pregnancy by inhibiting ovulation.

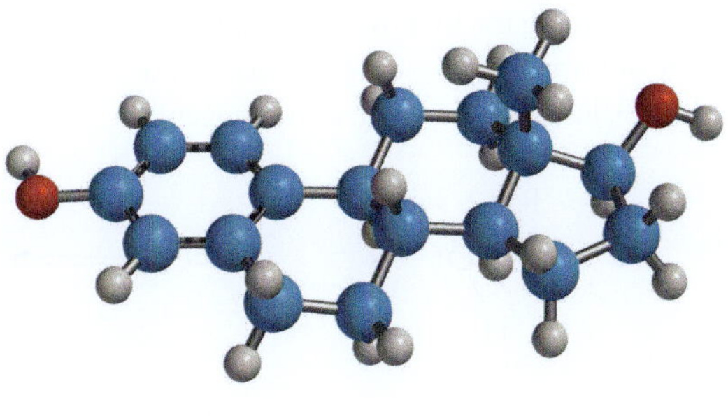

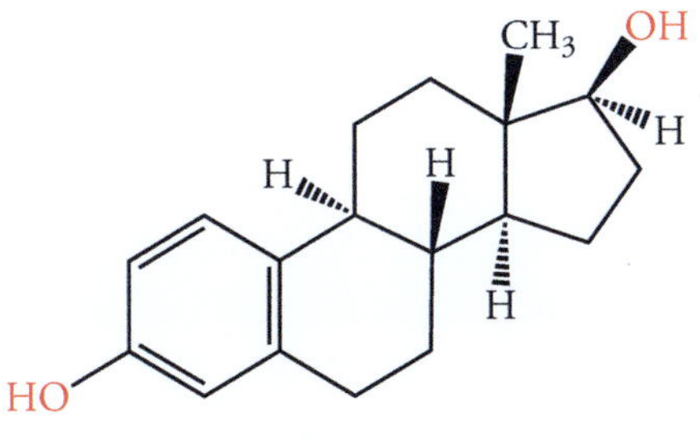

estradiol

17-ethynylestradiol

Estradiol itself is not an effective oral contraceptive because its C-17 hydroxyl group is rapidly oxidized to estrone in metabolic reactions. This oxidation product has greatly reduced estrogenic activity.

estrone

The hormonal action of estradiol is related to the C-17 hydroxyl group, which is located above the plane of the five-membered ring. Thus, it was decided to synthesize a compound structurally related to estradiol that is a tertiary alcohol with the correct stereochemistry. Such a compound would survive metabolic oxidation because tertiary alcohols cannot be oxidized. Based on the reactivity of Grignard reagents, one might propose to add methyl Grignard reagent to estrone to produce a tertiary alcohol. However, estrone has a phenolic hydroxyl group that would react with the Grignard reagent. This difficulty has been cleverly bypassed. Instead of making a tertiary alcohol by adding an alkyl group derived from a Grignard reagent, a tertiary alcohol is made by adding acetylide anion—the conjugate base of acetylene (Section 4.5). The reaction of the acetylide ion with a carbonyl compound produces an acetylenic alcohol. Sodium acetylide reacts with estrone to give ethynyl estradiol. Although the acetylide ion could potentially attack from either side of the C-17 carbonyl group of estrone, a methyl group extends above the plane of the ring in the vicinity of the carbonyl group. The acetylide anion thus approaches the "bottom" of the ring, so the product has its −OH group "up," which is the stereochemistry required for hormonal activity. This tertiary alcohol cannot be oxidized and is an effective oral contraceptive.

10.7 ADDITION REACTIONS OF OXYGEN COMPOUNDS

The nucleophilic oxygen atom of both water and alcohols can attack the carbonyl carbon atom of aldehydes and ketones to give addition products. Addition of water yields a hydrate; addition of an alcohol yields a hemiacetal or hemiketal. Because water and alcohols are weak nucleophiles, each of these reactions is reversible.

Addition of Water

Water adds to aldehydes and ketones to form **hydrates**. The proton of water bonds to the oxygen atom of the carbonyl group; the hydroxide ion adds to the carbon atom. Formaldehyde is over 99% hydrated. The hydrate of formaldehyde, called formalin, was once used to preserve biological specimens. It is no longer used because formaldehyde is carcinogenic.

hemiacetal

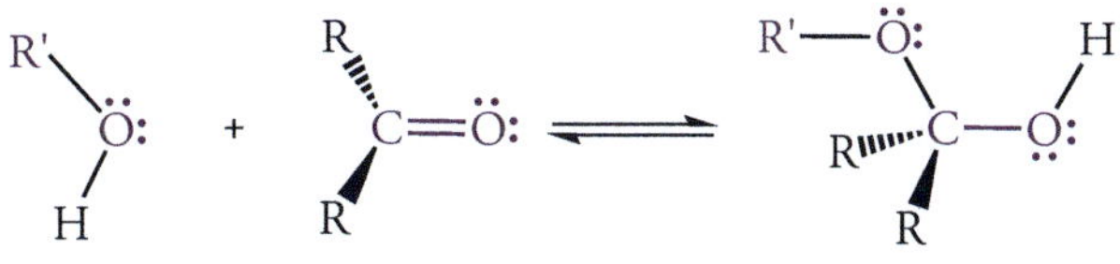

hemiketal

Hemiacetals and hemiketals are usually unstable compounds. That is, the equilibrium constant for formation of either a hemiacetal or a hemiketal is less than 1, and the equilibrium position for the preceding reactions lies to the left. However, when both the carbonyl group and the alcohol are part of the same molecule, the equilibrium constant is larger because a stable cyclic product forms. Cyclization is favorable because the two functional groups are close to each other.

Carbohydrates (Chapter 13), which contain both carbonyl and hydroxyl groups, exist to only a small extent as open-chain molecules. They exist largely as cyclic hemiacetals or hemiketals.

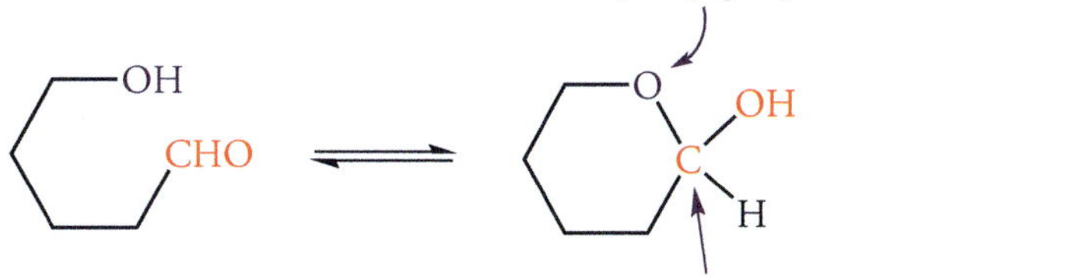

The ring oxygen atom is derived from the hydroxyl group

The ring carbon atom is derived from the carbonyl group

Mechanism of Addition of Alcohols

The first step in the acid-catalyzed addition of an alcohol to an aldehyde or ketone is protonation of the carbonyl oxygen atom (an electron pair donor, or Lewis base) to produce a resonance-stabilized carbocation.

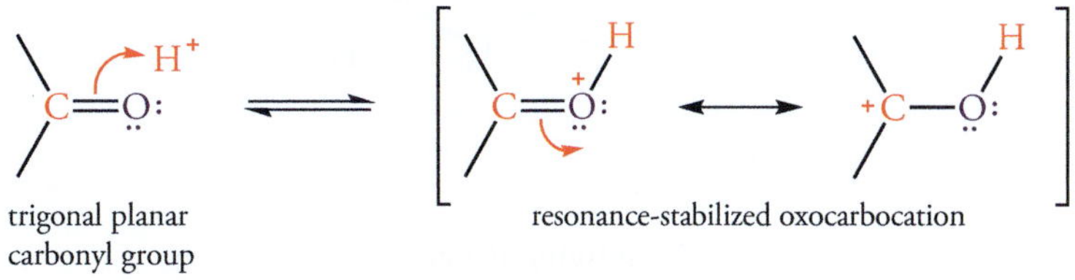

trigonal planar
carbonyl group

resonance-stabilized oxocarbocation

A proton is transferred from the conjugate acid of the hemiacetal in an acid-base reaction and becomes available again for the first step of the reaction sequence. Thus, the reaction is acid catalyzed.

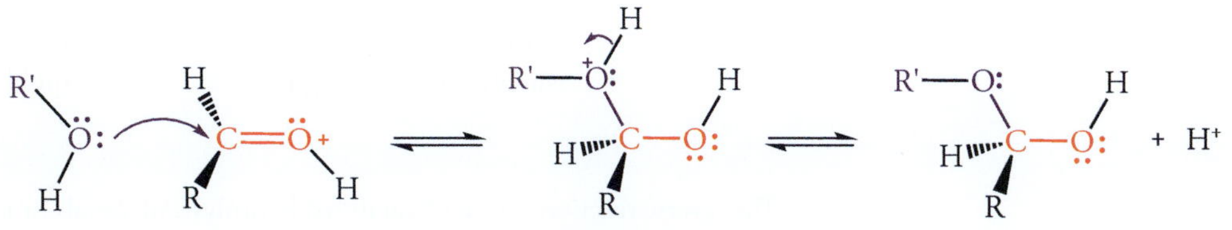

conjugate acid of
hemiacetal

hemiacetal

$+ H^+$

10.8 FORMATION OF ACETALS AND KETALS

The hydroxyl group in either a hemiacetal or a hemiketal can be replaced by substitution of the −OH by another alkoxy group, −OR′. This reversible reaction occurs readily in acidic solution to produce an acetal or ketal.

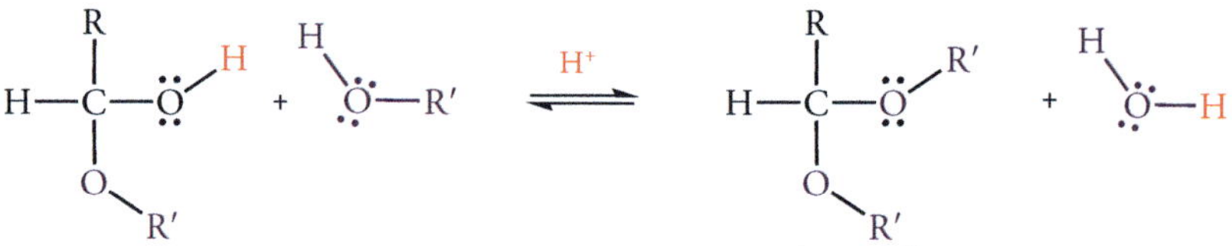

(an acetal)

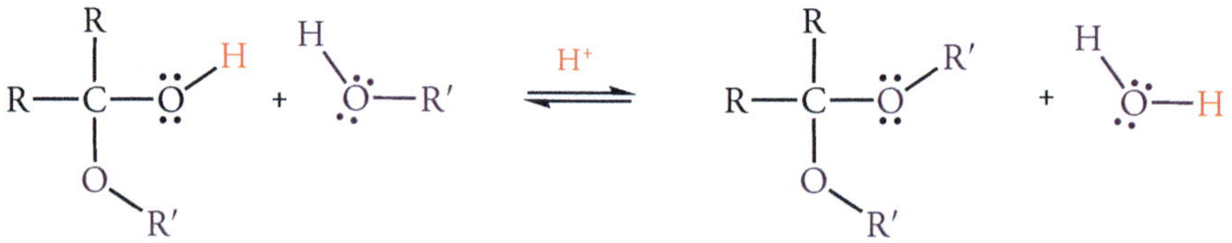

(a ketal)

Note that both acetals and ketals have two alkoxy groups (−OR′) attached to the same carbon atom. An acetal also has a hydrogen atom and an alkyl group attached to the carbon atom, whereas the ketal has two alkyl groups attached. The formation of an acetal or a ketal requires two molar equivalents of alcohol per mole of the original carbonyl compound.

Cyclic hemiacetals or hemiketals react with alcohols to produce cyclic acetals or ketals. Consider the cyclic hemiacetal of 5-hydroxypentanal. Its ring oxygen atom was originally the 5-hydroxyl oxygen atom. When this cyclic hemiacetal reacts with an alcohol, R′−OH, the product is a cyclic acetal. The oxygen atom in the −OR′ group of the acetal originated in the alcohol R′−OH. We will see this reaction again when we consider carbohydrates in Chapter 13.

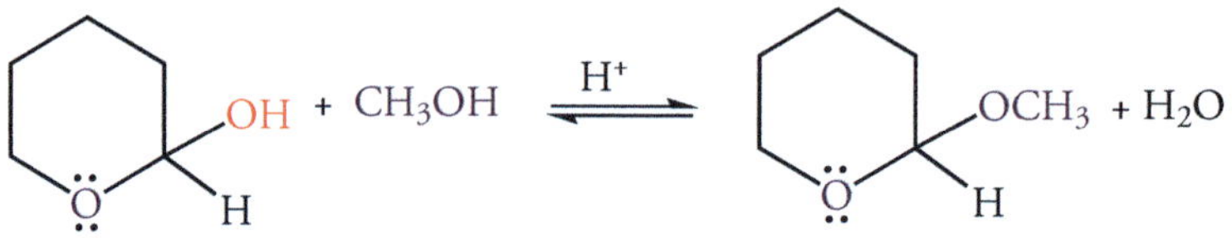

cyclic acetal

Reactivity of Acetals and Ketals

The conversion of a hemiacetal to an acetal and the conversion of a hemiketal to a ketal are reversible in acid solution. The position of the equilibrium can be shifted toward formation of an acetal or ketal by removing the water formed in the reaction or by increasing the concentration of the alcohol.

hemacetal + alcohol ⇌ acetal + water

Adding alcohol pushes the equilibrium to the right

Removing water pulls the equilibrium to the right

The reverse reaction, the acid-catalyzed hydrolysis of acetals or ketals, is favored when water is added. Acetals and ketals react with water in a hydrolysis reaction to give a carbonyl compound and the alcohol. However, acetals and ketals do not react in neutral or basic solution.

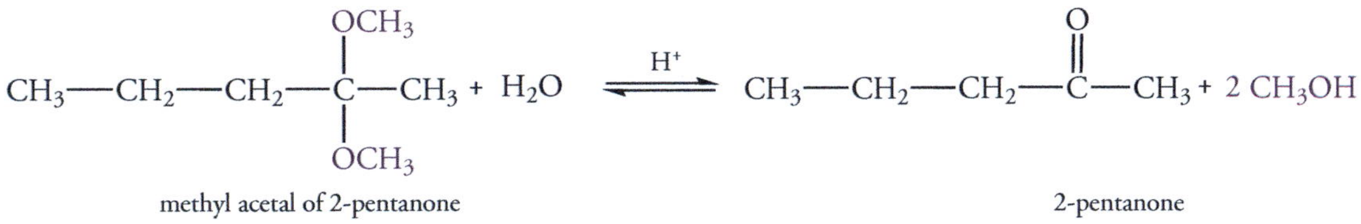

$$CH_3\text{—}CH_2\text{—}CH_2\text{—}\underset{\underset{OCH_3}{|}}{\overset{\overset{OCH_3}{|}}{C}}\text{—}CH_3 + H_2O \underset{}{\overset{H^+}{\rightleftharpoons}} CH_3\text{—}CH_2\text{—}CH_2\text{—}\overset{\overset{O}{||}}{C}\text{—}CH_3 + 2\ CH_3OH$$

methyl acetal of 2-pentanone   2-pentanone

Mechanism of Acetal and Ketal Formation

The conversion of hemiacetals and hemiketals to acetals and ketals occurs in four reversible, acid-catalyzed steps. These steps are shown here for the conversion of a hemiacetal to an acetal. In step 1, the acid protonates the oxygen atom of the hydroxyl group. In step 2, water leaves, and a resonance-stabilized oxocarbocation forms.

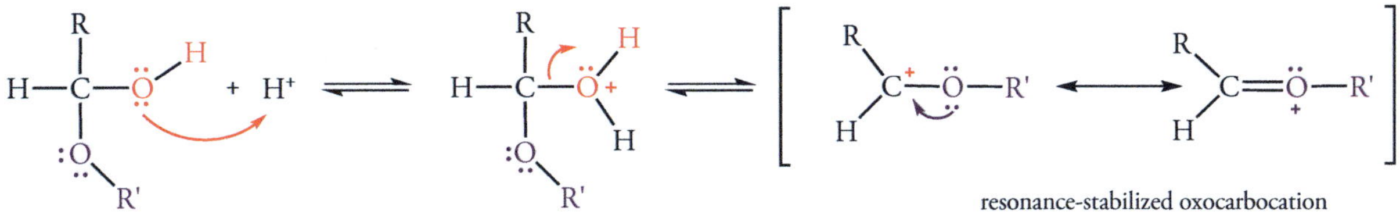

resonance-stabilized oxocarbocation

In step 3, the carbocation (a Lewis acid) combines with the alcohol (a Lewis base). In step 4, the proton bonded to the oxygen atom is lost to give an acetal. Note that H^+ is a catalyst: it starts the reaction by protonating the hemiacetal and is regenerated in the last step when the acetal forms.

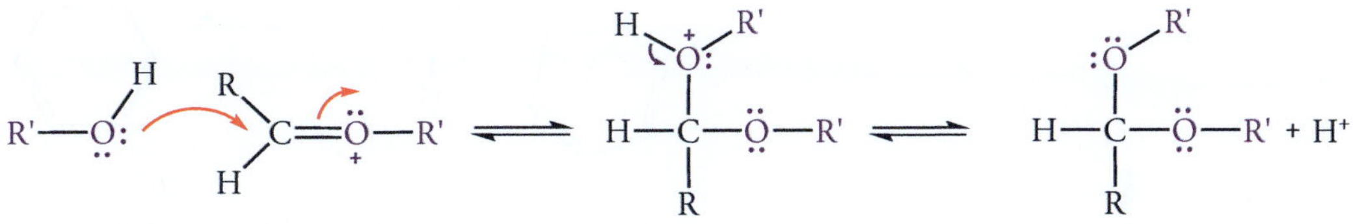

10.9 ADDITION OF NITROGEN COMPOUNDS

The carbonyl groups of aldehydes and ketones react with nucleophiles that contain nitrogen; that is, they react with ammonia, NH_3, and with amines of the general formula RNH_2 (Chapter 12). The reaction gives imines, which are compounds with the carbonyl carbon atom bonded to the nitrogen atom by a double bond. The reaction occurs in two steps called an **addition-elimination reaction**. In the addition step, the nitrogen atom bonds to the carbonyl carbon atom and a hydrogen atom bonds to the carbonyl oxygen atom. This step to give a hemiaminal resembles the addition of an alcohol to a carbonyl compound. The initial addition product loses a molecule of water in an elimination reaction to give an **imine**. The net result of reacting an aldehyde or ketone with GNH_2, where G represents any group, is the replacement of the carbonyl oxygen atom to give a C=N— group.

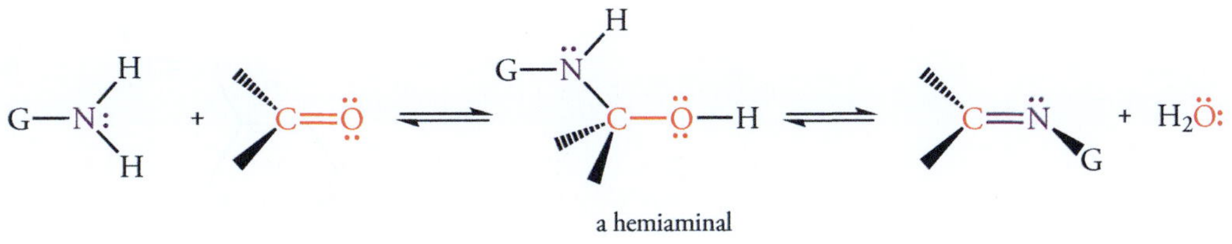

a hemiaminal

Note that the formation of an imine is accompanied by the release of water as a product. Imines can be isolated, but they react with water in the reverse reaction to produce the original carbonyl compound and the nitrogen compound. Thus, one way to isolate an imine is to remove water from the solution as the imine forms.

Problem 10.9

Write the structure of the intermediate hemiaminal produced by reaction of benzaldehyde and methylamine. Write the structure of the imine that results from dehydration of the hemiaminal.

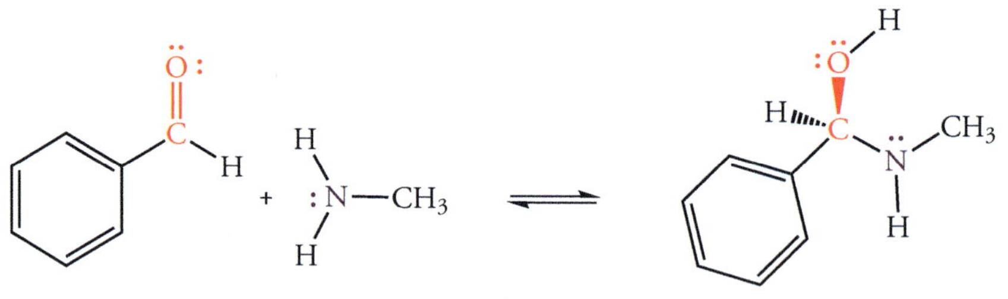

Solution

Form a bond between the nitrogen atom of the amine and the carbonyl carbon atom; add a hydrogen atom to the carbonyl oxygen atom and remove a hydrogen from the amine. This hemiaminal addition product is the result of nucleophilic attack of the nitrogen atom on the carbonyl carbon atom followed by proton transfer from the amine nitrogen to oxygen.

Dehydration occurs by elimination of the −OH group bonded to the carbon atom and the hydrogen atom bonded to the nitrogen atom to give the imine.

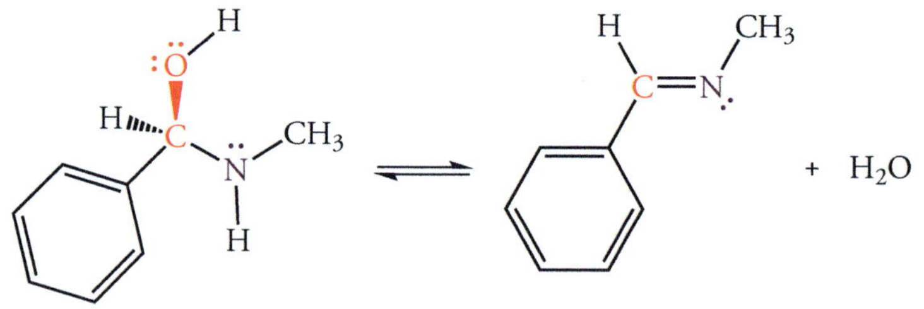

Note that two geometric isomers can result, because the nitrogen atom is sp²-hybridized. The *trans* isomer is shown. The *cis* isomer is less stable due to steric hindrance of the methyl and phenyl groups.

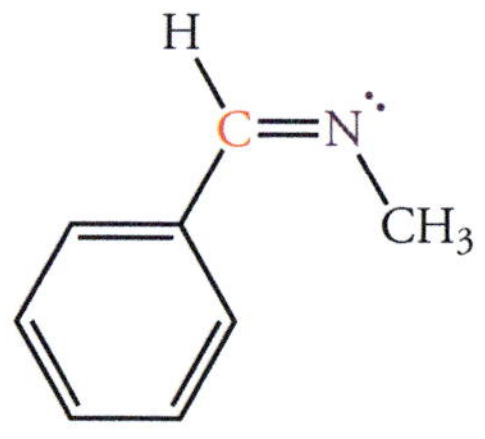

cis isomer

Problem 10.10

Write the structure of the reactants required to produce the following compound.

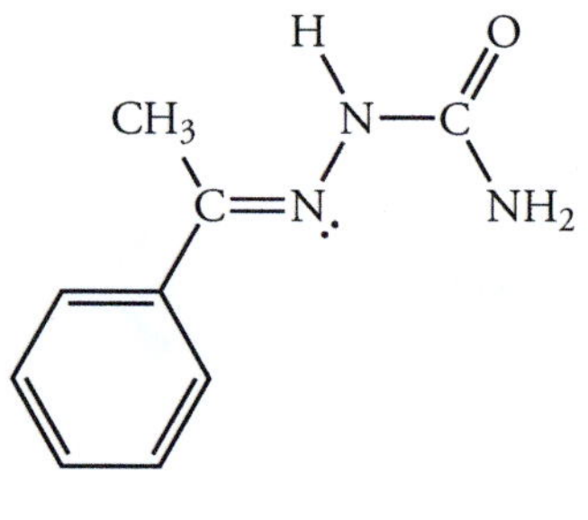

Addition Reactions of Nitrogen Compounds and Vision

We learned at our mother's knee that "carrots are good for us." This homely injunction is true because carrots contain β-carotene, which mammals require for vision. β-Carotene is a pigment that is largely responsible for the color of carrots. It is also available in egg yolk, liver, and various fruits and vegetables. β-Carotene has an all-*trans* configuration.

Persons who do not have adequate β-carotene in their diets suffer from *night blindness.* Mammals have a liver enzyme system that splits β-carotene in half to give two molecules of an aldehyde named retinal.

There are several isomers of retinal because geometric isomers can exist about each of the double bonds. The all-*trans* compound and the isomer with a *cis* orientation about one of the double bonds, known as *cis*-11-retinal, play an important role in vision.

11-*cis*-retinal undergoes an addition reaction with a protein in the retina called opsin to form a substance called rhodopsin. The aldehyde group of 11-*cis*-retinal reacts with a specific amino group in the protein to form an imine. The shape of the imine adduct of 11-*cis*-retinal allows it to bind the active site of the protein. Rhodopsin is a visual receptor in the retina that absorbs visible light. When light strikes rhodopsin, the *cis* double bond is isomerized to a *trans* double bond, a process called photo-isomerization. The resulting all-*trans* isomer no longer fits into the opsin binding site, the imine spontaneously hydrolyzes, and the all-*trans* retinal is released from opsin. This process occurs in about one millisecond. During that time a nerve impulse is generated and travels to the brain, where it is translated into a visual image.

If 11-*cis*-retinal cannot bind opsin to give rhodopsin, vision is impaired. We recall from earlier discussions that formaldehyde, which is produced by the oxidation of methyl alcohol, can cause blindness. Blindness occurs because formaldehyde competes with 11-*cis*-retinal for the active site amine group of opsin. If no rhodopsin is formed, then no "light-induced" messages will get to the brain.

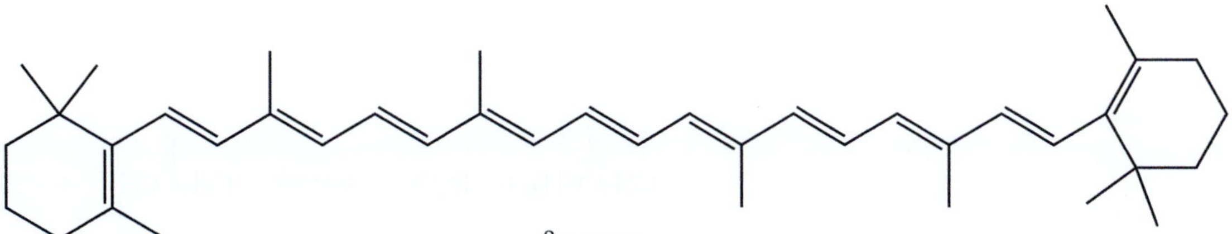

β-carotene

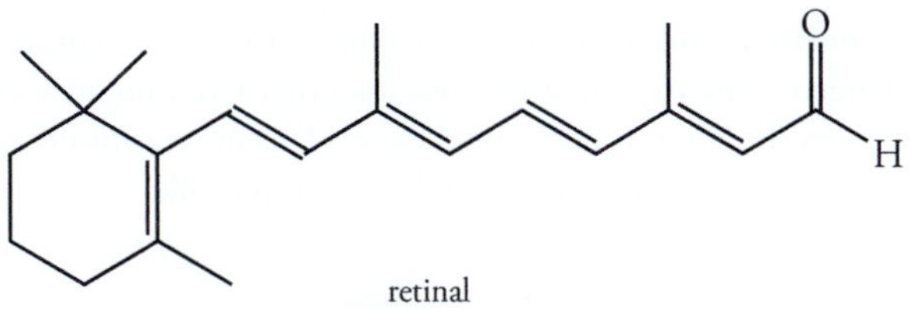

retinal

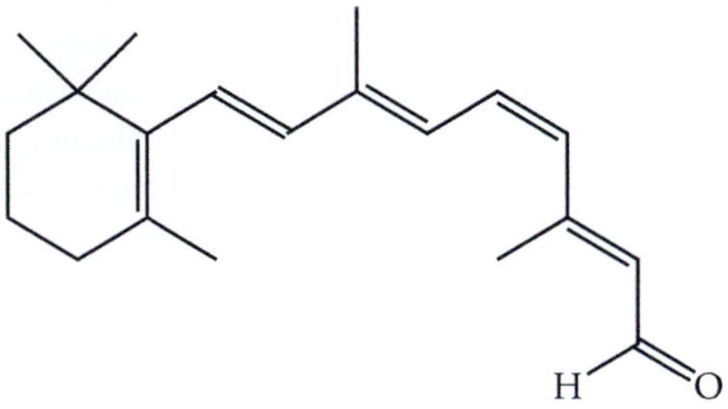

11-*cis*-retinal

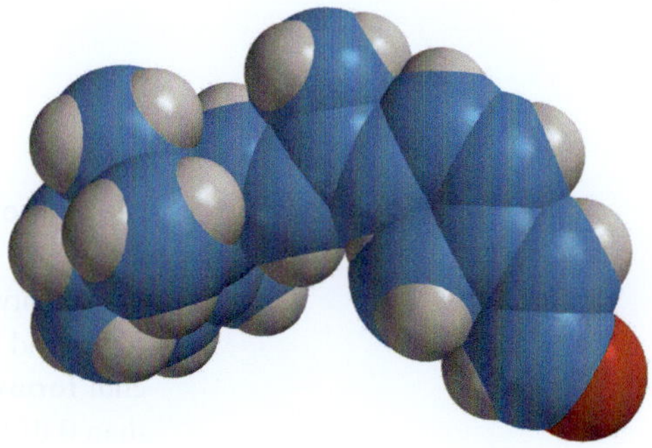

10.10 REACTIVITY OF THE α-CARBON ATOM

The carbonyl group itself is not the only reactive site of carbonyl compounds. Many important reactions occur at the carbon atom directly attached to the carbonyl carbon atom. This carbon atom is called the **α-carbon atom.**

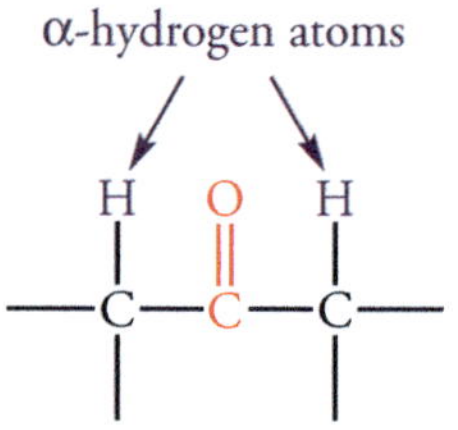

α-hydrogen atoms

Acidity of α Hydrogens

The carbonyl carbon atom has a partial positive charge and it attracts electrons in neighboring bonds by an inductive effect. As a result, the α-carbon atom loses electron density and acquires a partial positive charge. This effect is transmitted in turn to the bonds holding hydrogen atoms to the α-carbon atom. Thus, an **α-hydrogen atom** is more acidic than a hydrogen atom in a C—H bond of a hydrocarbon.

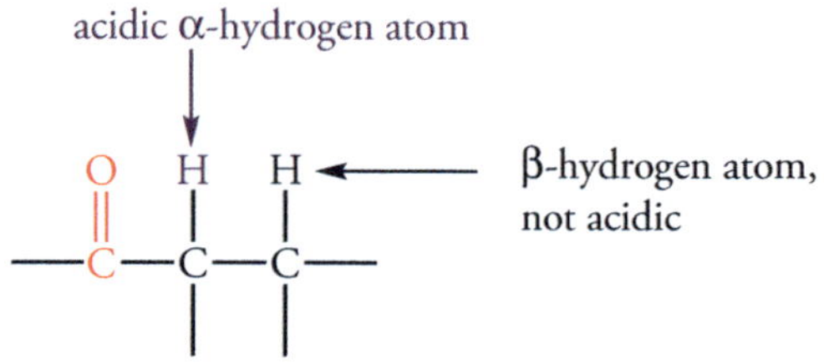

acidic α-hydrogen atom

β-hydrogen atom, not acidic

$$CH_3CH_3 + H_2O \rightleftharpoons CH_3CH_2^- + H_3O^+ \quad pK_a = 50$$

$$CH_3\overset{O}{\overset{\|}{C}}CH_3 + H_2O \rightleftharpoons CH_3\overset{O}{\overset{\|}{C}}CH_2^- + H_3O^+ \quad pK_a = 20$$

However, this tremendous difference of 30 powers of 10 in K_a is due to more than just the inductive effect of the carbonyl group. When a carbonyl compound loses its α-hydrogen atom, the resulting anion, called an **enolate anion**, is resonance-stabilized. One of its contributing resonance structures has a negative charge on the oxygen atom, and the other has a negative charge on the carbon atom. Because the charge on the anion is delocalized, an enolate anion is more stable than a carbanion, such as $CH_3CH_2^-$, in which no such resonance stabilization is possible.

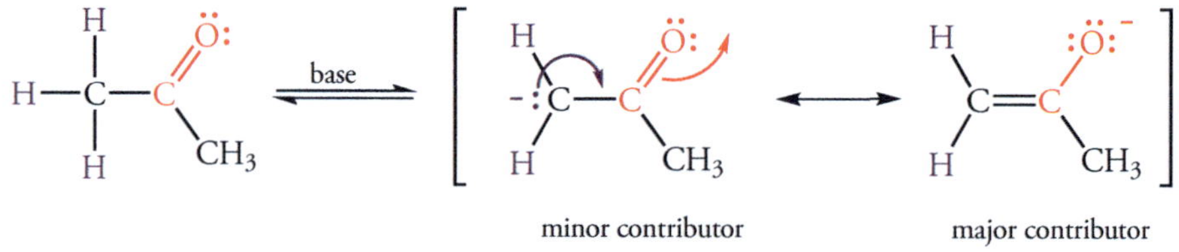

minor contributor   major contributor

resonance structures of the enolate anion

Keto-Enol Equilibria

The acidity of the α-hydrogen atoms of carbonyl compounds has another consequence: both aldehydes and ketones exist as an equilibrium mixture of isomeric compounds called the **keto** and the enol **forms**. Simple aldehydes and ketones exist predominantly in the keto form; acetone has less than 0.01% of the enol form.

stronger double bond / weaker double bond

acetone (keto form) / acetone (keto form)

This type of isomerism is called **tautomerism**, and the two isomeric forms are called **tautomers**. Note that tautomers differ in the location of a hydrogen atom and a double bond. Thus, they are structural isomers, not contributing resonance forms, as in the case of the enolate ion, where only the locations of the electrons differ. The tautomers equilibrate in a rearrangement reaction in which a hydrogen atom bonded to an α-carbon atom is transferred to the carbonyl oxygen atom. The carbon-oxygen double bond becomes a single bond, and the carbon-carbon single bond becomes a double bond.

Although the enol form is present only in small amounts at equilibrium, it is often responsible for the reactivity of carbonyl compounds. Tautomerism is important in the chemistry of carbohydrates (Chapter 13) and in the metabolism of these compounds. For example, one of the intermediates in carbohydrate metabolism is a three-carbon compound called dihydroxyacetone phosphate. It undergoes an enzyme-catalyzed isomerization reaction to form another three-carbon molecule called D-glyceraldehyde 3-phosphate. The isomerization reaction occurs by transfer of a hydrogen atom from the α-carbon atom of dihydroxyacetone phosphate to the carbonyl oxygen atom. Tautomerization in the first step of the reaction yields an enediol intermediate. The second tautomerization step yields D-glyceraldehyde 3-phosphate.

dihydroxyacetone phosphate / enediol intermediate / D-glycerladehyde-3-phosphate

Isomerization occurs because the enediol intermediate is in equilibrium with dihydroxyacetone phosphate and D-glyceraldehyde 3-phosphate. Similar isomerization reactions occur in many enzyme-catalyzed reactions of carbohydrates.

10.11 THE ALDOL CONDENSATION

Two molecules of an aldehyde can join in a reaction called an **aldol condensation**. This base-catalyzed reaction gives a product that is both an aldehyde and an alcohol, called an **aldol**. The aldol can be isolated, but it can also react further to form a conjugated unsaturated carbonyl compound and water.

The addition step of the aldol condensation occurs at room temperature when an aldehyde is treated with an aqueous solution of sodium hydroxide. The reaction occurs in a three-step process.

1. One aldehyde molecule reacts with base (OH⁻) at its α C—H bond to give a nucleophilic enolate anion.

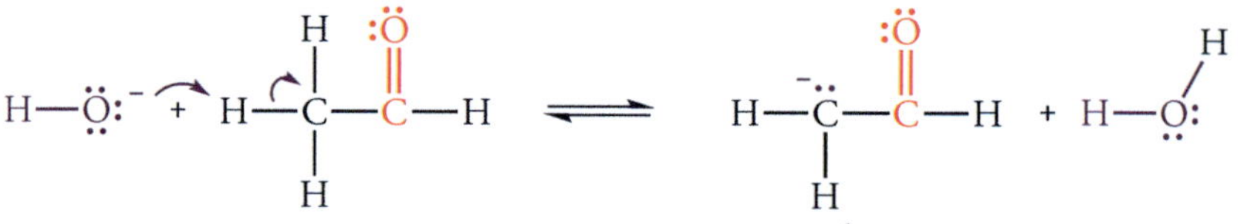

$$H-\overset{..}{\underset{..}{O}}:^- + H-\overset{\overset{H}{|}}{\underset{\underset{H}{|}}{C}}-\overset{:\overset{..}{O}:}{\underset{}{C}}-H \rightleftharpoons H-\overset{\overset{}{|}}{\underset{\underset{H}{|}}{\overset{..}{C}}}-\overset{:\overset{..}{O}:}{\underset{}{C}}-H + H-\overset{H}{\underset{..}{O}}:$$

enolate anion

2. The nucleophilic enolate anion reacts with the carbonyl carbon atom of another aldehyde molecule. The alkoxide ion product is the conjugate base of an aldol.

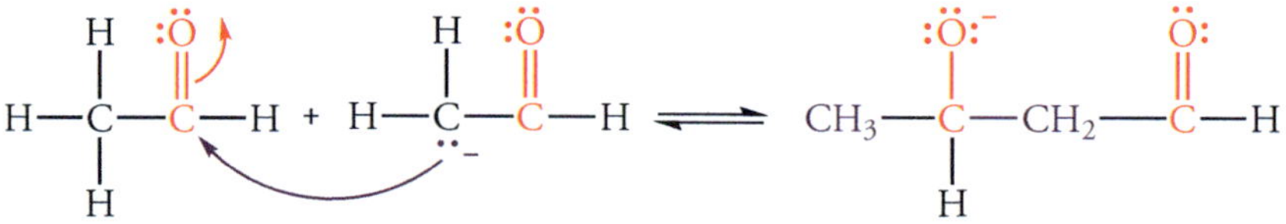

an alkoxide anion

3. The alkoxide anion extracts a proton from the solvent, water, which regenerates a hydroxide anion.

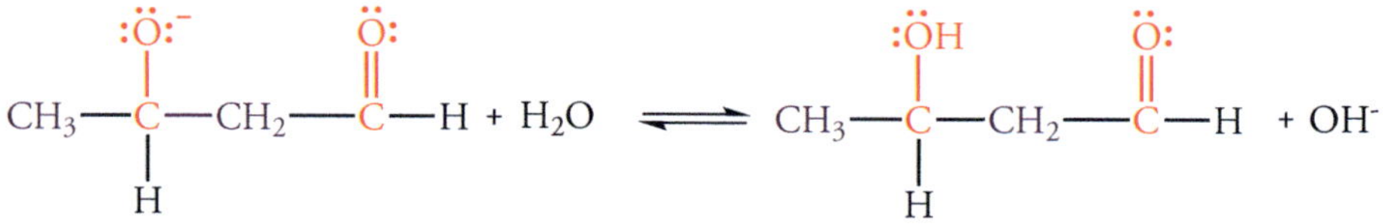

$$CH_3-\overset{:\overset{..}{O}:^-}{\underset{\underset{H}{|}}{C}}-CH_2-\overset{\overset{..}{O}:}{\underset{}{C}}-H + H_2O \rightleftharpoons CH_3-\overset{:\overset{..}{O}H}{\underset{\underset{H}{|}}{C}}-CH_2-\overset{\overset{..}{O}:}{\underset{}{C}}-H + OH^-$$

aldol product

Under the conditions of the reaction, little of the carbonyl compound is converted to the enolate. The aldehyde and the enolate are in equilibrium, and the enolate is replaced as it reacts. The enolate is present in an excess of aldehyde, so its nucleophilic carbon atom is surrounded by many electrophilic carbon atoms. The hydroxide ion is a catalyst, and its concentration does not change.

The aldol condensation can seem confusing at first sight. We can picture it as shown below.

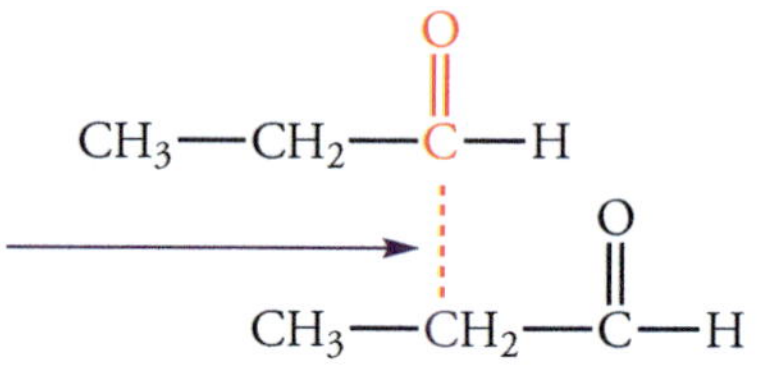

Form a new bond between the α-carbon atom and the carbonyl carbon ⟶

$$CH_3-CH_2-\overset{\overset{O}{||}}{C}-H$$

$$CH_3-CH_2-\overset{\overset{O}{||}}{C}-H$$

α-carbon atom

Next, draw a second structural formula of propanal with the α-carbon atom near the carbonyl carbon atom of the first structure. Now we have the aldol product.

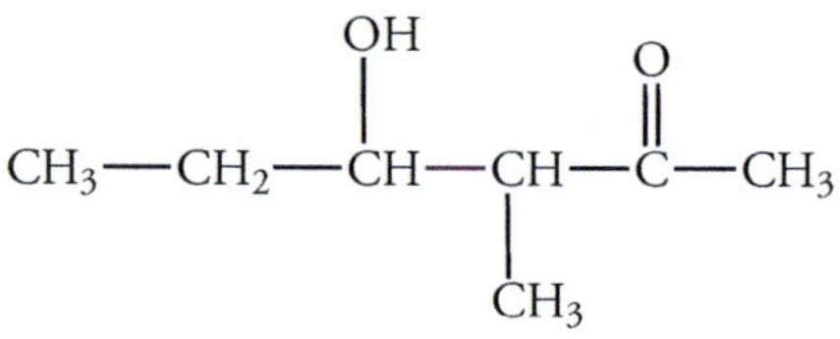

$$CH_3-CH_2-\overset{\overset{OH}{|}}{CH}-\overset{\overset{}{\underset{\underset{CH_3}{|}}{CH}}}{CH}-\overset{\overset{O}{||}}{C}-CH_3$$

Dehydration of Aldols

If the aldol reaction mixture is heated under basic conditions, the aldol loses water. The base-catalyzed dehydration reaction occurs in two steps.

1. The base removes an α-hydrogen atom to give a resonance-stabilized enolate ion.

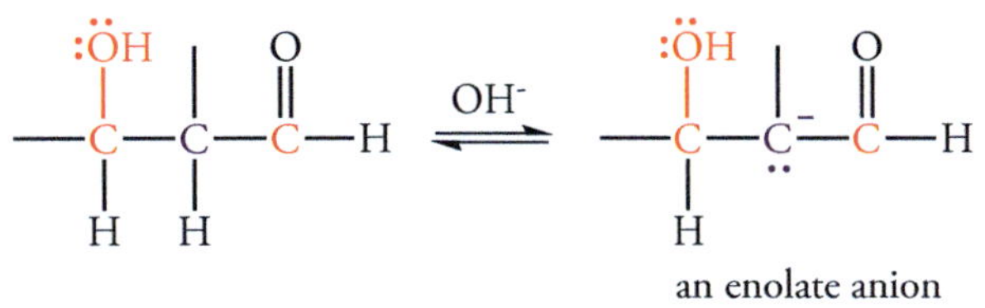

2. The hydroxide ion leaves in the second step of this E2 reaction.

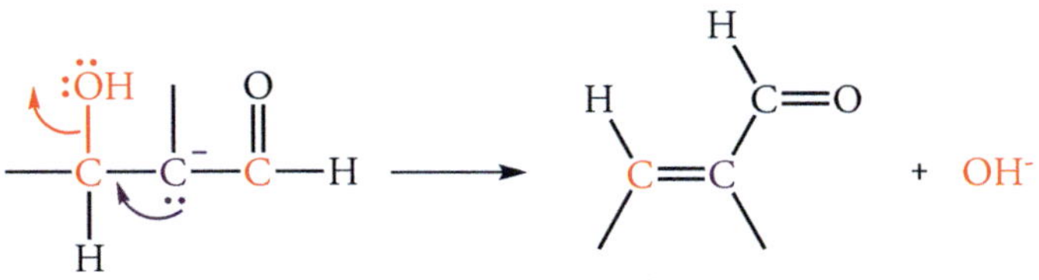

Problem 10.11
Without writing the individual steps of the mechanism, draw the product of the aldol condensation of propanal.

Solution
First draw the structural formula of propanal with the carbonyl group on the right.

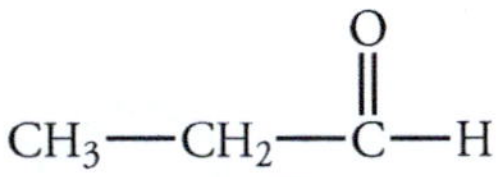

$$CH_3—CH_2—\overset{\overset{\displaystyle O}{\|}}{C}—H$$

Next, draw a second structural formula of propanal with the α-carbon atom near the carbonyl carbon atom of the first structure and connect them with a new bond.

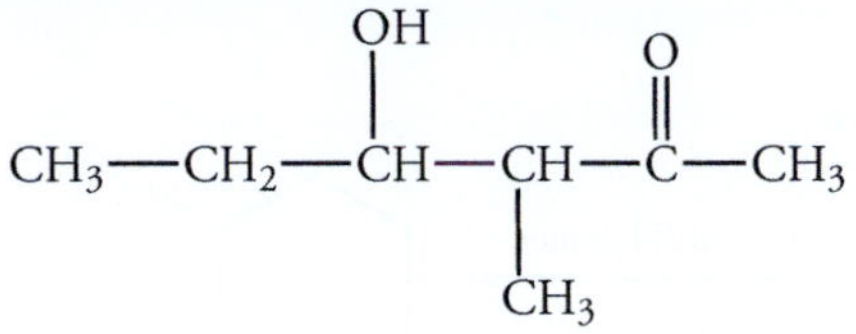

$$CH_3—CH_2—\overset{\overset{\displaystyle OH}{|}}{CH}—\underset{\underset{\displaystyle CH_3}{|}}{CH}—\overset{\overset{\displaystyle O}{\|}}{C}—CH_3$$

Problem 10.12
What compound is required to form the following unsaturated compound using an aldol condensation?

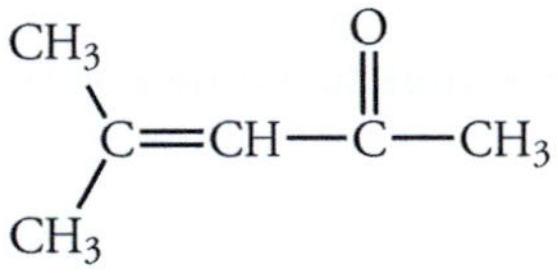

SUMMARY OF REACTIONS

1. Oxidation of Aldehydes (Section 10.4)

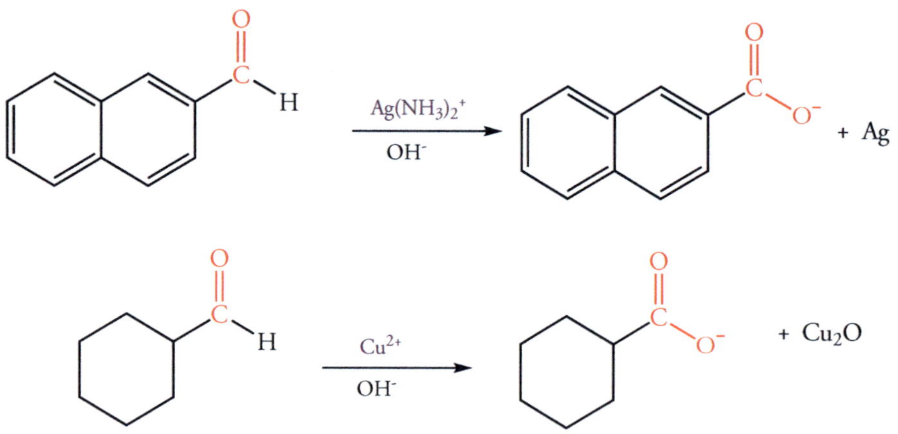

2. Reduction of Aldehydes and Ketones to Alcohols (Section 10.4)

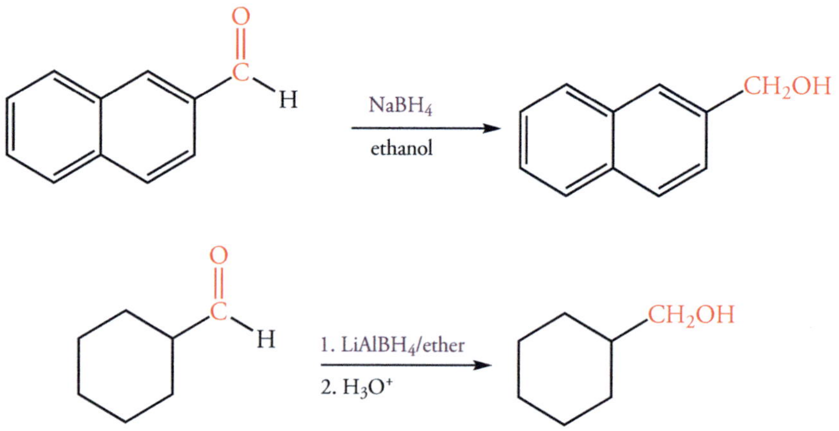

3. Reduction of Aldehydes and Ketones to Methylene Groups (Section 10.4)

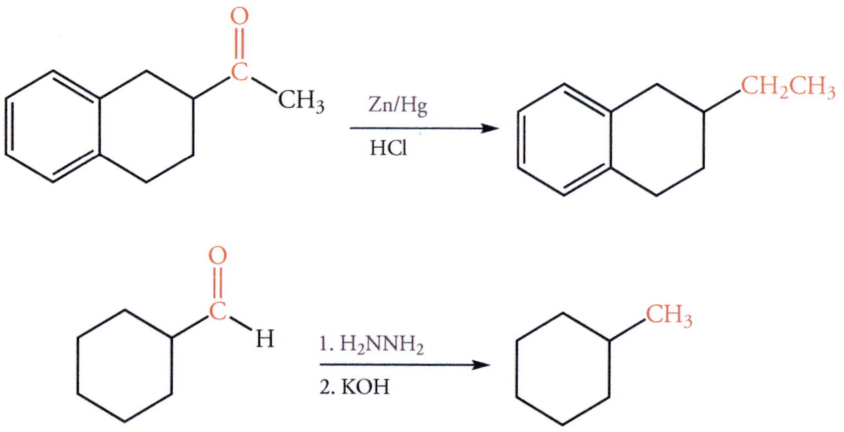

4. Reduction of Aldehydes and Ketones to Methylene Groups (Section 10.4)

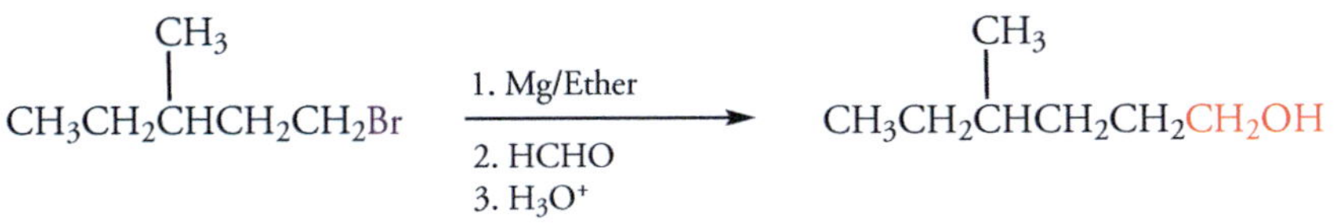

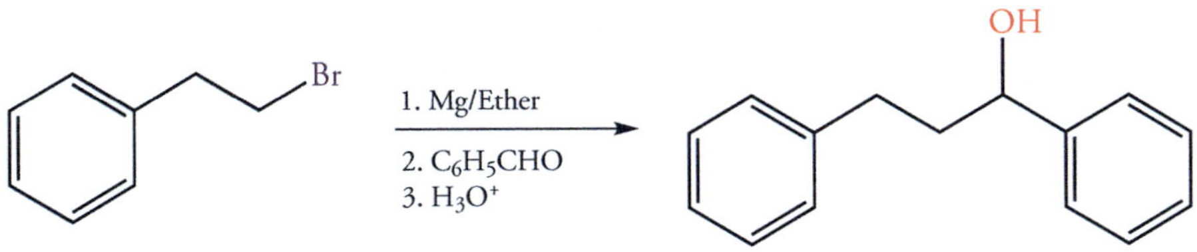

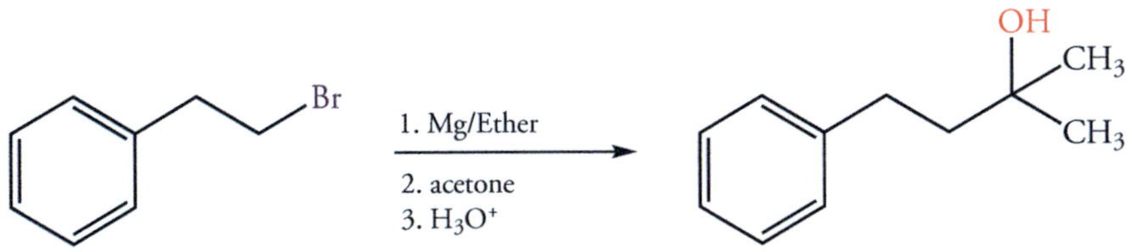

$$CH_3CH_2\underset{\underset{CH_3}{|}}{C}HCH_2CH_2Br \quad \xrightarrow[\substack{2.\ HCHO \\ 3.\ H_3O^+}]{1.\ Mg/Ether} \quad CH_3CH_2\underset{\underset{CH_3}{|}}{C}HCH_2CH_2CH_2OH$$

$$\xrightarrow[\substack{2.\ C_6H_5CHO \\ 3.\ H_3O^+}]{1.\ Mg/Ether}$$

$$\xrightarrow[\substack{2.\ acetone \\ 3.\ H_3O^+}]{1.\ Mg/Ether}$$

5. Formation of Acetals and Ketals (Section 10.8)

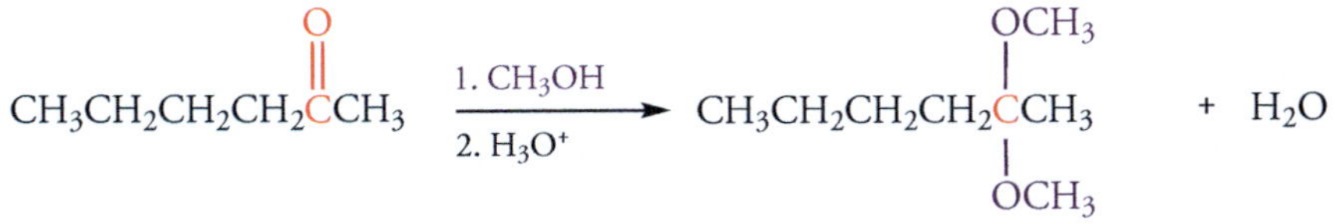

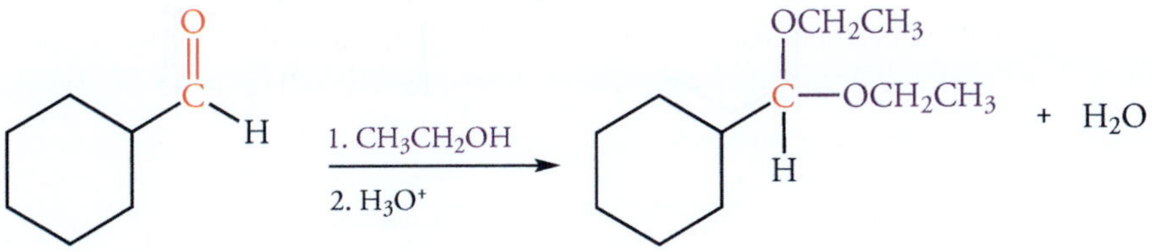

$$CH_3CH_2CH_2CH_2\overset{\overset{O}{\|}}{C}CH_3 \quad \xrightarrow[\substack{2.\ H_3O^+}]{1.\ CH_3OH} \quad CH_3CH_2CH_2CH_2\underset{\underset{OCH_3}{|}}{\overset{\overset{OCH_3}{|}}{C}}CH_3 \quad + \quad H_2O$$

$$\xrightarrow[\substack{2.\ H_3O^+}]{1.\ CH_3CH_2OH} \qquad + \quad H_2O$$

6. Addition of Nitrogen Compounds to Carbonyl Compounds (Section 10.9)

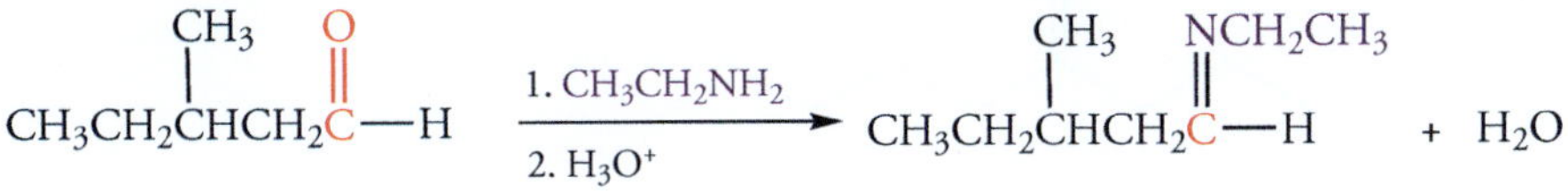

$$CH_3CH_2\underset{\underset{CH_3}{|}}{C}HCH_2\overset{\overset{O}{\|}}{C}-H \quad \xrightarrow[\substack{2.\ H_3O^+}]{1.\ CH_3CH_2NH_2} \quad CH_3CH_2\underset{\underset{CH_3}{|}}{C}HCH_2\overset{\overset{NCH_2CH_3}{\|}}{C}-H \quad + \quad H_2O$$

7. Aldol Condensation (Section 10.11)

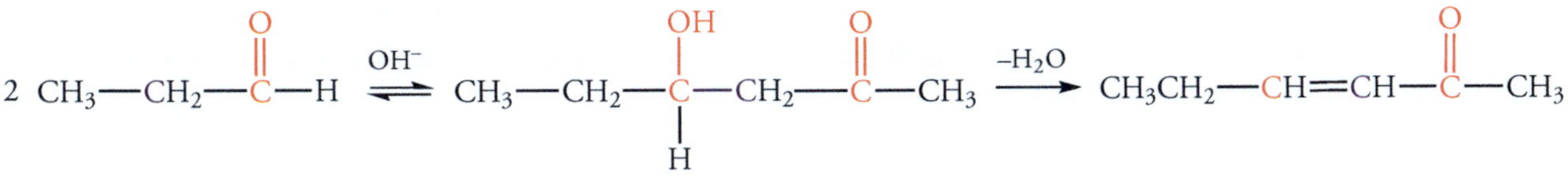

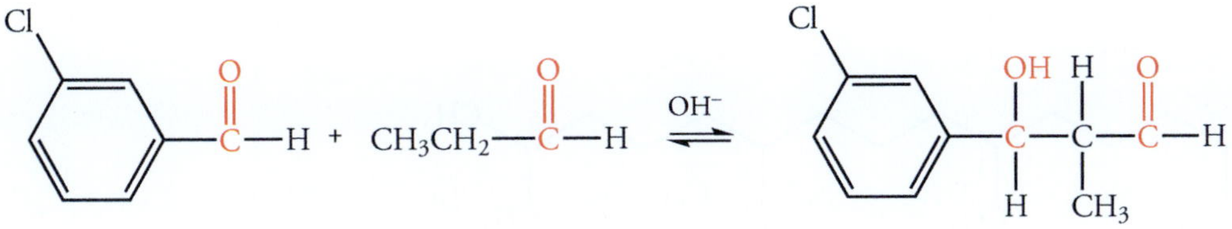

$$2\ CH_3-CH_2-\overset{\overset{O}{\|}}{C}-H \quad \rightleftharpoons \quad CH_3-CH_2-\underset{\underset{H}{|}}{\overset{\overset{OH}{|}}{C}}-CH_2-\overset{\overset{O}{\|}}{C}-CH_3 \quad \xrightarrow{-H_2O} \quad CH_3CH_2-CH=CH-\overset{\overset{O}{\|}}{C}-CH_3$$

EXERCISES

Nomenclature of Aldehydes and Ketones

10.1 Write the structure for each of the following compounds:
(a) 2-methylbutanal (b) 3-ethylpentanal (c) 2-bromopentanal

10.2 Write the structure of each of the following compounds:
(a) 3-bromo-2-pentanone (b) 2,4-dimethyl-3-pentanone (c) 4-methyl-2-pentanone

10.3 Give the IUPAC name of each of the following compounds:

(a) $CH_3CH_2CH_2CHO$

(b)
$$CH_3\overset{\overset{\displaystyle CH_3}{|}}{\underset{\underset{\displaystyle CH_3}{|}}{C}}-CH_2CHO$$

(c)
$$CH_3CH_2\overset{\overset{\displaystyle CH_3}{|}}{C}H\underset{\underset{\displaystyle CH_2H_3}{|}}{C}HCHO$$

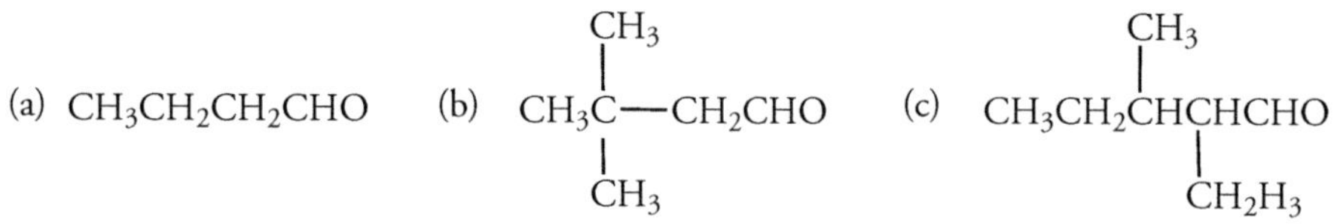

10.4 Give the IUPAC name of each of the following compounds:

(a)
$$CH_3CH_2\overset{\overset{\displaystyle O}{||}}{C}CH_2CH_3$$

(b)
$$CH_3\overset{\overset{\displaystyle CH_3}{|}}{C}H\underset{\underset{\displaystyle O}{||}}{C}CH_2CH_3$$

(c)
$$CH_3\overset{\overset{\displaystyle CH_3}{|}}{C}HCH_2\overset{\overset{\displaystyle O}{||}}{C}CH_3$$

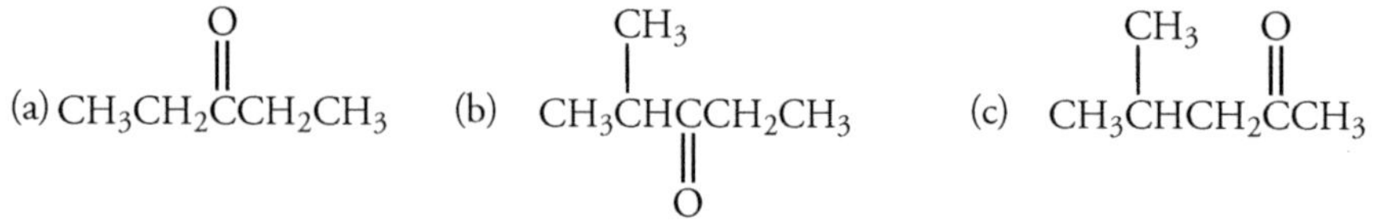

10.5 Give the IUPAC name of each of the following compounds:

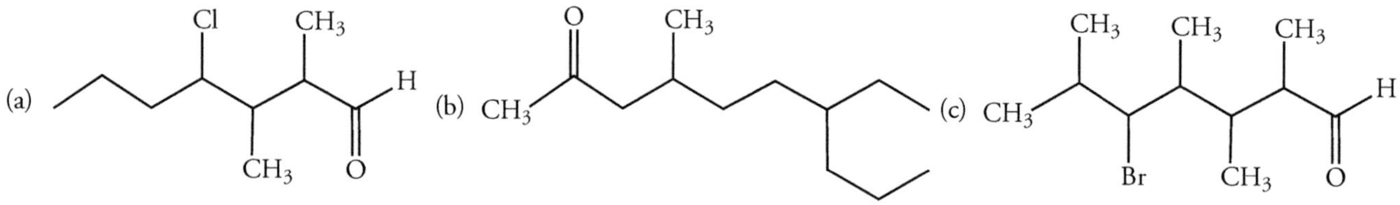

10.6 Give the IUPAC name of each of the following compounds:

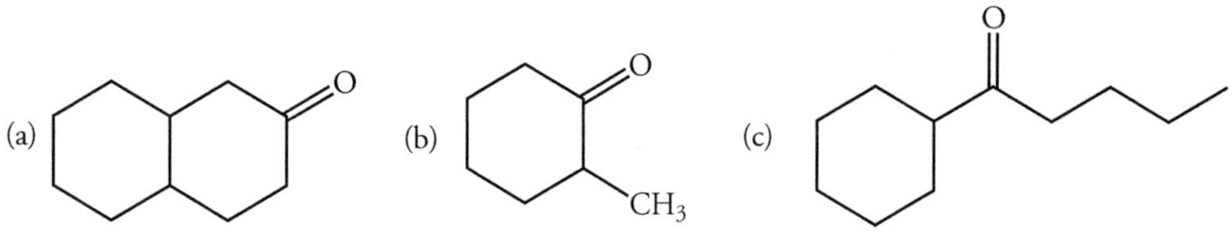

10.7 Draw the product of ozonolysis (Section 4.7) obtained from 1-methylcyclohexene. Name the compound.

10.8 Draw the products of ozonolysis of vitamin K1 assuming that the double bond in the quinone ring is unaffected. Name the carbonyl compound derived from the side chain.

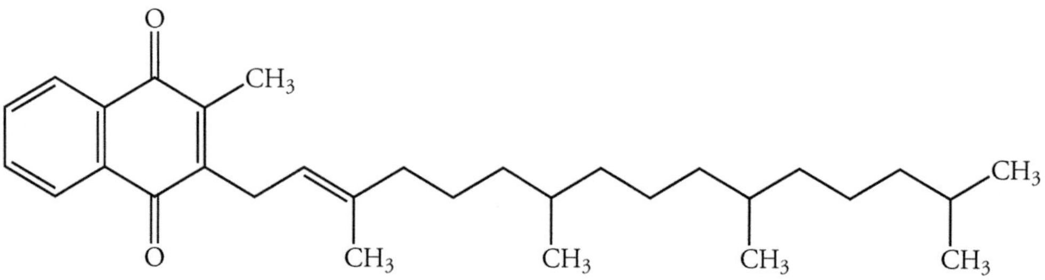

Properties of Aldehydes and Ketones

10.9 The dipole moments of 1-butene and butanal are 0.34 and 2.52 D, respectively. Explain this difference.

10.10 The dipole moments of acetone and isopropyl alcohol are 2.7 and 1.7 D, respectively. Explain this difference.

10.11 The boiling points of butanal and 2-methylpropanal are 75 and 61 °C, respectively. Explain this difference.

10.12 The boiling points of 2-heptanone, 3-heptanone, and 4-heptanone are 151, 147, and 144 °C, respectively. What is responsible for this trend?

10.13 The solubilities of butanal and 1-butanol in water are 7 and 9 g/100 mL, respectively. Explain this difference.

10.14 The solubilities of butanal and 2-methylpropanal in water are 7 and 11 g/100 mL, respectively. Explain this difference.

Oxidation and Reduction of Carbonyl Compounds

10.15 What is observed when an aldehyde reacts with Fehling's solution? What is observed when an aldehyde reacts with Tollens reagent?

10.16 Draw the structure of the product of each of the following reactions:

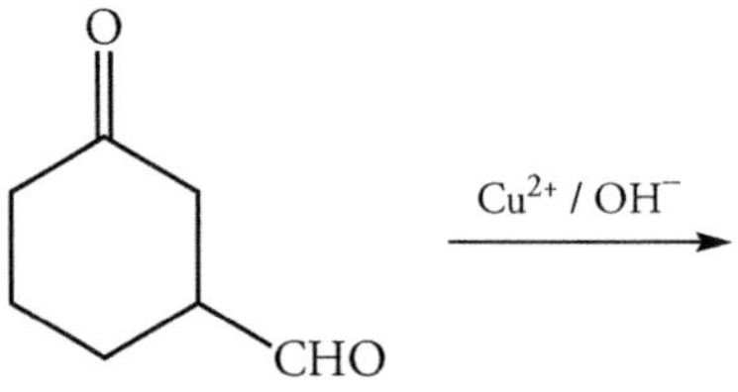

10.17 What is the product when each of the following reacts with lithium aluminum hydride?

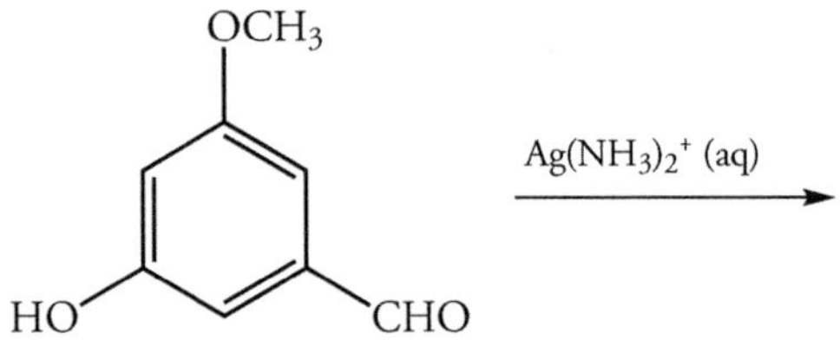

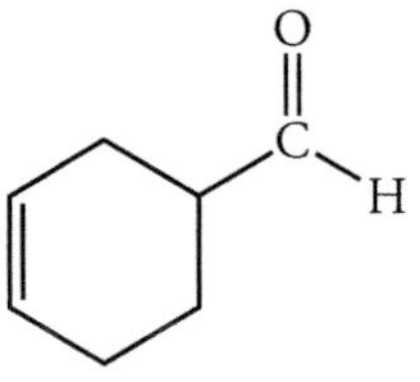

10.18 What is the product when each of the following reacts with sodium borohydride?

(a) 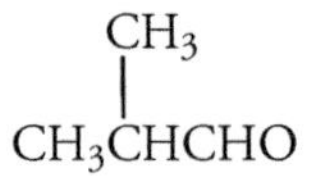 CH_3CHCHO with CH_3 substituent (b) 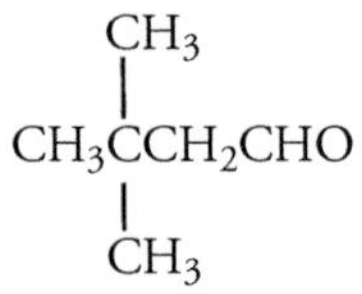 CH_3CCH_2CHO with CH_3 substituents (c) 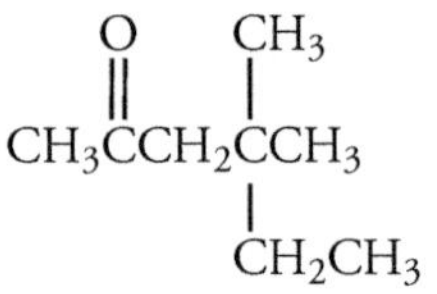 $CH_3CCH_2CCH_3$ with O, CH_3, and CH_2CH_3 substituents

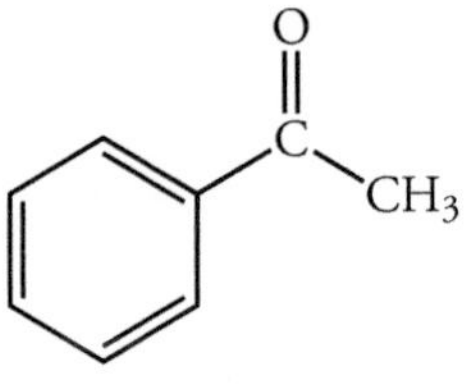

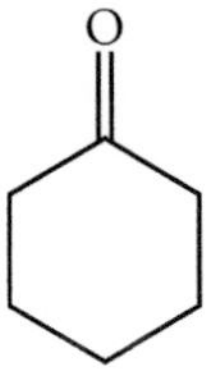

10.19 The reduction of carvone by lithium aluminum hydride yields two products. Explain why.

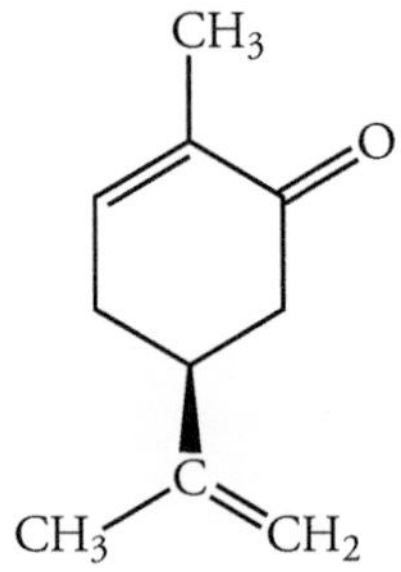

carvone

10.20 The reduction of the following compound by sodium borohydride yields two products. Explain why.

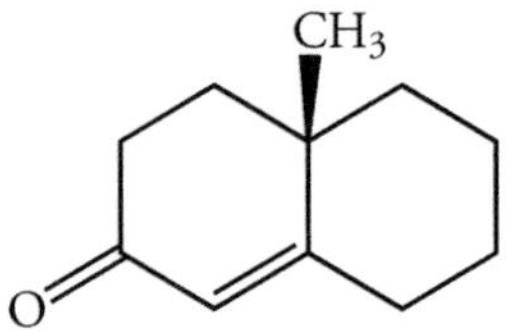

Addition Reactions

10.21 Formaldehyde has been used to disinfect rooms and surgical instruments. Why is this compound so effective compared to other carbonyl compounds?

10.22 Glutaraldehyde is used as a sterilizing solution for instruments that cannot be heated in an autoclave. Explain its sterilizing action.

$$\underset{H-\overset{\overset{\displaystyle O}{\|}}{C}-CH_2CH_2CH_2CH_2-\overset{\overset{\displaystyle O}{\|}}{C}-H}{}$$

glutaraldehyde

11

CARBOXYLIC ACIDS AND ESTERS

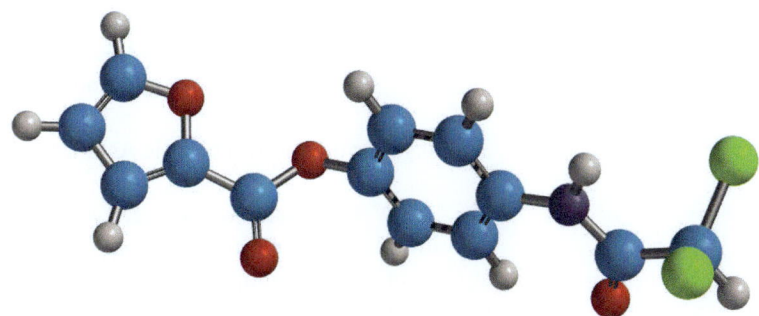

DILOXANIDE FURANOATE (AN ANTIBIOTIC)

11.1 CARBOXYLIC ACIDS AND ACYL GROUPS

In this chapter we will consider the structure, properties, and reactions of carboxylic acids. These compounds have a **carboxyl group** bonded to a hydrocarbon unit that can be saturated, unsaturated, aromatic, or heterocyclic.

carboxyl group   carboxylic acids

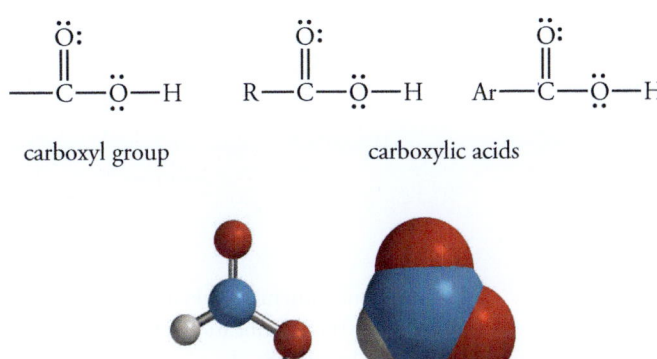

methanoic acid

The carboxyl carbon atom is sp^2-hybridized, and three of its valence electrons form three σ bonds at 120° angles to one another (Figure 11.1). One of the σ bonds is to a hydrogen atom or a carbon atom of an alkyl, aromatic, or heterocyclic group. The other two σ bonds are to oxygen atoms: one to the hydroxyl oxygen atom and the other to the carbonyl oxygen atom. The carbonyl carbon atom of the carboxyl group also has one electron in a 2p orbital forming a π bond with an electron in a 2p orbital of the carbonyl oxygen atom.

Although the bond angles at the carboxyl carbon atom are all approximately 120°, the carboxyl group is often represented with only vertical and horizontal lines. To save space, two condensed representations of the carboxyl group are commonly used. Unless required to account for the mechanism of a reaction, the nonbonding electrons are not shown.

R—C—Ö—H   or   R—COOH   or   R—CO$_2$H

Equivalent representations of carboxylic acids

Principles of Organic Chemistry. http://dx.doi.org/10.1016/B978-0-12-802444-7.00011-2

Figure 11.1
Bonding in Carboxylic
Acids

An electron of the sp² hybrid orbital of the carbonyl carbon atom and an electron of the sp³ hybrid orbital of the oxygen atom form a σ bond

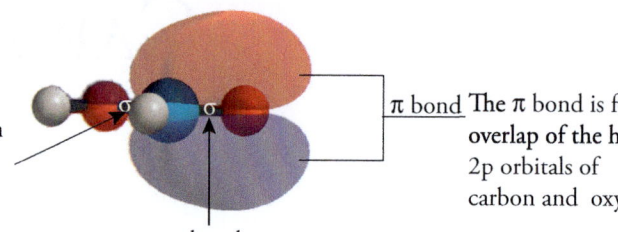

π bond The π bond is formed by **overlap of the half-filled** 2p orbitals of carbon and oxygen

σ bond
An electron of the sp² hybrid orbital of the carbonyl carbon atom and an electron of the sp² hybrid orbital of the oxygen atom form a σ bond

The Acyl Group and Carboxylic Acid Derivatives

The "RCO" unit contained in a carboxylic acid is called an **acyl group**. Several families of acid derivatives have oxygen-containing groups or electronegative atoms, such as nitrogen or a halogen, in place of the OH group of a carboxylic acid. This group, represented by L, can be replaced in a **nucleophilic acyl substitution** reaction (Section 11.6).

$$CH_3-\overset{\overset{\textstyle O}{\|}}{C}-L \;+\; Nu{:}^- \longrightarrow CH_3-\overset{\overset{\textstyle O}{\|}}{C}-Nu \;+\; L{:}^-$$

If an alkoxy (−OR) or phenoxy (−OAr) group is bonded to the acyl group, the derivative is an **ester**. Esters are formed from a carboxylic acid and an alcohol in a condensation reaction (Section 11.8). Esters are moderately reactive toward water in hydrolysis reactions and produce a carboxylic acid and an alcohol (Section 11.9).

$$R-\overset{\overset{\textstyle \ddot{O}:}{\|}}{C}-\ddot{O}-R' \qquad R-\overset{\overset{\textstyle \ddot{O}:}{\|}}{C}-\ddot{O}-Ar \qquad Ar-\overset{\overset{\textstyle \ddot{O}:}{\|}}{C}-\ddot{O}-Ar'$$

Examples of esters (acyl groups shown in red)

If the substituent is linked to the acyl group through a nitrogen atom, the compound is called an **amide**. The classification of amides depends on the number of carbon groups, including the acyl group, bonded to the nitrogen atom. These compounds are much less reactive than esters in hydrolysis reactions. The amide functional group is responsible for the structural stability of proteins (Chapter 14). We will discuss amides in Chapter 12.

$$R-\overset{\overset{\textstyle \ddot{O}:}{\|}}{C}-\underset{\underset{\textstyle H}{|}}{\ddot{N}}-H \qquad R-\overset{\overset{\textstyle \ddot{O}:}{\|}}{C}-\underset{\underset{\textstyle H}{|}}{\ddot{N}}-R' \qquad R-\overset{\overset{\textstyle \ddot{O}:}{\|}}{C}-\underset{\underset{\textstyle R'}{|}}{\ddot{N}}-R'$$

primary amide   secondary amide   tertiary amide

When the substituent attached to an acyl group is a chlorine atom, the derivative is called an **acid chloride** or **acyl chloride**. When two acyl groups are bonded to a common oxygen atom, the compound is an **acid anhydride**. Neither of these highly reactive compounds occurs in nature, but they are used in the laboratory synthesis of esters and amides.

$$R-\overset{\overset{\textstyle \ddot{O}:}{\|}}{C}-\ddot{C}\ddot{l}: \qquad\qquad R-\overset{\overset{\textstyle \ddot{O}:}{\|}}{C}-\ddot{O}-\overset{\overset{\textstyle \ddot{O}:}{\|}}{C}-R$$

acid chloride   acid anhydride

When a substituent is linked to an acyl group through a sulfur atom, the derivative is called a **thioester**. Thioesters are less reactive than acid chlorides and acid anhydrides, but they are sufficiently reactive to participate in many biochemical acyl transfer reactions.

$$R—\overset{\overset{\displaystyle \ddot{O}:}{\|}}{C}—\ddot{S}—R$$

thioester

Esters, amides, anhydrides, and thioesters may make up part of a cyclic structure. Cyclic esters are called **lactones**. Cyclic amides are called **lactams**. Cyclic acyl derivatives behave chemically like acyclic acyl compounds.

lactone
(a cyclic ester)

lactam
(a cyclic amide)

Common Names

The carboxylic acids and their derivatives are abundant in nature and were among the first organic substances to be isolated. Because they have been known for so long, the common acids, esters, and other acyl compounds are often referred to by their common names. Some common acids are formic acid (HCO_2H), acetic acid (CH_3CO_2H), and benzoic acid ($C_6H_5CO_2H$). Both the common and the IUPAC names of a few commonly encountered carboxylic acids are given in Table 11.1.

Table 11.1
Nomenclature of Carboxylic Acids

Formula	Common Name	IUPAC Name
HCO_2H	Formic acid	Methanoic acid
CH_3CO_2H	Acetic acid	Ethanoic acid
$CH_3CH_2CO_2H$	Propionic acid	Propanoic acid
$CH_3(CH_2)_2CO_2H$	Butyric acid	Butanoic acid
$CH_3(CH_2)_3CO_2H$	Valeric acid	Pentanoic acid
$CH_3(CH_2)_4CO_2H$	Caproic acid	Hexanoic acid
$CH_3(CH_2)_6CO_2H$	Caprylic acid	Octanoic acid
$CH_3(CH_2)_{10}CO_2H$	Lauric acid	Dodecanoic acid
$CH_3(CH_2)_{12}CO_2H$	Myristic	Tetradecanoic acid
$CH_3(CH_2)_{14}CO_2H$	Palmitic	Hexadecanoic acid
$CH_3(CH_2)_{16}CO_2H$	Stearic	Octadecanoic acid

In common names of carboxylic acids and their derivatives, the positions of groups attached to the parent chain are designated alpha (α), beta (β), gamma (γ), delta (δ), and so forth. The —COOH group itself is not designated by a Greek letter.

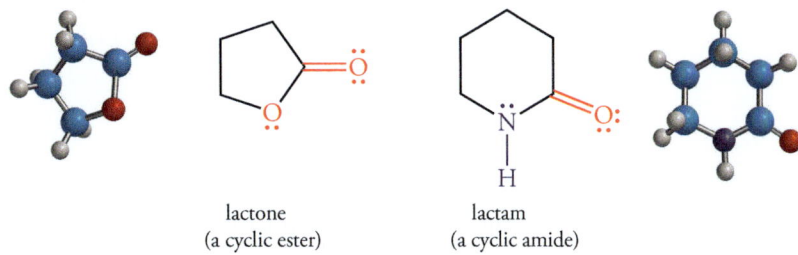

β-bromo-γ-ethylvaleric acid

Some unbranched carboxylic acids contain a —COOH group at each end of the chain. The common names and IUPAC names of some of these **dicarboxylic acids** are listed in Table 11.2.

Table 11.2
Nomenclature of Dicarboxylic Acids

Formula	IUPAC Name	Common Name
$HO_2C—CO_2H$	Ethanedioic acid	Oxalic acid
$HO_2C—CH_2—CO_2H$	Propanedioic acid	Malonic acid
$HO_2C—(CH_2)_2—CO_2H$	Butanedioic acid	Succinic acid
$HO_2C—(CH_2)_3—CO_2H$	Pentanedioic acid	Glutaric acid
$HO_2C—(CH_2)_4—CO_2H$	Hexanedioic acid	Adipic acid

IUPAC Names of Carboxylic Acids

The IUPAC rules to name carboxylic acids are similar to those for aldehydes. The final -e of the parent hydrocarbon is replaced by the ending -oic acid. The carboxyl group has a higher priority than aldehyde, ketone, halogen, hydroxyl, and alkoxy groups. The priority order for carbonyl groups is carboxylic acid > aldehyde > ketone. The carbonyl group of an aldehyde or ketone is indicated by the prefix oxo-. Compounds that have a —CO_2H group bonded to a cycloalkane ring are named as derivatives of the cycloalkane, and the suffix -carboxylic acid is added. The ring carbon atom to which the carboxyl carbon atom is bonded is assigned the number 1, but this number is not included in the name. The following examples illustrate the IUPAC rules for naming carboxylic acids.

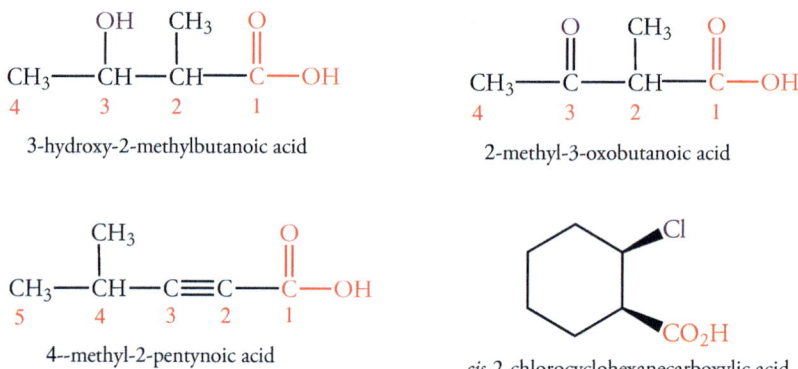

3-hydroxy-2-methylbutanoic acid

2-methyl-3-oxobutanoic acid

4--methyl-2-pentynoic acid

cis-2-chlorocyclohexanecarboxylic acid

Names of Carboxylic Acid Derivatives

The name of the conjugate base of a carboxylic acid, a **carboxylate** anion, is obtained by changing the -oic acid ending to -oate. For a salt of a carboxylic acid, the name of the carboxylate anion is preceded by the name of the metal ion.

sodium ethanoate
(sodium acetate)

potassium 3-phenylpropanoate
(potassium β-phenylpropionate)

An ester is named by first writing the name of the alkyl or aryl group bonded to the bridging oxygen atom, followed by the name of the acyl portion of the ester, which is derived from a carboxylic acid and named as a carboxylate.

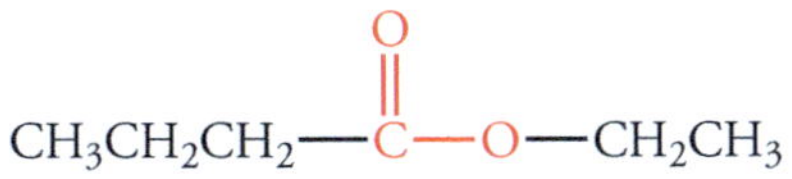

$CH_3CH_2CH_2\underset{\underset{O}{\parallel}}{C}\!-\!O\!-\!CH_2CH_3$

ethyl butanoate

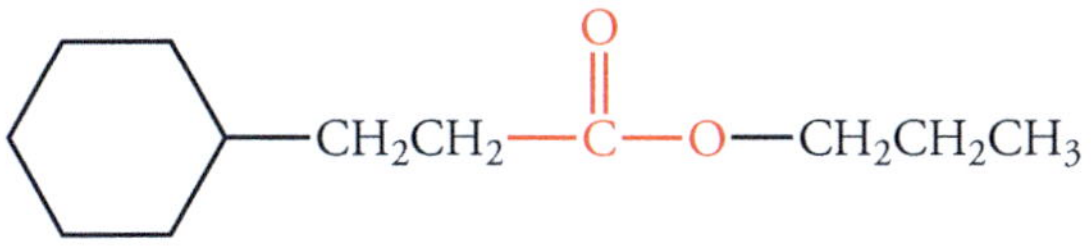

$CH_2CH_2\underset{\underset{O}{\parallel}}{C}\!-\!O\!-\!CH_2CH_2CH_3$

propyl-3-cyclohexylpropanoate
(propyl-β-cyclohexylpropionate)

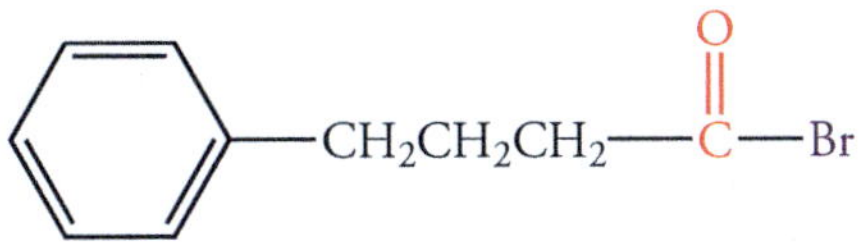

$-CH_2CH_2CH_2\underset{\underset{O}{\parallel}}{C}\!-\!Br$

4-phenylbutanoyl bromide

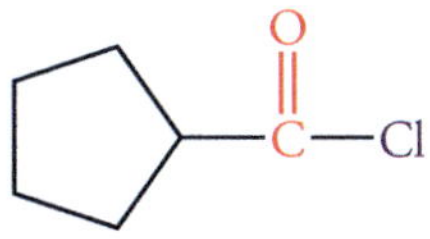

$\underset{\underset{O}{\parallel}}{C}\!-\!Cl$

cyclopentanecarbonyl chloride

An acid anhydride consists of two acyl groups bonded through a bridging oxygen atom. Although acid anhydrides can have two different acyl groups, compounds containing identical acyl groups are more common. They are named by replacing the suffix *-oic acid* with *-oic anhydride*.

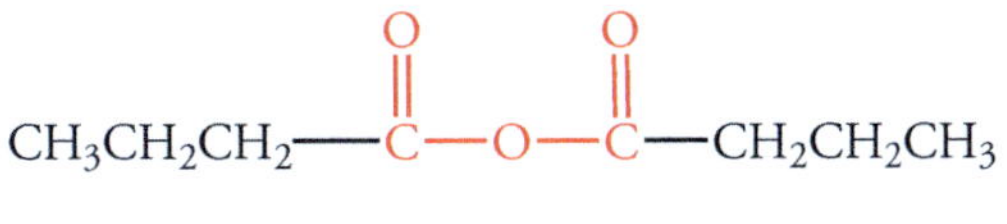

$CH_3CH_2CH_2\underset{\underset{O}{\parallel}}{C}\!-\!O\!-\!\underset{\underset{O}{\parallel}}{C}\!-\!CH_2CH_2CH_3$

butanoic anhydride

$\underset{\underset{O}{\parallel}}{C}\!-\!O\!-\!\underset{\underset{O}{\parallel}}{C}$

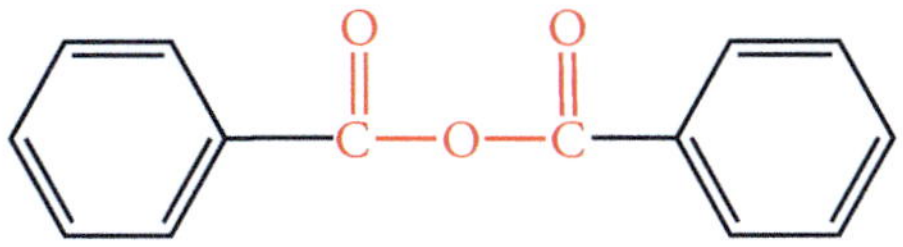

benzoic anhydride

Amides are named by replacing the suffix for the acid (*-oic acid*) with the suffix *-amide*. An amide functional group bonded to a cycloalkane ring is named as a *carboxamide*.

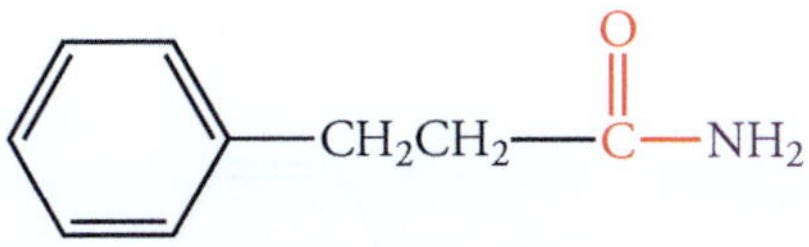

$-CH_2CH_2\underset{\underset{O}{\parallel}}{C}\!-\!NH_2$

3-phenylpropanamide

$\underset{\underset{O}{\parallel}}{C}\!-\!NH_2$

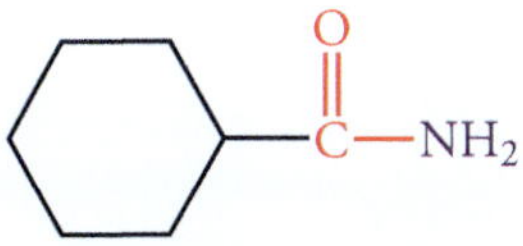

cyclohexanecarboxamide

In secondary and tertiary amides, the nitrogen atom is bonded to one or more alkyl or aryl groups instead of hydrogen atoms. We will discuss the names of amides with groups bonded to nitrogen along with the names of amines in Chapter 14.

Problem 11.1

The structure of oleic acid, an unsaturated carboxylic acid present as an ester in vegetable oils, is shown below. What is the IUPAC name of oleic acid?

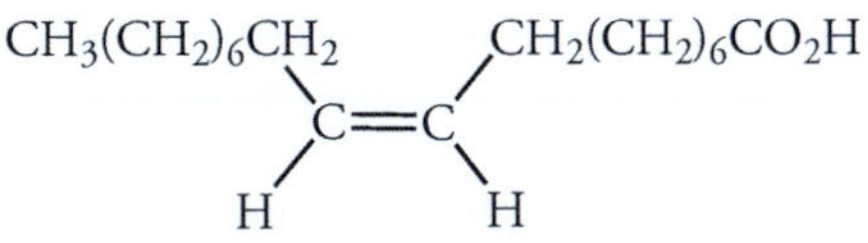

$$CH_3(CH_2)_6CH_2 \qquad CH_2(CH_2)_6CO_2H$$
$$\underset{H}{\overset{}{C}}=\underset{H}{\overset{}{C}}$$

Solution

First, determine the length of the continuous chain that contains the —COOH group. It contains 18 carbon atoms. The double bond is located at C-9, numbering from the carboxyl group on the right. Thus, the compound is a 9-octadecenoic acid. The configuration about the double bond is Z, and therefore the complete name is (Z)-9-octadecenoic acid.

Problem 11.2

Mevalonic acid is required to form isopentenyl pyrophosphate, an intermediate in terpene synthesis. It has the following structure. What is its IUPAC name?

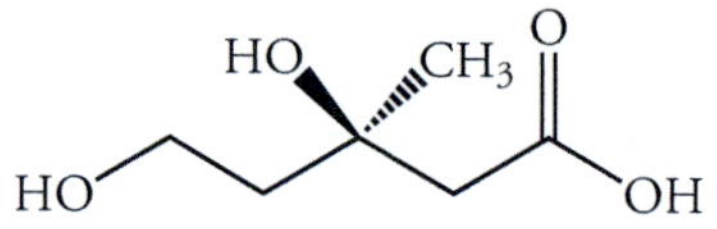

Problem 11.3
Assign the IUPAC name to clofibrate, a drug used to lower the concentration of blood triacylglycerols and cholesterol.

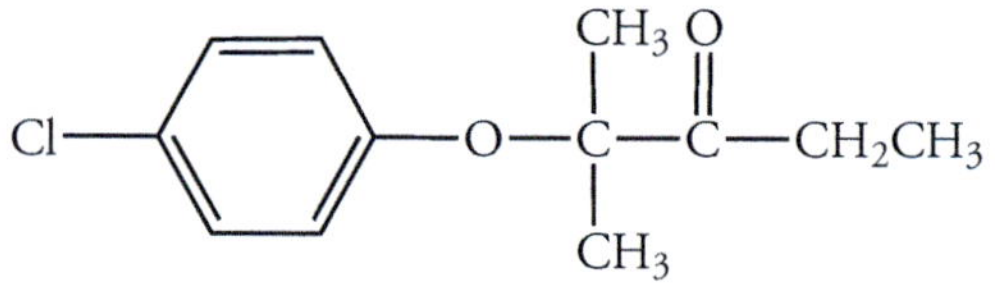

chlofibrate

Solution
First, identify the alcohol portion of the ester; it is located at the right of the molecule. The alcohol portion contains two carbon atoms, so the compound is an ethyl ester.

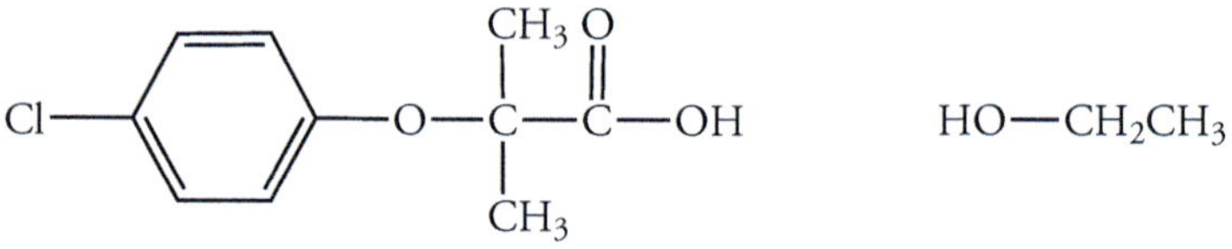

acyl portion of ester   ethanol

The acyl portion is a substituted propanoic acid with a methyl group and an aryl-containing group at C-2. Imagine removing the aryl-containing group from the acid and adding a hydrogen atom to its oxygen atom. The resulting compound is *p*-chlorophenol. The original group is *p*-chlorophenoxy.

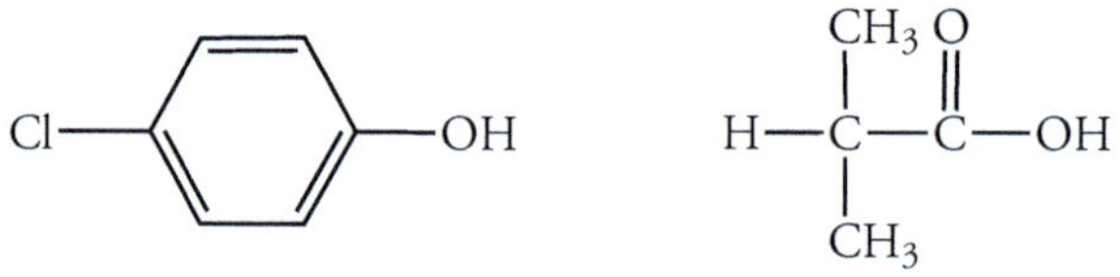

p-chlorophenol   2-methylpropanoic acid

The name of the acid is 2-(*p*-chlorophenoxy)-2-methylpropanoic acid. Now change the *-oic* ending of the acid to *-oate* and write the name of the alkyl group of the alcohol as a separate word in front of the modified acid name. The ester is named ethyl 2-(*p*-chlorophenoxy)-2-methylpropanoate.

Problem 11.4
Isobutyl formate has the odor of raspberries. Based on this common name, draw its structural formula. What is its IUPAC name?

11.3 PHYSICAL PROPERTIES OF CARBOXYLIC ACIDS

The physical properties of carboxylic acids and esters, such as boiling point and solubility, are distinctly different because of the different types of intermolecular interactions that are possible in each class of compounds. Differences in biological properties such as taste depend on a class of proteins in membranes called G-coupled protein receptors, a topic that is beyond the scope of this text.

Melting Points, Boiling Points, and Solubilities

The boiling points of carboxylic acids are high relative to the other organic compounds we have discussed (Table 11.3). Carboxylic acids interact very strongly by forming hydrogen-bonded dimers. These dimers have higher boiling points than substances of comparable molecular weight that do not form dimers.

Hydrogen-bonded dimer of acetic acid

$$CH_3-\overset{\overset{\displaystyle \ddot{O}:}{\|}}{C}-\ddot{O}-H \qquad CH_3-\overset{\overset{\displaystyle \ddot{O}:}{\|}}{C}-CH_3 \qquad CH_3-\overset{\overset{\displaystyle CH_2}{\|}}{C}-CH_3$$

bp 118 °C bp 56.5 °C bp –7 °C

Table 11.3
Physical Properties of Carboxylic Acids

IUPAC Name	Melting Point (°C)	Boiling Point (°C)	Solubility in Water (g/100 mL at 20 °C)
Methanoic acid		101	Miscible
Ethanoic acid		118	Miscible
Propanoic acid		141	Miscible
Butanoic acid		164	Miscible
Pentanoic acid		186	4.97
Hexanoic acid		205	0.960
Octanoic acid		239	0.068
Decanoic acid	32	270	0.015
Dodecanoic acid	44	299	0.0055

Carboxylic acids with low molecular weights are soluble in water because the carboxyl group forms several hydrogen bonds with water. A carboxylic acid acts both as a hydrogen bond donor through its hydroxyl hydrogen atom and as a hydrogen bond acceptor through the lone pair electrons of both oxygen atoms. The solubility of carboxylic acids, like that of alcohols, decreases with increasing chain length because the long, nonpolar hydrocarbon chain dominates the physical properties of the acid.

Esters are polar molecules, but their boiling points are lower than those of carboxylic acids and alcohols of similar molecular weight because there is no intermolecular hydrogen bonding between ester molecules.

Esters can form hydrogen bonds through their oxygen atoms to the hydrogen atoms of water molecules. Thus, esters are slightly soluble in water. However, because esters do not have a hydrogen atom to form a hydrogen bond to an oxygen atom of water, they are less soluble than carboxylic acids. Table 11.4 lists the solubilities and boiling points of some esters.

Odors of Acids and Esters

Liquid carboxylic acids have sharp, unpleasant odors. For example, butanoic acid occurs in rancid butter and aged cheese. Caproic, caprylic, and capric acids have the smell of goats. (The Latin word for goat, *caper*, is the source of the common names of these acids.)

In contrast to carboxylic acids, esters have pleasant fruity smells. In fact, the odors of many fruits are due to esters. For example, ethyl ethanoate is found in pineapples, 3-methylbutyl ethanoate in apples and bananas, 3-methylbutyl 3-methylbutanoate in apples, and octyl ethanoate in oranges.

Esters have low boiling points, and some fraction of them is driven off in the preparation of processed foods. Thus, esters are added back to the food to make them smell "natural." Nevertheless, government regulations require that the added esters be identified as additives on the label. The esters used in some products need not be the same as those in natural fruits, but of course they are selected to produce the same odor or taste. The choice of esters may be dictated by their cost and availability. Table 11.5 lists some of these flavoring agents.

Table 11.4
Physical Properties of Esters

IUPAC Name	Boiling Point (°C)	Solubility (g/100 mL H_2O)
Methyl methanoate	32	Miscible
Methyl ethanoate	57	24.4
Methyl propanoate	80	1.8
Methyl butanoate	102	0.5
Methyl pentanoate	126	0.2
Methyl hexanoate	151	0.06
Ethyl methanoate	54	Miscible
Ethyl ethanoate	77	7.4
Ethyl propanoate	99	1.7
Ethyl butanoate	120	0.5
Ethyl pentanoate	145	0.2
Propyl ethanoate	102	1.9
Butyl ethanoate	125	1.0
Methyl benzoate	199	0.1
Ethyl benzoate	213	0.08

Table 11.5
Esters Used As Flavoring Agents

IUPAC Name	Formula	Flavor
Methyl butanoate	$CH_3CH_2CH_2CO_2CH_3$	Apple
Pentyl butanoate	$CH_3CH_2CH_2CO_2CH_2(CH_2)_3CH_3$	Apricot
Pentyl ethanoate	$CH_3CO_2CH_2(CH_2)_3CH_3$	Banana
Octyl ethanoate	$CH_3CO_2CH_2(CH_2)_6CH_3$	Orange
Ethyl butanoate	$CH_3(CH_2)_2CO_2CH_2CH_3$	Pineapple
Ethyl methanoate	$HCO_2CH_2CH_3$	Rum

11.4 ACIDITY OF CARBOXYLIC ACIDS

Although acetic acid and other carboxylic acids are weak acids, they are much more acidic than alcohols or phenols. The acid ionization constant, K_a, for acetic acid is about 10^{11} times larger than the K_a for ethanol.

$$CH_3CH_2OH + H_2O \rightleftharpoons CH_3CH_2O^- + H_3O^+ \quad K_a = 1 \times 10^{-16}$$

$$CH_3CO_2H + H_2O \rightleftharpoons CH_3CO_2^- + H_3O^+ \quad K_a = 1.8 \times 10^{-5}$$

The greater acidity of acetic acid is the result of resonance stabilization of the negative charge in the conjugate base, acetate ion. In the ethoxide ion ($CH_3CH_2O^-$) the negative charge is concentrated on a single oxygen atom.

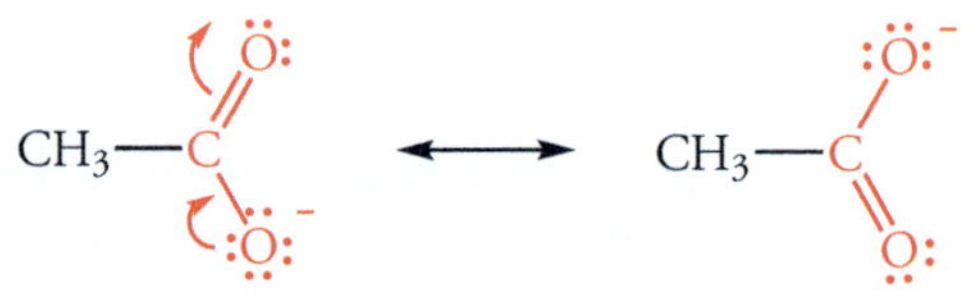

Resonance structures of acetate ion

The acidity of carboxylic acids is also partly the result of an inductive effect (Section 2.8). That is, the carbonyl group polarizes the H—O bond by attracting electron density through the σ bonds. The withdrawal of electron density from the H—O bond weakens it, and thus increases the acidity of the ionizable hydrogen atom.

The inductive effect of an alkyl or aryl group attached to the carbonyl carbon atom affects K_a. An alkyl group is electron-releasing with respect to hydrogen. This release of electron density to the carboxyl group stabilizes the acid and slightly destabilizes the conjugate base. Thus, acetic acid (pK_a = 4.74) is a weaker acid than formic acid (pK_a = 3.75). In contrast, the sp^2-hybridized carbon atom of an aryl group is electron-withdrawing relative to an alkyl group. Thus, benzoic acid (pK_a = 4.19) is a stronger acid than acetic acid.

Electron-withdrawing groups bonded to a carbon atom increase the acidity of carboxylic acids. For example, chloroacetic acid is a stronger acid than acetic acid. Other examples of the effect of electron-withdrawing groups on the acidity of carboxylic acids are given in Table 11.6.

Table 11.6
pK$_a$ Values of Carboxylic Acids

IUPAC Name	*Formula*	*pK$_a$*
Methanoic acid	HCO_2H	3.75
Ethanoic acid	CH_3CO_2H	4.72
Propanoic acid	$CH_3CH_2CO_2H$	4.87
Butanoic acid	$CH_3(CH_2)_2CO_2H$	4.82
2-Methylbutanoic acid	$(CH_3)_2CHCO_2H$	4.84
Pentanoic acid	$CH_3(CH_2)_3CO_2H$	4.81
2,2-Dimethylmethylpropanoic acid	$(CH_3)CCO_2H$	5.03
Fluoroethanoic acid	FCH_2CO_2H	2.59
Chloroethanoic acid	$ClCH_2CO_2H$	2.86
Bromoethanoic acid	$BrCH_2CO_2H$	2.90
Idoethanoic acid	ICH_2CO_2H	3.18
Dichloroethanoic acid	Cl_2CHCO_2H	1.24
Trichloroethanoic acid	Cl_3CCO_2H	0.64
Trifluoroethanoic acid	F_3CCO_2H	0.23
Methoxyethanoic acid	$CH_3OCH_2CO_2H$	3.55
Cyanoethanoic acid	$CNCH_2CO_2H$	2.46
Nitroethanoic acid	$NO_2CH_2CO_2H$	1.72

$$pK_a = 2.9 \qquad\qquad pK_a = 4.7$$

The electrons in the C—Cl bond are "pulled" toward the more electronegative chlorine atom and away from the carbon skeleton. As a result, the electrons of the oxygen atom are drawn away from the O—H bond, and the proton can therefore ionize more easily.

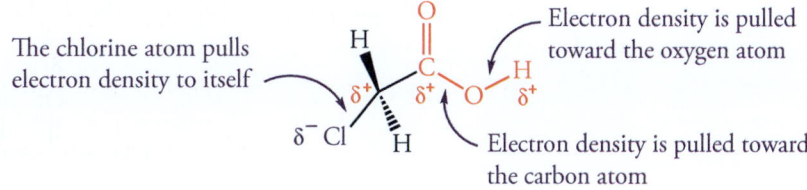

The chlorine atom pulls electron density to itself

Electron density is pulled toward the oxygen atom

Electron density is pulled toward the carbon atom

As the distance between the halogen atom and the carboxyl group increases, the inductive effect decreases dramatically. For β- and γ-substituted acids, the K_a value approaches that of an unsubstituted carboxylic acid.

$$\underset{\substack{| \\ \text{Cl}}}{\text{CH}_3\text{CH}_2\text{CHCO}_2\text{H}} \qquad \underset{\substack{| \\ \text{Cl}}}{\text{CH}_3\text{CHCH}_2\text{CO}_2\text{H}} \qquad \underset{\substack{| \\ \text{Cl}}}{\text{CH}_2\text{CH}_2\text{CH}_2\text{CO}_2\text{H}} \quad \text{CH}_3\text{CH}_2\text{CH}_2\text{CO}_2\text{H}$$

$$\text{p}K_a\ 2.84 \qquad\qquad \text{p}K_a\ 4.06 \qquad\qquad\qquad \text{p}K_a\ 4.52 \qquad\qquad\qquad \text{p}K_a\ 4.72$$

⬅ Increasing acidity

Salts of Carboxylic Acids

The reaction of carboxylic acids with hydroxide ions and the reaction of carboxylate salts with hydronium ions have some practical applications in separating carboxylic acids from mixtures. Because they are ionic, carboxylate salts are more soluble in water than their corresponding carboxylic acids. Carboxylic acids are often separated from other nonpolar organic compounds in the laboratory by adding a solution of sodium hydroxide to form the more soluble carboxylate salt. Consider, for example, a mixture of decanol and decanoic acid. Decanol is not soluble in water and does not react with sodium hydroxide. However, decanoic acid reacts with sodium hydroxide and thus dissolves in the basic solution, whereas the decanol remains undissolved.

$$\underset{\text{Insoluble in water}}{\text{CH}_3(\text{CH}_2)_8\text{CO}_2\text{H}} + \text{HO}^- \longrightarrow \underset{\text{Soluble in water}}{\text{CH}_3(\text{CH}_2)_8\text{CO}_2^-} + \text{H}_2\text{O}$$

Undissolved decanol is physically separated from the basic solution. Then HCl is added to neutralize the basic solution, and insoluble decanoic acid separates from the aqueous solution.

$$\underset{\text{Soluble in water}}{\text{CH}_3(\text{CH}_2)_8\text{CO}_2^-} + \text{H}_3\text{O}^+ \longrightarrow \underset{\text{Insoluble in water}}{\text{CH}_3(\text{CH}_2)_8\text{CO}_2\text{H}} + \text{H}_2\text{O}$$

This procedure is very useful in isolating acids from complex mixtures that are encountered in extracts from natural sources. It is also used to purify acids produced by chemical synthesis in the laboratory.

Problem 11.5

Pyruvic acid is a key metabolic intermediate in processes that provide energy for the growth and maintenance of cells. Explain why pyruvic acid ($\text{p}K_a$ 4.7) is about 100 times more acidic than propanoic acid ($\text{p}K_a$ 2.5).

$$\text{CH}_3 - \overset{\overset{\displaystyle \ddot{\text{O}}:}{\|}}{\text{C}} - \overset{\overset{\displaystyle \ddot{\text{O}}:}{\|}}{\text{C}} - \ddot{\text{O}} - \text{H}$$

pyruvic acid

Solution

Because the $\text{p}K_a$ of pyruvic acid is smaller than that of propanoic acid, pyruvic acid is a stronger acid. Thus the group attached to the carbonyl carbon atom pulls electron density away from the O—H bond. We recall that the carbonyl group is very polar and that the carbonyl carbon atom has a partial positive charge. As a result, the carbonyl carbon atom of the ketone group inductively pulls electron density from the carboxylic acid group, the polarity of the O—H bond increases, and the acidity increases.

$$\text{CH}_3 - \overset{\overset{\displaystyle \ddot{\text{O}}:}{\|}}{\underset{\delta^+}{\text{C}}} \leftarrow \overset{\overset{\displaystyle \overset{\delta^-}{\ddot{\text{O}}:}}{\|}}{\text{C}} \leftarrow \ddot{\text{O}} \leftarrow \text{H}$$

Problem 11.6

The pK_a values for the dissociation of the first of the two carboxyl groups of malic acid and oxalo-acetic acid are 3.41 and 1.70, respectively. Which compound is the stronger acid? Which of the two carboxyl groups in each compound is the more acidic?

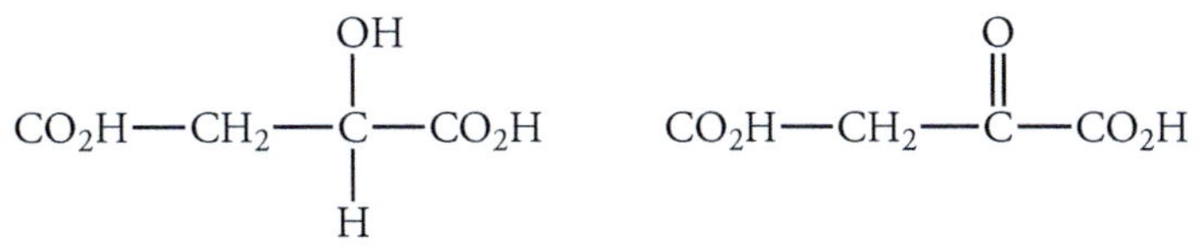

malic acid   oxaloacetic acid

11.5 SYNTHESIS OF CARBOXYLIC ACIDS

Carboxylic acids can be prepared by several oxidation reactions that we described in earlier chapters. Both aldehydes and alcohols are oxidized by Jones reagent (Section 8.7) to produce carboxylic acids. Aldehydes can be oxidized by Tollens reagent or Fehling's solution (Section 10.4).

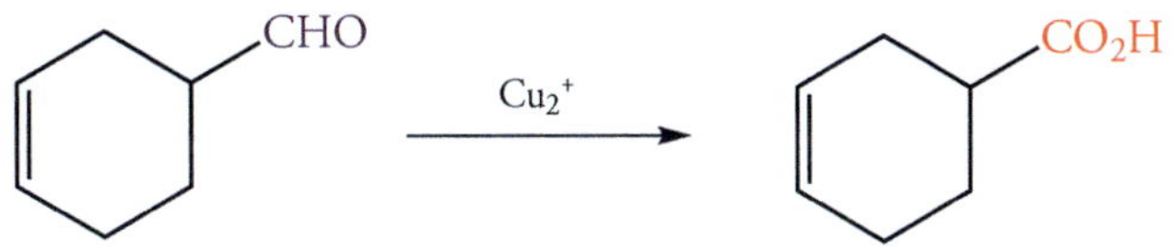

3-cyclohexenecarbaldehyde   3-cyclohexenecarboxylic acid

$$CH_3(CH_2)_8CH_2OH \xrightarrow[H_2SO_4/\text{acetone}]{CrO_3} CH_3(CH_2)_8CO_2H$$

1-decanol   decanoic acid

Alkylbenzenes are oxidized by potassium permanganate to give benzoic acids. The entire side chain is oxidized in this reaction (Section 5.9). The oxidation of tetralin to produce phthalic acid is an example.

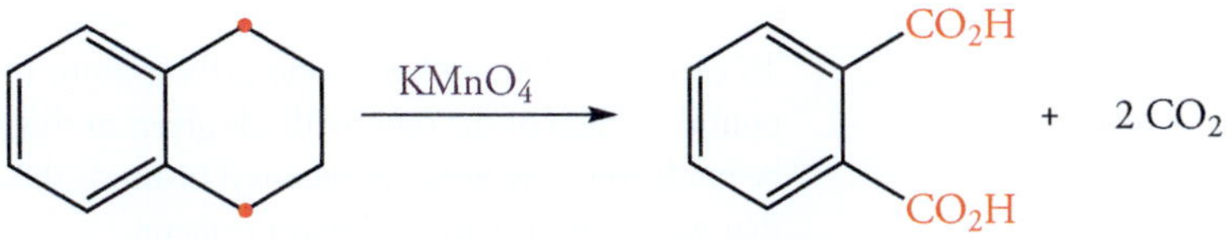

tetralin   phthalic acid

Carboxylic acids can also be made by two more general methods starting from haloalkanes. These methods provide structures that have one more carbon atom than the reactant haloalkane.

In Chapter 10 we saw that the Grignard reagent acts as a nucleophile, and that it reacts with the carbonyl group of aldehydes or ketones. A similar reaction occurs between a Grignard reagent and the carbon-oxygen double bond of carbon dioxide to yield the magnesium salt of a carboxylic acid. Adding aqueous acid to the solution of the conjugate base gives the carboxylic acid.

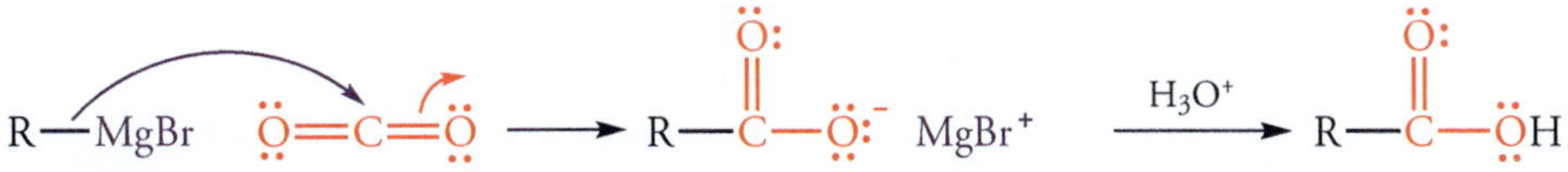

Starting from the haloalkane, the reaction sequence requires three steps. First, the haloalkane is converted to a Grignard reagent. Second, the ether solution is poured over solid carbon dioxide (dry ice). Finally, the reaction mixture is acidified. The reaction sequence can be presented using a single arrow with the three steps listed above and below the arrow.

Br
1. Mg/ether
2. CO2
3. H3O+
CH3
O
C
OH
CH3

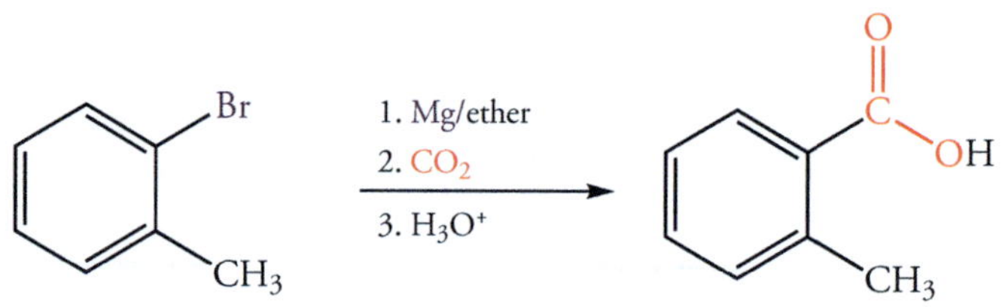

o-bromotoluene   o-methylbenzoic acid

The second synthesis that adds one carbon atom to the parent chain of the reacting haloalkane occurs by an S_N2 substitution reaction of halide ion by cyanide ion (Chapter 7). The resulting product, called a **nitrile** (RCN), can be hydrolyzed to produce a carboxylic acid.

$$R{-}\ddot{\underset{..}{Br}}\!: + {}^{-}\!:C{\equiv}N: \longrightarrow R{-}C{\equiv}N: + :\ddot{\underset{..}{Br}}\!:^{-}$$

$$R{-}C{\equiv}N: \xrightarrow{H_3O^+} R{-}\overset{\overset{\displaystyle O}{\|}}{C}{-}O{-}H + NH_4^+$$

CH2Br
1. KCN
2. H3O+
CH2CO2H

Problem 11.7

Suggest a synthetic sequence for the following transformation.

HO
CH2Br
HO
CH2CO2H

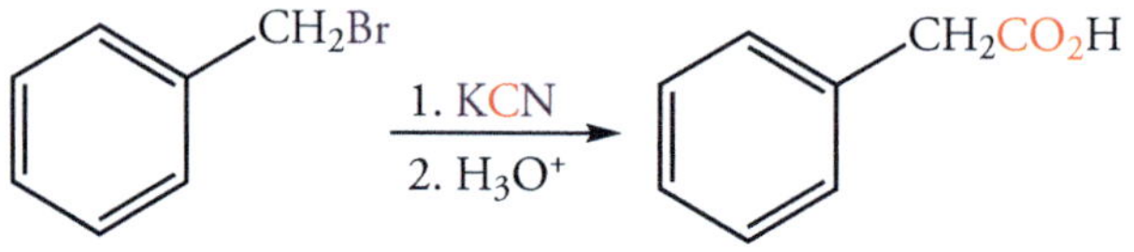

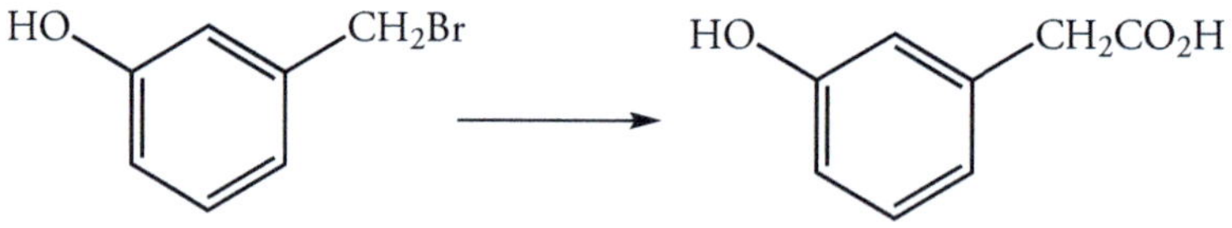

Solution

To carry out this synthesis, one carbon atom must be added to the side chain of the aromatic compound. Which of the two methods given in this section should be selected? A Grignard reagent cannot be made from the *m*-hydroxybenzyl bromide, because the phenolic hydroxyl group has an acidic proton that would destroy the Grignard reagent.

Consider the substitution reaction of the starting material with cyanide ion to give the nitrile. Because benzyl bromide is a primary haloalkane, it readily reacts in an S_N2 reaction. The resulting nitrile is hydrolyzed to form the desired acid.

HO
CH2Br
CN−
HO
CH2C≡N
H3O+
HO
CH2CO2H

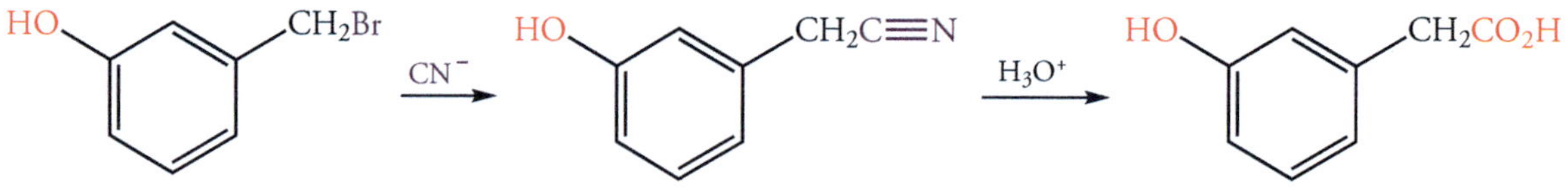

Problem 11.8

Outline a series of steps to prepare 2,2-dimethylheptanoic acid starting from 2-methyl-1-heptene.

$$CH_3CH_2CH_2CH_2CH_2{-}\overset{\overset{\displaystyle CH_3}{|}}{C}{=}CH_2 \longrightarrow CH_3CH_2CH_2CH_2CH_2{-}\overset{\overset{\displaystyle CH_3}{|}}{\underset{\underset{\displaystyle CH_3}{|}}{C}}{-}CO_2H$$

$$CaO + 3\ C \xrightarrow{2500\ °C} CaC_2 + CO$$

$$CaC_2 + 2\ H_2O \longrightarrow CH{\equiv}CH + Ca(OH)_2$$

The conversion of acetylene into acetic acid occurs in two steps, a hydration reaction followed by an oxidation reaction. Hydration of acetylene using mercuric sulfate and sulfuric acid gives acetaldehyde via an enol intermediate (Section 4.10). Oxidation of acetaldehyde by oxygen gas using cobalt(III) acetate as a catalyst gives acetic acid.

$$H{-}C{\equiv}C{-}H + 2\ H_2O \xrightarrow[\text{H}_2\text{SO}_4]{\text{HgSO}_4} CH_3{-}\underset{\displaystyle \overset{\displaystyle O}{\|}}{C}{-}H$$

$$CH_3{-}\underset{\displaystyle \overset{\displaystyle O}{\|}}{C}{-}H \xrightarrow{O_2 \;\; Co^{3+}} CH_3{-}\underset{\displaystyle \overset{\displaystyle O}{\|}}{C}{-}OH$$

Large-scale industrial syntheses are developed using low-cost feed stocks in efficient reactions that have low energy requirements. Although the reactants in the described synthesis of acetic acid are relatively inexpensive and the reactions occur in high yield, the process requires large amounts of energy. Over a half-century, the cost of energy increased so much that the process eventually became uneconomical.

Ethylene is inexpensive and is readily available from the refining of petroleum and natural gas. Chemical engineers devised a process with much lower energy requirements that catalytically oxidizes ethylene to acetaldehyde using molecular oxygen and a catalyst of palladium(II) and copper(II). This process, which became the primary source of acetaldehyde, was devised by Wacker-Chemie of Germany in 1959, and is known as the **Wacker process**.

Subsequent catalytic oxidation of acetaldehyde using molecular oxygen with cobalt(III) acetate as a catalyst is used to produce acetic acid. This process dominated the market for only 15 years, because in 1973 the Monsanto process was developed using methanol as a feedstock and reacting it with carbon monoxide. The formation of the carbon-carbon bond is catalyzed by rhodium(III), HI, and water.

$$CH_2{=}CH_2 + 2\ O_2 \xrightarrow[\text{Cu}^{2+}]{\text{Pd}^{2+}} 2\ CH_3{-}\underset{\displaystyle \overset{\displaystyle O}{\|}}{C}{-}H$$

$$CH_3{-}\underset{\displaystyle \overset{\displaystyle O}{\|}}{C}{-}H \xrightarrow[\text{Co}^{3+}]{O_2} CH_3{-}\underset{\displaystyle \overset{\displaystyle O}{\|}}{C}{-}OH$$

$$CH_3{-}OH + CO \xrightarrow[\text{HI/H}_2\text{O}]{\text{Rh}^{3+}} CH_3{-}\underset{\displaystyle \overset{\displaystyle O}{\|}}{C}{-}OH$$

Methanol is also an important commercial product. Using various catalysts, methane and water react to produce hydrogen and carbon monoxide in a 3:1 molar ratio, known as *water gas*.

$$CH_4 + H_2O \longrightarrow CO + 3\ H_2$$

The components of the water gas mixture can be adjusted to a 2:1 mixture of hydrogen gas to carbon monoxide, known as *synthesis gas*, and that mixture is converted to methanol in a catalyzed process.

$$CO + 2\ H_2 \longrightarrow CH_3OH$$

As a result, the water gas mixture is the source of both compounds required in the Monsanto process. Methane, obtained from natural gas, and water are the starting materials for the synthesis of acetic acid. Catalysts have been developed for all of the necessary reactions to make the various steps efficient and economical. Methods for producing these compounds are continuously evolving in response to environmental concerns.

11.6 NUCLEOPHILIC ACYL SUBSTITUTION

In Chapter 10 we discussed nucleophilic addition reactions in which the nucleophile attacks the electrophilic carbonyl carbon atom of aldehydes and ketones to give a tetrahedral product.

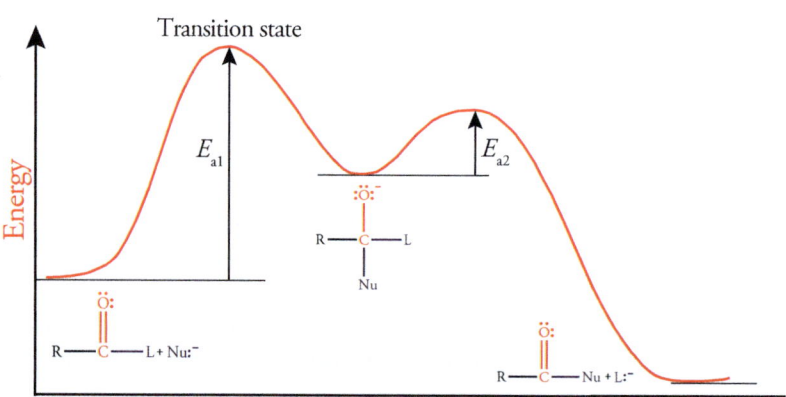

Acyl derivatives also react with nucleophiles at the carbonyl carbon atom to generate a tetrahedral intermediate. However, the tetrahedral intermediate is unstable, and a leaving group departs to form a different acyl derivative. The overall process is called **nucleophilic acyl substitution**. The process is also called an **acyl group transfer** reaction because it transfers an acyl group from one group (the leaving group) to another (the nucleophile).

tetrahedral intermediate

Why don't acyl derivatives behave like aldehydes and ketones and form stable tetrahedral products? The answer is that the intermediate formed from an acyl derivative has a good leaving group. In the case of an acid chloride, the leaving group is the chloride ion, which is a weak base. We recall that leaving group abilities are inversely related to base strength. An intermediate derived from a ketone does not have a good leaving group. A carbanion, the conjugate acid of a hydrocarbon, is an extremely strong base and therefore a very poor leaving group.

The stoichiometry of the nucleophilic acyl substitution reaction resembles that of an S_N2 substitution reaction of haloalkanes. However, the resemblance is only superficial. An S_N2 reaction is a one-step process in which the nucleophile bonds to the carbon atom as the leaving group leaves. Nucleophilic acyl substitution occurs in two steps (Figure 11.2). The rate-determining step is most commonly nucleophilic attack at the carbonyl carbon atom to form a tetrahedral intermediate. The loss of the leaving group occurs in a second, faster step.

**Figure 11.2
Mechanism of Nucleophilic
Acyl Substitution**

Relative Reactivity of Acyl Derivatives

The order of reactivity of acyl derivatives is acid chloride > acid anhydride > ester = acid > amide. This order of reactivity might appear to reflect the leaving group abilities, as reflected by their basicities. We know that HCl is a strong acid and NH_3 is a very weak acid. Thus, Cl^- is a weak base and NH_2^- is a strong base, meaning that Cl^- is a better leaving group than NH_2^-. However, *the rate-determining step is not loss of the leaving group but rather attack of the nucleophile.*

The order of reactivities reflects resonance stabilization of the reactant. Donation of an electron pair of the atom bonded to the acyl carbon atom decreases the partial positive charge on the carbon atom and decreases its electrophilicity. In the tetrahedral intermediate, no resonance stabilization is possible. Consider the general equation for the first step of the reaction.

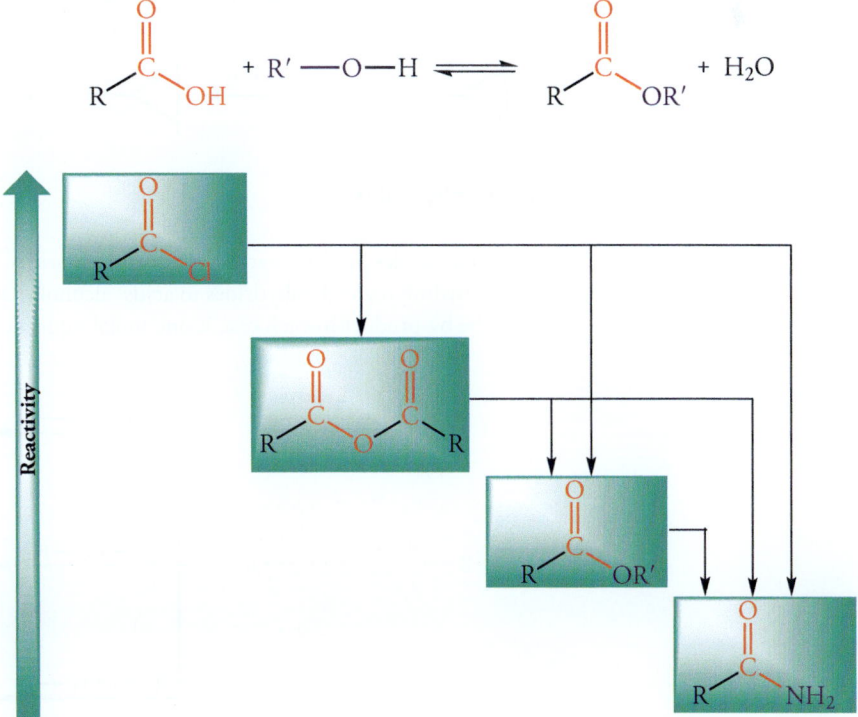

Resonance-stabilized reactant   No resonance stabilization

We recall from the discussion of substituent effects on electrophilic aromatic substitution that second-row elements such as oxygen and nitrogen effectively donate electrons by resonance. Furthermore, nitrogen is a better donor of electrons by resonance than oxygen because nitrogen is less electronegative. Thus, a nitrogen atom stabilizes amides more effectively than an oxygen atom stabilizes esters. Finally, we recall that chlorine, a third-row element, is not effective in donating electrons by resonance. Thus, acid chlorides have little resonance stabilization by the chlorine atom.

The order of reactivity means that a more reactive acyl derivative can be converted into a less reactive acyl derivative (Figure 11.3). The relative reactivity of acyl derivatives enables us to understand most of the chemical reactions presented in this chapter. Note that acids and esters both have an oxygen atom bonded to the carbonyl carbon atom. Thus, these two classes of compounds have similar reactivities and can be readily interconverted in equilibrium processes. We will discuss these important reactions in later sections.

Figure 11.3
Reactivity of Acyl Derivatives

The more reactive acyl derivatives can be converted spontaneously to the less reactive acyl derivatives by nucleophilic acyl substitution reactions.

Acid Chlorides

Acid chlorides are the most reactive acyl derivatives. They are prepared in the laboratory by treating carboxylic acids with thionyl chloride, $SOCl_2$. This reaction takes advantage of Le Châtelier's principle. Thionyl chloride reacts with carboxylic acids to form acid chlorides because the reaction is driven to the right by the formation of HCl and SO_2, which are released as gases and escape from the reaction mixture. Phosphorus trichloride, PCl_3, can also be used to prepare acyl chlorides.

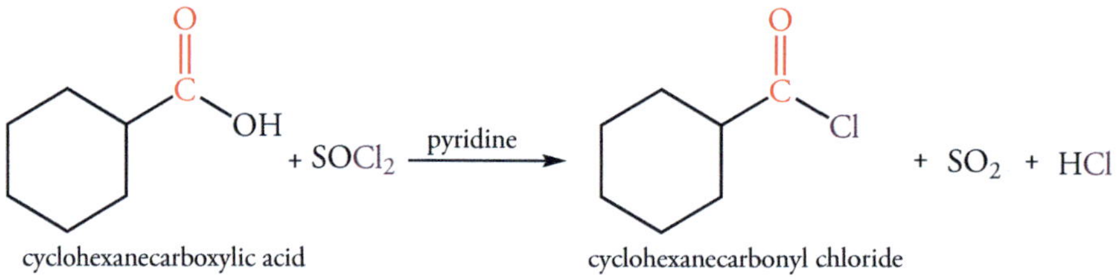

cyclohexanecarboxylic acid   cyclohexanecarbonyl chloride

Acid chlorides react with most nucleophiles and are hydrolyzed by the moisture in air. Reaction of an acid chloride with an alcohol gives an ester. Amines are easily converted into amides by reaction with acid chlorides.

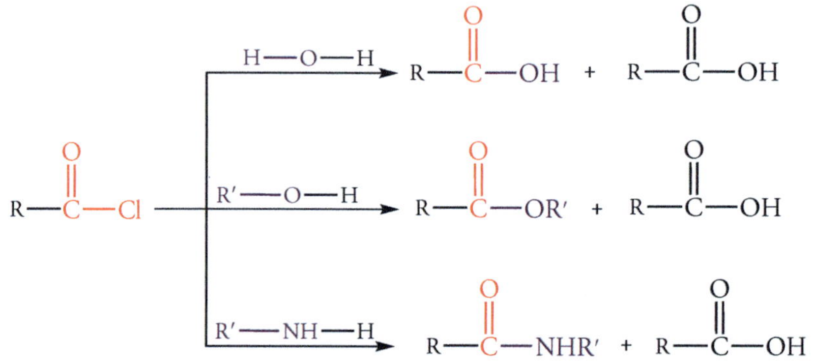

Esters are easily produced from acids by first converting the acid to an acid chloride. The acid chloride reacts with an alcohol to give excellent yields of an ester. A base such as pyridine is used to react with the HCl formed in the reaction.

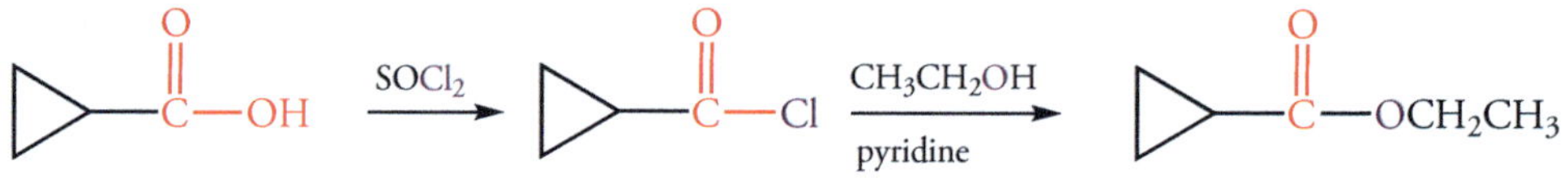

Acid Anhydrides

Acid anhydrides are less reactive than acid chlorides, but they are still very active acylating agents. Water hydrolyzes acid anhydrides to acids, alcohols react to give esters, and amines give amides. Note that the by-product in each case is one molar equivalent of a carboxylic acid.

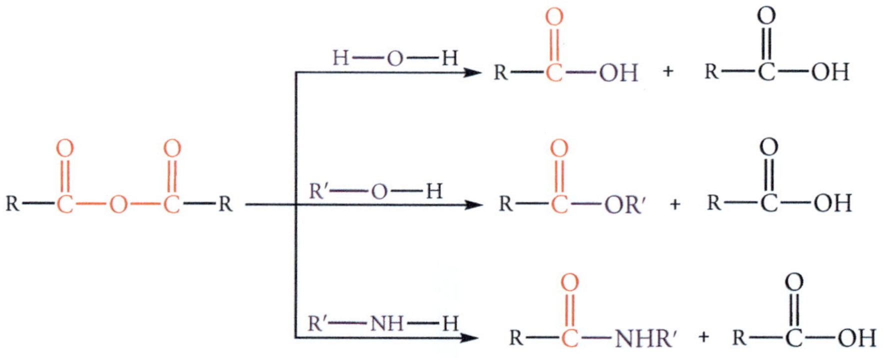

The most common acid anhydride is acetic anhydride. Reaction of acetic anhydride with the phenol functional group of salicylic acid yields an ester called acetyl salicylic acid, or aspirin.

$$CH_3-\overset{\displaystyle O}{\overset{\|}{C}}-O-\overset{\displaystyle O}{\overset{\|}{C}}-CH_3 \quad + \qquad \qquad \qquad \longrightarrow \qquad \qquad \qquad + \quad CH_3-\overset{\displaystyle O}{\overset{\|}{C}}-OH$$

acetic anhydride   salicylic acid   acetylsalicylic acid   acetic acid

Thioesters are Biochemical Acyl Transfer Reagents

The interconversion of acyl compounds in cells occurs by transferring acyl groups from one molecule to another. Esters are widely distributed in cells, but generally react too slowly to transfer acyl groups efficiently to other substrates.

Cells require an acyl transfer reagent that is sufficiently stable toward hydrolysis, but reactive enough toward nucleophiles to react at speeds sufficient to maintain life processes. These reagents are thioesters. Thioesters are more reactive than esters, and the thiol group is easily replaced by an alkoxy group from an alcohol. The increased reactivity of thioesters over esters results from the smaller ability of sulfur, a third-period element, to donate electrons by resonance to the carbonyl carbon atom and stabilize the thioester. Because thioesters are less stable than esters, the transfer of an acyl group from a thioester to an alcohol is a spontaneous, enzyme-catalyzed reaction.

$$CH_3-\overset{\displaystyle O}{\overset{\|}{C}}-SR \; + \; R'-O-H \quad \longrightarrow \quad CH_3-\overset{\displaystyle O}{\overset{\|}{C}}-OR' \; + \; H-S-R$$

The most important thioester is acetyl coenzyme A, which is formed from the thiol group of coenzyme A, a complex thiol that we will abbreviate as CoA—SH. Coenzyme A consists of three substructures—adenosine diphosphate (ADP), the vitamin pantothenic acid, and 2-aminoethanethiol. The thiol group of CoA—SH is bonded to an acyl group in acyl-CoA derivatives.

When an acetyl group is linked to CoA—SH, the adduct is acetyl coenzyme A. This extremely important metabolite is produced by the degradation of long-chain carboxylic acids contained in fats. Acetyl-CoA is also produced by metabolic degradation of many amino acids and carbohydrates. Acetyl-CoA is a donor of the two-carbon acetyl group in the biosynthesis of long-chain carboxylic acids.

Acetyl coenzyme A reacts with nucleophiles in enzyme-catalyzed biological reactions to give new acyl compounds.

$$CH_3-\overset{\displaystyle O}{\overset{\|}{C}}-S-CoA \; + \; HNu\text{:} \quad \longrightarrow \quad CH_3-\overset{\displaystyle O}{\overset{\|}{C}}-Nu \; + \; CoA-S-H$$

For example, acetyl coenzyme A provides the acetyl group in the biosynthesis of the neurotransmitter acetylcholine. Choline contains a hydroxyl group that is acetylated by acetyl coenzyme A to make acetylcholine. The reaction is catalyzed by an enzyme called choline acyltransferase.

$$CH_3-\overset{\displaystyle O}{\overset{\|}{C}}-S-CoA \; + \; (CH_3)_3\overset{+}{N}CH_2CH_2OH \; \xrightarrow[\text{acyltransferase}]{\text{choline}} \; CH_3-\overset{\displaystyle O}{\overset{\|}{C}}-O-CH_2CH_2\overset{+}{N}(CH_3)_3 \; + \; CoASH$$

acetylCoA   choline   acetylcholine

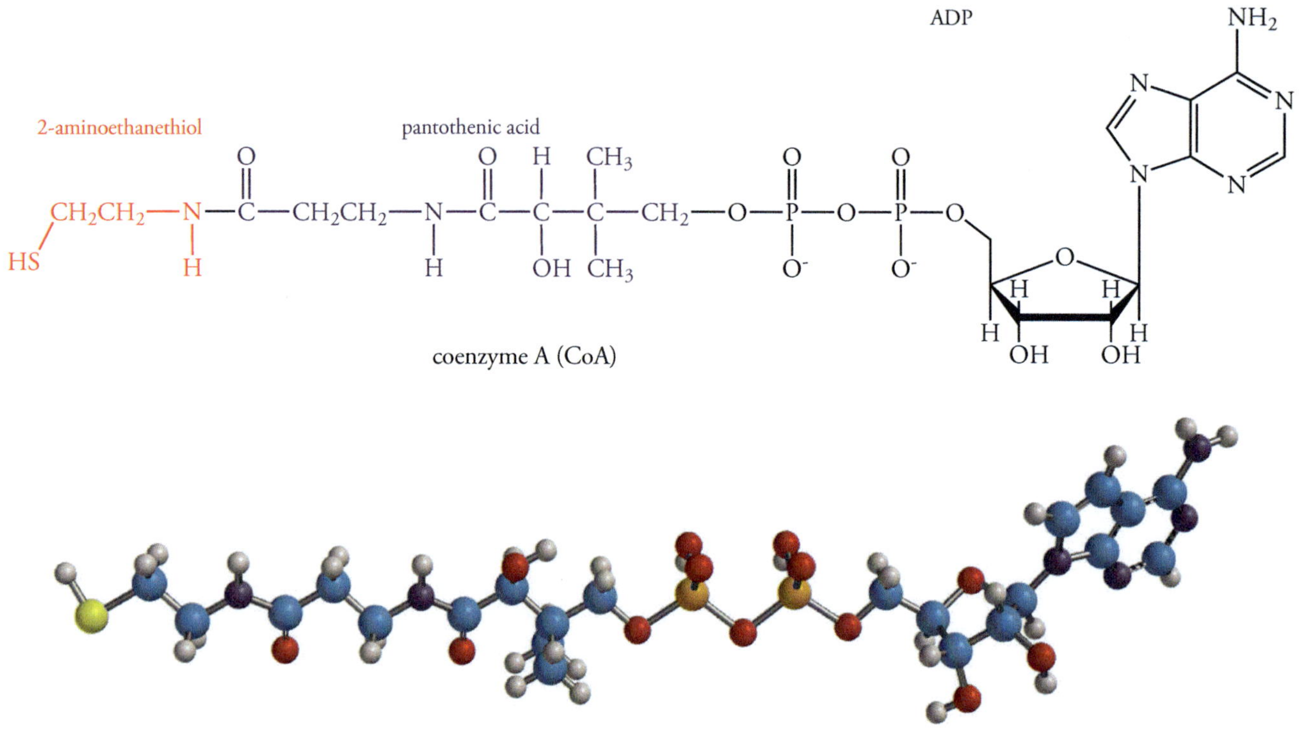

2-aminoethanethiol

pantothenic acid

coenzyme A (CoA)

11.7 REDUCTION OF ACYL DERIVATIVES

In Chapter 10 we saw that aldehydes (or ketones) can be reduced by either sodium borohydride or lithium aluminum hydride. These reactions involve nucleophilic attack of hydride ion on the carbonyl carbon atom. A similar process occurs with acyl derivatives. However, there are subsequent steps in the reaction mechanisms that occur after nucleophilic attack by the hydride ion. In this section, we will consider the reduction of esters, carboxylic acids, and acid chlorides. We will discuss the reduction of amides in Chapter 12.

Reduction of Esters

The reduction of esters requires the strong reducing agent lithium aluminum hydride. The milder reagent sodium borohydride does not reduce esters. We recall that lithium aluminum hydride does not reduce carbon-carbon double bonds.

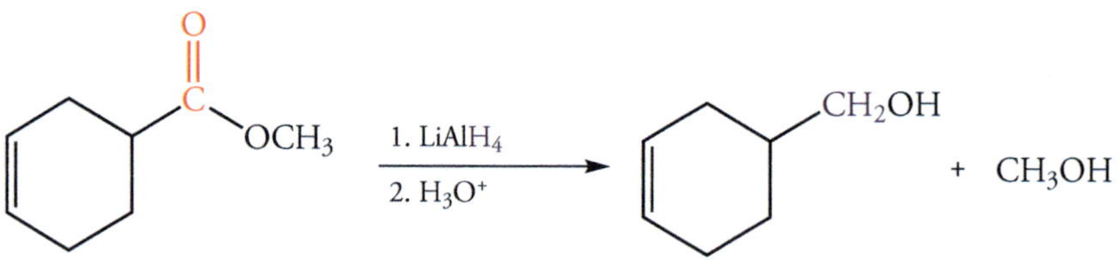

$$\text{1. LiAlH}_4 \quad \text{2. H}_3\text{O}^+$$

Note that the alcohol portion of the ester is a by-product of the reaction. The esters that are typically reduced by lithium aluminum hydride contain a low molecular weight alkyl group introduced in the conversion of an acid to an ester. The alcohol obtained by reduction of the acid portion of the ester is easily separated from the low molecular weight, water-soluble alcohol.

The mechanism of the reduction of an ester is pictured as nucleophilic attack of a one molar equivalent of a "hydride" ion on the carbonyl carbon atom. However, the aluminum atom participates in the reaction and is bonded to the oxygen atom. For simplicity, the structures shown have eliminated the aluminum group. Attack by the hydride ion produces a tetrahedral intermediate that loses an alkoxide ion. The resulting aldehyde is reduced even more rapidly than the original ester by a second molar equivalent of hydride ion. In total, two molar equivalents of hydride ion or one-half molar equivalent of lithium aluminum hydride are required for the reduction.

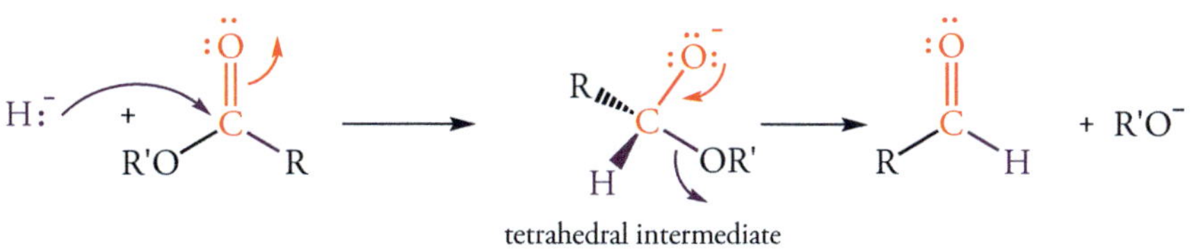

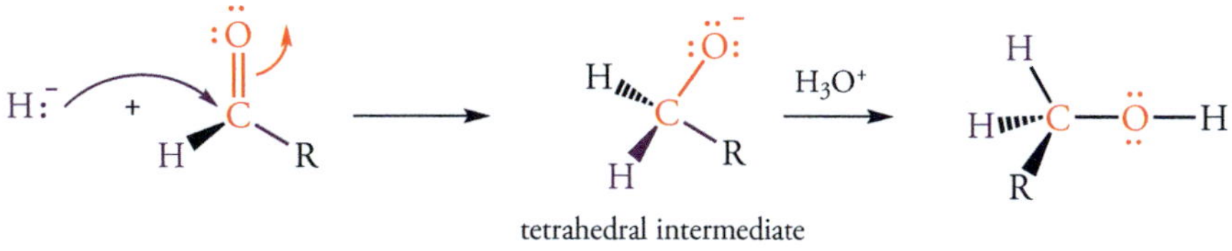

tetrahedral intermediate

tetrahedral intermediate

Reduction of Acids

Carboxylic acids are reduced by lithium aluminum hydride but not by the milder reagent sodium borohydride. As in the case of the reduction of esters, carboxylic acids are always completely reduced to alcohols; no intermediate aldehyde can be isolated.

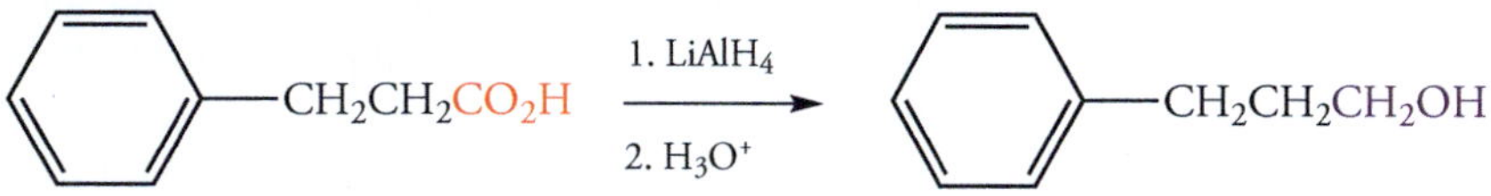

Lithium aluminum hydride is a strong base, and it reacts with the acidic proton of the carboxylic acid to give hydrogen gas. One molar equivalent of hydride ion is used in the reaction. The carboxylate salt is then reduced, requiring an additional two molar equivalents of hydride ion. Therefore, reduction of a carboxylic acid requires a total of three molar equivalents of hydride ion.

Reduction of Acid Chlorides

Esters or carboxylic acids cannot be reduced to aldehydes by hydride reducing agents because aldehydes are more reactive than esters or carboxylic acids. Acid chlorides are more reactive than esters toward nucleophilic hydride ion. As a result, acid chlorides are more rapidly reduced than esters. However, lithium aluminum hydride is such a strong reducing agent that acid chlorides are still reduced all the way to primary alcohols. The milder reducing agent lithium aluminum tri(*tert*-butoxy) hydride reacts with acid chlorides but much more slowly with aldehydes, which can therefore be isolated.

11.8 ESTERS AND ANHYDRIDES OF PHOSPHORIC ACID

Phosphoric acid, pyrophosphoric acid, and triphosphoric acid occur in living cells as esters. Just as alcohols react with carboxylic acids, they also react with phosphoric acid, pyrophosphoric acid, and triphosphoric acid.

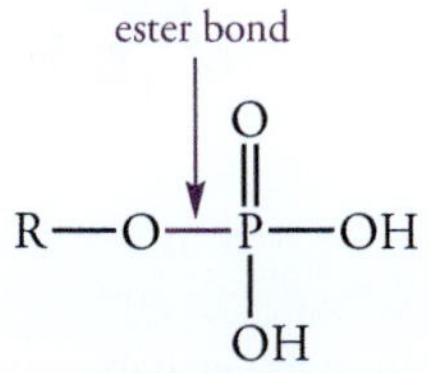

alkyl phosphate
(an ester of phosphoric acid)

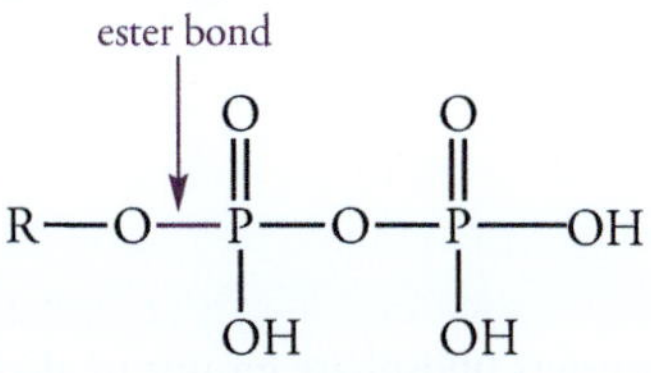

alkyl diphosphate
(an ester of pyrophosphoric acid)

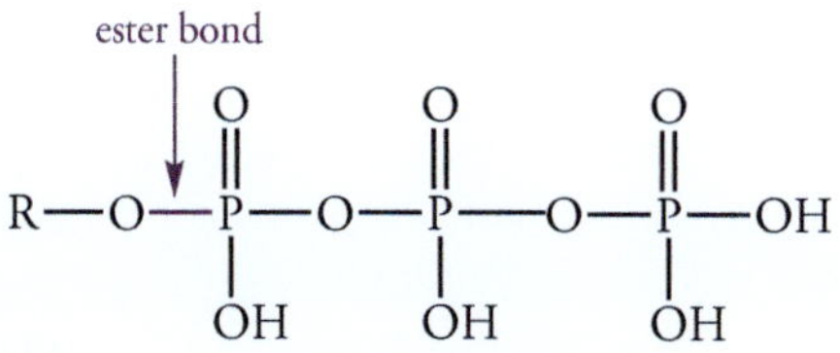

alkyl triphosphate
(an ester of triphosphoric acid)

Note that these esters are also acids. At physiological pH, the protons in the $-OH$ groups of these esters are ionized. That is why phosphoric acid derivatives are soluble in the aqueous environment of cells.

The esters are also acid anhydrides of phosphoric acid. The oxygen atom between the phosphorus atoms is part of the anhydride bond. These compounds are analogs of acid anhydrides of carboxylic acids.

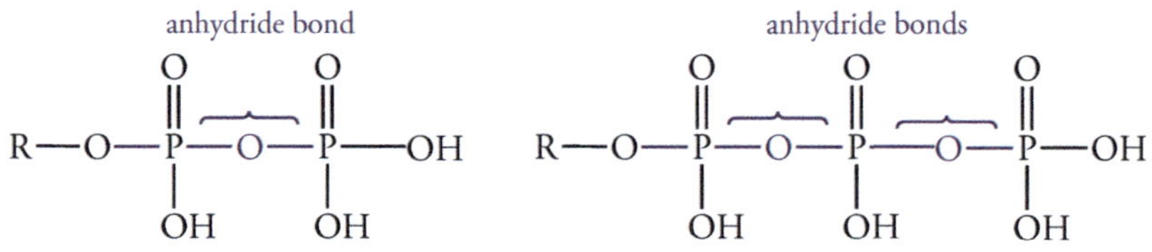

The anhydride bonds of diphosphates and triphosphates are hydrolyzed in many biological reactions. Adenosine triphosphate (ATP), a triphosphate ester of adenosine, stores some of the energy released in the degradation of carbohydrates, fats, and amino acids. When the terminal anhydride bond of ATP is hydrolyzed, adenosine diphosphate (ADP) is produced and 30.5 kJ of energy are released per mole of ADP produced. When the anhydride bond near the adenosine portion of the molecule is hydrolyzed, adenosine monophosphate (AMP) and pyrophosphoric acid are produced, and 36 kJ of energy are released per mole of AMP produced.

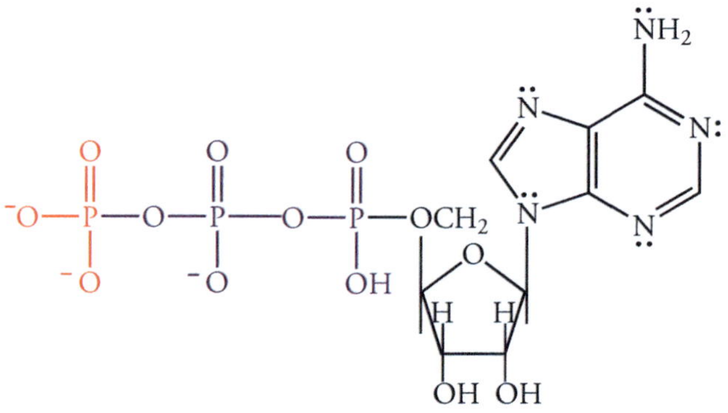

adenosine-5'-triphosphate (ATP^{3-})

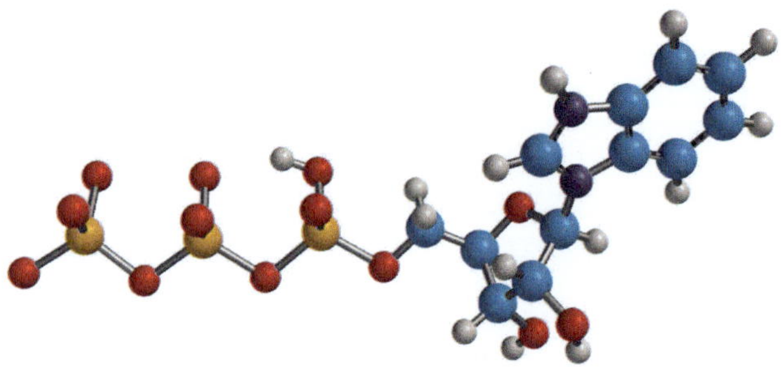

ATP transfers phosphate groups to alcohols in many enzyme-catalyzed metabolic reactions. For example, as part of the metabolism of glucose the hydroxyl group at C-6 is phosphorylated in a reaction with ATP to give glucose 6-phosphate and ADP.

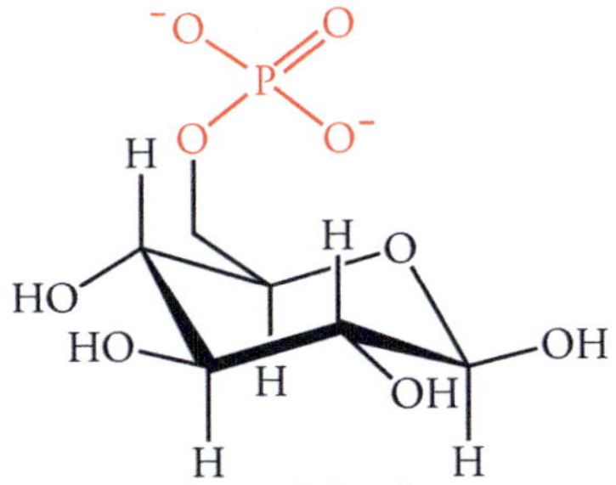

glucose 6-phosphate

Problem 11.9

The antibiotic chloramphenicol has a bitter taste. Its palatability for children is improved by using a suspension of the palmitate ester. The drug is orally administered, and the ester is enzymatically hydrolyzed in the intestine. Given the structure of the ester, write the structure of the antibiotic.

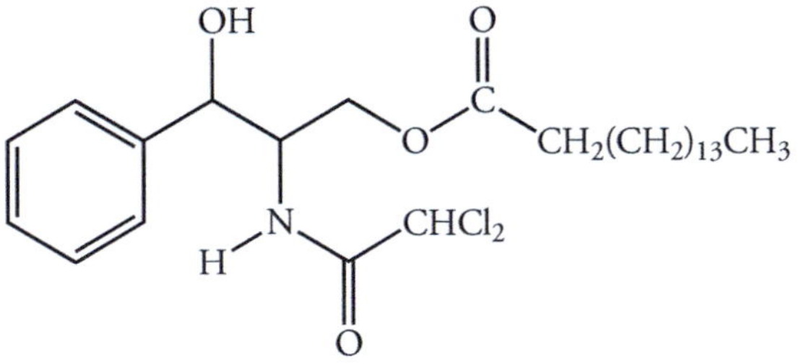

Solution

First, locate the ester functional group by examining the carbonyl carbon atoms. The carbonyl group at the bottom of the structure is bonded to a nitrogen atom. This is an amide group, a very stable acyl functional group. The carbonyl carbon atom at the top of the structure is bonded to an oxygen atom. This is the ester functional group. The carbon atom chain to the right of the carbonyl group is part of the acid. A total of 16 carbon atoms are contained in palmitic acid. Chloramphenicol is bonded in the ester through its primary hydroxyl group.

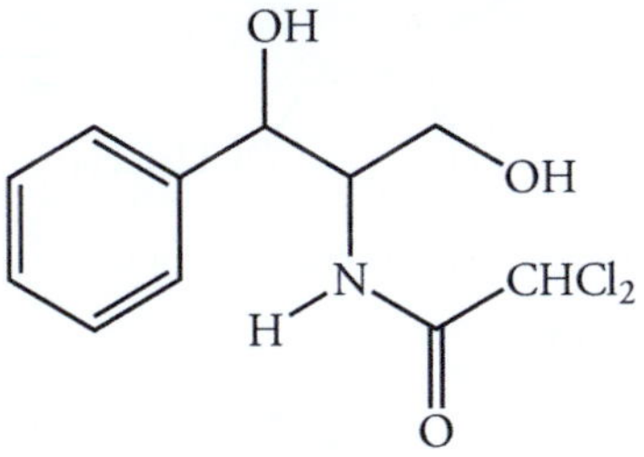

chloramphenicol

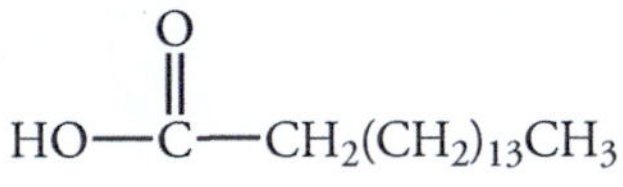

palmitic acid

Problem 11.10

Hydrolysis of diloxanide furanoate in the body is required for it to be effective against intestinal amebiasis. What is the acid component of the drug? Based on the name of the drug, what is the name of the acid?

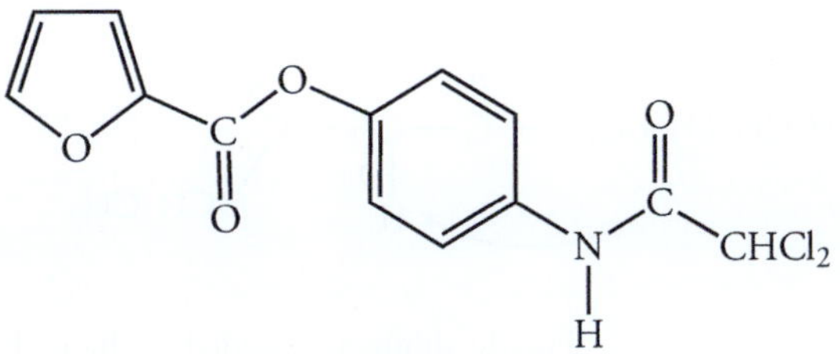

diloxanide

11.9 THE CLAISEN CONDENSATION

We recall that the hydrogen atom bonded to the α-carbon atom in an aldehyde or ketone is acidic (pK$_a$ = 20). As a result, carbonyl compounds with an α-carbon atom undergo condensation reactions in which the α-carbon atom of one molecule bonds to the carbonyl carbon atom of another molecule. The result is an aldol (Section 10.11).

Two molecules of an ester also react with each other in the presence of a base such as ethoxide ion to produce a condensation product. The reaction, which produces a β-keto ester, is called the **Claisen condensation**.

ethyl acetate   ethyl acetoacetate

The mechanism of the Claisen condensation resembles that of the aldol condensation, but there are important differences. A full equivalent of base is required for the Claisen condensation rather than the catalytic amount required for the aldol condensation. There are several equilibria in the sequence of reactions, and some are not favorable. The entire sequence is made favorable by driving the final step to completion by an acid-base reaction that requires a molar equivalent of base.

Like aldehydes, esters have an acidic α-hydrogen atom. The acid dissociation constants of esters (pK$_a$ = 25) are about 10^5 times smaller than those of aldehydes. The pK$_a$ of ethanol is 16. Thus, the reaction of sodium ethoxide with an ester (K$_a$ = 10^{-9}) produces only a small amount of the enolate at equilibrium.

weaker base   weaker acid   stronger base   stronger acid

The conjugate base of the ester reacts with another molecule of ester to form a carbon-carbon bond. The addition product is the conjugate base of a hemiacetal, which loses an alkoxide to give a β-keto ester.

β-Keto esters have pK$_a$ values around 11 because the negative charge of the conjugate base can be delocalized over both carbonyl groups. Ethanol is a weaker acid than β-keto esters, and, therefore, ethoxide ion essentially completely removes a proton from the product of the Claisen condensation. This final step drives the overall sequence of reactions to completion.

Finally, dilute acid added at the end of the reaction converts the conjugate base of the β-keto ester to the product.

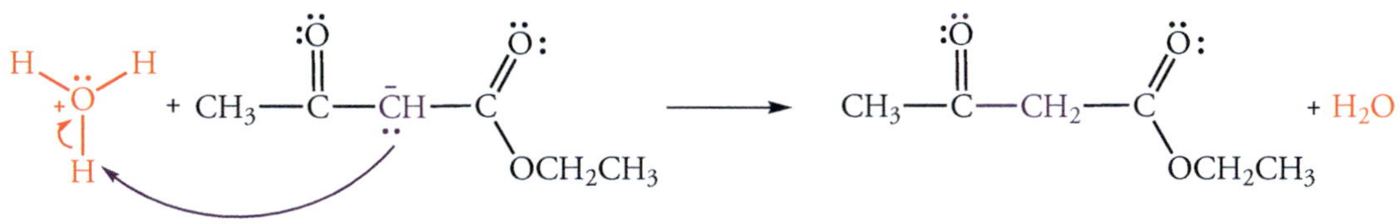

$$\text{H}-\overset{\text{H}}{\underset{\text{H}}{\overset{+}{\text{O}}}} \; + \; \text{CH}_3-\overset{:\overset{..}{\text{O}}:}{\text{C}}-\overset{..}{\text{CH}}-\overset{\overset{..}{\text{O}}:}{\text{C}}-\text{OCH}_2\text{CH}_3 \longrightarrow \text{CH}_3-\overset{:\overset{..}{\text{O}}:}{\text{C}}-\text{CH}_2-\overset{\overset{..}{\text{O}}:}{\text{C}}-\text{OCH}_2\text{CH}_3 \; + \; \text{H}_2\text{O}$$

Biochemical Claisen Condensations of Thioesters

A condensation of acetyl coenzyme A and oxaloacetic acid occurs in the citric acid cycle, in which an ester reacts with a ketone. The α-carbon atom of acetyl coenzyme A bonds to the carbonyl carbon atom of oxaloacetic acid in a Claisen condensation reaction. In this reaction, an acetyl group is transferred to oxaloacetic acid to form a compound that is subsequently hydrolyzed to citric acid. In the reactions shown below, the carboxylic acid groups are shown as carboxylate anions.

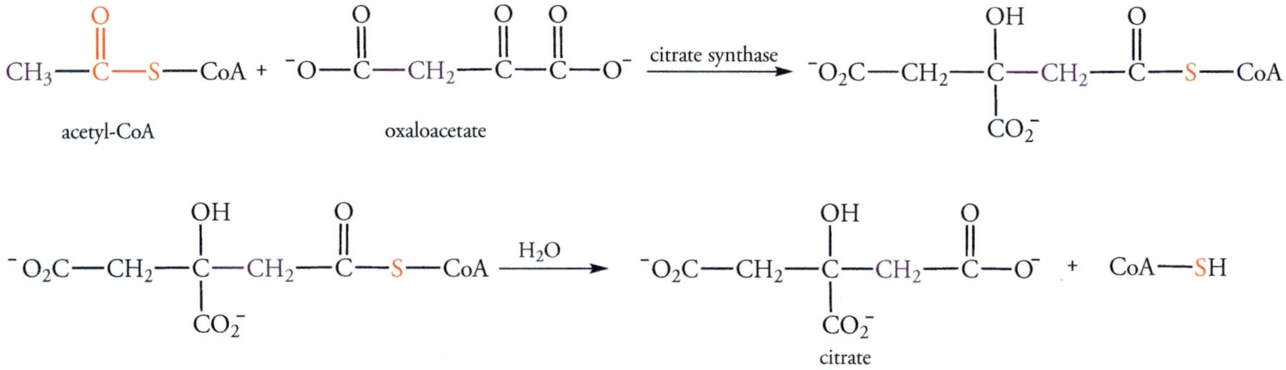

Most of the acetyl coenzyme A produced in this and other metabolic reactions reacts with oxaloacetic acid to form citric acid. However, in certain illnesses, such as diabetes, the metabolism of fats predominates over the metabolism of carbohydrates. When there is not enough oxaloacetic acid to react with all of the acetyl coenzyme A being produced, acetyl coenzyme A reacts with itself in a Claisen condensation.

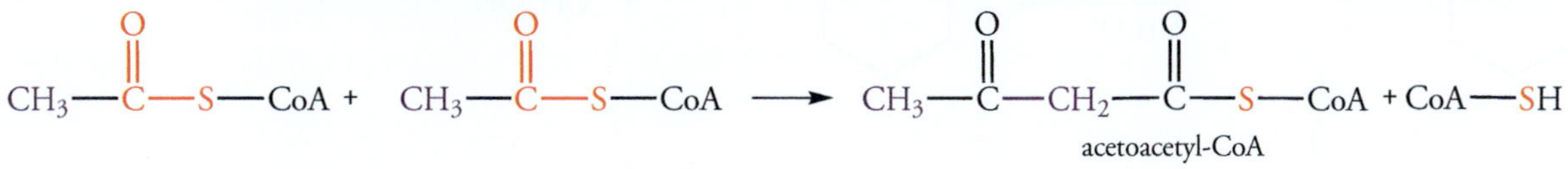

Hydrolysis of the resulting β-keto thioester yields acetoacetic acid (3-ketobutanoic acid). Subsequent reactions produce 3-hydroxybutanoic acid and acetone, which are collectively called ketone bodies. Detection of these compounds in the urine indicates diabetes.

SUMMARY OF REACTIONS

1. Synthesis of Carboxylic Acids by Oxidative Methods (Section 11.5)

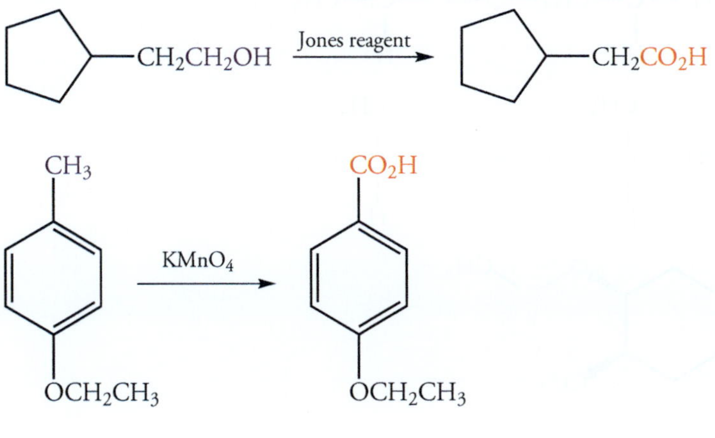

2. Synthesis of Carboxylic Acids from Haloalkanes (Section 11.5)

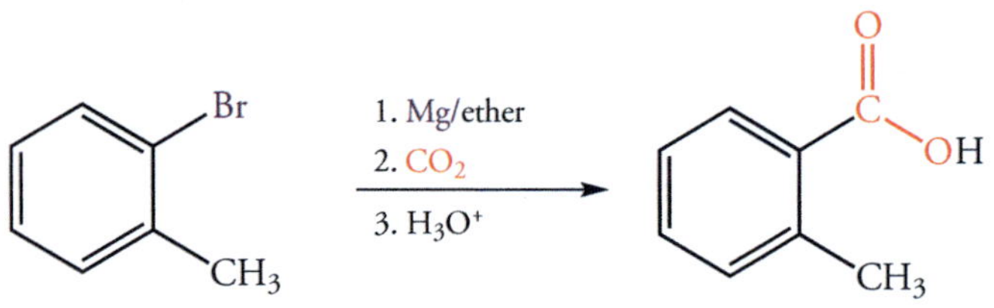
Br
CH3
1. Mg/ether
2. CO2
3. H3O+
O
C
OH
CH3

o-bromotoluene o-methylbenzoic acid

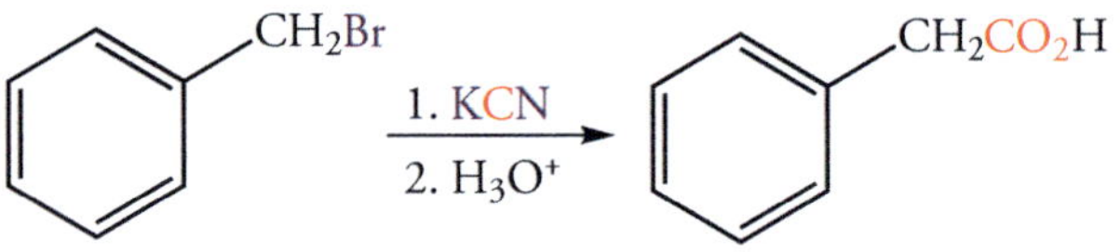
CH2Br
1. KCN
2. H3O+
CH2CO2H

3. Synthesis of Acyl Halides (Section 11.6)

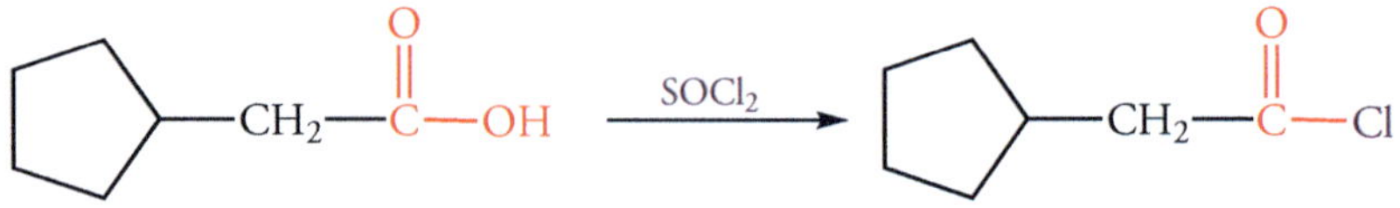
O
CH2—C—OH
SOCl2
O
CH2—C—Cl

4. Reduction of Carboxylic Acids and Acid Derivatives (Section 11.7)

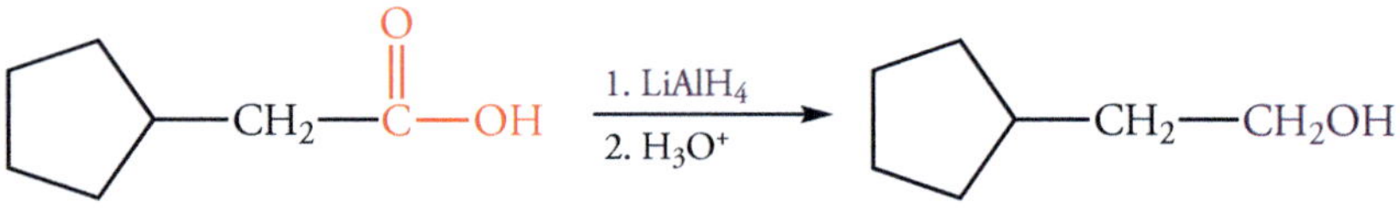
O
CH2—C—OH
1. LiAlH4
2. H3O+
CH2—CH2OH

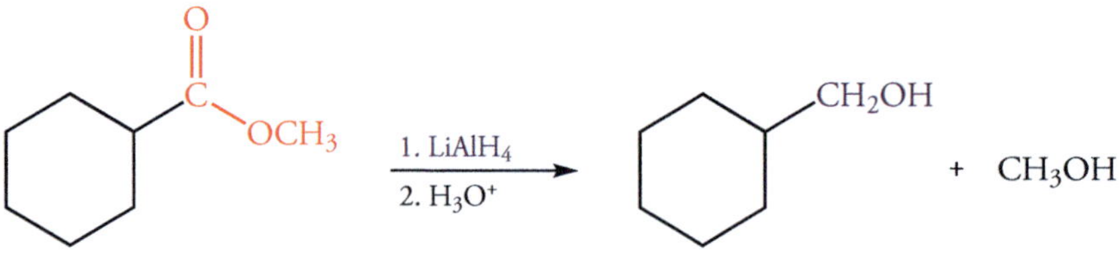
O
C
OCH3
1. LiAlH4
2. H3O+
CH2OH
+ CH3OH

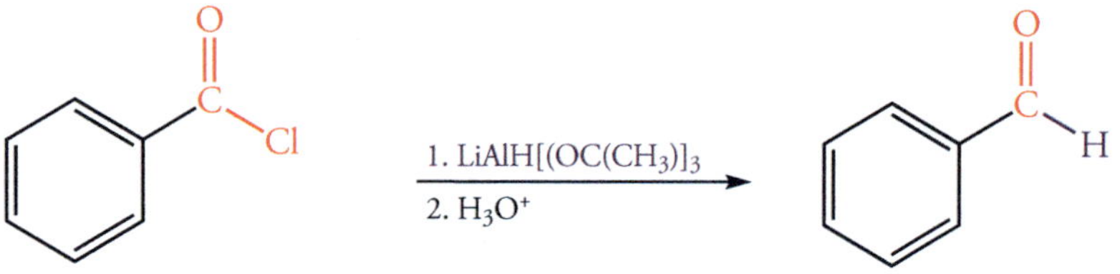
O
C
Cl
1. LiAlH[(OC(CH3)]3
2. H3O+
O
C
H

5. Synthesis of Esters (Sections 11.6 and 11.8)

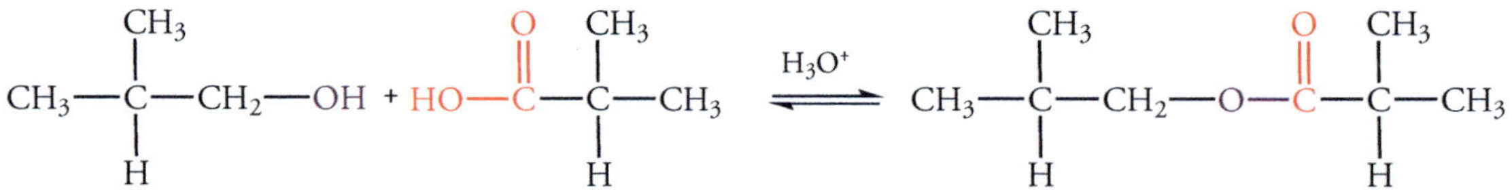
CH3
CH3—C—CH2—OH
H
+ HO—C—C—CH3
O
CH3
H
H3O+
CH3
CH3—C—CH2—O—C—C—CH3
H
O
CH3
H

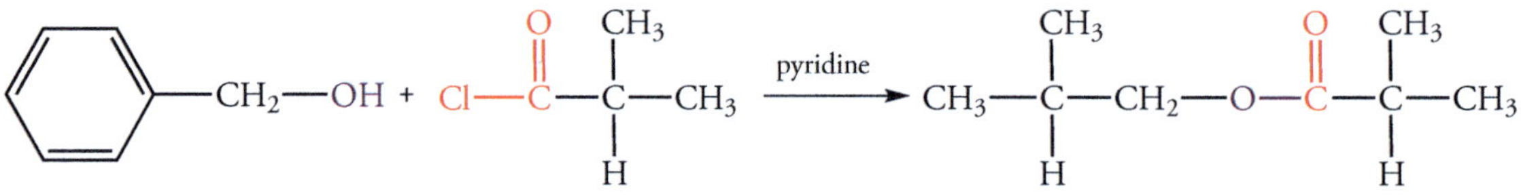
CH2—OH
+ Cl—C—C—CH3
O
CH3
H
pyridine
CH3
CH3—C—CH2—O—C—C—CH3
H
O
CH3
H

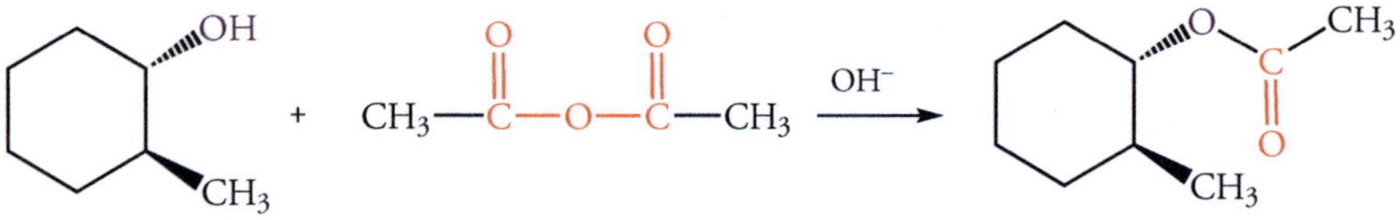
OH
CH3
+ CH3—C—O—C—CH3
O
O
OH–
O
C
O
CH3
CH3

6. Claisen Condensation (Sections 11.9)

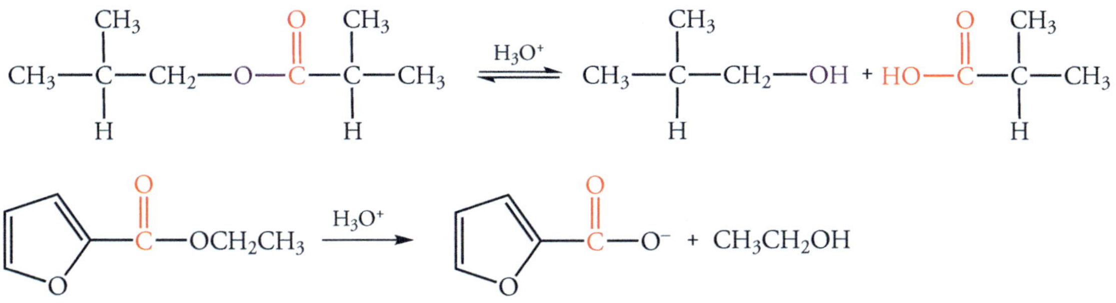

7. Claisen Condensation (Sections 11.9)

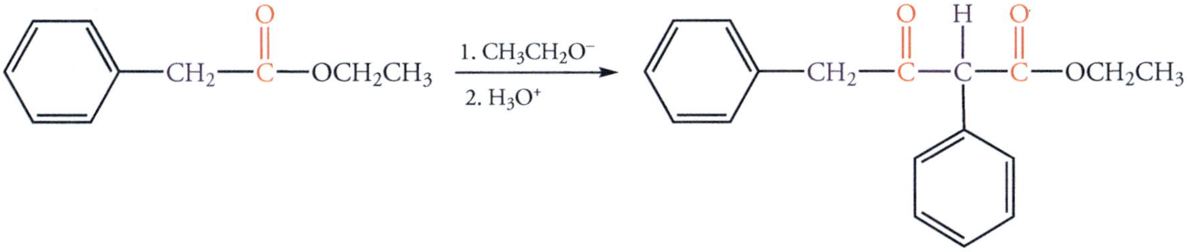

EXERCISES
Nomenclature

11.1 Give the common name for each of the following acids:
(a) $CH_3CH_2CO_2H$ (b) $CH_3(CH_2)_4CO_2H$ (c) $CH_3(CH_2)_{16}CO_2H$

11.2 Draw the structure of each of the following esters:
(a) octyl acetate (b) ethyl butyrate (c) propyl valerate

11.3 Give the IUPAC name for each of the following esters:

(a) $H-\overset{O}{\overset{\|}{C}}-O-CH_2CH_3$ (b) $CH_3CH_2CH_2-\overset{O}{\overset{\|}{C}}-O-CH_3$ (c) $CH_3-\overset{O}{\overset{\|}{C}}-O-CH_2(CH_2)_6CH_3$

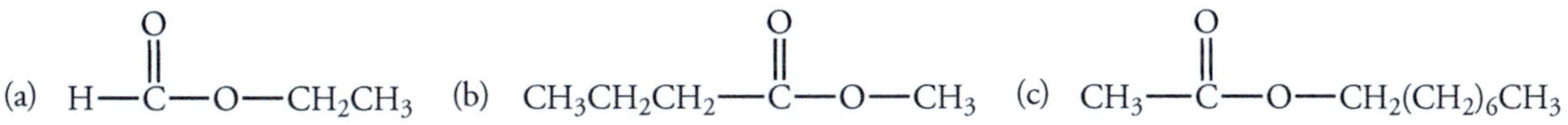

11.4 Give the IUPAC name for each of the following acids:

(a) $CH_3\overset{Cl}{\underset{|}{C}}HCH_2CO_2H$ (b) $Br\overset{CH_3}{\underset{|}{C}}HCO_2H$ (c) $CH_3\overset{CH_3}{\underset{|}{C}}HCHCH_2CO_2H$ with Br substituent

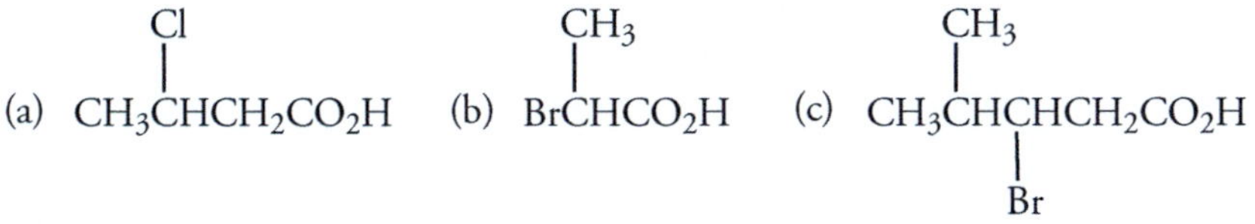

11.5 Give the IUPAC name for each of the following acids:

(a)

(b) $-CH_2CH_2CO_2H$ (c) CH_3O CO_2H

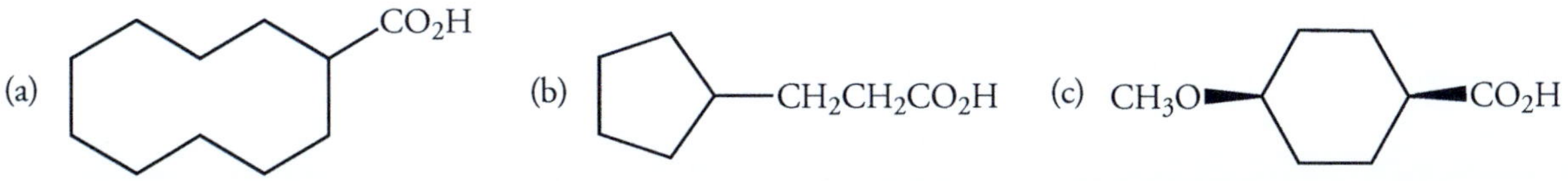

11.6 Give the IUPAC name for each of the following esters:

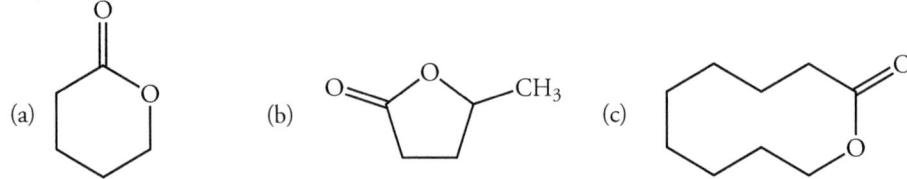

11.7 The IUPAC name of ibuprofen, the analgesic in Motrin, Advil, and Nuprin, is 2-(4-isobutylphenyl)propanoic acid. Draw its structure.

11.8 10-Undecenoic acid is the antifungal agent contained in Desenex® and Cruex®. Write the structure.

Cyclic Acyl Derivatives

11.9 The IUPAC names of lactones are derived by adding the term *lactone* at the end of the name of the parent hydroxy acid. Name each of the following compounds:

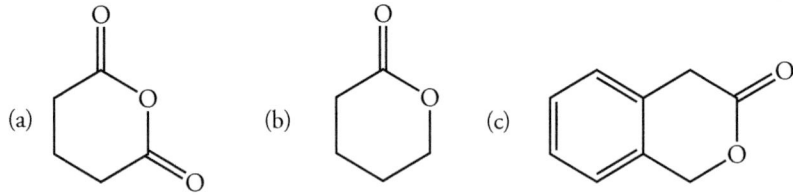

11.10 The IUPAC names of lactams are derived by adding the term lactam at the end of the name of the parent amino acid. Write the structure of each of the following lactams.
(a) 3-aminopropanoic acid lactam (b) 4-aminopentanoic acid lactam (c) 5-aminopentanoic acid lactam

11.11 Which of the following compounds are lactones?

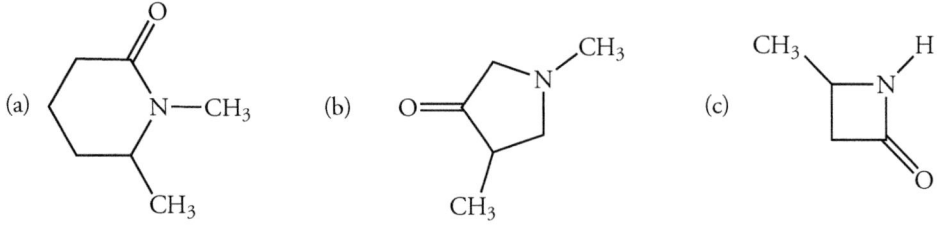

11.12 Which of the following compounds are lactams?

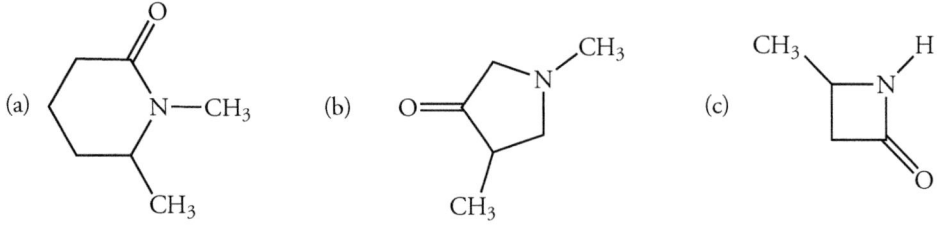

Molecular Formulas

11.13 What is the general molecular formula for a saturated carboxylic acid?

11.14 What is the general molecular formula for a saturated dicarboxylic acid?

11.15 How many isomeric acids have the molecular formula $C_4H_8O_2$?

11.16 How many isomeric esters have the molecular formula $C_4H_8O_2$?

Properties of Acids

11.17 Explain why 1-butanol is less soluble in water than butanoic acid.

11.18 Explain why adipic acid is much more soluble in water than hexanoic acid.

11.19 Explain why the boiling point of decanoic acid is higher than that of nonanoic acid.

11.20 Explain why the boiling point of 2,2-dimethylpropanoic acid (164 °C) is lower than that of pentanoic acid (186 °C).

11.21 The boiling points of methyl pentanoate and butyl ethanoate are 126 and 125 °C, respectively. Explain why the values are similar.

11.22 The boiling points of methyl pentanoate and methyl 2,2-dimethylpropanoate are 126 °C and 102 °C, respectively. Explain why these values differ.

Acidity of Carboxylic Acids

11.23 The K_a values of formic acid and acetic acid are 2.7×10^{-4} and 1.8×10^{-5}, respectively. Which compound is the stronger acid?

11.24 The pK_a values of acetic acid and benzoic acid are 4.74 and 4.19, respectively. Explain why these values differ.

11.25 The K_a of methoxyacetic acid is 2.7×10^{-4}. Explain why this value differs from the K_a of acetic acid (1.8×10^{-5}).

11.26 The pK_a values of benzoic acid and p-nitrobenzoic acid are 6.3×10^{-5} and 3.8×10^{-4}, respectively. Explain why these values differ.

11.27 The pK_a of benzoic acid is 4.2. The pK_a of probenecid, a drug used to treat gout, is 3.4. Explain why these values differ.

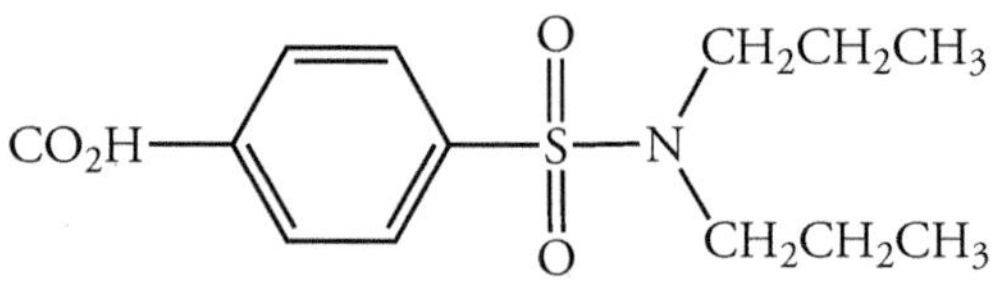

probenecid

11.28 Predict the pK_a of indomethacin, an anti-inflammatory agent.

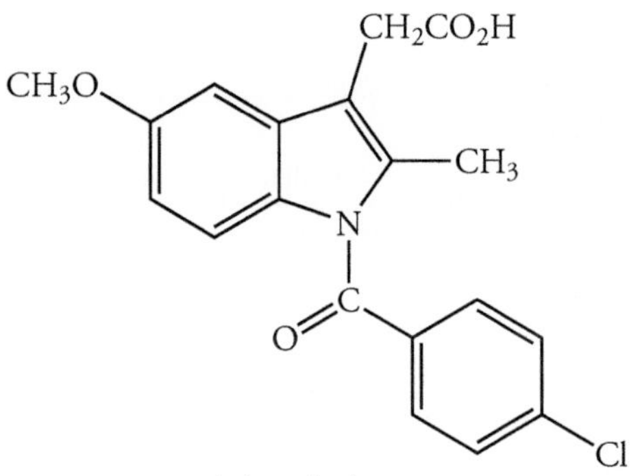

indomethacin

Nucleophilic Acyl Substitution

11.29 Indicate whether each of the following reactions will occur.

(a)

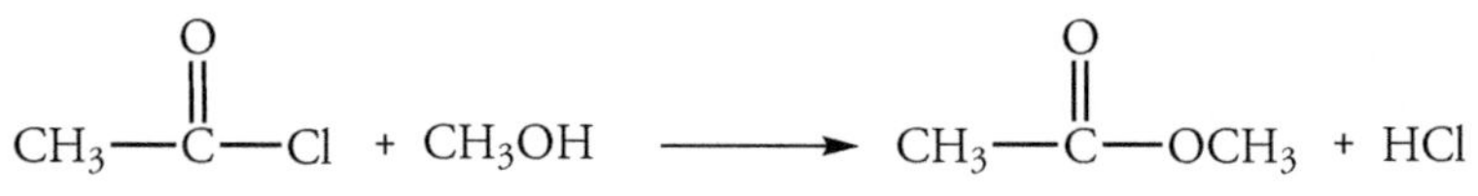

(b)

(c)

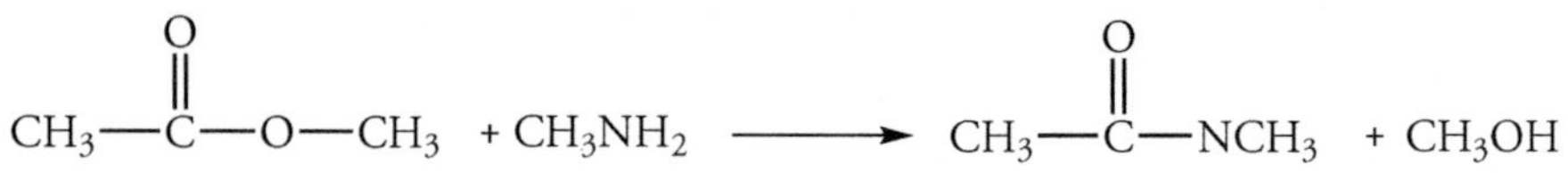

11.30 Indicate whether each of the following reactions will occur:

(a) 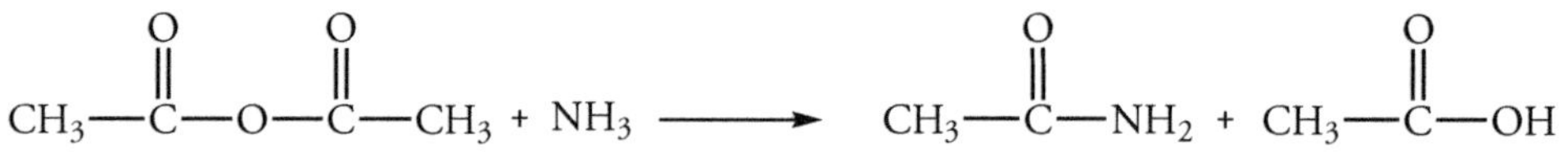 $\underset{\displaystyle CH_3}{} \overset{\displaystyle O}{\underset{\displaystyle \|}{C}} -O-\overset{\displaystyle O}{\underset{\displaystyle \|}{C}} -CH_3 + NH_3 \longrightarrow CH_3-\overset{\displaystyle O}{\underset{\displaystyle \|}{C}}-NH_2 + CH_3-\overset{\displaystyle O}{\underset{\displaystyle \|}{C}}-OH$

(b) 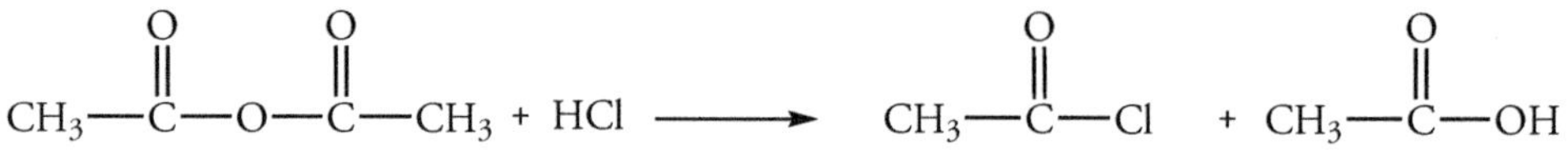 $CH_3-\overset{\displaystyle O}{\underset{\displaystyle \|}{C}}-O-\overset{\displaystyle O}{\underset{\displaystyle \|}{C}}-CH_3 + HCl \longrightarrow CH_3-\overset{\displaystyle O}{\underset{\displaystyle \|}{C}}-Cl + CH_3-\overset{\displaystyle O}{\underset{\displaystyle \|}{C}}-OH$

(c) $CH_3-\overset{\displaystyle O}{\underset{\displaystyle \|}{C}}-O-\overset{\displaystyle O}{\underset{\displaystyle \|}{C}}-CH_3 + CH_3OH \longrightarrow CH_3-\overset{\displaystyle O}{\underset{\displaystyle \|}{C}}-OCH_3 + CH_3-\overset{\displaystyle O}{\underset{\displaystyle \|}{C}}-OH$

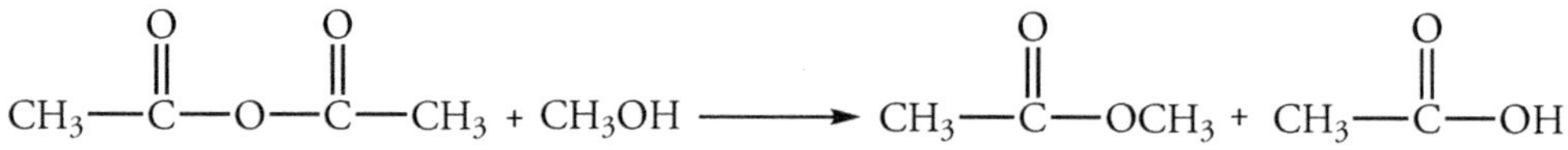

Reduction of Acyl Derivatives

11.31 Draw the structure of the product of each of the following reactions:

(a) 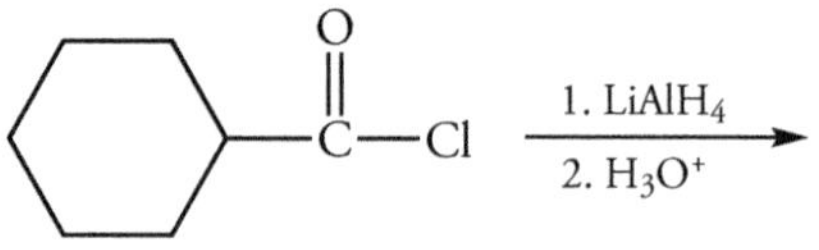 cyclohexyl—C(=O)—Cl $\xrightarrow[\text{2. } H_3O^+]{\text{1. LiAlH}_4}$

(b) 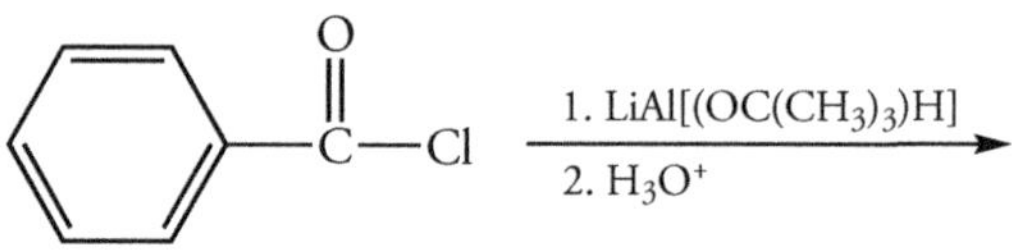 phenyl—C(=O)—Cl $\xrightarrow[\text{2. } H_3O^+]{\text{1. LiAl[(OC(CH}_3)_3)H]}}$

(c) $CH_3-\overset{\displaystyle CH_3}{\underset{\displaystyle H}{\overset{\displaystyle |}{\underset{\displaystyle |}{C}}}}-CH_2-\overset{\displaystyle O}{\underset{\displaystyle \|}{C}}-O-CH_3 \xrightarrow[\text{2. } H_3O^+]{\text{1. LiAlH}_4}$

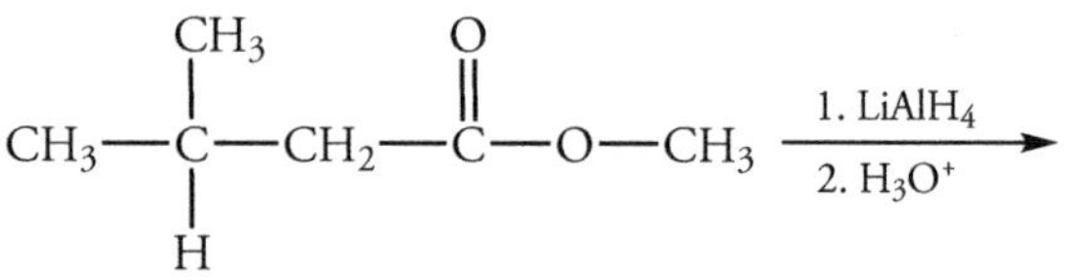

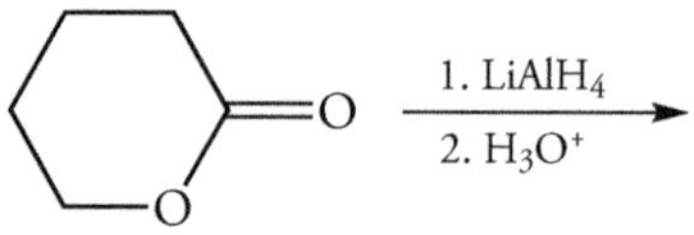

11.32 Draw the structure of the product of each of the following reactions:

(a) 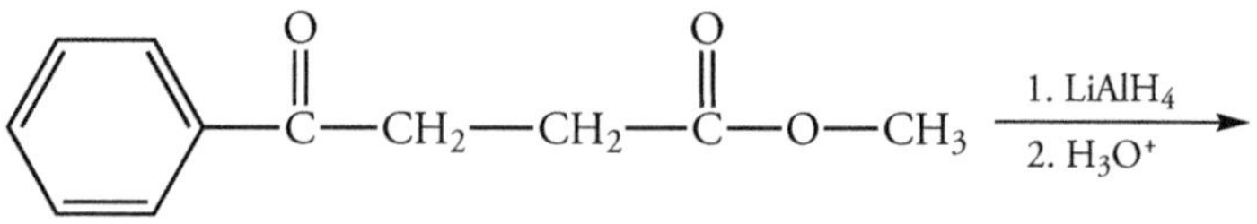 phenyl—C(=O)—CH₂—CH₂—C(=O)—O—CH₃ $\xrightarrow[\text{2. } H_3O^+]{\text{1. LiAlH}_4}$

(b) lactone (=O) $\xrightarrow[\text{2. } H_3O^+]{\text{1. LiAlH}_4}$

(c) phenyl—C(=O)—CH₂—CH₂—C(=O)—O—CH₃ $\xrightarrow[\text{CH}_3\text{CH}_2\text{OH}]{\text{NaBH}_4}$

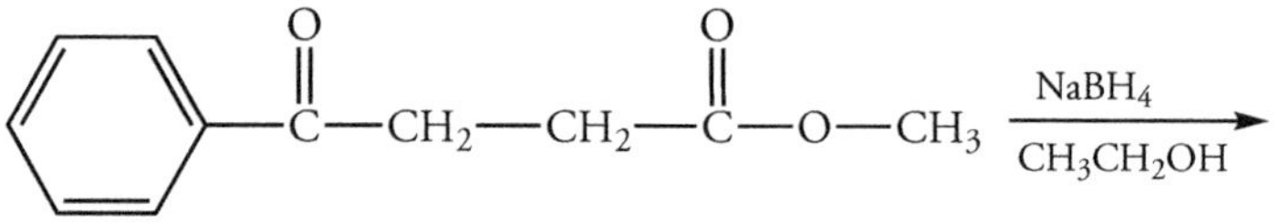

Chemical Reactions

11.33 Write the product of reaction of hexanoic acid with each of the following reagents:
(a) thionyl chloride   (b) methanol   (c) lithium aluminum hydride

11.34 Write the product of reaction of propanoyl chloride with each of the following compounds:
(a) 1-propanol   (b) 2-butanethiol   (c) benzylamine ($C_6H_5CH_2NH_2$)

11.35 Write the product of the reaction of butyrolactone with methylamine, CH_3NH_2.

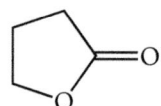

butyrolactone

12 AMINES AND AMIDES

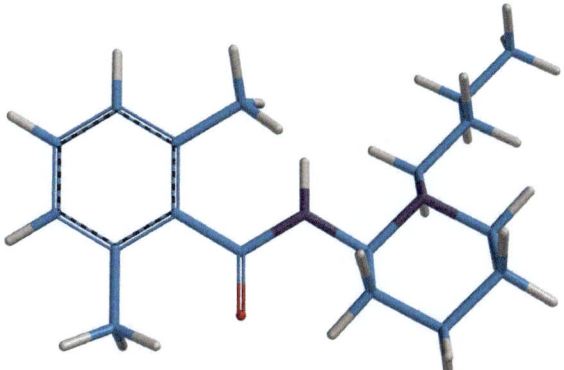

BUPIVACAINE (A LOCAL ANAESTHETIC)

12.1 ORGANIC NITROGEN COMPOUNDS

For most of this text we have concentrated on the compounds of carbon, hydrogen, and oxygen. We have paid less attention to compounds containing sulfur and nitrogen. Nitrogen is the fourth most common element in living systems after carbon, hydrogen, and oxygen. Organic compounds containing nitrogen are widely distributed in plants and animals and are necessary for life. Nitrogen is present in many vitamins and hormones. Nitrogen is essential in amino acids and proteins, in nucleotides and nucleic acids, and in scores of other cellular molecules. In addition, many nitrogen-containing compounds are important industrial products, including polymers such as nylon, many dyes, explosives, and pharmaceutical agents.

A nitrogen atom, which has five valence electrons, forms a total of three covalent bonds to carbon or hydrogen atoms in neutral compounds. A nitrogen atom in a functional group can form single, double, or triple bonds. We have discussed these functional groups in previous chapters. In this chapter we will focus on amines and amides, but we will also discuss other functional groups that are either the reactants required to form amines and amides or the products of their reactions.

$$R-\ddot{N}H_2 \qquad R-\overset{\overset{\displaystyle O}{\|}}{C}-\ddot{N}H_2 \qquad \overset{\displaystyle R}{\underset{\displaystyle R}{C}}=\ddot{N}-R \qquad R-C\equiv N:$$

an amine   an amide   an imine   a nitrile

Some amines affect the brain, spinal cord, and nervous system. These compounds include the neurotransmitters epinephrine, serotonin, and dopamine (Figure 12.1). Epinephrine, commonly called adrenaline, stimulates the conversion of stored glycogen into glucose. Serotonin is a hormone that causes sleep, and serotonin deficiency is responsible for some forms of mental depression. In Parkinson's disease, the dopamine concentration is low.

seratonin   dopamine   epinephrine

Principles of Organic Chemistry. http://dx.doi.org/10.1016/B978-0-12-802444-7.00012-4

Figure 12.1
Structures of
Neurotransmitters

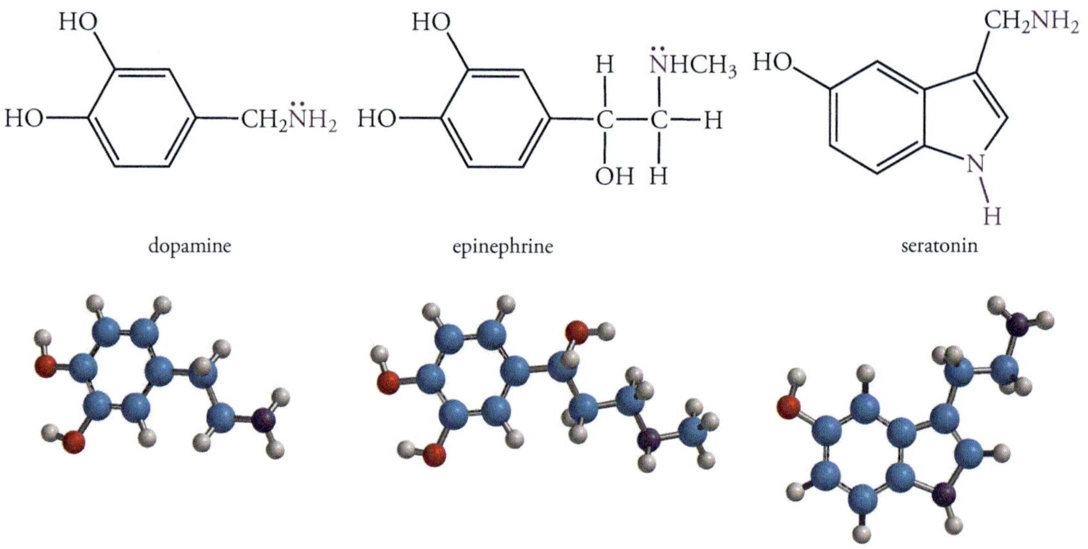

dopamine

epinephrine

seratonin

Proteins are made from nitrogen-containing molecules called α-amino acids. In proteins, each amine functional group of one α-amino acid is bonded to the carbonyl carbon of another α-amino acid in a chain of amino acyl groups that contains many amide bonds, or peptide bonds.

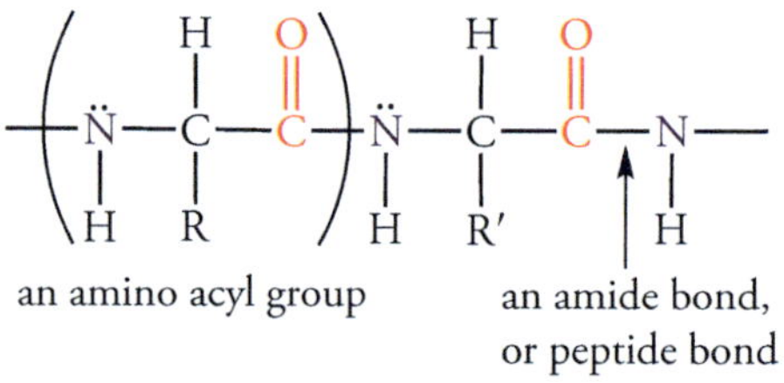

an amino acyl group

an amide bond, or peptide bond

Problem 12.1

Identify the nitrogen-containing functional groups in diazepam (Valium®).

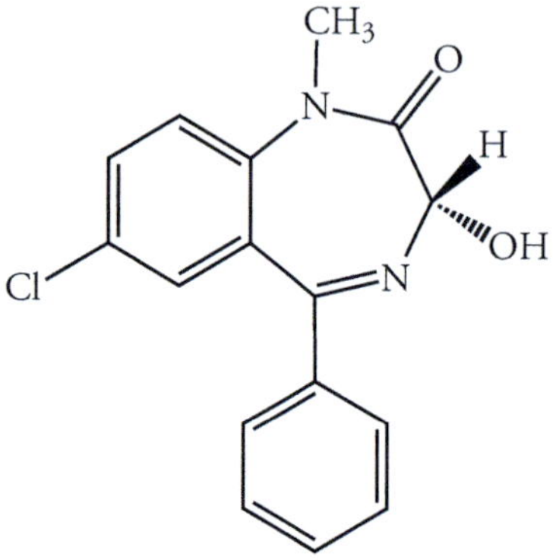

Solution

The nitrogen atom located at the top of this structure is bonded to a carbonyl carbon atom and is part of an amide group. The other nitrogen atom is bonded only to two carbon atoms—one by a double bond. This functional group is an imine.

12.2 BONDING AND STRUCTURE OF AMINES

We discussed the bonding and structure of simple amines in Chapter 1 (Figure 1.18). In the simplest amine, methylamine (CH_3NH_2), a methyl group has replaced one hydrogen atom of ammonia. The C—N—H and H—N—H bond, angles are approximately 112° and 106°, respectively, so methylamine has a pyramidal shape around the nitrogen atom. In methylamine and other amines, the nitrogen atom has five valence electrons in four sp³ hybrid orbitals. As expected from VSEPR theory, these orbitals point to the corners of a tetrahedron. Three are half-filled. They form three covalent bonds. The fourth orbital contains a pair of nonbonded electrons.

In the simplest amine, methylamine (CH_3NH_2), one hydrogen atom of ammonia has been replaced by a methyl group (Figure 12.2). The nitrogen atom of methylamine and other amines has five valence electrons in four sp^3 hybrid orbitals that are directed to the corners of a tetrahedron. Three of these orbitals are half-filled; the fourth contains a nonbonding pair of electrons that plays an important role in the chemical properties of amines.

**Figure 12.2
Structure of Methylamine**

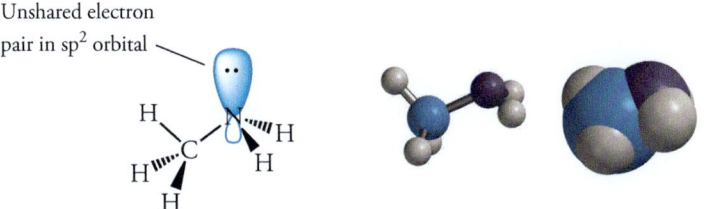

Unshared electron pair in sp^2 orbital

Perspective structural formula

Amines are classified by the number of alkyl (or aryl) groups attached to the nitrogen atom. Note that amines are not classified like alcohols. (The classification of alcohols is based on the number of groups attached to the carbon atom bearing the hydroxyl group.) For example, *tert*-butylamine has a *tert*-butyl group attached to an −NH_2 group. However, the amine is primary because only one alkyl group is bonded to the nitrogen atom. In contrast, *tert*-butyl alcohol is a tertiary alcohol because the carbon atom bonded to the −OH group is bonded to three alkyl groups. Trimethylamine is a tertiary amine because the nitrogen atom is bonded to three alkyl groups.

The nitrogen atom of an amine may be contained in a ring, a common feature of nitrogen compounds in nature. The simplest five- and six-membered nitrogen-containing heterocyclic compounds are pyrrolidine and piperidine.

Problem 12.2
Classify the nitrogen-containing functional groups in Mepivacaine®, a local anesthetic.

Mepivacaine ®

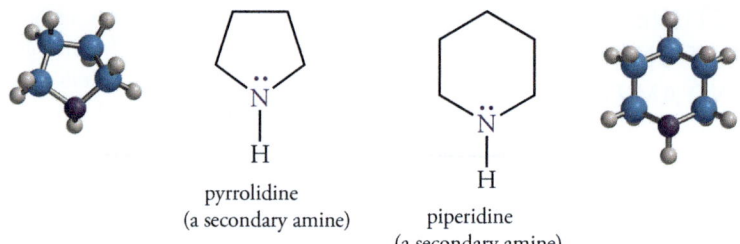

pyrrolidine
(a secondary amine)

piperidine
(a secondary amine)

Amides have an amino group or a substituted amino group bonded to a carbonyl carbon atom. The other two bonds of the nitrogen atom may be to hydrogen atoms, alkyl groups, or aryl groups. Amides are classified based on the number of carbon groups (including the acyl group) bonded to the nitrogen atom.

$$R-\overset{\overset{\ddot{O}:}{\parallel}}{C}-\ddot{N}H_2 \qquad R-\overset{\overset{\ddot{O}:}{\parallel}}{C}-\ddot{N}HR \qquad R-\overset{\overset{\ddot{O}:}{\parallel}}{C}-\ddot{N}R_2$$

primary amide   secondary amide   tertiary amide

The structures of amides resemble those of other carbonyl compounds: the three atoms bonded to carbon are in the same plane (Figure 12.3). The nitrogen atom of an amide has an unshared pair of electrons that is delocalized with the π electrons of the carbonyl group. An amide, such as formamide, can be represented by two resonance structures.

Contributing resonance structures of an amide

Thus, the bond between carbon and nitrogen has about 50% double-bond character, which restricts rotation about the carbon-nitrogen bond, which is therefore planar.

Figure 12.3
Bonding in an Amide

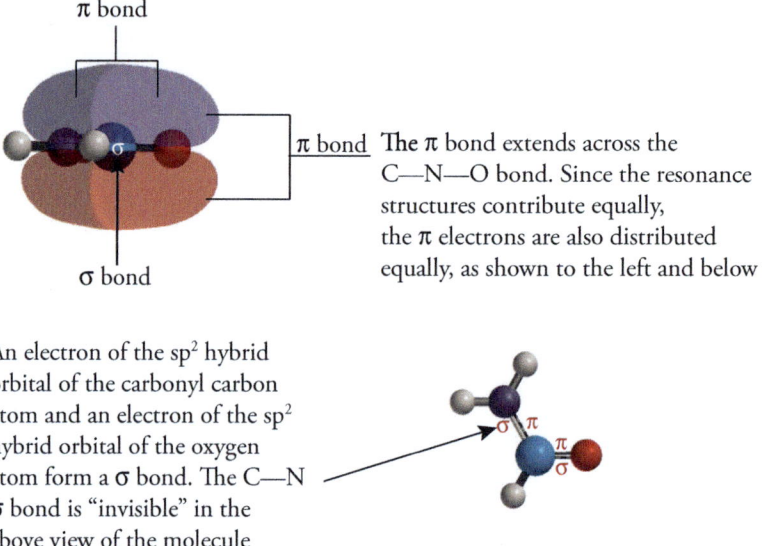

π bond

π bond The π bond extends across the C—N—O bond. Since the resonance structures contribute equally, the π electrons are also distributed equally, as shown to the left and below

σ bond

An electron of the sp² hybrid orbital of the carbonyl carbon atom and an electron of the sp² hybrid orbital of the oxygen atom form a σ bond. The C—N σ bond is "invisible" in the above view of the molecule

Problem 12.3
Classify Demerol®, a synthetic narcotic analgesic, as a primary, secondary, or tertiary amine.

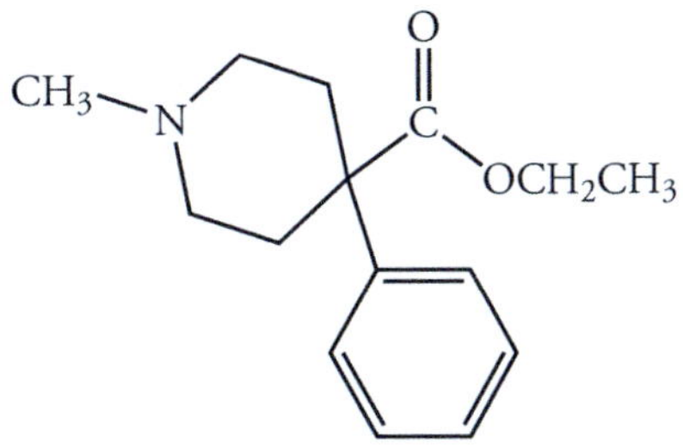

Demerol®

Solution

There are three carbon atoms bonded to the nitrogen atom. Two of the carbon atoms are in the heterocyclic ring. The third carbon atom bonded to the nitrogen atom is a methyl group. Demerol is a tertiary amine.

Problem 12.4

Classify the amine in the illicit drug methamphetamine (also colloquially known as "speed," and infamously as "crystal meth").

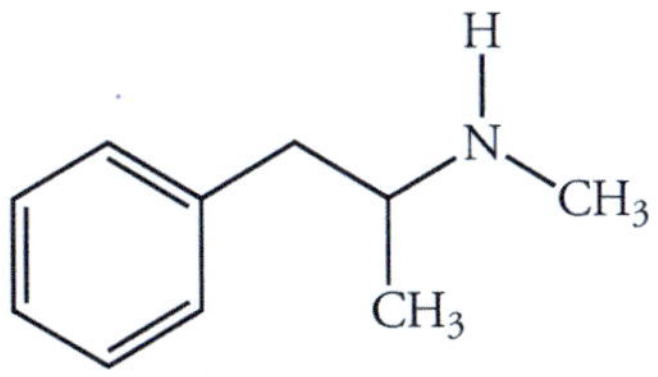

methamphetamine

12.4 NOMENCLATURE OF AMINES AND AMIDES

Like all organic compounds, many amines are known by their common names as well as by their IUPAC names. The IUPAC methods of naming amines and amides are related. The amine part of amides is named by the same method used for amines.

Amines

In common nomenclature, amines are described as *alkylamines*. The common name of a primary amine results from naming the alkyl group bonded to the amino group ($-NH_2$) and adding the suffix *-amine*. The entire name is written as one word. The common name for a secondary or tertiary amine is obtained by listing the alkyl groups alphabetically. When two or more identical alkyl groups are present, the prefixes *di-* and *tri-* are used.

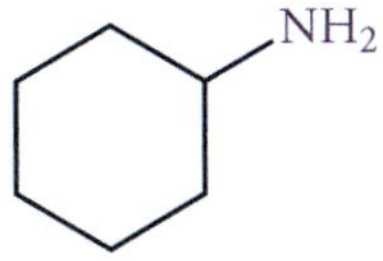

cyclohexylamine 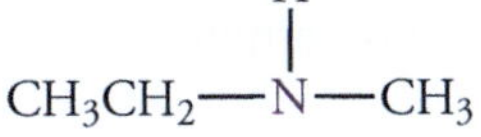   ethylmethylamine    diethylamine

γ-aminobutryic acid   β-(N,N-dimethylamino)caproic acid

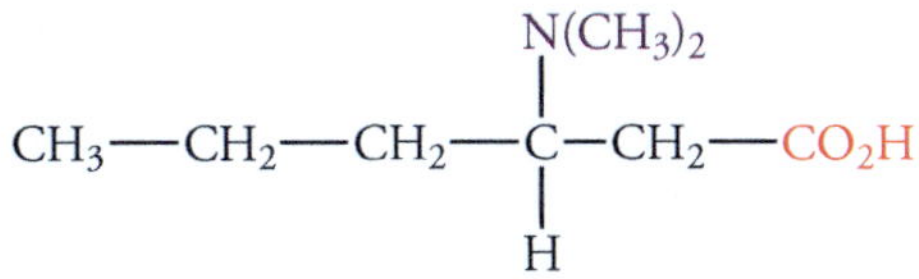

For more complex primary amines, the amino group is treated as a substituent. The nitrogen-containing substituent in complex secondary and tertiary amines is named as an N-alkylamino ($-NHR$) or N,N-dialkylamino ($-NRR'$) group. The upper case N indicates that the alkyl group is bonded to the nitrogen atom and not to the parent chain. The largest or most complicated group is used as the parent molecule.

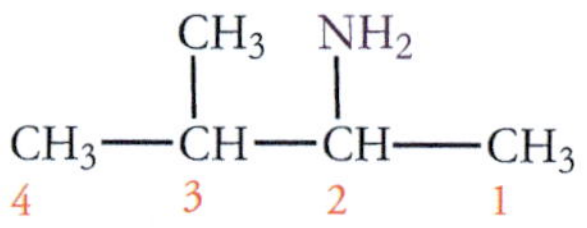

3-methyl-2-butanamine

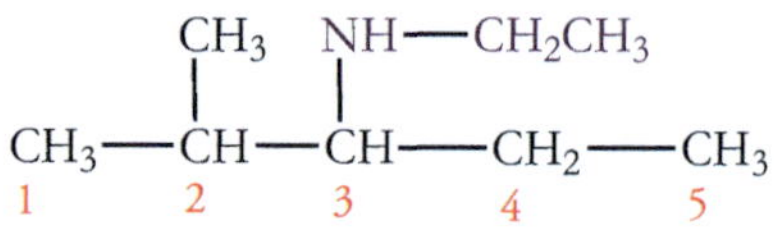

N-ethyl-2-methyl-3-pentanamine

Heterocyclic Aromatic Amines

Amines in which the nitrogen atom is part of an aromatic ring are called **heterocyclic aromatic amines**. In these compounds, the positions of substituents are established by using the numbering system indicated below. A nitrogen atom is assigned the number 1, and the direction of numbering provides the lowest possible numbers if the ring has more than one nitrogen atom.

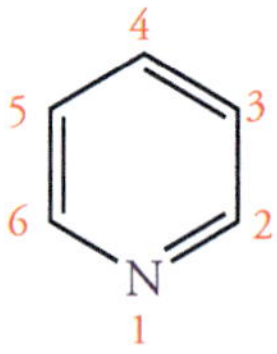

pyridine 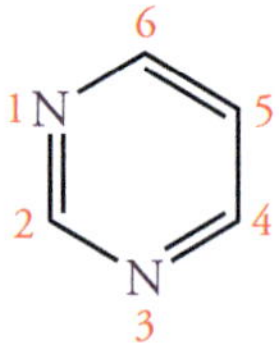   pyrimidine 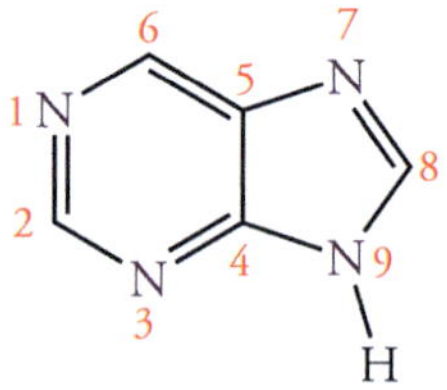   purine 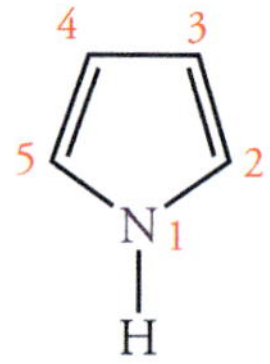   pyrrole 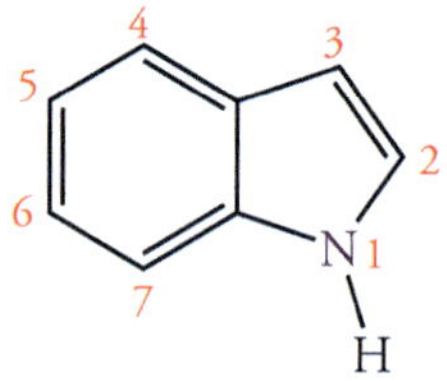   indole

Names of Amides

The common names of amides are formed by dropping the suffix -ic of the related acid and adding the suffix -amide. When there is a substituent on the nitrogen atom, the prefix N- followed by the name of the group bonded to nitrogen is attached to the name. Substituents on the acyl group are designated by Greek letters α, β, γ, and so on, as in the common names of carboxylic acids.

In the IUPAC system, the longest chain that contains the amide functional group is the parent. The final -e of the alkane is replaced by -amide. The substituents on nitrogen are indicated by the same method as in the common system. However, numbers are used for substituents on the parent chain.

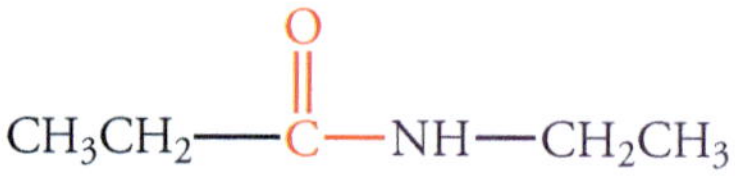

N-ethylpropanamide
(N-ethylpropionamide)

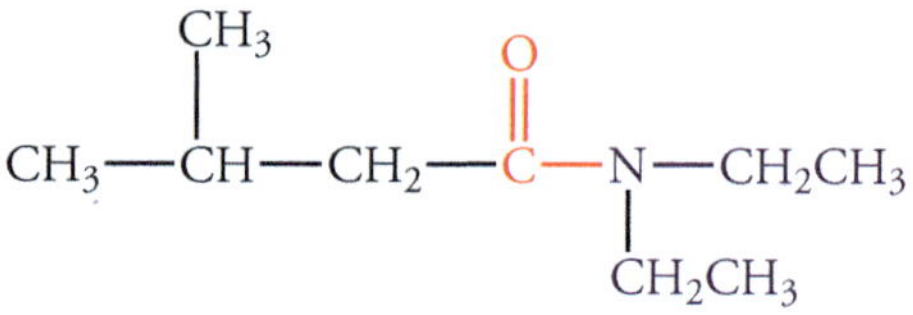

N,N-diethyl-3-methylbutanamide
(N,N-diethyl-β-methylbutyramide)

Amphetamines

There are recurring structures in groups of physiologically active compounds that contain nitrogen. One such structure is 2-phenylethanamine, which is required for binding at certain receptor sites that stimulate the central nervous system. The adrenal medulla produces the hormones epinephrine and norepinephrine, which contain a 2-phenylethanamine unit. Epinephrine, also known as adrenaline, makes glucose available to tissues under conditions of excitement. Norepinephrine maintains the muscle tone of blood vessels and hence controls blood pressure.

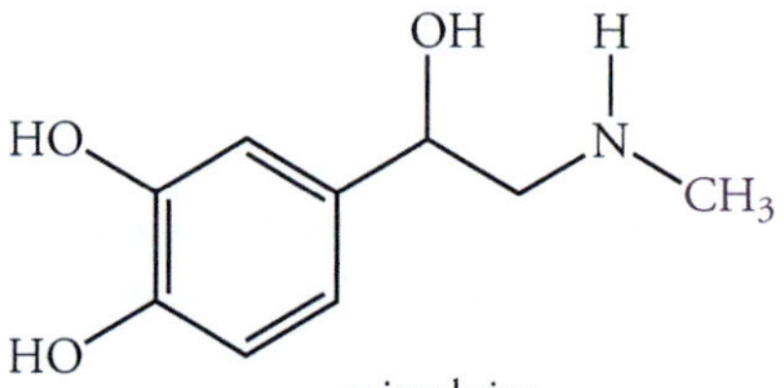

epinephrine

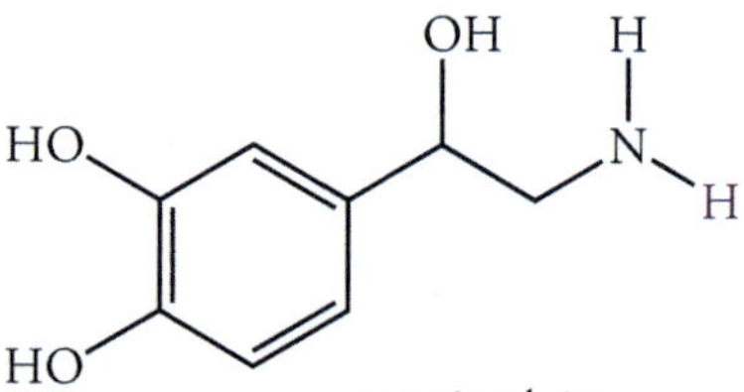

norepinephrine

Amphetamines are drugs known as "uppers" that mimic the action of the naturally occurring phenyl-lethanamines epinephrine and norepinephrine. Amphetamine (also known as benzedrine) has a structure similar to epinephrine. It is a moderate appetite suppressant and stimulates the cortex of the brain, which effectively counters fatigue. For that reason this illegal drug has been used by long-distance truck drivers and many other who want to ward off fatigue.

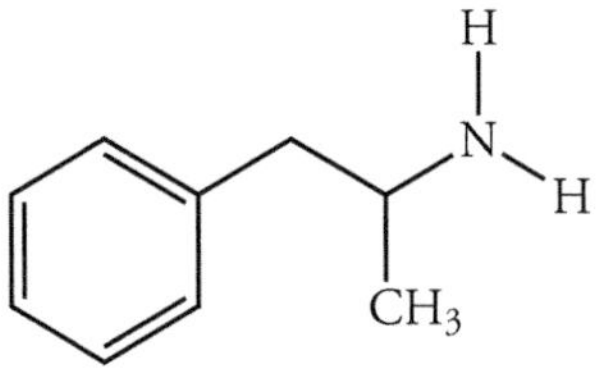

amphetamine

The structurally related illegal drugs are methamphetamine— "speed" or "crystal meth"— and methoxyamphetamine—known as "STP." These drugs produce severe psychological and physiological reactions. In addition, once the drug wears off, the user often "crashes" into a state of physical and mental exhaustion.

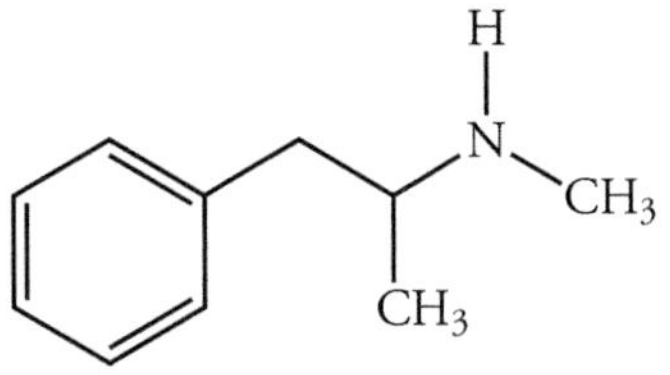

methamphetamine

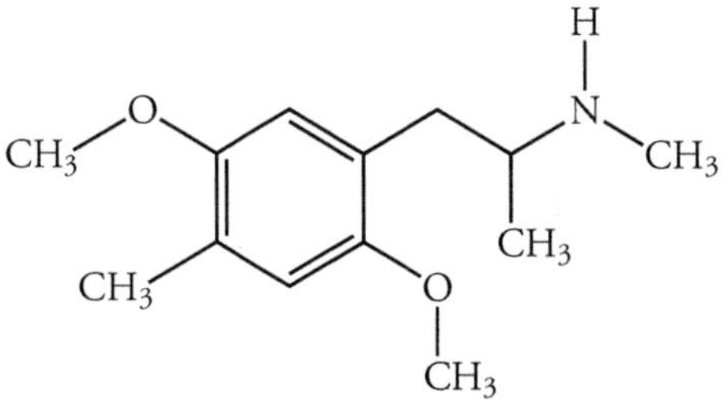

methoxyamphetamine

Phenylpropanolamine is a prescription drug used as an appetite suppressant by those on crash diets. Its use is not recommended by physicians, and the drug can cause health hazards for individuals suffering from hypertension. In young women it increases the risk of stroke.

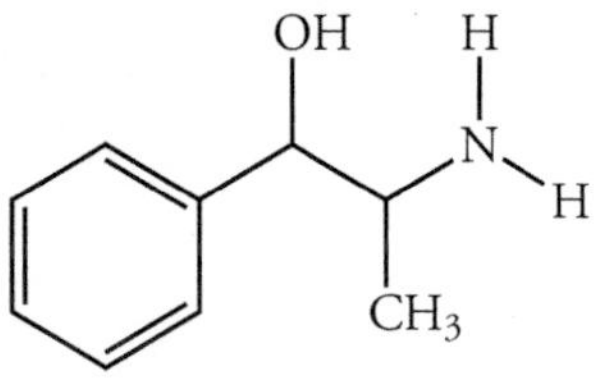

phenylpropanolamine

Although phenylethanamine derivatives generally act as stimulants, methylphenidate (Ritalin®) has a calming effect. This drug has been used in the treatment of hyperactivity in young boys. The use of this drug is controversial.

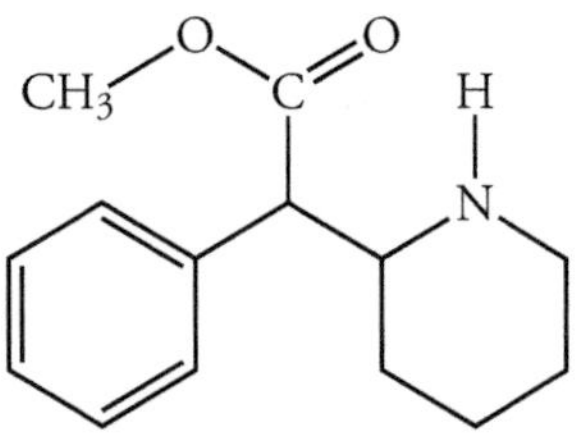

methylphenidate (Ritalin®)

Problem 12.5
Write the IUPAC name for baclofen, a muscle relaxant.

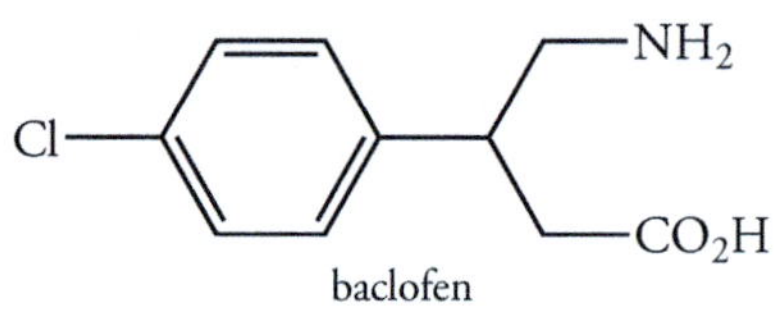

baclofen

Solution

The parent chain of baclofen is butanoic acid. It contains an amino group and an aryl group. The amino group is located at C-4. The aryl group bonded to C-3 is named *p*-chlorophenyl. Placing the groups in alphabetical order, the name is 4-amino-3-(*p*-chlorophenyl)butanoic acid. The name of the aryl group is written within parentheses to identify it clearly.

Problem 12.6

2-(3,4,5-Trimethoxyphenyl)ethanamine is the systematic name of mescaline, a hallucinogen. Write its structure.

Problem 12.7

2-(Diethylamino)-N-(2,6-dimethylphenyl)ethanamide is the IUPAC name for lidocaine, a local anesthetic. Write its structure.

Solution

The parent is the two-carbon amide, ethanamide. The group contained within parentheses and preceded by an N is 2,6-dimethylphenyl. Place this group on the nitrogen atom of ethanamide.

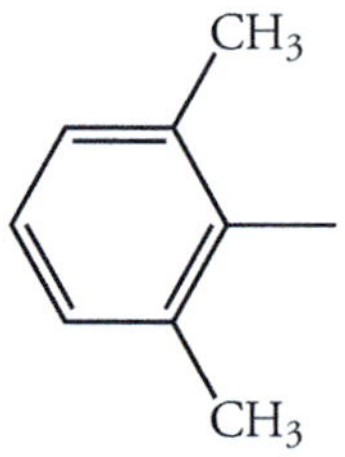

2,6-dimethylphenyl group 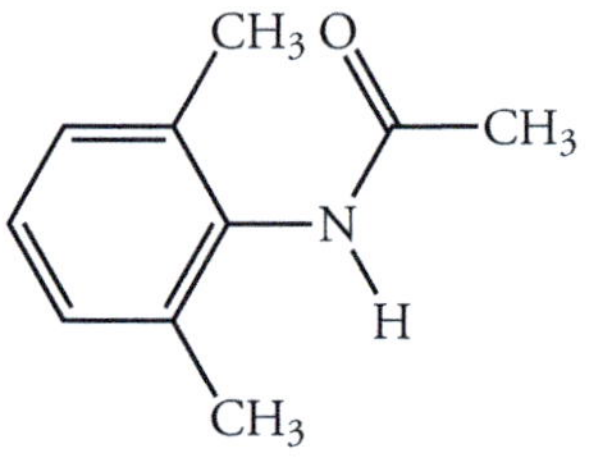 · (*N*-aryl substituted amide)

The compound is also an amine because of the diethylamino group bonded to the C-2 atom.

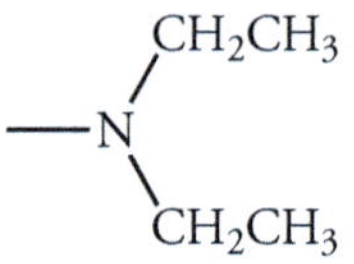

dimethylamino group 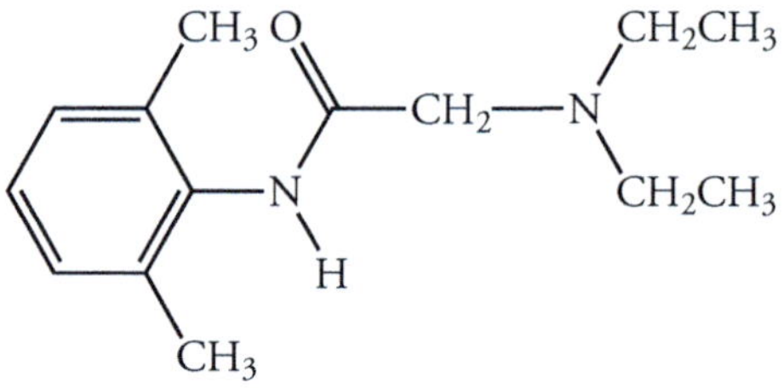 · lidocaine

Problem 12.8

Assign the IUPAC name of DEET, an insect repellent.

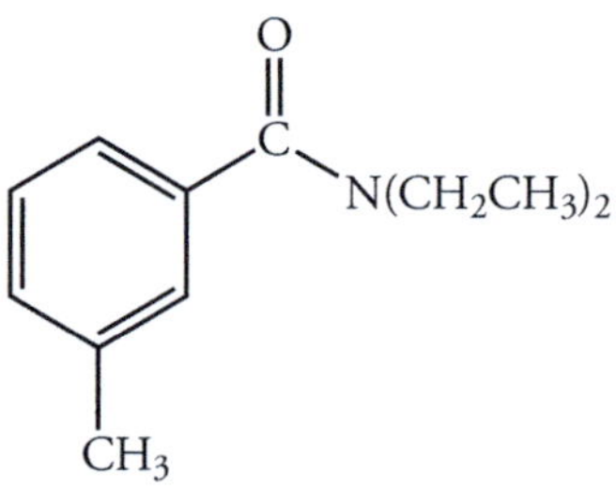

12.5 PHYSICAL PROPERTIES OF AMINES

Amines with low molecular weights are gases at room temperature, but amines with higher molecular weights are liquids or solids (Table 12.1). Amines have boiling points that are higher than those of alkanes of similar molecular weight, but lower than those of alcohols.

$$CH_3-CH_2-CH_3 \qquad CH_3-CH_2-NH_2 \qquad CH_3-CH_2-OH$$
$$\text{bp -42 °C} \qquad\qquad \text{bp 17 °C} \qquad\qquad \text{bp 78 °C}$$

Table 12.1
Boiling Points of Amines

Type of Amine	*Name*	*Boiling Point (°C)*
Primary amine	Methylamine	–7
	Ethylamine	17
	Propylamine	48
	Isopropylamine	33
	Butylamine	77
	Isobutylamine	68
	tert-Butylamine	45
	Cyclohexylamine	134
Secondary amine	Dimethylamine	7
	Ethylmethylamine	37
	Diethylamine	56
	Dipropylamine	111
Tertiary amine	Trimethylamine	3
	Triethylamine	90
	Tripropylamine	156
Aromatic amine	Aniline	184
	Pyridine	116

With the exception of formamide, primary amides are solids at room temperature. Substituted amides have lower melting points. All amides have high boiling points.

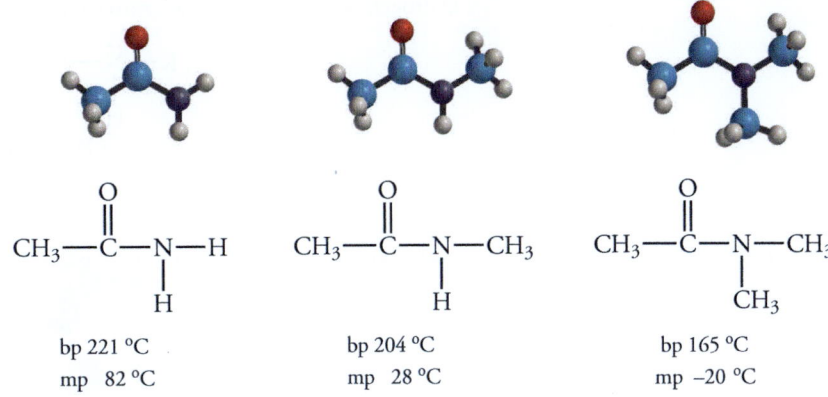

$$CH_3-\overset{\overset{\displaystyle O}{\|}}{C}-\overset{\overset{}{|}}{\underset{H}{N}}-H \qquad CH_3-\overset{\overset{\displaystyle O}{\|}}{C}-\overset{}{\underset{H}{N}}-CH_3 \qquad CH_3-\overset{\overset{\displaystyle O}{\|}}{C}-\overset{}{\underset{CH_3}{N}}-CH_3$$

bp 221 °C mp 82 °C   bp 204 °C mp 28 °C   bp 165 °C mp –20 °C

Hydrogen Bonding in Nitrogen Compounds

Primary and secondary amines have higher boiling points than hydrocarbons of comparable molecular weight because they can form intramolecular hydrogen bonds.

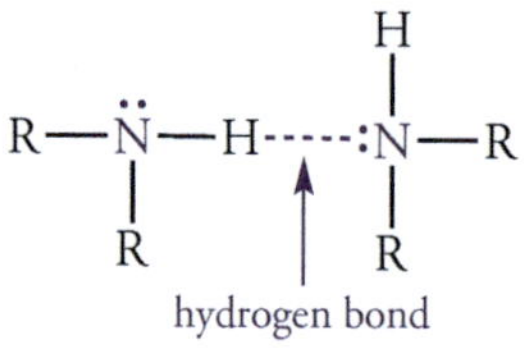

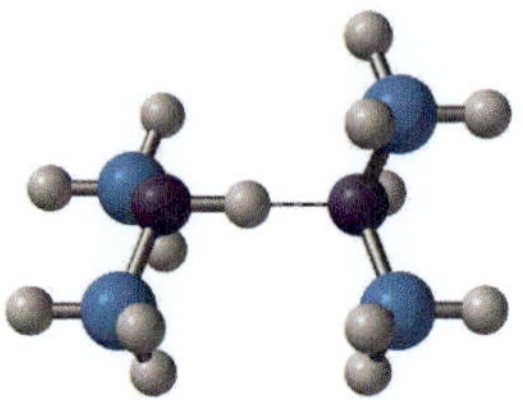

Tertiary amines have no hydrogen atoms bonded to the nitrogen atom and cannot serve as hydrogen bond donors. Thus, they have lower boiling points than primary and secondary amines of comparable molecular weight.

Amines have lower boiling points than alcohols because nitrogen is less electronegative than oxygen. As a result, the N—H bond is less polar than the O—H bond, and the N—H···N hydrogen bond in a mines is weaker than the O—H···O hydrogen bond in alcohols.

Amides form strong intermolecular hydrogen bonds between the amide hydrogen atom of one molecule and the carbonyl oxygen atom of a second molecule (C=O···H—N). This intermolecular interaction is responsible for the high melting and boiling points of primary amides. Substitution of the hydrogen atoms on the nitrogen atom by alkyl or aryl groups reduces the number of possible intermolecular hydrogen bonds and lowers the melting and boiling points. Tertiary amides cannot form intermolecular hydrogen bonds.

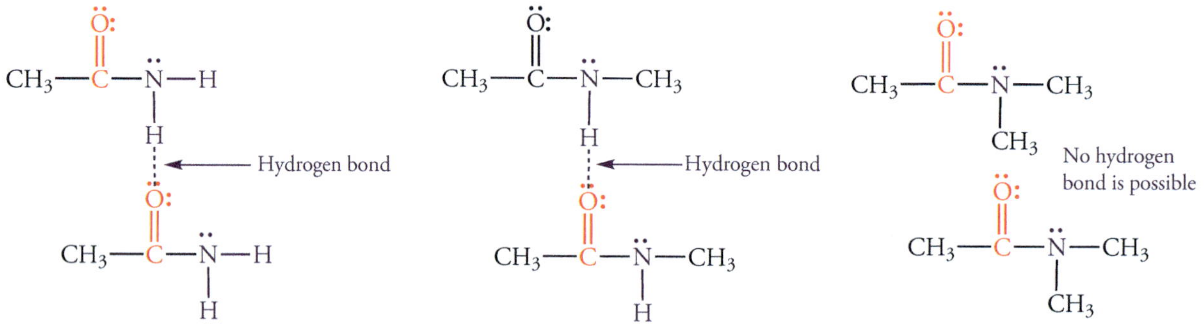

Solubility of Amines in Water

Primary and secondary amines function as both hydrogen bond donors and acceptors, and they readily form hydrogen bonds with water. Amines with five or fewer carbon atoms are miscible with water. Even tertiary amines are soluble in water because the nonbonding pair of electrons of the nitrogen atom is a hydrogen bond acceptor of a hydrogen atom of water.

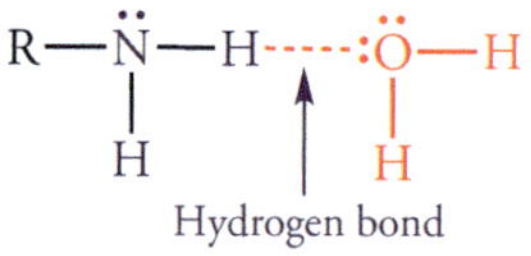

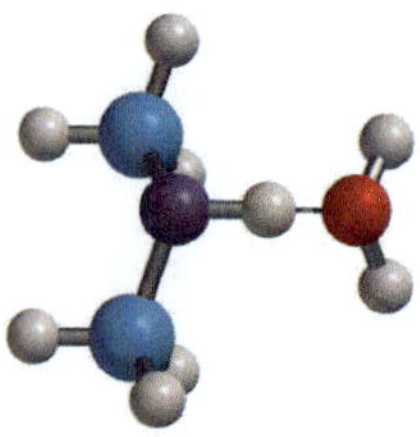

As we have seen for other types of compounds, the solubility of amines decreases with increasing molecular weight because the functional group is a less significant part of the structure. Amides having low molecular weights are soluble in water because hydrogen bonds form between the amide group and water. Even low molecular weight tertiary amides are water soluble because the carbonyl oxygen atom can form hydrogen bonds to the hydrogen atoms of water.

Odor and Toxicity of Amines

Amines with low molecular weights have sharp penetrating odors similar to ammonia. Amines with higher molecular weights smell like decaying fish. Two compounds responsible for the odor of decaying animal tissue have the common, graphic names putrescine and cadaverine.

$$NH_2CH_2CH_2CH_2CH_2NH_2 \qquad NH_2CH_2CH_2CH_2CH_2CH_2NH_2$$

putrescine

cadaverine

12.6 BASICITY OF NITROGEN COMPOUNDS

The nitrogen atom in both amines and amides has an unshared pair of electrons. However, there is a substantial difference in the chemistry of these two classes of compounds. In this section we consider the ability of these nitrogen compounds to donate an electron pair to a proton—that is, the basicity of the compounds. In Section 12.6 we will examine the donation of the electron pair to an electrophile other than a proton.

$$CH_3NH_2 + H_2O \rightleftharpoons CH_3NH_3^+ + OH^-$$

$$K_b = \frac{[OH^-][CH_3NH_3^+]}{[CH_3NH_2]}$$

Table 12.2 lists the base ionization constants for several amines. We recall that alkyl groups are electron-donating toward carbocations (Section 4.9), and that they are electron-donating in electrophilic aromatic substitution reactions (Section 5.7). Similarly, alkyl-substituted amines are slightly stronger bases than ammonia because the inductive donation of electrons to the nitrogen atom by alkyl groups makes the unshared pair of electrons more available to a proton.

Aryl-substituted amines are much weaker bases than ammonia and alkyl-substituted amines. Their K_b values are less than 10^{-9} (Table 12.2). For example, the K_b value of aniline is 10^{-6} times the K_b value of cyclohexylamine.

Table 12.2
Basicity of Amines and Acidity of Ammonium Ions

Name	K_b	K_a	pK_b	pK_a
Ammonia	1.8×10^{-5}	5.5×10^{-10}	4.74	9.26
Methylamine	4.3×10^{-4}	2.2×10^{-11}	3.37	10.66
Ethylamine	4.8×10^{-4}	2.1×10^{-11}	3.36	10.80
Dimethylamine	4.7×10^{-4}	2.1×10^{-11}	3.33	10.73
Diethylamine	3.1×10^{-4}	3.2×10^{-11}	3.40	10.49
Trimethylamine	1.0×10^{-3}	1.0×10^{-11}	3.22	11.00
Cyclohexylamine	4.6×10^{-4}	2.2×10^{-11}	3.33	10.66
Aniline	4.3×10^{-10}	2.5×10^{-5}	9.37	4.63

Aryl-substituted amines are weaker bases than ammonia because the unshared pair of electrons of the nitrogen atom is resonance delocalized over the π orbital system of the benzene ring. As a result, the unshared electron pair of nitrogen is less available for bonding with a proton.

Electron pair is localized on the nitrogen atom

Electron pair is resonance delocalized

cyclohexylamine

aniline

Heterocyclic Amines

The basicity of heterocyclic amines varies over a wide range and reflects both the hybridization of the nitrogen orbital containing the lone pair electrons and the effects of delocalization. Pyridine is a substantially weaker base than alkylamines. The electron pair of pyridine occupies an sp^2-hybridized orbital, and lies closer to the nitrogen nucleus than the electron pair in the sp^3-hybridized orbital of alkylamines. As a result, pyridine is a weaker base (larger pK_b) than an alkylamine.

sp^2-hybridized nitrogen
(less basic)

pyridine
pK_b 8.75

sp^3-hybridized nitrogen
(more basic)

piperidine
pK_b 2.88

Pyrrole is an exceedingly weak base. The pair of electrons of the nitrogen atom interacts with the four electrons of the two carbon-carbon double bonds to give an aromatic six-π-electron system similar to that of benzene. Thus, the electron pair is not readily available for protonation because it is required to maintain the sextet of electrons required for aromaticity in the ring. Pyrrolidine has a pK_b similar to acyclic amines.

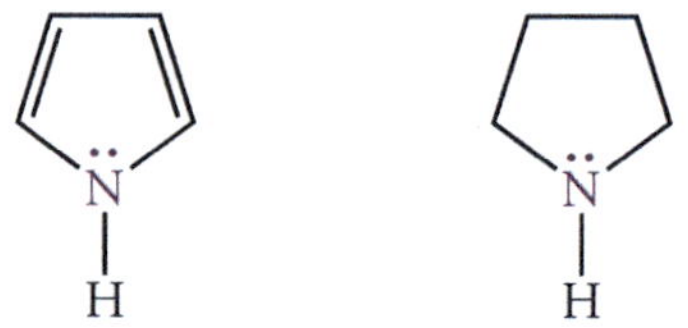

pyrrole, pK_b 15   pyrrolidine, pK_b 2.73

Imidazole is an aromatic ring found in many biological molecules. One of its nitrogen atoms resembles that of pyrrole, and is not basic. The second nitrogen atom, which is structurally similar to the nitrogen atom of pyridine, acts as a base. However, imidazole is about 100 times more basic than pyridine. The increased basicity results from resonance stabilization of the positive charge of the conjugate acid.

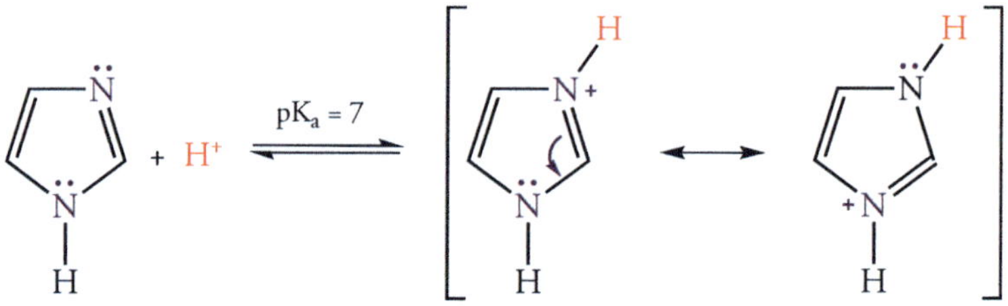

imidazole

$pK_a = 7$

Resonance-stabilized imidazolium ion

Amides

In contrast to amines, amides are extremely weak bases. (The K_b of ethanamide is approximately 10^{-15}!) This difference in basicity is due to the carbonyl group, which draws electron density away from the nitrogen atom. Thus, the unshared electron pair of the amide nitrogen atom is delocalized and not readily available to react with a proton. An amide is polar, planar, and resonance stabilized.

Delocalized electron pair of nitogen is not basic

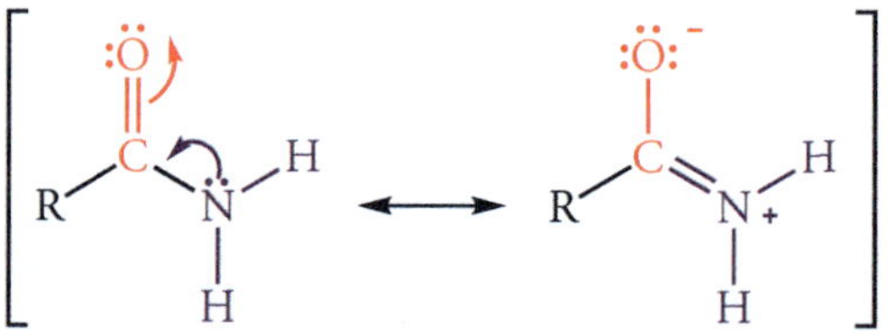

Contributing resonance structures of an amide

Acidity of Ammonium Ions

When an amine is protonated, the conjugate acid product is a positively charged, substituted ammonium ion. The ionization constant of the conjugate acid of methylamine, the methylammonium ion, is derived from the following reaction:

$$CH_3NH_3^+ + H_2O \rightleftharpoons CH_3NH_2 + H_3O^+$$

$$K_b = \frac{[H_3O^+]\left[CH_3NH_2\right]}{[CH_3NH_3^+]}$$

There is an inverse relationship between the K_a value of an acid and the K_b value of its related conjugate base. The K_a value of the methylammonium ion and the K_b value of methylamine illustrate this relationship. The K_a value for the methylammonium ion is relatively small; the K_b value for methylamine is relatively large. The values of K_a and K_b for a conjugate acid-base pair are related as follows:

$$(K_a)(K_b) = K_w = 1 \times 10^{-14}$$

pK$_a$ and pK$_b$

The basicity of an amine is usually listed as a pK_b, the negative logarithm of K_b. For an amine with $K_b = 10^{-4}$, the pK_b is 4. The pK_b values of strong bases are small. Thus, as pK_b increases, base strength decreases. It is also common practice to indicate the relative base strength of amines in terms of the pK_a of their conjugate acids. (Recall that $pK_a = -\log K_a$.) The sum of the pK_a and pK_b for a conjugate acid-base pair is 14. Thus, for an amine base with $pK_b = 5$, its conjugate ammonium ion has $pK_a = 9$.

Problem 12.9
Based on the K_b values in Table 12.2, estimate the K_b of phenindamine, an antihistamine.

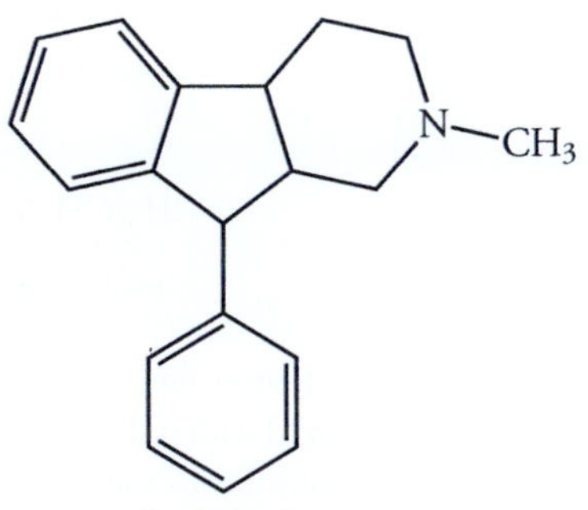

phenindamine

Solution
Phenindamine is a tertiary amine and the nitrogen atom is bonded only to alkyl groups. Thus, its pK_b should be similar to that of a simpler tertiary amine such as triethylamine, which is 3.0 from Table 12.2.

Problem 12.10
Which of the two nitrogen atoms of chlorpromazine (Thorazine®), an antipsychotic drug, is the more basic?

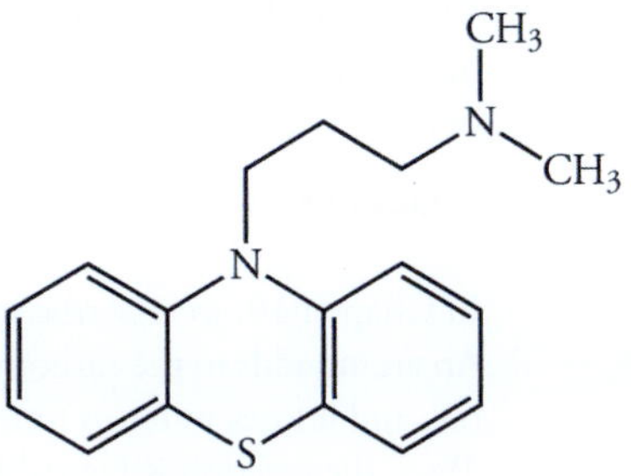

chlorpromazine

12.7 SOLUBILITY OF AMMONIUM SALTS

When an amine is added to a solution of a strong acid such as hydrochloric acid, the amine nitrogen atom is protonated to produce an ammonium salt.

$$RNH_2 \ + \ HCl \ \rightleftharpoons \ RNH_3^+Cl^-$$

Ammonium salts are more soluble than amines because the nitrogen atom of an ammonium salt has a positive charge. This property is used by drug companies to manufacture compounds that will be soluble in body fluids. Drugs containing an amino group are often prepared as ammonium salts to improve their solubility in body fluids. For example, the solubility of procaine (Novocain®) is only 0.5 g/100 mL, but the solubility of its ammonium salt is 100 g/100 mL.

procaine

The ammonium salts of many drugs are more stable and less prone to oxidation than the amine itself. Procaine is most stable at pH 3.6 and becomes less stable as the pH is increased. The ammonium salts have higher melting points than the amines and have virtually no odor. For example, ephedrine melts at 79 °C and has a fishy odor. Its hydrochloride salt, used in cold and allergy medications, melts at 217 °C and has no odor.

Amines can be separated from other substances by converting them to ammonium salts. Consider the separation of 1-chlorooctane from 1-octanamine, a mixture that can result from a synthesis of the amine by reaction of 1-chlorooctane with ammonia. Both compounds are insoluble in water. Adding HCl to a solution containing both compounds converts the 1-octanamine into its ammonium salt, whereas 1-chlorooctane is not affected.

$$CH_3(CH_2)_6CH_2NH_2 \ + \ HCl \ \longrightarrow \ CH_3(CH_2)_6CH_2NH_3^+Cl^-$$
(insoluble in water)   (soluble in water)

$$CH_3(CH_2)_6CH_2Cl \ + \ HCl \ \xrightarrow{\ \ X\ \ } \ \text{no reaction}$$
(insoluble in water)

The 1-chlorooctane is physically separated from the aqueous acid solution. Then the acid solution is neutralized with sodium hydroxide to form the free amine. The amine can then be physically separated from the aqueous solution.

$$CH_3(CH_2)_6CH_2NH_3^+ \ + \ OH^- \ \longrightarrow \ CH_3(CH_2)_6CH_2NH_2 \ + \ H_2O$$
(soluble in water)   (insoluble in water)

Amides are not sufficiently basic to be protonated in aqueous solutions. Thus, their solubility is unaffected by pH. They behave as neutral compounds.

12.8 NUCLEOPHILIC REACTIONS OF AMINES

We described some reactions of amines in earlier chapters. These reactions occur because the nonbonding electron pair of the nitrogen atom makes amines nucleophilic. We will review each type of reaction in this section.

Reactions of Amines with Carbonyl Compounds

In Chapter 10, we described the addition-elimination reaction of amines with carbonyl compounds. An amine adds to the carbonyl carbon atom to give a tetrahedral intermediate. This product is unstable, and it loses water to form an imine. In general, imines are less stable than carbonyl compounds. Thus, the reaction is favorable only if water is removed from the reaction mixture. Most imines are not stable; they rapidly hydrolyze in aqueous solution to give carbonyl compounds.

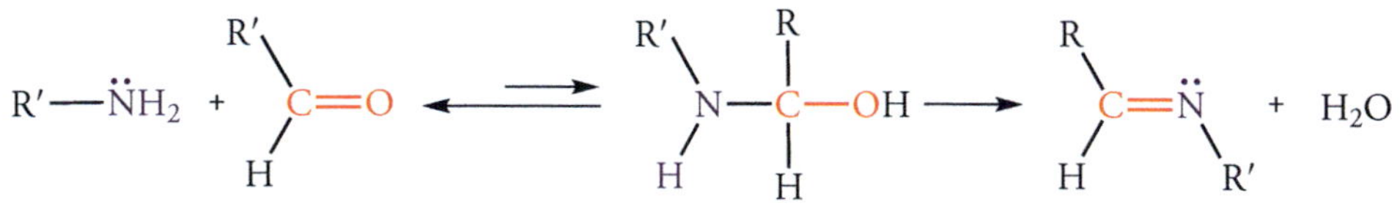

R'—N̈H₂ + (R'/H)C=O ⇌ tetrahedral intermediate R'—N(H)—C(R)(H)—OH → (R/H)C=N̈—R' + H₂O

amine   aldehyde   tetrahedral intermediate   imine

Reactions of Amines with Acyl Halides

In Chapter 11, we noted that an amide can be made by treating an amine with an acid halide. We recall that acid halides are very reactive acyl derivatives of acids; amides are very stable.

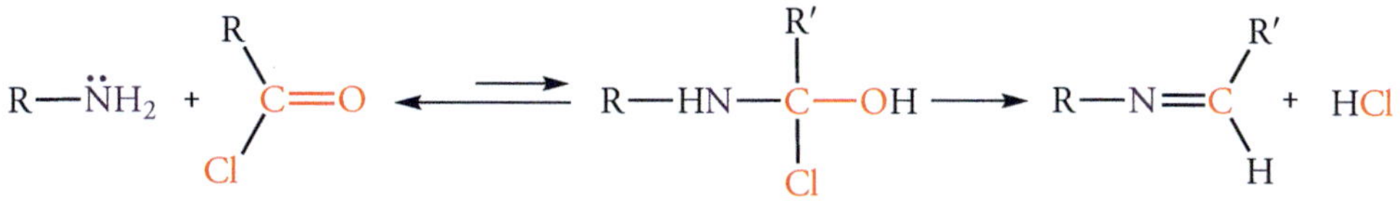

R—N̈H₂ + (R/Cl)C=O ⇌ R—HN—C(R')(Cl)—OH → R—N=C(R')(H) + HCl

amine   acyl chloride   tetrahedral intermediate   amide

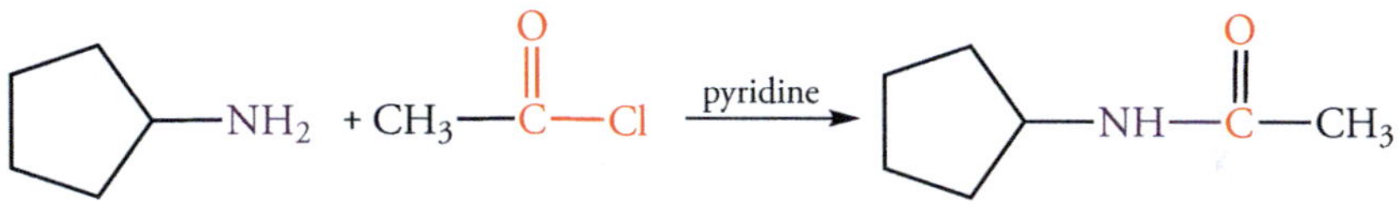

cyclopentyl—NH₂ + CH₃—C(=O)—Cl →(pyridine)→ cyclopentyl—NH—C(=O)—CH₃

Only ammonia and primary or secondary amines form amides. That is why pyridine, which cannot form an amide, is often used as a base to react with the HCl formed in the reaction.

Reactions of Amines with Alkyl Halides

We described nucleophilic substitution reactions of alkyl halides in Chapter 7. Primary and secondary alkyl halides react with nucleophiles by an S_N2 mechanism. Amines are nucleophiles that can displace a halide ion from a primary or secondary alkyl halide to form an ammonium halide salt that is subsequently neutralized.

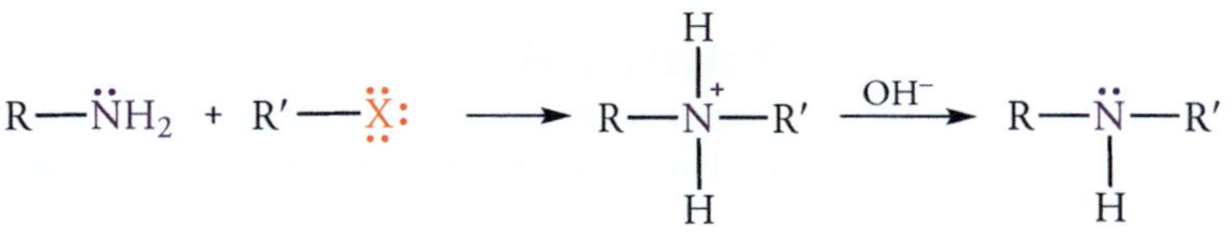

R—N̈H₂ + R'—Ẍ: → R—N⁺(H)(H)—R' →(OH⁻)→ R—N̈(H)—R'

The initial product of the nucleophilic substitution reaction is a secondary ammonium ion. It can lose a proton in an equilibrium reaction with the reactant primary amine.

R—N̈H₂ + R—N⁺(H)(H)—R' ⇌ R—N̈H₃⁺ + R—N̈(H)—R'

The secondary amine then can continue to react with the alkyl halide to give a tertiary amine and eventually a quaternary ammonium ion.

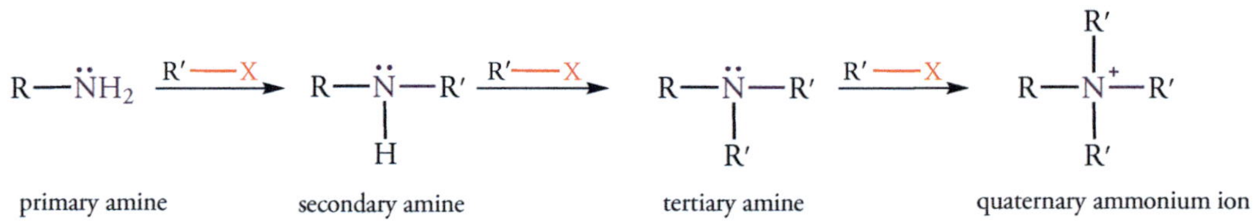

R—N̈H₂ $\xrightarrow{R'—X}$ R—N̈—R' $\xrightarrow{R'—X}$ R—N̈—R' $\xrightarrow{R'—X}$ R—N⁺—R'
                       | | |

primary amine   secondary amine   tertiary amine   quaternary ammonium ion

Quaternary ammonium salts are ammonium salts that have four alkyl or aryl groups bonded to a nitrogen atom. Some quaternary ammonium salts containing a long carbon chain are **invert soaps**.

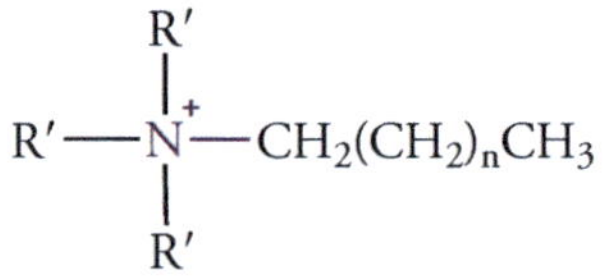

$$R'—\overset{\overset{\displaystyle R'}{|}}{\underset{\underset{\displaystyle R'}{|}}{N^{+}}}—CH_2(CH_2)_nCH_3$$

an invert soap

Invert soaps differ from soaps and detergents because the polar end of the ion in the micelle is positive rather than negative. Like soaps, the long hydrocarbon tail associates with nonpolar substances, and the polar head dissolves in water. Invert soaps are widely used in hospitals. They are active against bacteria, fungi, and protozoa, but they are not effective against spore-forming microorganisms. Benzalkonium chlorides are one type of invert soap. The alkyl groups of these compounds contain from 8 to 16 carbon atoms.

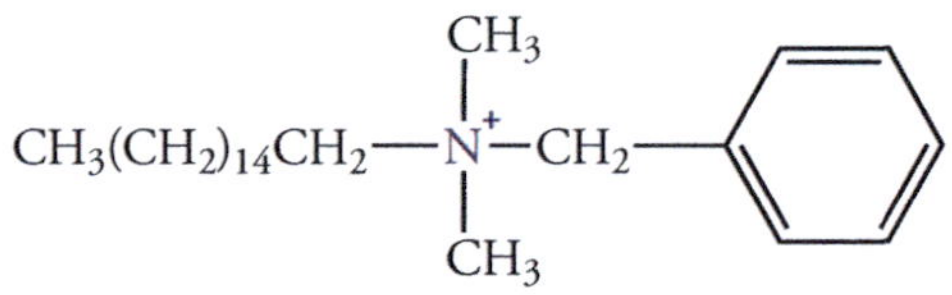

$$CH_3(CH_2)_{14}CH_2—\overset{\overset{\displaystyle CH_3}{|}}{\underset{\underset{\displaystyle CH_3}{|}}{N^{+}}}—CH_2—\bigcirc$$

a benzalkonium ion

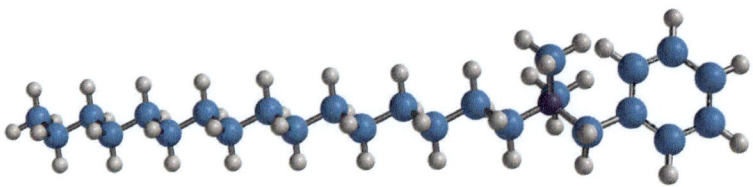

Problem 12.11

Flecainide, an antiarrhythmic drug, is an amide. Draw the structures of the compounds that could be used to produce the drug. What possible complications might occur with this combination of reactants?

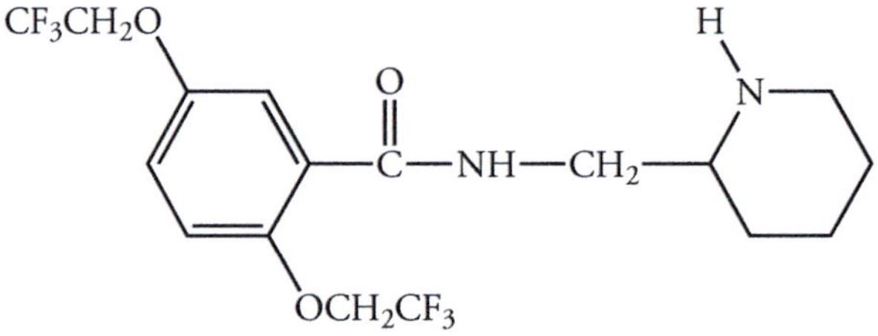

flecainide

Solution

Mentally separate the amide into two components by breaking the bond between the nitrogen atom and the carbonyl carbon atom. Place a hydrogen atom on the nitrogen atom. Place a chlorine atom on the carbonyl carbon atom.

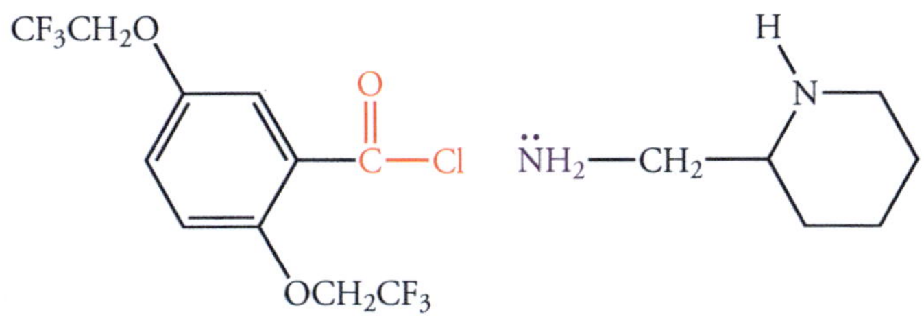

Note that the "amine" is actually a diamine: one part is a primary amine, the other part a secondary amine. Thus, the diamine could react at either nitrogen atom and form two isomeric amides. The primary amine is more reactive because the secondary amine is more sterically hindered.

Problem 12.12

Acetaminophen, the analgesic in many drugs, is an amide. Draw the structures of the compounds that could be used to produce it. What possible complications might occur with this combination of reactants?

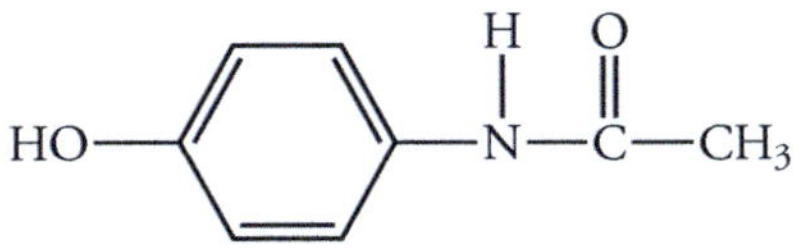

12.9 SYNTHESIS OF AMINES

We have seen many of the general methods to synthesize amines discussed in preceding sections and chapters. Except for the displacement reaction of an alkyl halide by ammonia or an amine, the remaining methods involve compounds that already have a nitrogen atom contained in a functional group, which is then transformed into an amine functional group.

Alkylation of Amines by Alkyl Halides

In Section 12.7 we saw that a nucleophilic substitution reaction of ammonia with an alkyl halide yields a mixture of products resulting from multiple alkylation. The chances for multiple alkylation can be diminished somewhat by selecting the proper reaction conditions. For example, if the reaction of an alkyl halide with ammonia is carried out with excess ammonia, an alkyl halide can be converted to a primary amine. When the concentration of ammonia is greater than the concentration of the primary amine product, the probability that the primary amine will continue to react with the alkyl halide decreases.

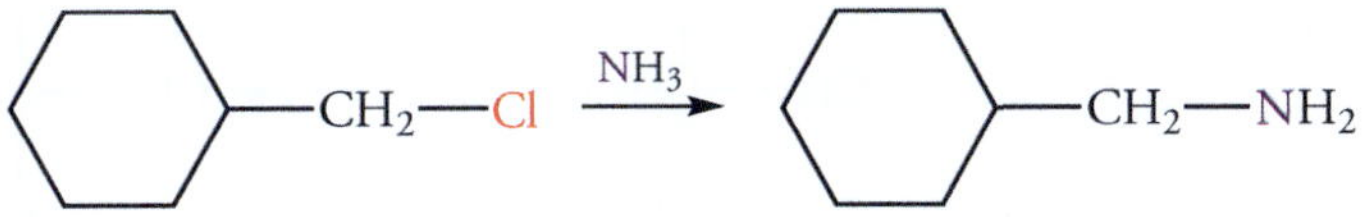

Reduction of Imines

We recall that the carbonyl group of aldehydes or ketones is reduced to an alcohol by either catalytic hydrogenation or metal hydrides. Imines are the nitrogen analogs of carbonyl compounds, and they are also reduced by the same reagents.

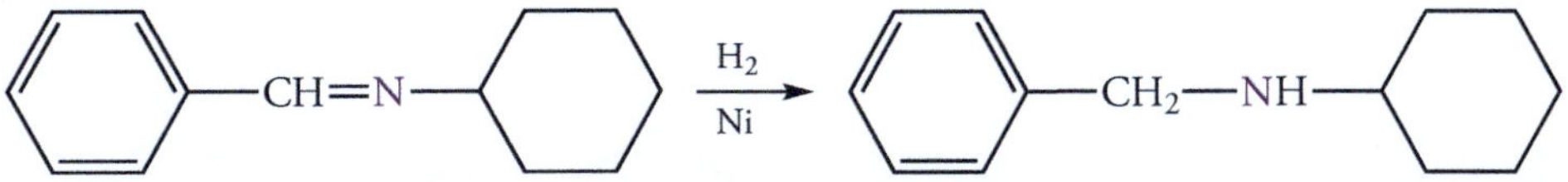

Imines do not have to be prepared and isolated for subsequent reduction. A mixture of a carbonyl compound and ammonia or the appropriate amine reacts in the presence of hydrogen gas and a metal catalyst. The imine that forms initially is reduced to an amine. The overall process is called **reductive amination**.

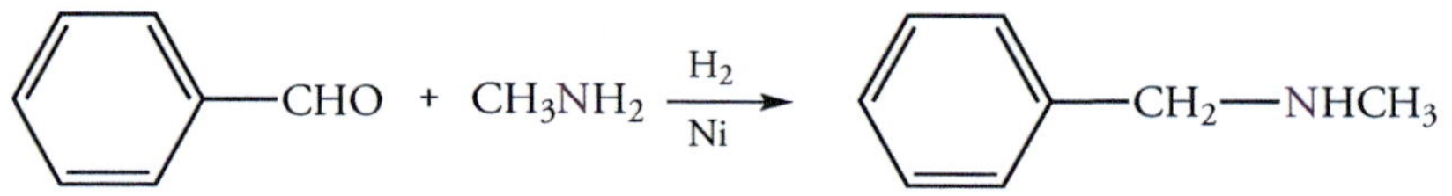

Reduction of Amides

Reduction of amides is one of the most frequently used methods of preparing amines. The method is very versatile because primary, secondary, and tertiary amines are easily prepared from the corresponding class of amide. Amides are prepared by acylation of amines using activated acyl derivatives such as acid chlorides or acid anhydrides (Section 12.6). Subsequent reduction of the amide with $LiAlH_4$ followed by acidic workup produces the amine.

(a primary amide) (a primary amine)

(a secondary amide) (a secondary amine)

(a tertiary amide) (a tertiary amine)

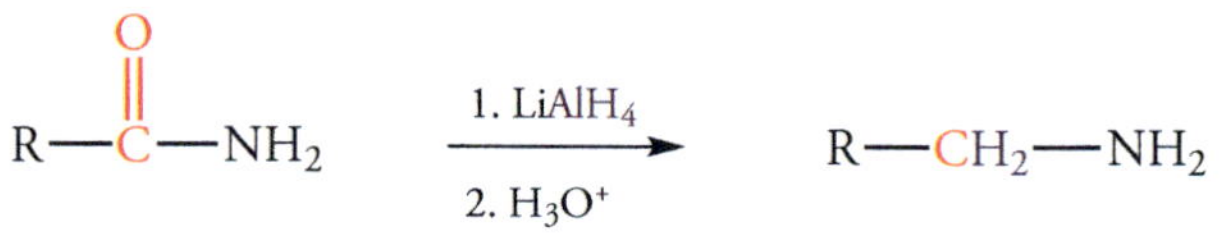

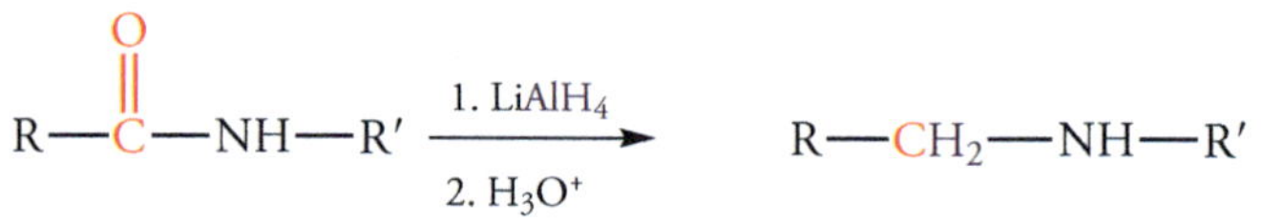

Reduction of Nitriles

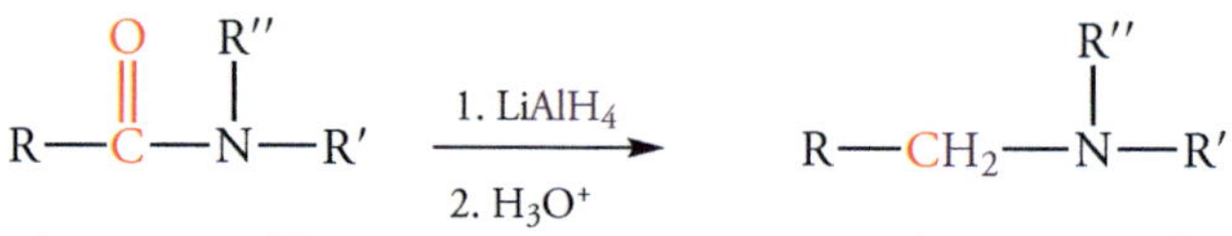

Nitriles can be prepared from primary alkyl halides by a direct S_N2 displacement reaction using sodium cyanide as the nucleophile (Section 7.6). Then the nitrile is reduced to a primary amine with lithium aluminum hydride.

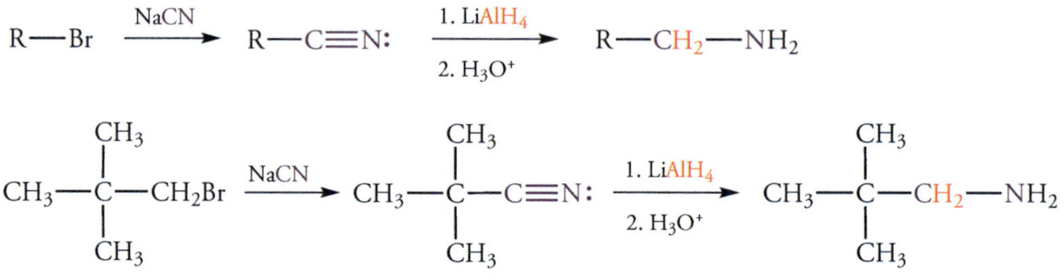

Reduction of Nitro Compounds

There is no synthetic procedure to introduce an amino group onto an aromatic ring in one step. However, it is possible to substitute an amino group onto an aromatic ring in two steps (Section 5.10). First the ring is nitrated. Then, the nitro group is reduced to an amino group.

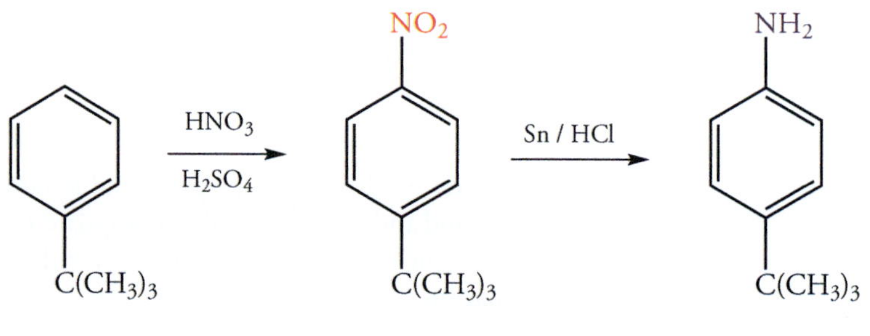

12.10 HYDROLYSIS OF AMIDES

Hydrolysis of an amide breaks the carbon-nitrogen bond and produces an acid and either ammonia or an amine. This reaction resembles the hydrolysis of esters, which we discussed in Chapter 11. There are, however, important differences. The hydrolysis of esters occurs relatively easily, whereas amides are very resistant to hydrolysis. Amides are hydrolyzed only by heating for hours with a strong acid or strong base. When amide hydrolysis is carried out in basic solution, the salt of the carboxylic acid forms; one mole of base is required per mole of amide. When amide hydrolysis is carried out under acidic conditions, the ammonium salt of the amine is formed, and one mole of acid is required per mole of amide.

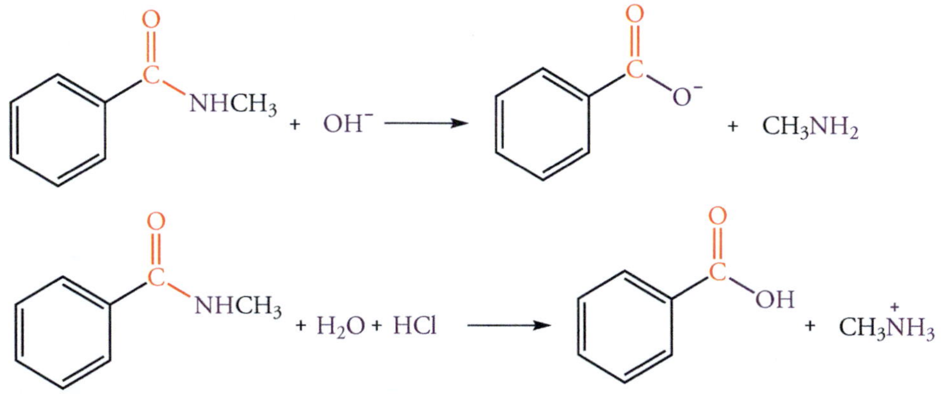

The great stability of amides toward hydrolysis has an important biological consequence, because amino acids in proteins are linked by amide bonds. Since amides are stable, proteins do not readily hydrolyze at physiological pH and at body temperature in the absence of a specific enzyme catalyst. However, in the presence of specific enzymes, the hydrolysis of amides is rapid. We will discuss these reactions in Chapter 13.

Problem 12.13

What are the products of the hydrolysis of phenacetin by a base? Phenacetin was once used in analgesic tablets, but it is no longer used because it has severe side effects.

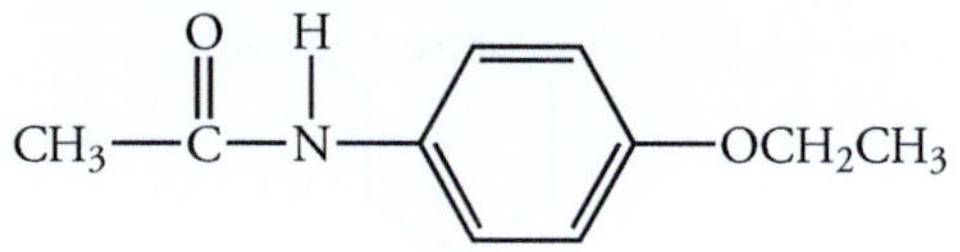

phenacetin

Solution

The functional group on the right side of the benzene ring is an ether, which does not react with base (Section 9.6). The functional group on the left is an amide.

Hydrolysis of an amide breaks the bond between the nitrogen atom and the carbonyl group. The acid fragment is acetic acid. The amine fragment is a substituted aniline containing an ether substituent.

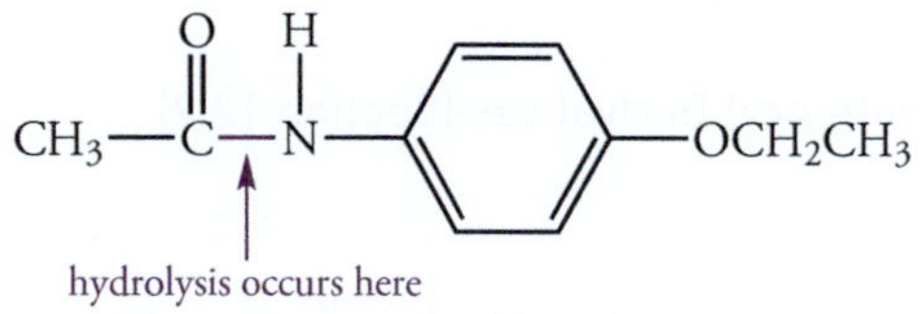

hydrolysis occurs here

Because a base is used in the hydrolysis, the acid product is present in the reaction mixture as the acetate ion. The amine is p-ethoxyaniline.

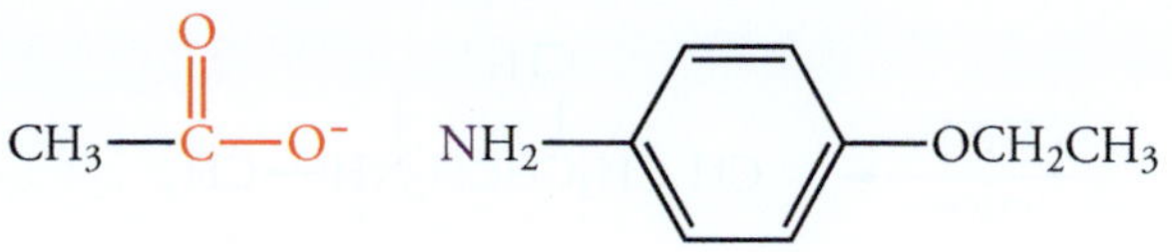

Problem 12.14

What are the products of the hydrolysis of nubucaine by an acid? Nubucaine is a local anesthetic.

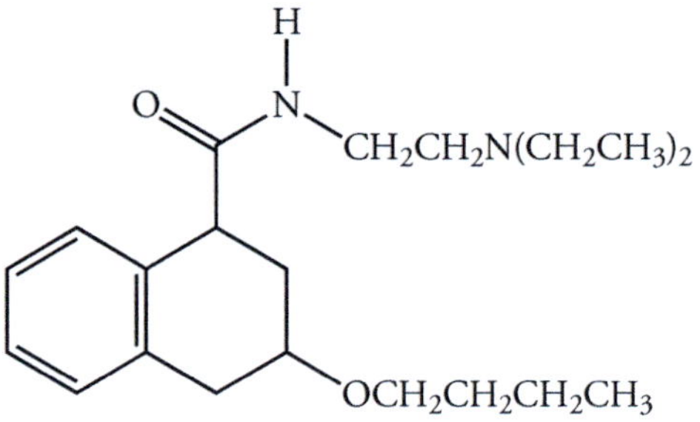

12.11 SYNTHESIS OF AMIDES

Carboxylic acids react to form an amide and water when heated to a high temperature with ammonia, a primary amine, or a secondary amine. Tertiary amines do not form amides because they have no hydrogen atom bonded to the nitrogen atom.

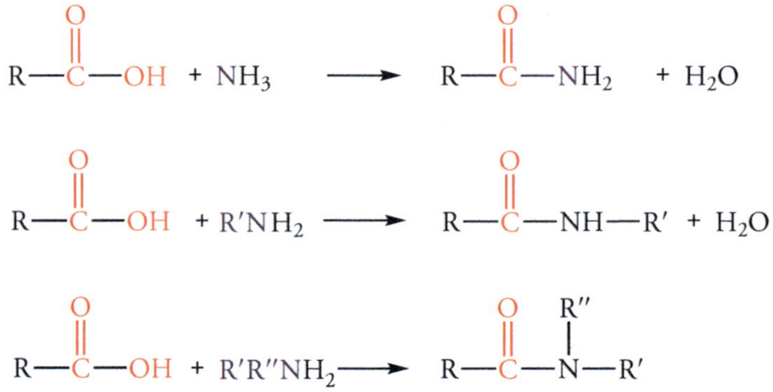

The high temperature of this direct reaction often affects other functional groups in the molecule. An amide can be synthesized at lower temperatures by the reaction of an acyl chloride with ammonia, a primary amine, or a secondary amine (Section 12.7).

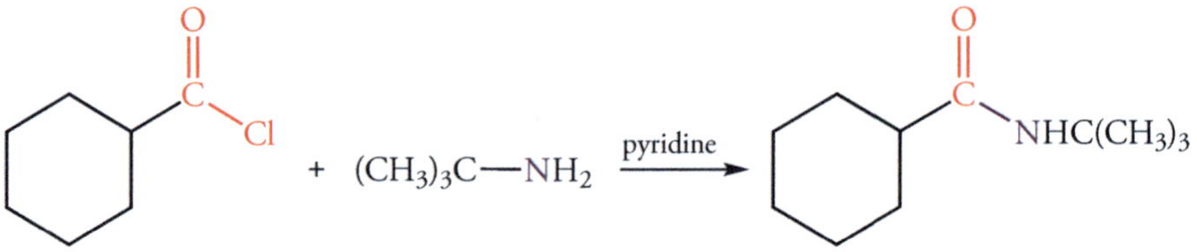

SUMMARY OF REACTIONS

1. Reaction of Amines with Carbonyl Compounds (Section 12.9)

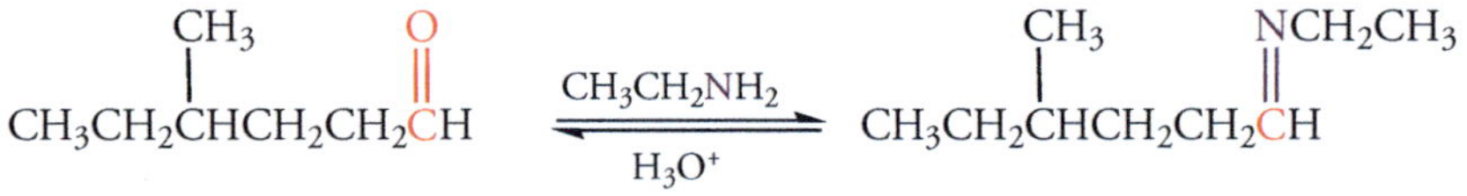

2. Reaction of Amines with Acyl Derivatives (Section 12.9)

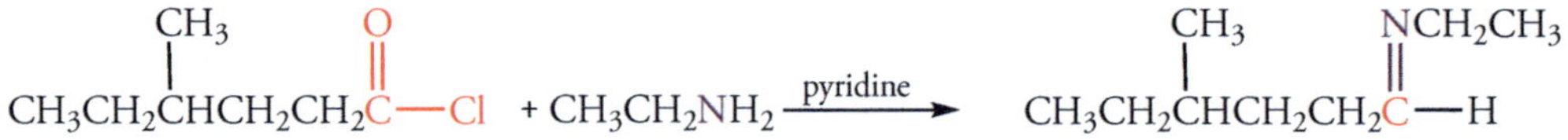

3. Reaction of Amines with Alkyl Halides (Section 12.9)

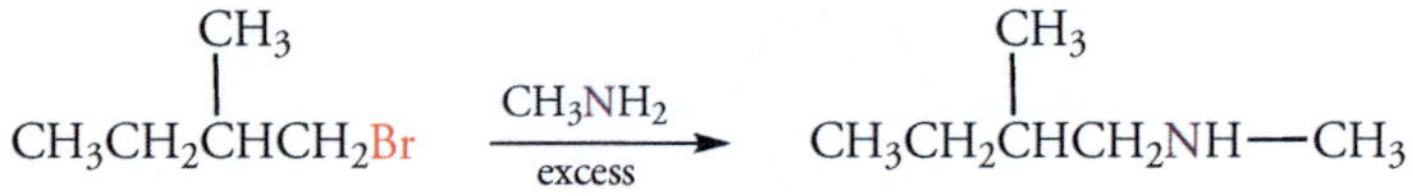

4. Synthesis of Amines (Section 12.9)

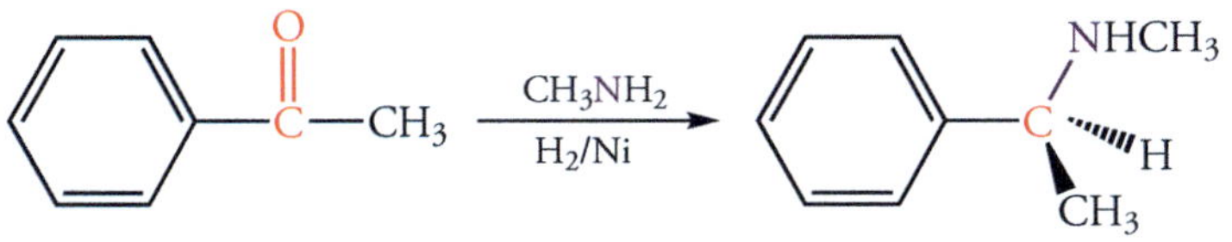

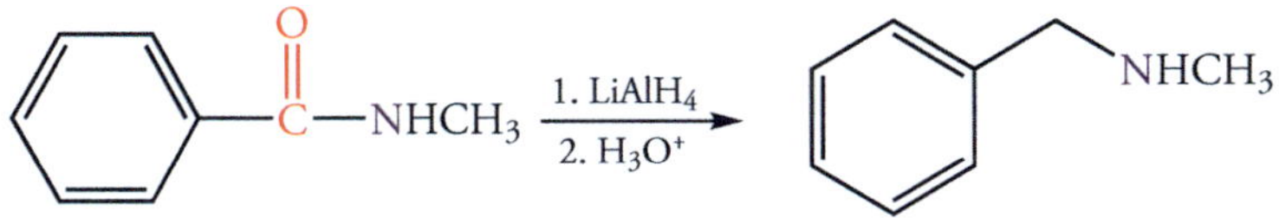

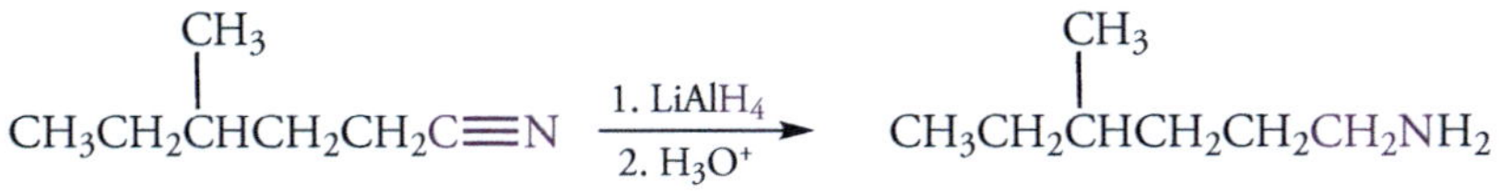

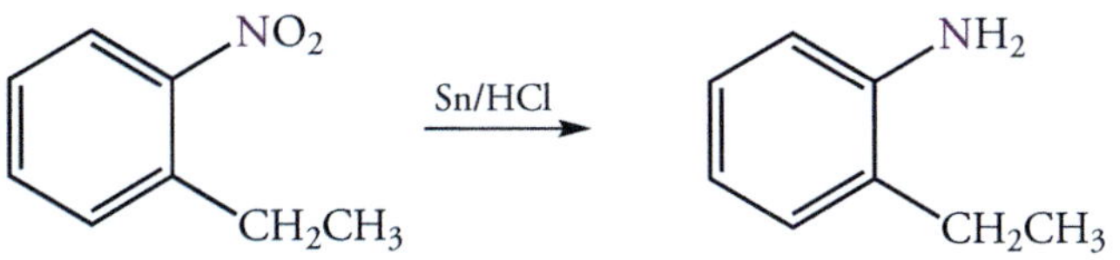

$$\text{C}_6\text{H}_5\text{-C(=O)-CH}_3 \xrightarrow[\text{H}_2/\text{Ni}]{\text{CH}_3\text{NH}_2} \text{C}_6\text{H}_5\text{-CH(NHCH}_3)\text{-CH}_3$$

$$\text{C}_6\text{H}_5\text{-C(=O)-NHCH}_3 \xrightarrow[\text{2. H}_3\text{O}^+]{\text{1. LiAlH}_4} \text{C}_6\text{H}_5\text{-CH}_2\text{-NHCH}_3$$

$$\text{CH}_3\text{CH}_2\overset{\overset{\text{CH}_3}{|}}{\text{C}}\text{HCH}_2\text{CH}_2\text{C}\equiv\text{N} \xrightarrow[\text{2. H}_3\text{O}^+]{\text{1. LiAlH}_4} \text{CH}_3\text{CH}_2\overset{\overset{\text{CH}_3}{|}}{\text{C}}\text{HCH}_2\text{CH}_2\text{CH}_2\text{NH}_2$$

o-nitroethylbenzene $\xrightarrow{\text{Sn/HCl}}$ o-aminoethylbenzene

5. Hydrolysis of Amides (Section 12.10)

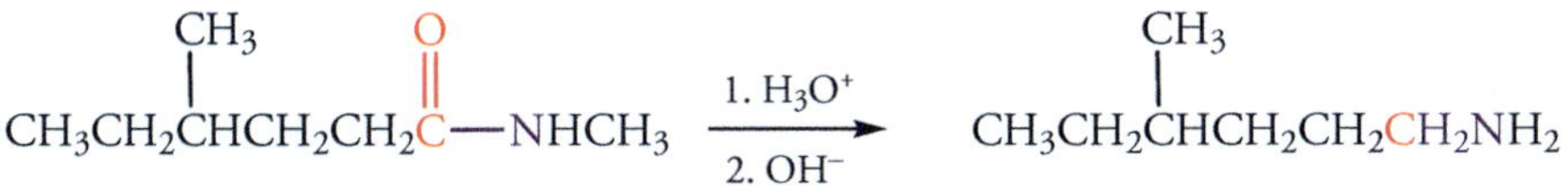

$$\text{CH}_3\text{CH}_2\overset{\overset{\text{CH}_3}{|}}{\text{C}}\text{HCH}_2\text{CH}_2\overset{\overset{\text{O}}{||}}{\text{C}}\text{-NHCH}_3 \xrightarrow[\text{2. OH}^-]{\text{1. H}_3\text{O}^+} \text{CH}_3\text{CH}_2\overset{\overset{\text{CH}_3}{|}}{\text{C}}\text{HCH}_2\text{CH}_2\text{CH}_2\text{NH}_2$$

6. Synthesis of Amides (Section 12.11)

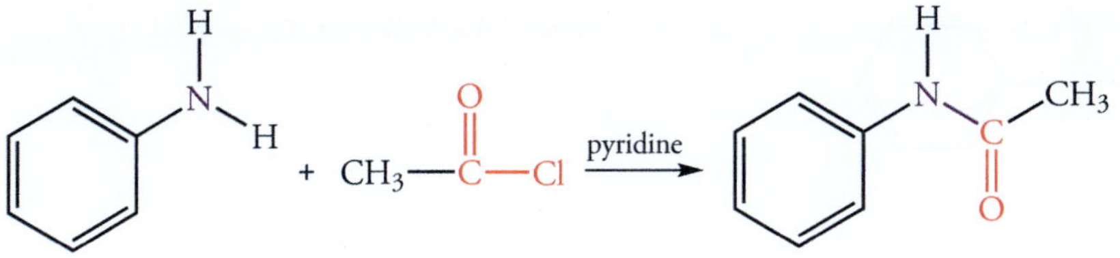

$$\text{C}_6\text{H}_5\text{-NH}_2 + \text{CH}_3\text{-C(=O)-Cl} \xrightarrow{\text{pyridine}} \text{C}_6\text{H}_5\text{-N(H)-C(=O)-CH}_3$$

EXERCISES

Classification of Amines and Amides

12.1 Classify each of the following amines or amides according to their degree of substitution:

(a) 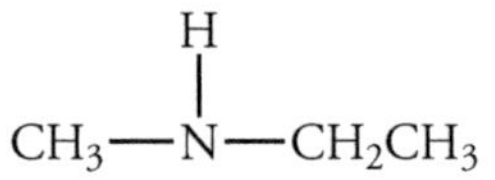
$$\underset{\text{H}}{\overset{\text{H}}{CH_3—N—CH_2CH_3}}$$

(b) 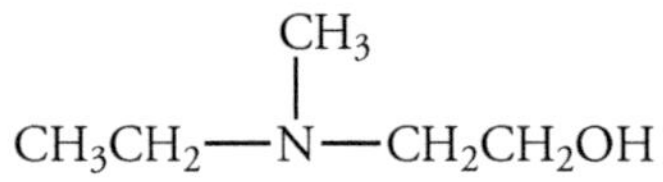
$$\underset{}{\overset{CH_3}{CH_3CH_2—N—CH_2CH_2OH}}$$

(c)

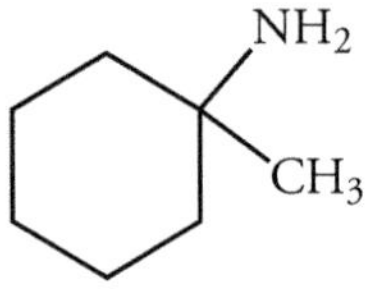

(d) 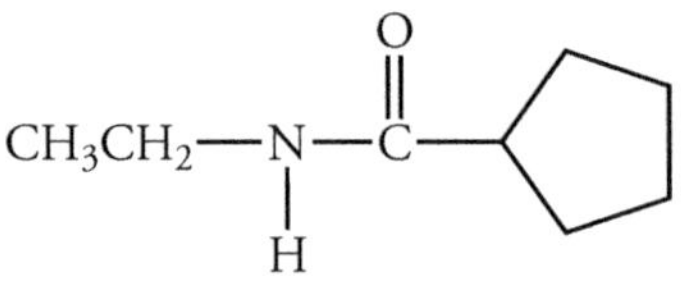 $CH_3CH_2—N—\overset{O}{\underset{H}{C}}$ cyclopentane

(e)

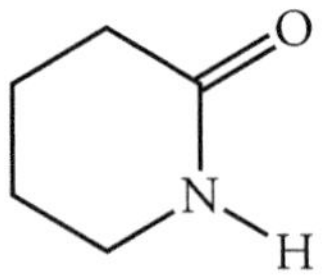

(f)

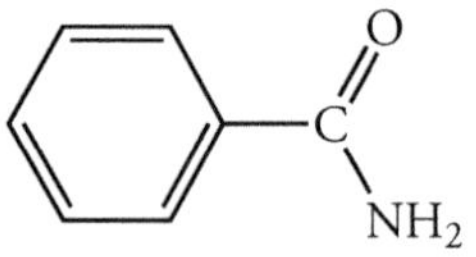

12.2 Classify each of the following amines or amides according to their degree of substitution:

(a) 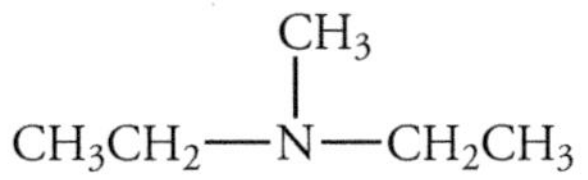
$$\underset{}{\overset{CH_3}{CH_3CH_2—N—CH_2CH_3}}$$

(b) 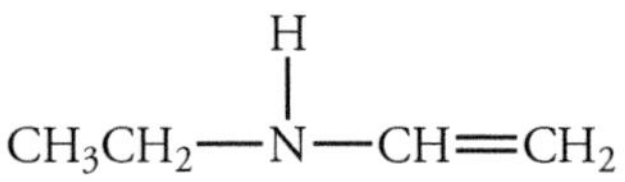
$$\underset{}{\overset{H}{CH_3CH_2—N—CH=CH_2}}$$

(c)

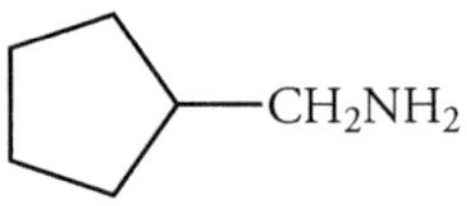

—CH_2NH_2

(d) 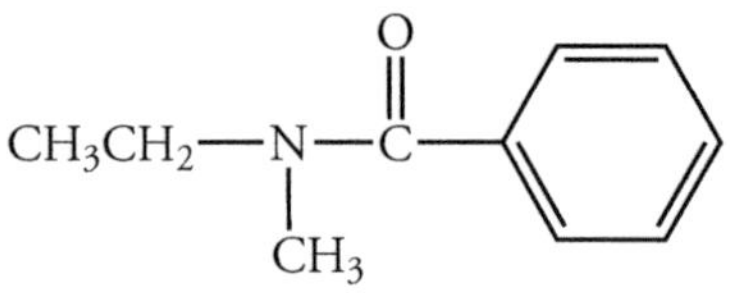 $CH_3CH_2—N—\overset{O}{\underset{CH_3}{C}}$

(e)

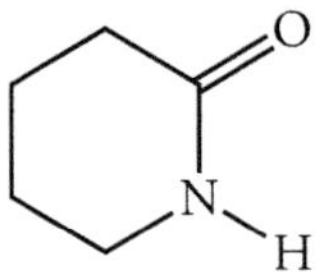

(f)

NHCH_2CH_3

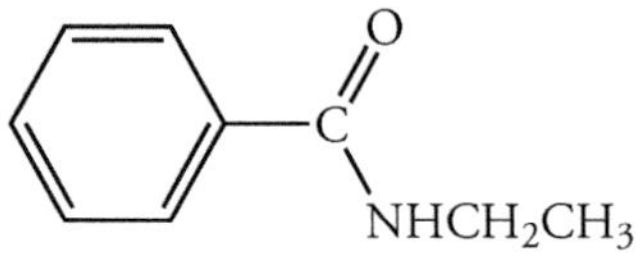

12.3 Classify the nitrogen-containing functional group in each of the following structures:

(a) acetaminophen, an analgesic in many drugs

(b) coniine, part of the hemlock poison drunk by Socrates

(a)

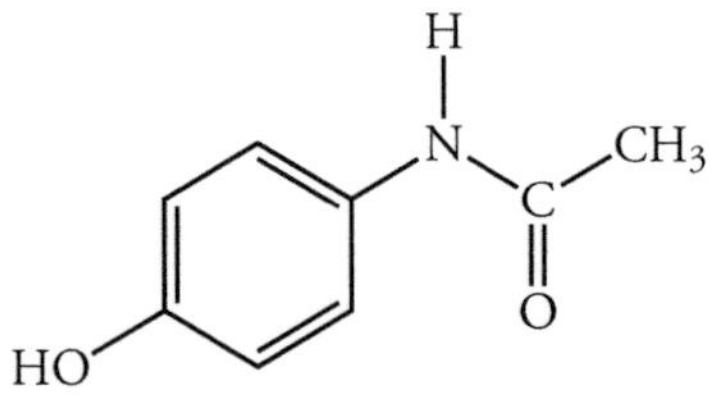

(b)

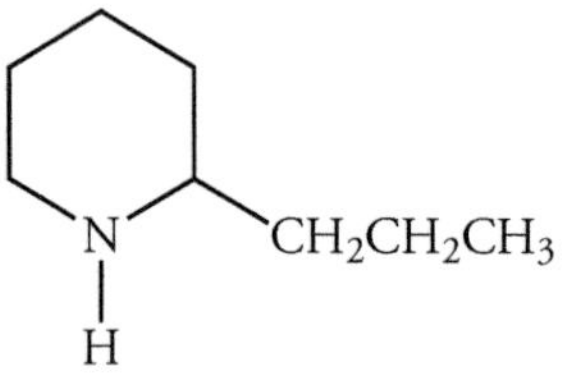

CH_2CH_2CH_3

12.4 Classify the nitrogen-containing functional group in each of the following structures:
(a) DEET, an insect repellent   (b) phencylidine, a hallucinogen

(a)

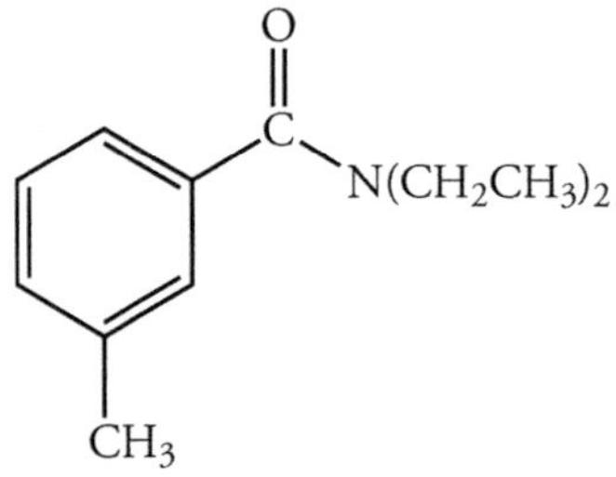

DEET

(b)

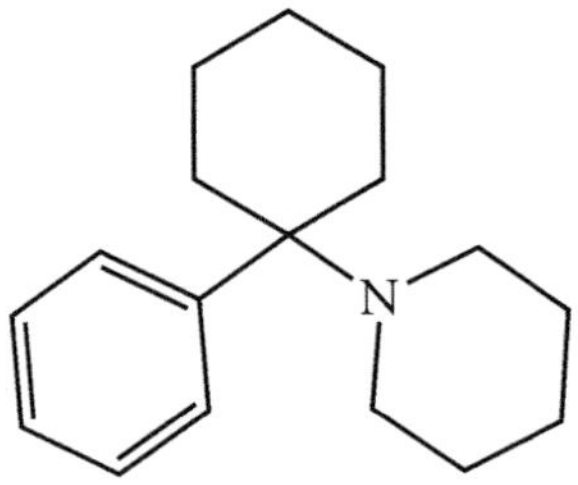

phencyclidine

12.5 Classify the nitrogen-containing functional group in encainide, an antiarrhythmic drug.

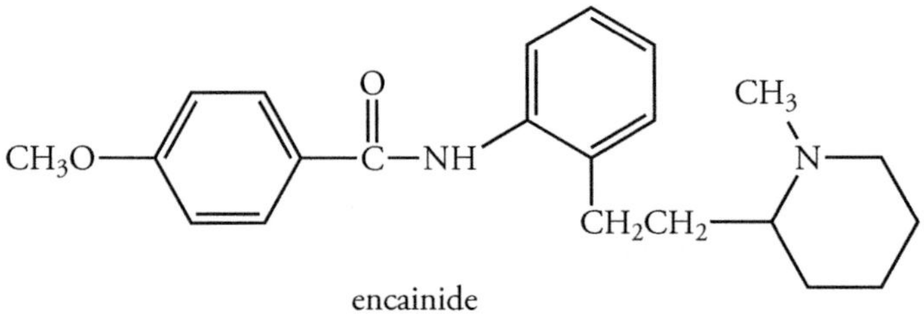

encainide

12.6 Classify the nitrogen-containing functional group in practolol, an antihypertensive drug.

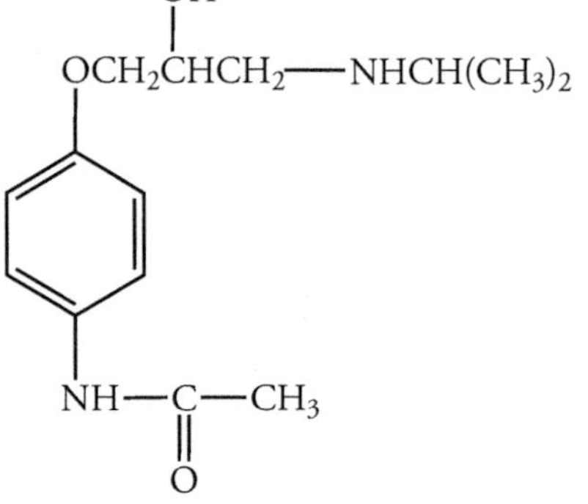

practolol

Nomenclature

12.7 Give the IUPAC name for each of the following compounds:

(a)
$$CH_3CH_2-\overset{\overset{\displaystyle CH_3}{|}}{N}-CH_2CH_2CH_3$$

(b)
$$CH_3CH_2-\overset{\overset{\displaystyle H}{|}}{N}-CH_2\overset{\overset{\displaystyle CH_3}{|}}{C}HCH_2CH_3$$

(c)

(d)

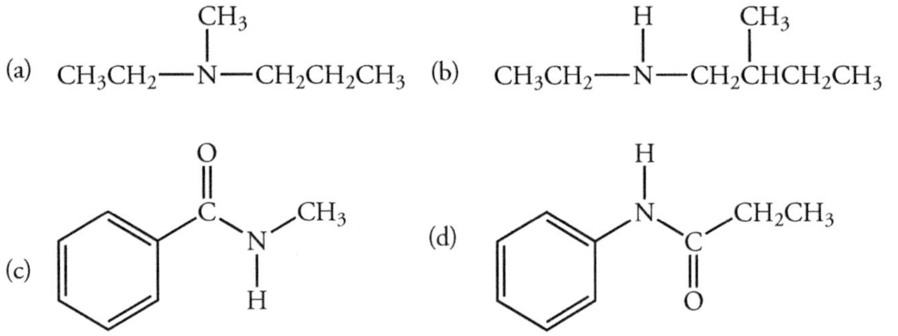

12.8 Classify the nitrogen-containing functional group in each of the following structures:

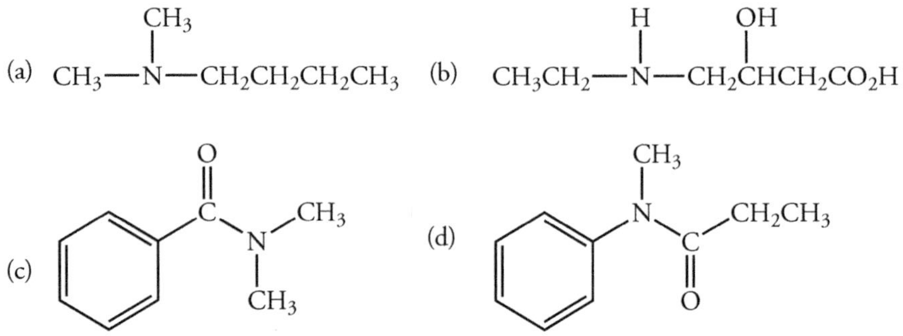

(a)
$$CH_3-\overset{\overset{\displaystyle CH_3}{|}}{N}-CH_2CH_2CH_2CH_3$$

(b)
$$CH_3CH_2-\overset{\overset{\displaystyle H}{|}}{N}-CH_2\overset{\overset{\displaystyle OH}{|}}{C}HCH_2CO_2H$$

(c)

(d)

12.9 An antidepressant drug is named *trans*-2-phenylcyclopropylamine. Draw its structure.

12.10 Tranexamic acid is a drug that aids blood clotting. Its IUPAC name is *trans*-4-(aminomethyl)cyclohexanecarboxylic acid. Draw its structure.

12.11 Name the following compound produced by the marine acorn worm:

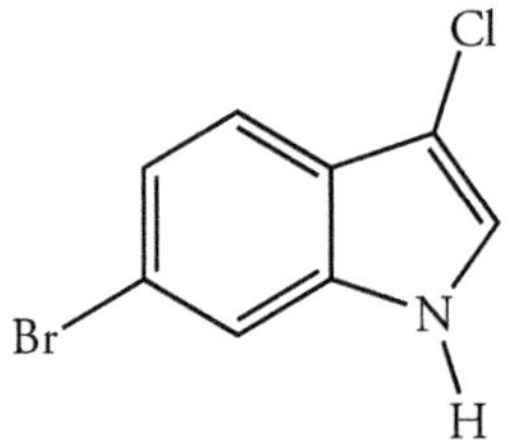

12.12 Draw the structure of each of the following compounds:
(a) 2-ethylpyrrole (b) 3-bromopyridine (c) 2,5-dimethylpyrimidine

Isomers of Amines

12.13 Draw and name all isomers with the molecular formula C_2H_7N.

12.14 Draw and name all isomers with the molecular formula C_3H_9N.

12.15 Draw and name all isomers for primary amines with the molecular formula $C_4H_{11}N$.

12.16 Draw and name all isomers for primary amines with the molecular formula $C_5H_{13}N$.

Properties of Amines

12.17 The boiling points of the isomeric compounds propylamine and trimethylamine are 49 and 3.5 °C, respectively. Explain this large difference.

12.18 The boiling point of 1,2-diaminoethane is 116 °C. Explain why this compound boils at a much higher temperature than propylamine (49 °C) although its molecular weight is similar.

Basicity of Amines

12.19 The pK_a values for cyclohexylamine and triethylamine are 3.34 and 2.99, respectively. Which compound is the stronger base?

12.20 The K_a values For dimethylamine and diethylamine are 4.7×10^{-4} and 3.1×10^{-4}, respectively. Which compound is the stronger base?

12.21 Based on the data in Table 12.2, estimate the K_b of each of the following molecules:

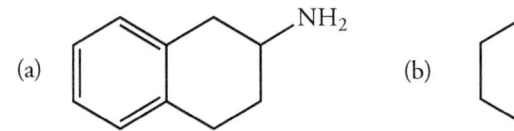

(a)
(b) 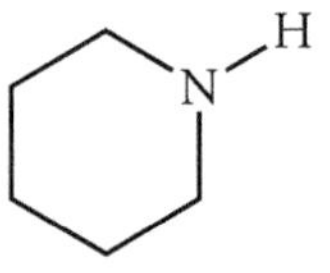
(c)

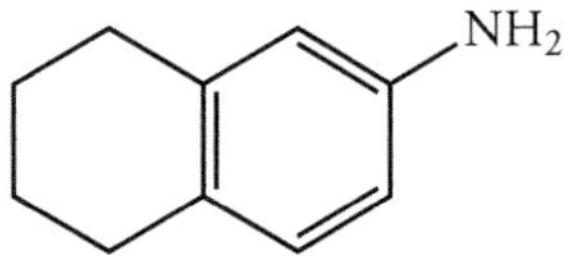

12.22 Based on the data in Table 12.2, estimate the K_b of each of the following molecules:

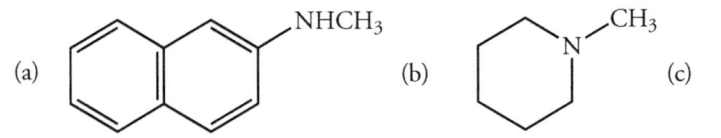

(a)
(b)
(c)

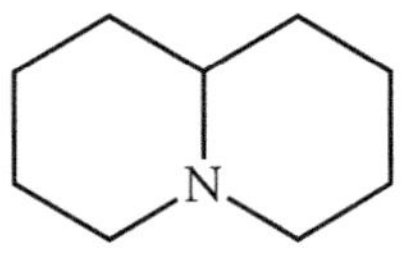

12.23 Explain why the pK_b of aniline (9.4) is different from the pK_b of p-nitroaniline (13.0).

12.24 Explain the difference in the pK_b of the following bases:

(a) $N\equiv C-CH_2CH_2NH_2$ (b) $N\equiv C-CH_2NH_2$

12.25 Physostigmine is used in 0.1 to 1.0% solutions to decrease the intraocular pressure in treatment of glaucoma. Rank the three nitrogen atoms in the molecule in order of increasing basicity.

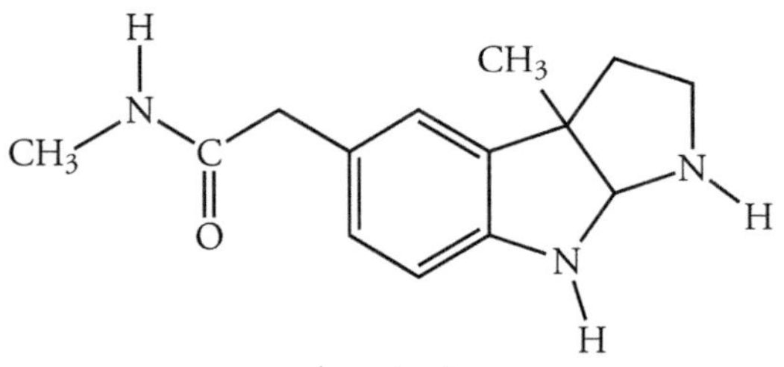

physostigmine

12.26 Nubucaine is a local anesthetic that is administered as the hydrochloride salt. Which nitrogen atom is protonated?

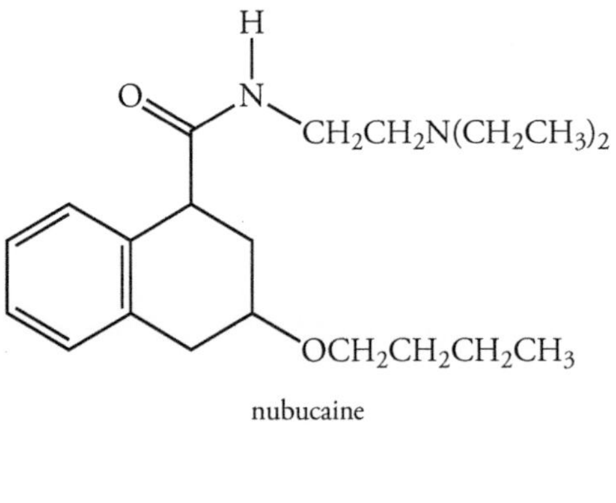

nubucaine

Reactions of Amines

12.27 Draw the structure of the compound formed when benzylmethylamine reacts with each of the following reagents:
(a) excess methyl iodide   (b) acetyl chloride   (c) hydrogen iodide

12.28 Draw the structure of the compound formed when piperidine reacts with each of the following reagents:
(a) allyl bromide   (b) benzoyl chloride   (c) acetic anhydride

Synthesis of Amines

12.29 Write the structure of the product of each of the following reactions:

(a)
$$CH_3-\overset{\displaystyle O}{\overset{\displaystyle \|}{C}}-CH_2CH_2CH_3 \ + \ CH_3CH_2NH_2 \quad \xrightarrow[\text{Ni}]{H_2}$$

(b) 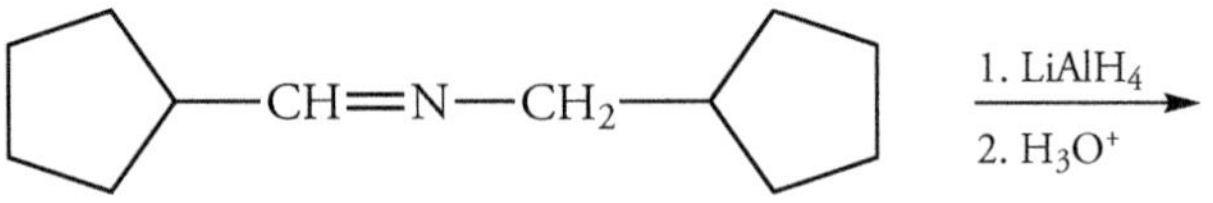
$\xrightarrow[\text{2. H}_3\text{O}^+]{\text{1. LiAlH}_4}$

(c) 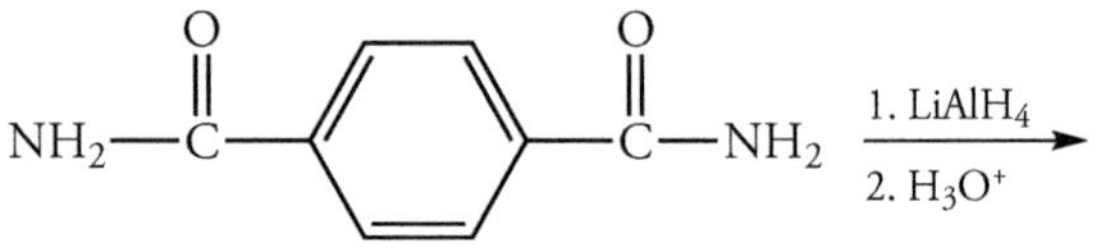
$$NH_2-\overset{\displaystyle O}{\overset{\displaystyle \|}{C}}-\!\!\bigcirc\!\!-\overset{\displaystyle O}{\overset{\displaystyle \|}{C}}-NH_2 \quad \xrightarrow[\text{2. H}_3\text{O}^+]{\text{1. LiAlH}_4}$$

12.30 Write the structure of the product of each of the following reactions:

(a)
$$CH_3-\overset{\displaystyle O}{\overset{\displaystyle \|}{C}}-CH_2CH_2CH_3 \ + \ CH_3NH_2 \quad \xrightarrow[\text{Ni}]{H_2}$$

(b)
$$CH_3-\overset{\displaystyle NCH_3}{\overset{\displaystyle \|}{C}}-CH_2CH_2CH_3 \quad \xrightarrow[\text{2. H}_3\text{O}^+]{\text{1. LiAlH}_4}$$

(c) 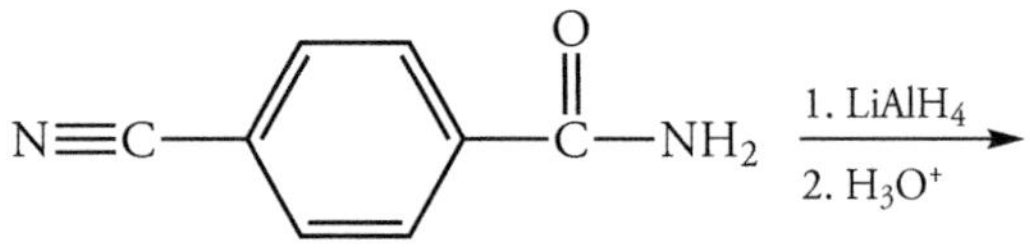
$$N\!\equiv\!C-\!\!\bigcirc\!\!-\overset{\displaystyle O}{\overset{\displaystyle \|}{C}}-NH_2 \quad \xrightarrow[\text{2. H}_3\text{O}^+]{\text{1. LiAlH}_4}$$

12.31 Write the structure of the final product of each of the following sequences of reactions:

(a)
$$CH_3(CH_2)_4CO_2H \quad \xrightarrow[\text{pyridine}]{\text{SOCl}_2} \quad \xrightarrow{\text{NH}_3} \quad \xrightarrow[\text{2. H}_3\text{O}^+]{\text{1. LiAlH}_4}$$

(b) 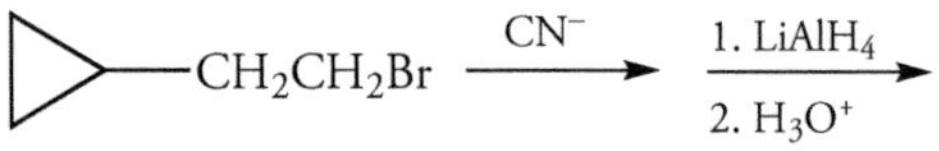
$$\triangleright\!-CH_2CH_2Br \quad \xrightarrow{\text{CN}^-} \quad \xrightarrow[\text{2. H}_3\text{O}^+]{\text{1. LiAlH}_4}$$

(c)
$$CH_3(CH_2)_3CO_2H \quad \xrightarrow[\text{2. H}_3\text{O}^+]{\text{1. LiAlH}_4} \quad \xrightarrow{\text{PBr}_3} \quad \xrightarrow[\text{excess}]{\text{NH}_3}$$

12.32 Write the structure of the final product of each of the following sequences of reactions:

(a)
$$CH_3\overset{\overset{\displaystyle OH}{|}}{C}HCH_2CH_2CH_3 \xrightarrow{PCC} \xrightarrow[CH_3NH_2]{H_2/Ni}$$

(b) 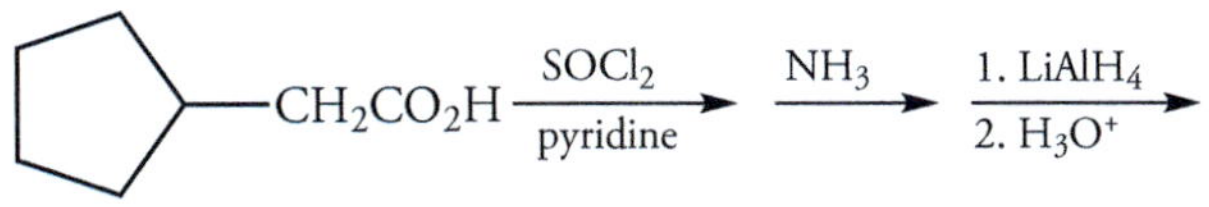
cyclopentyl—$CH_2CO_2H \xrightarrow[\text{pyridine}]{SOCl_2} \xrightarrow{NH_3} \xrightarrow[\text{2. } H_3O^+]{\text{1. LiAlH}_4}$

(c)
$$HOCH_2(CH_2)_4CH_2OH \xrightarrow[\text{excess}]{HBr} \xrightarrow[\text{excess}]{CN^-} \xrightarrow[\text{2. } H_3O^+]{\text{1. LiAlH}_4}$$

12.33 Outline the steps required to convert benzoic acid into N-ethylbenzylamine.

12.34 Outline the steps required to convert benzyl chloride into 2-phenylethanamine.

Synthesis of Amines

12.35 Write the products of the reaction of each of the following compounds with strong aqueous acid.

(a) 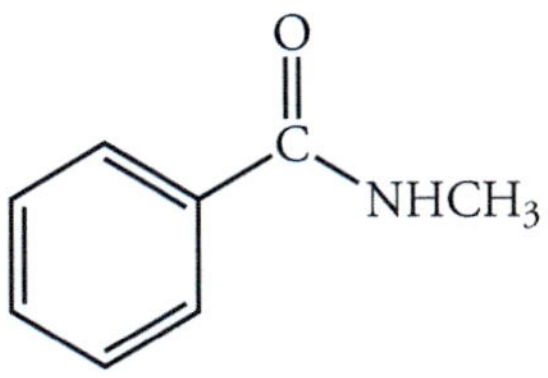 phenyl—C(=O)—NHCH₃

(b) $CH_3CH_2\overset{\overset{\displaystyle O}{\|}}{C}-N(CH_3)_2$

(c) 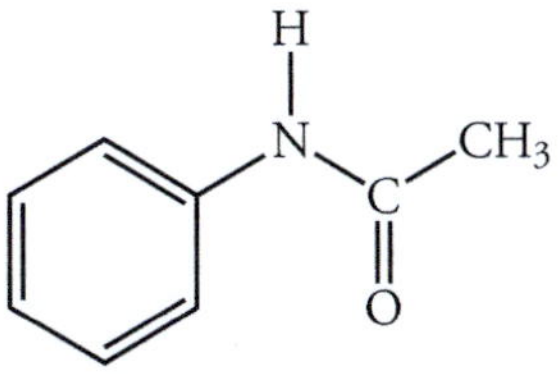 phenyl—N(H)—C(=O)—CH₃

12.36 Write the products of the reaction of each of the following compounds with strong aqueous base.

(a) 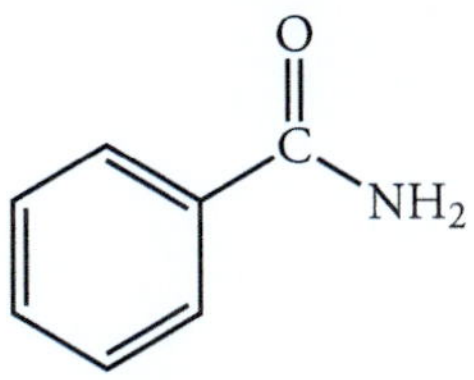 phenyl—C(=O)—NH₂

(b) $CH_3CH_2\overset{\overset{\displaystyle O}{\|}}{C}-N(CH_3CH_2)_2$

(c) 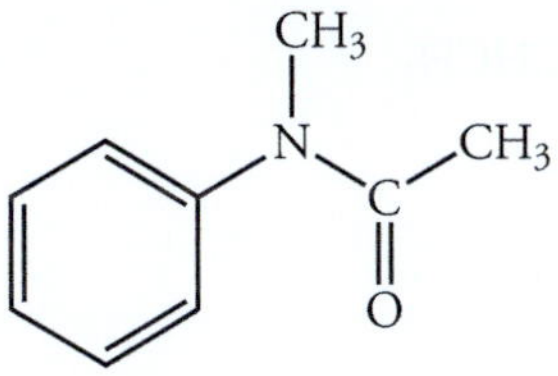 phenyl—N(CH₃)—C(=O)—CH₃

12.37 Write the product of reduction of each compound in Exercise 12.35 by lithium aluminum hydride followed by hydrolysis with aqueous acid.

12.38 Write the product of reduction of the following cyclic amides (lactams) by lithium aluminum hydride followed by hydrolysis with aqueous acid.

(a) 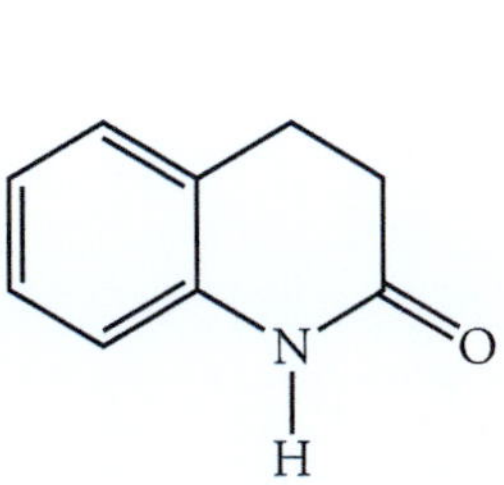 (b) 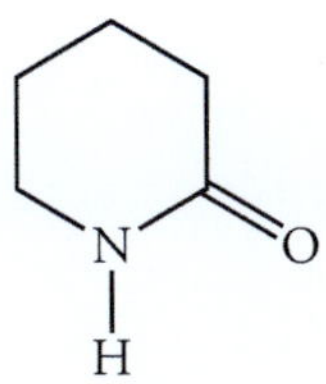 (c)

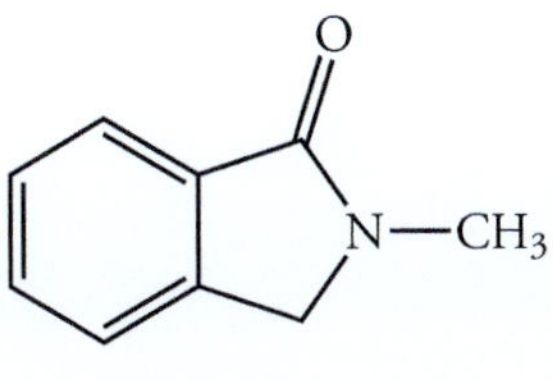

Synthesis of Amides

12.39 Write the structure of the product of each of the following reactions:

(a)

$$CH_3(CH_2)_3 \overset{\overset{\displaystyle O}{\|}}{C} Cl \ + \ CH_3NH_2 \ \xrightarrow{\text{pyridine}}$$

(b)

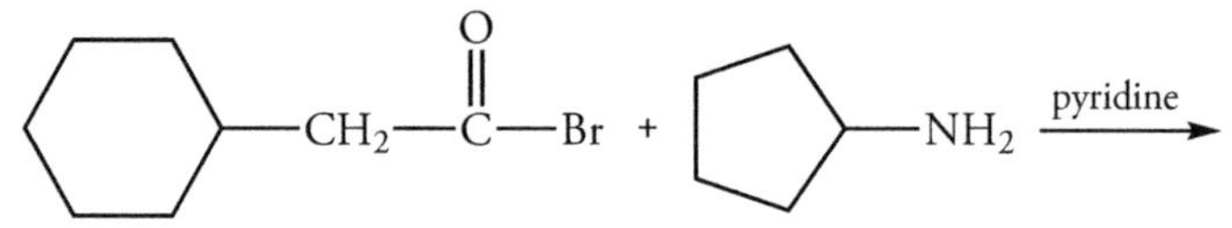

cyclohexyl–CH_2–$\overset{\overset{\displaystyle O}{\|}}{C}$–Br + cyclopentyl–$NH_2$ $\xrightarrow{\text{pyridine}}$

(c)

cyclopentyl–CH_2–$\overset{\overset{\displaystyle O}{\|}}{C}$–Br + NH_3 $\longrightarrow$

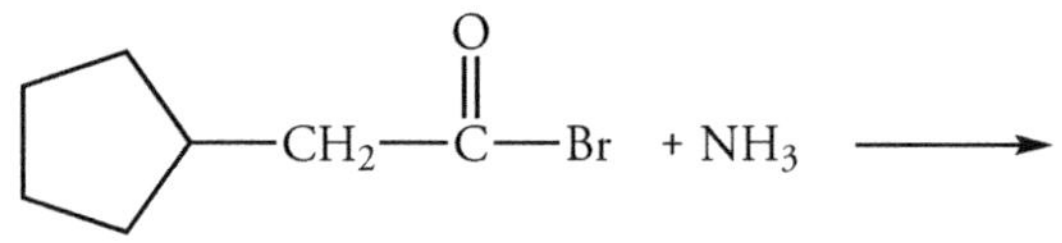

12.40 Write the structure of the product of each of the following reactions:

(a)

$$CH_3(CH_2)_3 \overset{\overset{\displaystyle O}{\|}}{C} OCH_2CH_3 \ + \ CH_3CH_2NH_2 \ \longrightarrow$$

(b)

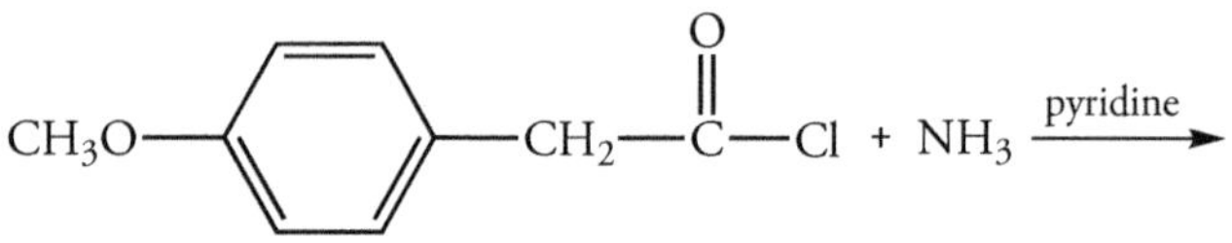

CH_3O–⟨benzene⟩–CH_2–$\overset{\overset{\displaystyle O}{\|}}{C}$–Cl + NH_3 $\xrightarrow{\text{pyridine}}$

(c)

cyclopentyl–CH_2–$\overset{\overset{\displaystyle O}{\|}}{C}$–$SCH_3$ + CH_3NH_2 $\longrightarrow$

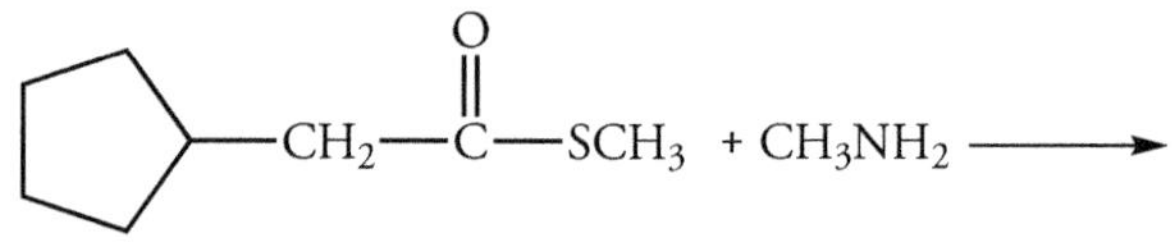

12.41 Select two reactants that could be used to prepare crotamiton, which is used to treat scabies.

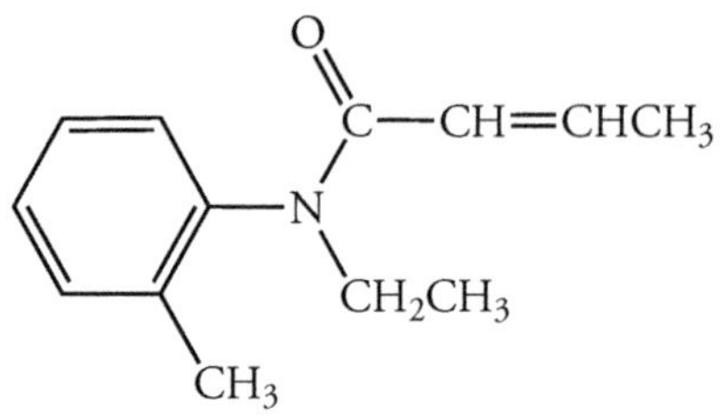

⟨structure: benzene ring with N(CH₂CH₃) bearing C(=O)–CH=CHCH₃, ortho CH₃⟩

crotamiton

12.42 Select two reactants that could be used to prepare bupivacaine, a local anesthetic.

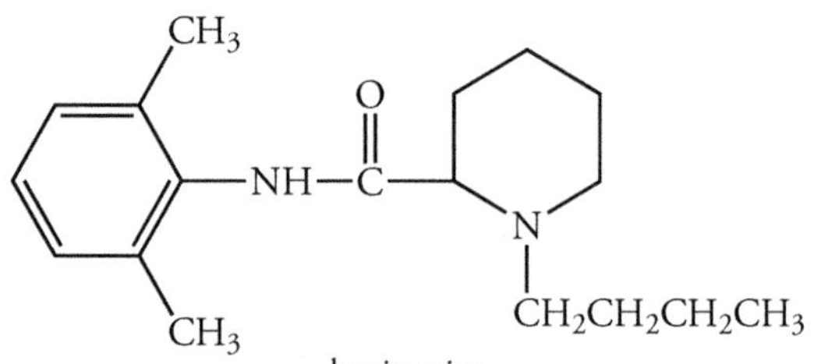

⟨structure: 2,6-dimethylphenyl–NH–C(=O)–piperidine with N–CH₂CH₂CH₂CH₃⟩

bupivacaine

13

CARBOHYDRATES

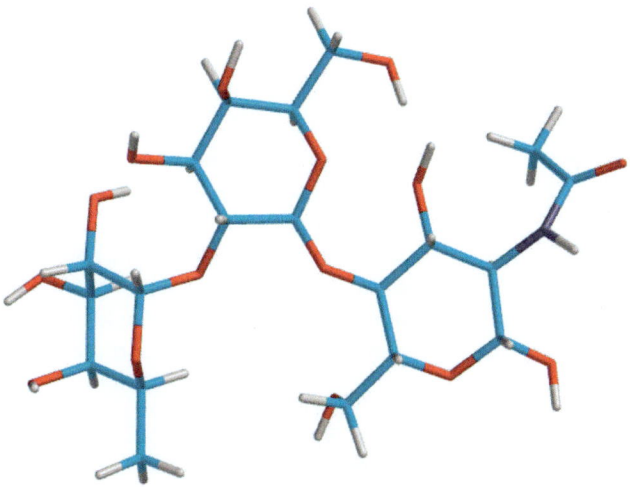

TYPE O BLOOD GROUP ANTIGEN

13.1 CLASSIFICATION OF CARBOHYDRATES

If importance were measured by abundance, the carbohydrates would hold first prize, for they are far and away the most abundant molecules in the biological world (excepting only water). The functions of carbohydrates are as varied as their structures. They are a major source of metabolic energy, and are important structural components in plants and in animal cells. **Carbohydrates** are polyhydroxy aldehydes or ketones or compounds that can be hydrolyzed to form them. They span a wide range of structures, from molecules containing three carbon atoms to giant molecules containing thousands of carbon atoms.

Carbohydrates are divided into three large structural classes—monosaccharides, oligosaccharides, and polysaccharides. Carbohydrates such as glucose and fructose that cannot be hydrolyzed into smaller molecules are **monosaccharides**.

Carbohydrates consisting of 2-10 or so monosaccharides are **oligosaccharides**. Hydrolysis of oligosaccharides may yield identical monosaccharides or two or more different monosaccharides. Oligosaccharides are called **disaccharides**, **trisaccharides**, and so forth, depending on the number of linked monosaccharide units. The disaccharide lactose, also called "milk sugar," contains one molecule of glucose and one of galactose. Maltose, another disaccharide, contains two glucose units.

Polysaccharides contain thousands of covalently linked monosaccharides. Those that contain only one type of monosaccharide are **homopolysaccharides**. Examples include starch and cellulose that are made by plants and contain only glucose. Glycogen, found in animals, is a homopolysaccharide containing glucose. Polysaccharides that contain more than one type of monosaccharide are called **heteropolysaccharides**.

The monosaccharides in oligo- and polysaccharides are linked by acetal or ketal bonds, which are called **glycosidic bonds** in carbohydrate chemistry. Glycosidic bonds link the aldehyde or ketone site of one monosaccharide and a hydroxyl group of another monosaccharide. Hydrolysis of the glycosidic bonds yields the component monosaccharides.

Monosaccharides are further classified by their most highly oxidized functional group. Monosaccharides are called **aldoses** if their most highly oxidized functional group is an aldehyde. They are **ketoses** if their most highly oxidized functional group is a ketone. The suffix *-ose* indicates that a compound is a carbohydrate. The prefix *aldo-* or *keto-* indicates that the compound is an aldehyde or ketone. The number of carbon atoms in an aldose or ketose is indicated by the prefix *tri-*, *tetr-*, *pent-*, and *hex-*. Aldoses are numbered from the carbonyl carbon atom. Ketoses are numbered from the end of the carbon chain closest to the carbonyl carbon atom.

Principles of Organic Chemistry. http://dx.doi.org/10.1016/B978-0-12-802444-7.00013-6

a ketotetrose

an aldopentose

13.2 CHIRALITY OF CARBOHYDRATES

Monosaccharides are conveniently represented by their Fischer projection formulas (Section 6.4). We recall that in a Fischer projection formula the carbon chain is represented by a vertical line. Groups attached at the top and bottom of the vertical line represent bonds going into the page, and horizontal lines represent bonds coming out of the page. By convention, the carbonyl carbon atom. The most oxidized carbon atom in these compounds, is placed near the "top" in the Fischer projection formula. The simplest aldose, glyceraldehyde, has three carbon atoms, one of which is a stereogenic center. This aldotriose can exist in two enantiomeric forms.

(D)-glyceraldehyde   Fischer projection   (L)-glyceraldehyde   Fischer projection

Monosaccharides with multiple stereogenic centers are represented with the carbon backbone continually pushed back behind the plane of the page. A curve of atoms in a C shape results, with the attached hydrogen atoms and hydroxyl groups pointing out from the backbone of the carbon chain.

D-galactose   Fischer projection   D-fructose   Fischer projection

D- and L- Series of Monosaccharides

The stereochemistry of each stereogenic center of a monosaccharide can be assigned by the R,S notation. However, late in the nineteenth century, the German chemist Emil Fischer devised a stereochemical nomenclature that preceded the R,S notation. It is still in common use for carbohydrates and amino acids (Chapter 14). In the Fischer stereochemical system, the configurations of all stereogenic centers are based on their relationship to the naturally occurring stereoisomer of glyceraldehyde. We recall that when this aldotriose has the hydroxyl group on C-2, which is located on the right in the projection formula, its configuration is symbolized as D. Its enantiomer, called L-glyceraldehyde, has the hydroxyl group on the left at C-2 in the Fischer projection formula.

Monosaccharides are made in cells from the "building block" D-glyceraldehyde, so nearly all naturally occurring monosaccharides have the same configuration as D-glyceraldehyde at the stereogenic carbon atom farthest from the carbonyl group. For example, the hydroxyl groups on the C-4 atom of ribose and on the C-5 atoms of glucose and fructose are on the right side in the following projection formulas. Each compound is a member of the D series of aldoses.

D-glyceraldehyde   D-ribose   D-glucose   D-fructose

The configuration of the hydroxyl groups of a monosaccharide relative to one another determines its chemical identity. Thus, in ribose the hydroxyl groups at the stereogenic carbon atoms are all on the same side in the Fischer projection formula. The name D-ribose has two components: the term D defines the configuration at the stereogenic carbon atom farthest from the aldehyde; the term *ribo-* by itself defines the relative configuration of the three stereogenic centers at C-2, C-3, and C-4. Taken together, the name D-ribose defines the absolute configuration at every stereogenic center in the molecule. Thus, in D-ribose, the hydroxyl groups at C-2, C-3, and C-4 are all on the right in the Fischer projection formula. In the enantiomer, L-ribose, the hydroxyl groups at C-2, C-3, and C-4 are all on the left because the entire structure is the mirror image of the D isomer.

The OH group is on the left in the Fischer projection formula for the L-isomer

The OH group is on the right in the Fischer projection formula for the D-isomer

L-ribose   D-ribose

The Fischer projection formulas of the aldotetroses, aldopentoses, and aldohexoses of the D series are shown in Figure 13.1. D-Glyceraldehyde, at the top of the "tree," is the parent aldose. When we insert a new stereogenic center (H—C—OH) between the carbonyl carbon atom and the stereogenic center below it, the resulting molecules are D-aldotetroses. Because the new H—C—OH group can have its group can have its −OH group on the right or left, two aldotetroses, D-erythrose and D-threose, are possible. Note that aldotetroses contain two nonequivalent stereogenic centers, so $2^2 = 4$ stereoisomers are possible. The two L-aldotetroses are not shown in Figure 13.1. D-Erythrose and L-erythrose are enantiomers, as are D-threose and L-threose.

Inserting another stereogenic center (H—C—OH), which can have either of two configurations, between the carbonyl carbon atom and the chiral center at C-2 in D-erythrose leads to two D-aldopentoses: D-ribose and D-arabinose. Similarly, inserting a new stereogenic center (H—C—OH) between the carbonyl carbon atom and the stereogenic center at C-2 in D-threose gives D-xylose and D-lyxose. Repeating the process one more time in each of the four D-aldopentoses gives a total of eight D-aldohexoses. D-Glucose and D-galactose are the most widely found in nature; D-mannose and D-talose occur in smaller amounts. The others are extremely rare.

Any of the members of a group of isomeric monosaccharides shown in Figure 13.1 are related as diastereomers. They are not enantiomers because they are not mirror images. The enantiomer of each compound is the L-monosaccharide of the same name.

Epimers

We recall that diastereomers are stereoisomers, but not enantiomers. Diastereomers that contain two or more stereogenic carbon atoms but differ in configuration at only one stereogenic center are called epimers. Thus, the diastereomers D-glucose and D-galactose are epimers because they differ in configuration only at C-4. D-Glucose and D-mannose are epimers that differ in configuration at C-2.

Figure 13.1
Structures of D-Aldoses

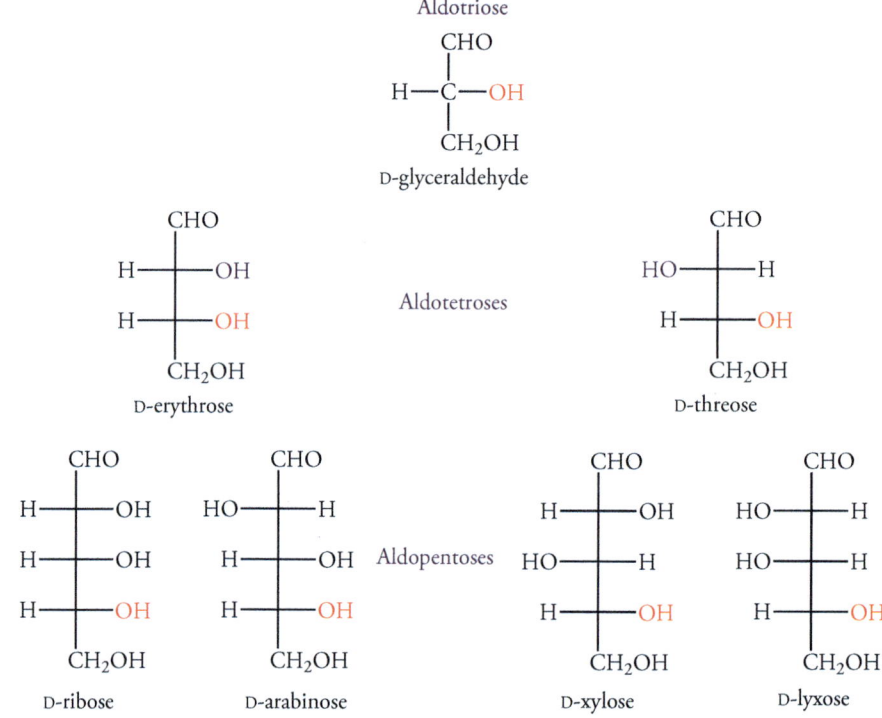

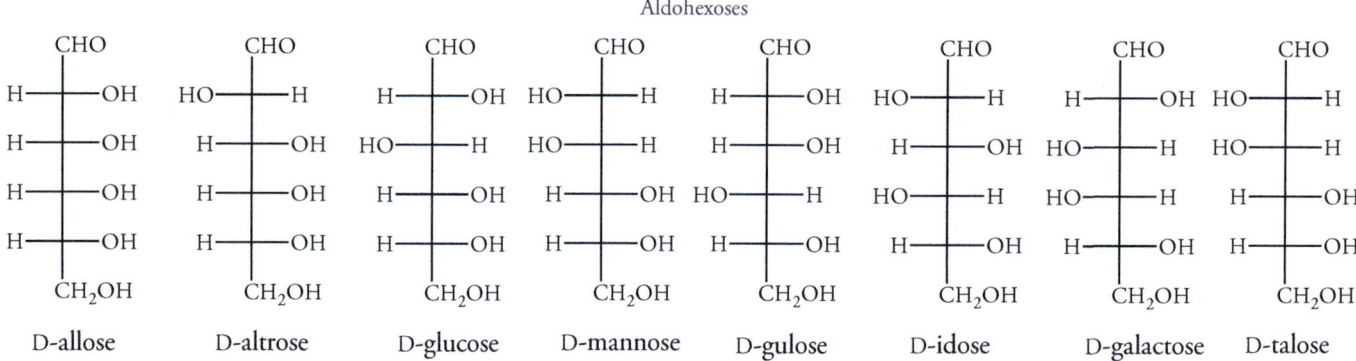

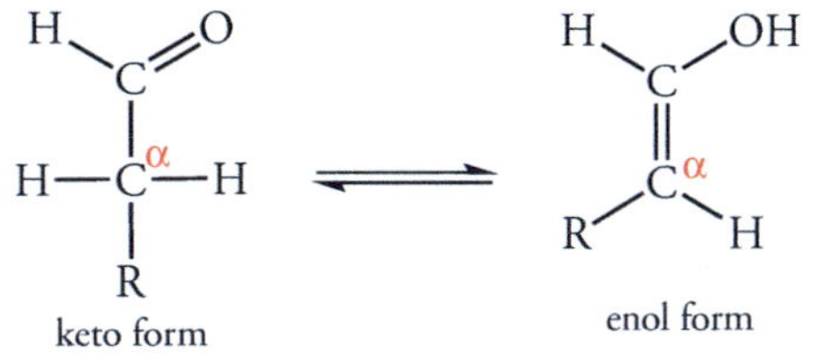

keto form   enol form

The interconversion of epimers such as D-glucose and D-mannose at C-2 illustrates the tautomerization reaction we discussed in Chapter 10. The α-hydrogen atom of an aldehyde is involved in a tautomerization reaction, and is responsible for an equilibrium between keto and enol forms of carbonyl compounds.

In an aldose the α-carbon atom is a stereogenic center with a hydroxyl group bonded to it. Tautomerization yields an enediol in which the α-carbon atom is not a stereogenic center. In the reverse reaction to regenerate the aldose, a stereogenic center is formed again at the α-carbon atom, which can have either of two configurations. The resultant compounds are C-2 epimers.

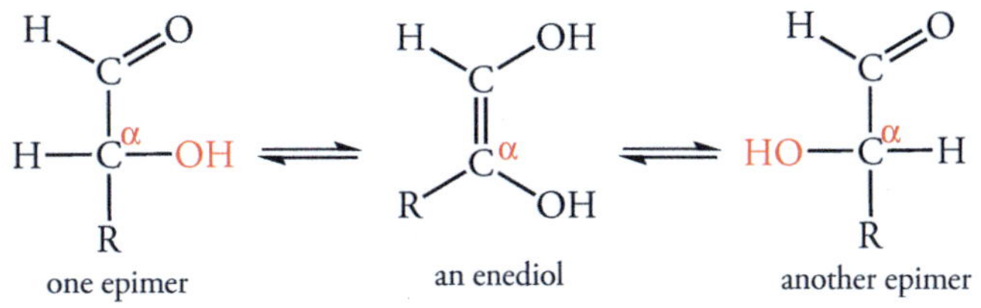

one epimer   an enediol   another epimer

Ketoses

Our focus to this point has been the aldoses, which play an important role in many biological processes. However, several ketoses also play a pivotal role in metabolism. The Fischer projection formulas of the ketotetroses, ketopentoses, and ketohexoses of the D series are shown in Figure 13.2. The "parent" ketose is the ketotriose called dihydroxyacetone. We can construct ketoses from dihydroxyacetone by consecutively inserting stereogenic centers (H—C—OH) between the ketone carbonyl carbon atom and the carbon atom directly below it.

The simplest ketose, dihydroxyacetone, does not contain any stereogenic carbon atoms. This ketose is produced in the metabolism of glucose as a phosphate ester at the C-3 hydroxyl group. Fructose is produced by isomerization of glucose in glycolysis—a metabolic pathway that all cells use to degrade glucose and produce energy. The ketopentoses ribulose and xylulose are both intermediates in the pentose phosphate pathway, another important metabolic pathway that produces the ribose necessary for ribonucleic acids.

Problem 13.1
What stereochemical relationships exist between (a) and (b) and between (a) and (c)?

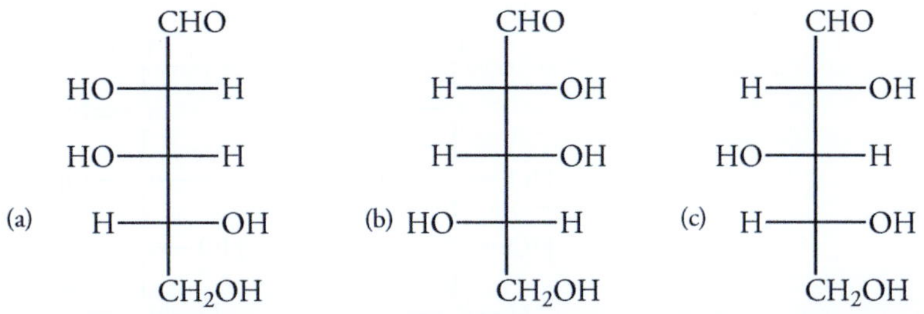

Solution
Structures (a) and (b) are enantiomers because the configuration at every center in (b) is opposite that of corresponding centers in (a). Structures (a) and (c) are diastereomers. Both are D-monosaccharides and differ in configuration only at C-2, so they are epimers.

Problem 13.2

What relationship exists between D-allose and D-talose (see Figure 13.1)?

Problem 13.3

What is the structure of L-arabinose, which is present in some antiviral drugs?

Solution

L-Arabinose is the enantiomer of D-arabinose (see Figure 13.1). Therefore, the L-isomer can be drawn by reflecting the planar projection formula in an imagined mirror perpendicular to the plane of the page and parallel to the carbon chain. Each hydroxyl group that is on the right in D-arabinose is on the left in L-arabinose, and vice versa.

Figure 13.2
Structures of D-2-Ketoses

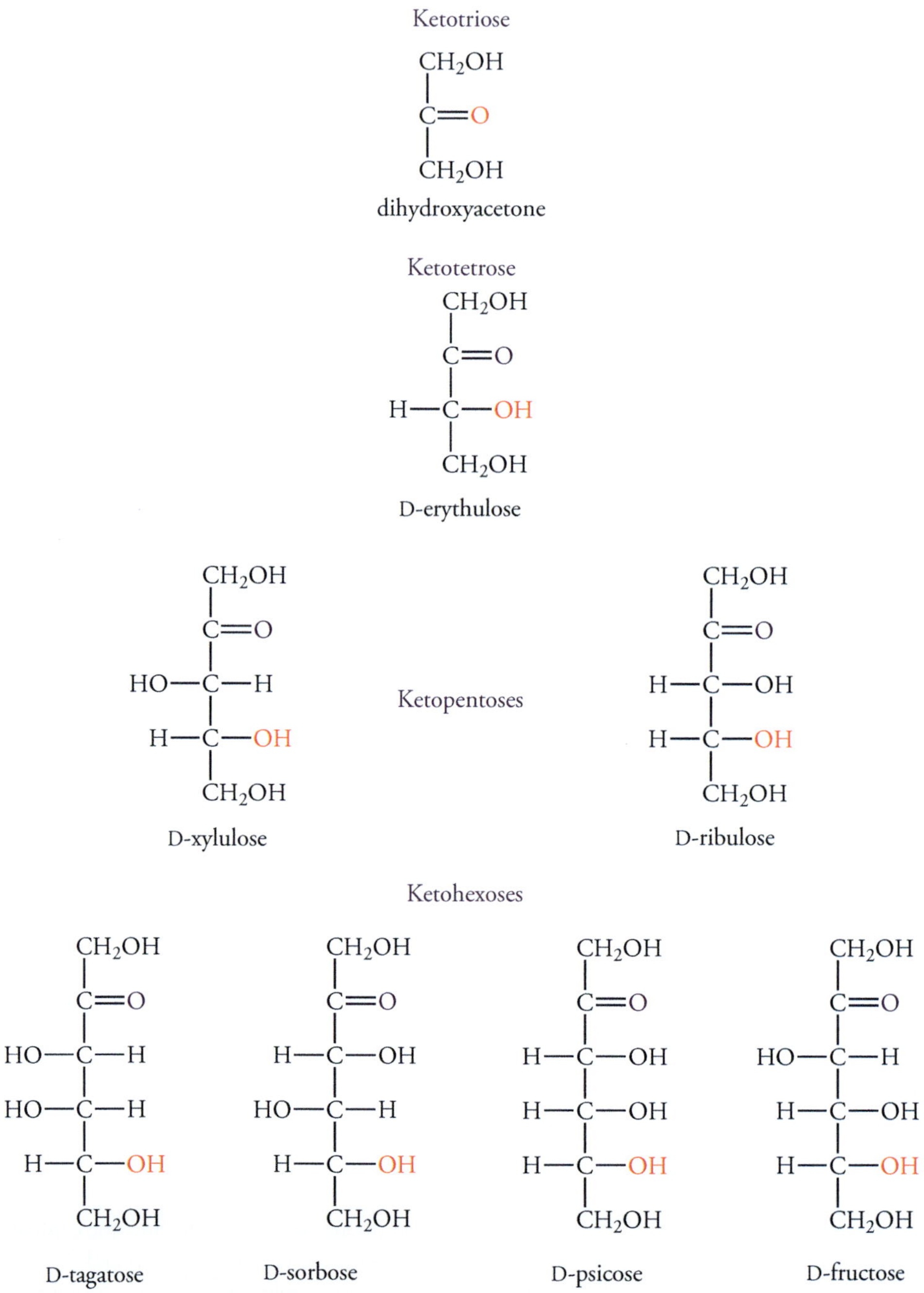

Ketotriose

CH_2OH
|
$C{=}O$
|
CH_2OH

dihydroxyacetone

Ketotetrose

CH_2OH
|
$C{=}O$
|
$H{-}C{-}OH$
|
CH_2OH

D-erythulose

Ketopentoses

CH_2OH
|
$C{=}O$
|
$HO{-}C{-}H$
|
$H{-}C{-}OH$
|
CH_2OH

D-xylulose

CH_2OH
|
$C{=}O$
|
$H{-}C{-}OH$
|
$H{-}C{-}OH$
|
CH_2OH

D-ribulose

Ketohexoses

CH_2OH
|
$C{=}O$
|
$HO{-}C{-}H$
|
$HO{-}C{-}H$
|
$H{-}C{-}OH$
|
CH_2OH

D-tagatose

CH_2OH
|
$C{=}O$
|
$H{-}C{-}OH$
|
$HO{-}C{-}H$
|
$H{-}C{-}OH$
|
CH_2OH

D-sorbose

CH_2OH
|
$C{=}O$
|
$H{-}C{-}OH$
|
$H{-}C{-}OH$
|
$H{-}C{-}OH$
|
CH_2OH

D-psicose

CH_2OH
|
$C{=}O$
|
$HO{-}C{-}H$
|
$H{-}C{-}OH$
|
$H{-}C{-}OH$
|
CH_2OH

D-fructose

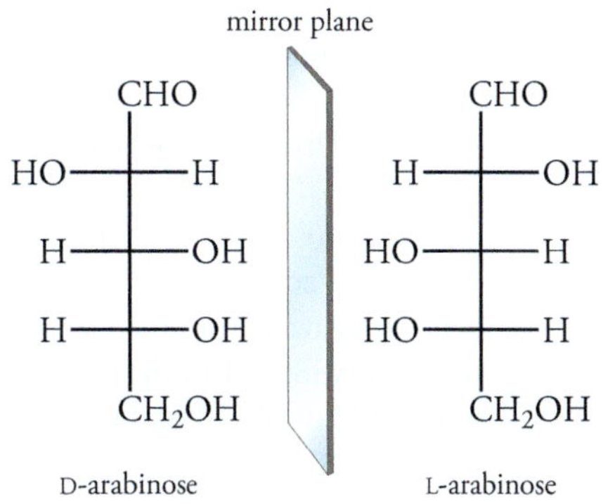

mirror plane

D-arabinose   L-arabinose

Problem 13.4
Draw the structure of the C-3 epimer of D-ribulose (see Figure 13.2).

13.3 HEMIACETALS AND HEMIKETALS

We recall that aldehydes and ketones react reversibly with alcohols to form hemiacetals and hemiketals, respectively (Chapter 10).

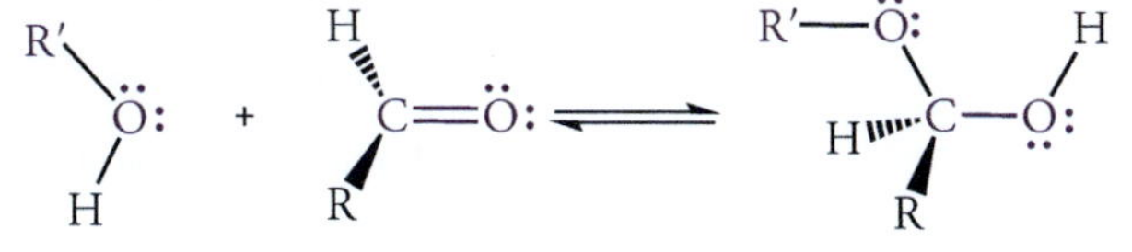

hemiacetal

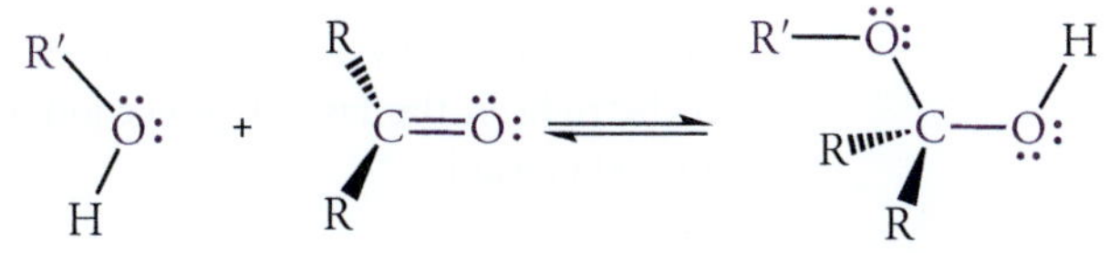

hemiketal

When the hydroxyl group and the carbonyl group are part of the same molecule, a cyclic hemiacetal is formed in an intramolecular reaction. Cyclic hemiacetals containing five or six atoms in the ring form readily because of the proximity of the two functional groups.

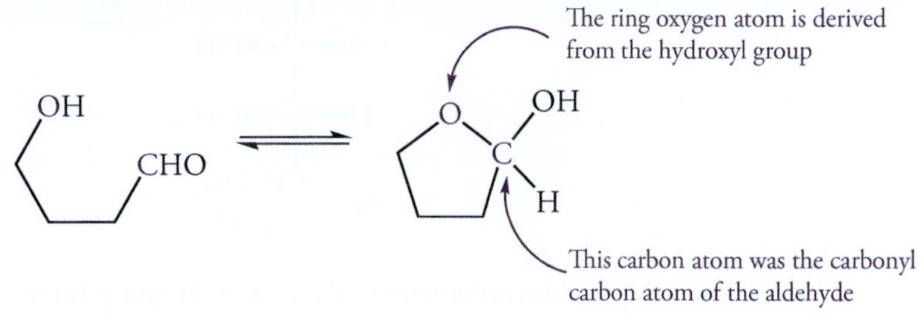

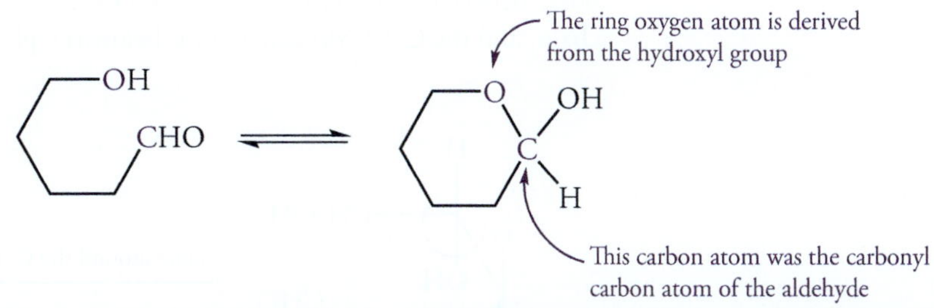

The ring oxygen atom is derived from the hydroxyl group

This carbon atom was the carbonyl carbon atom of the aldehyde

The ring oxygen atom is derived from the hydroxyl group

This carbon atom was the carbonyl carbon atom of the aldehyde

The cyclic hemiacetal or hemiketal forms of aldo- and ketohexoses and pentoses are the predominant forms of monosaccharides rather than the open-chain structures we have discussed to this point. Cyclic hemiacetals and hemiketals of carbohydrates that contain five-membered rings are called

furanoses; cyclic hemiacetals and hemiketals that contain six-membered rings are called **pyranoses**. These structures are usually represented by planar structures called Haworth projection formulas.

Haworth Projection Formulas

In a Haworth projection formula, a cyclic hemiacetal or hemiketal is represented as a planar structure and viewed edge-on. Bond lines representing atoms toward the viewer are written as heavy wedges; bond lines away from the viewer are written as ordinary bond lines. The carbon atoms are arranged clockwise with C-1 of the pyranose of an aldohexose on the right. For the furanose of the ketohexose shown, C-2 is placed on the right side of the structure.

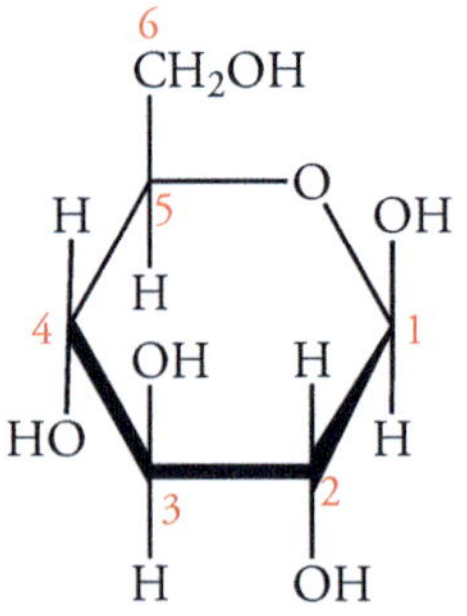

Haworth projection of
a pyranose

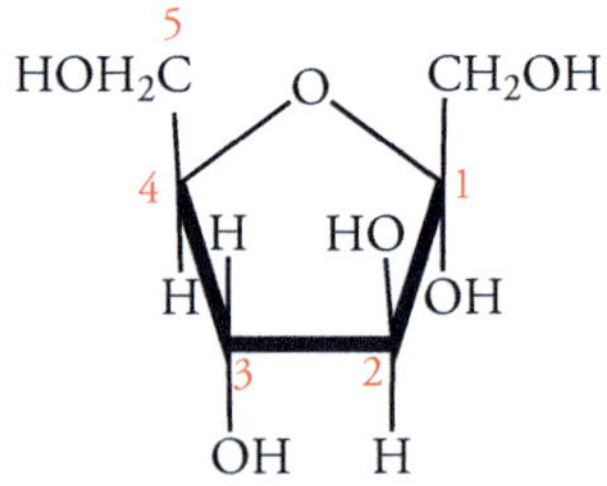

Haworth projection of
a furanose

We can convert the Fischer projection formula of D-glucose into a hemiacetal written as a Haworth projection formula in the following way: The open-chain form of D-glucose is shaped like a "C" that is arranged vertically on the page. Turn this curved chain to the right so that it is horizontal. Groups on the right in the Fischer projection are then directed downward, whereas groups on the left are directed upward.

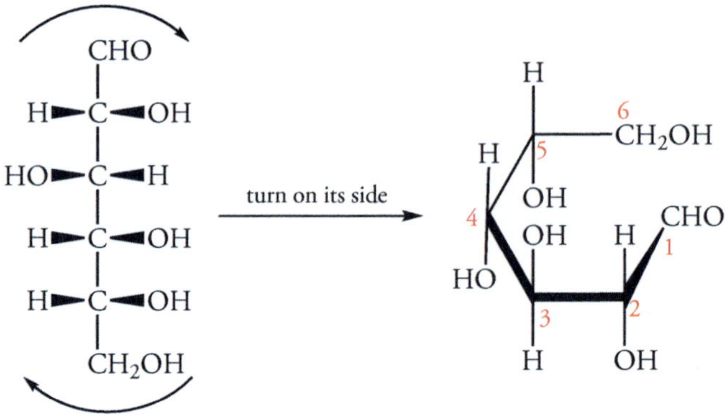

turn on its side

In this conformation, the C-5 −OH group is not near enough to the carbonyl carbon atom to form a ring. To bring the C-5 —OH group near the carbonyl carbon atom, rotate the structure about the bond between C-4 and C-5. The —CH₂OH group is now above the plane of the curved carbon chain, and the C-5 hydrogen atom is below the plane.

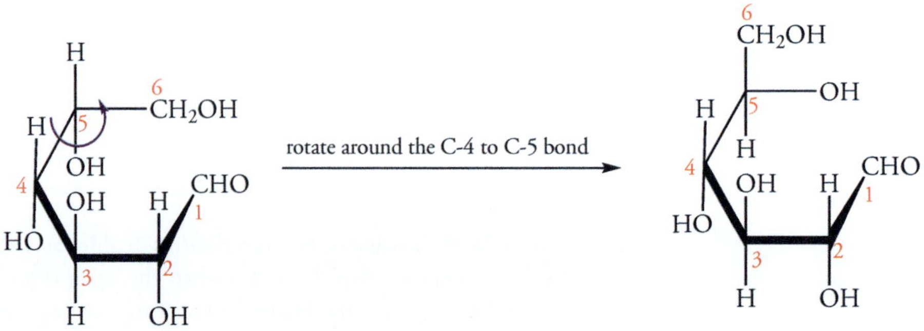

rotate around the C-4 to C-5 bond

Now add the oxygen atom of the C-5 —OH group to the carbonyl carbon atom, and add a hydrogen atom to the carbonyl oxygen atom. A six-membered ring that contains five carbon atoms and one oxygen atom results. There are now four different groups bonded to C-1 in this cyclic hemiacetal. Thus, a new stereogenic center is formed at the original carbonyl carbon atom, and two configurations are possible at C-1. If the hydroxyl group of the hemiacetal is directed below the plane, the compound is α-D-glucopyranose; if it is above the plane, the compound is β-D-glucopyranose.

α-D-glucopyranose   β-D-glucopyranose

Note that in both forms of glucose, as well as in all carbohydrates with a D configuration, the —CH$_2$OH group is located above the ring in a Haworth projection.

The α and β forms of D-glucose are diastereomers that differ in configuration at one center. Hence, they are epimers. Compounds whose configurations differ only at the hemiacetal center are a special type of epimer called **anomers**. The chiral carbon atom at the hemiacetal center that forms in the cyclization reaction is called the **anomeric carbon atom**. Now let's consider the cyclic form of the ketohexose D-fructose. D-Fructose cyclizes in aqueous solution to give a mixture that contains 20% α- and β-D-fructofuranose and 80% α- and β-D-fructopyranose. The furanose isomers are formed when the C-5 —OH group adds to the carbonyl carbon atom of the C-2 keto group. A ring of four carbon atoms and one oxygen atom results. The pyranose isomers form when the C-6 hydroxyl group adds to the C-2 carbonyl carbon atom. Again, α and β designate the configuration at the anomeric carbon atom.

α-D-fructofuranose   α-D-fructofuranose
(a hemiketal)   (a hemiketal)

Mutarotation

Evidence for the reversible closure of an open-chain monosaccharide to form two anomeric cyclic forms is provided by a phenomenon known as **mutarotation**. When D-glucose is crystallized from methanol, α-D-glucopyranose, which melts at 146 °C, is obtained. It has $[\alpha]_D$ = +112.2°. When α-glucose is crystallized from acetic acid, the β-anomer, which melts at 150 °C, is obtained. It has $[\alpha]_D$ = +18.7°. We recall that diastereomers have different chemical and physical properties, so these data are not surprising.

When α-D-glucopyranose is dissolved in water, the rotation of the solution slowly changes from the initial value of +112.2° to an equilibrium value of +54°. When β-D-glucopyranose is dissolved in water, the rotation of the solution slowly changes from the initial value of +18.7° to the same equilibrium value of +54°. The gradual changes in rotation to an equilibrium point for either of a set of anomeric carbohydrates is known as **mutarotation**. Mutarotation results from the interconversion of the cyclic hemiacetals with the open-chain form in solution.

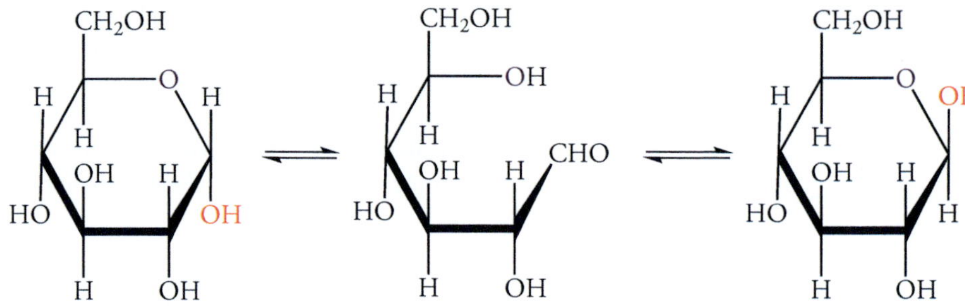

α-D-glucopyranose
36%

open chain

β-D-glucopyranose
64%

Problem 13.5

Draw the Haworth projection of the α-anomer of the pyranose form of D-galactose, that is, α-D-galactopyranose.

Solution

First, write galactose in the Fischer projection. Because the pyranose form is a six-membered ring, draw the ring consisting of five carbon atoms and one oxygen atom. Place the −CH$_2$OH group at C-6 above the plane of the ring.

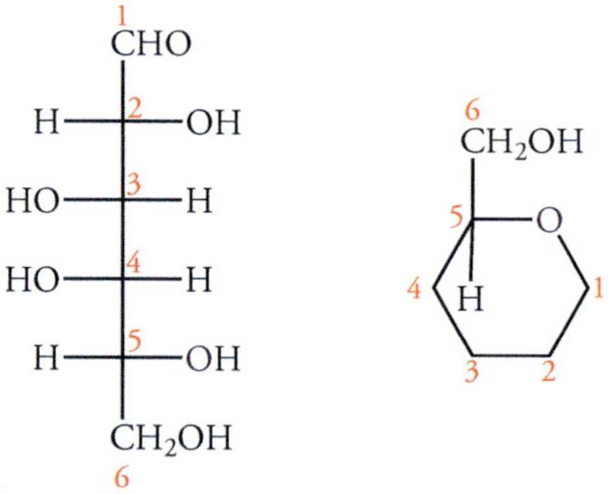

D-galactose

Now enter the hydroxyl groups and hydrogen atoms at C-2, C-3, and C-4. An atom or group on the right in the Fischer projection is below the ring of the Haworth projection, and an atom or group on the left is above the ring.

Finally, the α-anomer must have a hydroxyl group below the plane of the ring at the anomeric carbon atom, C-1. The hydrogen atom at C-1 is above the plane.

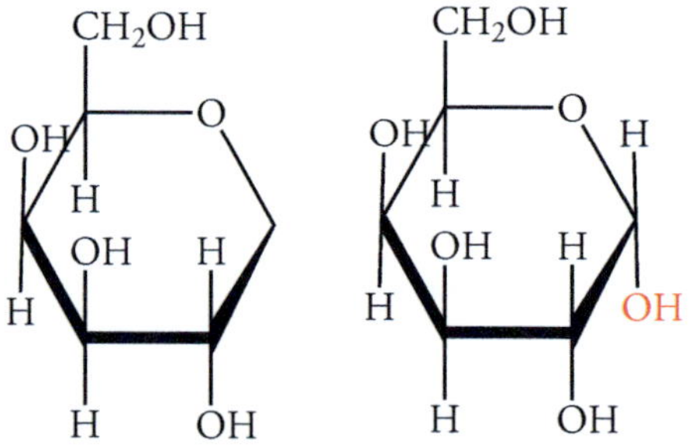

α-D-galactopyranose

Problem 13.6

Draw the Haworth projection for the pyranose form of D-galactose with the β configuration, that is, β-D-galactopyranose.

Solution

Although the chair conformation could be derived from the Haworth projection—which can, in turn, be derived from the open-chain formula—there is an easier way to obtain the structure. We need only recall that the β-anomer of glucose has all of its hydroxyl groups in equatorial positions. The α-anomer of galactose must therefore have the hydroxyl group at C-1 in an axial position. We also recall that galactose is the C-4 epimer of glucose. Therefore, the hydroxyl group at the C-4 atom must also be axial.

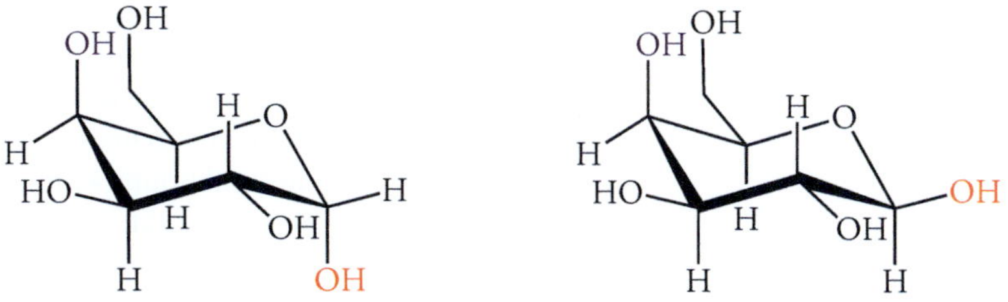

α-D-galactopyranose   β-D-galactopyranose

Problem 13.7
Draw the Haworth projection formula of the pyranose form of D-mannose with the β configuration; that is, β-D-mannopyranose.

13.4 CONFORMATIONS OF MONOSACCHARIDES

Haworth projection formulas do not give an accurate three-dimensional representation of carbohydrates, because the six atoms of pyranose rings exist in chair conformations just like cyclohexane (Section 3.6). Any hydroxyl group (or other group) that is up in the Haworth projection is also up in the chair conformation. However, "up" and "down" do not correspond to axial and equatorial, respectively. Each carbon atom must be individually examined. On one set of alternating carbon atoms, an "up" substituent is axial; on the intervening carbon atoms, an "up" substituent is equatorial.

Haworth projection formulas are converted into chair representations by "moving" two carbon atoms. The anomeric carbon atom is lowered below the plane of the ring, and the C-4 atom is raised above the plane of the ring. The remaining four atoms—three carbon atoms and the ring oxygen atom—are unchanged. This process is shown in Figure 13.3 for both α-D-glucopyranose and β-D-glucopyranose. Both the hydrogen atoms and the hydroxyl groups can be shown. However, a more condensed form that eliminates the C—H bonds is often used.

Note the changes in the locations of the hydroxyl groups in the Haworth projection compared to those in the chair conformation. Although the hydroxyl groups were both up and down in the Haworth projection formula, all hydroxyl groups are equatorial in β-D-glucopyranose. This anomer is the more stable based on the position of equilibrium observed in mutarotation experiments. β-D-Glucopyranose, the most abundant aldohexose, is the only aldohexose that has all of its hydroxyl groups in equatorial positions.

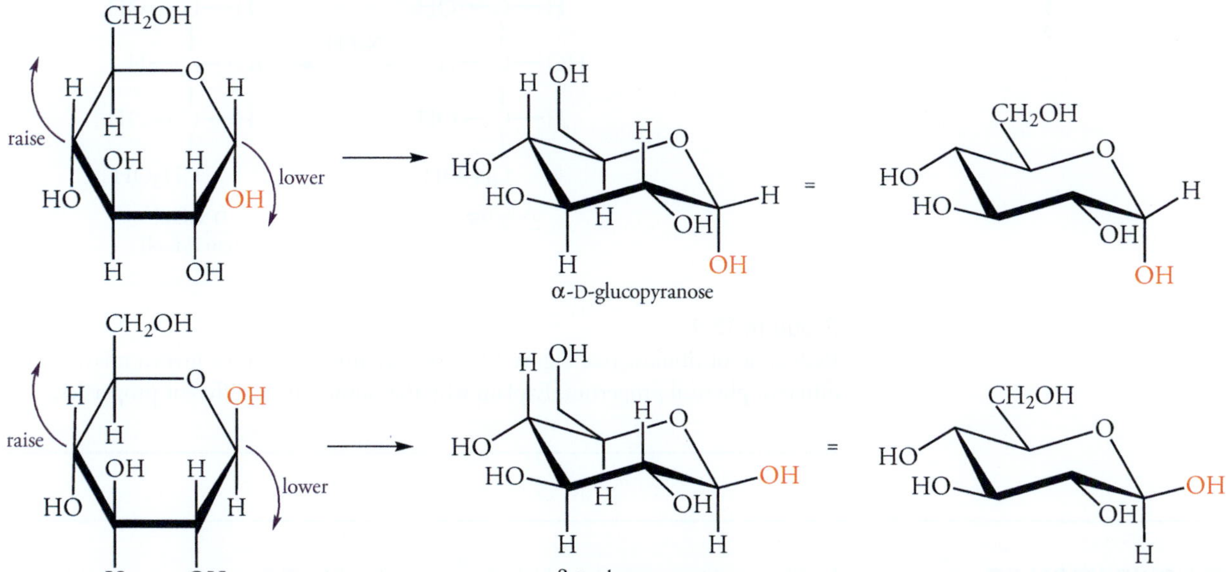

Figure 13.3
Conversion of Haworth Projections to Chair Conformations

13.5 REDUCTION OF MONOSACCHARIDES

Although five- and six-carbon monosaccharides exist predominantly as hemiacetals and hemiketals, they undergo the characteristic reactions of simple aldehydes and ketones. One such reaction is reduction. Treating an aldose or ketose with sodium borohydride reduces it to a polyalcohol called an **alditol**. The reduction reaction occurs via the aldehyde group in the small amount of the open-chain form of the aldose in equilibrium with its cyclic hemiacetal. As the aldehyde is reduced, the equilibrium shifts to produce more aldehyde until eventually all of the monosaccharide is reduced. The alditol derived from D-glucose is called D-glucitol. D-Glucitol occurs in some fruits and berries. Produced and sold commercially as a sugar substitute, it is also called sorbitol.

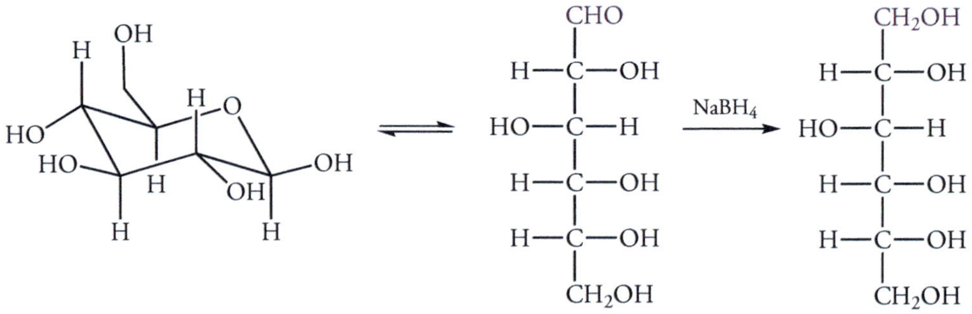

D-glucose

D-glucitol
(an alditol)

Problem 13.8

D-Xylitol is used as a sweetener in some chewing gums that are said to have a lower probability of causing cavities in teeth compared to those containing glucose and fructose. Deduce the structure of D-xylitol from its name.

Solution

The name indicates that it is derived from D-xylose, which can be reduced by sodium borohydride to an alditol. The configurations of the carbon atoms bearing the hydroxyl groups in D-xylitol are the same as those in D-xylose.

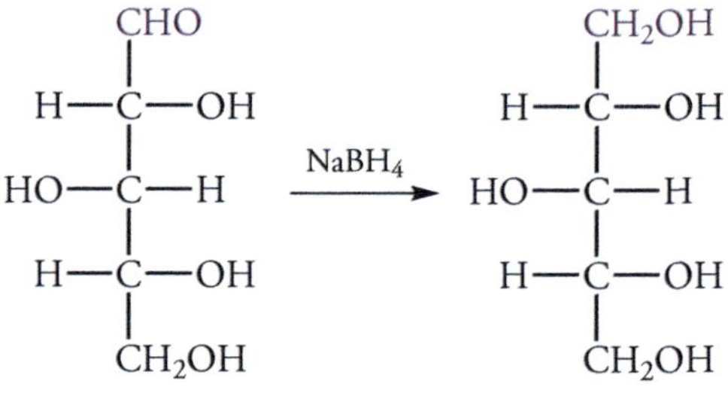

D-xylose

D-xylitol
(an alditol)

Problem 13.9

Reduction of ribulose (see Figure 13.2) by sodium borohydride gives a mixture of two isomers with different physical properties. Explain why the isomers have different properties.

13.6 OXIDATION OF MONOSACCHARIDES

In Chapter 10 we saw that aldehydes are oxidized by Tollens reagent and Fehling's solution. These reagents also oxidize open-chain aldoses that exist in equilibrium with the cyclic hemiacetal form. When some of the open-chain form reacts, the equilibrium shifts to form more compound for subsequent oxidation, and eventually all of the aldose is oxidized. Oxidation yields a product with a carboxylic acid at the original C-1 atom. This product is called an **aldonic acid**.

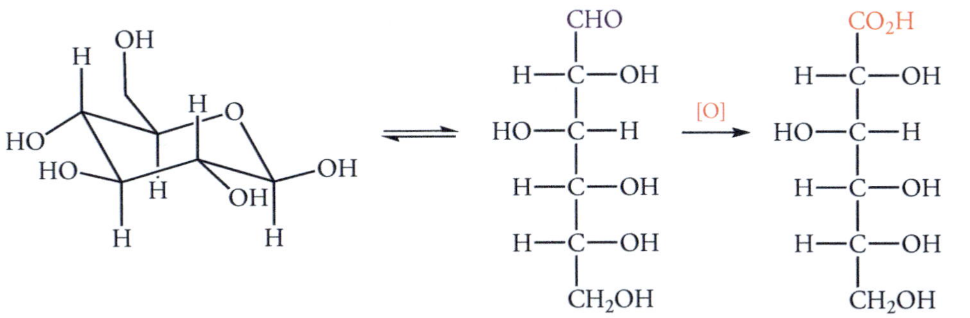

D-glucose

D-gluconic acid
(an aldonic acid)

If Tollens reagent is used as the oxidizing agent, metallic silver forms a mirror on the walls of the test tube. If Fehling's solution is used, a red precipitate of Cu_2O indicates that a reaction has occurred. Carbohydrates that react with Fehling's solution are called **reducing sugars**. The term *reducing* refers to the effect of the carbohydrate on the reagent that is reduced in the reaction.

Fehling's solution also oxidizes ketoses. We certainly do not expect this, because ketones are not oxidized by Fehling's solution. However, we recall that α-hydroxy ketones tautomerize in basic solution. Because Fehling's solution is basic, the tautomer of a ketose is an enediol that not only reverts to the α-hydroxy ketone, but also forms an isomeric α-hydroxy aldehyde.

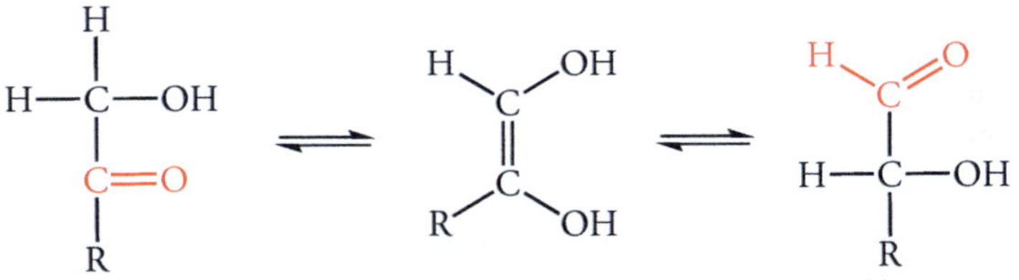

ketose   an enediol   aldose

Shifting a hydrogen atom from the C-2 hydroxyl group to C-1 regenerates the original ketose. However, shifting a hydrogen atom from the C-1 hydroxyl group to the C-2 atom forms an aldose. In basic solution, then, a ketose is in equilibrium with an aldose. The aldose reacts with Fehling's solution, and more ketose is converted into aldose. The equilibrium shifts as predicted by Le Châtelier's principle, and eventually all of the ketose is oxidized.

Other hydroxyl groups of aldoses may be oxidized if stronger oxidizing agents are used. For example, dilute nitric acid oxidizes both the aldehyde group and the primary alcohol of aldoses to give **aldaric acids**.

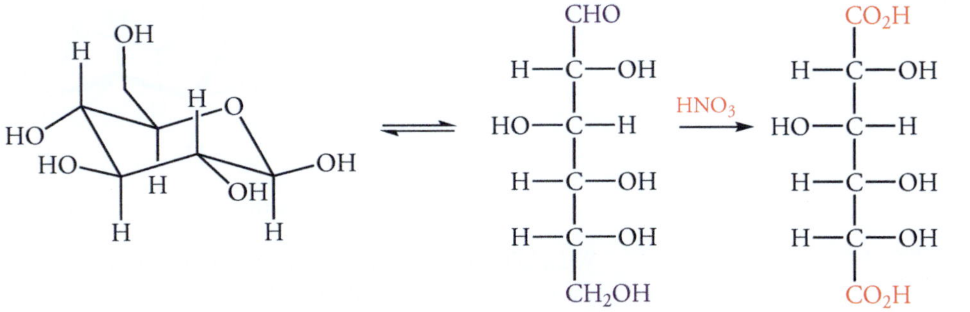

D-glucose

D-glucaric acid
(an aldaric acid)

The terminal $-CH_2OH$ group of an aldose can be oxidized enzymatically in cells without oxidation of the aldehyde group. The product is a **uronic acid**. The enzyme responsible for this reaction uses $NADP^+$ (nicotinamide adenine dinucleotide phosphate, which has a structure similar to NAD^+) as the oxidizing agent. An example of this reaction is the oxidation of D-glucose to give D-glucuronic acid, a component in the polysaccharide hyaluronic acid, which is found in many tissues, including the vitreous humor of the eye.

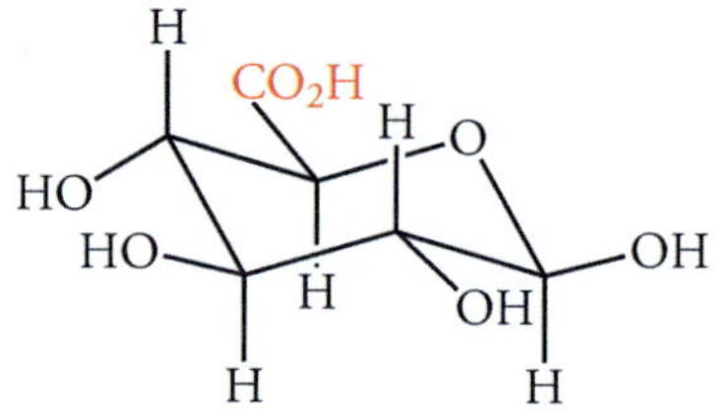

β-D-glururonic acid

13.7 GLYCOSIDES

In Chapter 10 we saw that hemiacetals and hemiketals react with alcohols to yield acetals and ketals, respectively. Acid catalyzes the reaction, shifting the equilibrium to the right if there is excess alcohol or the water that forms is removed. In this substitution reaction, an −OR group replaces the −OH group.

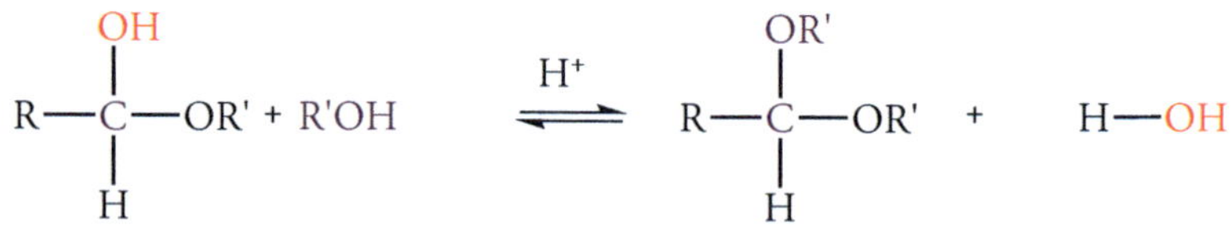

(a hemiacetal)   (an acetal)

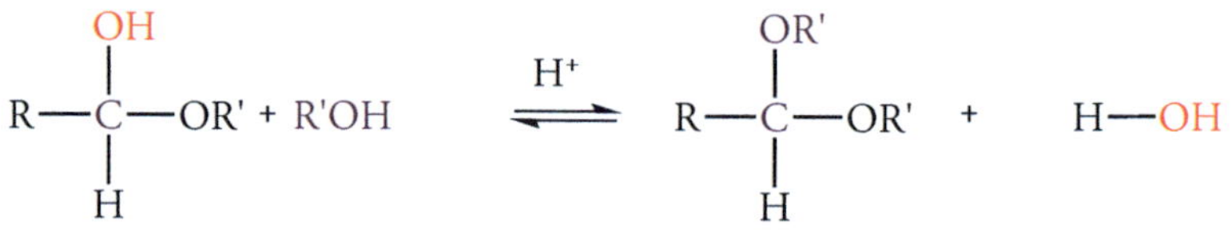

(a hemiketal)   (a ketal)

The cyclic hemiacetal and hemiketal forms of monosaccharides also react with alcohols to form acetals and ketals. These acetals and ketals are called **glycosides**, and the carbon-oxygen bond formed is called a **glycosidic bond**. The group bonded to the anomeric carbon atom of a glycoside is an **aglycone**. In aglycones, an oxygen atom from an alcohol or phenol is linked to the anomeric carbon atom.

Glycosides are named by citing the aglycone group first and then replacing the -*ose* ending of the carbohydrate with -*oside*. The configuration at the glycosidic carbon atom must be indicated.

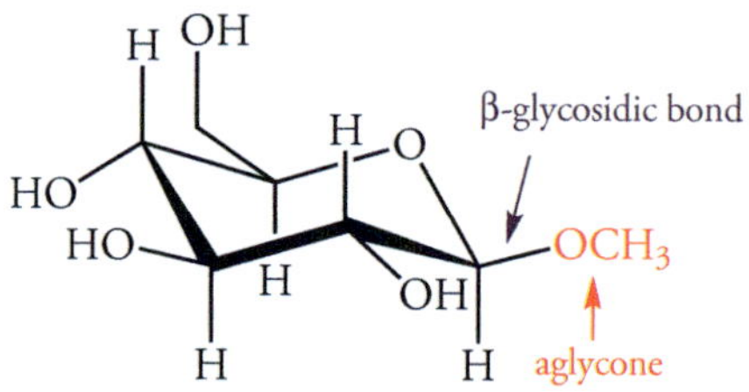

methyl β-D-glucopyranoside

We recall that the conversion of hemiacetals or hemiketals into acetals or ketals, respectively, occurs via a resonance-stabilized oxocarbocation (Section 10.8) formed by protonation of the −OH group followed by loss of water.

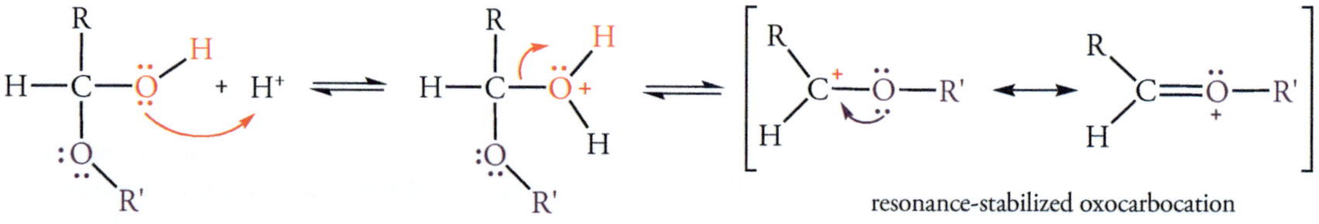

Subsequently, the intermediate oxocarbocation reacts with an alcohol that acts as a nucleophile.

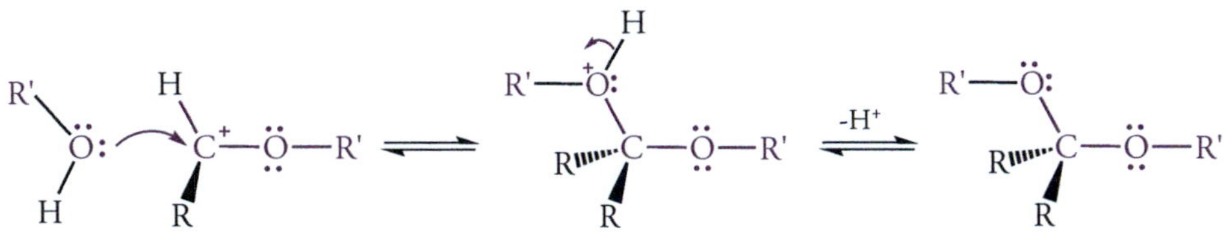

Attack of the alcohol is when the intermediate formed from a monosaccharide can occur from either of two directions. As a result, a mixture of α- and β-anomers forms (Figure 13.4).

Figure 13.4
Formation of α- and β-Glycosides

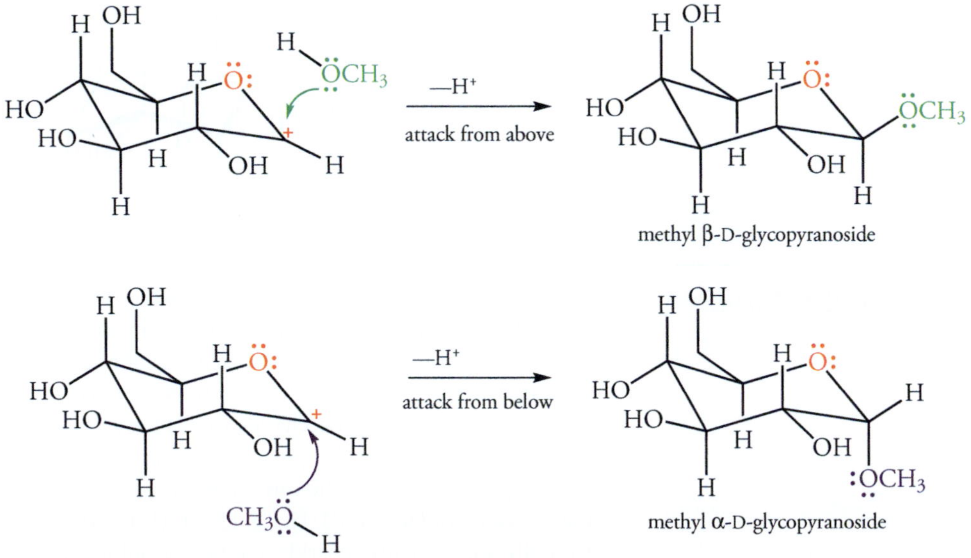

methyl β-D-glycopyranoside

methyl α-D-glycopyranoside

Glycosides are hydrolyzed in acid solution. However, in neutral or basic solution, glycosides are stable compounds, and the anomers have different physical properties because they are diastereomers. Glycosides are not reducing sugars because they do not hydrolyze to form a free aldehyde group in Fehling's solution, which is basic.

Problem 13.10

Examine the following molecule to determine its component functional groups. From what compounds can the substance be formed?

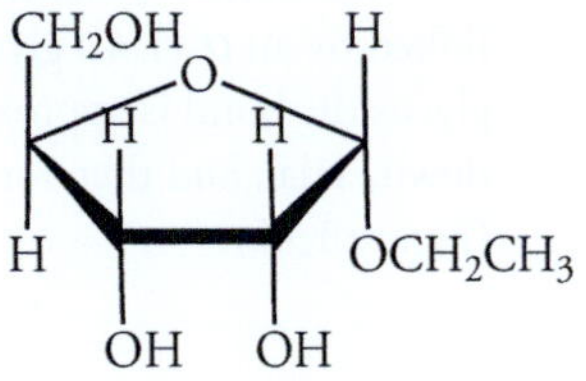

Solution

The compound is a furanose form of a monosaccharide, because there are four carbon atoms and one oxygen atom in a five-membered ring. The ring carbon atom on the right is an acetal center because there are two −OR groups and one hydrogen atom bonded to it.

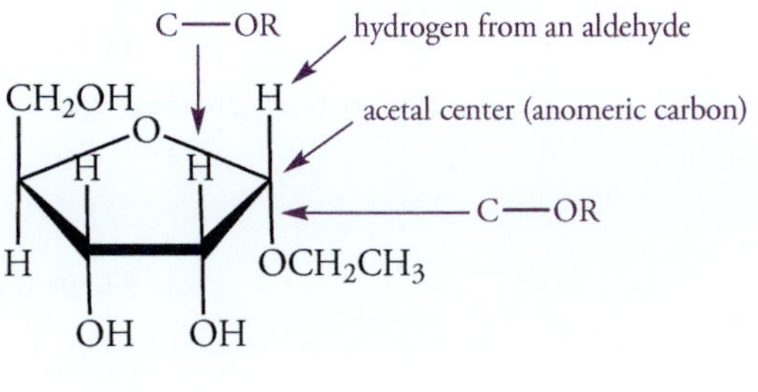

13.7 Glycosides   357

The acetal has the α configuration, and the alcohol used to form the acetal is ethanol. The carbohydrate has the D configuration because the —CH$_2$OH group is "up" in the Haworth projection. The other two stereogenic carbon atoms of the pentose have hydroxyl groups "down," which corresponds to the right in the Fischer projection. The carbohydrate component of the compound is ribose. The name of the compound is ethyl α-D-ribofuranoside.

Problem 13.11
Examine the following molecule to determine its component functional groups. From what compounds can the substance be formed?

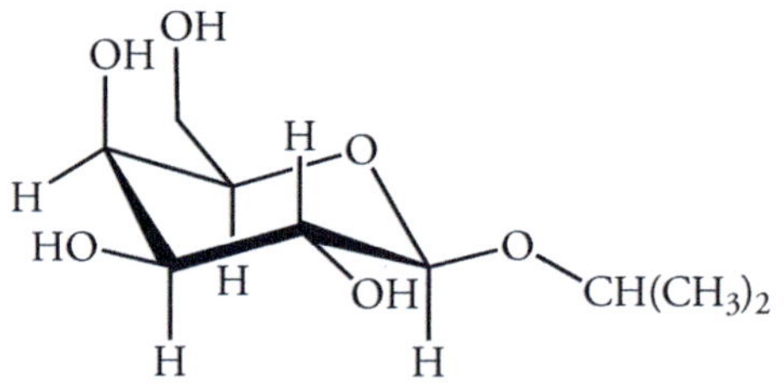

13.8 DISACCHARIDES

Disaccharides have glycosidic linkages between the anomeric center of one monosaccharide unit and the hydroxyl oxygen atom of the second monosaccharide unit, which is the aglycone. A glycosidic bond between the C-1 atom of the hemiacetal of an aldose and the C-4 atom of the second monosaccharide is very common. Such bonds are designated (1,4'). The prime superscript indicates which carbon atom of the monosaccharide that is the aglycone provides the oxygen atom. Maltose, cellobiose, and lactose all have (1,4') glycosidic bonds (Figure 13.5).

In principle, any of the carbon atoms of a monosaccharide could provide the oxygen atom of the aglycone. In fact, (1,1'), (1,2'), (1,3'), (1,4'), and (1,6') glycosidic bonds have all been found in naturally occurring disaccharides containing aldohexoses. Note that a (1,1') glycosidic bond connects both anomeric carbon atoms. Sucrose has a (1,2') glycosidic bond connecting the anomeric carbon atoms of glucose and fructose.

Maltose

In maltose the glycosidic oxygen atom of one glucose is α and is bonded to the C-4 atom of another glucose unit that is the aglycone. Therefore, maltose is an α-1,4'-glycoside. Maltose is produced by the enzymatic hydrolysis of starch (a homopolysaccharide) catalyzed by the enzyme amylase. Maltose is further hydrolyzed by the enzyme maltase to produce two molecules of D-glucose.

The monosaccharide unit on the left is the hemiacetal of the α-D-glucopyranosyl unit. It is linked by an α-(1,4') glycosidic bond to β-D-glucopyranose, the aglycone. The oxygen atom of the glycosidic bond is approximately in the center of the structure, between the two rings. It is projected down, axial, and therefore α. It is linked to C-4 of the aglycone, and so the link is axial-equatorial (Figure 13.5).

Figure 13.5
Maltose
(a) Bond-line structure of maltose. (b) Molecular model of maltose.

(a)

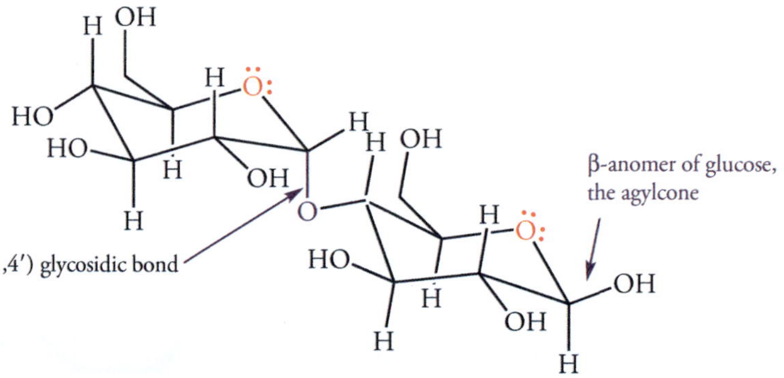

α-(1,4') glycosidic bond

β-anomer of glucose, the agylcone

4-O-(α-D-glucopyranosyl)-β-D-glucopyranose
(maltose)

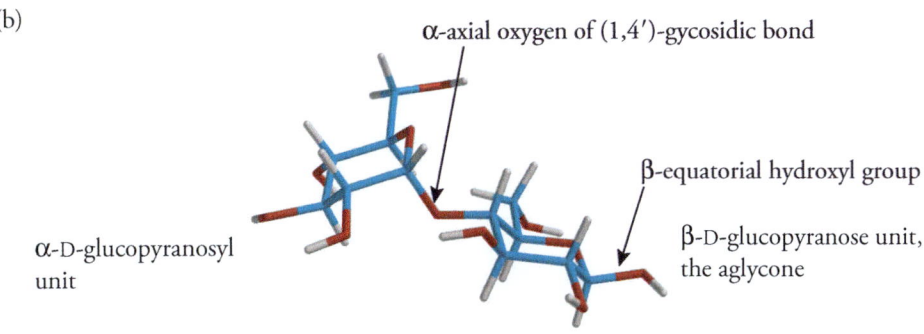

(b)

α-axial oxygen of (1,4′)-gycosidic bond

β-equatorial hydroxyl group

β-D-glucopyranose unit, the aglycone

α-D-glucopyranosyl unit

4-O-(α-D-glucopyranosyl)- β-D-glucopyranose
(maltose)

Maltose has a more formal, IUPAC name: 4-O-(α-D-glucopyranosyl)-β-D-glucopyranose. This rather forbidding name is not quite as bad as it looks. The term in parentheses refers to the glucose unit on the left, which contributes the acetal portion of the glycosidic bond. The term *-pyrano* tells us that this part of the structure is a six-membered ring, and the suffix *-osyl* indicates that the ring is linked to a partner by a glycosidic bond. The prefix 4-O- refers to the position of the oxygen atom on the aglycone, the right-hand ring. The term β-D-glucopyranose describes the aglycone.

Because the aglycone is a hemiacetal, maltose undergoes mutarotation. For the same reason maltose is a reducing sugar. The free aldehyde formed by ring opening can react with Benedict's solution. The acetal part of the structure is called the "nonreducing end" of the disaccharide. If we do not want to specify the configuration of the aglycone, we use the name 4-O-(α-D-glucopyranosyl)-D-glucopyranose.

Maltose undergoes mutarotation at its hemiacetal anomeric center. Recall that the process occurs via an open-chain structure containing an aldehyde. The free aldehyde formed by ring opening can react with Fehling's solution, so maltose is a reducing sugar.

Cellobiose

Cellobiose is a disaccharide that has two molecules of D-glucose linked by a β-1,4′-glycosidic bond. Thus, cellulose differs from maltose in the configuration of its glycosidic bond (Figure 13.6).

As in maltose, the aglycone of cellobiose is a hemiacetal, which can be either α or β. In solution, the two forms of cellobiose exist in equilibrium. Thus, cellobiose mutarotates, and is a reducing sugar. Again, do not confuse the configuration of the hemiacetal center with that of the glycosidic bond, which is β in cellobiose.

Cellobiose is produced by hydrolysis of cellulose, a homopolysaccharide of glucose in which all units are linked by β-1,4′-glycosidic bonds. Humans do not have an enzyme to hydrolyze cellobiose. Small differences in configuration at the (1,4′) linkage result in remarkable differences in the chemical reactivity of these biomolecules. Glycosidic bonds are hydrolyzed by enzymes called glycosidases. A glycosidase that hydrolyzes α-1,4′-glycosidic bonds does not hydrolyze molecules that have β-1,4′-glycosidic bonds (and vice versa).

Figure 13.6
Structure of Cellobiose
(a) Bond-line structure of cellobiose. (b) Molecular model of cellobiose.

(a)   β-D-glucopyranosyl unit

H OH

β-(1,4′) glycosidic bond

HO

HO

H

H   Ö:

OH

H

H

O

H

OH

H

HO

H

Ö:

β-anomer of glucose, the agylcone

H   H

HO

OH

H

4-O-(β-D-glucopyranosyl)-β-D-glucopyranose
(cellobiose)

(b)

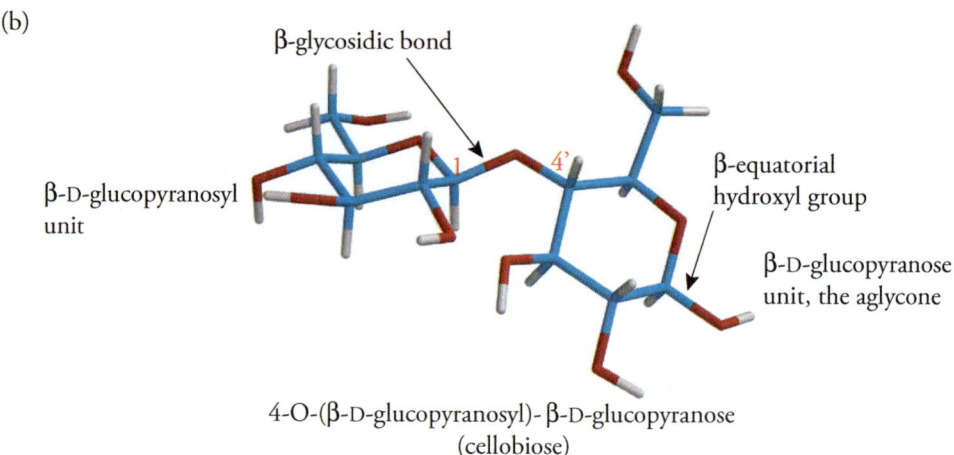

β-glycosidic bond

β-D-glucopyranosyl unit

β-equatorial hydroxyl group

β-D-glucopyranose unit, the aglycone

4-O-(β-D-glucopyranosyl)- β-D-glucopyranose
(cellobiose)

Lactose

Lactose, a disaccharide found in different concentrations in the milk of mammals, including both humans and cows, is often called milk sugar. The IUPAC name of lactose is 4-O-(β-D-galactopyrano-syl)-D-glucopyranose (Figure 13.7).

Figure 13.7
Lactose

β-D-galactopyranosyl unit

C-4 epimer of glucose

β-(1,4') glycosidic bond

β-anomer of glucose, the agylcone

4-O-(β-D-galactopyranosyl)- β-D-glucopyranose
(lactose)

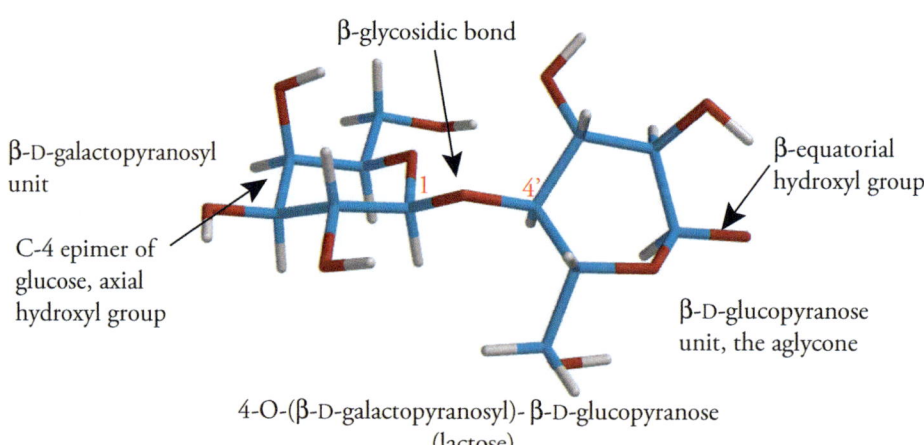

β-glycosidic bond

β-D-galactopyranosyl unit

C-4 epimer of glucose, axial hydroxyl group

β-equatorial hydroxyl group

β-D-glucopyranose unit, the aglycone

4-O-(β-D-galactopyranosyl)- β-D-glucopyranose
(lactose)

As in maltose and cellobiose, the aglycone of lactose is a hemiacetal, and it can be either α or β. Com-pare this structure to that of cellobiose. The monosaccharide shown on the left in the structure has an axial hydroxyl group at the C-4 atom, so it is galactose. The oxygen atom at the C-4 atom of the ring on the right is equatorial, so it is glucose. Both lactose and cellobiose are linked by a β-glycosidic linkage to the C-4 atom of a D-glucopyranose ring on the right. In both lactose and cellobiose, the glycosidic bond is β-1,4′.

Compare the structure of lactose to that of cellobiose. The monosaccharide shown on the left in the structure has an axial hydroxyl group at the C-4 atom, so it is galactose. The oxygen atom at the C-4 atom of the ring on the right is equatorial, so it is glucose. Both lactose and cellobiose are linked by a β-glycosidic linkage to the C-4 atom of a D-glucopyranose ring on the right. In both lactose and cellobiose, the glycosidic bond is β-1,4′. Humans have an enzyme called β-galactosidase (also known as lactase) that catalyzes the hydrolysis of the β-1,4′-galactosidic linkage. However, β-galactosidase does not catalyze the hydrolysis of the β-1,4′-glucosidic linkage of cellobiose.

As in the case of cellobiose and maltose, the aglycone component of lactose is a hemiacetal, which can be either α or β. In solution, the two forms of lactose exist in equilibrium. Thus, lactose undergoes mutarotation, and is a reducing sugar. The lactose content of milk varies with species; cow's milk contains about 5% lactose, whereas human milk contains about 7%. The enzyme lactase, which is present in the small intestine, catalyzes hydrolysis of lactose to form glucose and galactose. Galactose is then isomerized into glucose in a reaction catalyzed by the enzyme UDP-galactose-4-epimerase.

Sucrose

Some disaccharides have a glycosidic linkage between both anomeric centers. Sucrose, common table sugar (Figure 13.8), is a disaccharide in which the anomeric centers are linked 1,2′.

Figure 13.8 Sucrose

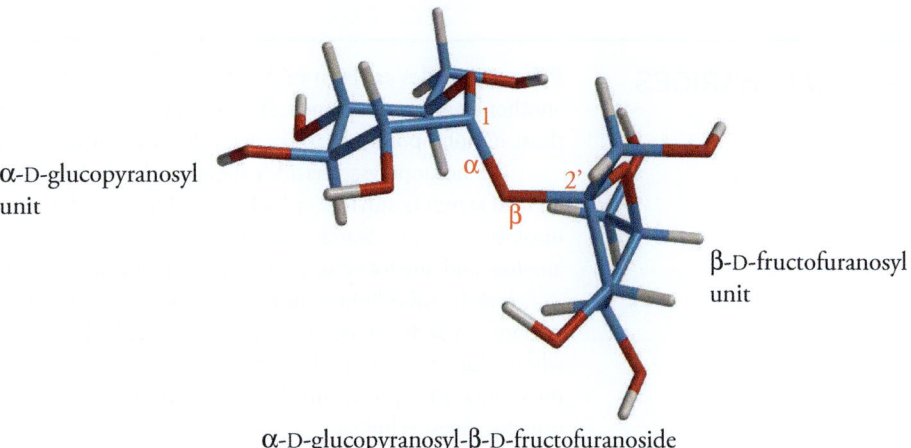

α-D-glucopyranosyl-β-D-fructofuranoside
(sucrose)

α-D-glucopyranosyl unit

β-D-fructofuranosyl unit

α-D-glucopyranosyl-β-D-fructofuranoside
(sucrose)

Sucrose has both an acetal and a ketal functional group. Neither ring can exist in equilibrium with either an aldehyde or ketone. As a result, sucrose cannot mutarotate, nor is it a reducing sugar. The systematic name, α-D-glucopyranosyl-β-D-fructofuranoside, ends in the suffix *-oside*, which indicates that sucrose exists as a glycoside, and so is not a reducing sugar.

Problem 13.12

Describe the structure of the following disaccharide.

Solution

The hemiacetal center located on the aglycone ring (at the right) has a hydroxyl group in the β-configuration. The glycosidic bond is from C-1 of the acetal ring (on the left) to C-3 of the aglycone ring (on the right). Furthermore, the oxygen bridge is formed through a β-glycosidic bond. Thus, the bridge is β-1,3'. Next, we examine both rings to determine the identity of the monosaccharides. The ring on the left is β-D-galactopyranose: all of its hydroxyl groups are equatorial except the one at C-4, which is axial. The ring on the right is β-D-glucopyranose. The compound is 3-O-(β-D-galactopyranosyl)-β-D-glucopyranoside.

Problem 13.13

Describe the structure of the following disaccharide.

13.9 POLYSACCHARIDES

Polysaccharides are high molecular weight substances consisting of monosaccharides linked to one another by glycosidic bonds. Because the structures of heteropolysaccharides are more complex than those of homopolysaccharides, we will only consider homopolysaccharides in this section.

The homopolysaccharides starch and cellulose, found in plants, contain only glucose. About 20% of starch is amylose, which is soluble in cold water; the remaining 80%, called amylopectin, is insoluble in water. Starch is present in potatoes, rice, wheat, and other cereal grains. The amount of amylose and amylopectin in starch is variable and depends on its source.

Starch and cellulose differ by one structural feature, but this difference has great biological importance. Starch, whose glucosyl units are linked α-1,4', can be digested by most animals. Cellulose, whose glucosyl units are linked β-1,4', can be digested only by cattle and other herbivores, because microorganisms present in their digestive tracts have enzymes that hydrolyze β-glucosides. Termites can also digest cellulose.

Amylose is a linear polymer with 200 to 2000 α-linked glucose units that serves as a major source of food for some animals. The molecular weight of amylose ranges from 40,000 to 400,000 amu. Cellulose is a β-linked polymer of glucose (Figure 13.9) that can contain 5,000 to 10,000 glucose units. Certain algae produce cellulose molecules that contain more than 20,000 glucose units. Amylopectin contains chains similar to those in amylose, but only about 25 glucose units occur per chain. Amylopectin has branches of glucose-containing chains interconnected by a glycosidic linkage between the C-6 hydroxyl group of one chain and the C-1 atom of another glucose chain (Figure 13.5). The molecular weight of amylopectin may be as high as 1 million. Because each chain has an average molecular weight of 3000, there may be as many as 300 interconnected chains.

The structure of glycogen is similar to that of amylopectin, but glycogen has more branches, and the branches are shorter than those of amylopectin. The average chain length in glycogen is 12 glucose units. Glycogen has a molecular weight greater than 3 million. Glycogen is synthesized by animals as a storage form of glucose. Although glycogen is found throughout the body, the largest amounts are in the liver. An average human adult has enough glycogen for about 15 hours of normal activity.

Human Blood Groups

Complex carbohydrates coat the surfaces of nearly all human cells, acting as markers that identify the cell. Human blood cells contain surface markers that divide blood into three major classes, designated A, B, and O. The classification of blood groups relies on differences in the structures of oligosaccharides bonded to a protein called glycophorin, which is embedded in the membrane of red blood cells. The blood-group oligosaccharides contain several different monomers: galactose (Gal), N-acetyl-galactosamine (GalNAc), and N-acetylglucosamine (GlcNAc).

They also contain the rather unusual sugar 6-deoxy-α-L-galactose. This sugar has the common name α-L-fucose.

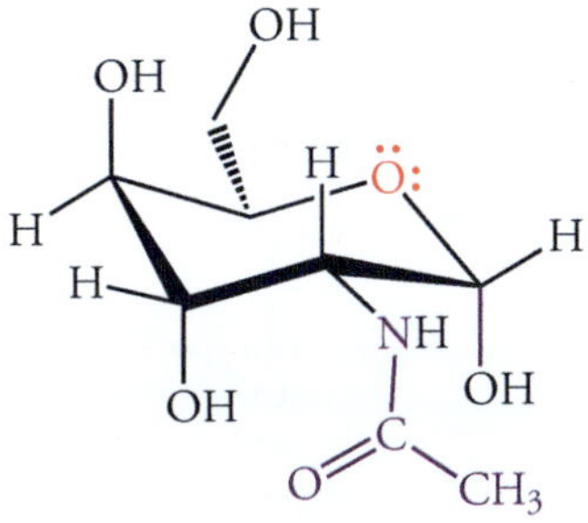

N-acetylgalactosamine
(GalNAc)

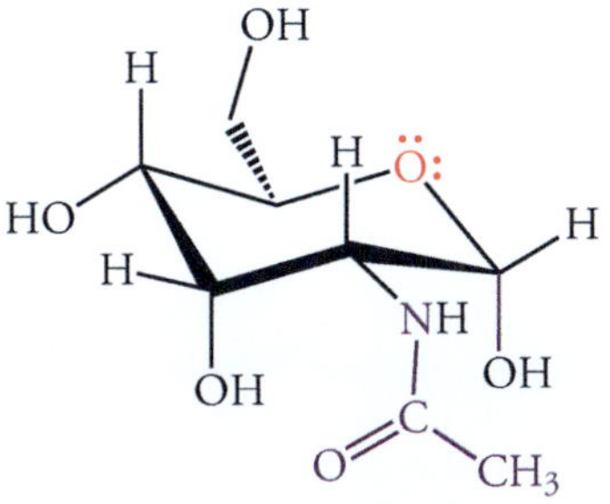

N-acetylglucosamine
(GlcNAc)

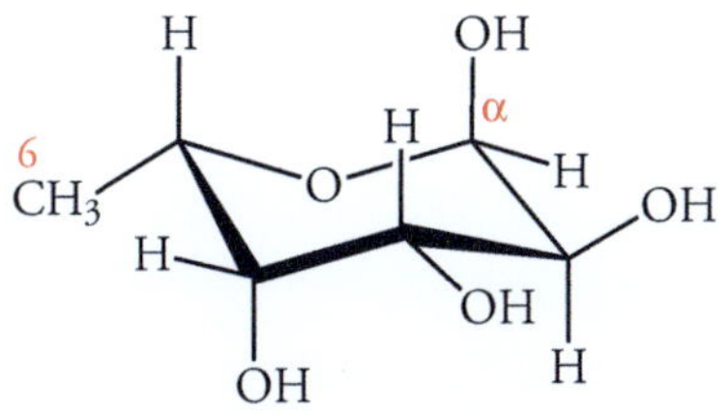

6-deoxy-α-L-galactose
(α-L-fucose)

The α-L-fucose moiety of each blood group is attached to a trisaccharide in blood groups A and B and to a disaccharide in blood group O (see structures in the drawings below). Type B, which is not shown, differs from type A by a galactose on the left side of the structure rather than N-acetylgalactosamine. In each oligosaccharide the β-galactose residue is attached to α-L-fucose by a 1,2′-glycosidic bond. The β-N-acetylglucosamine portion of the oligosaccharide shown is bonded to a "spacer unit" that consists of an oligosaccharide chain that may contain from 2 to as many as 50 monosaccharide units. This chain is bonded to a protein in a variety of possible ways including a glycosidic bond to the hydroxyl group of a serine residue in the protein.

Each blood group is further subdivided into two types of chains that differ in their glycosidic linkages. In a type 1 chain, the central β-Gal moiety is linked to β-GlcNAc by a 1,4′-glycosidic bond. In a type 2 chain, the β-Gal moiety is linked to β-GlcNAc by a 1,3′-glycosidic bond. The structures below show a type 1 linkage.

These carbohydrates are the antigenic determinants of their groups. A person with type A blood makes antibodies that "attack" type B blood, forming clumps of type B cells. Similarly, a person with type B blood makes antibodies that "attack" type A blood. However, persons who are type A or type B do not make antibodies against type O blood, so type O persons are called universal donors. They are not, however, universal acceptors, because they produce antibodies against both type A and type B blood.

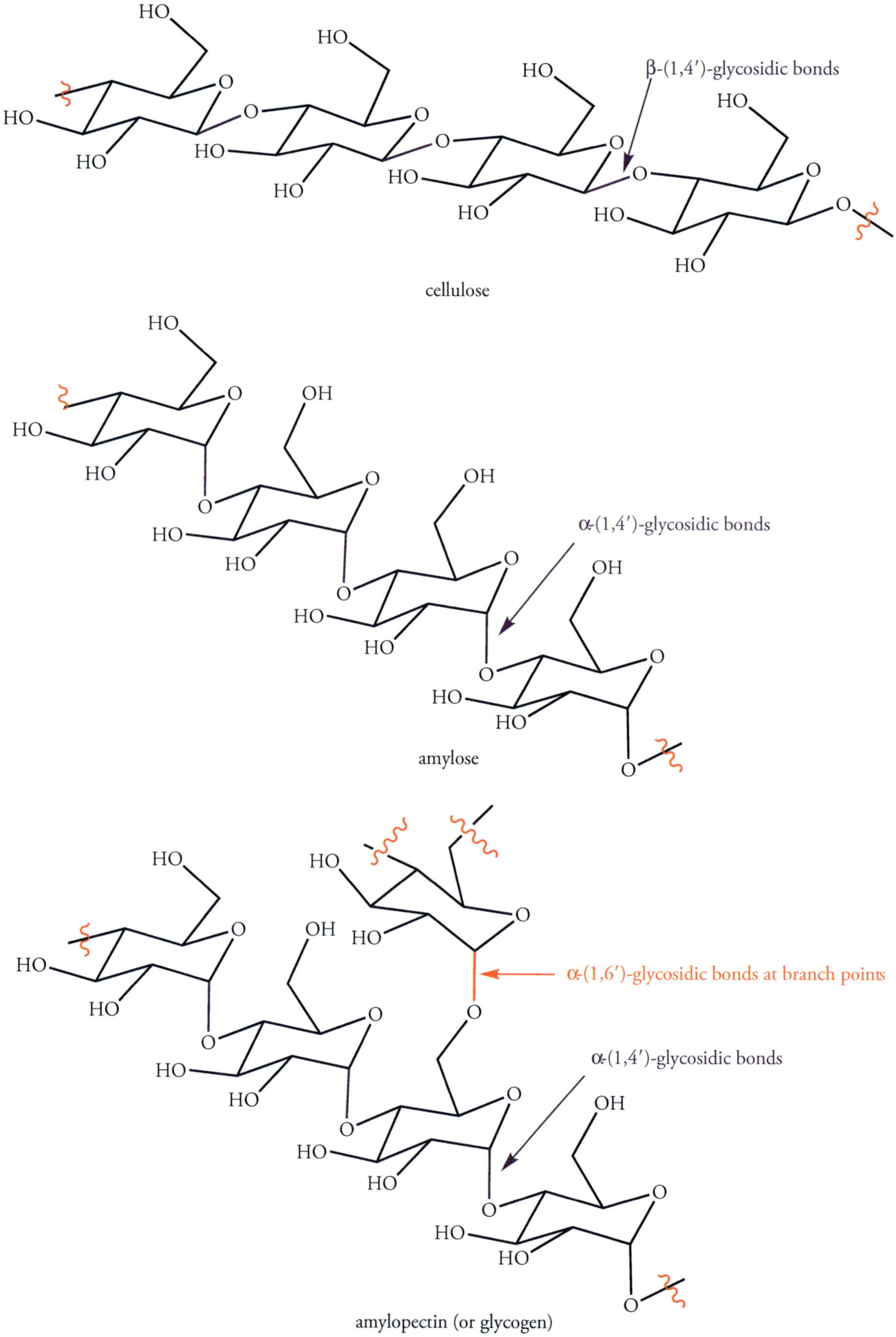

cellulose

amylose

α-(1,4')-glycosidic bonds

β-(1,4')-glycosidic bonds

α-(1,6')-glycosidic bonds at branch points

α-(1,4')-glycosidic bonds

amylopectin (or glycogen)

Figure 13.9 Structures of Polysaccharides

Type A

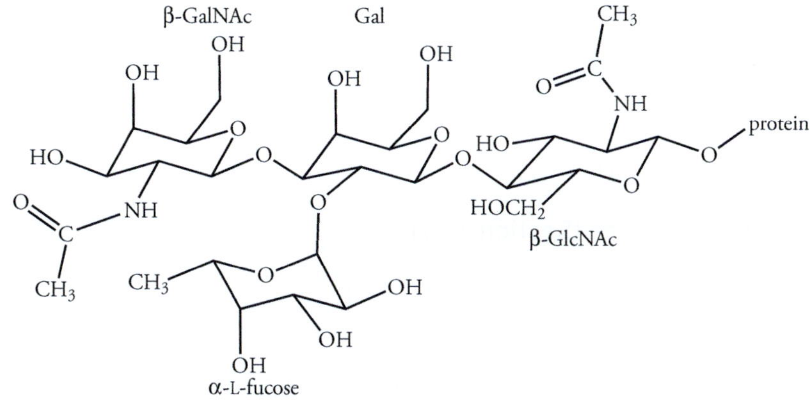

β-GalNAc   Gal

OH   OH   OH   OH   CH₃

HO   O   O   HO   NH   protein

O   NH   O   HOCH₂   O

CH₃   CH₃   O   OH   β-GlcNAc

OH   OH

OH

α-L-fucose

Type B

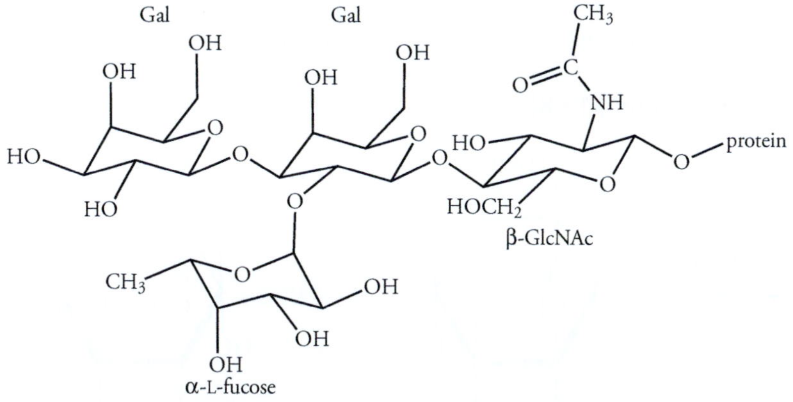

Gal   Gal

OH   OH   OH   OH   CH₃

HO   O   O   HO   NH   protein

HO   O   HOCH₂   O

CH₃   OH   β-GlcNAc

OH

OH

α-L-fucose

Type O

OH   OH   CH₃

galactose   HO   O   HO   NH   protein

HO   O   HOCH₂   O   β-GlcNAc

CH₃   O   OH

OH

OH

α-L-fucose

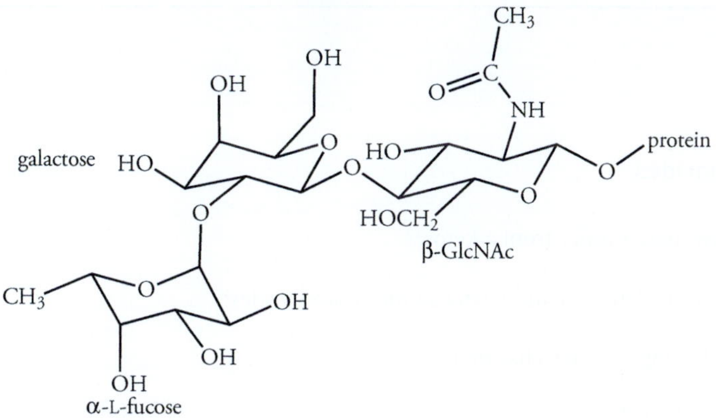

SUMMARY OF REACTIONS

1. Reduction of Monosaccharides (Section 13.5)

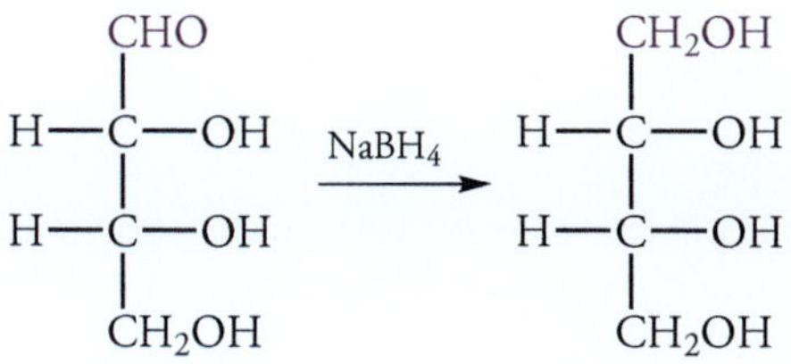

$$
\begin{array}{c}
\text{CHO} \\
\text{H}-\text{C}-\text{OH} \\
\text{H}-\text{C}-\text{OH} \\
\text{CH}_2\text{OH}
\end{array}
\xrightarrow{\text{NaBH}_4}
\begin{array}{c}
\text{CH}_2\text{OH} \\
\text{H}-\text{C}-\text{OH} \\
\text{H}-\text{C}-\text{OH} \\
\text{CH}_2\text{OH}
\end{array}
$$

2. Oxidation of Monosaccharides (Section 13.6)

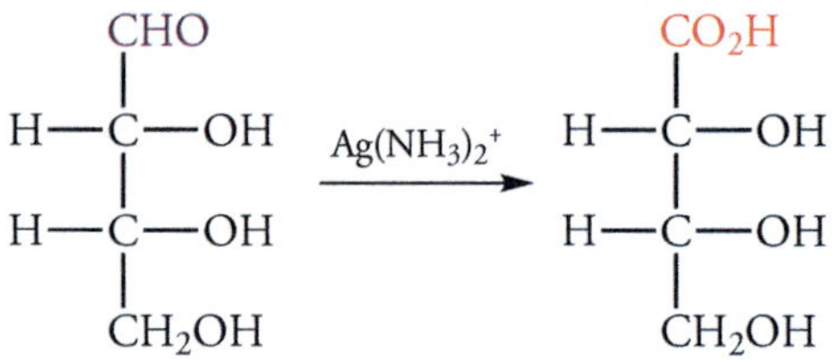

$$\begin{array}{c} \text{CHO} \\ | \\ \text{H--C--OH} \\ | \\ \text{H--C--OH} \\ | \\ \text{CH}_2\text{OH} \end{array} \quad \xrightarrow{\text{Ag(NH}_3\text{)}_2{}^+} \quad \begin{array}{c} \text{CO}_2\text{H} \\ | \\ \text{H--C--OH} \\ | \\ \text{H--C--OH} \\ | \\ \text{CH}_2\text{OH} \end{array}$$

3. Isomerization of Monosaccharides (Section 13.7)

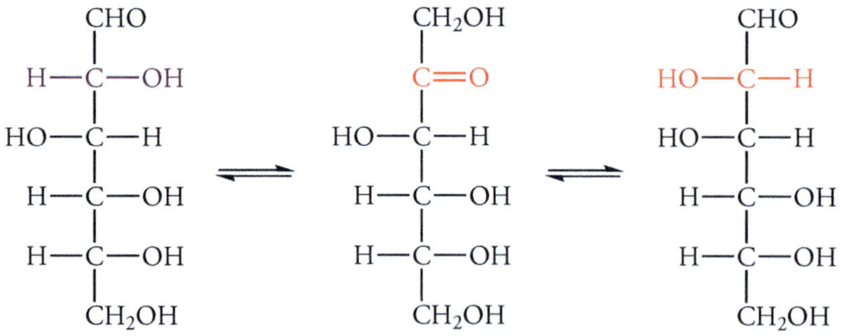

$$\begin{array}{c} \text{CHO} \\ | \\ \text{H--C--OH} \\ | \\ \text{HO--C--H} \\ | \\ \text{H--C--OH} \\ | \\ \text{H--C--OH} \\ | \\ \text{CH}_2\text{OH} \end{array} \rightleftharpoons \begin{array}{c} \text{CH}_2\text{OH} \\ | \\ \text{C}{=}\text{O} \\ | \\ \text{HO--C--H} \\ | \\ \text{H--C--OH} \\ | \\ \text{H--C--OH} \\ | \\ \text{CH}_2\text{OH} \end{array} \rightleftharpoons \begin{array}{c} \text{CHO} \\ | \\ \text{HO--C--H} \\ | \\ \text{HO--C--H} \\ | \\ \text{H--C--OH} \\ | \\ \text{H--C--OH} \\ | \\ \text{CH}_2\text{OH} \end{array}$$

4. Formation of Glycosides (Section 13.8)

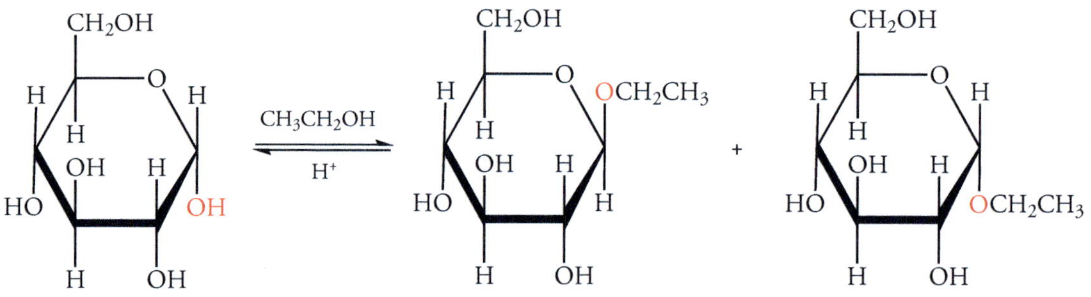

EXERCISES

Classification of Monosaccharides

13.1   What is an aldose? How does it differ from a ketose?

13.2   To what carbon atom do the letters D and L refer in monosaccharides?

13.3   Classify each of the following monosaccharides:

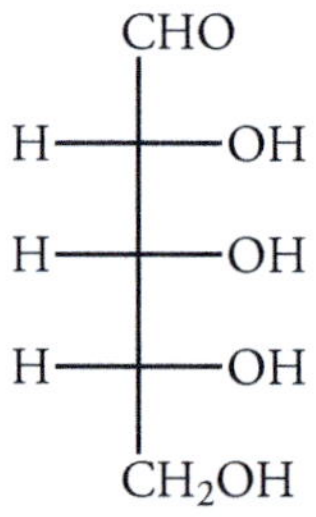

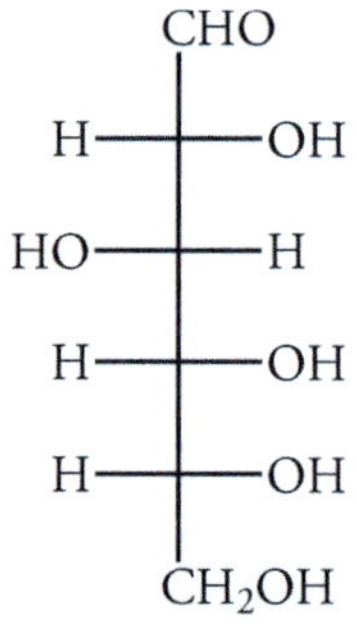

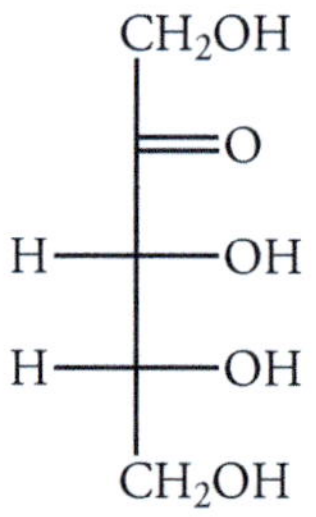

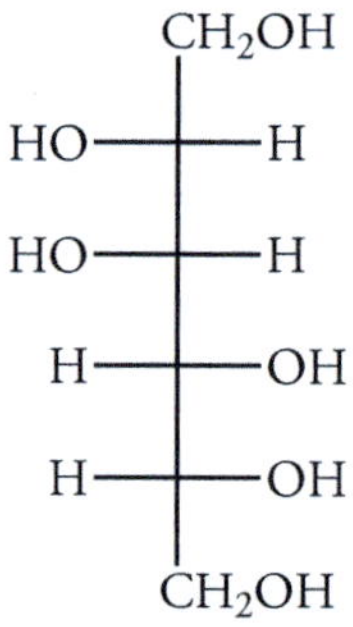

13.4   Classify each of the following monosaccharides:

(a) 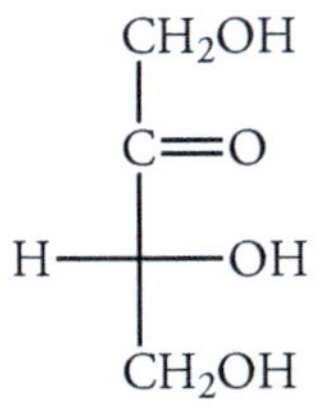
$$\begin{array}{c} CH_2OH \\ | \\ C{=}O \\ H\!-\!\!\!-\!OH \\ | \\ CH_2OH \end{array}$$

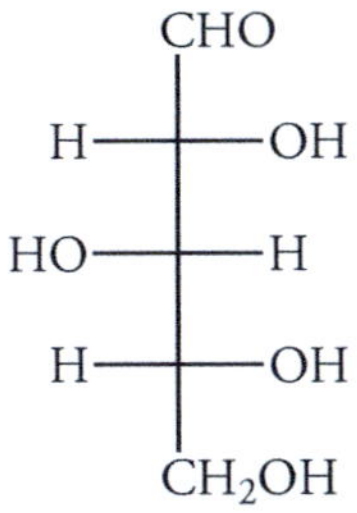

(b)
$$\begin{array}{c} CHO \\ H\!-\!\!\!-\!OH \\ HO\!-\!\!\!-\!H \\ H\!-\!\!\!-\!OH \\ CH_2OH \end{array}$$

(c)
$$\begin{array}{c} CHO \\ H\!-\!C\!-\!OH \\ HO\!-\!\!\!-\!H \\ CH_2OH \end{array}$$

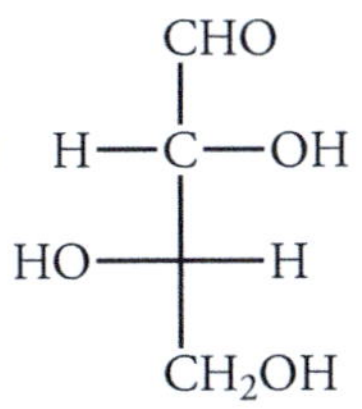

(d) 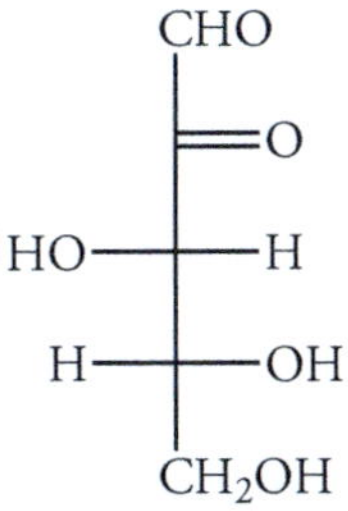
$$\begin{array}{c} CHO \\ {=}O \\ HO\!-\!\!\!-\!H \\ H\!-\!\!\!-\!OH \\ CH_2OH \end{array}$$

13.5   Classify each of the monosaccharides in Exercise 13.3 as D or L.

13.6   Classify each of the monosaccharides in Exercise 13.4 as D or L.

13.7   Draw the Fischer projection formulas of the isomeric D-ketopentoses with the carbonyl group at the C-2 atom.

13.8   Draw the Fischer projection formulas of the isomeric 3-ketopentoses. Indicate whether each isomer is chiral or achiral.

13.9   Draw the Fischer projection formula of each of the following:
(a) L-xylose   (b) L-erythrose   (c) L-galactose

13.10   Draw the Fischer projection formula of each of the following:
(a) D-xylose   (b) D-erythrose   (c) D-galactose

13.11   Draw the Fischer projection formula of the C-2 epimer of each of the following:
(a) L-ribose   (b) L-threose   (c) L-mannose

Haworth Projection Formulas

13.12   Draw the Haworth projection formula of the pyranose form of each of the following compounds:
(a) α-D-mannose   (b) β-D-galactose   (c) α-D-glucose   (d) α-D-galactose

13.13   Draw the Haworth projection formula of the furanose form of each of the following compounds:
(a) α-D-fructose   (b) β-D-fructose   (c) α-D-ribulose   (d) β-D-xylulose

13.14   Identify the monosaccharide represented by each of the following structures:

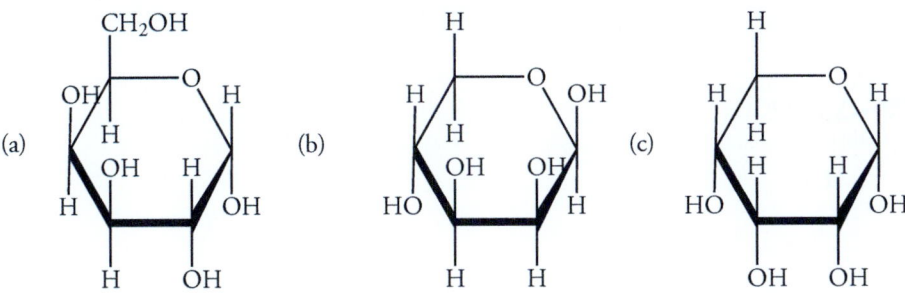

13.15   Identify the monosaccharide represented by each of the following structures:

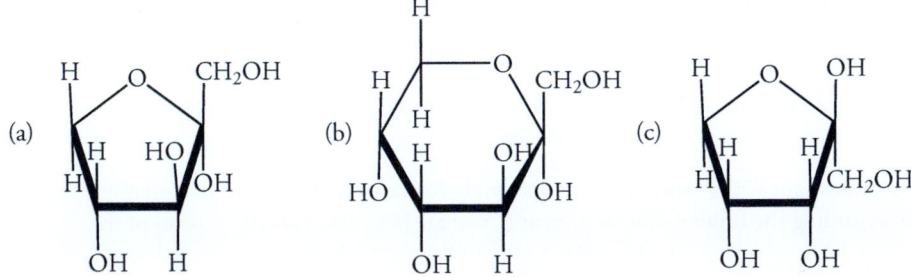

13.16 Name each compound in Exercise 13.15, indicating the type of ring and configuration of the anomeric center.

13.17 Name each compound in Exercise 13.16, indicating the type of ring and configuration of the anomeric center.

Conformations of Monosaccharides

13.18 Draw the chair conformation of β-galactopyranose and β-mannopyranose, and compare the number of axial hydroxyl groups in each compound.

13.19 Draw the standard chair conformation of β-talopyranose and β-allopyranose, and compare the number of axial hydroxyl groups in each compound.

Mutarotation

13.20 Can all aldopentoses mutarotate? Why?

13.21 Can L-glucose mutarotate?

13.22 Which of the following compounds can mutarotate?

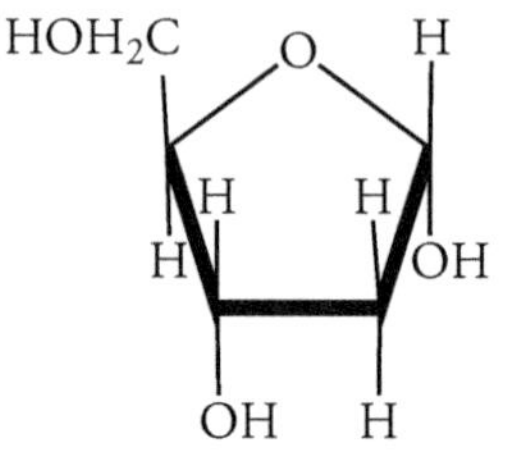

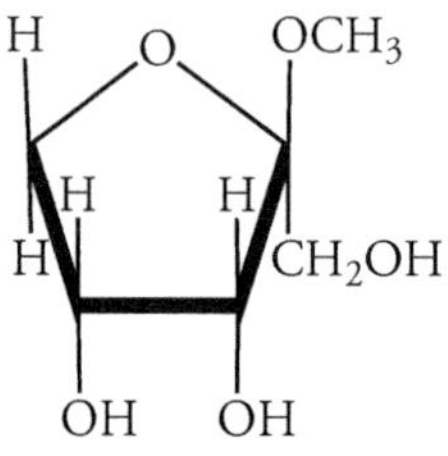

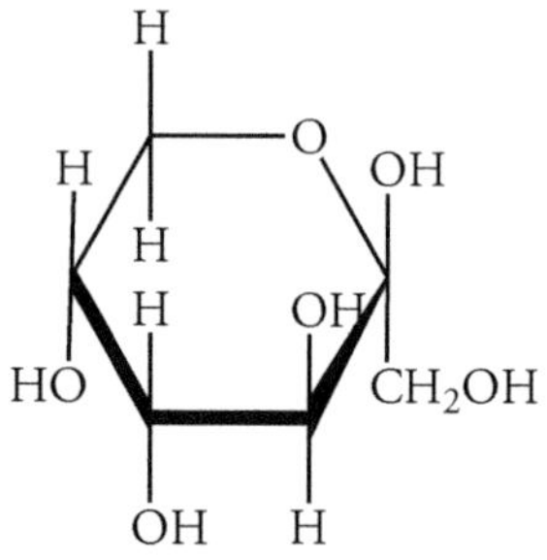

13.23 Which of the following compounds can mutarotate?

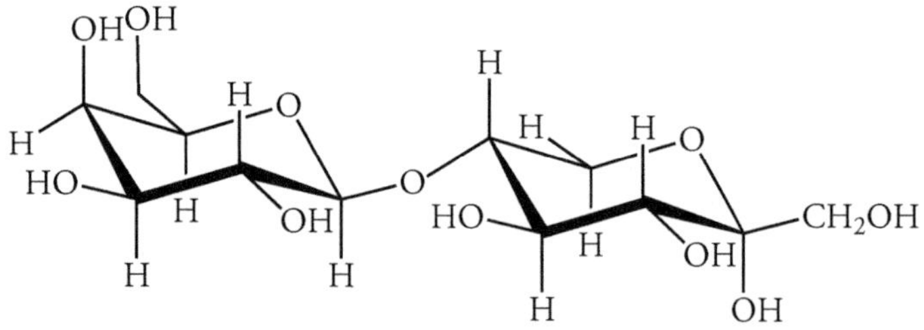

13.24 The $[\alpha]_D$ values of the α and β anomers of D-galactose are +150.7° and +52.8°, respectively. In water, mutarotation of D-galactose results in a specific rotation of +80.2°. Which anomer predominates?

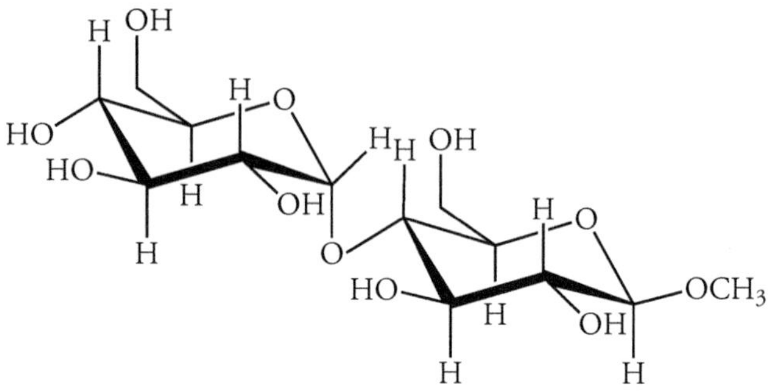

13.25 The $[\alpha]_D$ of the α and β anomers of D-mannose are +20.3° and –17.0°, respectively In water, mutarotation of D-mannose results in a specific rotation of +14.2. Disregarding the furanose forms present (less than 1%), calculate the percent of the α anomer.

Reduction of Monosaccharides

13.26 Draw the Fischer projections of the alditols of D-erythrose and D-threose. One compound is optically active, and the other is a meso compound. Explain why.

13.27 Which of the alditols of the D-pentoses are optically inactive? Explain why.

13.28 Reduction of D-fructose with sodium borohydride yields a mixture of two alditols. Explain why. Name the two alditols.

13.29 Reduction of D-tagatose with sodium borohydride yields a mixture of galactitol and talitol. What is the structure of D-tagatose?

Oxidation of Monosaccharides

13.30 Draw the structures of each of the following compounds:
(a) D-mannonic acid   (b) D-galactonic acid   (c) D-ribonic acid

13.31 Draw the structures of each of the following compounds:
(a) D-allonic acid   (b) D-talonic acid   (c) D-xylonic acid

13.32 Oxidation of D-erythrose and D-threose with nitric acid yields aldaric acids, one of which is optically inactive. Which one? Explain why.

13.33 Which of the D-aldopentoses will yield optically inactive aldaric acids when oxidized with nitric acid?

Isomerization of Monosaccharides

13.34 Draw the structures of the aldose and ketose that can exist in equilibrium with D-allose in basic solution.

13.35 Draw the structures of two aldoses that can exist in equilibrium with D-xylulose in basic solution.

Glycosides

13.36 Draw the Haworth projection formulas of the two glycosides formed from the pyranose forms of mannose and methyl alcohol.

13.37 Draw the Haworth projection formulas of the two glycosides formed from the furanose forms of fructose ethyl alcohol.

13.38 Vanillin is found as the β-glycoside of D-glucose. Draw the structure of the glycoside.

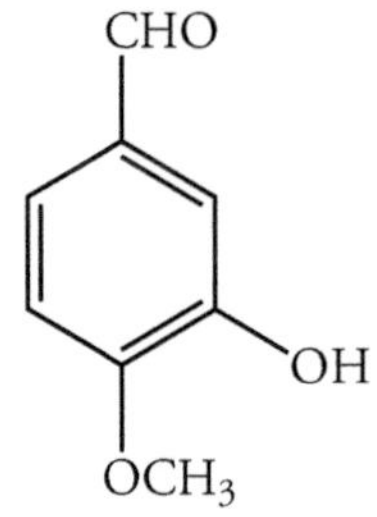

vanillin

13.39 Salicin is found in the bark of several species of fruit trees. What are the hydrolysis products of salicin?

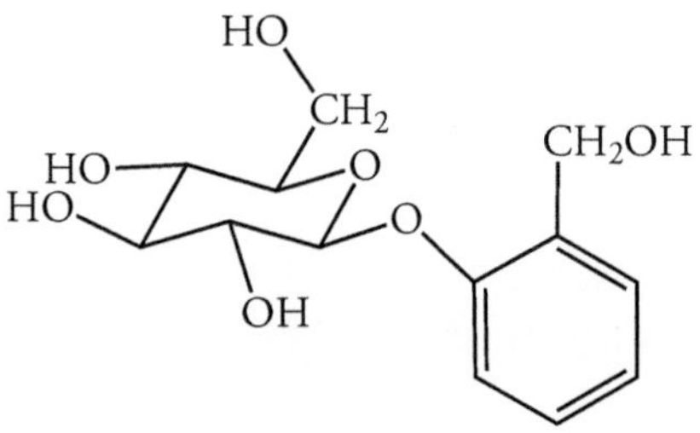

salicin

Disaccharides

13.40 Identify the component monosaccharides of each of the following compounds and describe the type of glycosidic linkage in each.

(a)

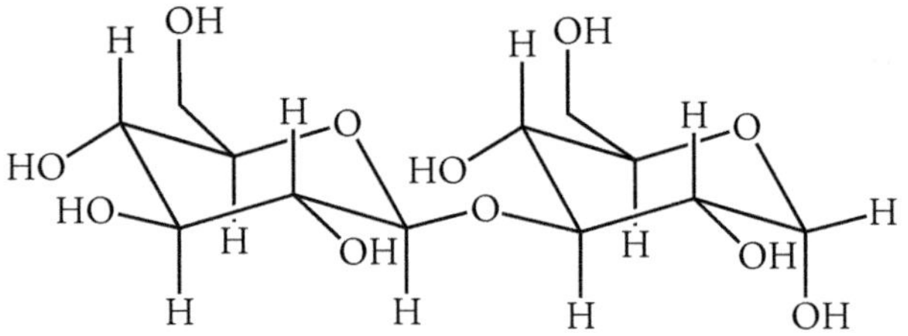

(b)

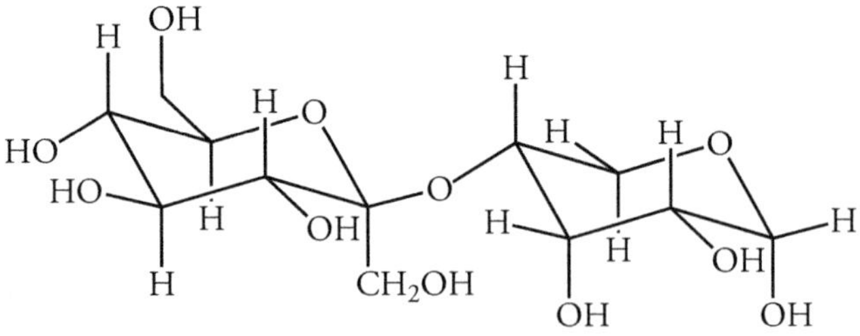

13.41 Identify the component monosaccharides of each of the following compounds and describe the type of glycosidic linkage in each.

(a)

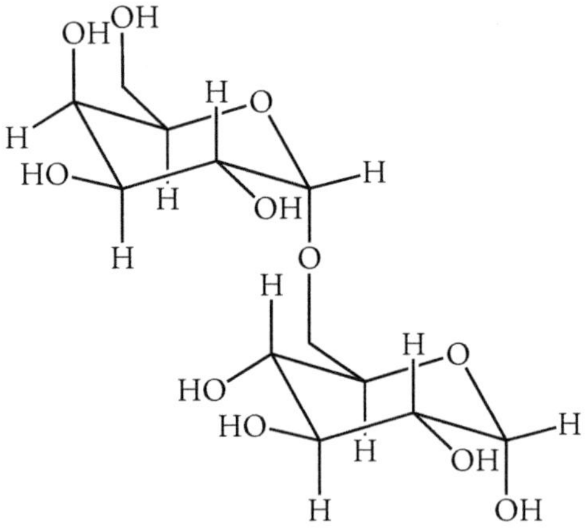

(b)

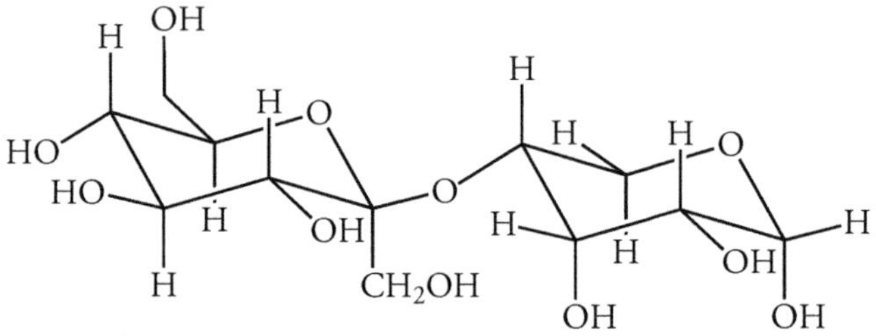

14 AMINO ACIDS, PEPTIDES, AND PROTEINS

The highest operation in nature and in art is the attainment of significant form.
Goethe

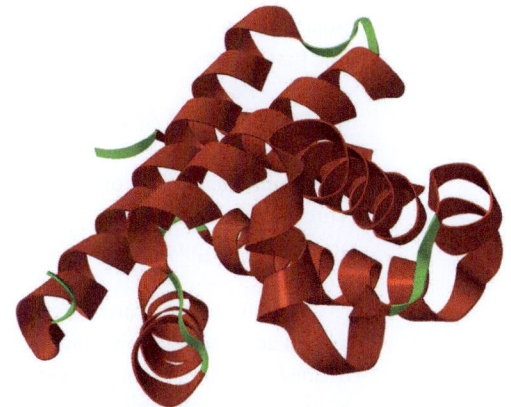

RIBBON MODEL OF HUMAN MYOGLOBIN

14.1 PROTEINS AND POLYPEPTIDES

From amoebas to zebras, the proteins of all organisms contain α-amino acids linked by amide bonds between amino and carboxylic acid functional groups. In proteins, amide bonds are called **peptide bonds**.

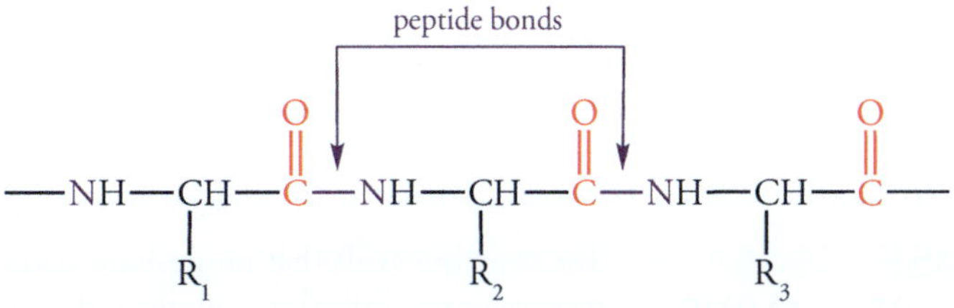

The name protein is derived from the Greek *proteios*, meaning "pre-eminence" or "holding first place," which reflects the crucial role that proteins play in virtually all cellular processes. The name was suggested in 1839 by the Dutch chemist Gerardus Johannes Mulder, who could not have known how prophetic his suggested name would be. Proteins have an extraordinary range of functions. Proteins called enzymes catalyze nearly all of the chemical reactions in cells. Proteins are required for the transport of most substances across cell membranes. They are the major structural substances of skin, blood, muscle, hair, and other tissues of the body. Proteins in the immune system, called antibodies, resist the effects of foreign substances that enter the body.

Proteins are polymers of 50 or more α-amino acids; some proteins contain more than 8000 amino acid units. Polypeptides are smaller molecules that contain fewer than about 50 amino acids. Some important hormonal polypeptides with physiological functions such as pain relief and control of blood pressure contain as few as nine amino acid units.

We begin this chapter by describing the structure and properties of the 20 amino acids isolated from proteins. Then we will consider the structure and properties of polypeptides and proteins. We will also describe the chemical synthesis of polypeptides and analytical methods to determine their structure.

14.2 AMINO ACIDS

Amino acids contain both an amino group and a carboxylic acid group. About 250 have been found in natural sources; however, only 20 of them occur in large amounts in proteins (Table 14.1). The amino acids of proteins in all cells are α-amino acids. They have an amino group bonded to the α-carbon atom of a carboxylic acid. The Fischer projection formula for an α-amino acid is shown below.

Principles of Organic Chemistry. http://dx.doi.org/10.1016/B978-0-12-802444-7.00014-8

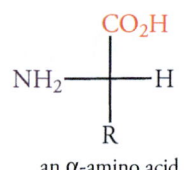

an α-amino acid

In this structure, the R group is called the **side chain**. There are 20 different R groups in amino acids isolated from proteins. Of these 20 α-amino acids, 19 are chiral; they have the L configuration. The one that is not chiral is glycine: its R group is hydrogen.

Classification of Amino Acids

The amino acids in proteins are primary amines except for proline, which is a secondary amine. Three-letter abbreviations of the amino acids, used as a short-hand to describe protein structure, are given in Figure 14.1. (An alternate one-letter shorthand method exists, but we will not use it in this text.)

The amino acids are classified by their side-chain R groups as neutral, basic, and acidic. **Neutral amino acids** contain one amino group and one carboxyl group. The neutral amino acids are further divided according to the polarity of the R group. Serine, threonine, and tyrosine are neutral amino acids that are also alcohols. Phenylalanine, tyrosine, and tryptophan contain aromatic rings. Cysteine and methionine contain a sulfur atom. The remaining neutral amino acids have hydrocarbon side chains.

The three **basic amino acids**, lysine, arginine, and histidine, have a basic nitrogen-containing functional group in the side chain. The two **acidic amino acids**, aspartic acid and glutamic acid, have carboxylic acid side chains. The acidic amino acids also have close relatives, asparagine and glutamine, that have neutral amides in the side chains.

Amino acids are also classified by the tendency of their side chains to interact favorably or unfavorably with water. Those amino acids with polar side chains are said to be **hydrophilic**; that is, water-loving. Those whose side chains are nonpolar are said to be **hydrophobic**. Hydrophobic amino acids have alkyl or aromatic groups that do not form hydrogen bonds to water.

14.3 ACID-BASE PROPERTIES OF α-AMINO ACIDS

The α-amino acids that do no have acidic or basic side chains have no *net* charge. However, their properties resemble those of salts rather than uncharged molecules. Amino acids have low solubilities in organic solvents, but are moderately soluble in water, unlike most organic compounds of comparable molecular weight. The physical states of amino acids also differ from those of comparable carboxylic acids and amines. For example, ethyl amine is a gas, and acetic acid is a liquid at room temperature. In contrast, glycine is a solid.

CH_3CH_2—NH_2 CH_3—CO_2H NH_2—CH_2—CO_2H

ethyl amine   acetic acid   glycine

mp –84 °C   mp 16 °C   mp 232 °C

Ionic Form of Amino Acids

When an amino acid dissolves in an aqueous buffer at pH 7 (Figure 14.1), its α-carboxyl group ionizes to give a carboxylate ion, and its α-amino group ionizes to give an ammonium ion. Thus, it exists as a dipolar ion, sometimes called a **zwitterion** (German, *zwitter*, hybrid). The dipolar ion acts both as an acid and a base. That is, it is **amphoteric**.

$$^+NH_3—\overset{\displaystyle CO_2^-}{\underset{\displaystyle R}{C}}—H$$

Structure of a dipolar ion (zwitterion)

When an amino acid dissolves in basic solution, the carboxylate group exists as an anion and the ammonium ion exists as an unprotonated amino group. This species is the *conjugate base* of the original amino acid. It has a net charge of –1.

Figure 14.1
Structures of the α-Amino Acids at pH 7
At pH 7 the α-amino and α-carboxyl groups are both ionized.

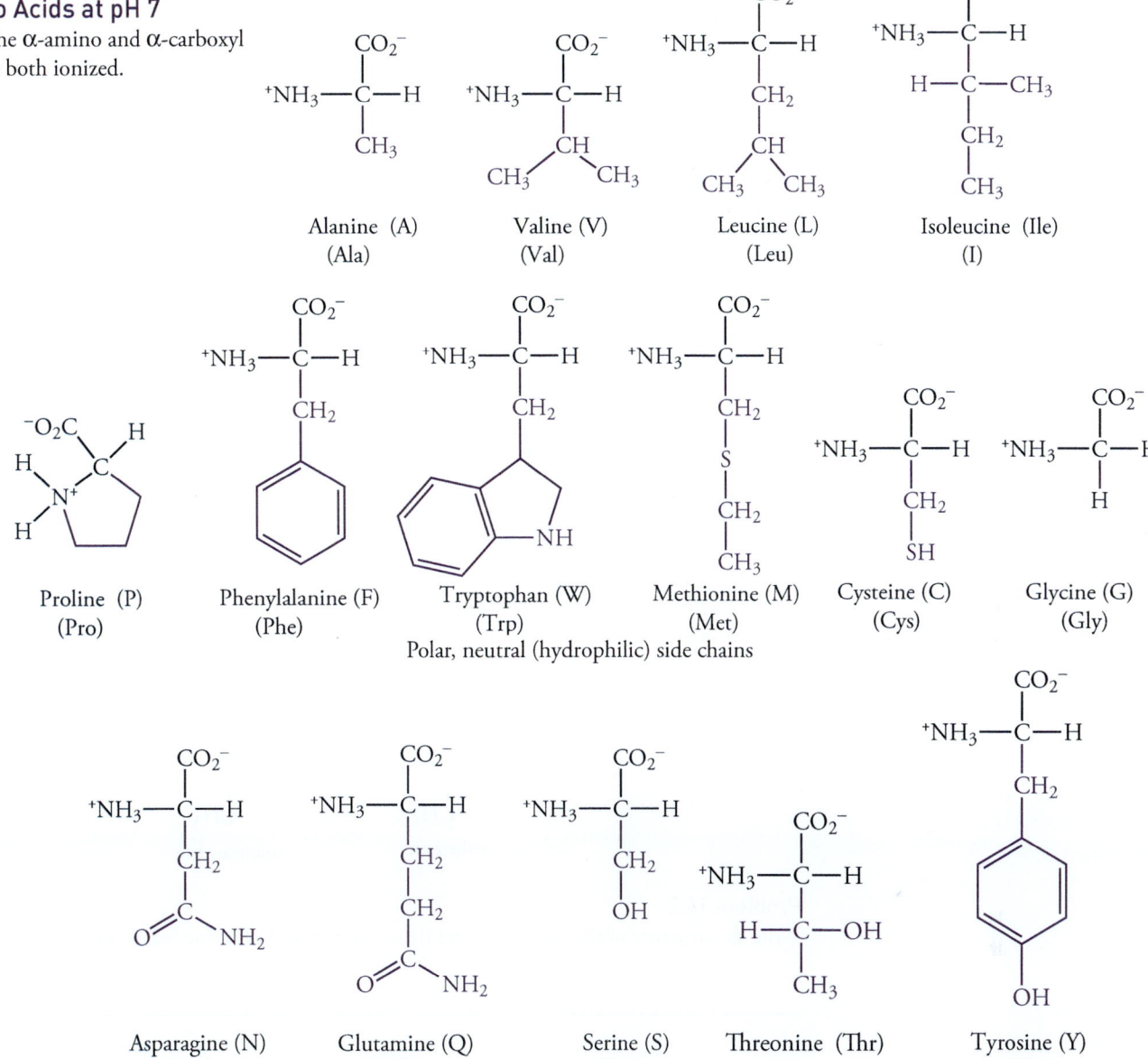

Nonpolar (hydrophobic) side chains

Alanine (A)
(Ala)

Valine (V)
(Val)

Leucine (L)
(Leu)

Isoleucine (Ile)
(I)

Proline (P)
(Pro)

Phenylalanine (F)
(Phe)

Tryptophan (W)
(Trp)

Methionine (M)
(Met)

Cysteine (C)
(Cys)

Glycine (G)
(Gly)

Polar, neutral (hydrophilic) side chains

Asparagine (N)
(Asn)

Glutamine (Q)
(Gln)

Serine (S)
(Ser)

Threonine (Thr)
(T)

Tyrosine (Y)
(Tyr)

Basic Amino Acids

Acidic Amino Acids

Lysine (Lys)
(K)

Arginine (Arg)
(R)

Histidine (His)
(H)

Aspartate (Asp)
(D)

Glutamate (Glu)
(E)

$$\overset{\displaystyle CO_2^-}{\underset{\displaystyle R}{^+NH_3-\overset{\textstyle |}{\underset{\textstyle |}{C}}-H}} \;+\; OH^- \;\longrightarrow\; \overset{\displaystyle CO_2^-}{\underset{\displaystyle R}{NH_2-\overset{\textstyle |}{\underset{\textstyle |}{C}}-H}} \;+\; HOH$$

zwitterion conjugate base

When an amino acid dissolves in acid solution, the carboxylate group exists as carboxylic acid group and the amino group exists as an ammonium ion. This species is the *conjugate acid* of the original amino acid. It has a net charge of +1.

$$\overset{\displaystyle CO_2^-}{\underset{\displaystyle R}{^+NH_3-\overset{\textstyle |}{\underset{\textstyle |}{C}}-H}} \;+\; H_3O^+ \longrightarrow\; \overset{\displaystyle CO_2H}{\underset{\displaystyle R}{NH_2-\overset{\textstyle |}{\underset{\textstyle |}{C}}-H}} \;+\; H_2O$$

zwitterion conjugate acid

Problem 14.1

What are the structures of the dipolar ion and conjugate base of alanine?

Solution

The zwitterion is written by removing a proton from the carboxyl group and adding a proton to the nitrogen atom. The conjugate base is written by removing a proton from the ammonium ion site of the dipolar ion. The nitrogen atom becomes neutral, whereas the carboxylate ion retains a negative charge.

$$\overset{\displaystyle CO_2^-}{\underset{\displaystyle CH_3}{^+NH_3-\overset{\textstyle |}{\underset{\textstyle |}{C}}-H}} \qquad\qquad \overset{\displaystyle CO_2^-}{\underset{\displaystyle CH_3}{NH_2-\overset{\textstyle |}{\underset{\textstyle |}{C}}-H}}$$

zwitterion conjugate base

Problem 14.2

Write the structure of the zwitterion and the conjugate acid of serine (see serine in Table 14.1).

---

pK$_a$ Values of α-Amino Acids

The pK$_a$ values of the amino acids depend upon the structure of the amino acid. The pK$_a$ values of glycine are 2.35 and 9.78. Table 14.1 lists the pK$_a$ values of all 20 α-amino acids.

$$\overset{\displaystyle CO_2H}{\underset{\displaystyle H}{^+NH_3-\overset{\textstyle |}{\underset{\textstyle |}{C}}-H}} \;+\; H_2O \;\rightleftharpoons\; \overset{\displaystyle CO_2^-}{\underset{\displaystyle H}{^+NH_3-\overset{\textstyle |}{\underset{\textstyle |}{C}}-H}} \;+\; H_3O^+ \qquad \begin{array}{l} K_a = 5 \times 10^{-3} \\ pK_a = 2.35 \end{array}$$

conjugate acid of glycine zwitterion

$$\overset{\displaystyle CO_2^-}{\underset{\displaystyle H}{^+NH_3-\overset{\textstyle |}{\underset{\textstyle |}{C}}-H}} \;+\; H_2O \;\rightleftharpoons\; \overset{\displaystyle CO_2^-}{\underset{\displaystyle H}{NH_2-\overset{\textstyle |}{\underset{\textstyle |}{C}}-H}} \;+\; H_3O^+ \qquad \begin{array}{l} K_a = 1.6 \times 10^{-10} \\ pK_a = 9.78 \end{array}$$

zwitterion conjugate base of glycine

When an amino acid dissolves in solution, several species usually exist. When the pH of the solution equals the pK_a of the ionizing group, the concentrations of the conjugate acid and the dipolar form are equal. For example, the pK_a of the carboxyl group of glycine is 2.35, and at pH 2.35 the concentrations of the conjugate acid and the dipolar ion are equal. The pK_a of the ammonium ion of glycine is 9.78, and at pH 9.78 the concentrations of the conjugate base and the dipolar ion are equal. At pH values between 2.35 and 9.78, the dipolar ion is the major ionic form of glycine in solution.

Table 14.1
pK_a Values of Acidic and Basic Groups in α-Amino Acids

Amino Acid	α-CO₂H group	α-NH₃⁺ group	Side chain
Glycine	2.35	9.78	
Alanine	2.35	9.87	
Valine	2.29	9.72	
Leucine	2.33	9.74	
Isoleucine	2.32	9.76	
Methionine	2.17	9.27	
Proline	1.95	10.64	
Phenylalanine	2.58	9.24	
Tryptophan	2.43	9.44	
Serine	2.19	9.44	
Threonine	2.09	9.10	
Cysteine	1.89	10.78	8.53
Tyrosine	2.20	9.11	10.11
Asparagine	2.02	8.80	
Glutamine	2.17	9.13	
Aspartate	1.99	10.00	3.96
Glutamate	2.13	9.95	4.32
Lysine	2.16	9.20	10.80
Arginine	1.82	8.99	12.48
Histidine	1.81	9.15	6.00

Problem 14.3
In what ionic form does serine exist in 0.1 M HCl (see Table 14.1)?

Solution
A 0.1 M solution of HCl has a pH of 1.0. The pK_a values of serine are 2.19 and 9.44. Thus, at pH = 1, serine will exist as the conjugate acid.

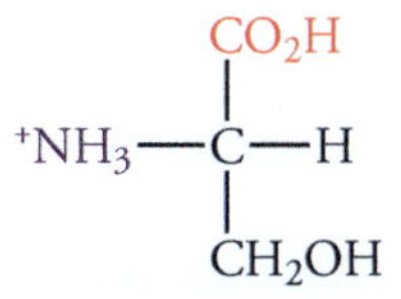

$$^+NH_3 - \underset{\underset{CH_2OH}{|}}{\overset{\overset{CO_2H}{|}}{C}} - H$$

conjugate acid of serine

Problem 14.4
In what ionic form does alanine exist in 0.01 M NaOH (see Table 14.1)?

14.4 ISOIONIC POINT

Table 14.2
Isoionic Points

Amino Acid	pH_i
Glycine	5.97
Alanine	6.10
Valine	5.96
Leucine	5.98
Isoleucine	6.02
Methionine	5.74
Proline	6.30
Phenylalanine	5/48
Tryptophan	5.89
Serine	5.68
Threonine	5.60
Cysteine	5.07
Tyrosine	5.66
Asparagine	5.41
Glutamine	5.65
Aspartic acid	2.77
Glutamic acid	3.22
Lysine	9.74
Arginine	10.76
Histidine	7.59

The pH at which the concentration of the zwitterion is at a maximum is the **isoionic point**, abbreviated pH_i. At this pH the amino acid has no net charge. In a more basic solution, when the pH is greater than the pH_i, the conjugate base predominates. In a more acidic solution, when the pH is less than the pH_i, the conjugate acid predominates. The isoionic points of some amino acids are given in Table 14.2. The isoionic points of the neutral amino acids are close to 7. Those of acidic and basic amino acids are significantly less than and greater than 7, respectively.

Titration of Amino Acids

The pK_a values of the carboxyl group and the ammonium group of an amino acid, as well as the pH_i, can be determined by titrating the conjugate acid with base. Figure 14.2 shows a typical titration curve for glycine. As base is added, some of the conjugate acid is converted to the zwitterion, and the pH increases. The pH at which the $-CO_2H$ group is one-half neutralized is equal to pK_1. After one equivalent of base has been added, the zwitterion is the major form in solution, and the pH at this point is pH_i. Addition of a second equivalent of base starts to convert the α-NH_3^+ group to the conjugate base of the amino acid. The pH at which half-neutralization has occurred is equal to pK_2.

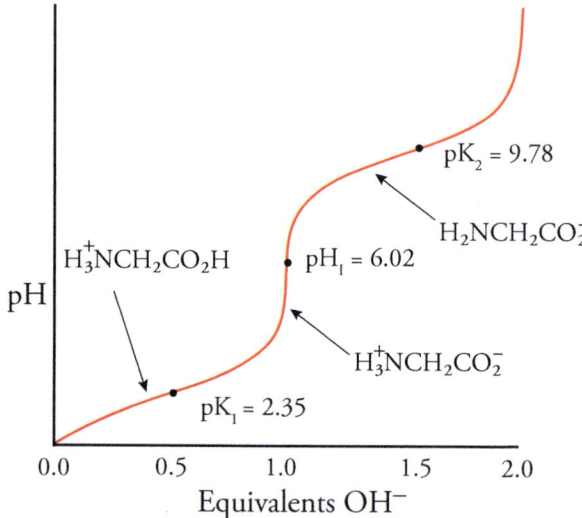

Figure 14.2 Titration Curve of Glycine

Isoionic Points of Proteins

A protein has an isoionic point that depends on its amino acid composition. At its isoionic point, a protein has no net charge, and its solubility is at a minimum. As a result, a protein tends to precipitate from solution at its isoionic pH. For example, casein, a protein in milk, has a net negative charge at the pH of milk, which is 6.3. The isoionic point of casein is 4.6 because it has many glutamic acid and aspartic acid residues. If milk is made more acidic, casein precipitates because the carboxylate ions of the side chains of glutamic acid and aspartic acid become protonated. Casein, which is used in making cheese, is obtained by adding an acid to milk or by adding bacteria that produce lactic acid.

Essential Amino Acids and Dietary Protein

Adequate amounts of about 10 of the 20 amino acids are synthesized in the body. The remaining amino acids must be obtained from food because either there is no biochemical pathway to produce them or the pathway produces inadequate amounts. Amino acids that must be obtained from food are called **essential amino acids**. Of course, all amino acids are necessary, but the term *essential* is reserved for those amino acids that must be obtained in the diet. The essential amino acids and their estimated minimum daily requirements are listed in Table 14.3.

Table 14.3
Essential Amino Acids and
Minimum Daily Requirements

Amino Acid	Minimum Daily Requirement (g)	
	Women	Men
Isoleucine	0.45	0.7
Leucine	0.6	1.1
Lysine	0.5	0.8
Methionine	0.55	1.0
Phenylalanine	1.1	1.4
Threonine	0.3	0.5
Tryptophan	0.15	0.25
Valine	0.65	0.8

Table 14.4
Biological Value of Dietary Protein

Food	Biological (%) Value
Whole hen's egg	94
Whole cow's milk	84
Fish	83
Beef	73
Soybeans	73
White potato	67
Whole grain wheat	65
Whole grain corn	59
Wry beans	58

Tyrosine is not listed as an essential amino acid because phenylalanine is converted into tyrosine by the body. Histidine is essential for growth in infants and may be essential for adults as well. The rate of synthesis of histidine cannot meet the needs of a growing body, so histidine may or may not be considered an essential amino acid depending on the age and state of health of the individual. The composition of dietary protein must provide the proper mix of essential amino acids and other amino acids to supply those processes that form essential body protein. If one or more necessary amino acids are not available at the time of synthesis of a vital protein, then the protein is not made.

Dietary proteins are rated in terms of biological value on a percentage scale (see Table 14.4). Complete proteins have a high biological value—they supply all of the amino acids in the amounts required for normal growth. Note that hen's eggs, cow's milk, and fish provide proteins of high biological value. Plant proteins vary more in biological value than animal proteins. However, not all plant proteins are deficient in the same amino acids. Gliadin, a wheat protein, is low in lysine; zein, a corn protein, is low in both lysine and tryptophan. Societies that eat large amounts of corn or wheat products must have other sources of lysine.

Vegetarians must carefully choose their food so that all the essential amino acids are available every day. For example, wheat is low in lysine, but beans are high in lysine as well as tryptophan. On the other hand, wheat is high in cysteine and methionine, whereas beans are low in these two amino acids. By eating both beans and wheat, a vegetarian increases the percentage of usable proteins. Some societies and ethnic groups have developed diets that provide a good nutritional supply of proteins even without the benefit of nutritionists. The Native American thrived on a diet that included both corn and beans—a mixture that we call succotash. Rice and black-eyed peas of the South as well as corn tortillas and beans in Mexico provide a reasonable balance of amino acids.

The diets of people living in some areas of the world today fall below the minimum daily requirement of protein for many reasons. As income decreases, the more costly animal protein is replaced by cereal grains and other incomplete protein sources. If a variety of plant proteins are not available, a number of diseases in young children result. Kwashiorkor is a protein deficiency disease that develops in young children after weaning, when their diet is changed to starches. The disease is characterized by bloated bellies and patchy skin. After a certain point, death is inevitable. Some forms of impaired mental development also result from incomplete nutrition.

14.5 PEPTIDES

When the α-amino group of one amino acid is linked to the carboxyl group of a second amino acid by an amide bond, the product is called a **peptide**. If the peptide contains two amino acid units, it is a **dipeptide**; a peptide that contains three amino acids is called a **tripeptide**. In general, a prefix *di-*, *tri-*, etc., indicates the number of amino acids in a peptide. But a peptide that contains, say, 14 amino acids is more likely called a 14-peptide than a tetradecapeptide. Peptides that contain only a "few" amino acids are called **oligopeptides**.

A peptide has two ends: the end with a free a-amino group is called the **N-terminal amino acid residue**; the end with the free carboxyl group is called the **C-terminal amino acid residue**. Peptides are named from the N-terminal amino acid to the C-terminal amino acid. The two isomeric dipeptides containing alanine and serine (Figure 14.3) illustrate the importance of designating the order of linkage of the amino acids.

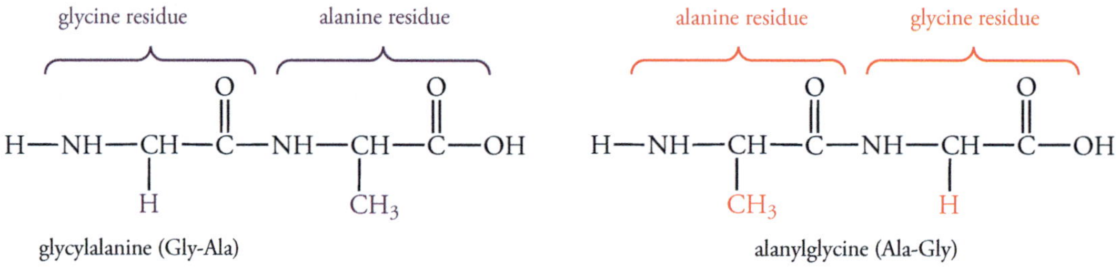

Figure 14.3
Peptide Nomenclature
(a) Structure of glycylalanine.
(b) Structure of alanylglycine.

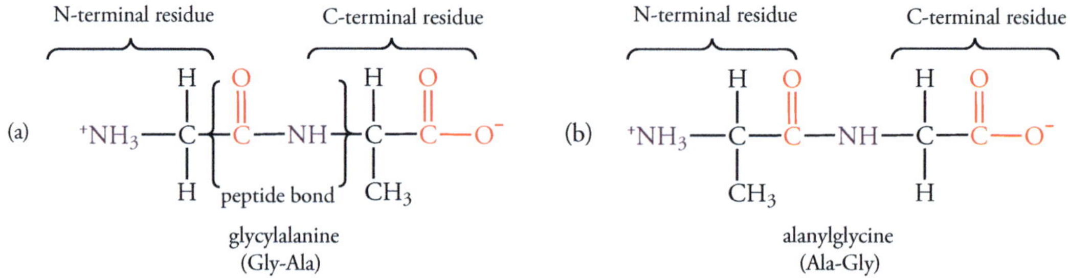

The number of isomers of peptides containing one each of n different amino acids is equal to $n!$, where

$$n! = 1 \times 2 \times 4 \ldots (n-1) \times n$$

There are thus six possible isomers in a tripeptide that contains three different amino acids. The isomeric tripeptides containing glycine, alanine, and valine are Gly-Ala-Val, Gly-Val-Ala, Val-Ala-Gly, Ala-Gly-Val, and Ala-Val-Gly. For a peptide that contains one each of 20 different amino acids, there are 2,432,902,008,176,640,000 isomers! (That is, on the order of 2×10^{18} isomers.) Proteins often contain hundreds of amino acid residues, and they contain two or more amino acids of the same kind, but the number of possibilities is still astronomically large. Viewed from an evolutionary standpoint, we can say that nature has just begun to experiment.

Biological Functions of Peptides

Cells contain many relatively small peptides that have diverse functions. Some are hormones with physiological functions such as pain relief and control of blood pressure (Table 14.5). These oligopeptides are produced and released in small amounts. They are rapidly metabolized, but their physiological action is necessary for only a short time. For example, the 14-peptide somatostatin, which inhibits the release of other hormones such as insulin, glucagon, and secretin, has a biological half-life of less than 4 min.

Enkephalins are peptides that bind specific receptor sites in the brain to reduce pain. The enkephalin receptor sites have a high affinity for opiates, including morphine, heroin, and other structurally similar substances. Hence, enkephalin receptors are commonly called opiate receptors. Opiates mimic the enkephalins that are normally present in the body to mitigate pain.

Peptides are produced in many tissues. For example, angiotensin II is made in the kidneys. It causes constriction of the blood vessels and thus increases blood pressure. Angiotensin II is the most potent vasoconstrictor known, and the production of excess angiotensin II is responsible for some forms of hypertension.

Oxytocin and vasopressin are structurally similar nonapeptides formed by the pituitary gland. Oxytocin causes the contraction of smooth muscle, such as that of the uterus. It is used to induce delivery or to increase the effectiveness of uterine contractions. Vasopressin is one of the hormones that regulate the excretion of water by the kidneys, and it affects blood pressure. The structures of oxytocin and vasopressin differ by only two amino acids. They are both cyclic peptides that result from a disulfide bond between what would otherwise be the N-terminal amino acid cysteine and another cysteine five amino acid residues away. The C-terminal amino acid exists as an amide in both compounds.

$$S \text{————} S$$
$$\bar{O}_2C \text{——} Cys\text{-}Tyr\text{-}Ile\text{-}Gln\text{-}Asn\text{-}Cys\text{-}Pro\text{-}Leu\text{-}Gly\text{-}NH_3^+$$

oxytocin

↑ neutral, nonpolar side chain

$$S \text{————} S$$
$$\bar{O}_2C \text{——} Cys\text{-}Tyr\text{-}Phe\text{-}Gln\text{-}Asn\text{-}Cys\text{-}Pro\text{-}Arg\text{-}Gly\text{-}NH_3^+$$

vasopressin

↑ positively charged, polar side chain

The structural difference between oxytocin and vasopressin may seem small at first glance. But, in fact, the difference is enormous. Residue 3 in oxytocin is isoleucine and residue 3 in vasopressin is phenylalanine. This is a relatively small difference because both residues are nonpolar and about the same size. However, residue 8 in oxytocin is leucine, a nonpolar amino acid with a *sec*-butyl side chain, whereas residue 8 in vasopressin is arginine, an amino acid with a basic side chain that has a positive charge at pH 7. Because of this difference in charge, the receptor for oxytocin has a weak affinity for vasopressin, and the receptor for vasopressin has a very low affinity for oxytocin.

Table 14.5
Peptide Hormones

Hormone	Amino Acid Residues	Function
Tuftsin	4	Stimulates phagocytosis
Met-enkephalin	5	Analgesic activity
Angiotensin II	8	Vasoconstriction, increased vasopressin secretion
Oxytocin	8	Affects uterine contractions
Vasopressin	8	An antidiuretic
Bradykinin	9	Produced in response to tissue injury
Somatostatin	14	Inhibits release of other hormones
Gastrin	17	Leads to pepsin secretion
Secretin	27	Stimulates pancreatic secretions
Glucagon	29	Stimulates glucose production from glycogen
Calcitonin	32	Decreases calcium level in blood
Relaxin	48	Relaxation of pubic joints
Insulin	51	Affects blood sugar level

Problem 14.5
(a) Determine the number of isomeric tripeptides containing one alanine and two glycine residues.
(b) Write representations of the isomers using three-letter abbreviations.

Problem 14.6

Identify the terminal amino acids of tuftsin, a tetrapeptide that stimulates phagocytosis and promotes the destruction of tumor cells. Write the amino acid sequence using three-letter abbreviations for the amino acids. Also write the complete name without abbreviations.

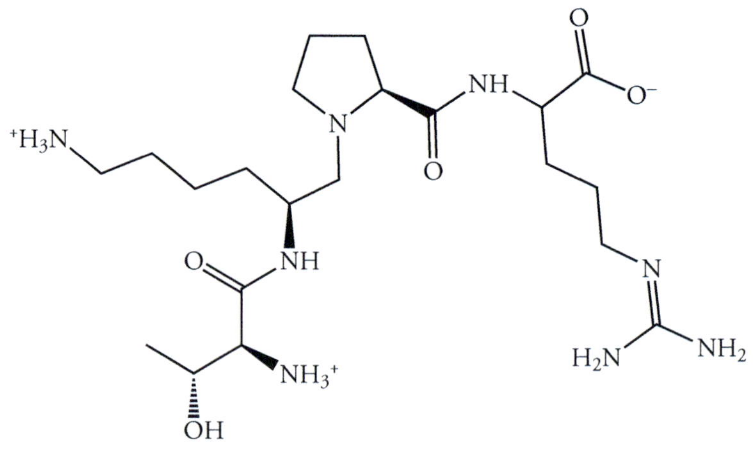

tuftsin

14.6 PEPTIDE SYNTHESIS

The synthesis of peptides and polypeptides is an important aspect of research in biochemistry and in the biotechnology industry. A highly specialized set of reagents has been developed for such syntheses to give high yields. Several steps are required because two amino acids cannot be simply combined to give a desired dipeptide. For example, two amino acids, such as alanine and glycine, yield a mixture of dipeptides. Each amino acid could form bonds with its own kind to form Gly-Gly and Ala-Ala or to a different amino acid to give Gly-Ala and Ala-Gly. Also, the amino acids in the reaction mixture can continue to react with the dipeptide products to yield oligopeptide.

The synthesis of a specific dipeptide requires modification of both amino acids. One amino acid is protected at its carboxyl group, by a reagent we will call P_C, and has the amino group available for peptide bond formation. The second amino acid is protected at the amino group, by a reagent we will call P_N, and has the carboxyl group available for peptide bond formation. Then only one condensation reaction is possible.

The carboxyl group is protected by converting it to a benzyl ester. Because esters are more reactive toward nucleophiles such as hydroxide ion than are amides, the C-terminus is easily "unprotected" at the end of the synthesis.

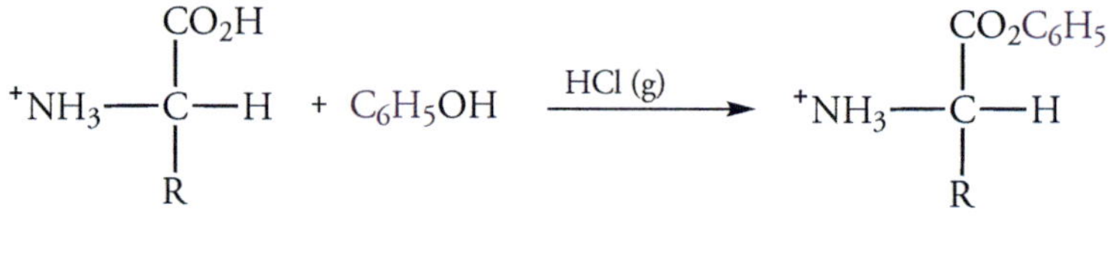

Several protecting groups have been developed to protect the amino terminus of an amino acid; the *tert*-butoxycarbonyl (Boc) derivative is typical. Reaction of an amino acid with di-*tert*-butyl dicarbonate gives a Boc-amino acid.

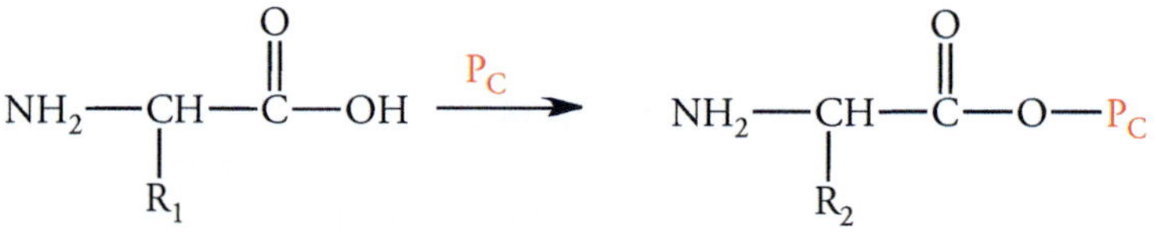

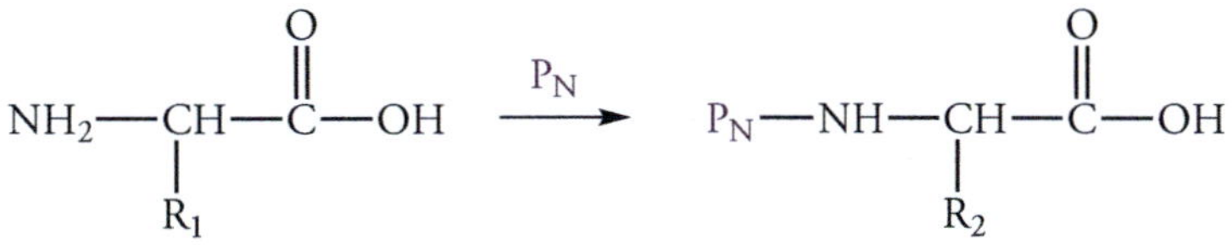

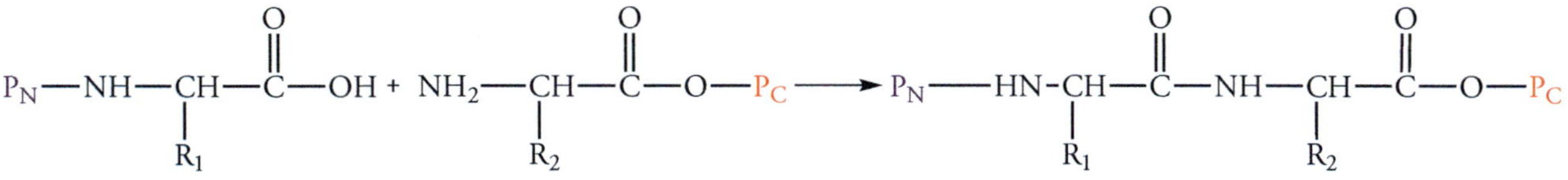

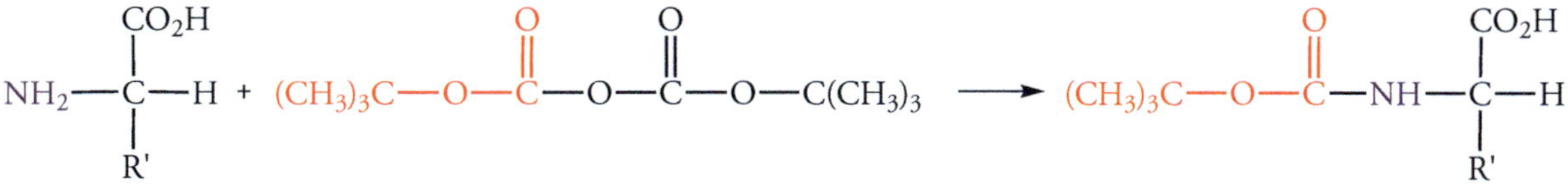

$$NH_2-CH(R_1)-C(=O)-OH \xrightarrow{P_N} P_N-NH-CH(R_2)-C(=O)-OH$$

$$P_N-NH-CH(R_1)-C(=O)-OH \; + \; NH_2-CH(R_2)-C(=O)-O-P_C \longrightarrow P_N-HN-CH(R_1)-C(=O)-NH-CH(R_2)-C(=O)-O-P_C$$

$$NH_2-C(R')(H)-CO_2H \; + \; (CH_3)_3C-O-C(=O)-O-C(=O)-O-C(CH_3)_3 \longrightarrow (CH_3)_3C-O-C(=O)-NH-C(R')(H)-CO_2H$$

N-*tert*-butoxycarbonyl amino acid
(a Boc-amino acid)

Note that the carbonyl group of the Boc group is bonded to both an oxygen atom and a nitrogen atom. This functional group is a *tert*-butyl carbamate that is easily hydrolyzed without affecting the amide or even the ester groups. The Boc group can be removed with trifluoroacetic acid. Both the amide groups and the ester of a protected carboxyl group are unaffected by the reaction conditions. The by-products of the reaction, CO_2 and 2-methylpropene, are gases.

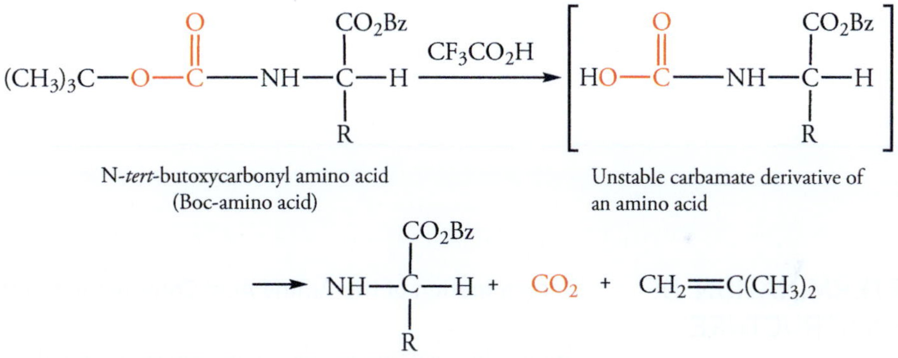

$$(CH_3)_3C-O-C(=O)-NH-C(CO_2Bz)(H)-R \xrightarrow{CF_3CO_2H} \left[HO-C(=O)-NH-C(CO_2Bz)(H)-R \right]$$

N-*tert*-butoxycarbonyl amino acid
(Boc-amino acid)

Unstable carbamate derivative of
an amino acid

$$\longrightarrow NH-C(CO_2Bz)(H)-R \; + \; CO_2 \; + \; CH_2{=}C(CH_3)_2$$

The protecting groups of both the amino and the carboxyl group are sensitive to acids and bases, so the condensation of two protected amino acids to form a peptide bond must be carried out under neutral conditions. It turns out that a reagent called dicyclohexylcarbodiimide (DCCI) causes condensation of two amino acids by removing the elements of water. The reaction has a very high yield and no other functional groups are modified. The by-product of the reaction is dicyclohexylurea.

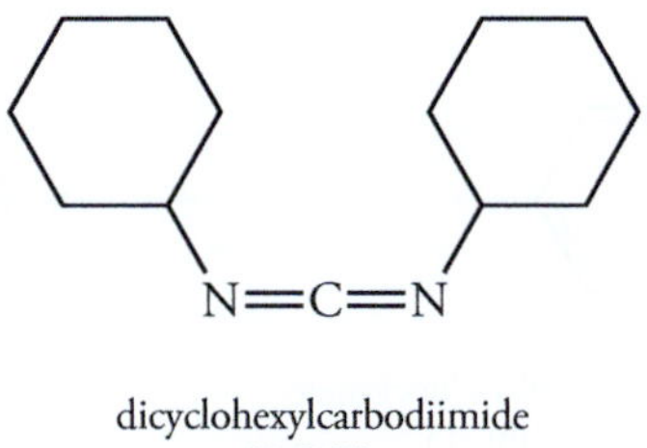

dicyclohexylcarbodiimide
(DCCI)

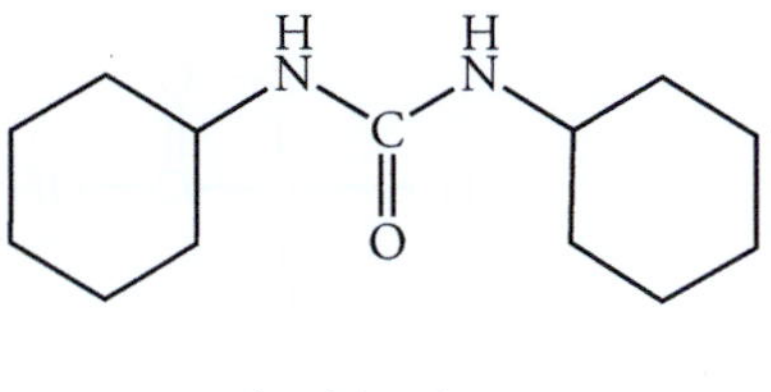

dicyclohexylurea

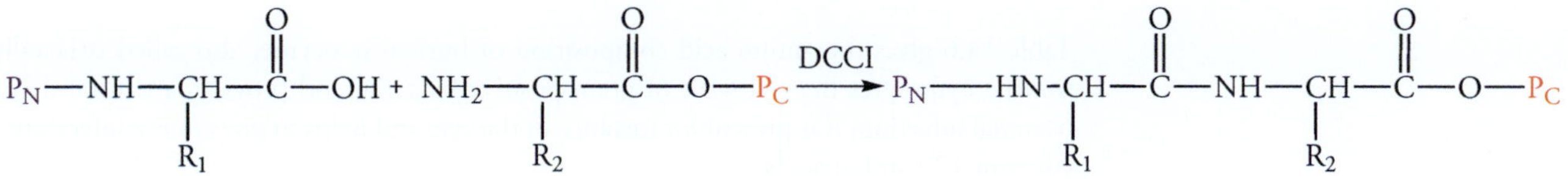

$$P_N-NH-CH(R_1)-C(=O)-OH \; + \; NH_2-CH(R_2)-C(=O)-O-P_C \xrightarrow{DCCI} P_N-HN-CH(R_1)-C(=O)-NH-CH(R_2)-C(=O)-O-P_C$$

$(CH_3)_3C-\overset{\overset{\displaystyle O}{\|}}{C}-O-HN-CH-\overset{\overset{\displaystyle O}{\|}}{C}-NH-CH-\overset{\overset{\displaystyle O}{\|}}{C}-O-Bz \quad \xrightarrow{CF_3CO_2H}$

with R_1, R_2

$NH_2-CH-\overset{\overset{\displaystyle O}{\|}}{C}-NH-CH-\overset{\overset{\displaystyle O}{\|}}{C}-O-Bz \quad \xrightarrow[DCCI]{Boc\text{-amino acid}} \quad Boc\text{-tripeptide-}CO_2Bz$

with R_1, R_2

This dipeptide is still protected at the carboxyl group and can only react at the free amino group. Reaction with another Boc-amino acid and DCCI yields a tripeptide. Ultimately, after the proper number of reaction sequences, the final polypeptide is liberated by hydrolysis with a base.

$Boc-NH-\underset{R_1}{\overset{H}{C}}-\overset{O}{C}-NH-\underset{R_2}{\overset{H}{C}}-\overset{O}{C}-NH-\underset{R_3}{\overset{H}{C}}-\overset{O}{C}-OBz \quad \xrightarrow{6\ M\ HCl}$

protected tripeptide

$\overset{+}{NH_3}-\underset{R_1}{\overset{H}{C}}-\overset{O}{C}-HN-\underset{R_2}{\overset{H}{C}}-\overset{O}{C}-NH-\underset{R_3}{\overset{H}{C}}-\overset{O}{C}-OH$

tripeptide

14.7 DETERMINATION OF PROTEIN STRUCTURE

Determination of the Amino Acid Composition of Proteins by Chemical Methods

At one time, determining the amino acid composition was a difficult and time consuming process. This analysis is now performed automatically in an instrument called an amino acid analyzer. About 10 μg of protein are required. Modern instruments can detect about 5 nmol of a given amino acid.

The first step in the analysis is hydrolysis of the protein in HCl. The amino acids released by hydrolysis of the protein are converted to phenylthiocarbamyl (PTC) amino acids by reaction with phenylisothiocyanate (PICT). The mixture of PTC amino acids is separated by chromatography, and the per cent composition of each amino acid in the mixture is then determined automatically.

$NH_2-\underset{H}{\overset{R_1}{C}}-\overset{O}{C}-OH \ + \ \langle\text{phenyl}\rangle-N\!=\!C\!=\!S \ \longrightarrow \ \langle\text{phenyl}\rangle-HN-\overset{S}{C}-NH-\underset{H}{\overset{R_1}{C}}-CO_2H$

phenylisothiocyanate (PICT)   phenylisothiocarbamyl amino acid (PCT amino acid)

Table 14.6 gives the amino acid composition of human lysozyme, also called α-lactalbumin. Lysozyme hydrolyzes the cell walls of gram-positive bacteria, and provides a natural defense against bacterial infection. It is present for instance in the eye, and helps to prevent eye infections. Lysozyme contains 120 amino acids.

Table 14.6
Amino Acid Composition of Human Lysozyme

Amino Acid	Number of Amino Acids	Percent Composition
Ala	5	4.1
Arg	1	0.8
Asn	4	3.3
Asp	12	9.8
Cys	8	6.5
Gln	7	4.9
Glu	8	6.5
Gly	6	9.8
His	2	11.4
Ile	12	9.8
Leu	14	11.4
Lys	12	9.8
Met	2	1.6
Phe	4	3.3
Pro	2	1.6
Ser	8	6.5
Thr	7	5.7
Trp	3	2.4
Tyr	4	3.3
Val	2	1.6

Chemical End Group Analysis

The identity of the N-terminal amino acid of a polypeptide can be determined by a method invented by Pehr Edman that is called the Edman degradation. In the Edman degradation, the polypeptide is treated with phenyl isothiocyanate—the Edman reagent—which reacts with the N-terminal amino acid to give an N-phenylthiourea derivative. This derivative forms by addition of the terminal N—H bond across the C=N of the phenyl isothiocyanate. After the adduct has formed, anhydrous trifluoroacetic acid is added to the reaction mixture. This reagent cleaves the polypeptide at the N-terminal residue. Under these conditions, the peptide bonds in the protein do not break (Figure 14.4). A complex cyclization reaction occurs to give a substituted phenylthiohydantoin. This ring contains the carbonyl carbon atom, the α-carbon atom, and the amino-nitrogen atom. The R group of the amino acid is attached to the ring. Comparison of the product with the phenylthiohydantoin of known amino acids establishes the identity of the amino acid. This entire process can be carried out automatically by an instrument called an automatic sequenator.

Because the Edman degradation does not cleave the peptide bonds in the protein, it can be repeated to identify the amino acids sequentially from the N-terminal amino acid end of the molecule. The yield of the Edman degradation approaches 100%, and sequences of 30 residues of a polypeptide can be determined from 5 pmol (5×10^{-12} mol) samples. This means that the sequence of a peptide with 30 amino acid residues, with a molecular weight of about 3000, can be determined from a 15 ng sample!

Figure 14.4
Edman Degradation

First, the peptide is converted to its N-terminal PCT derivative by treatment with phenylisothiocyanate. Next, the PCT protein is treated with trifluoroacetic acid, then with water to give the phenylthiohydantoin derivative. The N-terminal amino acid is released in this step. The other peptide bonds are not affected.

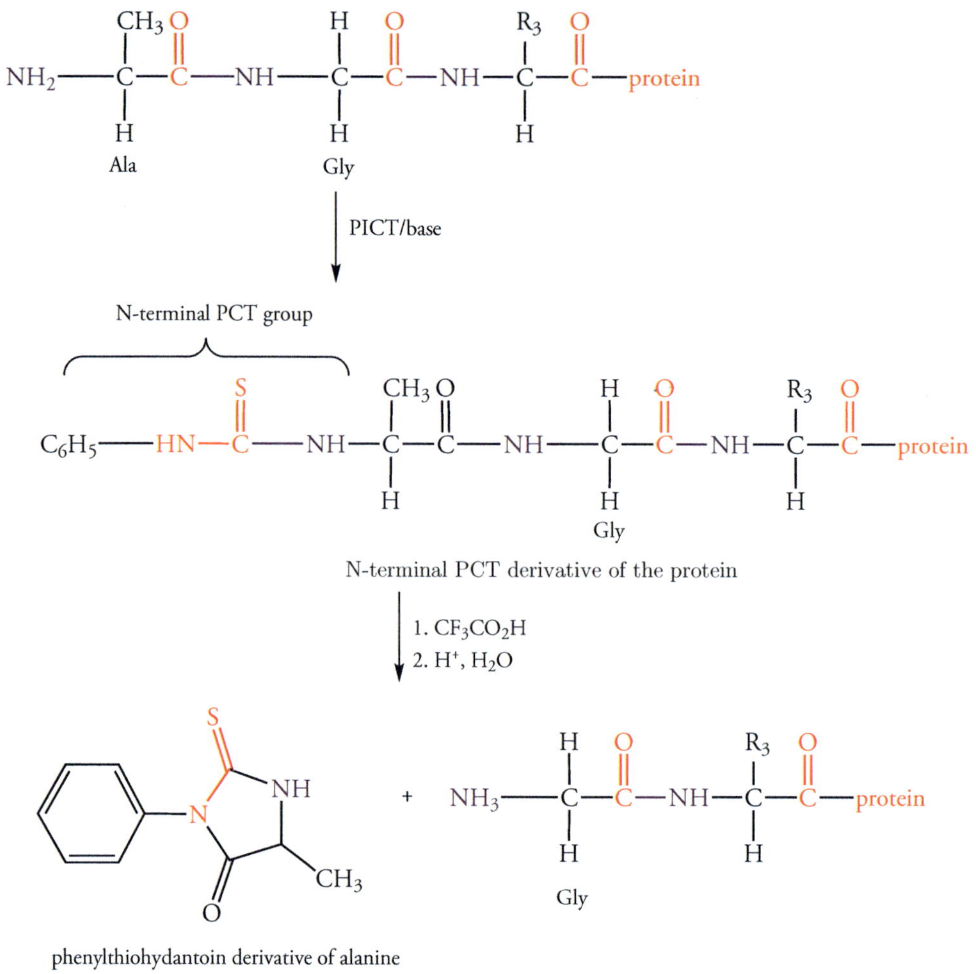

phenylthiohydantoin derivative of alanine

Enzymatic Cleavage of Polypeptide Chains

Many proteins contain hundreds of amino acids. To determine their sequences other reactions are required to provide sequences short enough to be determined by Edman degradation. Enzymatic cleavage by two enzymes, trypsin and chymotrypsin, are used to produce smaller peptides. Trypsin cleaves polypeptide chains on the C-terminal side of basic residues such as arginine and lysine. Chymotrypsin cleaves the polypeptide on the C-terminal side of aromatic residues.

The sequences each oligopeptide fragment produce in these enzymatic reactions are determined by Edman degradation. Then, in the final step the fragments are aligned to provide the entire sequence.

Primary Structures and Evolutionary Relationships

The primary structures of thousands of protein are known. Comparing the primary structures of proteins that are common to many species reveals evolutionary relationships. As organisms evolve, their genes change through mutation. Because the primary structure of a protein reflects the gene coding for it, differences among primary structures are a record of evolutionary change. Comparing the amino acid sequences of proteins found in different species thus opens a window to the past. In a sense, then, proteins can be regarded as living fossils.

In closely related species, the primary structures of common proteins are similar. Counting the number of differences in amino acid sequences among these proteins gives some idea of how far various species have diverged in the course of evolution. For example, the protein cytochrome c is an excellent protein for evolutionary comparisons because it is found in the respiratory electron transport system, which is present in all aerobic organisms (Figure 14.5).

Figure 14.5 shows that as evolutionary lines diverge, the number of sequence variations increases, so that closely related species have few differences and distantly related species have many difference in primary structure. Thus, human and chimpanzees have identical cytochrome c sequences. The primary structures of cytochrome c molecules from California gray whales differs from that of pigs, cows, and sheep by only two residues. We conclude that the whale has evolved from land animals related to modern hoofed animals. Gray whale cytochrome c differs from human cytochrome c 10 residues.

Peking duck and penguins also have cytochrome c sequences that differ by only three residues, but they differ by 11 residues from bullfrogs. Thus, these species are closely related to each other but distantly to bullfrogs.

The difference between human cytochrome c and baker's yeast cytochrome c is 45 residues, which we do not find particularly surprising since these are distantly related species. However, 59 of the 104 residues in cytochrome c are identical. Identical residues are essential for the structure and function of the protein.

Figure 14.5 Evolutionary Family Tree for Cytochrome c

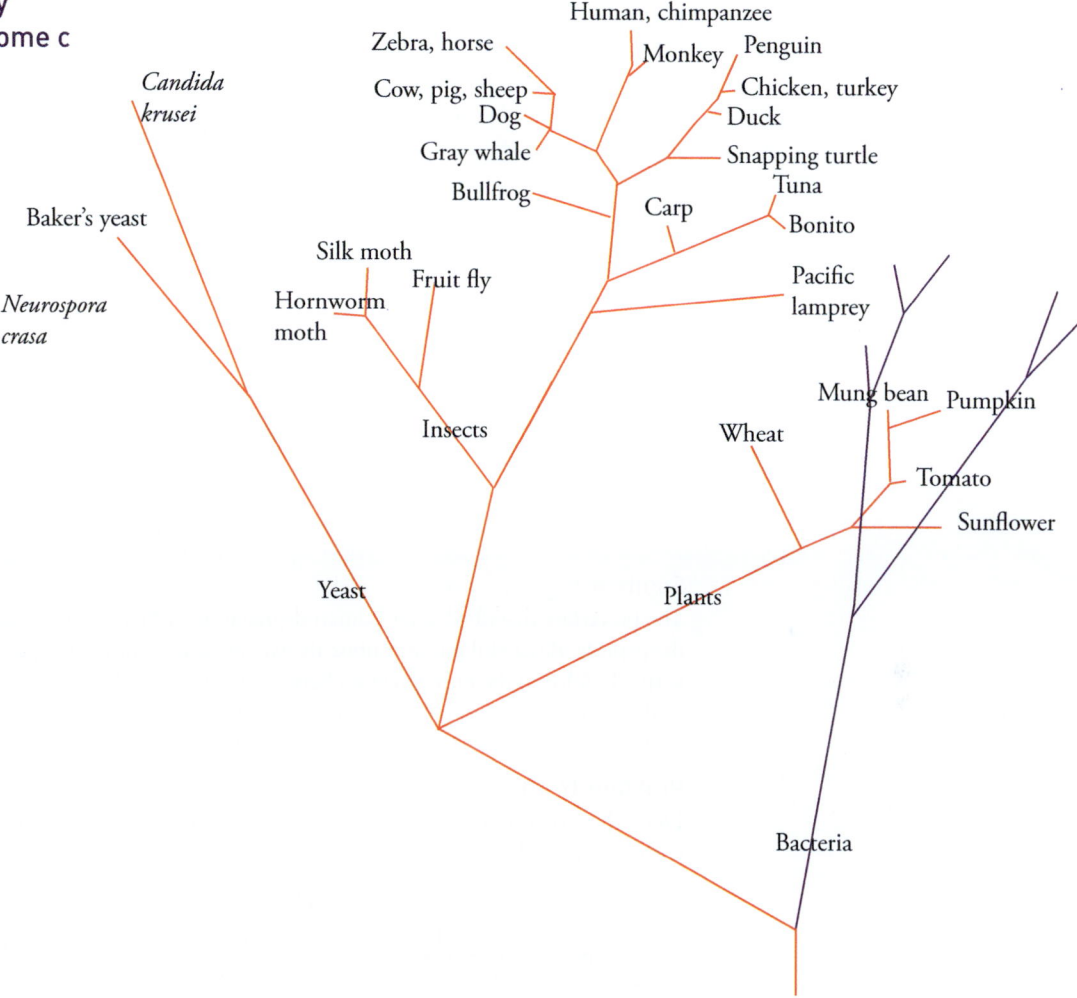

Problem 14.7

There are several enkephalins. Predict the products of the chymotrypsin-catalyzed hydrolysis of the following enkephalin.

Tyr-Giy-Giy-Phe-Leu

Solution

Chymotrypsin catalyzes hydrolysis of peptide bonds on the C-terminal side of aromatic amino acids in peptides and proteins. The enkephalin contains both phenylalanine and tyrosine, so chymotrypsin cleaves the peptide in two places. Consider each step separately. Tyrosine is the N-terminal amino acid, and hydrolysis at its carboxyl end results in free tyrosine.

Tyr—Gly—Gly—Phe—Leu $\xrightarrow{\text{chymotrypsin}}$ Tyr + Gly—Gly—Phe—Leu

The phenylalanine in the tetrapeptide is bonded to the C-terminal amino acid, leucine. Hydrolysis at the carboxyl group of phenylalanine frees leucine. A tripeptide results.

Gly—Gly—Phe—Leu $\xrightarrow{\text{chymotrypsin}}$ Gly—Gly—Phe + Leu

The products of the reaction are tyrosine, leucine, and the tripeptide Gly—Gly—Phe.

Problem 14.8
Predict the products of the trypsin-catalyzed hydrolysis of the following pentapeptide.

Ala—Lys—Gly—Arg—Leu

Problem 14.9
β-Endorphin, a 31-peptide, has analgesic effects and promotes the release of growth hormone and prolactin. Treating β-endorphin with phenyl isothiocyanate followed by hydrolysis with anhydrous trifluoroacetic acid yields the following product. What does this information reveal about the structure of the peptide?

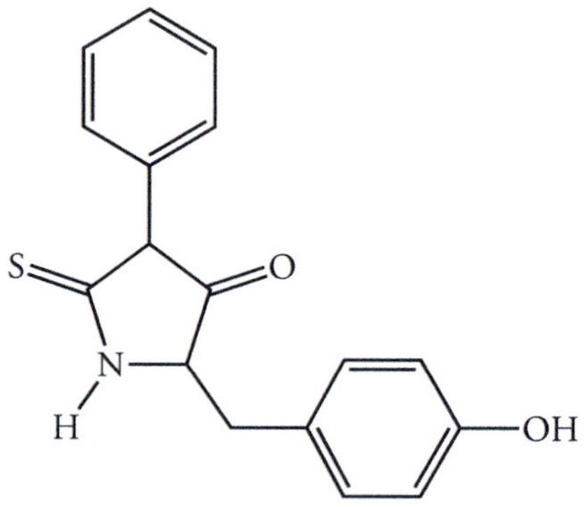

Solution
The procedure described is an Edman degradation, which removes the N-terminal amino acid from the peptide. Be careful not to confuse the two aromatic rings in the phenylthiohydantoin. The aromatic ring bonded to the nitrogen atom between the C=O and the C=S is the phenyl group of the phenyl isothiocyanate. The group bonded to the ring between the nitrogen atom and the C=O is the R group of the amino acid. The N-terminal amino acid of β-endorphin is tyrosine.

Problem 14.10
Draw the structure of the phenylthiohydantoin obtained by reaction of the following tetrapeptide with phenyl isothiocyanate.

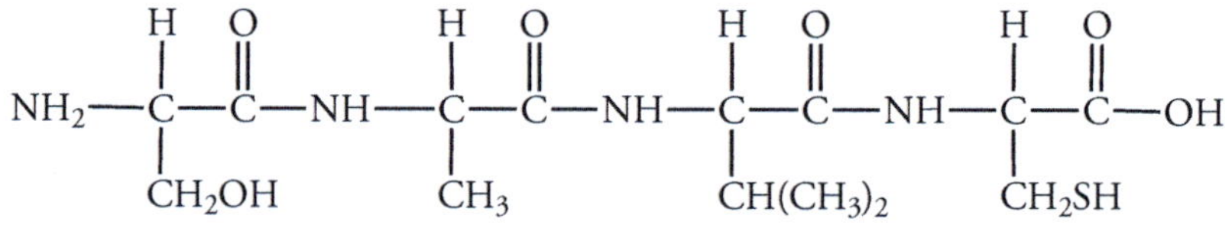

14.8 PROTEIN STRUCTURE

The biological activity of a protein depends on the three-dimensional shape of the molecule, called its **native state** or **native conformation**. Protein structure is described at four levels: primary, secondary, tertiary, and quaternary. Each of these divisions is somewhat arbitrary because it is the total structure of the protein that controls function. Nevertheless, it is useful to consider the levels of structure individually, so we will proceed by describing each level of structure in turn.

Primary Structure

The linear sequence of amino acids in a protein and the location of disulfide bonds is called its **primary structure**. The peptide bond, which is largely responsible for the primary structure of a protein, is a very strong bond. The peptide bond is a resonance hybrid of two contributing structures, and the carbon-nitrogen bond has partial double bond character. The peptide bond is planar, and there is restricted rotation about the carbon-nitrogen bond.

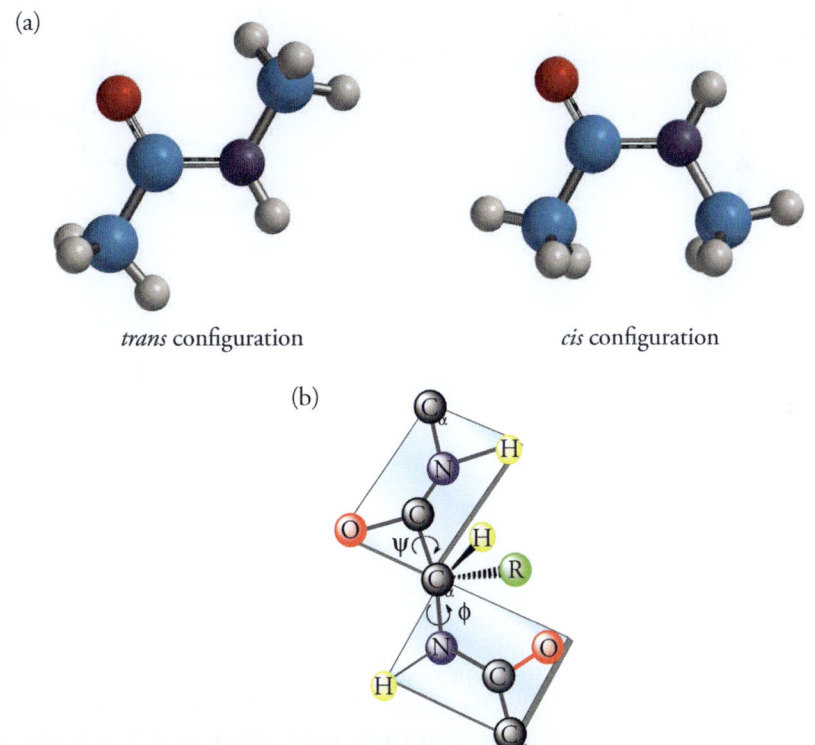

contributing structure 1   contributing structure 2   resonance hybrid

Two conformations are possible about the planar peptide bond. The *trans* conformation shown for the amide bond of N-methylacetamide is more stable than the alternate *cis* conformation due to steric effects (Figure 14.6). The amino acid residues in proteins usually exist in such *trans* conformations as well for the same reason.

The bond between the α-carbon atom of peptides and the carbonyl carbon atom is a rotationally free single bond. Similarly, the single bond between the nitrogen atom and the a-carbon atom of the next amino acid is also rotationally free. There is, in addition, free rotation about the bonds between the α-carbon atoms and the R groups. Thus, a protein chain consists of rigid peptide units connected to one another by freely rotating single bonds.

Figure 14.6
Structure of the Peptide Bond

(a) Rotation around the C—N bond, which has 50% double bond character, does not occur at room temperature. However, rotation around the N—C_a bond (φ) and the C—C_a bond (ψ) is possible, and many conformations are possible in peptides and proteins.
(b) We can think of the α-carbon as a "hinge" between two planar peptide bonds. If one takes two note cards, and links them with a swivel, it is easy to see that many arrangements are possible. However, some φ and ψ are not possible because of steric interference of the side chain R group. Glycine, for example, can assume many more conformations than amino acids like proline and tryptophan.

(a)

trans configuration   *cis* configuration

(b)

The disulfide bond, which is a covalent bond between two sulfur atoms, is also considered part of the primary structure of a protein. A disulfide bond results from the oxidation of the —SH (sulfhydryl) groups of two cysteine molecules to form cystine.

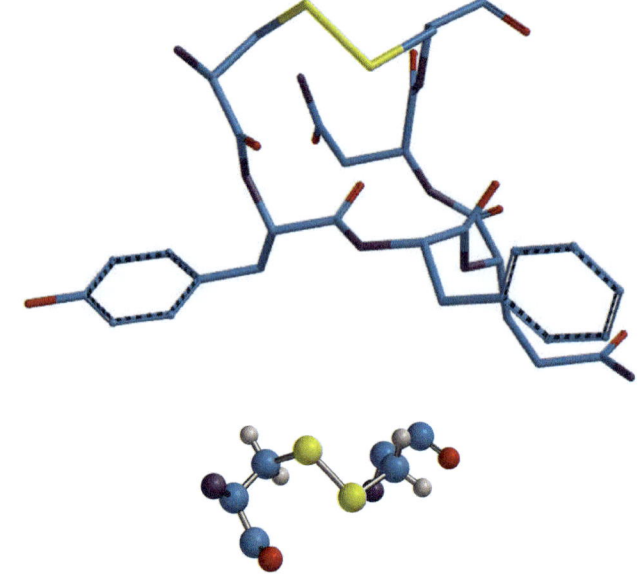

2 cysteine residues

cystine residue

Many proteins contain cysteine residues. Many of these cysteine residues exist in an oxidized form, linking them by intramolecular bonds to other cysteine residues. As a result, the protein conformation is much less flexible. Intrachain disulfide bonds occur in oxytocin and vasopressin (Figure 14.7). Disulfide bonds can also link a cysteine residue in one polypeptide chain with a cysteine residue in another polypeptide chain such as occurs in insulin, which consists of two polypeptide chains linked by two disulfide bonds.

Figure 14.7
Structure of Vasopressin
A disulfide bond, shown in yellow, forms between cysteine residues 1 and 6. An expanded view of a disulfide bond is shown below the peptide structure. It has the least sterically hindered s-trans conformation.

Secondary Structure

The specific spatial arrangement of the amino acid residues close to one another in the polypeptide chain into a regularly repeating conformation is called the **secondary structure** of the protein (Figure 14.8). Because the bonds separating the planar and rigid peptide units can rotate freely, these peptide units can be oriented at angles to each other and can exist in many conformations. Intramolecular hydrogen bonding between the amide hydrogen atom of one peptide unit and the carbonyl oxygen atom of another peptide unit is very common. These hydrogen bonds lead to structures called α-helices and β-pleated sheets, which we will discuss below.

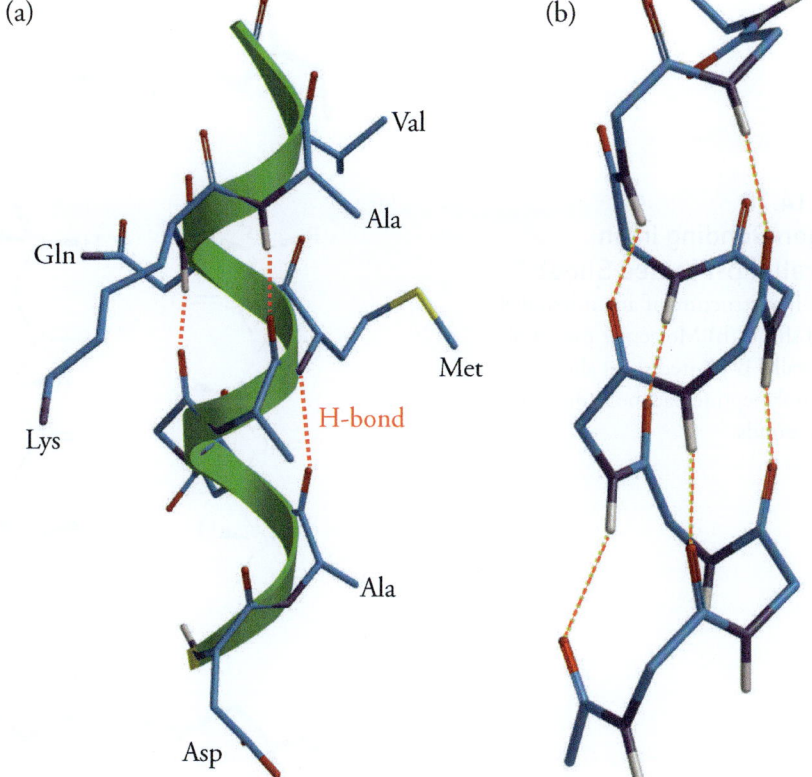

Many proteins regions in which the polypeptide chain is coiled into a spiral known as an α-helix. The helix could be either right- or left-handed, but for proteins consisting of L-amino acids, the right-handed α-helix is the only one observed. The α-helix is stabilized by hydrogen bonds between the proton of the N—H group of one amino acid and the oxygen atom of the C=O group of another amino acid in the next turn of the helix (Figure 14.8).

Figure 14.8
Structure of the α-Helix
Hydrogen bonds between the peptide bond N—H hydrogen and carbonyl oxygen atoms are approximately parallel to the long axis of the α-helix. (a) The α-helix is shown as a ribbon, which partly obscures hydrogen bonds. (b) Side chains are omitted to show the pattern of hydrogen bonds in the α-helix.

Many proteins contain a type of secondary structure in which the polypeptide chain has a completely extended conformation. Two adjacent regions of fully extended structures form hydrogen bonds that link the chains at approximately right angles to the long axis of the chain. This type of secondary structure is called a β-**pleated sheet**. There are two kinds of β-pleated sheets, called **parallel** and **antiparallel**. In a parallel β-pleated sheet, the C— and N— termini of the sheet are together (Figure 14.9); in the antiparallel β-pleated sheet, they are opposed (Figure 14.10).

**Figure 14.9
Hydrogen Bonding in a Parallel
β-Pleated Sheet**

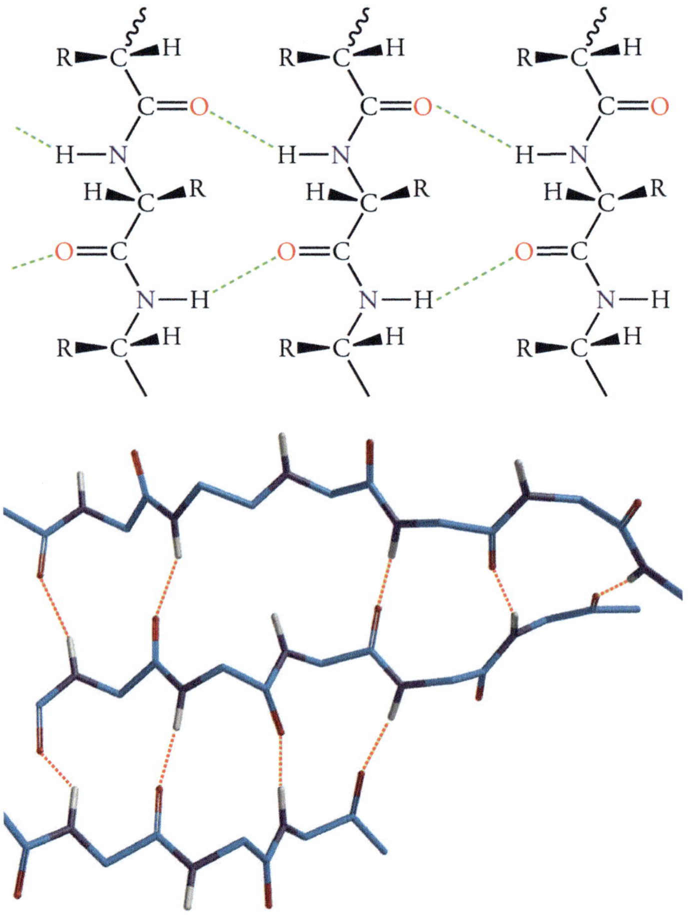

**Figure 14.10
Hydrogen Bonding in an
Antiparallel β-Pleated Sheet**
(a) Bond-line structure of an antiparallel
β pleated sheet. (b) Molecular model of
an antiparallel β-pleated sheet showing
only the polypeptide backbone and the
hydrogen bonds.

(a)

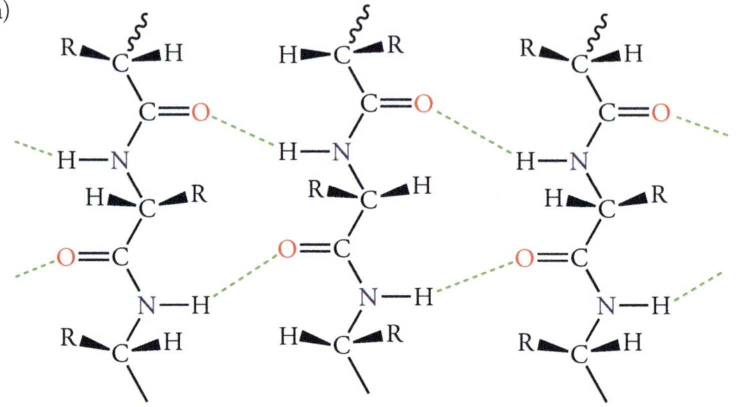

(b)

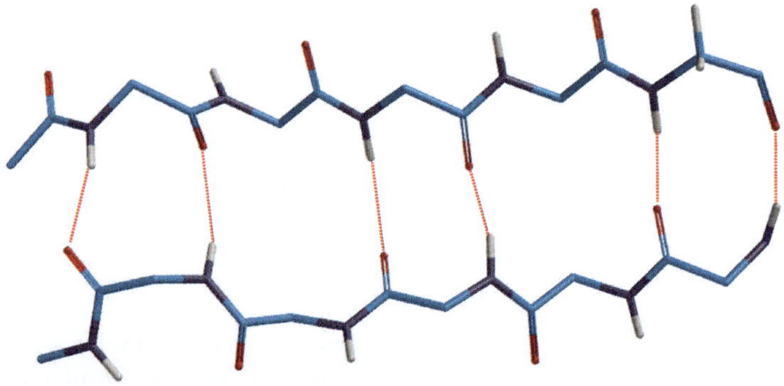

Many proteins have regions of all three types of secondary structure. These regions are part of the tertiary structure of proteins, which we will consider next.

Tertiary Structure

The three-dimensional shape of the protein is its **tertiary structure**. This spatial arrangement brings many amino acid residues that are far apart in the polypeptide chain close to one another. The proximity of amino acids in the tertiary structure is responsible for the activity of many enzymes. The three-dimensional folded shapes of proteins depends on their primary and secondary structures, which together make long-range interaction possible between amino acids. The forces of attraction between amino acids include ionic bonds, hydrogen bonds, and hydrophobic interactions. Hydrogen bonds form between various amino acid side chains, such as the —OH groups of serine, threonine, and tyrosine. The side chains of acidic and basic amino acids can also form hydrogen bonds. However, these amino acids may also form an intrachain ionic bond called a **salt bridge**. These ionic attractive forces between the carboxylate groups and the ammonium groups of side chains pull portions of chains together. These interactions are weak unless they occur "inside" the folded polypeptide chain, out of contact with water.

Aspartate residue

Lysine residue

Proteins contain many nonpolar side chains that are repelled by water. As a result, they tend to associate with one another on the "inside" of a folded protein molecule, out of contact with water. The tendency of nonpolar side chains to collect out of contact with the solvent is called the **hydrophobic effect**. Hydrophobic interactions among nonpolar side chains in proteins are weak, but abundant, and are primarily responsible for maintaining the folded conformation of a protein. The hydrophobic portions of the protein associate within the interior of the folded structure. Polar or charged (hydrophilic) groups are located at the surface near water molecules.

A small protein called 1GB1, which contains the ligand (antigen) binding site at the N-terminus of an immunoglobulin is shown in Figure 14.11. This region, which has 56 amino acid residues, contains parallel and antiparallel β-strands. An α-helix interacts with the β-strands by intrachain hydrogen bonds and nonpolar, London forces.

Figure 14.11 Structure of 1GB1

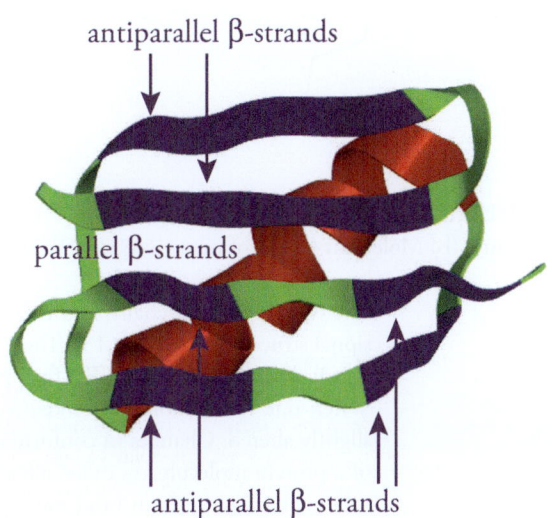

antiparallel β-strands

parallel β-strands

antiparallel β-strands

Quaternary Structure

The **quaternary structure** of a protein is the association of several protein chains or subunits into a closely packed arrangement. Each of the subunits has its own primary, secondary, and tertiary structure. The subunits are held together by hydrogen bonds and van der Waals forces between nonpolar side chains.

The subunits in a quaternary structure must be specifically arranged for the entire protein to function properly. Any alteration in the structure of the subunits or how they are associated causes marked changes in biological activity. Table 14.7 lists a few of the proteins that have quaternary structure.

Table 14.7
Examples of Proteins Having Quaternary Structure

Protein	Molecular Weight	Number of Subunits	Function
Alcohol dehydrogenase	80,000	4	Enzymatic reaction in fermentation
Aldolase	150,000	4	Enzymatic reaction in glycolysis
Fumarase	194,000	4	Enzymatic reaction in citric acid cycle
Hemoglobin	65,000	4	Oxygen transport in blood
Insulin	11,500	2	Hormone that regulates metabolism of glucose

Hemoglobin consists of two pairs of different proteins, each protein is bound to a molecule of heme. The heme is known as a **prosthetic group**. An Fe(II) ion at the center of the heme is the site of oxygen binding. Each subunit of hemoglobin has a histidine residue that forms a covalent bond to the heme (Figure 14.12).

(a)   (b)

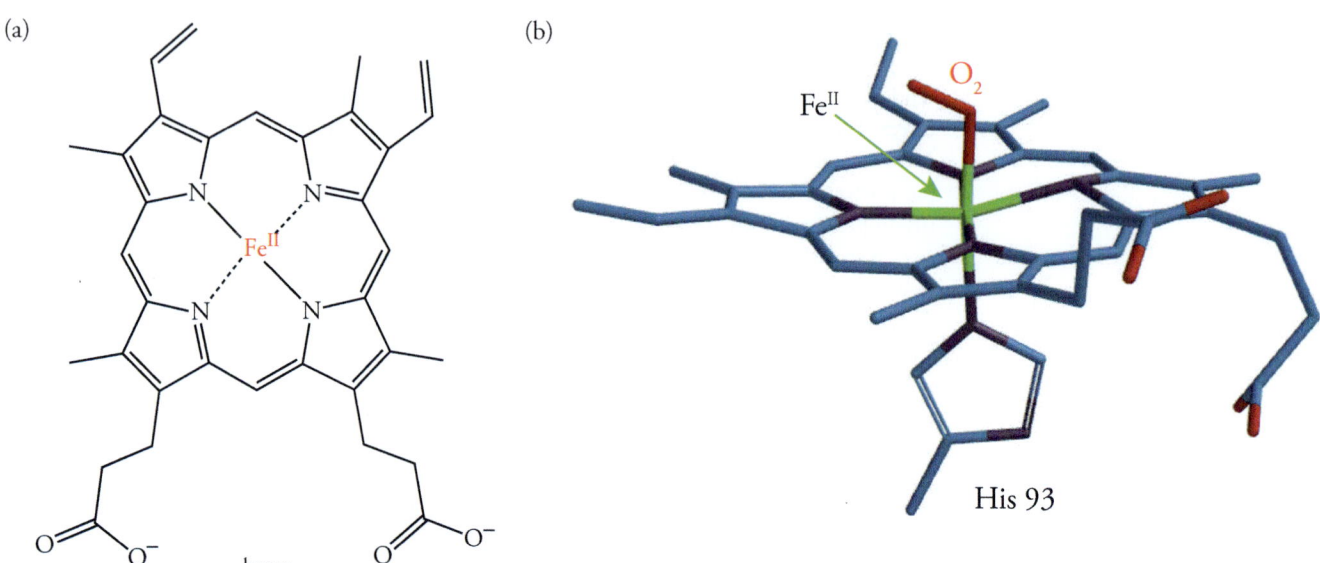

Figure 14.12 Structure of Heme
(a) Bond-line structure. (b) Molecular model of heme oxygen complex.

The two identical α chains and two identical β chains are arranged tetrahedrally in a three-dimensional structure (Figure 14.13). These units are held together by hydrophobic interactions, hydrogen bonding, and salt bridges. The four protein subunits of hemoglobin do not behave independently. When one heme molecule binds O_2, the conformation of the surrounding polypeptide chain is slightly altered. Changes in conformation at one site caused by a change at a spatially separated site of a protein molecule are called allosteric effects. As a result of **allosteric effects**, each heme in the other subunits then can bind more easily to additional oxygen molecules. As each oxygen binds, there are further conformational changes in the other protein chains that enhance their binding capability. As a result, once oxygenation occurs at one heme, there is cooperation at all other sites in hemoglobin, which then can carry four oxygen molecules.

Figure 14.13
Structure of Deoxyhemoglobin
The α and β subunits of hemoglobin interact cooperatively, and when one heme binds O_2, each of the others rapidly binds O_2.

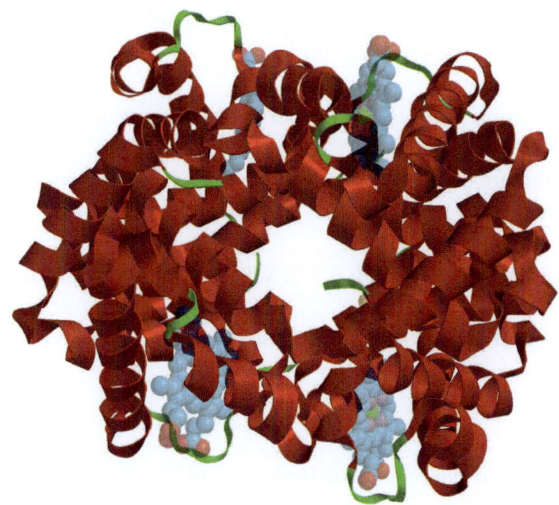

Exercises

Amino Acids

14.1 A D-glutamic acid residue is found in some bacterial cell walls. Draw its Fischer projection formula.

14.2 Gramicidin S is a cyclic peptide antibiotic that contains a D-phenylalanine residue. Draw the projection formula of D-phenylalanine.

14.3 The following amino acid is present in collagen. From what amino acid is it derived?

$$NH_2-CH_2-\underset{\underset{HO}{|}}{CH}-CH_2-CH_2-\underset{\underset{NH_2}{|}}{\overset{\overset{H}{|}}{C}}-\overset{\overset{O}{||}}{C}-OH$$

14.4 The following antibacterial agent, called alliin, is present in garlic. From what amino acid might it be derived?

$$CH_2{=}CH-CH_2-\overset{\overset{O}{||}}{S}-CH_2-\underset{\underset{NH_2}{|}}{\overset{\overset{H}{|}}{C}}-\overset{\overset{O}{||}}{C}-OH$$

Alliin

Acid-Base Properties of Amino Acids

14.5 Draw the structures of alanine and glutamic acid at pH = 1 and pH = 12.

14.6 Draw the structures for the zwitterions of cysteine and valine.

14.7 How could you distinguish between aqueous solutions of asparagine and aspartic acid?

14.8 Would you expect an aqueous solution of lysine at pH 7 to be neutral, acidic, or basic? Explain.

Isoionic Points

14.9 Estimate the isoionic points of the following tripeptides:
(a) Ala-Val-Gly (b) Ser-Val-Asp (c) Lys-Ala-Val

14.10 Estimate the isoionic points of the following tripeptides:
(a) Glu-Val-Ala (b) Arg-Val-Gly (c) His-Ala-Val

14.11 Examine the structures of oxytocin and vasopressin in Section 14.5. Which one has the higher isoionic point?

14.12 Examine the structure of the enkephalin whose sequence is shown below and estimate its isoionic point.
Tyr-Gly-Gly-Phe-Glu

14.13 The isoionic point of chymotrypsin is 9.5. What does this value indicate about its amino acid composition?

14.14 The isoionic point of pepsin is 1.1. What does this value indicate about its amino acid composition?

Peptides

14.15 Write the bond-line structure for alanylserine at pH 7.

14.16 How does glycylserine differ from serylglycine?

14.17 Identify the amino acids contained in the following tripeptide. Name the compound.

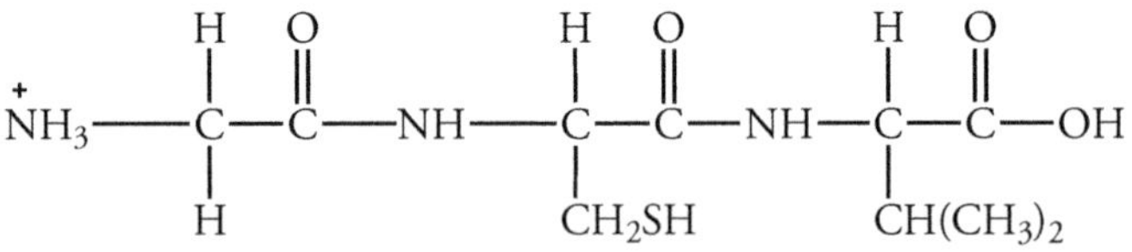

14.18 Identify the amino acids contained in the following tripeptide. Name the compound.

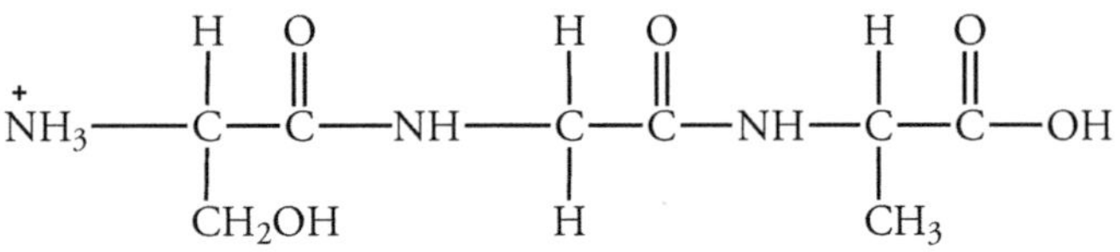

14.19 Thyrotropin-releasing hormone (TRH) causes the release of thyrotropin from the pituitary gland, which then stimulates the thyroid gland. Examine its structure and comment on one unusual structural feature.

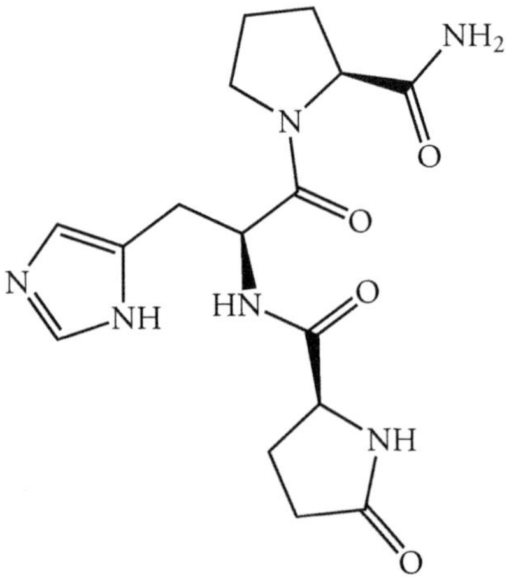

14.20 The tripeptide glutathione, which is important in detoxifying metabolites, has an unusual structural feature. Identify it.

glutathione

14.21 How peptide many isomers with the composition Gly$_2$, Ala$_2$ are possible?

14.22 How peptide many isomers with the composition Gly$_2$, Ala, Leu are possible?

Hydrolysis and Structure Determination

14.23 Assuming that only dipeptides are formed by partial hydrolysis, what is the minimum number that must be identified to establish the amino acid sequence of a pentapeptide?

14.24 Assuming that only tripeptides are formed by partial hydrolysis, what is the minimum number that must be identified to establish the amino acid sequence of an octapeptide?

14.25 The tetrapeptide tuftsin is hydrolyzed to produce Pro-Arg and Thr-Lys. Does this information establish the structure of tuftsin?

14.26 Partial hydrolysis of the octapeptide angiotensin II produces Pro-Phe, Val-Tyr-Ile, Asp-Arg-Val, and Ile-His-Pro. What is its amino acid sequence?

Enzymatic Hydrolysis

14.27 Which of the following tripeptides will be cleaved by trypsin? If cleavage occurs, name the products.
(a) Arg-Gly-Tyr   (b) Glu-Asp-Gly   (c) Phe-Trp-Ser (d) Ser-Phe-Asp

14.28 Which of the following tripeptides will be cleaved by trypsin? If cleavage occurs, name the products.
(a) Asp-Lys-Ser   (b) Lys-Tyr-Cys   (c) Asp-Gly-Lys   (d) Arg-Glu-Ser

14.29 Indicate which of the tripeptides in Exercise 14.27 will be cleaved by chymotrypsin and name the products.

14.30 Indicate which of the tripeptides in Exercise 14.28 will be cleaved by chymotrypsin and name the products.

14.31 The tetrapeptide tuftsin is hydrolyzed by trypsin to produce Pro-Arg and Thr-Lys. Does this information establish the amino acid sequence of tuftsin?

14.32 The pentapeptide met-enkephalin is hydrolyzed by chymotrypsin to give Met, Tyr, and Gly-Gly-Phe. Does this information establish the amino acid sequence of met-enkephalin?

14.33 The nonapeptide known as the sleep peptide is hydrolyzed by chymotrypsin to produce Ala-Ser-Gly-Glu and Ala-Arg-Gly-Tyr and Trp. What two amino acid sequences are possible for the sleep peptide?

14.34 The sleep peptide is hydrolyzed by trypsin to produce Gly-Tyr-Ala-Ser-Gly-Glu and Trp-Ala-Arg. What is the amino acid sequence of the sleep peptide?

End Group Analysis

14.35 Explain why structure determination of insulin using the Edman method yields two phenylthiohydantoin products.

14.36 Cholecystokinin, a peptide that contains 33 amino acids, plays a role in reducing the desire for food, and its production is stimulated by food intake. Its N-terminal amino acid is lysine. Draw the structure of the phenylthiohydantoin product.

14.37 Reaction of angiotensin II with the Edman reagent yields the following product. What information has been established?

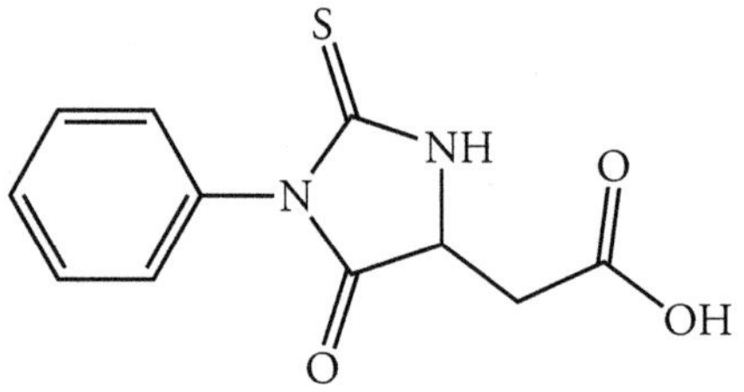

14.38 Reaction of corticotropin with the Edman reagent yields the following product. What information has been established?

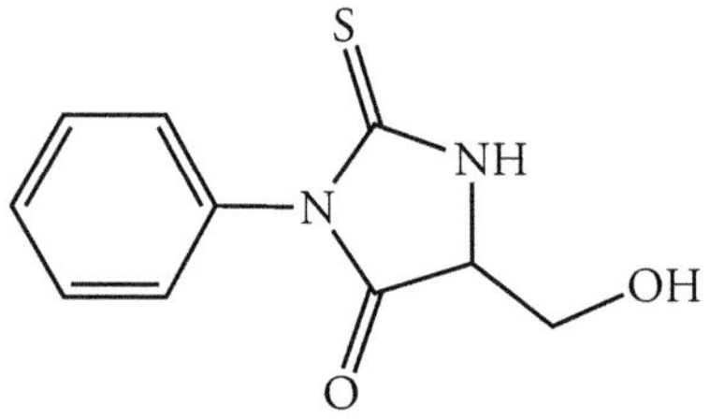

Proteins

14.39 Which amino acids can form salt bridges in proteins?

14.40 Which amino acids have R groups that form hydrogen bonds?

14.41 Which of the following amino acids are likely to exist in the interior of a protein dissolved in an aqueous solution?
(a) glycine   (b) phenylalanine   (c) glutamic acid

14.42 Which of the following amino acids are likely to exist in the interior of a protein dissolved in an aqueous solution?
(a) lysine   (b) cysteine   (c) glutamine

14.43 Noting that proline is a secondary amine, explain how proline can disrupt the α helix of a protein.

14.44 Examine the structures of valine and glutamic acid and suggest a reason why human hemoglobin is affected by the substitution of valine for glutamic acid at position 6 in the β chain.

14.45 Which of the following amino acids are likely to exist in the interior of a protein dissolved in an aqueous solution?
(a) glycine   (b) phenylalanine   (c) glutamic acid   (d) arginine

14.46 Which of the following amino acids are likely to exist in the interior of a protein dissolved in an aqueous solution?
(a) proline   (b) cysteine   (c) glutamine   (d) aspartic acid

14.47 If a protein is embedded in a hydrophobic lipid bilayer of a biological membrane, which of the amino acids listed in Exercise 14.45 will be in contact with the interior of the bilayer?

14.48 Noting that proline is a secondary amine, explain how proline can disrupt the α helix of a protein.

14.49 Examine the structures of valine and glutamic acid and suggest a reason why human hemoglobin is affected by the substitution of valine for glutamic acid at position 6 in the β chain.

15

SYNTHETIC POLYMERS

NYLON 66

15.1 NATURAL AND SYNTHETIC MACROMOLECULES

In most of our study of organic chemistry, we have focused on the chemical reactions, structure, and physical properties of "small" molecules. We examined very large molecules, or macromolecules, in only two chapters. Repeated condensation reactions of small molecules called monomers form polysaccharides and proteins (polyamides), two classes of macromolecules found in living organisms.

Some naturally occurring macromolecules are important commercial products. For example, wood and cotton are carbohydrates; wool and silk are proteins. But synthetic polymers far outstrip natural polymers in commercial importance. The chemical industry has developed many synthetic macromolecules with diverse properties and a wide variety of uses. These synthetic macromolecules are indispensable in a modern society.

Synthetic polymers have a wide range of properties. For example, certain transparent polymers can be molded into precise shapes in the manufacture of corrective lenses. The polymer rubber used in tires must be flexible enough to be distorted from one shape to another. Synthetic fibers used for clothing have to feel good against the body and be able to hold a dye.

The physical properties of synthetic macromolecules result from the number and kind of monomer units, as well as resulting intermolecular interactions and intramolecular interactions, such as London forces, dipole-dipole forces, and hydrogen bonding. We recall that the properties of proteins and other naturally occurring macromolecules result from intermolecular interactions. For example, the strength of structural proteins such as collagen and cellulose is due to the many intermolecular hydrogen bonds. Hydrogen bonding can also be an important feature to consider in designing synthetic macromolecules.

15.2 STRUCTURE AND PROPERTIES OF POLYMERS

By synthesizing appropriate monomers and learning how to polymerize them, organic chemists can prepare macromolecules to meet required specifications. The number of monomer units—a few hundred to several thousand—in a polymer affects its physical properties. However, no synthetic polymerization process can be stopped precisely after a specific number of monomers have been incorporated into the polymer. All polymerization reactions give mixtures of polymer molecules with a range of molecular weights. Therefore, we refer to the average molecular weight of a polymer. The average molecular weight of synthetic polymers is in the 10^5 to 10^6 range.

The types of monomers incorporated in a polymer strongly influence the flexibility and shape of the polymer. For example, polymers whose monomers contain aromatic rings are less flexible than those whose monomers are acyclic. In some polymers, the chains are cross-linked by covalent bonds. Cross-linking creates larger macromolecules with more rigid structures. Cross-links are also important in naturally occurring polymers. For example, the proteins in wool fibers are cross-linked by many disulfide bonds.

London Forces in Polymers

As in the case of proteins, the inter- and intramolecular interactions of polymer chains are extremely important in determining physical properties. London forces are largely responsible for the folded or coiled conformations of a polymer. Intermolecular London forces between individual chains help to hold them together. Both intra- and intermolecular London forces increase with the polarizability of functional groups in the polymer.

Only London forces affect the properties of polyethylene. This molecule, which can be produced as a linear polymer, resembles a giant alkane (Figure 15.1). The attractive forces between pairs of hydrogen atoms on adjacent chains of a polymer are very small. However, thousands of those interactions exist in a polymer, producing a large total interaction energy.

Principles of Organic Chemistry. http://dx.doi.org/10.1016/B978-0-12-802444-7.00015-X

Figure 15.1 London Forces in Polyethylene

The closely packed, all-anti conformation of the alkyl chains results in many London attractive forces.

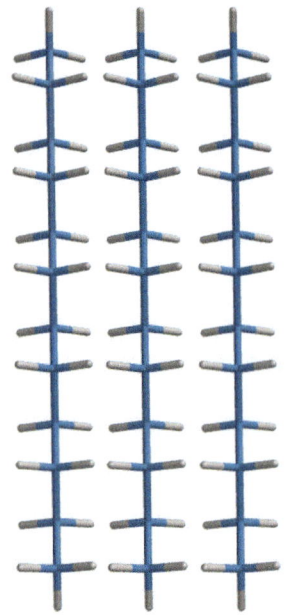

The linear polymer of ethylene is called high-density polyethylene (HDPE). It has a high density and is high melting (135 °C) because parts of the molecules "line up" in a tightly packed, orderly array. HDPE is used to make materials as simple as bottle caps for milk containers and as complex as cabinets for televisions and computers.

Ethylene can also be polymerized with branches off the main chain. The resulting polymer is called low-density polyethylene (LDPE). LDPE is less dense than HDPE because the branches prevent the main chains from packing closely (Figure 15.2). The more open structure not only is less dense but also has smaller London forces. Because the intermolecular forces between chains are smaller, LDPE has a lower melting point (120 °C) and is a more flexible material. It is used to make plastic bags and flexible bottles for consumer products such as soft drinks and bleach. Containers made of LDPE are also used for windshield wiper fluid, antifreeze, and engine oil.

Structural units such as aromatic rings have polarizable electrons that create strong London forces. Hence, polymers with aromatic rings have higher tensile strength than polymers with acyclic units, because there are strong London forces between aromatic rings in neighboring polymer chains. Furthermore, the aromatic rings reduce the flexibility of the polymer chain and the number of possible conformations. Chains of sp^3-hybridized carbon atoms in polymers such as polyethylene are more flexible. They can exist in gauche and anti conformations around each carbon-carbon bond. The allowed motions of the chains affect the properties of the polymer.

Hydrogen Bonding in Polymers

We noted that intermolecular hydrogen bonding dramatically affects the properties of naturally occurring macromolecules. Some synthetic polymers also have extensive hydrogen bonding between polymer chains. The substantial strength of polyamides such as nylon 66 and Kevlar is due to hydrogen bonding. The amount of hydrogen bonding in nylon 66 is affected by the flexibility of the chain and the conformation around the carbon-carbon bonds. The maximum number of hydrogen bonds is formed only if nylon 66 is in the all-anti conformation (Figure 15.3). Kevlar, used in bulletproof vests, is extensively hydrogen-bonded because the aromatic ring and the amide bond restrict the conformation to the one best suited to form the maximum number of hydrogen bonds (Figure 15.4).

Figure 15.2 Branching in Low-Density Polyethylene

The branched chains of the alkanes prevent close packing and results in a lower density polymer than polyethylene.

Figure 15.3 Hydrogen Bonding in Nylon 66

In the all-anti conformation, hydrogen bonds form between amide hydrogen atoms and carbonyl oxygen atoms.

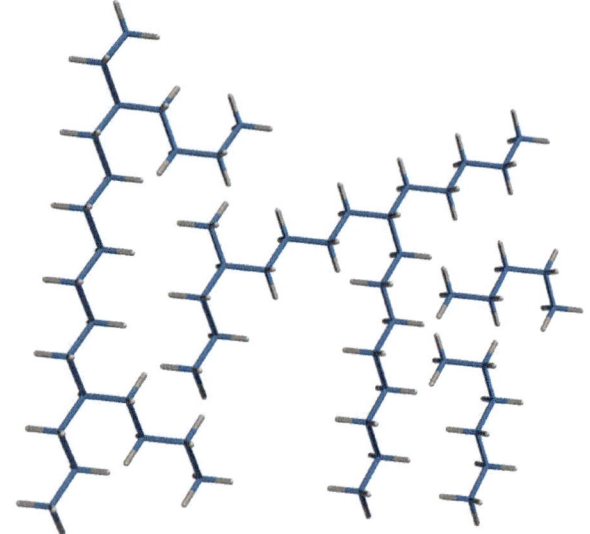

Many gauche and anti conformations around σ bonds of methylene groups

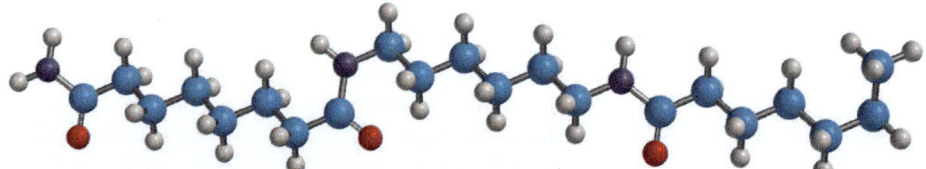

Rigid conformation around each planar amide bond

Figure 15.4 Hydrogen
Bonding in Kevlar

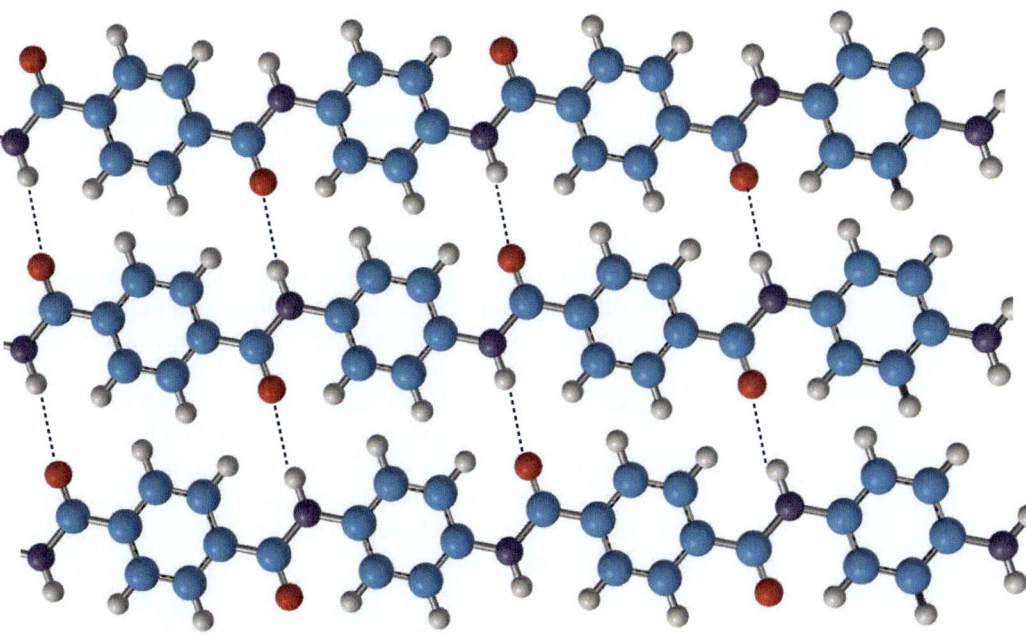

Problem 15.1

How would the properties of an addition polymer formed from 3-methyl-1-pentene differ from those of a polymer formed from propene?

Sample Solution

The polymer of 3-methyl-1-pentene has large branched *sec*-butyl groups off the main chain, whereas propene has relatively small methyl groups as branches. As a result, the polymer of 3-methyl-1-pentene has a more open structure. The bulkier chains cannot pack closely, and the intermolecular forces between chains are smaller. The polymer of 3-methyl-1-pentene should have a lower melting point and be a more flexible material.

Problem 15.2

There are three isomeric benzenedicarboxylic acids. Each reacts with ethylene glycol to produce a polyester. Which one should produce the polyester that packs most efficiently and hence has the largest intermolecular attractive forces?

15.3 CLASSIFICATION OF POLYMERS

Polymers can be classified by their macroscopic physical properties, by the method of polymerization, or by a number of structural features, such as cross-linking and stereochemistry. The three major classes based on physical properties are elastomers, plastics, and fibers.

Elastomers

Elastomers are elastic materials that regain their original shape if they are distorted. Some common elastomers are rubber, a naturally occurring polymer of isoprene, and neoprene, a synthetic polymer of 2-chloro-1,3-butadiene. These elastomers contain carbon-carbon double bonds separated by intervening units containing two sp^3-hybridized carbon atoms.

repeating unit

polyisoprene

An elastomer's properties depend on both the groups bonded to the sp³-hybridized carbon atoms and the geometry of the polymer chain around the double bond. Elastomers are amorphous materials. The individual chains of the polymers are random coils that are tangled in an irregular way. The coils "straighten out" when they are stretched. When the force is released, the elastomer returns to its coiled state because the intermolecular forces are greatest in this arrangement.

The flexibility of an individual chain of an elastomer depends on the structure of the intervening unit of sp³-hybridized carbon atoms. Some rotation can occur around the σ bonds. The double bonds provide rigidity because they restrict rotation around π bonds. Polymer chains of both E and Z configurations are known. For example, natural rubber is polyisoprene with a Z configuration, whereas gutta-percha, an industrial polymer of isoprene, has an E configuration. This stereochemical difference is reflected in the properties of the two polymers. Rubber is an elastomer, but gutta-percha is less flexible. In gutta-percha the *trans* double bonds give a zigzag arrangement resembling that of saturated fatty acids. We recall that the regularity of the zigzag chains of fatty acids allows adjacent chains to nestle together, resulting in large London forces. Similar large forces exist in gutta-percha. Natural rubber has a "bent" chain similar to that of an unsaturated fatty acid. This arrangement of atoms gives a more open, less regular relationship among polymer chains. As a result, the polymer chains are not closely packed, and they can slide past one another as the elastomer is distorted (Figure 15.5).

Figure 15.5 Polyisoprene

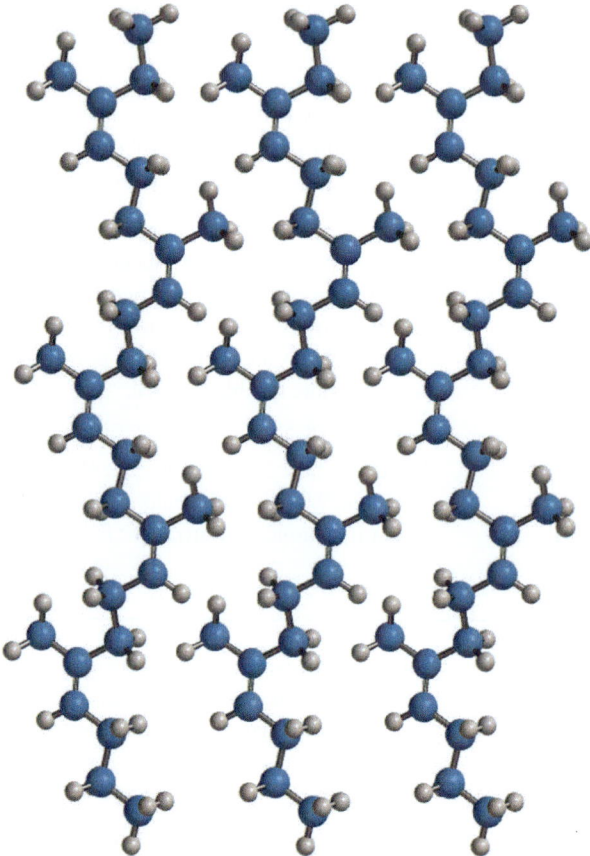

Plastics

In polymer chemistry the term plastic is used for those polymers that harden upon cooling. They can be molded or extruded into shapes that remain after cooling (Gr. *plastikos* "fit for molding"). **Thermoplastics** are polymers that reversibly soften when heated, becoming sufficiently fluid to be molded. **Thermosetting polymers** can be molded when they are first prepared. However, after being heated they "set," hardening irreversibly. If heated to a high temperature, thermosetting polymers decompose rather than melt. The difference between thermoplastics and thermosetting polymers is related to cross-linking. The polymer chains of thermoplastics are not cross-linked.

When a thermoplastic is heated, the kinetic energy of the polymer chains increases, overcoming the intermolecular forces, and causing the polymer to melt. Polyethylene is a thermoplastic in which the London forces between hydrocarbon chains are the only intermolecular forces. Thermosetting polymers have extensive cross-links between polymer "chains" that result in much larger polymer molecules. Bakelite is a thermosetting polymer (Section 15.6). The only way that the structure of this material can be disrupted is by cleaving covalent bonds. This process irreversibly decomposes the material.

Fibers

Some thermoplastics are prepared as thin filaments that can be spun into fibers similar to natural fibers. The length of the polymer molecule must be at least 500 nm, which corresponds to a minimum average molecular weight of 10^4. The structure of the polymer chains must also provide sufficiently strong intermolecular forces to give the fiber an adequate tensile strength.

Filaments of thermoplastics are prepared by two methods. If the thermoplastic is stable in the molten state, it may be passed through tiny pores in a die called a spinneret and then cooled. For less stable thermoplastics the polymer is dissolved in a volatile solvent and forced through the spinneret. The solvent evaporates and a filament precipitates. Regardless of the method of formation, the fiber is then drawn out to several times its length after it has cooled. Cold-drawing orients the molecules along the axis of the fiber. The resultant intermolecular forces between polymer molecules increase the tensile strength of the fiber.

Problem 15.3
Assign the polymer represented by the following structure to one of the three classes of polymers. Identify a compound whose physical properties it most closely resembles.

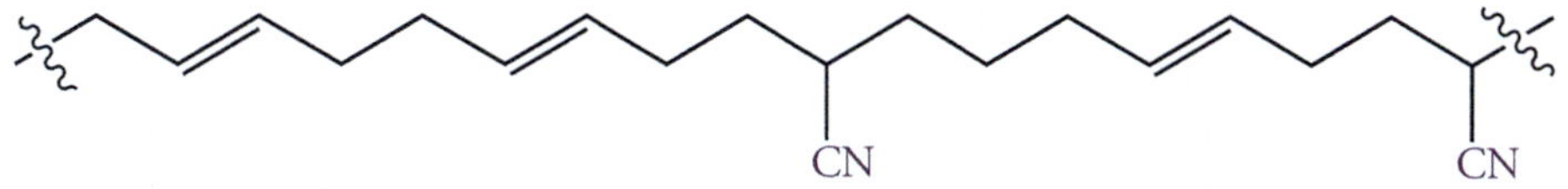

Solution
The polymer is an elastomer. The sp³-hybridized carbon atoms between double bonds provide some flexibility to the elastomer. However, the *trans* arrangement of the double bond, which resembles that of gutta-percha, allows chains to pack efficiently and leads to less flexibility in the elastomer.

Problem 15.4
(a) What type of plastic is best suited to make the handles for cooking utensils for the home?
(b) What type of plastic is most likely to be used for the frames of eyeglasses?

15.4 METHODS OF POLYMERIZATION

Polymers can be divided into two broad classes called **addition polymers** and **condensation polymers**. Addition polymers result from the successive addition reactions of one alkene or a mixture of alkenes by radical, cationic, or anionic mechanisms. Condensation polymers result from condensation reactions of monomers that contain two or more functional groups such as an alcohol and a carboxylic acid or an amine and a carboxylic acid. These functional groups react in condensation reactions to eliminate a small molecule such as water. Condensation polymerization reactions are often carried out at high temperatures so that the eliminated molecule evaporates, helping drive the reaction to completion.

Addition polymers are also called **chain growth polymers**. The polymer chain grows when the reactive intermediate formed in an initiation step adds to another monomer unit. The initiating species may be a radical, carbocation, or carbanion. For example, dibenzoyl peroxide yields a benzoyl radical.

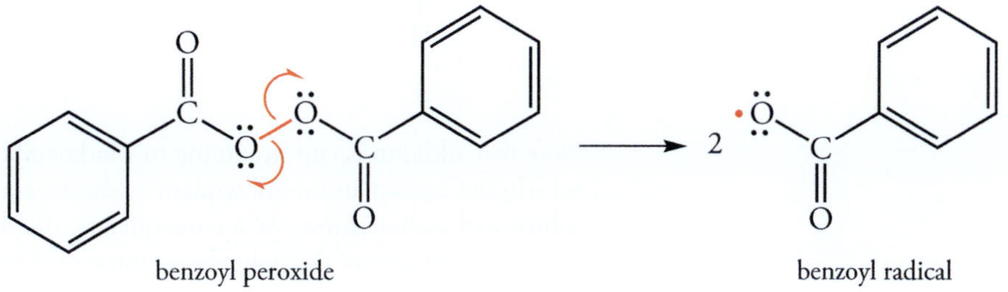

benzoyl peroxide   benzoyl radical

This radical reacts with a monomer to give another radical that then reacts with another unit of monomer.

The successive additions of monomers give a growing chain that always has a reactive end. The number of polymer chains formed therefore depends on the concentration of intermediates initially formed. A monomer cannot react until it encounters one of the growing chains with a reactive site.

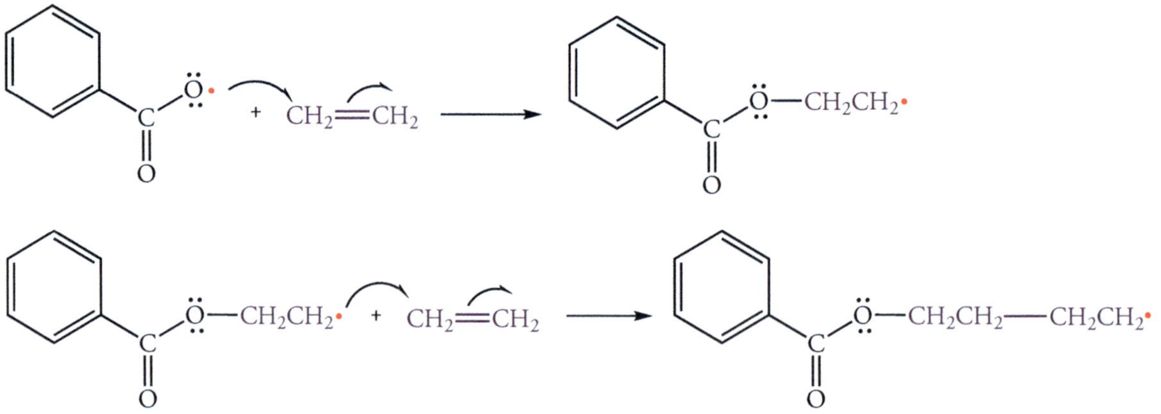

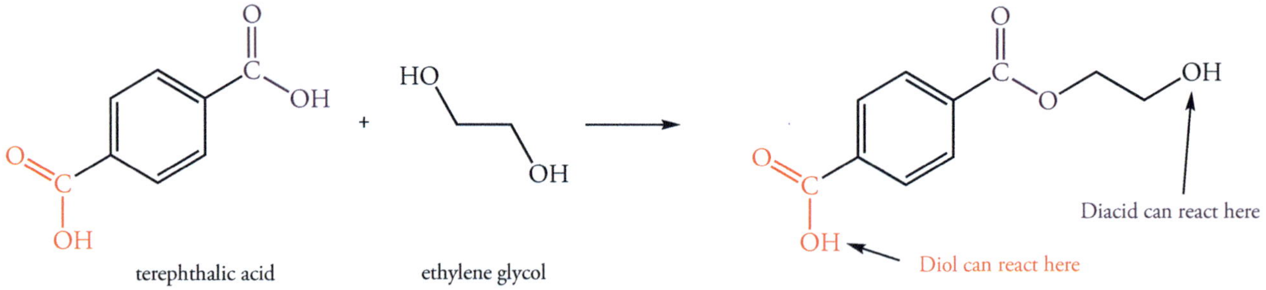

Cationic polymerization proceeds through carbocation intermediates rather than radicals. A Lewis acid such as BF$_3$, Al(CH$_2$CH$_3$)$_3$, TiCl$_4$, or SnCl$_4$ is used to react with the alkene to form a carbocation, which in turn reacts with another alkene molecule to form another cation. Consider the reaction with 2-methylpropene (isobutylene) as the monomer. The Lewis acid that acts as an electrophile is represented by E$^+$.

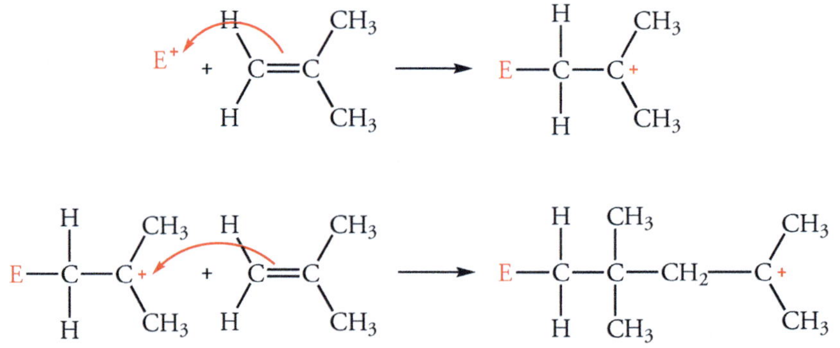

Note that addition occurs according to Markovnikov's rule, with the electrophile adding to the less substituted carbon atom. Subsequent reactions continue, with the carbocation adding to the less substituted carbon atom. As a consequence, the more stable tertiary carbocation is formed each time. The structure of the polymer is represented as shown below.

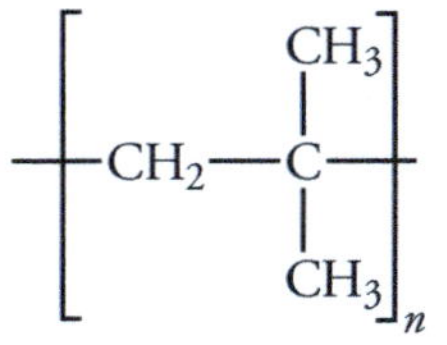

Low molecular weight polyisobutylene is used in lubricating oil and in adhesives for removable paper labels. Higher molecular weight polyisobutylene is used to produce inner tubes for bicycle and truck tires.

Anionic polymerization is initiated by a carbanion that behaves as a nucleophile. One example is the butyl anion, which is provided by butyllithium. The lithium compound has a very polar bond, and the carbon atom has a partial negative charge. In the following reactions, the butyl group is represented as Bu⁻. The monomer is acrylonitrile.

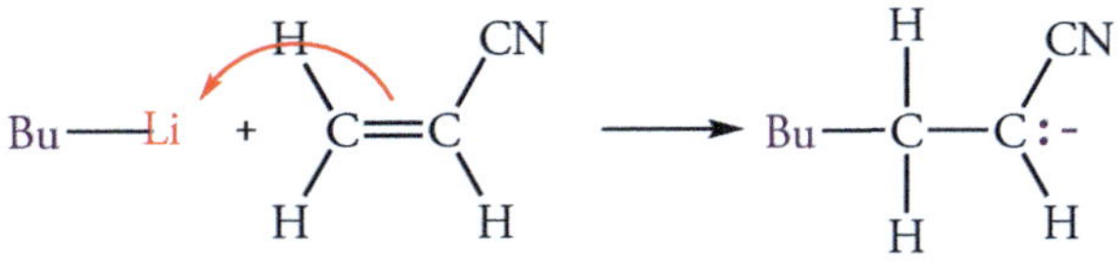

Addition occurs at the less substituted carbon atom because the resulting carbanion is resonance stabilized.

Continued reaction of the initial carbanion product—the new reactant—give another resonance stabilized carbanion as the process continues.

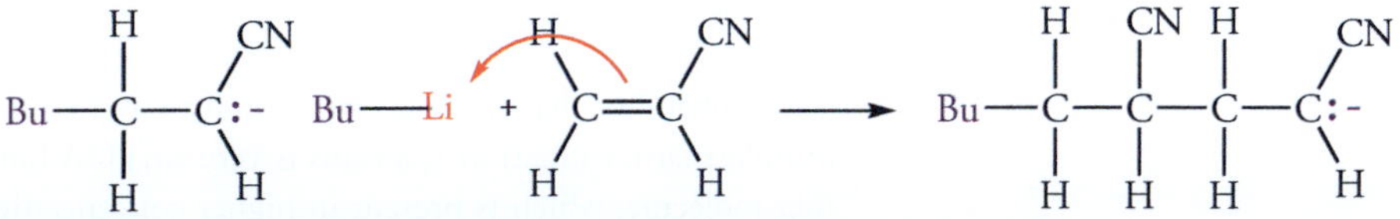

The structure of the polymer is represented as shown below.

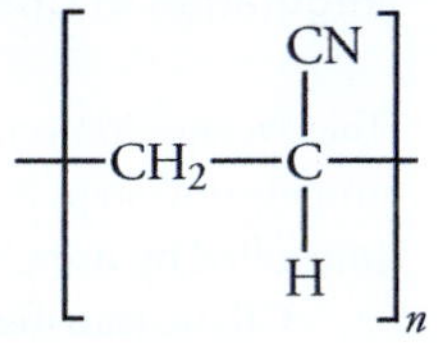

Polyacrylonitrile is used in fibers that can be spun to give the textiles Orion® or Acrilan®. Some rugs are also produced using this polymer. Condensation polymers are also called step-growth polymers. In reactions between two units, such as a diol and a diacid, an ester forms between one alcohol site and one acid site. This ester also has an alcohol site and an acid site at the ends of the molecule. Monomers can continue to react in condensation reactions with this product.

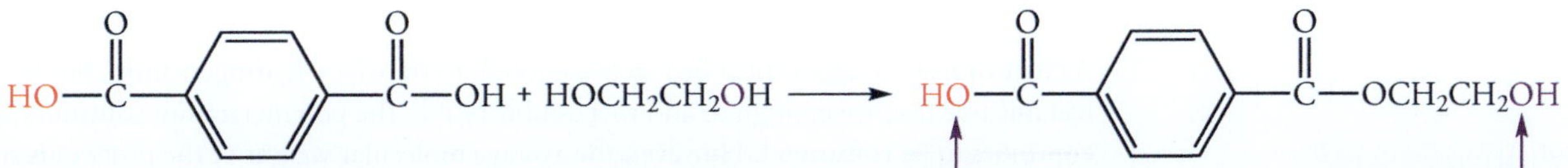

terephthalic acid   ethylene glycol   This hydroxyl group can react with ethylene glycol   This hydroxyl group can react with terephthalic acid

However, subsequent condensation reactions are not restricted to the reaction of the monomers with the ends of the growing polymer chain. The monomers in the reaction mixture can continue to react with each other to start additional chains. So the monomers in a step-growth polymerization generate many low molecular weight oligomers rather than a smaller number of steadily growing, high molecular weight chains. Formation of true polymers occurs only after the monomer is used up. At this point large increases in the chain length result from the reaction of the ends of the oligomers with each other. Thus, in step-growth polymerization, the polymer is formed in "blocks" that result in a substantially higher molecular weight product.

15.5 ADDITION POLYMERIZATION

Addition polymerization occurs by a chain reaction in which one carbon-carbon double bond adds to another. Monomers continue to react with the end of the growing polymer chain in an addition polymerization reaction until the reactive intermediate is destroyed in a termination reaction.

Disproportionation and dimerization are two possible termination reactions. In disproportionation, a hydrogen atom at a carbon atom α to the radical center is abstracted by a radical in another chain. This produces a double bond in one polymer molecule, and the other polymer molecule becomes saturated. Because no new radical intermediates are formed, the next propagation step cannot occur.

$$2\,R\!-\!\!(CH_2CH_2)_nCH_2CH_2 \cdot \longrightarrow R\!-\!\!(CH_2CH_2)_nCH_2\!=\!\!CH_2 + R\!-\!\!(CH_2CH_2)_nCH_2\!-\!H$$

In the dimerization reaction, two radicals combine to form an even longer polymer chain. Again, the destruction of radicals prevents propagation.

$$2\,R\!-\!\!(CH_2CH_2)_nCH_2CH_2 \cdot \longrightarrow R\!-\!\!(CH_2CH_2)_nCH_2CH_2\!-\!\!(CH_2CH_2)_nCH_2\!-\!R$$

The probability that the reactive sites of two growing polymer chains will react in either of these bimolecular termination reactions is very small. A bimolecular reaction of one chain with a monomer molecule, which is present in higher concentration and consumed throughout the reaction, is more likely.

Regulation of Chain Length

The average molecular weight of an addition polymer is controlled by the number of times the propagation steps occur before the chain is terminated. However, the length of the chain can also be controlled by using either chain-transfer agents or inhibitors.

Chain-transfer agents control the chain length of a polymer by interrupting the growth of one chain and then initiating the formation of another chain. Thiols are common chain-transfer agents.

$$2\,R\!-\!CH_2CH_2 \cdot + R'\!-\!S\!-\!H \longrightarrow R\!-\!CH_2CH_2\!-\!H + R'\!-\!S \cdot$$
a terminated chain

$$R'\!-\!S \cdot + CH_2\!=\!\!CH_2 \longrightarrow R'\!-\!S\!-\!CH_2CH_2 \cdot$$
initation of another chain

A chain-transfer reagent must be reactive enough to transfer a hydrogen atom, but the resulting radical must be reactive enough to add to a double bond. The polymerization continues, and monomer continues to be consumed. However, the average molecular weight of the product is smaller because more chains are formed by the chain-transfer process.

Inhibitors react with the radical site of a growing polymer chain to give a less reactive radical. Benzoquinone is a typical inhibitor used in free radical polymerization reactions.

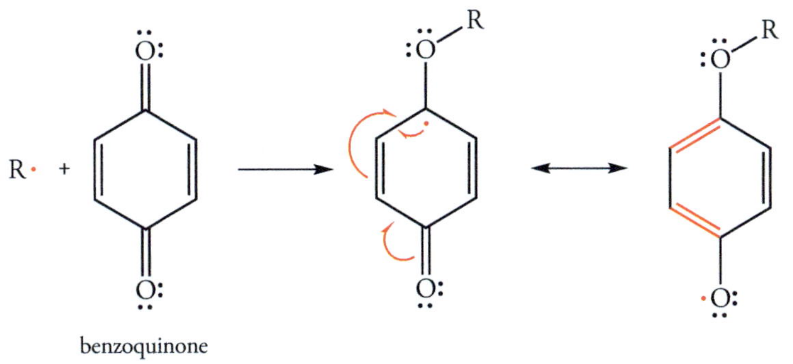

benzoquinone

Chain Branching

In practice, the linear polymer we might expect for alkenes is nor the major product of the free radical process. (Cationic polymerization is generally used to prepare linear addition polymers of alkenes.) The product chains have many alkyl branches, which most often are the four-carbon-atom butyl groups produced by **short chain branching**. These products are the result of intramolecular hydrogen abstraction by way of a six-membered transition state that generates a secondary radical from a primary radical.

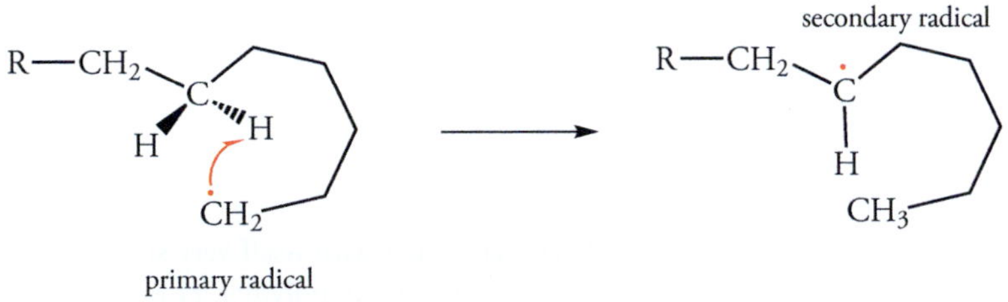

The polymerization continues at the new radical site, and a butyl group branch is located on the chain.

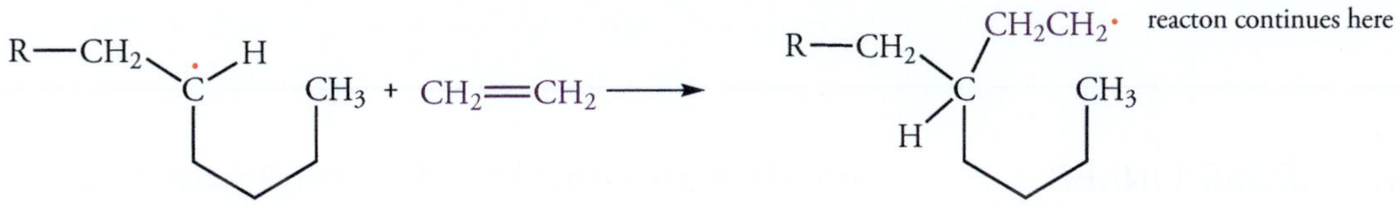

Large chain branching occurs by a random process. Intermolecular hydrogen atom abstraction can occur between the terminal radical of one chain and any of the hydrogen atoms located in another chain. In this case, one chain is terminated and the polymerization continues at a site within the other chain. The length of the resulting branch depends on the site of hydrogen abstraction. Short chain branching is more common than long chain branching because intramolecular reactions are more probable than intermolecular reactions.

15.6 COPOLYMERIZATION OF ALKENES

The addition polymers that we have discussed are homopolymers, made up of repeating units derived from a single unsaturated monomer. **Copolymers** incorporate two different monomers in the polymer chain. They are formed in reactions of a mixture of two monomers. Copolymerization of various combinations of monomers provides many more possible structures and a greater variety of materials that might have desirable physical properties than homopolymerization.

Two monomers can react randomly to give a random copolymer. The exact composition depends on the reaction conditions and on the concentrations of the two monomers.

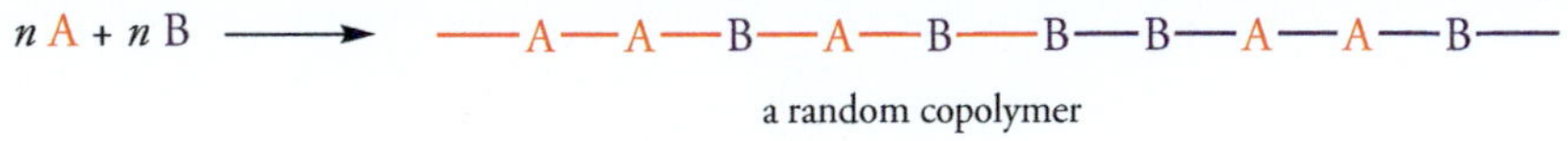

a random copolymer

Few pairs of monomers give totally random copolymers. In fact, monomers are usually selected to avoid random copolymers. Monomers are chosen so that one monomer at the end of the growing polymer chain prefers to react with the other monomer in the mixture, and vice versa. In short, it is desirable to have a monomer at the end of a chain that reacts preferentially with the other monomer in the reaction mixture. Such polymers are called alternating copolymers.

$$n \, A + n \, B \longrightarrow \quad —B—A—B—A—B—A—B—A—B—A—$$

an alternating copolymer

It is difficult to form perfect alternating copolymers. However, the reaction of styrene with maleic anhydride produces a nearly perfect alternating copolymer.

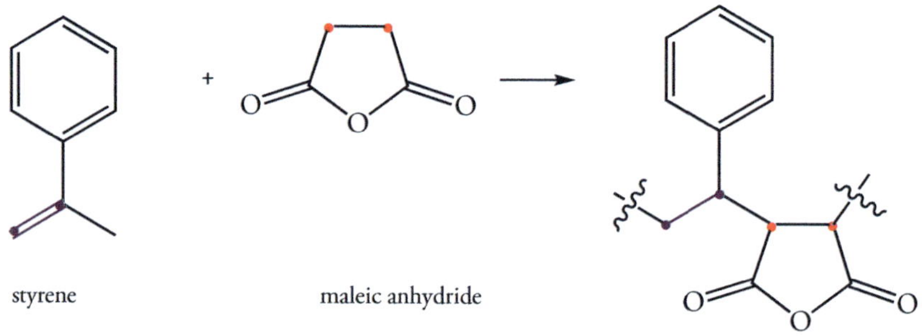

styrene   maleic anhydride

Maleic acid reacts with itself very slowly, and its homopolymer is difficult to form. Styrene readily reacts to form a homopolymer. However, a styrene group at the end of a growing polymer chain reacts faster with maleic anhydride than with itself. After the addition of styrene to maleic anhydride, a radical is produced that does not react with maleic anhydride. As a result, the next alkene that is added is styrene. Monomers that provide perfect alternating copolymers are highly desirable because the product can be reproduced.

15.7 CROSS-LINKED POLYMERS

Atoms bonded between polymer chains are called **cross-links**. They form during polymerization of the monomers or in separate reactions after formation of the polymer.

p-Divinylbenzene has two alkene functional groups, each of which can become part of a different polymer chain in a copolymerization reaction with styrene. One alkene group of p-divinylbenzene is incorporated in a chain whose major components are styrene units.

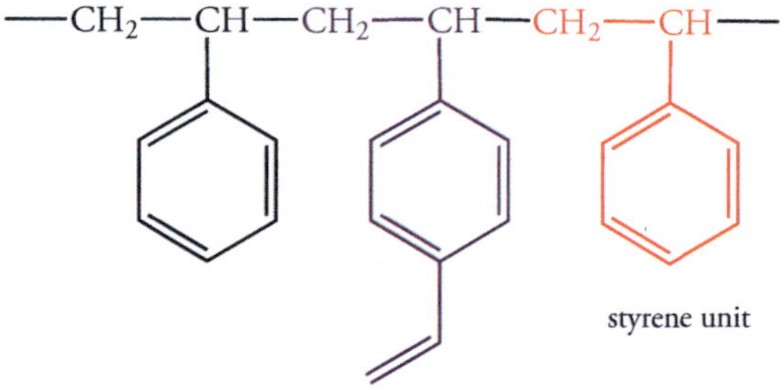

divinylbenzene unit

styrene unit

At some point in the reaction, the other alkene group reacts in a chain propagation process that develops a second chain. Thus, divinylbenzene becomes part of each polymer chain and forms a link between the two chains (Figure 15.6). The degree of cross-linking and the space between the divinylbenzene units depend on the amounts of two monomers used.

The importance of cross-links in determining the properties of a polymer was accidentally discovered by Charles Goodyear in his study of the properties of rubber. Natural and synthetic rubbers can be used to make rubber bands, but are too soft and tacky for many applications such as tires. The resilience of rubber decreases when it is heated because the polyisoprene chains slide past each other more easily when stretched at higher temperatures. When tension is released, natural rubber does not regain its original structure.

Figure 15.6 Cross-Links in Addition Polymerization

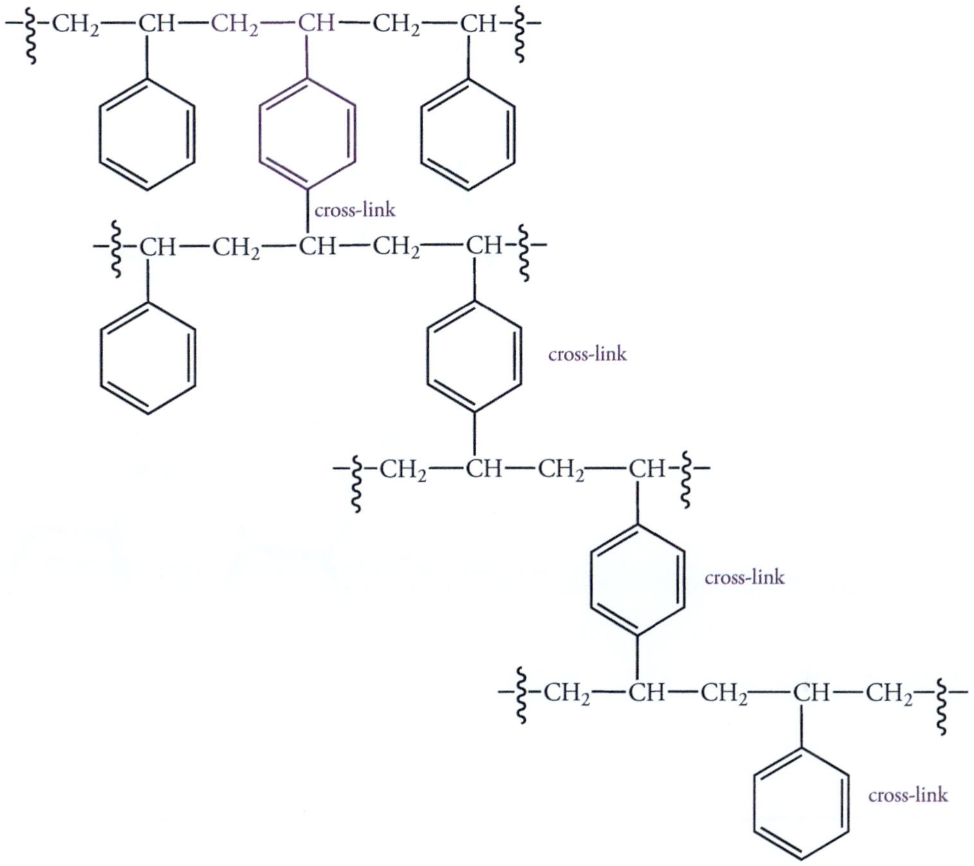

In 1839 Charles Goodyear found that heating natural rubber with a small amount of sulfur produces a material with new properties. He called this process **vulcanization**. (Vulcan is the fire good in ancient Roman mythology—not to mention the extraterrestrial creatures in Star Trek who dwell on the planet Vulcan.)

The sulfur reacts with polyisoprene to replace some C—H bonds with disulfide bonds. As a result, the polymer chains become connected by cross-links that may contain one, two, or more sulfur atoms (Figure 15.7). These cross-links increase the rigidity of the rubber because more of the chains are linked into a larger molecule. The freedom of movement of one chain relative to another is diminished. After distortion, the vulcanized rubber returns to its original molded shape. The amount of sulfur—3% to 10% by weight—controls the flexibility of the rubber.

Figure 15.7 Cross-Links in Vulcanized Polyisoprene

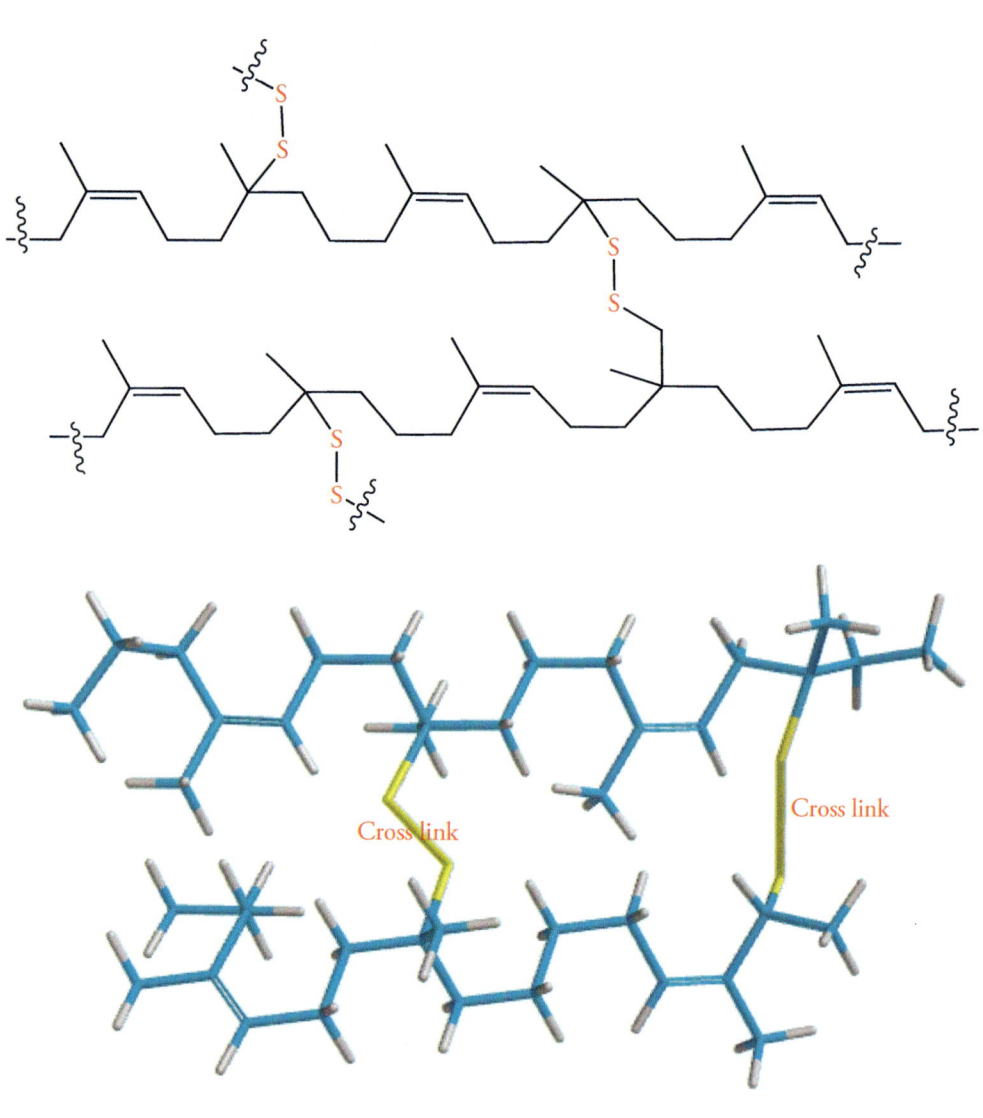

Cross link

Cross link

15.8 STEREOCHEMISTRY OF ADDITION POLYMERIZATION

Addition polymerization of some alkenes generates stereogenic centers along the entire backbone of the polymer. The relationship of these centers to one another affects the physical properties of the polymer. Consider the polymer formed from propene. If the methyl groups are all on the same side of the backbone of the zigzag chain, the polymer is **isotactic**. If the methyl groups are in a regular alternating sequence on opposite sides of the backbone, the polymer is **syndiotactic**. If the methyl groups are randomly oriented, the polymer is **atactic**. The three forms of polypropylene are shown in Figure 15.8.

The regularity of structure in isotactic and syndiotactic polymers is responsible for their higher melting points, so they can be used to manufacture objects that will be exposed to boiling water. Atactic polymers formed in radical chain polymerization (which also have branches that result from hydrogen abstraction processes) have lower melting points. Both isotactic and syndiotactic forms of polymers are produced with catalysts designed by K. Ziegler, a German, and G. Natta, an Italian. These catalysts yield polymers with no chain branching. The development of methods to form stereochemically regular linear polymers revolutionized polymer science.

The Ziegler-Natta catalysts are organometallic compounds that contain a transition metal. For example, triethylaluminum and titanium(III) chloride combine to give such a catalyst. The structure of the catalyst and its function in the polymerization process are beyond the scope of this text. However, each catalyst coordinates alkene monomers and allows them to react stereoselectively.

Figure 15.8 Atactic, Syntactic, and Syndiotactic Polymers

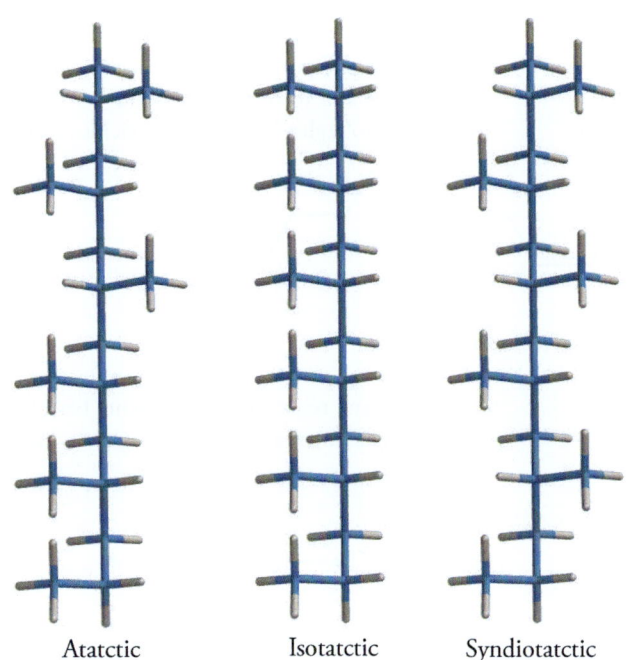

| Atatctic | Isotatctic | Syndiotatctic |

Diene Polymers

Conjugated dienes can form addition polymers by a 1,4-addition reaction. The remaining double bond of each monomer unit occurs at every fourth carbon atom along the chain. Natural rubber, for example, is a polymer of 2-methyl-1,3-butadiene (isoprene) with *cis* geometry at all of the double bonds. The polymer is obtained from the latex synthesized under the bark of some trees that grow in southeast Asia. The isomeric gutta-percha is a *trans* isomer of natural rubber that is produced by trees of a different genus.

natural rubber (polyisoprene, all *cis*)

gutta-percha (all *trans*)

The different properties of natural rubber and gutta-percha reflect both the geometries around the double bonds and their molecular weights. The molecular weight of natural rubber is about 100,000, whereas that of gutta-percha is less than 10,000. As a result of the *cis* arrangement of the chain in natural rubber, the adjacent molecules cannot fit close to one another. Natural rubber has random coils that can be stretched out when the material is pulled. After the tension is released, the material returns to its original structure. Gutta-percha molecules, on the other hand, can pack closer because the *trans* arrangement of the double bonds and the favored anti conformation around the saturated carbon atoms provide a chain with a regular zigzag arrangement. So gutta-percha is a highly crystalline, hard, inflexible material. It was once used in golf ball covers.

Early attempts to polymerize isoprene in industrial processes to prepare synthetic rubber were not successful because the reactions were not stereospecific. However, a variety of Ziegler-Natta catalysts are now available. One catalyst that contains titanium stereospecifically gives polyisoprene with *cis* double bonds, and another catalyst containing vanadium gives polyisoprene with *trans* double bonds.

The polymerization of 2-chloro-1,3-butadiene was one of the reactions considered by US industry to replace rubber made from natural sources located in areas of the world that could be cut off in a crisis such as war. This diene structurally resembles isoprene, with a chlorine atom replacing the methyl group of isoprene. Free radical polymerization gives a mixture of *cis* and *trans* double bonds as well as a mixture of 1,2-and 1,4-addition products. Polymerization of 2-chloro-1,3-butadiene using a Ziegler-Natta catalyst yields Neoprene®, a compound with *trans* double bonds.

cis-poly(1,3-butadiene)   *trans*-poly(1,3-butadiene)

Neoprene resists oxidizing agents better than natural rubber does. Neoprene is therefore used to manufacture materials such as gaskets and industrial hoses.

neoprene

Problem 15.5

The free radical polymerization of 1.3-butadiene yields some sections of the polymer that contain a vinyl group. Explain the origin of this group.

Solution

The presence of a vinyl group bonded to the main chain means than the other vinyl group of 1,3-butadiene is incorporated in the chain. Thus, the polymerization of this unit occurs by a 1,2-addition reaction similar to that of a simple alkene.

Problem 15.6

Draw the structure of the product of ozonolysis of natural rubber. Would this structure differ from the ozonolysis product of gutta-percha?

15.9 CONDENSATION POLYMERS

A condensation reaction is a reaction between two reactants that yields one larger product and a second, smaller product such as water. This type of reaction has been illustrated in the reactions of many functional groups containing oxygen or nitrogen. Products of condensation reactions include ethers, acetals, esters, imines, and amides.

We now consider condensation reactions that yield polymers. Two functional groups are required in a monomer so that after one functional group reacts, the other is available to link to another monomer. The functional groups in monomers may be arranged in two ways for condensation polymerization.

A single compound can contain two different functional groups such as an amino group and a carboxylic acid group. Reaction of the amino group of one molecule with the carboxylic acid of another molecule gives an amide that still has a free amino group and a free carboxylic acid group, which can continue to react to form a polymer. The general reaction is shown below.

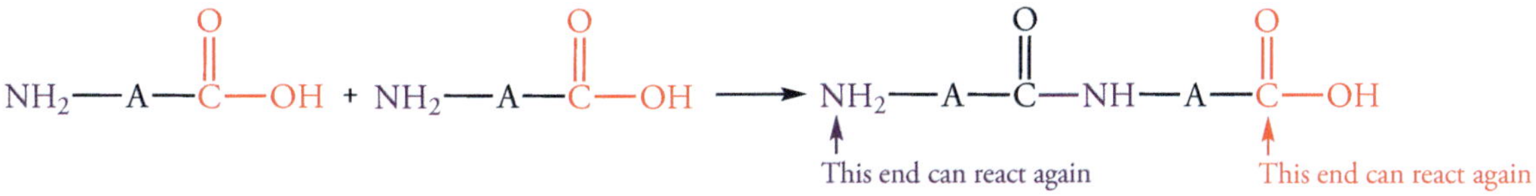

$$NH_2-A-\overset{\overset{\displaystyle O}{\|}}{C}-OH \ + \ NH_2-A-\overset{\overset{\displaystyle O}{\|}}{C}-OH \longrightarrow NH_2-A-\overset{\overset{\displaystyle O}{\|}}{C}-NH-A-\overset{\overset{\displaystyle O}{\|}}{C}-OH$$

This end can react again   This end can react again

Continued reaction of the carboxylic acid end with the amino group of another monomer or of the amino group end with the carboxylic acid group of another monomer yields a homopolymer.

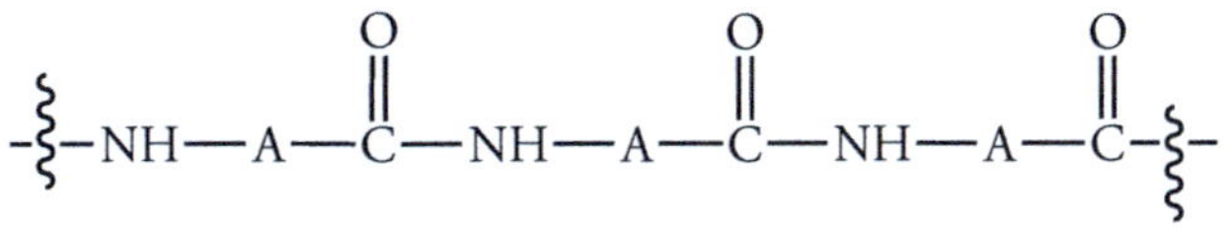

$$-NH-A-\overset{\overset{\displaystyle O}{\|}}{C}-NH-A-\overset{\overset{\displaystyle O}{\|}}{C}-NH-A-\overset{\overset{\displaystyle O}{\|}}{C}-$$

homopolymer of an amino acid

Condensation reactions also result from the copolymerization of two monomers. Each monomer contains two of the same functional group. Examples include the reaction of a monomer that is a dicarboxylic acid with a monomer that is a diol. The functional groups on one monomer can only react with the functional groups on the other monomer. The general reaction is shown below.

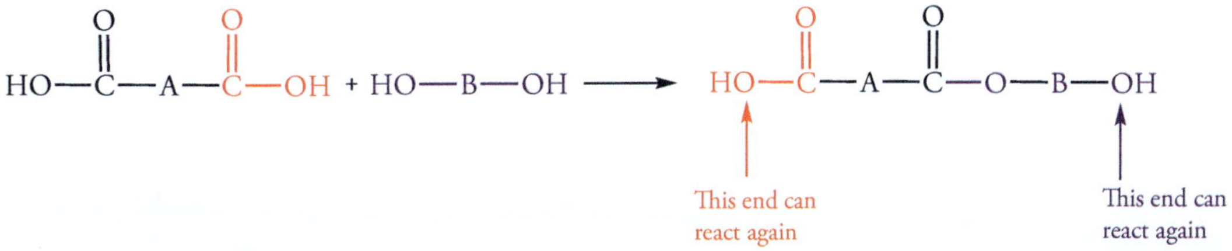

$$HO-\overset{\overset{\displaystyle O}{\|}}{C}-A-\overset{\overset{\displaystyle O}{\|}}{C}-OH \ + \ HO-B-OH \longrightarrow HO-\overset{\overset{\displaystyle O}{\|}}{C}-A-\overset{\overset{\displaystyle O}{\|}}{C}-O-B-OH$$

This end can   This end can
react again   react again

Continued reaction of the carboxylic acid end with the hydroxyl group of the diol monomer or of the hydroxyl group end with the carboxylic acid group of the dicarboxylic acid monomer yields a copolymer.

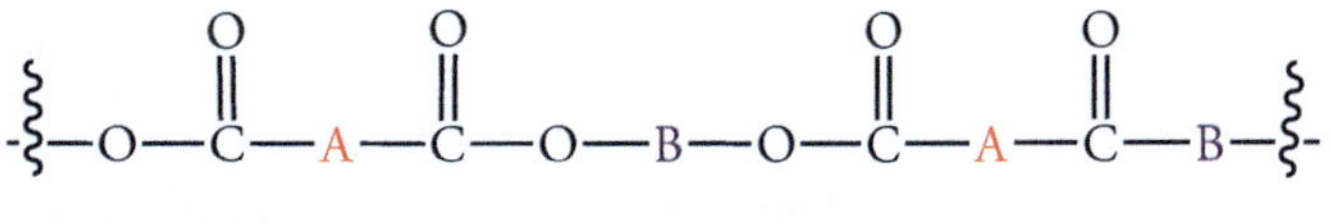

$$-O-\overset{\overset{\displaystyle O}{\|}}{C}-A-\overset{\overset{\displaystyle O}{\|}}{C}-O-B-O-\overset{\overset{\displaystyle O}{\|}}{C}-A-\overset{\overset{\displaystyle O}{\|}}{C}-B-$$

an alternating copolymer

A monomer can contain two different functional groups, but such monomers are not widely used. First, these monomers are more difficult to prepare without uncontrolled polymerization during their synthesis. Second, the monomer can only be used in one possible polymerization reaction. Condensation polymers formed from two different monomers are more common. The synthesis of each monomer is usually straightforward and less expensive. Each monomer can be used in reactions with other monomers. For example, any of a series of dicarboxylic acids can react with any of another series of diols.

15.10 POLYESTERS

Polyesters account for approximately 40% of the synthetic fibers produced in the United States. Poly(ethylene terephthalate), also known as PET, is the major polyester. It is a copolymer of ethylene glycol and terephthalic acid.

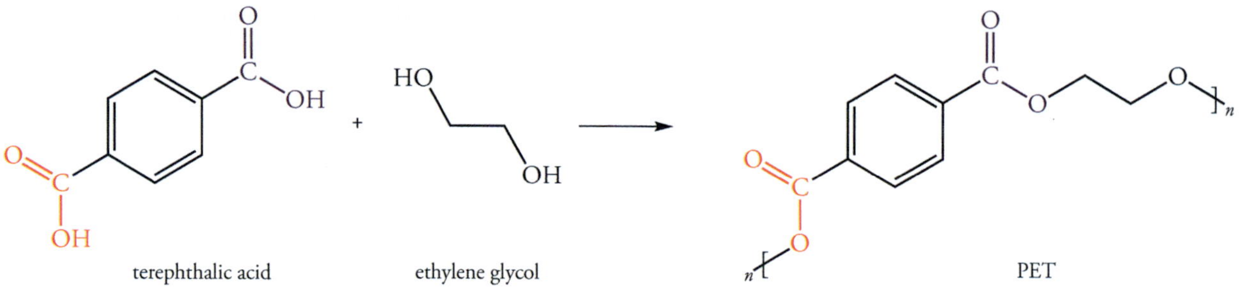

terephthalic acid   ethylene glycol   PET

PET and many other polyesters are produced industrially by transesterification reactions. A transesterification reaction involves exchange of an alkoxy group of an ester for a second alkoxy group provided by an alcohol. The reaction has an equilibrium constant close to 1, and the position of the equilibrium can be controlled by changing the experimental conditions. Ethylene glycol can serve as the nucleophile displacing alcohol units from esters. PET is prepared by the reaction of dimethyl terephthalate with ethylene glycol at 150 °C. Neither reactant is volatile at this temperature, but the second product is methanol, which boils at 65 °C. As methanol forms, it is continuously vaporized from the reaction mixture, driving the polymerization reaction to completion.

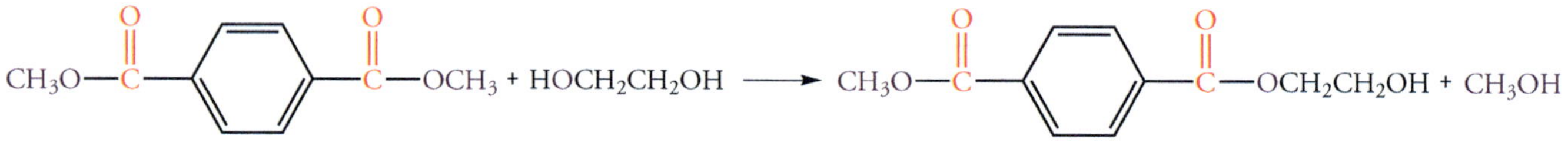

Cyclic anhydrides such as phthalic anhydride and maleic anhydride also react with glycols to form polyesters. The anhydride is a bifunctional molecule that reacts with the bifunctional glycol to give linear, alternating copolymers.

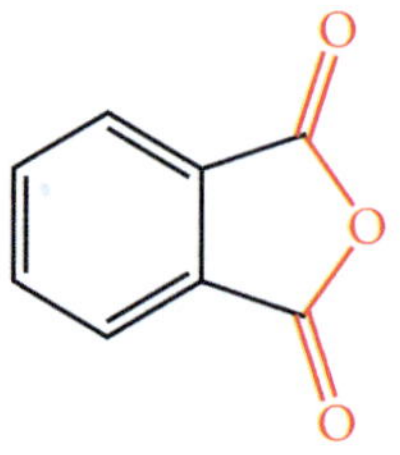

phthalic anhydride

However, when a triol reacts with an anhydride, a cross-linked polymer results. For example, the reaction of phthalic anhydride with 1,2,3-propanetriol (glycerol) initially occurs selectively with the primary hydroxyl groups to give a linear polymer.

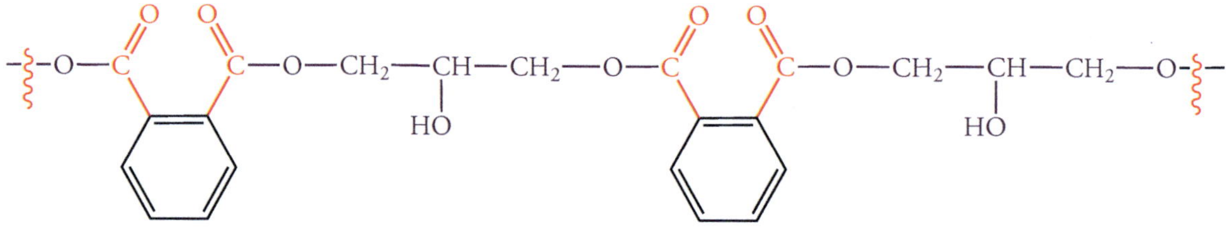

Reaction of phthalic anhydride with the secondary hydroxyl groups is so slow that continued polymerization can be carried our as a second step. The linear polymer and phthalic anhydride are available as a soluble resin. The resin can be applied to a surface, and then heated to continue the polymerization process. The resulting cross-linked polymer is an insoluble, hard, thermosetting plastic called glyptal (Figure 15.9).

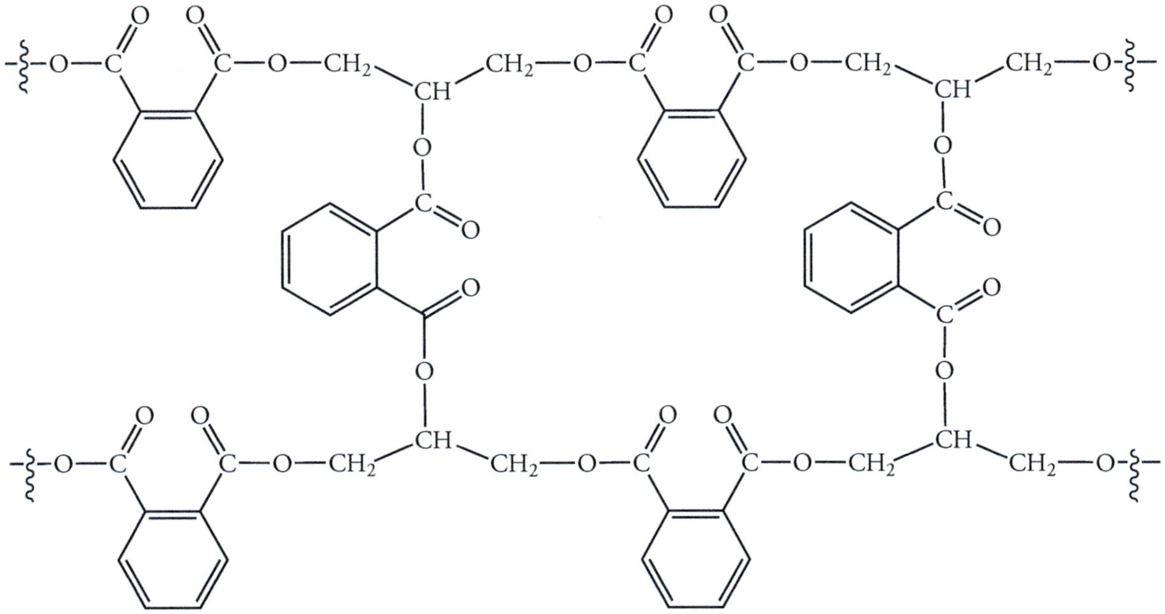

Figure 15.9 Cross-Links in a Condensation Polymer
The reaction of 2 moles of 1.2,3-propanetriol and 3 moles of phthalic anhydride gives a cross-linked polymer called a glyptal.

15.11 POLYCARBONATES

Carbonates are esters of carbonic acid. However, because carbonic acid is unstable, carbonates cannot be produced from carbonic acid and an alcohol.

carbonic acid   (a dialkyl carbonate)

Dialkyl carbonates can be made from the reaction of alcohols with phosgene, a highly toxic gas. The second chlorine atom of phosgene increases the electrophilicity of the carbonyl carbon atom. As in the reaction of an alcohol with an acid chloride, a base is required to neutralize the HCl by-product.

phosgene   (a dialkyl carbonate)

Although a polymeric carbonate could be produced in the reaction of a diol with phosgene, these products are usually obtained by a transesterification reaction with a dialkyl (or diaryl) carbonate. The reaction of diethyl carbon with a phenol called bisphenol A gives a polycarbonate known as Lexan.

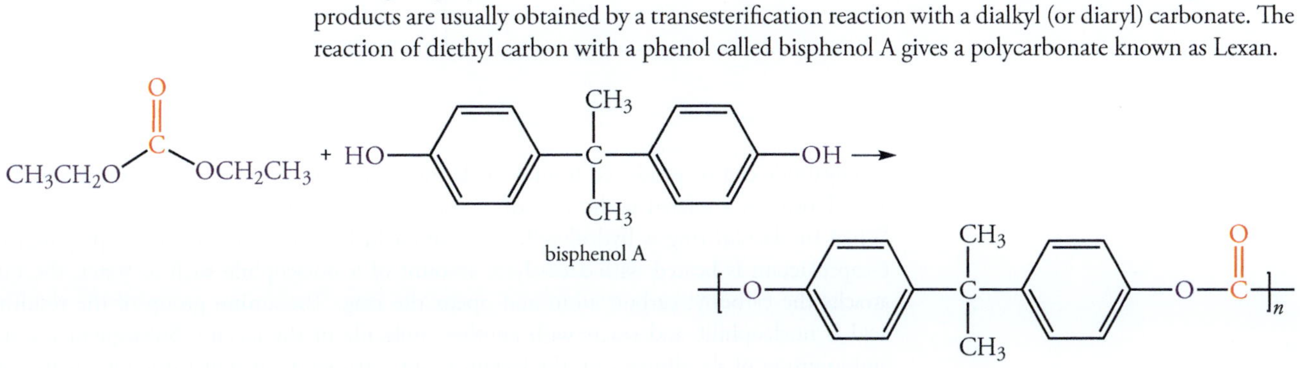

bisphenol A

Lexan

Lexan has very high impact strength and is strong enough to be used in crash helmets. It is also used to manufacture telephone housings. Because Lexan can be produced as a clear colorless polymer, it is used in bulletproof windshields and in the visors of astronauts' helmets.

15.12 POLYAMIDES

We recall that amides are best made by the reaction of acid chlorides and amines. Therefore, polyamides can be made by reaction of a monomer with two acid chloride functional groups and a monomer with two amine groups. However, the high reactivity of acid chlorides with nucleophiles such as water requires special precautions to preserve this reagent. Thus, these compounds are not much used in industrial laboratories.

An alternate method for the synthesis of amides is the direct heating of an amine with a carboxylic acid. The first product is an ammonium salt, which loses water when heated to form the amide.

$$\text{R'} - CO_2H \ + \ \text{R'} - NH_2 \ \longrightarrow \ \left[\text{R'} - CO_2^- \ \ \text{R'} - NH_3^+ \right] \xrightarrow{\text{heat}} \ \text{R'} - NH - \overset{\overset{\displaystyle O}{\|}}{C} - NH - R \ + H_2O$$

(an ammonium salt
of a carboxylic acid)

Polyamides can be made from the reaction of diacids with diamines. A diammonium salt is formed by proton transfer reactions. When the salt is heated to 250 °C, water is driven off and a polyamide forms. Nylon is a common name for polyamides. The most common polyamide is formed by the reaction of adipic acid, a six-carbon diacid, and 1,6-hexanediamine (hexamethylene diamine), a six-carbon diamine. The ammonium salt formed is called a nylon salt.

$$HO - \overset{\overset{\displaystyle O}{\|}}{C} - (CH_2)_4 - \overset{\overset{\displaystyle O}{\|}}{C} - OH \ + \ H_2N - (CH_2)_6 - NH_2 \ \longrightarrow$$

$$\left[{}^-O - \overset{\overset{\displaystyle O}{\|}}{C} - (CH_2)_4 - \overset{\overset{\displaystyle O}{\|}}{C} - O^- \ \ {}^+H_3N - (CH_2)_6 - NH_3^+ \right] \xrightarrow{250 \ ^\circ C}$$

nylon salt

$$\left[O - \overset{\overset{\displaystyle O}{\|}}{C} - (CH_2)_4 - \overset{\overset{\displaystyle O}{\|}}{C} - NH - (CH_2)_6 \right]_n$$

nylon 6,6

A polyamide can be produced from a single monomer containing both an amine and a carboxylic acid. However, a related cyclic structure called a lactam can also be converted into a polyamide. When the lactam ring is hydrolyzed, an amino acid is produced that can be polymerized. When ε-caprolactam is heated with a catalytic amount of a nucleophile such as water, the nucleophile attacks the carbonyl carbon atom and opens the ring. The amino group of the resulting amino acid is nucleophilic and reacts with another molecule of the lactam. Subsequent reaction of the amino group of the dimer with the lactam yields a trimer. Continued reaction yields a six-carbon homopolymer called nylon 6.

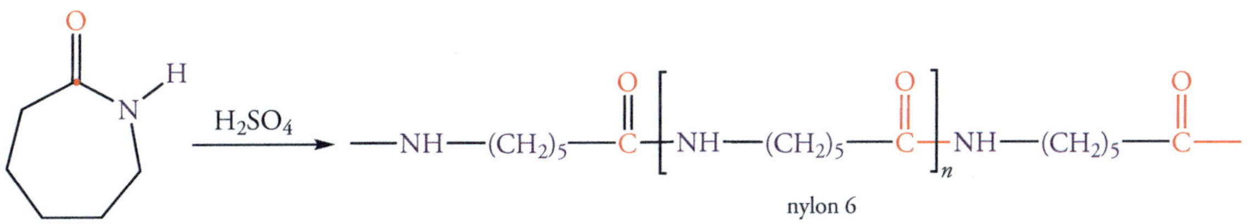

Nylons are used in many products. As a fiber, nylon is used in clothing, rope, tire cord, and parachutes. Because nylon has a high impact strength and resistance to abrasion, it can even be used to make bearings and gears.

15.13 POLYURETHANES

A urethane is an ester of a carbamic acid. Like carbonic acid, carbamic acids are unstable. They decompose to an amine and carbon dioxide. Therefore, urethanes cannot be made by esterification of a carbamic acid.

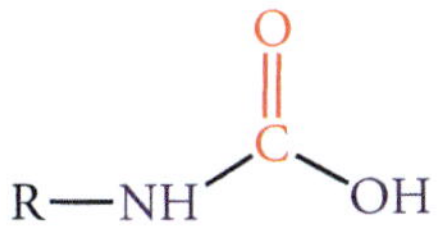

carbamic acid 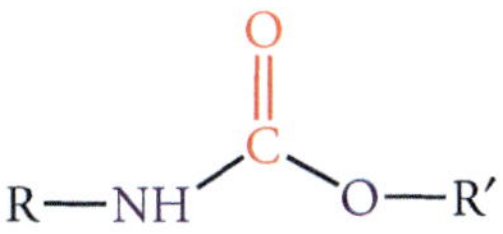   (a urethane)

However, carbamates can be made by addition of an alcohol to an isocyanate. The reaction formally results in addition of the alcohol across the carbon-nitrogen double bond.

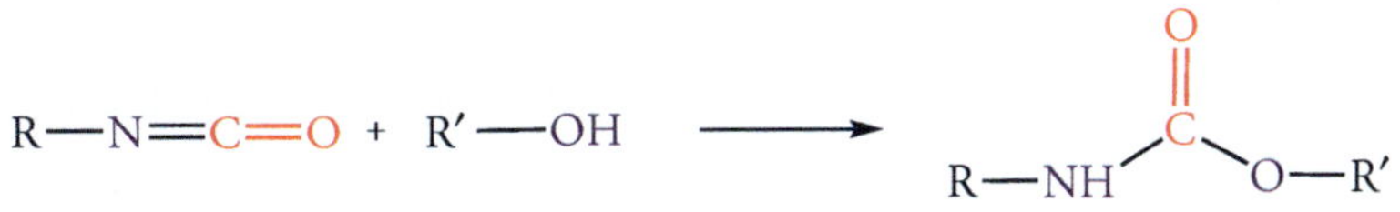

(an isocyanate)   (a urethane)

Polyurethane can be prepared by the reaction of a diisocyanate with a diol. The major diisocyanate used is toluene diisocyanate, which has the isocyanate groups at positions ortho and para to the methyl group. When ethylene glycol is added to the diisocyanate, a typical condensation polymerization occurs to give a polyurethane.

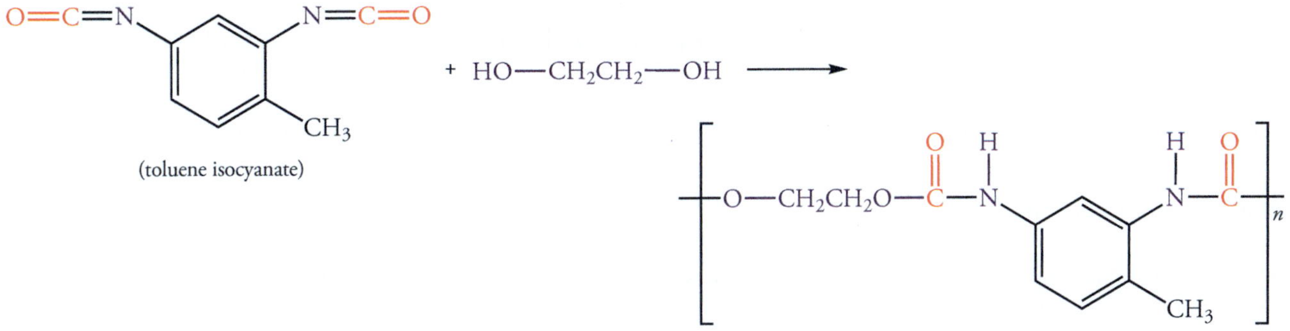

(toluene isocyanate)

(a polyurethane)

The major use of polyurethanes is in foams. Gases are blown into the liquid polymer to produce bubbles that are trapped as the material cools. When the resulting material is spongy, it is used for cushions. If monomers are selected to give cross-links, the more rigid foams that form are used for thermal insulation in building construction.

EXERCISES

Properties of Polymers

15.1   Explain why the polymer of 2-methylpropene is a sticky elastomer with a low melting point.

15.2   How would the properties of the polymer of the following diamine and adipic acid differ from those of nylon 66?

$$CH_3-NH-(CH_2)_6-NH-CH_3$$

15.3   Explain how 1,2,4,5-benzenetetracarboxylic acid dianhydride could be used to make a thermosetting polyester.

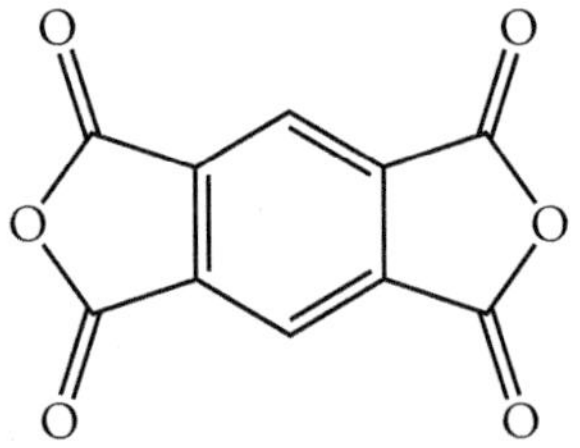

1,2,4,5-benzenetetracarboxylic acid dianhydride

15.4   How would the properties of the copolymer of l,4-butanediol with terephthalic acid differ from those of PET?

Addition Polymers

15.5   Vinyl acetate is used to make a polymer used in chewing gum. Draw a bond-line representation of the polymer.

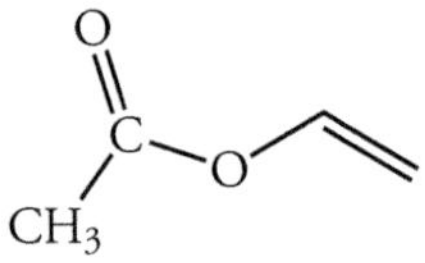

vinyl acetate

15.6   Draw a bond-line structure of polyvinyl alcohol. Explain why the polymer is prepared by the hydrolysis of polyvinyl acetate.

15.7   What monomer is required to prepare the following polymer?

$$-CFCl-CF_2-CFCl-CF_2-CFCl-CF_2-CFCl-CF_2-$$

15.8   Hexafluoropropene is a monomer used to prepare a polymer called Yiton. Draw a representation of the polymer.

Chain Transfer Reactions

15.9   Draw the structure of the branch formed by a short-chain transfer reaction in the formation of polystyrene.

15.10   Explain why formation of a polymer of l-hexene under free radical conditions would produce some molecules with methyl groups bonded to the main chain.

Copolymers

15.11   Draw a representation of an alternating polymer of isoprene and 2-methylpropene.

15.12   Styrene and 1,3-butadiene form a random polymer. What is the probability that a 1,3-butadiene unit will react with a growing polymer chain with styrene at its end?

15.13   Some hair sprays contain a solution of a copolymer made from the following monomers. Draw a representation of the polymer. Why does the copolymer hold hair in place?

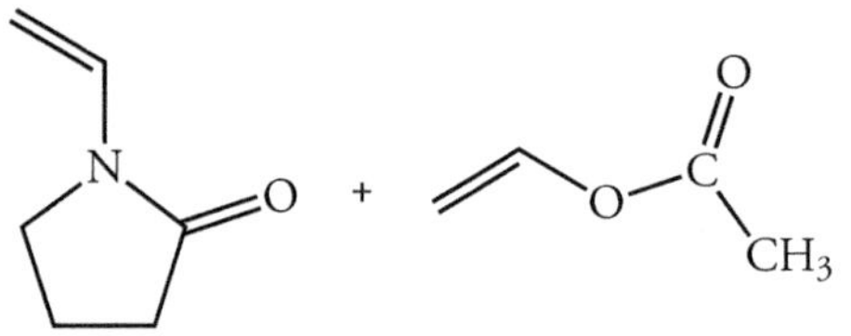

15.14   Saran is a copolymer of vinylidene chloride ($CH_2=CCl_2$) and a smaller amount of vinyl chloride. Draw a representation of the polymer.

Cross-Linked Polymers

15.15   What is the difference between the number of cross-links in the rubber used in tires and the rubber used in gloves?

15.16   Draw a representation of the polyester formed from butenedioic anhydride (maleic anhydride) and 1,2-propanediol. Explain how this polymer could be cross-linked by reacting it with styrene.

Stereochemistry of Polymerization

15.17   Which of the following alkenes can be polymerized to give isotactic and syndiotactic structures?
(a) 1-chloroethene   (b) 1,1-dichloroethene   (c) 2-methylpropene   (d) styrene

15.18   Are syndiotactic or isotactic forms of polypropylene optically active?

15.19   S-Methyl-1-pentene reacts with a Ziegler-Natta catalyst to give an isotactic polymer. What relationship exists between the alkyl groups on the polymer chain?

15.20   Ethylene and *cis*-2-butene form a syndiotactic copolymer in a reaction catalyzed by a vanadium catalyst. Draw a representation of the polymer.

Condensation Polymers

15.21   What monomers are required to prepare the following polymers?

(a)

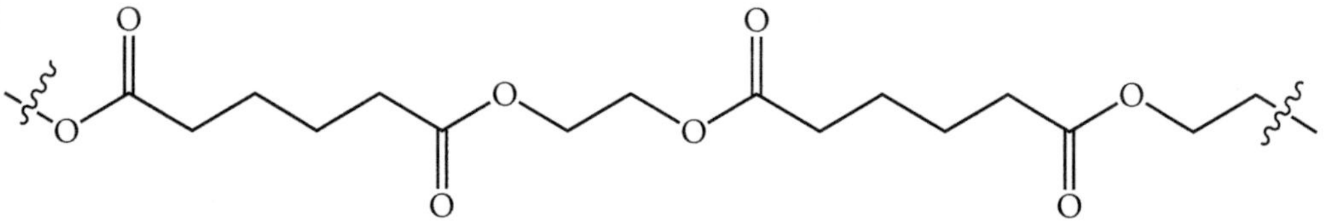

(b)

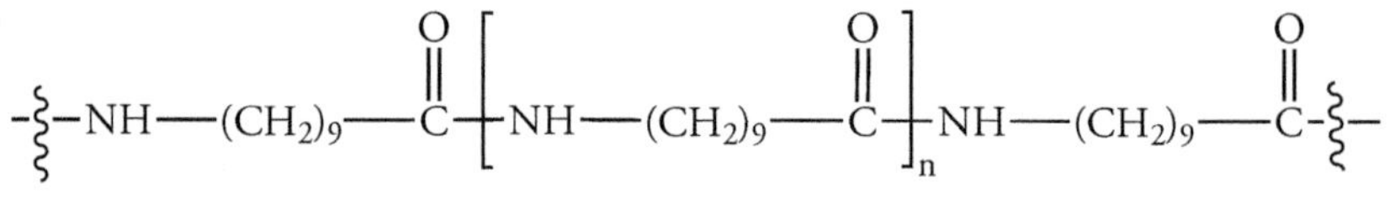

(c)

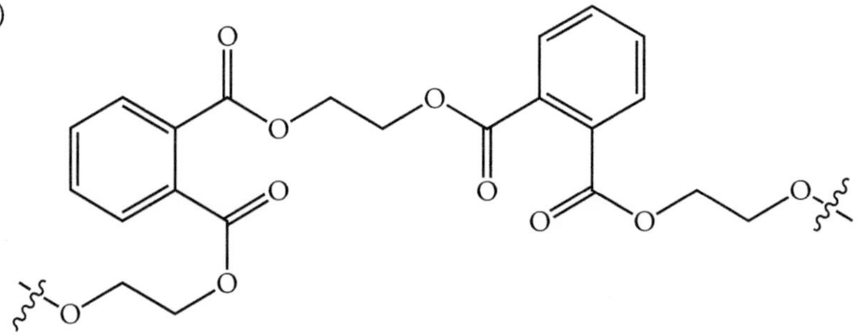

15.22 What monomers are required to prepare the following polymers?

(a)

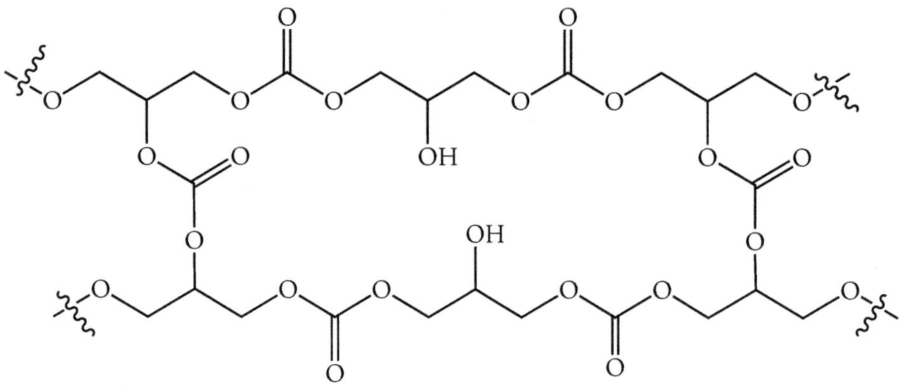

(b)

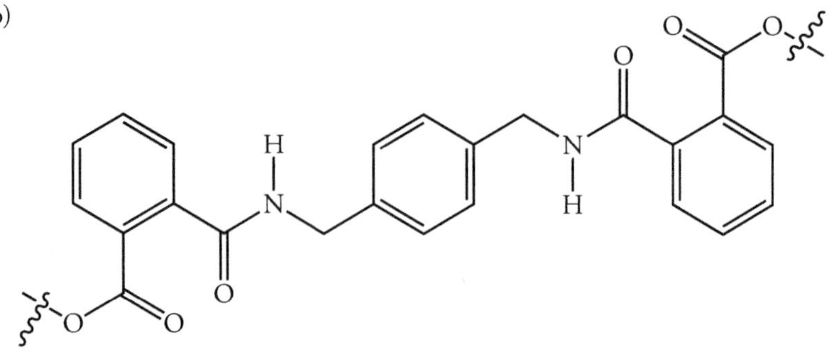

(c)

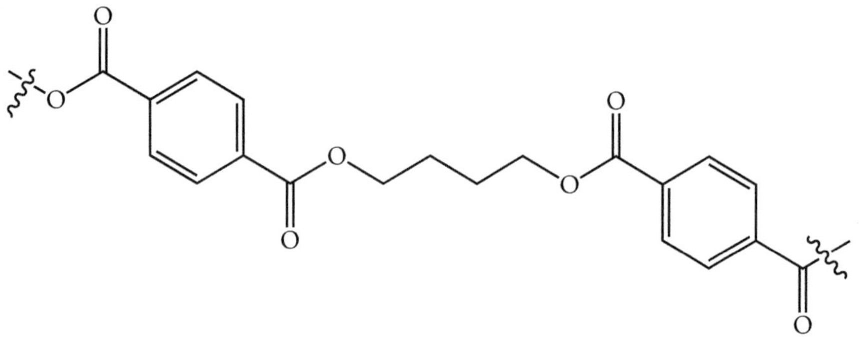

Polyesters

15.23 A homopolymer of lactic acid has been used to make body implants. Write a bond-line representation of the polymer.

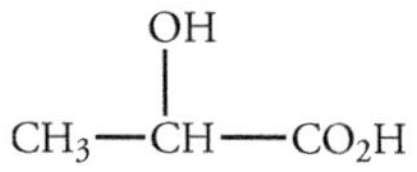

lactic acid

15.24 A polymer of β-propiolactone is obtained by using a catalytic amount of hydroxide ion. Draw the structure of the polymer. Why does the polymerization reaction continue?

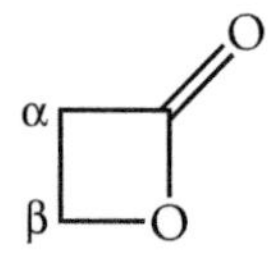

β-propiolactone

15.25 Kodel is a polymer of terephthalic acid and *trans*-di-1,4-(hydroxymethyl)cyclohexane. Draw a representation of the polymer.

15.26 What monomers are used to prepare the following polyester? Identify an unusual feature of this polyester.

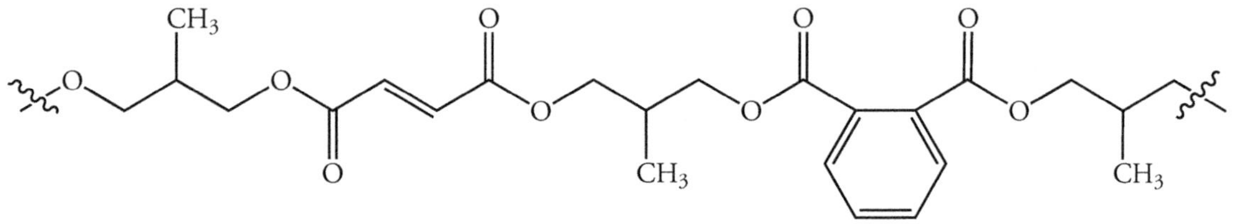

Polyamides

15.27 The following structure represents a group of polyamides called Qiana. The value of x is 8, 10, or 12. What are the component monomers? What is the significance of the value of x?

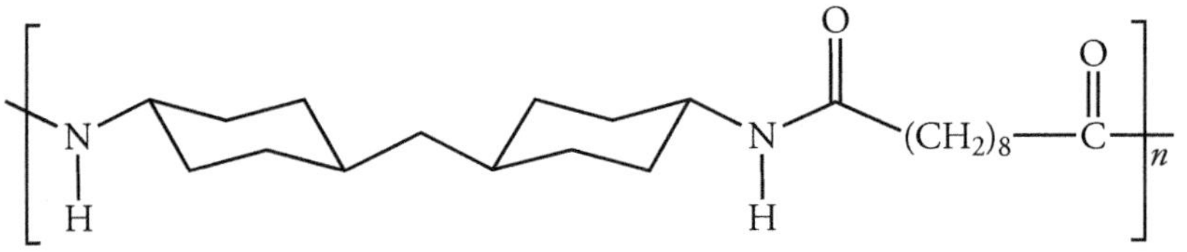

15.28 A polyamide contains the following structural unit, which is prepared from the reaction of a lactam. Draw the structure of the lactam.

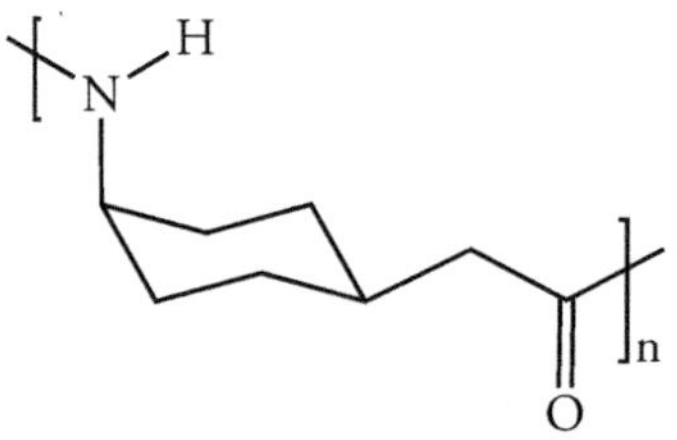

Polyurethanes

15.29 Explain why the addition of glycerol to the polymerization of toluene diisocyanate and ethylene glycol produces a stiffer foam.

15.30 An oligomer of tetramethylene glycol reacts with toluene diisocyanate to form a polyurethane called Lycra. Draw a representation of the polyurethane.

tetramethylene glycol

16

SPECTROSCOPY

16.1 SPECTROSCOPIC STRUCTURE DETERMINATION

The determination of the molecular structure of a compound obtained either from a natural source or as the product of a chemical reaction is an important part of experimental chemistry. It is of paramount importance because without a knowledge of structure we cannot explain either the physical or the chemical properties of molecules. At one time, the structure of an organic compound was determined by chemical reactions that related the unknown compound to other known compounds. To determine the structures of complex molecules, it was often necessary to use reactions that systematically degraded the large molecule into smaller molecules. Then it was possible to "reason backward" to postulate what the structure of the original compound must have been to yield the observed products. Thus, structure determination by chemical reactions is a time-consuming process. For example, consider the problem of determining the structure of a relatively simple compound with molecular formula $C_5H_{10}O$. Eighty-eight isomers are possible, including ethers, alcohols, aldehydes, and ketones. Many chemical reactions would be required to identify the functional group and the hydrocarbon skeleton. Structure determination by chemical reactions also has another severe limitation: each reaction destroys part of the sample of the unknown compound.

Spectroscopic determination requires only minuscule amounts of a compound compared to chemical methods, and spectroscopic methods require much less time than chemical methods. Chemists now determine structure by nondestructive spectroscopic techniques and recover the sample unchanged after they determine its spectrum. In this chapter we examine ultraviolet spectroscopy, infrared spectroscopy, and nuclear magnetic resonance spectroscopy. Each type of spectroscopy provides a different kind of information.

Ultraviolet spectroscopy provides information about the π system in a compound, so it allows us to distinguish between conjugated and nonconjugated compounds. Infrared spectroscopy reveals the functional groups in a compound. For example, consider again the determination of the structure of a compound with molecular formula $C_5H_{10}O$. **Infrared spectroscopy** can eliminate some of the 88 possible structures by identifying the functional group containing the oxygen atom. If infrared spectroscopy reveals that the compound is a ketone, the number of possible isomers drops to a more manageable three compounds.

$$CH_3-CH_2-\overset{\overset{\displaystyle O}{\|}}{C}-CH_2-CH_3$$

I

$$CH_3-\overset{\overset{\displaystyle O}{\|}}{C}-CH_2-CH_2-CH_3$$

II

$$CH_3-\overset{\overset{\displaystyle O}{\|}}{C}-\underset{\underset{\displaystyle CH_3}{|}}{CH}-CH_3$$

III

Nuclear magnetic resonance spectroscopy (NMR) provides information about the carbon-hydrogen framework of a compound. To further characterize the $C_5H_{10}O$ compound known to be a ketone by infrared spectroscopy, we must determine how the carbon atoms are distributed on either side of the carbonyl carbon atom. To determine the structure of the ketone, we need to "see" either the structurally nonequivalent carbon nuclei or the nuclei of hydrogen atoms bonded to them. Nuclear magnetic resonance spectroscopy provides this information. In this chapter we will first consider how to determine the molecular structure using information about structurally different hydrogen nuclei. With information about the hydrogen atoms, we can deduce how the carbon atoms are arranged in the structure. Then we will see how NMR spectroscopy can detect structural differences among the carbon atoms themselves.

16.2 SPECTROSCOPIC PRINCIPLES

Spectroscopy is a study of the interaction of electromagnetic radiation with molecules. Electromagnetic radiation encompasses X-rays, ultraviolet, visible, infrared, microwaves, and radio waves. Electromagnetic radiation is described as a wave that travels at the speed of light (3×10^8 m/s). Waves are characterized by a wavelength (λ, Greek lambda) and a frequency (ν, Greek nu). The wavelength is the length of one wave cycle, such as from trough to trough (Figure 16.1). The wavelength is expressed in the metric unit convenient for each type of electromagnetic radiation. The frequency is the number of waves that move past a given point in a unit of time. Frequency is expressed in hertz (Hz). Wavelength and frequency are inversely proportional, and are related by $\lambda = c/\nu$, where c is the speed of light. Thus, as the wavelength of the electromagnetic radiation increases, the corresponding frequency decreases.

The energy, E associated with electromagnetic radiation is quantized. The relationship is given by the following equation

$$E = h\nu \tag{16.1}$$

In this equation h is Planck's constant. The energy of electromagnetic radiation is therefore directly proportional to its frequency. Since wavelength is inversely proportional to frequency, $\lambda = c/\nu$, we can rewrite this as Equation (16.2).

$$E = \frac{hc}{\lambda} \tag{16.2}$$

The energy of electromagnetic radiation is also directly proportional to the quantity $1/\lambda$. This quantity the **wavenumber**, Equation (16.3).

$$E = hc\left(\frac{1}{\lambda}\right) \tag{16.3}$$

The frequency of ultraviolet radiation is higher than that of infrared radiation. Alternatively expressed, the wavelength of ultraviolet radiation is shorter than the wavelength of infrared radiation. Because ultraviolet radiation has a high frequency, it has higher energy than infrared radiation (Figure 16.2).

Figure 16.1 Electromagnetic Radiation

The wavelength, λ, of electromagnetic radiation is the distance between any two peaks or troughs of the wave.

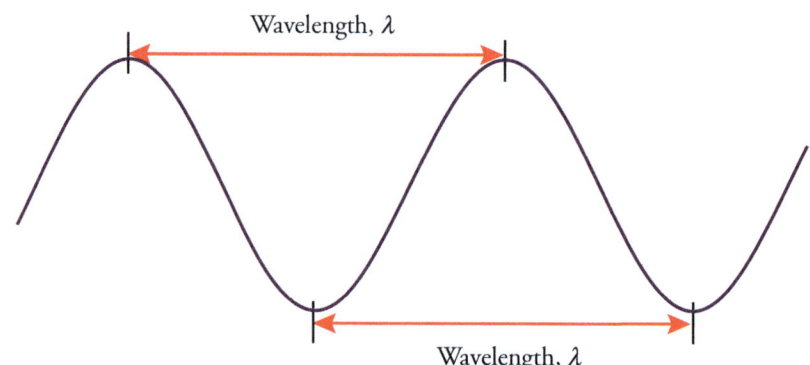

Wavelength, λ

Wavelength, λ

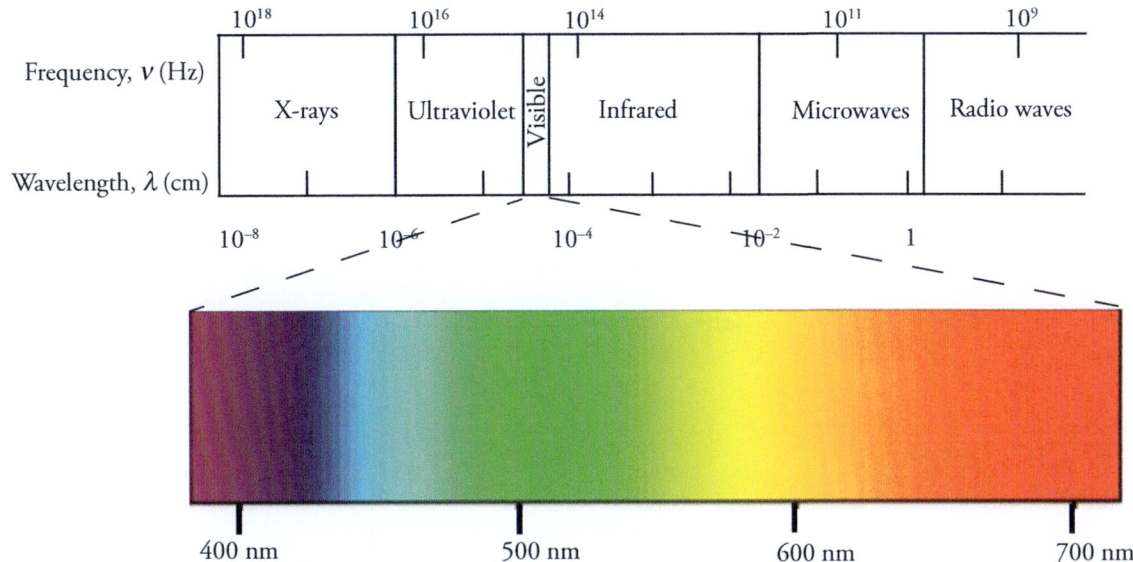

Figure 16.2 Electromagnetic Spectrum
The regions of the spectrum are characterized by a frequency and a wavelength. Usually the wavelength or the reciprocal of the wavelength, the wavenumber, is used to identify absorptions by organic molecules. The relationship of the visible spectrum to other spectral regions is shown within the expansion of that region.

Molecules can absorb only certain discrete amounts of energy. To change the energy of a molecule from E_1 to E_2, the energy difference, $(E_2 - E_1)$, is provided by a characteristic electromagnetic radiation with a specific frequency (or wavelength). The energy absorbed by the molecule can change its electronic or vibrational energy. For example, ultraviolet radiation causes changes in the electron distribution in π bonds; infrared radiation causes bonds to stretch and bond angles to bend.

In various types of spectroscopy, radiation is passed from a source through a sample that may or may not absorb certain wavelengths of the radiation. The wavelength is systematically changed, and a detector determines which wavelengths of light are absorbed (Figure 16.3). At a wavelength that corresponds to the energy $E_2 - E_1$ necessary for a molecular change; the radiation emitted by the source is absorbed by the molecule. The amount of light absorbed by the molecule (absorbance) is plotted as a function of wavelength. At most wavelengths the amount of radiation detected by the detector is equal to that emitted by the source—that is, the molecule does not absorb radiation. At such wavelengths a plot of absorbance on the vertical axis versus wavelength yields a horizontal line (Figure 16.3). When the molecule absorbs radiation of a specific wavelength, the amount of radiation that arrives at the detector is less than that emitted by the source. This difference is recorded as an absorbance.

Figure 16.3 Features of a Spectrum
The portion of the spectrum where no absorption occurs is the base line. This horizontal line may be located at the top or bottom of a graph. Absorption then is recorded as a "peak" extending down from the base line. In an infrared spectrum (a) the base line is at top of the spectrum. In an NMR spectrum (b) the base line is at the bottom of the spectrum.

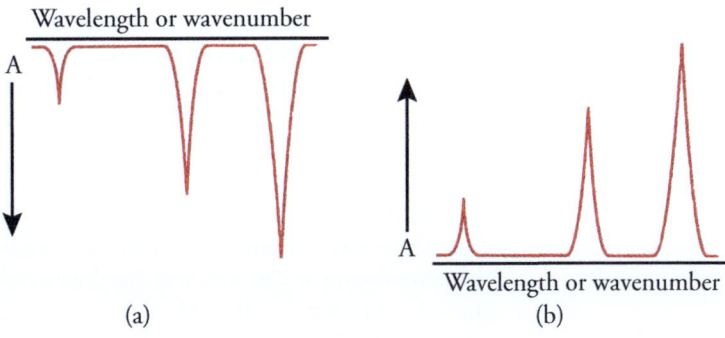

16.3 ULTRAVIOLET SPECTROSCOPY

The ultraviolet region of the electromagnetic spectrum spans wavelengths from 200 to 400 nm (1 nm = 10^{-9} m). In the ultraviolet region of the electromagnetic spectrum a molecule with conjugated double bonds absorbs energy. Both sigma bonds and isolated carbon-carbon double bonds require electromagnetic radiation of higher frequency to absorb energy. Ultraviolet (UV) spectra have a simple appearance. A UV spectrum is a plot of the absorbance of light on the vertical axis and the wavelength of light in nanometers (run) on the horizontal axis (Figure 16.4). The wavelength corresponding to the UV "peak" is called the λ_{max}. The intensity of the absorption depends on the structure of the compound and the concentration of the sample in the solution. Concentrations in the 10^{-3} to 10^{-5} M range are typically used to obtain a spectrum. Thus UV spectroscopy requires very small samples.

Figure 16.4 UV Spectrum of Isoprene

The ultraviolet spectrum of isoprene dissolved in methanol is typical of the spectra of conjugated dienes. The position of maximum absorption, λ_{max}, occurs at 222 nm.

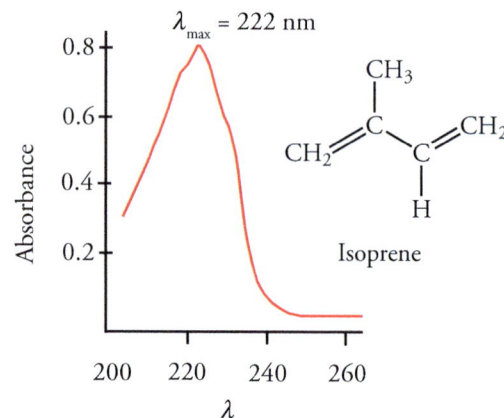

The energy absorbed by conjugated systems moves bonding π electrons into higher energy levels. The specific wavelength of ultraviolet light required to "excite" the electrons in a conjugated molecule depends on the structure of the compound. Ultraviolet spectroscopy provides information about the extent of the conjugation: as the number of double bonds increases, the wavelength of absorbed light also increases. For 1,3-butadiene, λ_{max} = 217 nm. As the number of conjugated double bonds increases, compounds absorb at longer wavelengths; that is, a lower energy is required for electronic excitation. For example 1,3,5-hexatriene and 1,3,5,7-octatetraene have λ_{max} values of 268 and 304 nm, respectively.

Although the extent of the conjugation is the primary feature in determining λ_{max}, the degree of substitution causes changes that are useful in structure determination. For example, the presence of an alkyl group such as methyl bonded to one of the carbon atoms of the conjugated system causes approximately a 5 nm shift to longer wavelengths. Thus, 2-methyl-l,3-butadiene (isoprene) absorbs at 222 nm, compared to 217 nm for l,3-butadiene.

Some naturally occurring compounds with extensively conjugated double bonds absorb at such long wavelengths that λ_{max} occurs in the visible region (400-800 nm) of the spectrum. β-Carotene, which is contained in carrots, absorbs at 455 nm in the blue-green region of the spectrum. Because blue-green light is absorbed, the light that is transmitted to our eyes is yellow-orange; that is, we see the complement of the absorbed light. Thus, the color of a compound provides qualitative information about its λ_{max} (Table 16.1). A compound is colored only if absorption occurs in some portion of the visible spectrum. Compounds that absorb only in the ultraviolet region are colorless because no "visible" light is absorbed.

Other kinds of conjugated molecules besides polyenes have ultraviolet absorptions. For example, benzene absorbs at 254 nm, and substituents on the aromatic ring affect the position of the absorption. By studying the effect of substituents on aromatic rings, as well as other classes of conjugated compounds, chemists have compiled information relating structural effects on ultraviolet spectra that make it possible to establish the structures of unknown compounds.

Table 16.1
Absorbed Light and Reflected Color

Absorbed wavelength (nm)	Reflected color
400 (violet)	Yellow-green
450 (blue)	Orange
510 (green)	Purple
590 (orange)	Blue
640 (red)	Blue-green
7300 (purple)	Green

Problem 16.1

Predict the λ_{max} of 2,4-dimethyl-1,3-pentadiene.

Solution

The compound contains two conjugated double bonds. Thus, the compound should absorb near 217 nm, as in butadiene. However, there are two branching methyl groups as well as the C-5 methyl group bonded to the unsaturated carbon atoms of the butadiene-type system. Thus, the compound should absorb at 217 + 3(5) = 232 nm.

Problem 16.2

Naphthalene and azulene are isomeric compounds that have extensively conjugated π systems. Naphthalene is a colorless compound, but azulene is blue. Deduce information about the absorption of electromagnetic radiation by these two compounds.

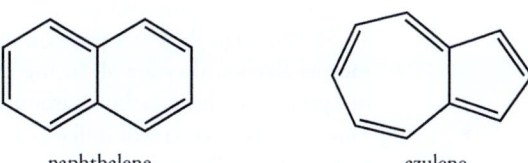

naphthalene   azulene

16.4 INFRARED SPECTROSCOPY

Bonded atoms in a molecule do not remain at fixed positions with respect to each other. Molecules vibrate at various frequencies that depend on molecular structure. Similarly, the angle between two atoms bonded to a common central atom expands and contracts by a small amount at a frequency that depends on molecular structure. These vibrational and bending frequencies correspond to the frequencies of light in the **infrared** region in the electromagnetic spectrum. Every type of bond or bond angle in a molecule absorbs infrared radiation at a specific wavelength. However, the number of different absorptions is large for even the simplest of organic molecules.

The infrared spectrum of 1-methylcyclopentene is shown in Figure 16.5. The wavelength, given on the bottom of the graph, is plotted against percent transmittance of light by the sample. An absorption corresponds to a "peak" pointed toward the bottom of the graph. Because the wavelength of absorbed light is inversely proportional to its energy, absorptions that occur at high wavelength (toward the right of the graph) represent molecular vibrations that require low energy. The plot also gives the corresponding value of the wavenumber, $1/\lambda$, for the absorption. The energy of absorbed light is directly proportional to the wavenumber. Thus, absorptions that occur at higher wavenumbers (toward the left of the graph) represent molecular vibrations that require high energy. The infrared spectrum of an organic molecule is complex, and a peak-by-peak analysis is very difficult. However, the total spectrum is characteristic of the compound and can be used to establish the identity of a compound. If the spectrum of an unknown compound has all of the same absorption peaks—both wavelength and intensity—as a compound of known structure, then the two samples are identical. If the unknown has one or more peaks that differ from the spectrum of a known, then the two compounds are not identical or some impurity in the unknown sample causes the extra absorptions. On the other hand, if the unknown lacks even one absorption peak that is present in the known structure, the unknown has a different structure than the known.

Figure 16.5 Infrared Spectrum of 1-Methylcylcopentene

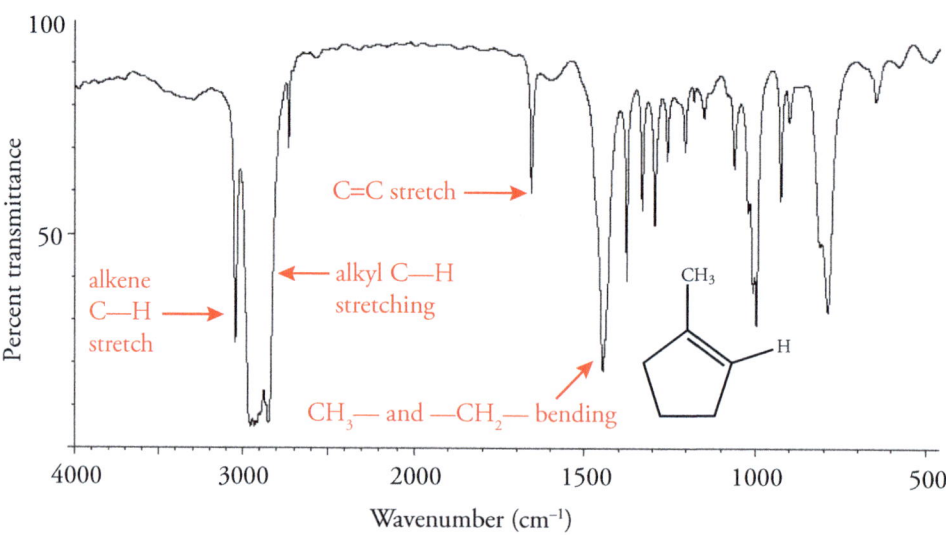

Characteristic Group Vibrations

Although the infrared spectrum of an organic compound is complex, distinctive bands appear in spectra of compounds with common functional groups. Distinctive absorptions corresponding to the vibration of specific bonds or functional groups are called **group vibrations**. Thus, we can use the presence or absence of these absorptions to characterize a compound. For example, the absorption at 1650 cm^{-1} in 1-methylcyclopentene is due to the stretching of a carbon-carbon double bond (Figure 16.5). Although the exact position of the carbon-carbon double bond absorption varies slightly for various alkenes, they are all in the 1630-1670 cm^{-1} region. Isomers such as 3-methylcyclopentene or cyclohexene have carbon-carbon double bonds that have absorptions in the 1650 cm^{-1} region. However, the spectra will differ in some other areas, meaning that the compounds are not identical to 1-methylcyclopentene.

In the following sections, we will consider a few characteristic group vibrations. We will not discuss the small differences in group vibrations that result from small differences in structure in this text.

Identifying Hydrocarbons

In the early chapters of this text, we saw that hydrocarbons are classified as saturated and unsaturated based on the absence or presence of multiple bonds. Multiple bonds decrease the number of hydrogen atoms in a molecular formula below the number given in C_nH_{2n+2}, the molecular formula for a saturated acyclic hydrocarbon. However, the molecular formula of hydrocarbons do not unambiguously indicate the presence or absence of a multiple bond. Both 1-octene and cyclooctane have the same molecular formula, C_8H_{16}.

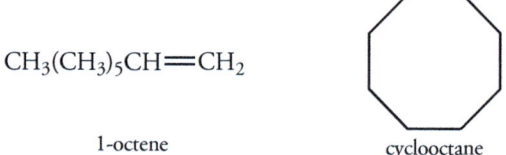

1-octene   cyclooctane

The structural features present in 1-octene that are absent in cyclooctane are a carbon-carbon double bond and sp^2-hybridized C—H bonds. Thus, the characteristic group absorptions of these features will be present in the spectrum of 1-octene and absent in the spectrum of cyclooctane. Now also consider the difference between 1-octyne and an isomeric bicyclic hydrocarbon.

$$CH_3(CH_3)_5CH\equiv CH$$

1-octyne

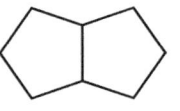

(a bicyclic hydrocarbon)

The structural features present in 1-octyne and absent in the bicyclic hydrocarbon are a carbon-carbon triple bond and an sp-hybridized C—H bond. Thus, the characteristic group absorptions of these features will be present in the spectrum of 1-octyne and absent in the spectrum of the bicyclic hydrocarbon. The energy of the infrared radiation absorbed by a C—H bond depends on the hybridization of the hybrid orbital (Table 16.2). The bond strengths of carbon-hydrogen bonds are in the order of $sp^3 < sp^2 < sp$, because the increased s character of the hybrid orbital keeps the bonding electrons closer to the carbon atom. Thus, the energy required to stretch the bond also increases. The sp^3-hybridized C—H bonds in saturated hydrocarbons like octane (Figure 16.6a) absorb infrared radiation in the 2850-3000 cm^{-1} region. The sp^2-hybridized C—H bonds in alkenes such as 1-octene absorb energy at 3080 cm^{-1}. This peak is well separated from the absorptions associated with the sp^3-hybridized C—H bond in this molecule (Figure 16.6b). An sp-hybridized C—H bond in a molecule such as 1-octyne (Figure 16.6c) absorbs infrared radiation at 3320 cm^{-1}.

Hydrocarbons can also be classified based on absorptions due to the carbon-carbon bond. Carbon-carbon bond strength increases in the order single < double < triple. Thus, the wavenumber position (cm^{-1}) of the absorption corresponding to stretching these bonds increases in the same order. Saturated hydrocarbons all contain many carbon-carbon single bonds that absorb in the 800-1000 cm^{-1} region, but the intensity is very low. Carbon-carbon single bonds present in unsaturated compounds also absorb in the same region. Many other molecular bond stretching vibrations and bond angle bending modes occur in the same region and are much more intense. Thus, this region has limited diagnostic value. Moreover, we already know that most organic compounds have carbon-carbon single bonds.

Unsaturated hydrocarbons are identified by the absorption for the carbon-carbon double bond, which occurs in the 1630-1670 cm^{-1} region. The intensity of the absorption decreases with increased substitution. Terminal alkenes have the most intense absorptions. The double bond in 1-octene absorbs at 1645 cm^{-1} (Figure 16.6b).

The absorption for carbon-carbon triple bonds occurs in the 2100-2140 cm^{-1} region. Terminal alkynes have the most intense absorptions; internal (disubstituted) alkynes have lower intensity absorptions. The triple bond in 1-octyne absorbs at 2120 cm^{-1} (Figure 16.6c).

Table 16.2
Characteristic Infrared Group Frequencies

Class	Group	Wavenumber (cm^{-1})
Alkane	C—H	2850-3000
Alkene	C—H	3080-3140
	C=C	1630-1670
Alkyne	C—H	3300-3320
	C≡C	2100-2140
Alcohol	O—H	3400-3600
	C—O	1050-1200
Ether	C—O	1070-1150
Aldehyde	C=O	1725
Ketone	C=O	1700-1780

Identifying Oxygen-Containing Compounds

Many functional groups contain oxygen. These functional groups have the characteristic infrared absorptions given in Table 16.2. The characteristic group frequencies of aldehydes and ketones are from 1700 to 1780 cm^{-1}. The carbon-oxygen double bond of carbonyl compounds requires more energy to stretch than does the carbon-oxygen single bond of ethers and alcohols. Thus, aldehydes and ketones absorb infrared radiation at higher wavenumber positions (1700-1780 cm^{-1}) than alcohols and ethers (1050-1200 cm^{-1}).

Figure 16.6a Infrared Spectrum of n-Octane

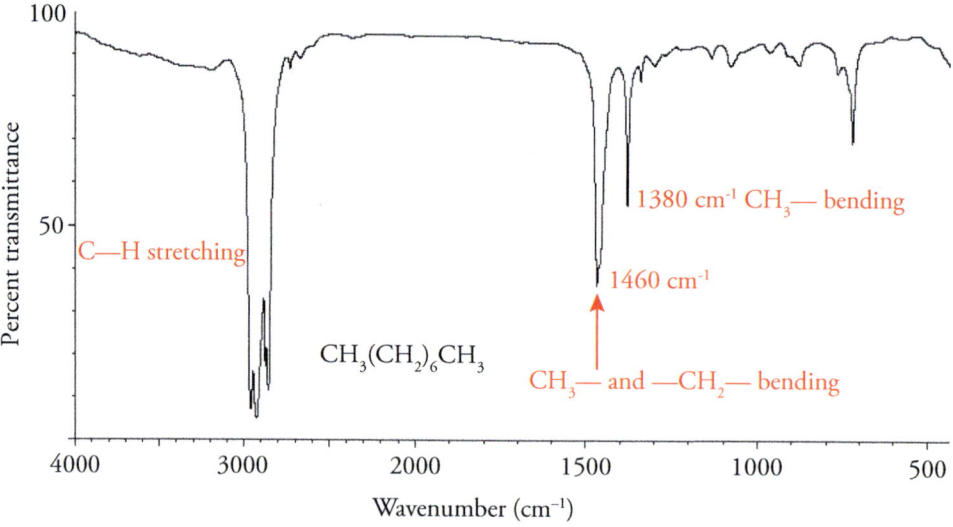

Figure 16.6b Infrared Spectrum of 1-Octene

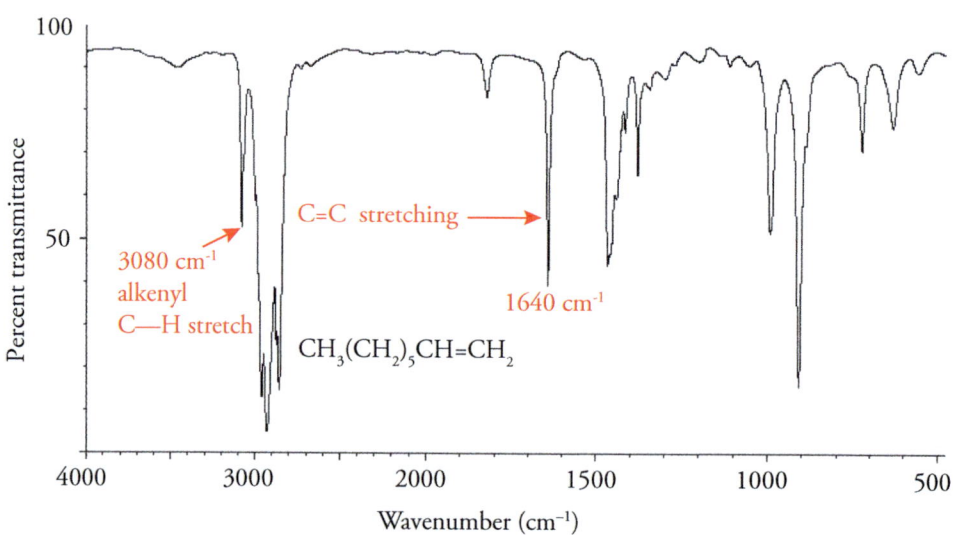

Figure 16.6c Infrared Spectrum of 1-Octyne

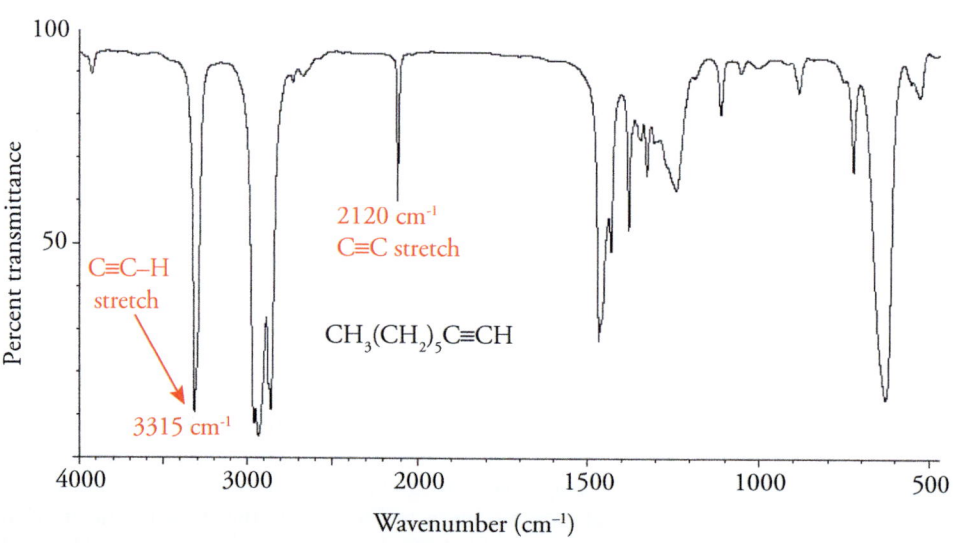

The absorption for a carbonyl group is extremely intense and is easily detected because it is in a region or the infrared spectrum that is devoid of conflicting absorptions. Note that carbon-carbon double bond stretching vibrations are at a lower wavenumber position than those of carbonyl compounds. A typical spectrum of a ketone is shown for 2-heptanone in Figure 16.7. The carbonyl stretching vibration occurs al 1712 cm⁻¹.

The position of the carbonyl group absorption of acyl derivatives depends on the inductive and resonance effects of atoms bonded to the carbonyl carbon atom. We recall that a carbonyl group is represented by two contributing resonance forms.

contributing resonance structures of a carbonyl group

Figure 16.7 Infrared Spectrum of 2-Heptanone

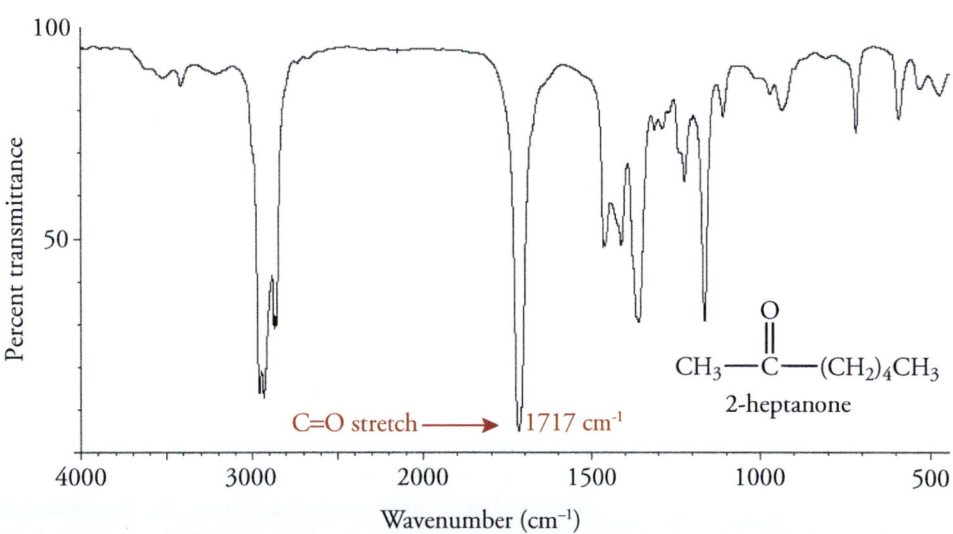

Because less energy is required to stretch a single bond than a double bond, any structural feature that stabilizes the contributing polar resonance form with a carbon-oxygen single bond will cause the infrared absorption to occur at lower wavenumber position. Thus, any group that donates electrons by resonance causes a shift in the absorption to lower wavenumbers. We recall that the nitrogen atom of amides is very effective in donation of electrons to the carbonyl carbon atom (Section 12.5).

contributing resonance structures of an amide

Thus, the double bond character of the carbonyl group decreases. As a result, an amide carbonyl group absorbs in the 1650-1690 cm⁻¹ region, which is at a lower wavenumber than for ketones. The carbon-oxygen single bond stretching vibration of alcohols and ethers appears in a region complicated by many other absorptions. However, the absorption of a carbon-oxygen single bond is more intense than the absorption of carbon-carbon single bonds. The presence of a hydroxyl group is better established by the oxygen-hydrogen stretching vibration that occurs as an intense broad peak in the 3360 cm⁻¹ region. The absorption is illustrated in the spectrum of 1-butanol (Figure 16.8).

Ethers can be identified by a process of elimination. If a compound contains an oxygen atom and the infrared spectrum lacks absorptions characteristic of a carbonyl group or a hydroxyl group, we can conclude that the compound is an ether.

Figure 16.8 Infrared Spectrum
of 1-Butanol

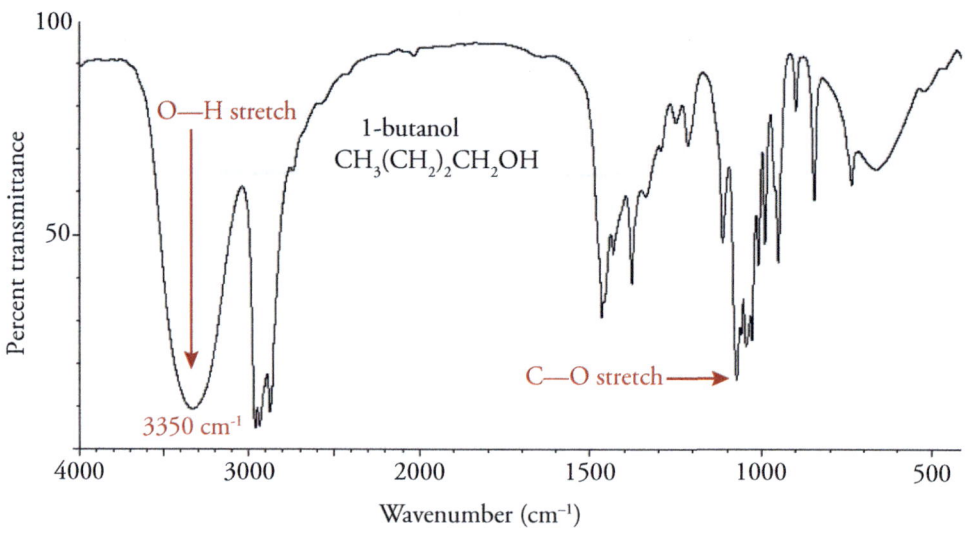

Problem 16.3

Explain how you could distinguish between the following two compounds by infrared spectroscopy.

I   II

Solution

The compounds are isomers. The compound on the left is a diene, which has two differently substituted double bonds. Thus, this compound may have two absorptions in the 1630-1670 cm⁻¹ region corresponding to the two carbon-carbon double bond stretching vibrations. There should also be a C—H stretching absorption in the 3080-3140 cm⁻¹ region for the hydrogen atoms bonded to sp²-hybridized carbon atoms. The compound on the right is an alkyne. Thus, this compound has an absorption in the 2100-2140 cm⁻¹ region that corresponds to the carbon-carbon triple bond stretching vibration. There is also a C—H stretching absorption in the 3300-3320 cm⁻¹ region for the hydrogen atom bonded to the sp-hybridized carbon atom.

Problem 16.4

Explain how you could distinguish between the following two compounds by infrared spectroscopy.

I   II

Problem 16.5

The carbonyl group of an acid chloride absorbs at 1800 cm⁻¹. Explain why this value is at a higher wavenumber than for an aldehyde or ketone.

$$R-C-Cl$$
an acid chloride

Solution

Chlorine does not effectively donate electrons by resonance via its 3p orbitals. Thus, the dipolar resonance form of a carbonyl group is less important, and the carbonyl group has more double bond character than an aldehyde or ketone. The chlorine atom also inductively withdraws electrons, destabilizing the dipolar resonance form. As a result, the carbonyl infrared absorption for an acid chloride requires more energy than for an aldehyde or ketone and occurs at a higher wavenumber position.

Problem 16.6
The carbonyl stretching vibration of ketones is at a longer wavelength than the carbonyl stretching vibration of aldehydes. Suggest a reason for this difference.

16.5 NUCLEAR MAGNETIC RESONANCE SPECTROSCOPY

Many atomic nuclei behave as if they are spinning—that is, they have a **nuclear spin**. For example, the magnetic field of an ^1H nucleus spins about its axis in either of two directions, described as clockwise and counterclockwise, or alternatively as α and β. Because the nucleus is charged, a magnetic moment results from the spinning nucleus. Thus, the hydrogen nucleus is a tiny magnet, with two possible orientations in the presence of an external magnetic field. The magnetic moment of the nucleus may be aligned with the external magnetic field or against it, with the former being of lower energy. If a spinning hydrogen nucleus with its magnetic moment aligned with the external field is irradiated with electromagnetic radiation in the radio-frequency range, the absorbed energy causes the nuclear field to "flip" and spin in the opposite direction (Figure 16.9). Thus, absorption of energy results in a higher energy state for the hydrogen nucleus. The process is called **nuclear magnetic resonance** (NMR). Some nuclei, such as ^{12}C, do not have a nuclear spin, and cannot be detected by NMR. However, the ^{13}C isotope has a nuclear spin, and can be detected by NMR (Section 16.7).

NMR spectroscopy uses electromagnetic radiation in the radio-frequency range. The energy associated with this radiation is very small. The energy required for absorption of energy depends on the strength of the external magnetic field. Increasing the strength of the external magnetic field increases the energy difference between the two spin states. An NMR experiment is done by selecting a magnetic field strength and then varying the radio frequency to find the proper frequency to make the hydrogen nucleus absorb energy to that its spin "flips."

Figure 16.9 Absorption of Electromagnetic Radiation by a Nucleus
When the magnetic moment of a spinning nucleus (H′) is aligned with the magnetic field of an NMR spectrometer (H_0) a low energy state results. Absorption of a specific frequency causes a change in the spin of the nucleus and produces a magnetic moment opposed to the magnetic field of the instrument.

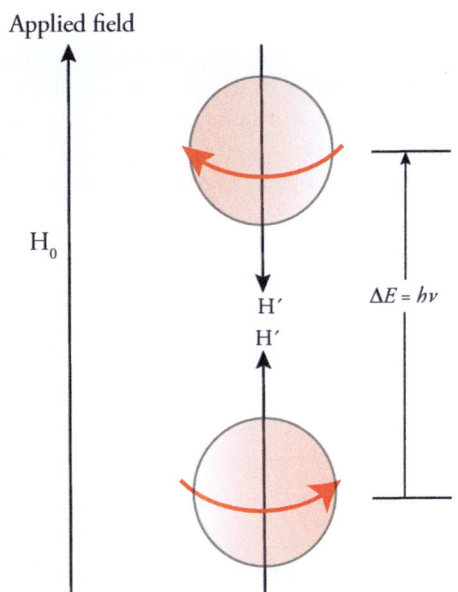

Chemical Shift

The magnetic field strength required to "flip" the spin of various hydrogen nuclei within a molecule differs. If all hydrogen nuclei absorbed the same electromagnetic radiation in an NMR experiment at the same magnetic field strength, then only a single absorption would be observed. As a consequence we would only know that the molecule contained hydrogen atoms.

The hydrogen nuclei in organic molecules are surrounded by electrons that also have spins. The electrons thus set up small local magnetic fields that are opposed to the applied external magnetic field. The local fields affect the magnetic environment of the hydrogen nuclei. When a local field opposes the external magnetic field, we say that the nucleus is **shielded**. The effective field felt by the nucleus is the applied magnetic field minus the local magnetic field generated by the electrons.

$$H_{effective} = H_{applied} - H_{local}$$

The local magnetic fields due to the spins of the electrons differ throughout the molecule because the bonding characteristics differ. Thus, the degree of shielding of each hydrogen nucleus is unique, and distinct resonances are obtained for each structurally nonequivalent hydrogen atom in a molecule. At a constant radio frequency, the external magnetic field required is larger for the more shielded nucleus.

The strengths of the local magnetic fields for various hydrogen atoms are about 10^{-6} times that of the applied magnetic field. Thus, the magnetic fields required to flip various structurally different hydrogen nuclei in a molecule differ on the order of parts per million (ppm). Rather than using absolute values of the field strength, a relative scale is used. An NMR chart is labeled on the horizontal axis with a **delta scale**, in which one delta unit (δ) is 1 ppm of the magnetic field used. The resonance for the hydrogen atoms of tetramethylsilane, $(CH_3)_4Si$, is used as a reference for the magnetic resonance spectra of hydrogen compounds. This resonance is defined as 0.0 δ. By convention, an absorption that occurs at lower field (higher energy) than tetramethylsilane (TMS) appears to the left of the TMS absorption, and is assigned a positive δ value.

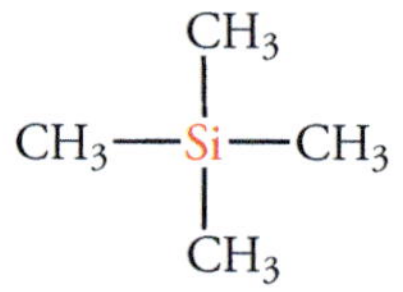

tetramethylsilane
(TMS)

By examining the NMR spectrum, we can tell at a glance how many sets of structurally nonequivalent hydrogen atoms a molecule contains. Consider the NMR spectrum of 1,2,2-trichloropropane shown in Figure 16.10. The spectrum consists of two peaks pointed to the top of the graph. The resonances occur at 2.2 and 4.0 δ. There are two different sets of hydrogen atoms in 1,2,2-trichloropropane. Each set of hydrogen atoms gives rise to one peak.

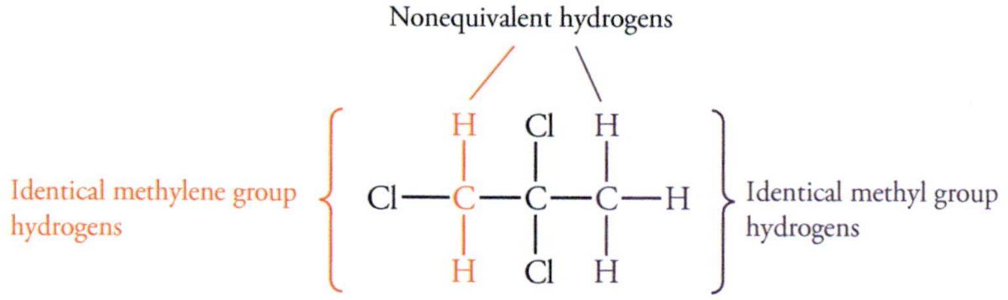

For most organic molecules, the various hydrogen resonances appear between 0 and 10 δ. This range is conveniently divided into regions that reflect certain structural characteristics. Hydrogen atoms bonded to sp^2-hybridized carbon atoms absorb at lower fields than hydrogen atoms bonded to saturated sp^3 carbon atoms. For example, hydrogen atoms bonded to sp^2-hybridized carbon atoms absorb in the 5.0-6.5 δ region. Hydrogen atoms bonded to saturated carbon atoms without any directly bonded substituents absorb in the 0.7-1.7 δ range. The hydrogen atoms bonded to an aromatic ring occur in the general region of 6.5-8.0 δ. The exact position of absorption for hydrogen atoms bonded to either sp^2-hybridized or sp^3-hybridized carbon atoms also depends on the number and type of substituents also bonded to the atom. Hydrogen atoms bonded to carbon atoms that are also bonded to electronegative atoms such as oxygen, nitrogen, or halogens absorb at lower fields. Examples of typical chemical shifts are listed in Table 16.3.

Table 16.3
Chemical Shifts of Hydrogen Atoms

Partial Structural Formula	Chemical Shift (ppm)	Partial Structural Formula	Chemical Shift (ppm)
$-CH_3$	0.7-1.3	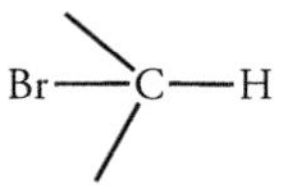 Br—C—H	2.5-4.0
$-CH_2-$	1.2-1.4	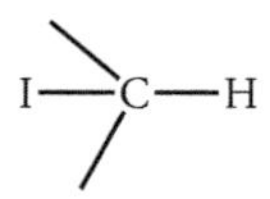 I—C—H	2.0-4.0
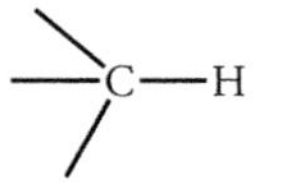 $\rangle$C—H	1.4-1.7	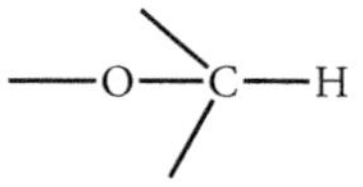 —O—C—H	3.3-4.0
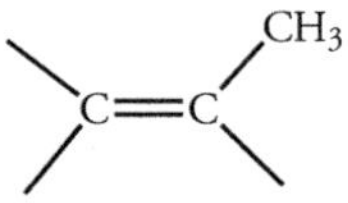 C=C(CH$_3$)	1.6-1.9	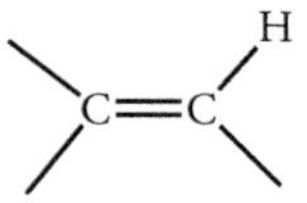 C=C(H)	5.0-6.5
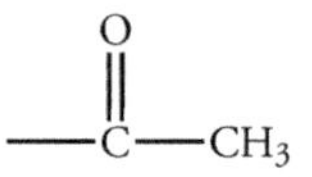 —C(=O)—CH$_3$	2.1-2.4	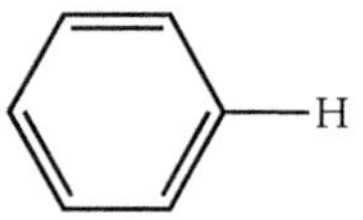 (benzene ring)—H	6.5-8.0
—C≡C—H	2.5-2.7	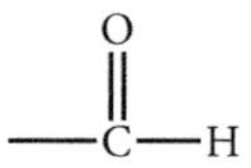 —C(=O)—H	9.7-10.0
 Cl—C—H	3.0-3.4	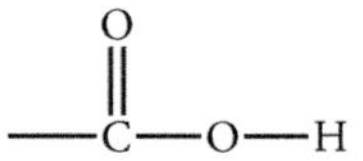 —C(=O)—O—H	10.5-13.0

Relative Peak Areas

The set of hydrogen atoms bonded to C-1 of 1,2,2-trichloropropane has an absorption at 4.0 δ, and the set of hydrogen atoms bonded to the C-3 atom has its absorption at 2.2 δ. This assignment is made based on the generalization that hydrogen atoms bonded to a carbon atom that is also bonded to an electronegative atom absorb at a lower field (higher δ). However, there is another method to confirm this assignment, namely "proton counting." The area under each resonance peak is proportional to the relative number of hydrogen atoms of each kind.

The relative area of a resonance peak is obtained from an electronic integrator used after the spectrum has been recorded. The area is proportional to the vertical displacement of a "stair step" superimposed on the resonance peak. These vertical distances have a ratio equal to the ratio of the number of hydrogen atoms. Thus, the ratio of the integrated intensities of the two resonances of 1,2,2-trichloropropane is 3:2 (Figure 16.11).

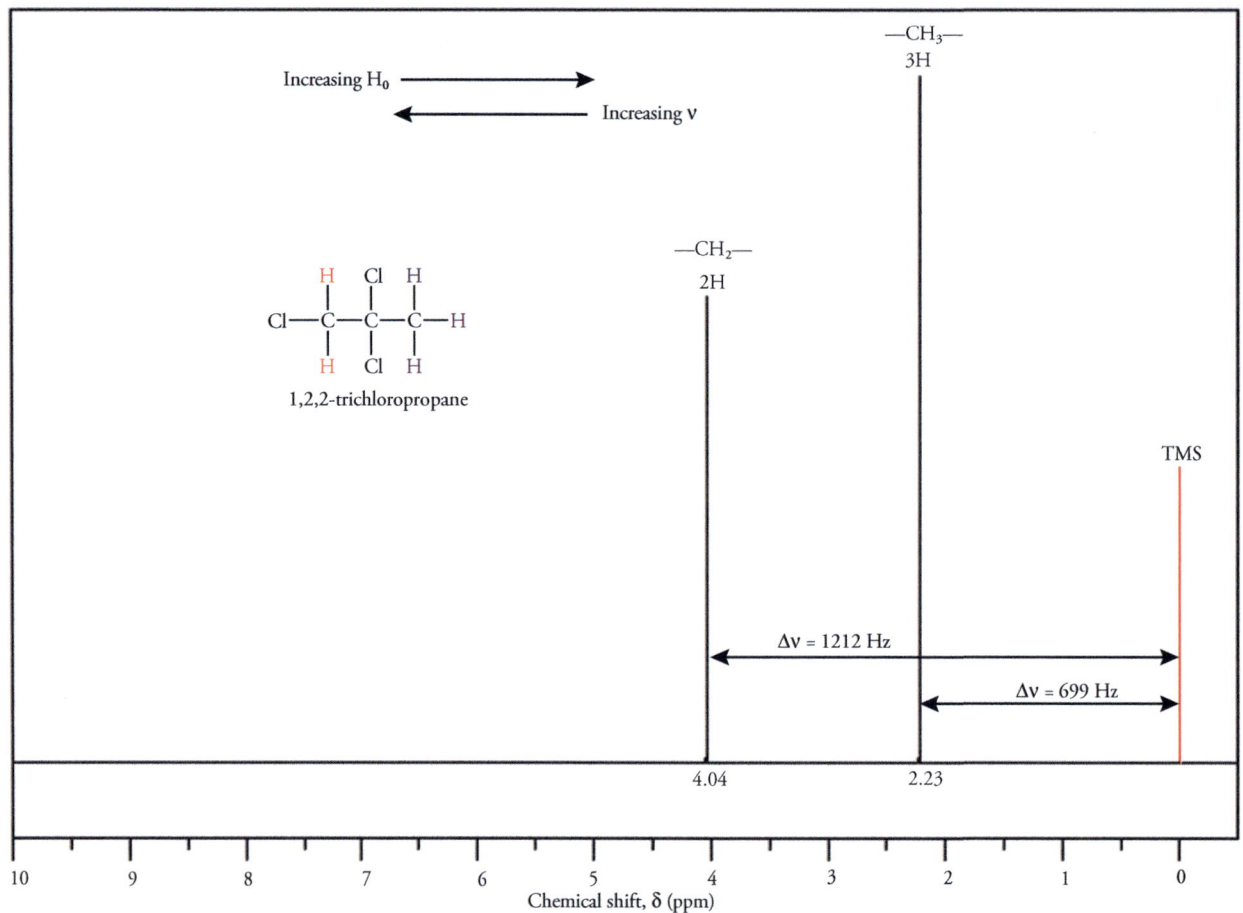

Figure 16.10 NMR Spectrum of 1,2,2-Trichloropropane
The three equivalent hydrogen atoms bonded to C-3 have a chemical shift of 2.23 δ. The two equivalent hydrogen atoms bonded to C-1 have a chemical shift of 4.04 δ.

Problem 16.7
Predict the chemical shift of the resonances of 1,2-dichloro-2-methylpropane. What are the relative intensities of the absorptions?

Solution
Write the structure of the compound to determine the number of sets of nonequivalent hydrogen atoms.

$$
\begin{array}{c}
\quad\ \ CH_3\ \ H \\
\quad\ \ \ |\quad\ | \\
CH_3-C-C-Cl \\
\quad\ \ \ |\quad\ | \\
\quad\ \ CH_3\ \ H
\end{array}
$$

One set of hydrogen atoms is located at C-1, which has one chlorine atom bonded to it. These hydrogen atoms should have a resonance near 4 δ. The six hydrogen atoms of the two equivalent methyl groups should have a resonance near 1 δ. The ratio of the two sets of hydrogen atoms is 6:2. The ratio of the relative areas at 1 and 4 δ are 3:1.

How many sets of nonequivalent hydrogen atoms are contained in each of the following ketones?

(a) $CH_3-CH_2-\overset{\overset{\displaystyle O}{\displaystyle \|}}{C}-CH_2-CH_3$ (b) $CH_3-\overset{\overset{\displaystyle O}{\displaystyle \|}}{C}-CH_2-CH_2-CH_3$ (c) $CH_3-\overset{\overset{\displaystyle O}{\displaystyle \|}}{C}-\underset{\underset{\displaystyle CH_3}{|}}{CH}-CH_3$

Figure 16.11 Integrated Intensities of an NMR Spectrum

The area of each resonance is proportional to the number of hydrogen atoms. The vertical distances of the "stair steps" shown is a measure those areas. The ratio of peak heights, represented by the green arrows, 3:2, corresponds to the three hydrogen atoms bonded to C-3 and the two hydrogens bonded to C-1 in 1,2,2-trichloropropane.

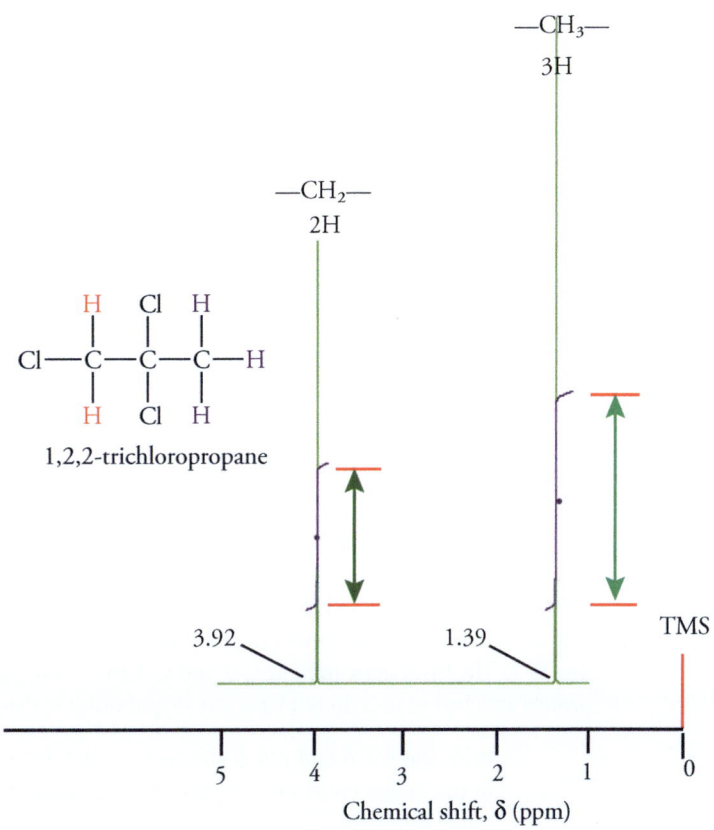

16.6 SPIN-SPIN SPLITTING

In 1,2,2-trichloropropane, each of the two sets of hydrogen atoms is responsible for a single peak. Now consider the spectrum of 1,1,2-tribromo-3,3-dimethylbutane (Figure 16.12).

Nonequivalent hydrogens

Identical methyl group hydrogens

$CH_3-\overset{\overset{\displaystyle CH_3}{|}}{\underset{\underset{\displaystyle CH_3}{|}}{C}}-\overset{\overset{\displaystyle H}{|}}{\underset{\underset{\displaystyle Br}{|}}{C}}-\overset{\overset{\displaystyle H}{|}}{\underset{\underset{\displaystyle Br}{|}}{C}}-Br$

1,1,2-tribromo-3,3-dimethylbutane

The nine equivalent hydrogen atoms of the three equivalent methyl groups give rise to the intense peak at 1.2 δ. The single hydrogen atoms bonded to C-1 and C-2 are nonequivalent. The resonance of the hydrogen atom at the C-1 atom is located at 6.4 δ as a consequence of the two electronegative bromine atoms bonded to that carbon atom. The resonance of the hydrogen atom at the C-2 atom is located at 4.4 δ because only one bromine atom is bonded to the carbon atom. Thus, the intensities of the absorptions and the chemical shifts of the hydrogen atoms are as expected based on molecular structure.

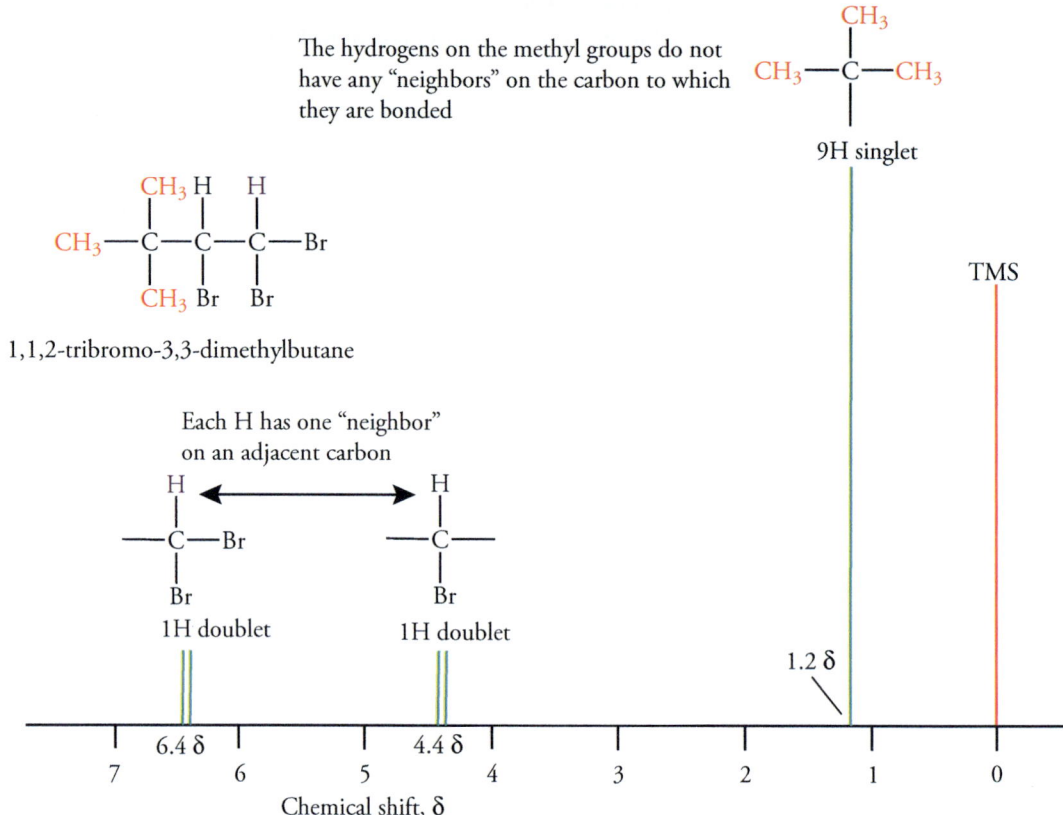

The hydrogens on the methyl groups do not have any "neighbors" on the carbon to which they are bonded

9H singlet

TMS

1,1,2-tribromo-3,3-dimethylbutane

Each H has one "neighbor" on an adjacent carbon

1H doublet   1H doublet

1.2 δ

6.4 δ   4.4 δ

7   6   5   4   3   2   1   0

Chemical shift, δ

Figure 16.12
NMR Spectrum of 1,1,2-Tribromo-3,3-dimethylbutane

The areas under the peaks are in the ratio of 9:1:1, which corresponds to the ratio of hydrogens at C-3, C-2, and C-1. The hydrogens atom at C-1 and C-2 each have one "neighbor." The nine hydrogens in the methyl groups attached to C-3 do not have any neighboring hydrogen atoms on the quaternary carbon.

Both the 4.4 δ and 6.4 δ absorptions of 1,1,2-tribromo-3,3-dimethylbutane are "split" as shown in the inserts containing expanded representations of the resonances. Each area contains two peaks called **doublets**. The phenomenon of multiple peaks is common in NMR spectroscopy. Other common multiplets include **triplets** and **quartets**, meaning that resonances are split into three and four peaks, respectively. This characteristic of a set of peaks associated with a given type of nucleus is called its **multiplicity**.

Multiple absorptions for a set of equivalent hydrogen atoms is known as **spin-spin splitting**. It results from the interaction of the nuclear spin(s) of one or more nearby "neighboring" hydrogen atom(s) with that set of equivalent hydrogen atoms. The small magnetic field of nearby hydrogen nuclei affects the magnetic field "felt" by other hydrogen nuclei. Consider the hydrogen atom on the C-1 atom of 1,1,2-tribromo-3,3-dimethylbutane. It has a neighboring hydrogen nucleus at C-2 that can be spinning in either of two possible directions. In those molecules where the hydrogen nucleus at C-2 is spinning clockwise, the hydrogen atom at C-1 experiences a small magnetic field that differs from that felt in molecules where the hydrogen nucleus at C-2 is spinning counterclockwise. Thus, two slightly different external magnetic fields are needed for the C-1 hydrogen nuclei in various molecules to absorb electromagnetic radiation. A doublet results. The same explanation accounts for the doublet for the hydrogen atom at the C-2 atom. In this case, the hydrogen nucleus at C-1 is the neighboring atom and its spin affects the magnetic field experienced by the hydrogen atom at C-2. Thus, in general, sets of hydrogen atoms on neighboring carbon atoms couple with each other. If hydrogen atom A couples and causes splitting of the resonance for hydrogen atom B, the resonance for hydrogen atom B is also split by hydrogen atom A. The resonance for the nine methyl hydrogen atoms of 1,1,2-tribromo-3,3-dimethylbutane is not "split" because the neighboring quaternary carbon atom has no hydrogen atoms.

A set of one or more hydrogen atoms that has n equivalent neighboring hydrogen atoms has $n + 1$ peaks in the NMR spectrum. The appearance of several sets of multiplets resulting from $n = 1$ through $n = 4$ is shown in Figure 16.13 for some common structures. The areas of the component peaks of a doublet are equal; the areas of the component peaks of other multiplets are not. To understand the relative peak areas of multiplets resulting from more than one neighboring hydrogen atom, let's consider the spectrum of 1,1,2-trichloroethane (Figure 16.14).

Figure 16.13 NMR Splitting Pattern for Neighboring Hydrogens

The resonance of one hydrogen atom (H_A) with neighboring hydrogens is shown. The number of equivalent neighboring hydrogen atoms is responsible for the multiplicity of the resonance. The number of peaks for H_A equals the number (n) of neighboring hydrogens +1.

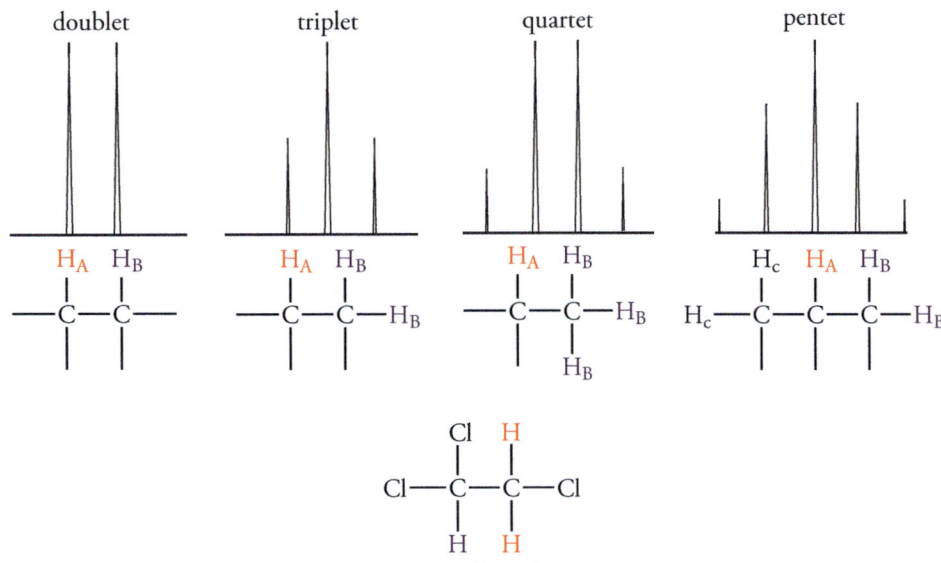

1,1,2-trichloroethane

The doublet near 4 δ corresponds to the two hydrogen atoms bonded to the C-2 atom. These hydrogen atoms have one neighboring hydrogen nucleus that spins in either of two directions we can designate as α or β. As a result, the C-2 hydrogen atoms in various molecules experience two different magnetic fields, resulting in a doublet absorption. Now let's consider the triplet resonance for the C-1 hydrogen atom. The spins of the neighboring two hydrogen nuclei at the C-2 atom can be $\alpha\alpha$, $\alpha\beta$, $\beta\alpha$, and $\beta\beta$. Because the magnetic fields generated by either $\alpha\beta$ or $\beta\alpha$ sets of spins are equivalent, the hydrogen atom at the C-1 atom experiences three different magnetic fields in the ratio of 1:2:1. This ratio is the same as the ratio of the component peaks of the observed triplet.

Extension of the possible combinations of spins of n neighboring equivalent hydrogen atom accounts for the area ratios listed in Table 16.4.

Table 16.4
Number of Peaks of Multiplets and Area Ratios

Number of equivalent adjacent hydrogens	Total Number of Peaks	Area Ratios
0	1	1
1	2	1:1
2	3	1:2:1
3	4	1:3:3:1
4	5	1:4:6:4:1
5	6	1:5:10:10:5:1
6	7	1:6:15:20:15:6:1

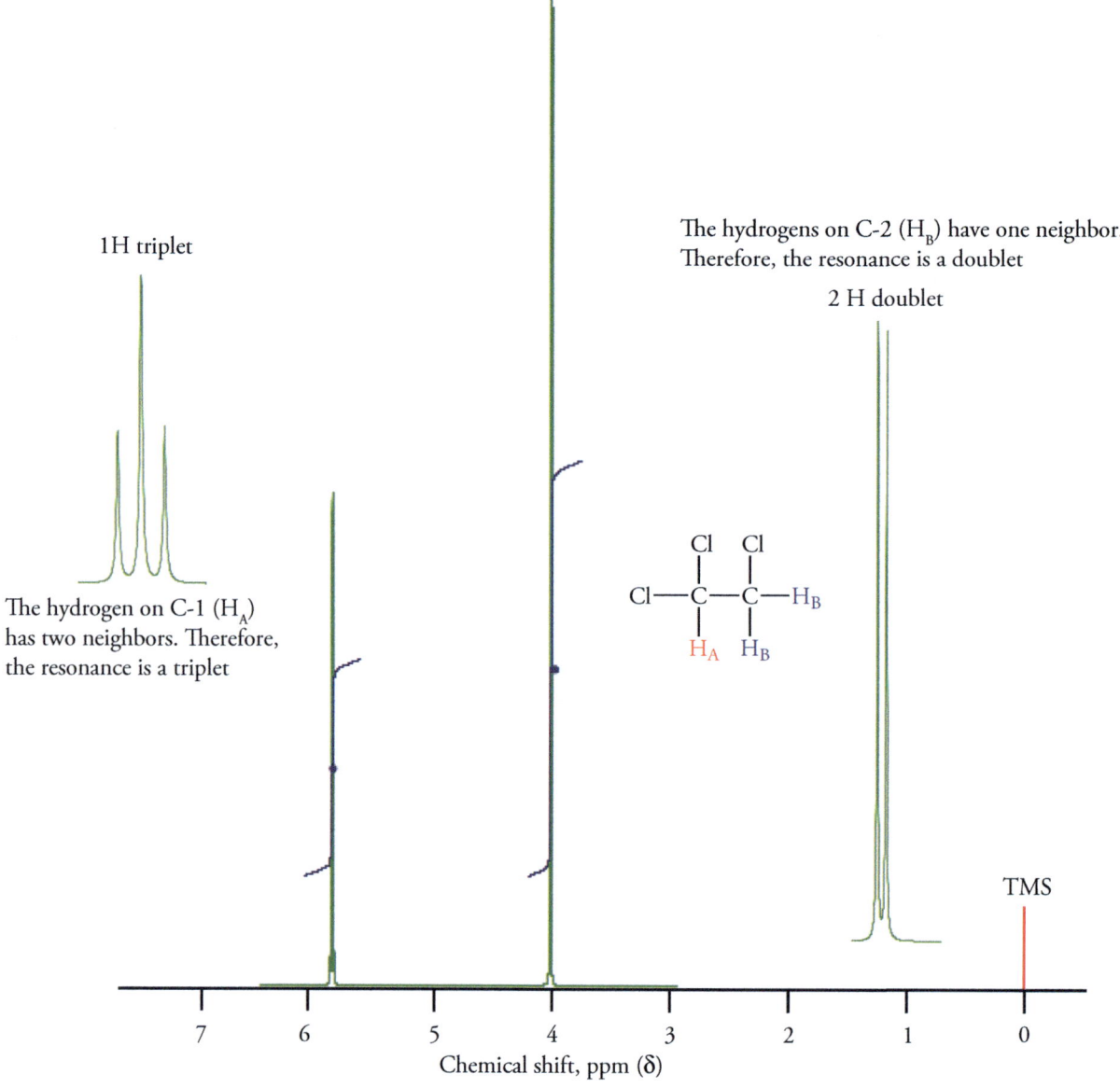

Figure 16.14 NMR Spectrum of 1,1,2-Trichloroethane

The insets show the doublet for the 4.0 δ resonance of the hydrogen atoms bonded to C-2, and the triplet for the 5.8 δ resonance of the hydrogen atom bonded to C-1. Note that the integrated intensities of the 4.0 and 5.8 δ resonances are in the ratio of 2:1.

Problem 16.9

Describe the NMR spectrum of 1,3-dichloropropane.

Solution

First draw the structure of 2-chloropropane to determine the number of sets of nonequivalent hydrogen atoms and the number of neighboring hydrogen atoms that can couple with each set.

2-chloropropane

The six hydrogen atoms located on the equivalent C-1 and C-3 atoms are equivalent. They have a resonance in the 1 δ region. The single hydrogen atom at C-2 has a resonance near 4 δ because a chlorine atom is bonded to that carbon atom. The relative areas of the 1 and 4 δ peaks are 6:1. The 4 δ resonance is a septet because the C-2 hydrogen atom has six neighboring hydrogen atoms that couple with it. Each of the two equivalent methyl groups has only one neighboring hydrogen atom—the C-2 hydrogen atom. Thus, the 1 δ resonance is a doublet.

Problem 16.10
Describe the NMR spectrum of 1,3-dichloropropane citing the number of resonances, their δ values, and their multiplicities.

16.7 ^{13}C NMR SPECTROSCOPY

^{13}C NMR spectroscopy allows us to detect the structural environment of carbon atoms. This is often an advantage, especially for carbon atoms that are not bonded to hydrogen atoms and thus cannot be detected by hydrogen NMR spectroscopy. NMR spectra can be easily obtained for many isotopes with half-integer spins, such as ^{19}F and ^{31}P, because their natural abundance is 100%. The detection of the isotope ^{13}C is more difficult because it has an abundance of only 1%. Therefore, the signals are weaker.

Let's consider the location of the ^{13}C isotope in a compound such as 2-butanol. Most of the carbon atoms are ^{12}C, which has no nuclear spin. The probability is equal for the location of ^{13}C at any of the positions in a molecule. The probability of finding a ^{13}C at C-1 of a molecule is 1%. The probability of finding a ^{13}C at C-2 is also 1%, and so on. The probability of finding two or more ^{13}C in the same molecule and simultaneously bonded to each other is very low. For example, the probability of finding ^{13}C in the same molecule at both C-1 and C-2 is only 0.01%. As a result, a ^{13}C NMR spectrum shows a sum of the signals generated by individual atoms at all of the possible sites in a collection of isotopically substituted molecules. The observed spectrum therefore resembles that expected for a molecule with ^{13}C located at every position, but without coupling between the ^{13}C isotopes.

Characteristics of ^{13}C Spectra

The ^{13}C spectra of organic compounds is shown using a δ scale relative to the resonance of ^{13}C in TMS. The chemical shift of ^{13}C shows many of the same trends as hydrogen chemical shifts. However, the range of chemical shifts for ^{13}C is much larger, on the order of 200 ppm (Table 16.5). Thus, the chemical shifts of ^{13}C are very sensitive to changes in structural environment. As a result, it is usually possible to "see" distinct signals for every nonequivalent ^{13}C in a molecule.

The resonances for ^{13}C are split by hydrogen atoms. However, by using specialized experimental methods, all splitting by hydrogen atoms can be eliminated. This greatly simplifies the spectrum since *each chemically unique* carbon atom appears as a single line in the spectrum.

The ^{13}C spectrum of 2-butanol is shown in Figure 16.15. There are four nonequivalent carbon atoms. The signal at lowest field is assigned to C-2 because it is deshielded by an oxygen atom. The line at 32.0 δ is assigned to C-1 because it is bonded to C-1, but not to an alkyl group. The line at 10.0 δ is assigned to C-4 because it is furthest from C-2, and thus the least deshielded by the oxygen atom. We know that a methyl group is electron releasing relative to hydrogen. Therefore, the line at 22.8 δ is assigned to C-3. The line at 32 δ is assigned to C-1.

Table 16.5
Chemical Shifts of ¹³C Atoms

¹³C Atom	Chemical Shift (ppm)	¹³C Atom	Chemical Shift (ppm)
RCH₂CH₃	12-15	RCH=CH₂	115-120
R₂CHCH₃	16-25	RCH=CH₂	125-140
R₃CH	12-35	$\underset{\text{R—C—OR}}{\overset{\text{O}}{\parallel}}$	170-175
$\underset{\text{R—C—CH}_3}{\overset{\text{O}}{\parallel}}$	30	$\underset{\text{R—C—H}}{\overset{\text{O}}{\parallel}}$	190-200
RCH₂Cl	40-45	$\underset{\text{R—C—R}}{\overset{\text{O}}{\parallel}}$	205-220
RCH₂Br	27-35	$\underset{\text{R—C—OH}}{\overset{\text{O}}{\parallel}}$	180
RCH₂OH	50-65	benzene	128

Figure 16.15 ¹³C NMR Spectrum of 2-Butanol

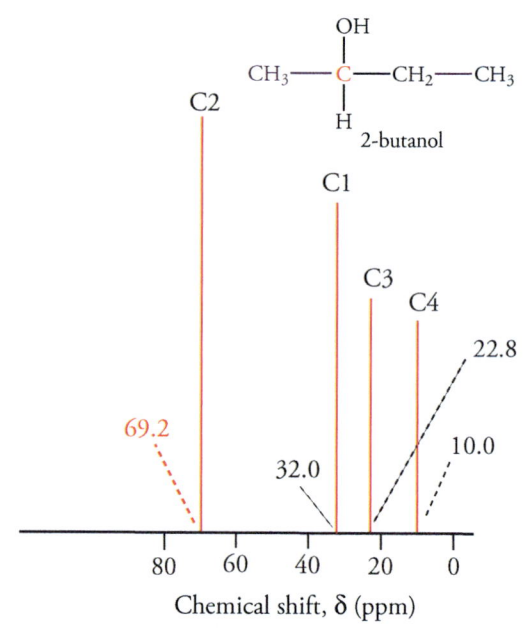

Problem 16.11

The isomeric alcohols 3-heptanol and 4-heptanol cannot be easily distinguished by hydrogen NMR spectroscopy. Describe how ¹³C NMR spectroscopy can be used to distinguish between these isomers.

Solution

Write the structure of the compound to determine the number of equivalent and nonequivalent carbon atoms.

$$\underset{\text{4-heptanol}}{\overset{\overset{\displaystyle OH}{|}}{CH_3CH_2CH_2CHCH_2CH_2CH_3}} \qquad \underset{\text{3-heptanol}}{\overset{\overset{\displaystyle OH}{|}}{CH_3CH_2CHCH_2CH_2CH_2CH_3}}$$

4-Heptanol has only four sets of nonequivalent carbon atoms as a result of the symmetry of the molecule. The C-1 and C-7 atoms are equivalent; the C-2 and C-6 atoms are equivalent; the C-3 and C-5 atoms are equivalent. The C-4 atom is unique. For 3-heptanol, all carbon atoms are nonequivalent. Thus, the two compounds can be distinguished based on the number of resonances in proton-decoupled spectra of the two compounds, four for 4-heptanol and seven for 3-heptanol.

Problem 16.12

How can a compound of molecular formula $C_4H_{10}O$ be established as an ether or an alcohol using ^{13}C NMR spectroscopy?

EXERCISES

UV Spectroscopy

16.1 The λ_{max} values of naphthalene, anthracene, and tetracene are 314, 380, and 480 nm, respectively. Suggest a reason for this order of the wavelength of maximum absorption. Are any of the compounds colored?

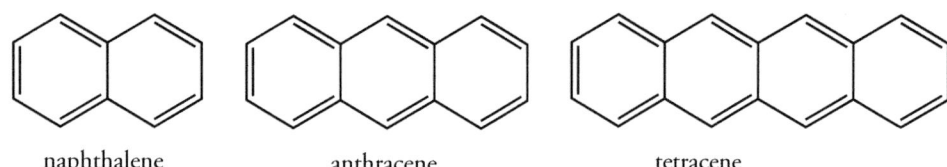

naphthalene   anthracene   tetracene

16.2 How many conjugated double bonds are contained in lycopene? Compare the conjugation in this compound to that of β-carotene. Using this information, predict the color of lycopene.

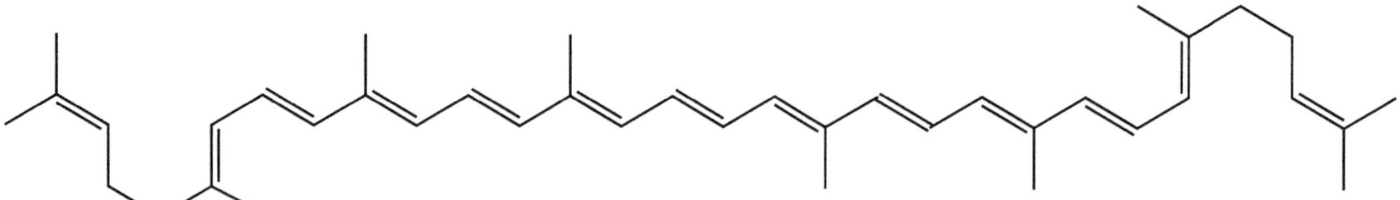

lycopene

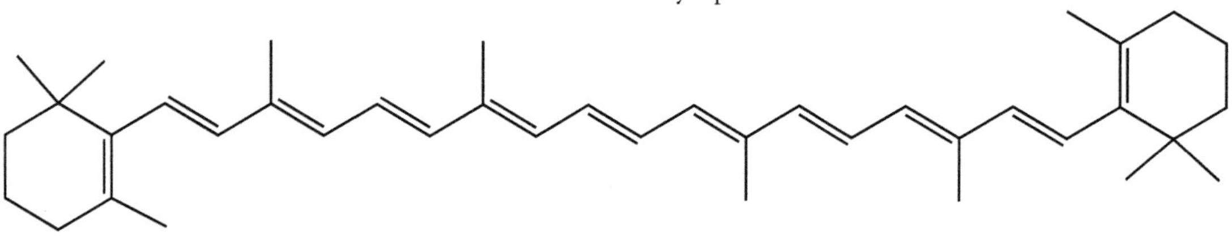

β-carotene

16.3 How might 2,4-hexadiyne be distinguished from 2,5-hexadiyne by ultraviolet spectroscopy?

CH_3—C≡C—C≡C—CH_3 CH_3—C≡C—CH_2—C≡CH
    2,4-hexadiyne   1,4-hexadiyne

16.4 One of the following unsaturated ketones has λ_{max}= 225 nm and the other has λ_{max}= 252 nm. Assign each value to the proper structure.

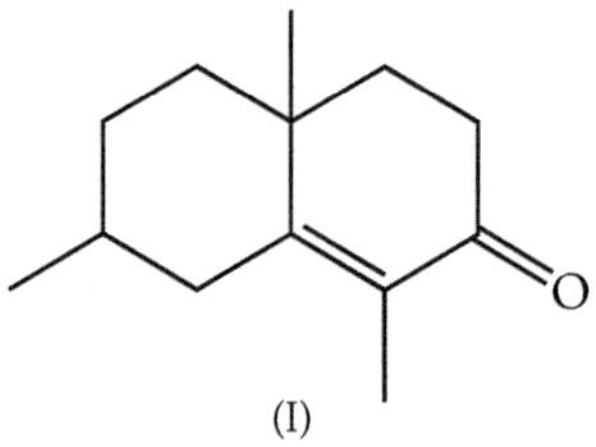

(I) 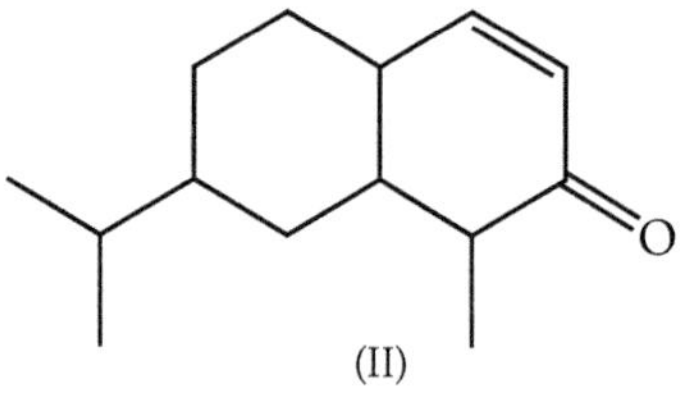   (II)

Infrared Spectroscopy

16.5 How can infrared spectroscopy be used to distinguish between propanone and 2-propen-l-ol?

CH_3—C—CH_3 (with O double bonded to C) CH_2=CH—CH_2OH
    propanone   2-propene-1-ol

16.6 How can infrared spectroscopy be used to distinguish between 1-pentyne and 2-pentyne?

$$CH_3—CH_2—CH_2—C≡CH \qquad CH_3—C≡C—CH_2—CH_3$$

1-pentyne   2-pentyne

16.7 How could infrared spectroscopy be used to distinguish between the following isomers?

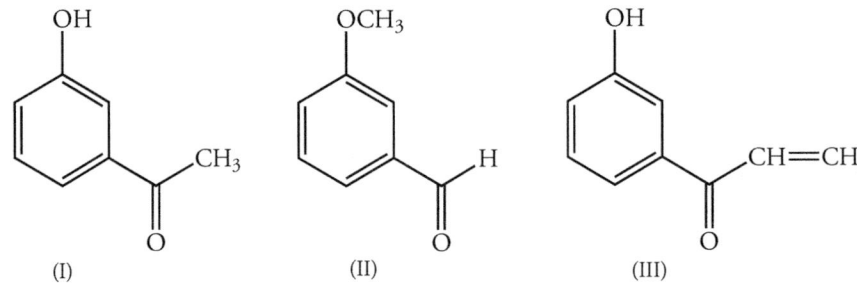

(I)   (II)   (III)   (IV)

16.8 How could infrared spectroscopy be used to distinguish between the following isomers?

(I)   (II)   (III)

16.9 An infrared spectrum of a compound with molecular formula $C_4H_6O_2$ has a intense broad band between 3500 and 3000 cm⁻¹ and an intense peak at 1710 cm⁻¹. Which of the following compounds best fits these data?

     (I) $CH_3CH_2CO_2CH_3$ (II) $CH_3CO_2CH_2CH_3$ (III) $CH_3CH_2CH_2CO_2H$

16.10 Dehydration of 1-methylcyclohexanol gives two isomeric alkenes. The minor product has a more intense C=C stretching vibration than the major product. Assign the structures of these compounds.

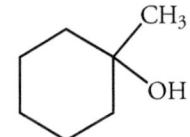

1-methylcyclohexanol

Proton Magnetic Resonance Spectroscopy

16.11 The NMR spectrum of a compound with molecular formula $C_3H_6Cl_2$ has only a singlet at 2.8 δ. What is the structure of the compound?

16.12 The NMR spectrum of a compound with molecular formula C_4H_9Br has only a singlet at 1.8 δ. What is the structure of the compound?

16.13 The NMR spectrum of a compound with the formula $C_5H_{11}Br$ consists of a singlet at 1.1 δ (9 H atoms) and a singlet at 3.2 δ (2 H atoms). What is the structure of the compound?

16.14 The NMR spectrum of a compound with the formula $C_7H_{15}Cl$ consists of a singlet at 1.1 δ (9 H atoms) and a singlet at 1.6 δ (6 H atoms). What is the structure of the compound?

16.15 The NMR spectrum of iodoethane has resonances at 1.9 δ and 3.1 δ. What are the intensities of the two resonances? What are the multiplicities?

16.16 The NMR spectrum of 1-chloropropane has resonances at 1.0 δ, 1.8 δ, and 3.5 δ. What are the intensities of the resonances? What are the multiplicities?

16.17 The NMR spectrum of an ether with molecular formula $C_5H_{12}O$ has singlets at 1.1 and 3.1 δ whose intensities are 3:1. What is the structure of the compound?

16.18 The NMR spectrum of an ether with molecular formula $C_6H_{14}O$ consists of an intense doublet at 1.0 δ and a septet at 3.6 δ. What is the structure of the ether?

16.19 Describe the differences in the NMR spectra of the following two esters:
(I) $CH_3CH_2CO_2CH_3$ (II) $CH_3CO_2CH_2CH_3$

16.20 Two isomeric esters having the molecular formula $C_5H_{10}O_2$ have different NMR spectra. One has the following characteristics: 1.0 δ (doublet, 6 H atoms), 2.0 δ (multiplet, 1 H atom), 4.0 δ (doublet, 2 H atoms), and 8.1 δ (singlet, 1 H atom). The other has the following characteristics: 1.2 δ (triplet, 3 H atoms), 1.4 δ (triplet, 3 H atoms), 2.5 δ (quartet, 2 H atoms), and 4.1 δ (quartet, 2 H atoms). What is the structure of each compound?

16.21 Determine the structure of a compound having the molecular formula C_4H_9Cl based on the following ^1H NMR spectrum.

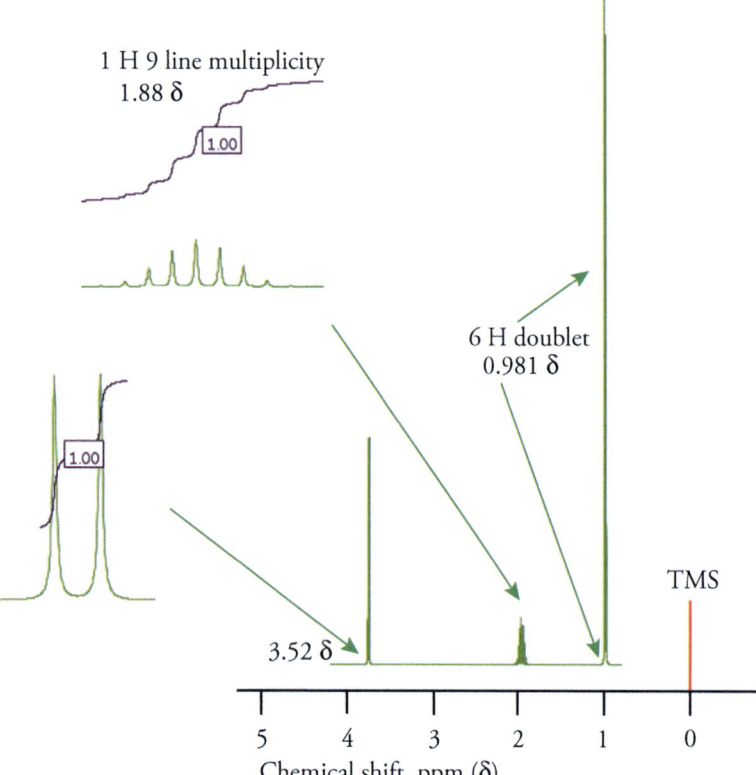

16.22 Determine the structure of a compound having the molecular formula $C_3H_6Cl_2$ based on the following ^1H NMR spectrum.

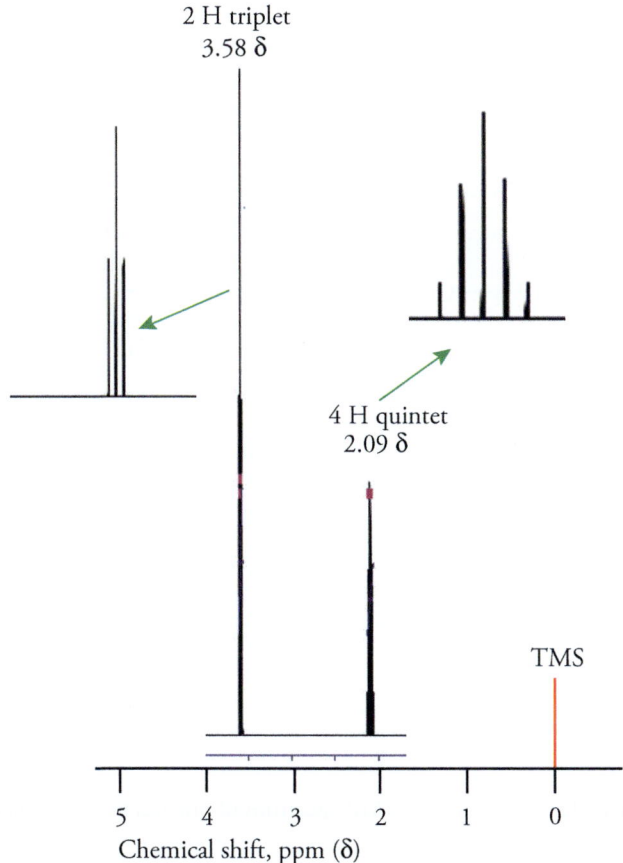

2 H triplet
3.58 δ

4 H quintet
2.09 δ

TMS

5   4   3   2   1   0
Chemical shift, ppm (δ)

Carbon-13 NMR

16.23 Determine the number of signals in the ^{13}C NMR spectrum of each of the following aromatic compounds:

(a) 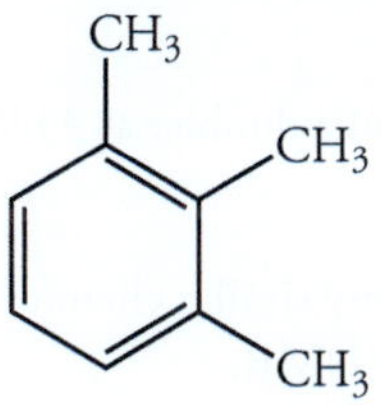 CH₃ CH₃ CH₃

(b) 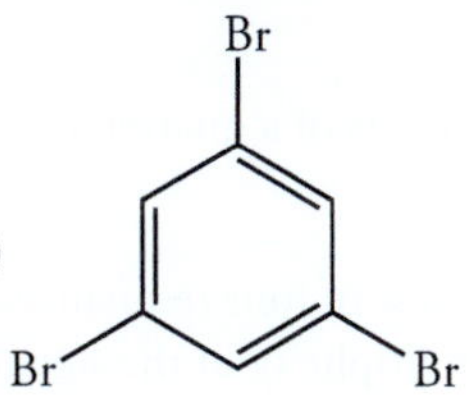 Br Br Br

(c) 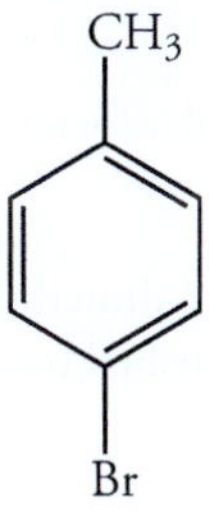 CH₃ Br

16.24 Consider the following isomeric hydrocarbons. One compound has a ^{13}C NMR spectrum with absorptions at 20.8, 30.7, and 31.4 δ. A second compound has a ^{13}C NMR spectrum with absorptions at 16.0, 23.6, 31.3, and 34.2 δ. A third compound has a ^{13}C NMR spectrum with absorptions at 23.0, 26.5, 32.9, 35.1, and 44.6 δ. Assign these features to the appropriate compound.

(I) 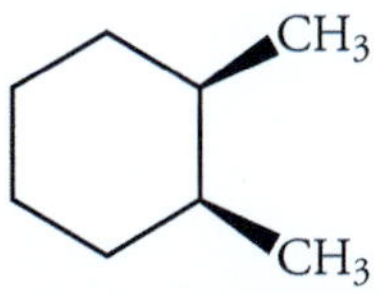 CH₃ CH₃

(II) 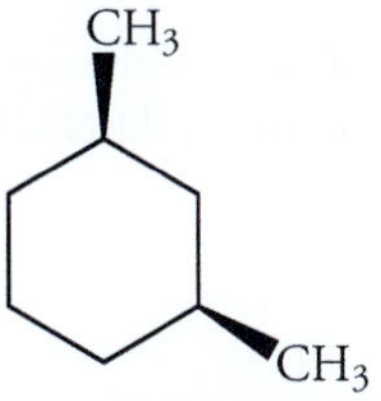 CH₃ CH₃

(III)  CH₃ Br

16.25 Determine the number of signals in the ^{13}C NMR spectrum of each of the following esters:

(a) $CH_3-CH_2-\overset{\displaystyle O}{\overset{\displaystyle \|}{C}}-O-CH_2-CH_3$ (b) $CH_3-\overset{\displaystyle O}{\overset{\displaystyle \|}{C}}-O-CH_2-CH_2-CH_3$ (c) $CH_3-\overset{\displaystyle O}{\overset{\displaystyle \|}{C}}-O-\overset{\displaystyle H}{\underset{\displaystyle CH_3}{\overset{\displaystyle |}{\underset{\displaystyle |}{C}}}}-CH_3$

16.26 Determine the number of signals in the ^{13}C NMR spectrum of each of the following alcohols:

(a) $CH_3-CH_2-\overset{\displaystyle OH}{\underset{\displaystyle H}{\overset{\displaystyle |}{\underset{\displaystyle |}{C}}}}-CH_2-CH_3$ (b) $CH_3-\overset{\displaystyle OH}{\underset{\displaystyle H}{\overset{\displaystyle |}{\underset{\displaystyle |}{C}}}}-CH_2-CH_2-CH_2-CH_3$ (c) $CH_3-\overset{\displaystyle H}{\underset{\displaystyle CH_3}{\overset{\displaystyle |}{\underset{\displaystyle |}{C}}}}-\overset{\displaystyle OH}{\underset{\displaystyle H}{\overset{\displaystyle |}{\underset{\displaystyle |}{C}}}}-\overset{\displaystyle H}{\underset{\displaystyle CH_3}{\overset{\displaystyle |}{\underset{\displaystyle |}{C}}}}-CH_3$

16.27 Dehydration of 1-methylcyclohexanol gives two isomeric alkenes. The ^{13}C NMR spectrum of one has five resonances and the other has seven resonances. Assign the structures of these compounds.

1-methylcyclohexanol

16.28 Hydration of 2-pentyne gives a mixture of two isomeric ketones. The ^{13}C NMR spectrum of one has five resonances and the other has three resonances. Assign the structures of these compounds.

$CH_3-C\equiv C-CH_2-CH_3$

2-pentyne

16.29 The ^{13}C NMR spectrum of a hydrocarbon with molecular formula C_4H_6 consists of a triplet at 30.2 δ and a doublet at 1.36 δ. What is its structure?

16.30 The ^{13}C NMR spectrum of a hydrocarbon with molecular formula C_6H_{14} consists of a quartet at 19.1 δ and a doublet at 33.9 δ. What is its structure?

16.31 The ^{13}C NMR spectra of both 3-methylpentane and 2,2-dimethylbutane consist of four resonances with very similar chemical shifts. Explain how the two compounds can be distinguished based on the multiplicity of the signals in ^{1}H NMR.

$CH_3-CH_2-\overset{\displaystyle CH_3}{\underset{\displaystyle H}{\overset{\displaystyle |}{\underset{\displaystyle |}{C}}}}-CH_2-CH_3$ $CH_3-\overset{\displaystyle CH_3}{\underset{\displaystyle CH_3}{\overset{\displaystyle |}{\underset{\displaystyle |}{C}}}}-CH_2-CH_3$

3-methylpentane   2,2-dimethylbutane

16.32 The ^{13}C NMR spectra of both 1-butanol and 2-butanol consist of four resonances with very similar chemical shifts. Explain how the two compounds can be distinguished based on the multiplicity of the signals in ^{1}H NMR.

$CH_3-CH_2-CH_2-CH_2OH$ $CH_3-CH_2-\overset{\displaystyle OH}{\underset{\displaystyle H}{\overset{\displaystyle |}{\underset{\displaystyle |}{C}}}}-CH_3$

1-butanol   2-butanol

Solutions to In-Chapter Problems

Chapter 1

1.5 Resonance Structures

Problem 1.2

Nitrites (NO_2^-) are added as antioxidants in some processed meats. Write resonance structures for the nitrite ion.

1.6 Predicting the Shapes of Simple Molecules

Problem 1.4

Using one of the resonance forms for the nitrite ion (NO_2^-), determine the shape of this ion.

Solution

trigonal planar geometry

1.9 Structural Formulas

Problem 1.6

Hexamethylenediamine, a compound used to produce nylon, has the following structural formula. Write three condensed structural formulas for it.

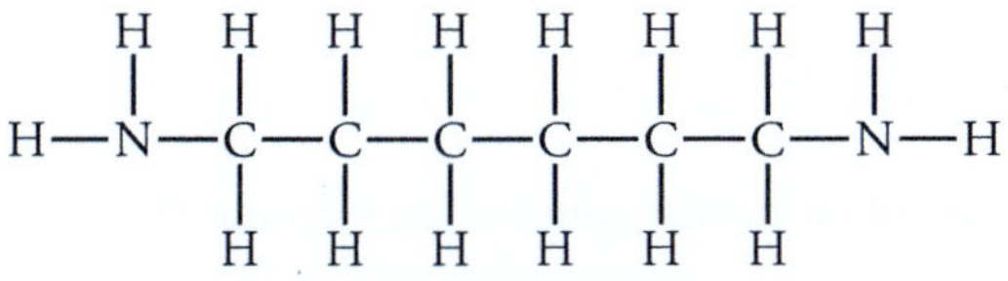

hexamethylenediamine

Solution

With the C—H bonds understood, we write only the C—C bonds.

$$NH_2—CH_2—CH_2—CH_2—CH_2—CH_2—CH_2—CH_2—NH_2$$

With both the C—H and C—C bonds understood, we write,

$$NH_2CH_2CH_2CH_2CH_2CH_2CH_2CH_2NH_2$$

The above structure can be condensed still further to give the fully condensed formula.

$$NH_2(CH_2)_6NH_2$$

Problem 1.8

What is the molecular formula of indoleacetic acid, a plant growth hormone that promotes shoot growth?

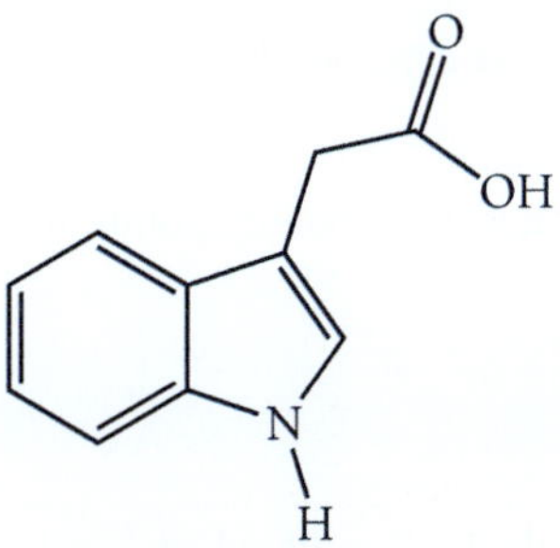

indoleacetic acid

Principles of Organic Chemistry. http://dx.doi.org/10.1016/B978-0-12-802444-7.10000-X

Solution

When we draw the complete structure we see that there are 10 carbon atoms, 1 nitrogen atom, 2 oxygen atoms, and 10 hydrogen atoms. The formula of indoleacetic acid is $C_{10}H_9NO_2$.

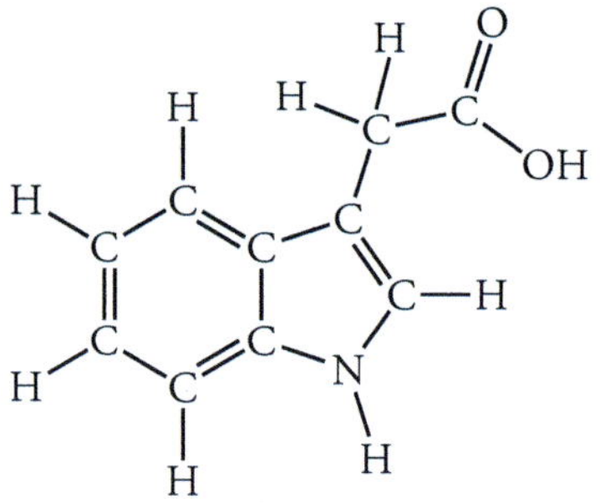

1.10 Isomers

Problem 1.10

Compare the following structures of two intermediates in the metabolism of glucose. Are they isomers? How do they differ?

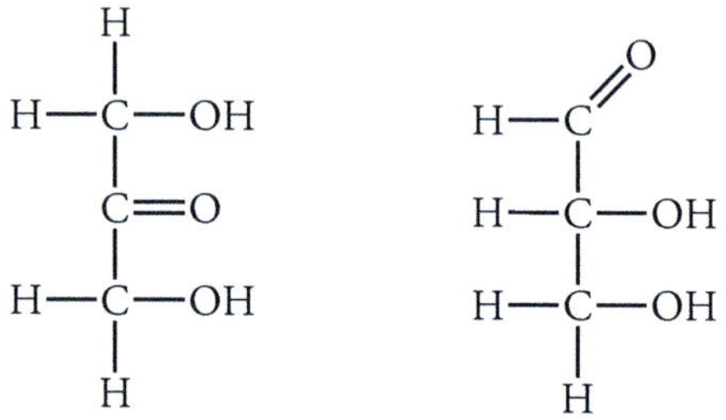

Solution

The compounds are isomers. Although both compounds have carbonyl groups, the compound on the left contains keto- group; the compound on the right contains an aldehyde group, abbreviated (CHO).

CHAPTER 2

2.1 Structure and Physical Properties

Problem 2.2

The boiling points of CCl_4 and $CHCl_3$ are 77 and 62 °C, respectively. Which compound is more polar? Is the polarity consistent with the boiling points? Why or why not?

Solution

(a) $CHCl_3$ is more polar than CCl_4. (b-c) Based upon polarity we might expect $CHCl_3$ to be more polar, but its molecular weight is much higher than that of CCl_4, and that is why it has a higher boiling point.

Problem 2.4

Explain why the boiling points of ethanethiol and dimethyl sulfide are very similar.

$$CH_3-CH_2-SH \qquad\qquad CH_3-O-CH_3$$

   ethanethiol, bp 35 °C   dimethyl sullfide, bp 37 °C

Solution

The molecular weights of these compounds are similar and neither can form intermolecular hydrogen bonds. Thus, their boiling points are similar.

2.3 Acid-Base Reactions

Problem 2.6

Consider the reaction of Br⁻ with ethene to give a charged intermediate called a carbocation. Which reactant is the Lewis acid, which one is the Lewis base?

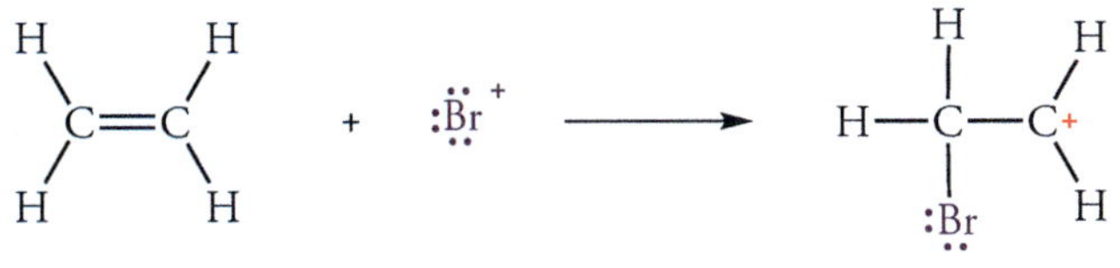

a carbocation

Solution

The positively charged bromine atom acts as a Lewis acid and the π bond of the alkene acts as a Lewis base in this reaction.

2.4 Oxidation-Reduction Reactions

Problem 2.8

Is the following process an oxidation or a reduction reaction?

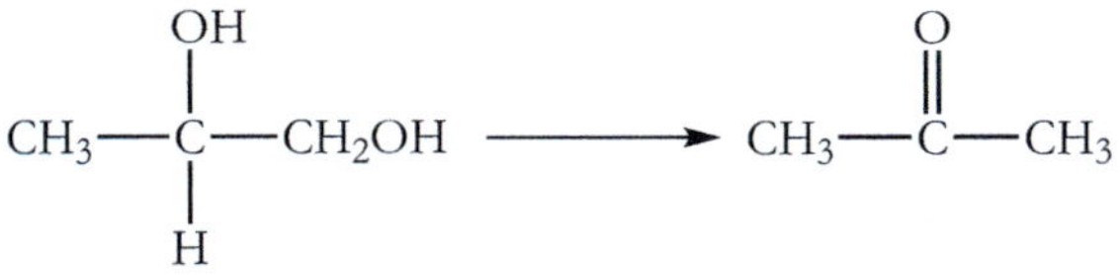

Solution

The reactant contains two hydroxyl groups. The one on the left is oxidized to a keto group; the one on the right is reduced to a methyl group. Therefore, there is no *net* oxidation or reduction in this reaction.

2.5 Classification of Organic Reactions

Problem 2.10

Classify the following reaction.

Solution

This is a molecular rearrangement in which the alkene double bond becomes a single bond, and the hydroxyl group becomes an carbonyl group.

2.8 Effect of Structure on Acidity

Problem 2.12

The pK_a value of the C—H bond of nitromethane is 10.2, whereas the pK_a value of methane is approximately 49. Explain why nitromethane is so much more acidic.

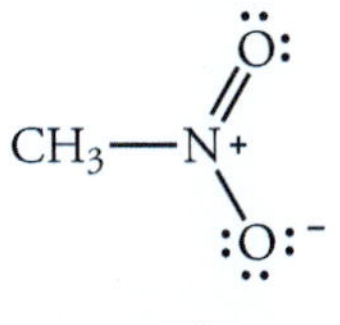

nitromethane

Solution

The nitro group is much more electron-withdrawing than a hydrogen atom, but even more importantly, the anion that results from the ionization of nitromethane is resonance-stabilized.

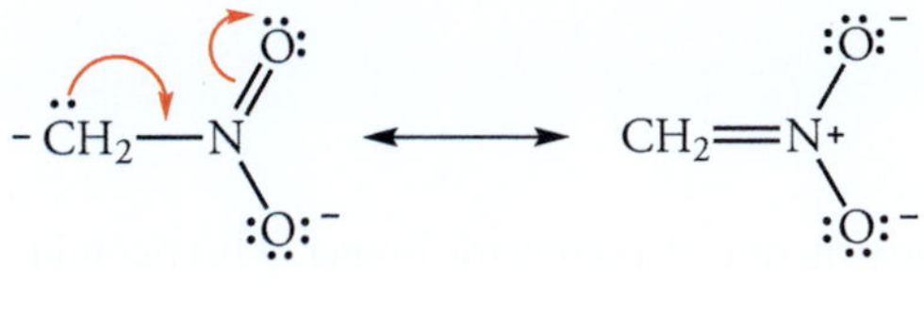

CHAPTER 3

3.2 Alkanes

Problem 3.2

Pentaerythritol tetranitrate is a drug used to reduce the frequency and severity of angina attacks. Classify the carbon atoms in this compound.

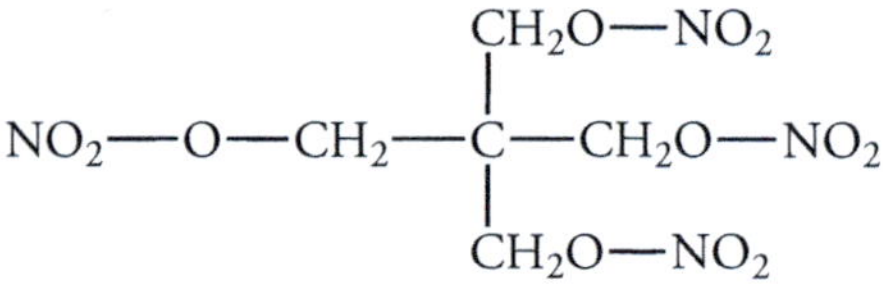

$$CH_2O—NO_2$$
$$NO_2—O—CH_2—\overset{\displaystyle CH_2O—NO_2}{\underset{\displaystyle CH_2O—NO_2}{C}}—CH_2O—NO_2$$

pentaerythritol tetranitrate

Solution

The central carbon is quaternary; all the others are primary.

3.3 Nomenclature of Alkanes

Problem 3.4

Name the following compound.

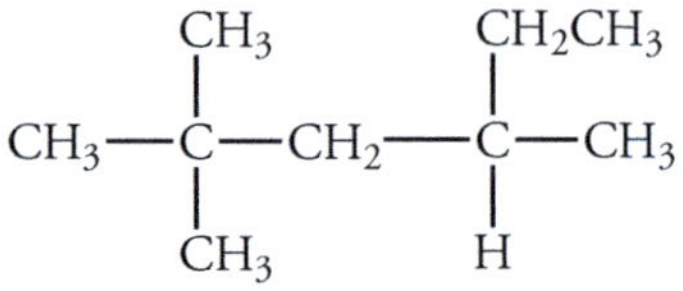

$$CH_3—\overset{\displaystyle CH_3}{\underset{\displaystyle CH_3}{C}}—CH_2—\overset{\displaystyle CH_2CH_3}{\underset{\displaystyle H}{C}}—CH_3$$

Solution

The parent compound contains six carbons, so the molecule is a methyl-substituted hexane. There are two methyl groups at the first branch point, one on the left, at C-2. Numbering from the left, we find another methyl group at C-4. The compound is 2,2,4-trimethylhexane.

3.5 Cycloalkanes

Problem 3.7

Determine the molecular formula of menthol based on the bond-line structure shown below.

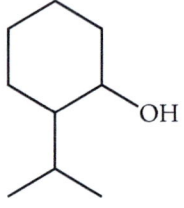

menthol

Solution

The molecular formula of menthol is $C_9H_{18}O$.

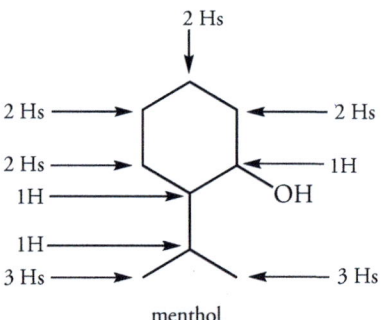

menthol

Problem 3.9

Brevicomin, the sex attractant of a species of pine beetle, has the following structure. Write the structure of a geometric isomer of brevicomin.

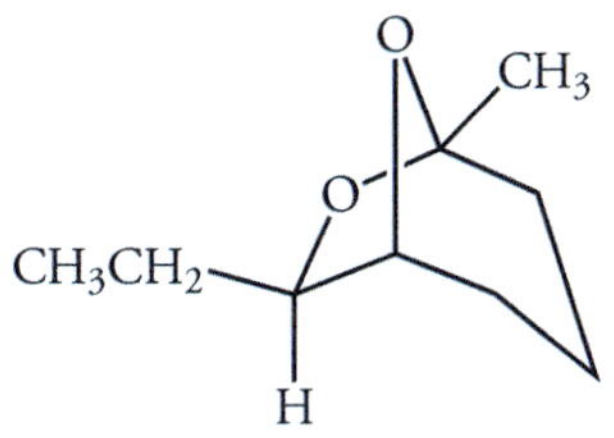

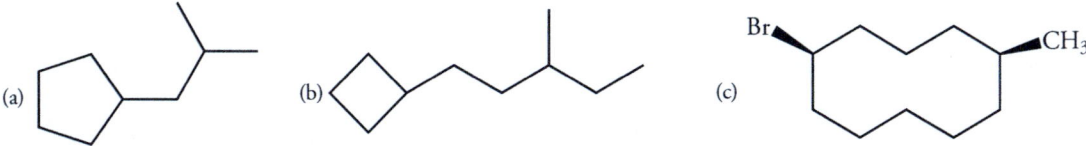

brevicomin

Solution

Brevicomin has two alkyl substituents. The methyl group is at a "bridge-head," and no isomers can exist at this point. However, the ethyl group can "exchange places" with the hydrogen atom, giving a different geometrical isomer, as shown below.

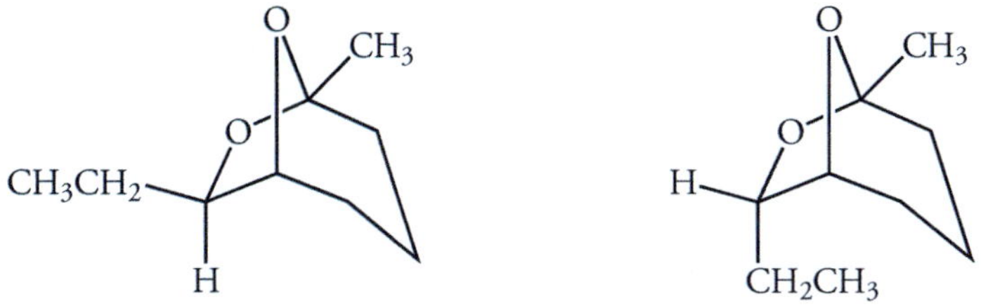

geomerical isomers of brevicomin

Problem 3.11

What are the names of the following compounds?

(a) (b) (c)

Solution

Compound (a) is isobutylcyclopentane; compound (b) is 1-cyclobutyl-3-methylpentane; compound (c) is *cis*-1-bromo-5-methylcyclodecane.

CHAPTER 4

4.1 Unsaturated Hydrocarbons

Problem 4.2

β-Carotene, found in carrots, has 40 carbon atoms and contains two rings and 11 double bonds. What is its molecular formula?

Solution

For n = 40, the number of hydrogen atoms for a saturated compound without rings is 82. Each ring and each double bond results in a reduction of two hydrogen atoms. Thus the total number of hydrogen atoms is:

Number of hydrogen atoms = 82 – [2(no. of rings)] – [2(no. of double bonds)] = 82 – [2(2)] – [2(11)] = 58

The molecular formula is $C_{40}H_{58}$.

Problem 4.4

Tremorine is used to treat Parkinson's disease. Classify this alkyne.

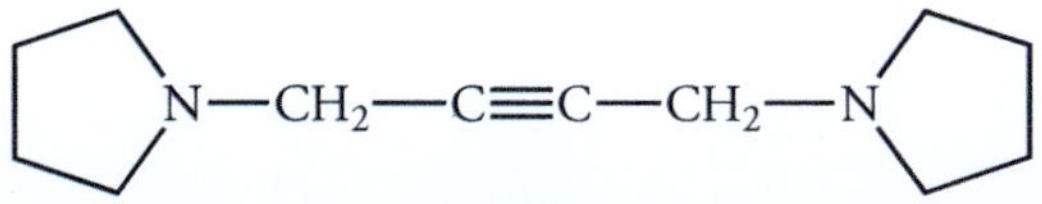

tremorine

Solution

Tremorine is a disubstituted alkyne.

4.2 Geometric Isomerism

Problem 4.6
Is *cis-trans* isomerism possible about either of the double bonds of bombykol, a pheromone secreted by the female silkworm moth?

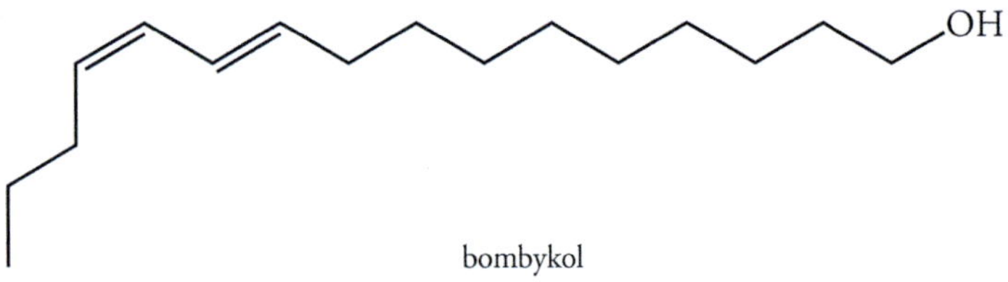

bombykol

Solution
The two double bonds of have different substituents, and each can have a *cis-* or a *trans-*configuration. Therefore, four geometrical isomers are possible. Of these, the one shown above is naturally occurring. The IUPAC name of bombykol is (10E,12Z)-hexadeca-10,12-dien-1-ol.

4.3 E,Z Nomenclature of Geometrical Isomers

Problem 4.8
Assign the E or Z configuration to tamoxifen, a drug used in the treatment of breast cancer.

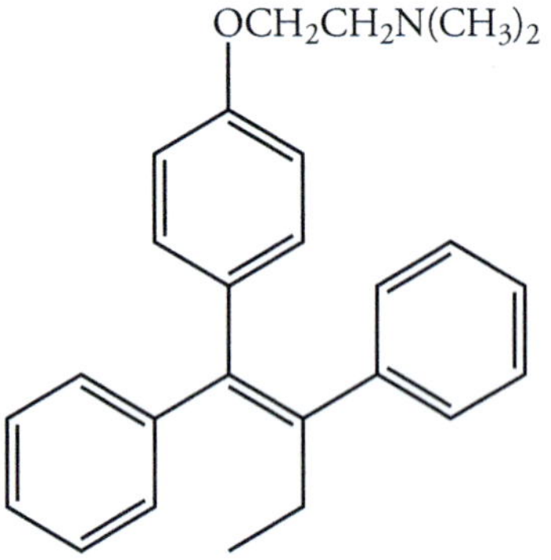

$OCH_2CH_2N(CH_3)_2$

tamoxifen

Solution
Since, the highest priority groups are on the same side of the double bond, tamoxifen has a Z configuration.

4.4 Nomenclature of Alkenes and Alkynes

Problem 4.10
Draw the structures of the following isomeric compounds.

(a) 5-methyl-1,3-cyclohexadiene   (b) 3-methyl-1,4-cyclohexadiene

Solution

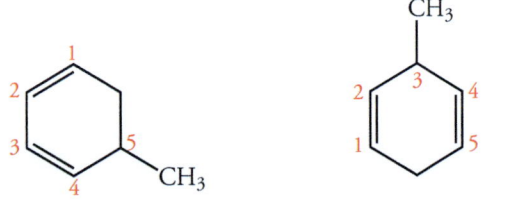

(a) 5-methyl-1,3-cyclohexadiene   (b) 3-methyl-1,4-cyclohexadiene

Problem 4.12

(3E,11E)-1,3,11-Tridecatriene-5,7,9-triyne is a compound found in safflowers that is a chemical defense against nematode infestations. Write the structure of the compound.

Solution

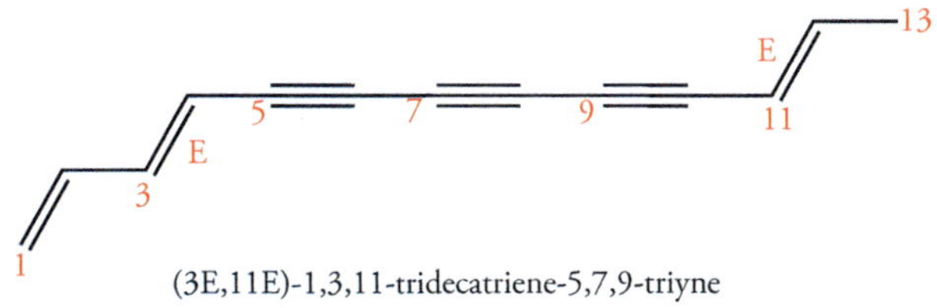

1
3
E
5
7
9
11
E
13

(3E,11E)-1,3,11-tridecatriene-5,7,9-triyne

4.6 Hydrogenation of Alkenes and Alkynes

Problem 4.14

Write the structure obtained by complete hydrogenation of ipsdienol, a pheromone of the Norwegian spruce beetle.

HO

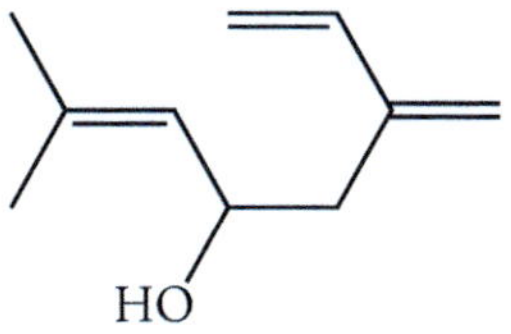

ispdienol

Solution

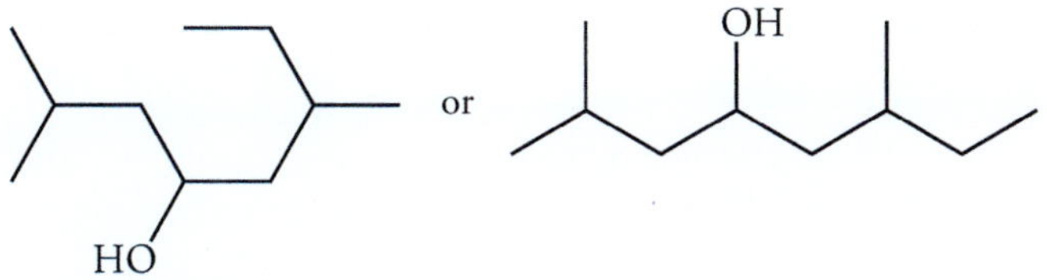

HO
or
OH

Problem 4.16

(E)-11-Tetradecen-1-ol is an intermediate required to synthesize the sex attractant of the spruce bud-worm moth. How can this compound be prepared from an alkyne? Suggest a name for the alkyne. (The *-ol* ending refers to the hydroxyl group, and the alkyne name must contain the *-ol*.)

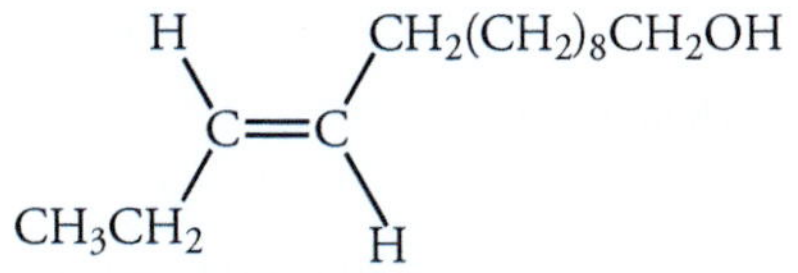

H
CH2(CH2)8CH2OH
C=C
CH3CH2
H

(E)-11-tetradecen-1-ol

Solution

A *trans*-double bond can be produced from an alkyne by treating the alkyne with sodium amide in liquid ammonia in the presence of sodium metal. The name of the starting alkyne is 11-tetradecyn-1-ol.

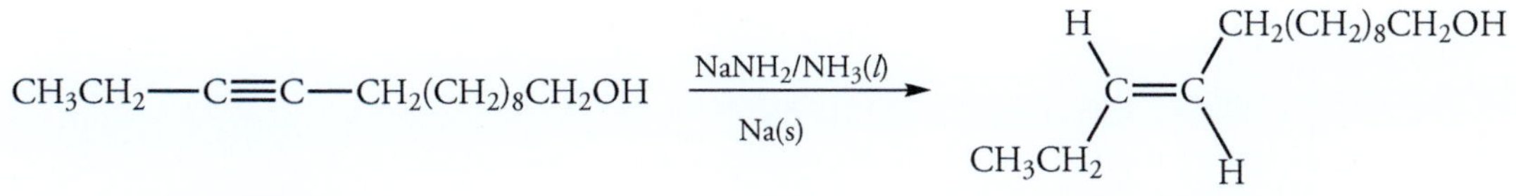

CH3CH2—C≡C—CH2(CH2)8CH2OH
NaNH2/NH3(l)
Na(s)
H
CH2(CH2)8CH2OH
C=C
CH3CH2
H

11-tetradecyn-1-ol

(E)-11-tetradecen-1-ol

4.8 Addition Reactions of Alkenes and Alkynes

Problem 4.18
Predict the product of the addition of HCl to 1-methylcyclohexene.

Solution
The hydrogen adds to the carbon atom with the most directly bonded hydrogens.

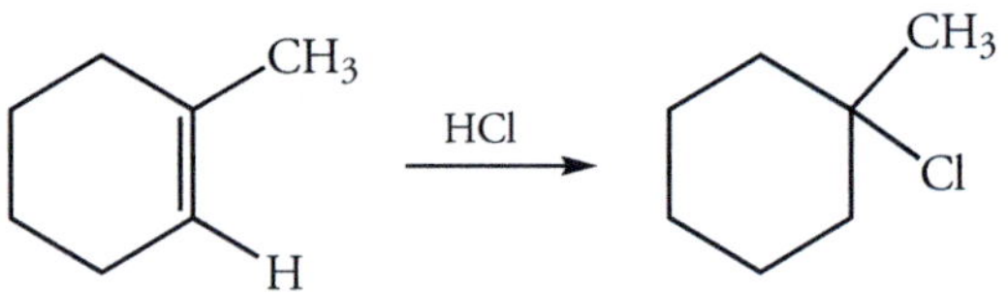

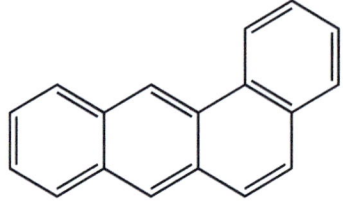

4.10 Hydration of Alkenes and Alkynes

Problem 4.20
What product results from the hydration of 1-methylcyclopentene?

Solution
The hydrogen adds to the carbon atom with the most directly bonded hydrogens.

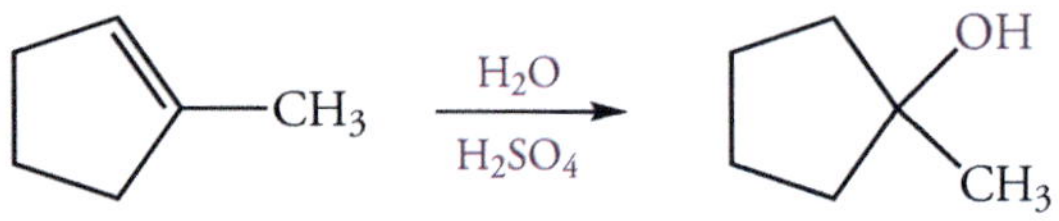

CHAPTER 5

5.2 Aromaticity

Problem 5.2
1,2-Benzanthracene is a carcinogen present as a combustion product in tobacco smoke. What is its molecular formula? Is the compound aromatic based on the Hückel rule?

benzanthracene

Solution
The molecular formula of 1,2-benzanthracene is $C_{12}H_{18}$. It is aromatic based on the $4n+2$ Hückel rule, where $n = 4$.

Problem 5.4
Adenine is present in the nucleic acids (DNA and RNA). Which nitrogen atoms in the rings have nonbonding electrons in sp^2 hybrid orbitals? How many electrons does each nitrogen atom contribute to the aromatic π system?

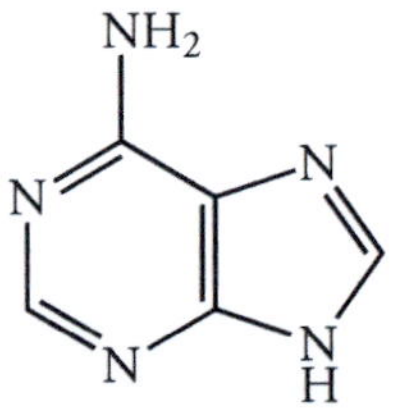

adenine
(6-aminopurine)

Solution

The structure shown below indicates which nitrogen atoms contribute to the aromatic system:

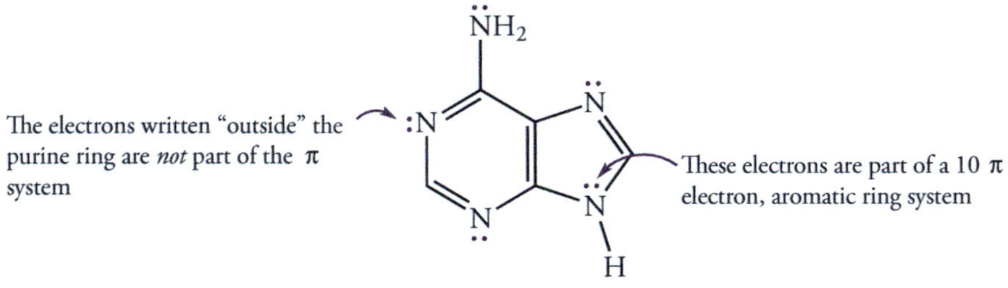

The electrons written "outside" the purine ring are *not* part of the π system

These electrons are part of a 10 π electron, aromatic ring system

adenine
(6-aminopurine)

5.3 Nomenclature of Aromatic Compounds

Problem 5.6
Name the following compound, which is used to make a local anesthetic.

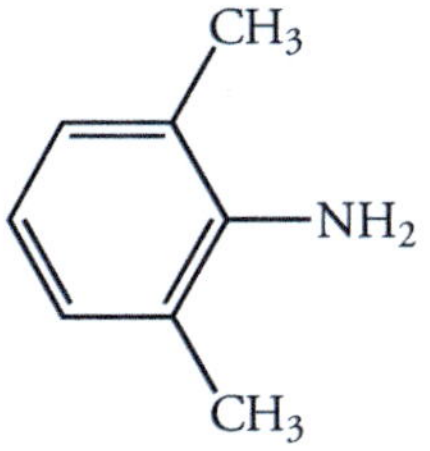

Solution
The parent compound is aniline; the name of the substituted aniline is 2,6-dimethylaniline.

Problem 5.8
Name the following compound as a substituted pyridine.

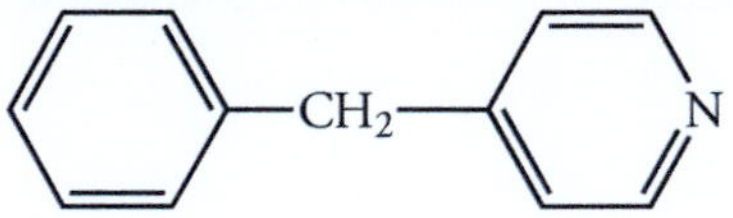

Solution
The compound has a benzyl substituent at the para position of the pyridine ring. The name of the compound is *p*-benzylpyridine.

5.5 Structural Effects in Electrophilic Aromatic Substitution

Problem 5.10
Predict whether the following compounds react faster or slower than benzene in a reaction with bromine and $FeBr_3$.

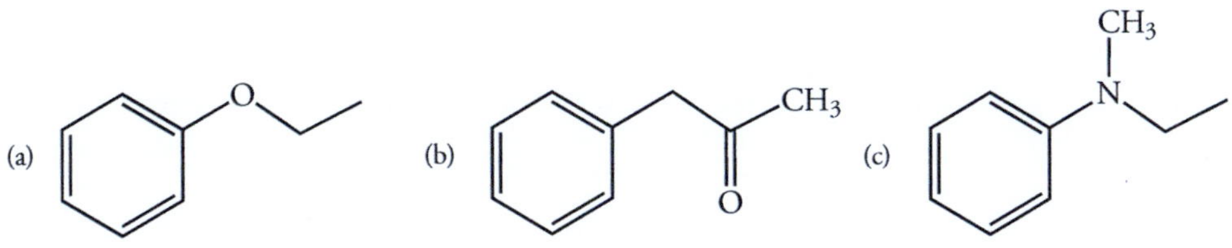

(a)   (b)   (c)

Solution
Compound (a) has an electron releasing substituent, thus the rate of electrophilic aromatic substitution is faster than that of benzene. Compound (b) has a methylene group bound to the ring, but the carbonyl group is electron withdrawing substituent, and electrophilic aromatic substitution occurs at a slightly slower rate than that of benzene. Compound (c) has an electron releasing substituent, so electrophilic aromatic substitution occurs at a faster rate than that of benzene.

Problem 5.12
Predict the structure of the product(s) formed in the bromination of each of the following compounds.

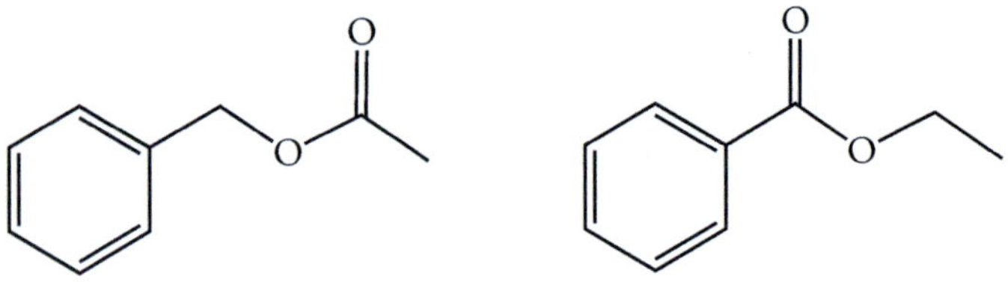

Solution
Compound (c) has a methylene group bound to the ring, but the carboxylate group is electron withdrawing substituent, so electrophilic aromatic substitution occurs at a slightly slower rate than that of benzene. Compound (c) has an electron withdrawing substituent, and the rate of electrophilic aromatic substitution is slower than that of benzene.

5.8 Reactions of Side Chains

Problem 5.14
A compound with molecular formula $C_{10}H_{14}$ is oxidized to give the following dicarboxylic acid. What are the structures of two possible compounds that would give this result?

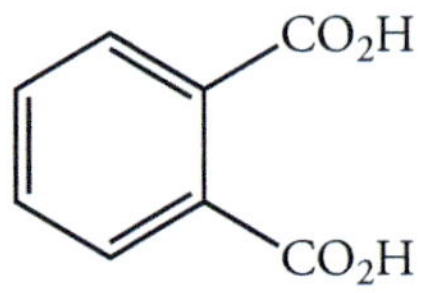

phthalic acid

Solution
Compounds with alkyl groups ortho to each other or with a fused ring bound to benzene will give this product. Two examples are given below.

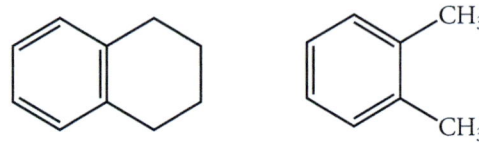

5.10 Synthesis of Substituted Aromatic Compounds

Problem 5.16
Propose a method to synthesize *p*-nitrobenzoic acid starting from benzene.

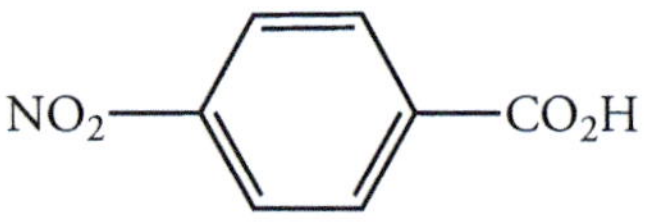

NO_2—⟨benzene ring⟩—CO_2H

p-nitrobenzoic acid

Solution
Both substituents in *p*-nitrobenzoic acid are *meta*- directors, but they are *para*- to each other! We can circumvent is apparent difficulty by making toluene first by Friedel-Crafts alkylation. Nitration of toluene gives p-nitrotoluene. But don't let this reaction go for too long or the explosive compound 2,4,6-trinitrotoluene (TNT) will form. Oxidation of *p*-nitrotoluene with $KMnO_4$ gives the desired product.

CHAPTER 6

6.2 Mirror Images and Chirality

Problem 6.2

The structure of nicotine is shown below. Is nicotine chiral?

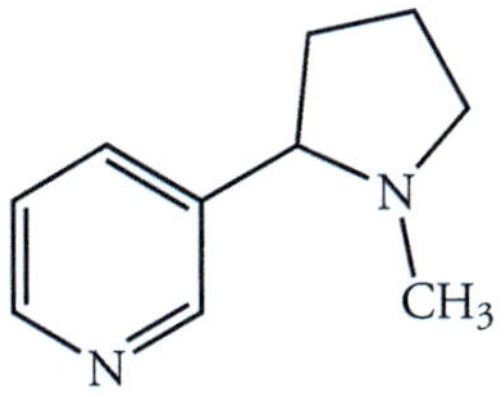

nicotine

Solution

Nicotine is a chiral compound with one stereogenic center. It can exist as either of two enantiomers, as shown below. The compound on the right is the naturally occurring, biologically active stereoisomer.

nicotine

6.5 Absolute Configuration

Problem 6.4

Arrange the groups at the stereogenic center of baclofen, an antispastic drug, in order from low to high priority.

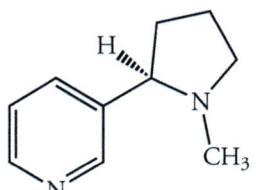

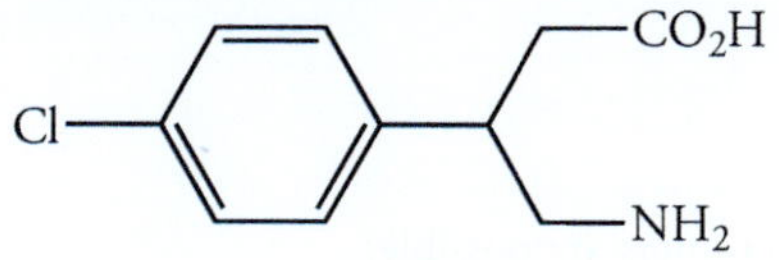

baclofen

Solution

The priority rankings of the groups bound to the stereogenic center are shown below.

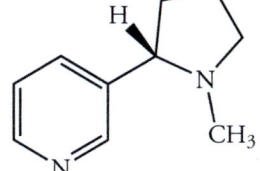

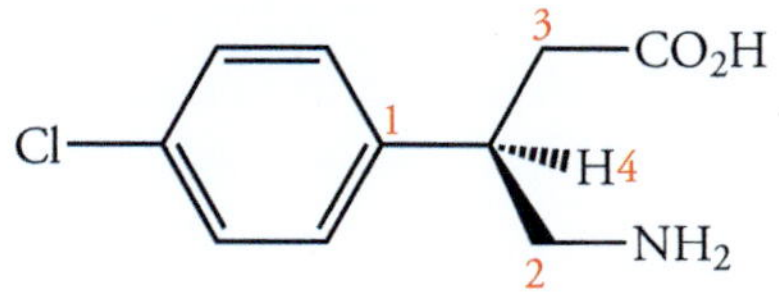

baclofen

Problem 6.6

What is the configuration of epinephrine, commonly known as adrenaline?

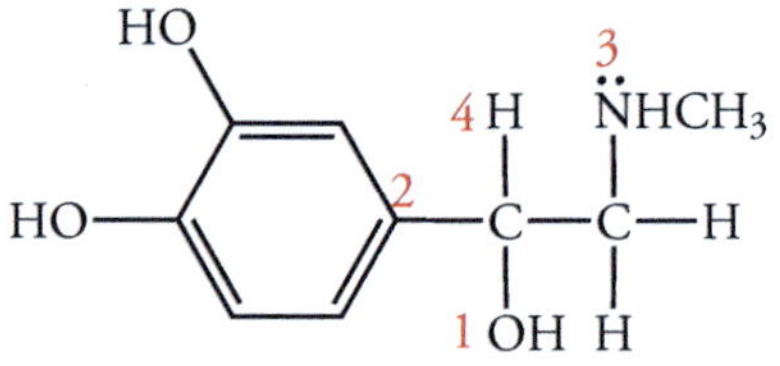

epinephrine

Solution

The priority rankings of the groups bound to the stereogenic center are shown above. The phenyl group has a higher priority ranking than the methylene group. The configuration of epinephrine is R.

6.6 Molecules with Multiple Stereogenic Centers

Problem 6.8

Write the Fischer projection formulas of the stereoisomeric 2,3-dibromobutanes. What relationships should exist between the optical activities of these isomers?

Solution

The meso compound has a plane of symmetry and is optically inactive. The (2R,3S) and (2S,3R) isomers are enantiomers with equal and opposite optical rotations.

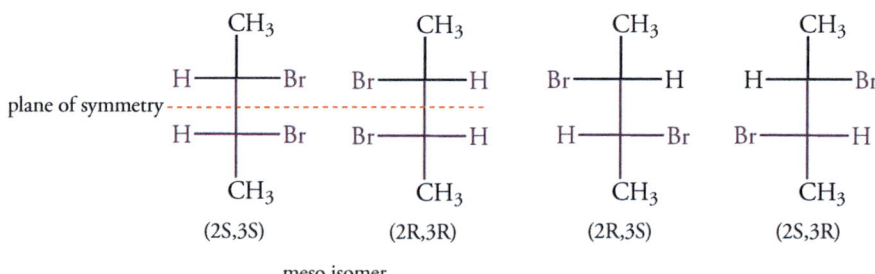

plane of symmetry

(2S,3S)   (2R,3R)   (2R,3S)   (2S,3R)

meso isomer

Problem 6.10

Determine the number of stereogenic centers in nootkatone, found in grapefruit oil. How many stereoisomers are possible?

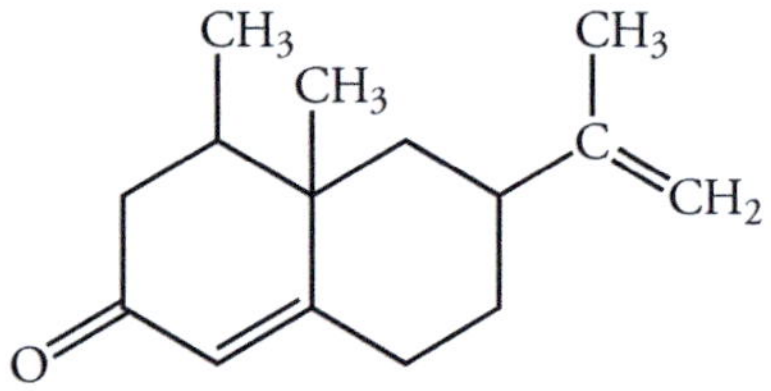

Solution

Nootkatone has three nonequivalent stereogenic centers. Therefore, eight diastereomers are possible. The naturally occurring isomer is shown below.

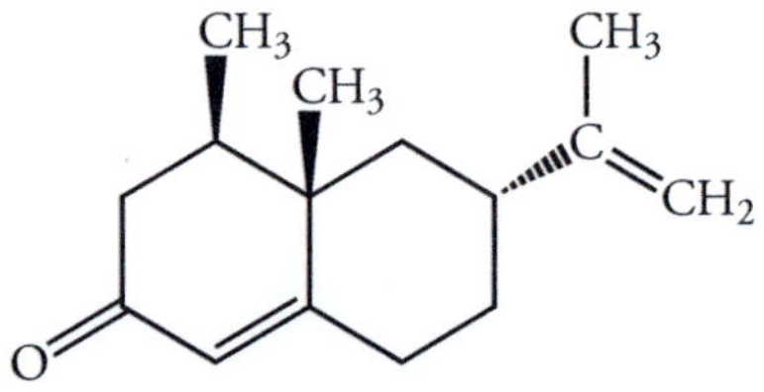

(+)-nootkatone

6.7 Synthesis of Stereoisomers

Problem 6.12

Free radical chlorination of (S)-2-bromobutane yields a mixture of compounds with chlorine substituted at any of the four carbon atoms. Write the structure of the 2-bromo-1-chlorobutane formed. Determine the configuration(s) of the stereogenic center(s). Is the product optically active?

Solution

Free radical chlorination of (S)-2-bromobutane at C-1 yields a mixture of two compounds. Only C-3 is a stereogenic center. Therefore, a racemic mixture of two products forms.

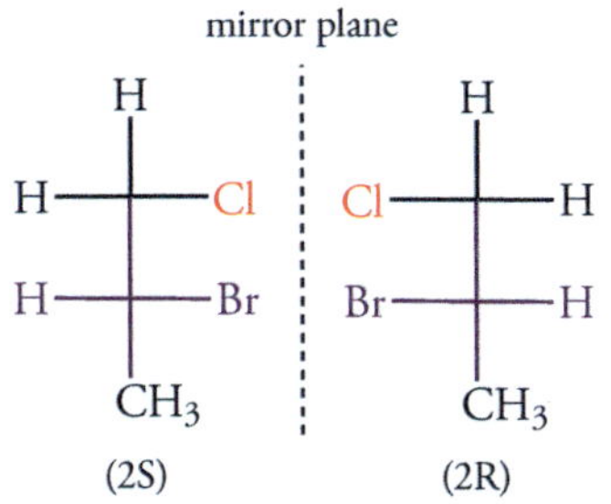

6.8 Reactions That Produce Stereogenic Centers

Problem 6.13

Sodium borohydride ($NaBH_4$) reacts with the C-2 carbonyl carbon atom of pyruvic acid to give lactic acid. What is the optical rotation of the product(s)?

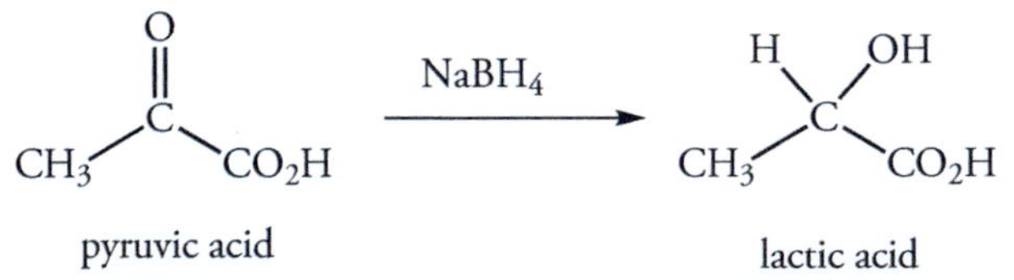

Solution

The carbonyl group of pyruvic acid is sp^2-hybridized, and is planar. Borohydride can attack from either the top or the bottom with equal probability. Thus an optically inactive racemic mixture is produced.

Problem 6.14

Reduction of pyruvic acid by NADH using the liver enzyme lactate dehydrogenase yields exclusively (S)-lactic acid. Write the Fischer projection of this product. Why does only a single product form?

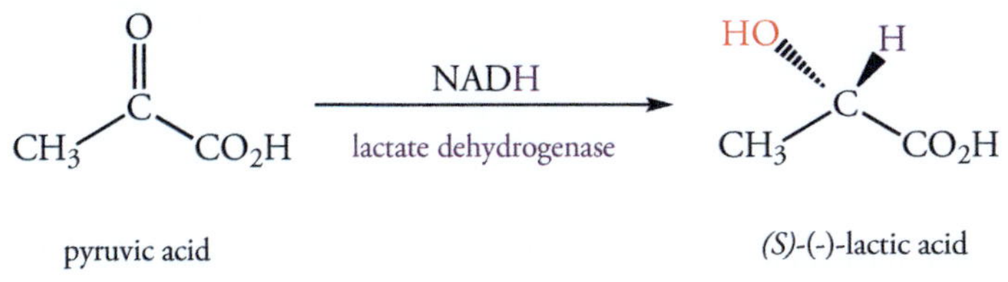

Solution

The active site of lactate dehydrogenase is chiral, and the planer substrate, pyruvic acid, can only bind in one way, and the hydride ion transfer reaction can only occur from one side. As a result, only one stereoisomer can form, and the product is chiral,

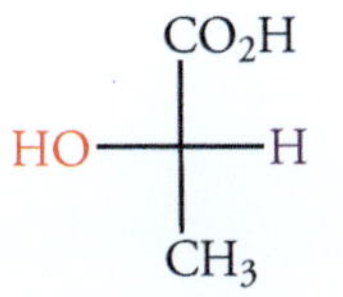

(S)-(-)-lactic acid

6.9 Reactions That Form Diastereomers

Problem 6.16

Write the structure of the oxirane (epoxide) that forms when (Z)-2-butene reacts with *m*-chloroperbenzoic acid (mCPBA). Assign the configurations of the stereogenic centers.

Solution

The epoxidation reaction occurs with retention of configuration, so the product is an optically inactive meso compound,

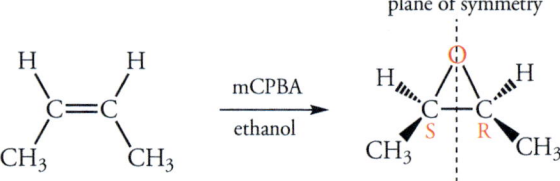

plane of symmetry

CHAPTER 7

7.1 Reaction Mechanisms and Haloalkanes

Problem 7.2

Name the following compound.

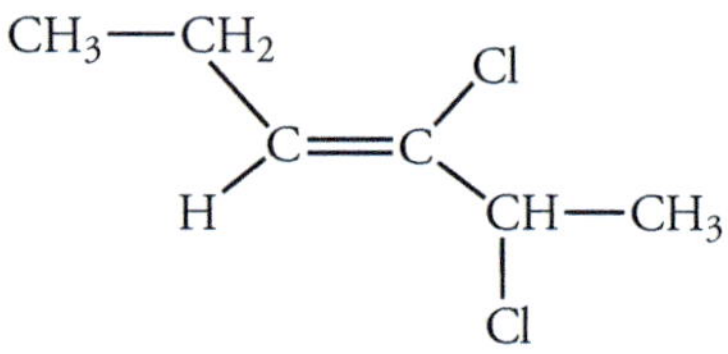

$$CH_3—CH_2 \quad Cl$$
$$C=C$$
$$H \quad CH—CH_3$$
$$\qquad\qquad Cl$$

Solution

The compound is (Z)-2,3-dichloro-3-hexene. Note that Cl— has a higher priority than —CHClCH₃.

7.2 Nucleophilic Substitution Reactions

Problem 7.4

Propose two ways to prepare 2-pentyne using nucleophilic substitution reactions.

Solution

2-Pentyne can be made in the two ways shown below.

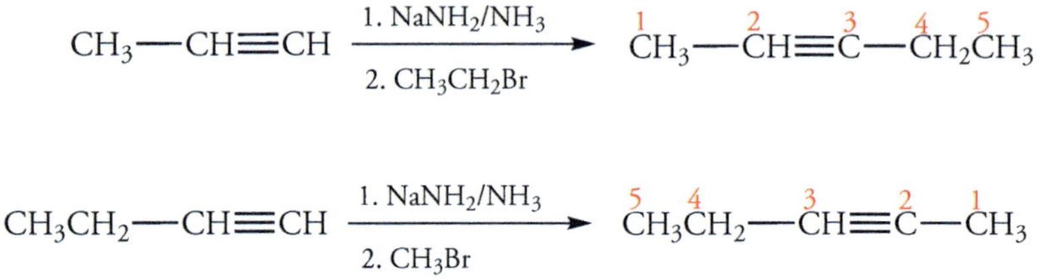

$$CH_3—CH\equiv CH \xrightarrow[\text{2. } CH_3CH_2Br]{\text{1. } NaNH_2/NH_3} \overset{1}{CH_3}—\overset{2}{CH}\equiv\overset{3}{C}—\overset{4}{CH_2}\overset{5}{CH_3}$$

$$CH_3CH_2—CH\equiv CH \xrightarrow[\text{2. } CH_3Br]{\text{1. } NaNH_2/NH_3} \overset{5}{CH_3}\overset{4}{CH_2}—\overset{3}{CH}\equiv\overset{2}{C}—\overset{1}{CH_3}$$

7.5 S$_N$2 Versus S$_N$1 Reactions

Problem 7.6

The relative rates of reaction of 1-iodobutane with a chloride ion in methanol, formamide, and dimethylformamide are 1, 12, and 1.2×10^6, respectively. Explain the relatively small difference in rate between methanol and formamide and the large difference in rate between formamide and dimethylformamide.

methanol formamide dimethylformamide

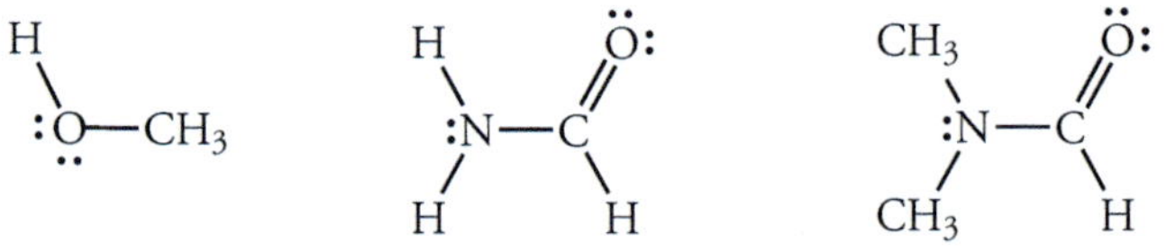

Solution

This is an S$_N$2 reaction, which is strongly favored in polar, aprotic solvents such as N,N-dimethylformamide.

7.7 Effect of Structure on Competing Reactions

Problem 7.8

The amount of elimination product for the reaction of 1-bromodecane with an alkoxide in the corresponding alcohol solvent is about 1% for the methoxide ion and 85% for the *tert*-butoxide ion. Explain these data.

Solution

Since 1-bromdecane is a primary haloalkane, it will react with methoxide ion by an S_N2 mechanism. However, the sterically hindered, strongly basic *tert*-butoxide ion abstracts a β hydrogen instead of displacing bromide, and an E2 reaction occurs.

CHAPTER 8

8.1 The Hydroxyl Group

Problem 8.2

Classify the alcohol functional groups in riboflavin (vitamin B₂).

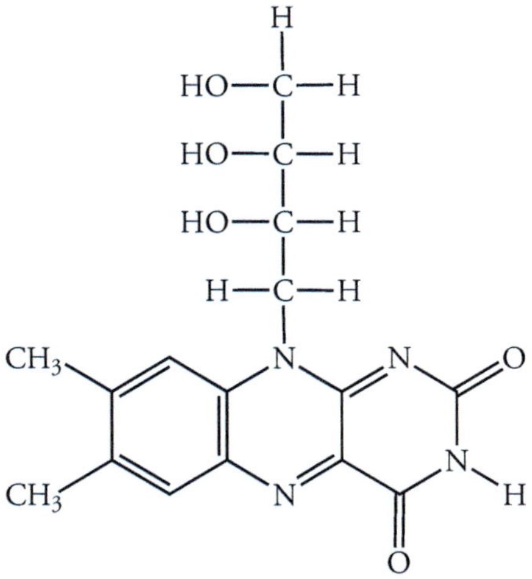

riboflavin (vitamin B₂)

Solution

The —OH group at the top of the above structure is primary; the other two are secondary.

Problem 8.4

Assign the IUPAC name for citronellol, a compound found in geranium oil that is used in perfumes.

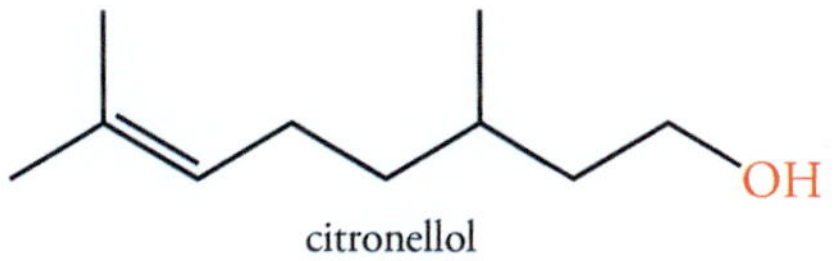

citronellol

Solution

The parent name of citronellol contains both an alcohol group and an alkenyl group. The alcohol takes precedence in the IUPAC numbering system, so the name is 3,7-dimethy-6-octen-1-ol. The structure of the S isomer of this compound, which is a terpene, is shown below,

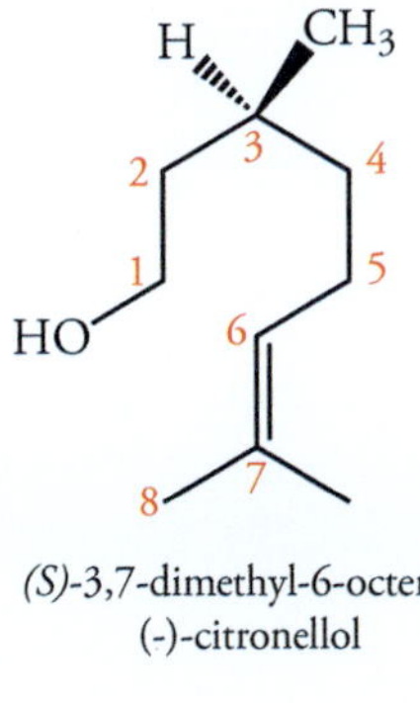

(S)-3,7-dimethyl-6-octene-1-ol
(-)-citronellol

8.5 Dehydration of Alcohols

Problem 8.6

Write the structures of the products of dehydration of 4-methyl-2-pentanol. Which of the isomeric alkenes should be the major product?

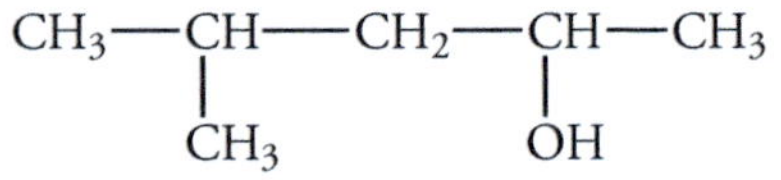

$$CH_3-CH-CH_2-CH-CH_3$$
$$\quad\quad\;\;|\quad\quad\quad\quad\;|$$
$$\quad\quad\;CH_3\quad\quad\;OH$$

Solution

Dehydration can occur with loss of water between C-2 and C-1 or between C-2 and C-3. The latter possibility gives the more stable disubstituted alkene. Of the two possible geometric isomers of 2-pentene, the *trans* isomer is more stable.

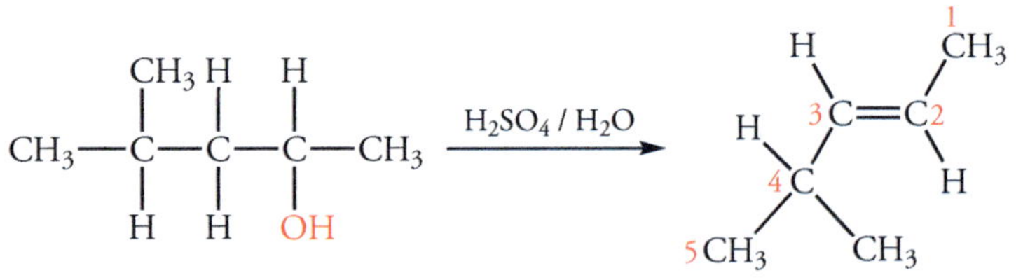

trans-4-methyl-2-pentene

8.6 Oxidation of Alcohols

Problem 8.8

Which of the isomeric $C_5H_{12}O$ alcohols react with PCC to produce a ketone, $C_5H_{10}O$?

Solution

The two $C_5H_{12}O$ alcohols shown below react with PCC to produce isomeric $C_5H_{10}O$ ketones.

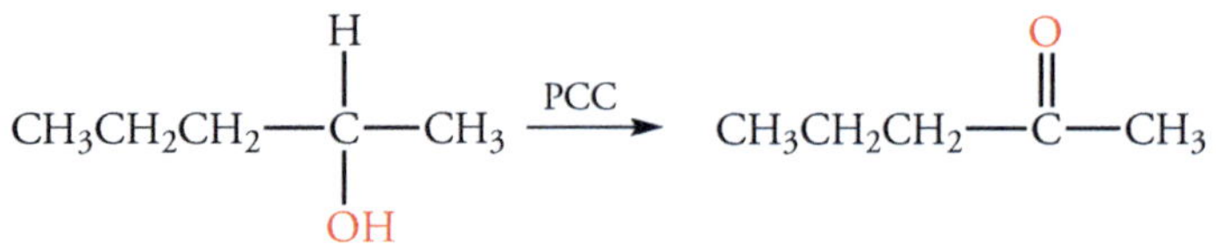

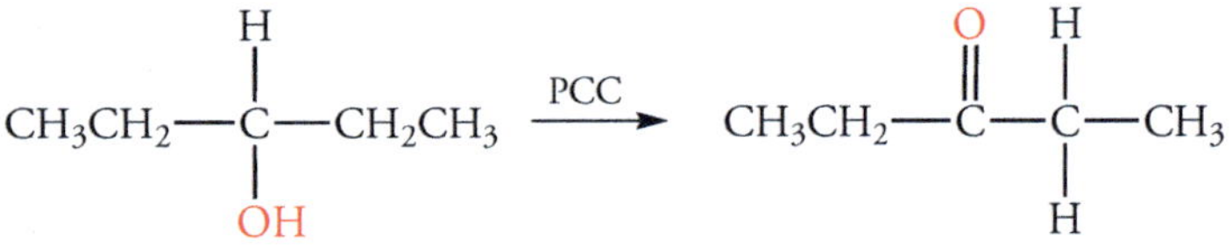

8.7 Synthesis of Alcohols

Problem 8.10

Write the product of (a) oxymercuration-demercuration and (b) hydroboration-oxidation of 3,3-dimethyl-1-butene.

Solution

(a) Oxymercuration-demercuration produces 3,3-dimethylbutane-2-ol as shown below.

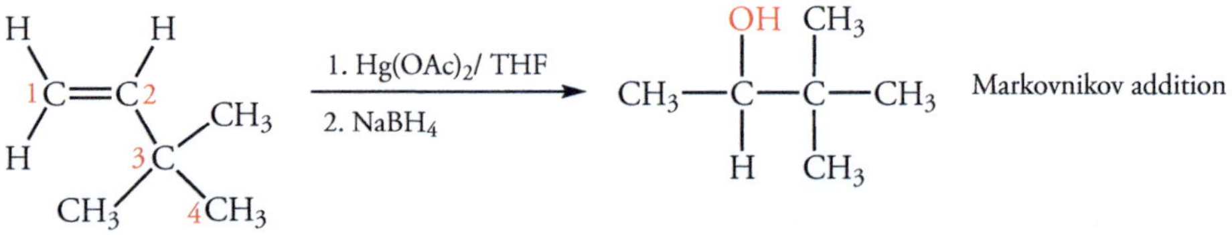

3,3-dimethyl-1-butene   3,3-dimethyl-butane-2-ol

(b) Hydroboration-oxidation produces 3,3-dimethylbutane-1-ol as shown below.

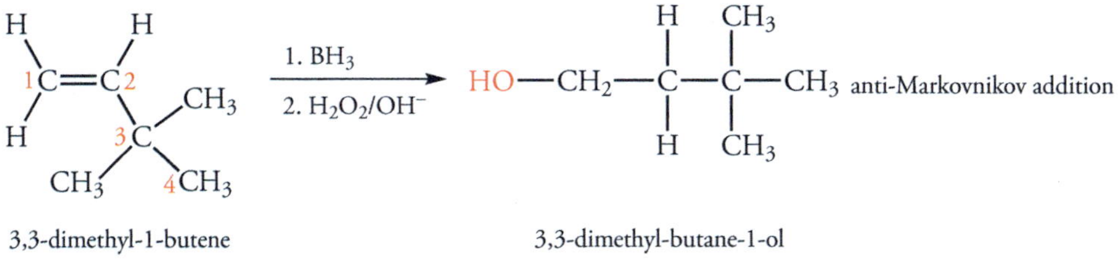

3,3-dimethyl-1-butene   3,3-dimethyl-butane-1-ol

CHAPTER 9

9.2 Nomenclature of Ethers

Problem 9.2
What are the IUPAC names of the following compounds?

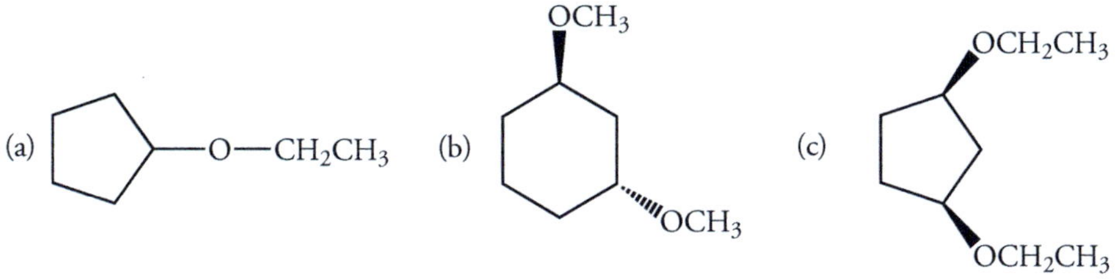

Solution
(a) Ethoxycyclopentane   (b) *trans*-1,3-dimethoxycyclohexane   (c) *cis*-2,4-diethoxycyclopentane

9.4 The Grignard Reagent and Ethers

Problem 9.4
Devise a synthesis of the following compound starting from benzene.

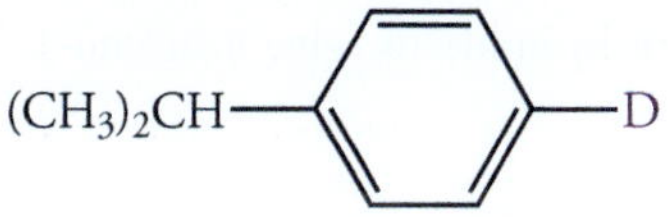

Solution

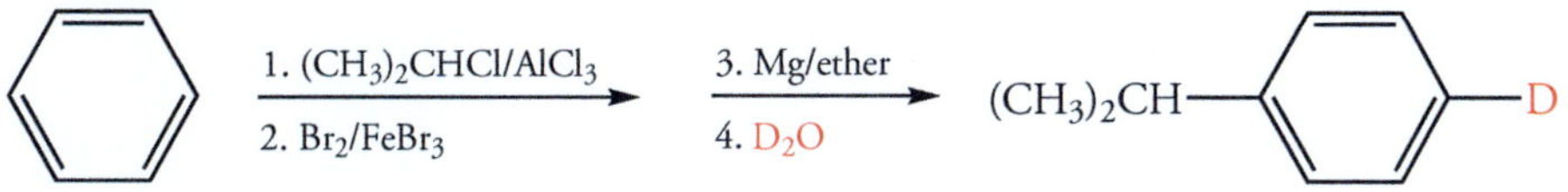

9.5 Synthesis of Ethers

Problem 9.6
Propose a synthesis of benzyl *tert*-butyl ether using the Williamson synthesis.

Solution

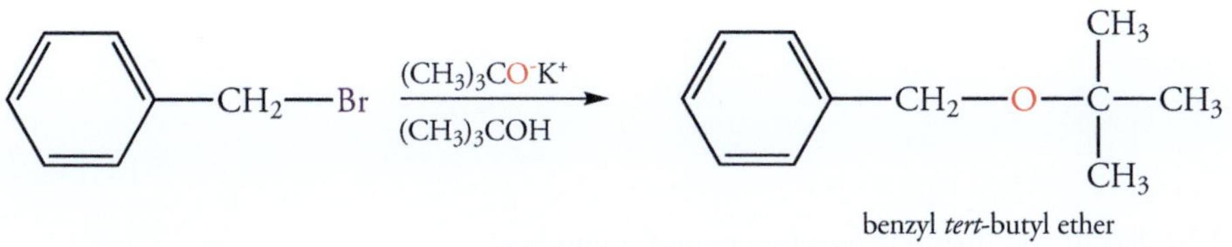

benzyl *tert*-butyl ether

9.6 Reactions of Ethers

Problem 9.8
Based on the mechanism of ether cleavage, predict the structure of the bromo alcohol that forms in the first step of the reaction of the following ether with HBr.

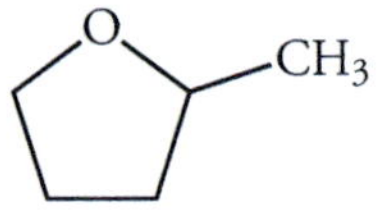

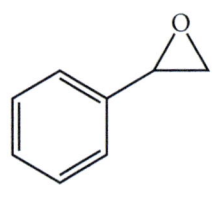

Solution
Protonation of the ether oxygen, followed by ring opening gives a secondary carbocation that reacts with bromide to give 4-bromo-1-pentanol, as shown below.

4-bromo-1-pentanol

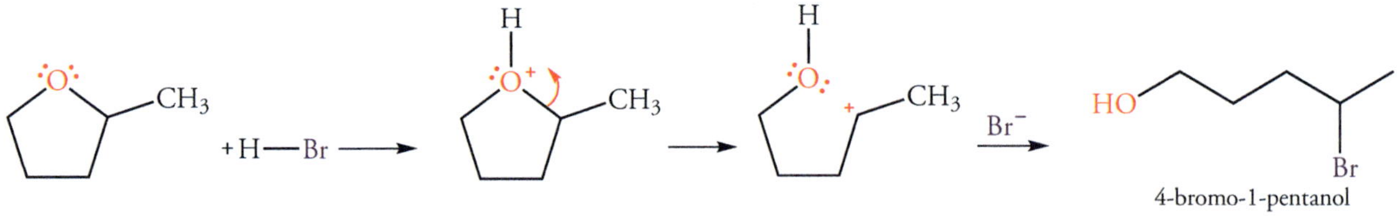

9.8 Reactions of Epoxides

Problem 9.10
Predict the product of the reaction of styrene oxide (phenyloxirane) in an acid-catalyzed reaction with methanol.

styrene oxide

Solution
Protonation of the ether oxygen, followed by ring opening gives a secondary carbocation that reacts with bromide to give 4-bromo-1-pentanol, as shown below.

CHAPTER 10

10.2 Nomenclature of Aldehydes and Ketones

Problem 10.2
What is the IUPAC name for the following compound, which is an alarm pheromone in some species of ants.

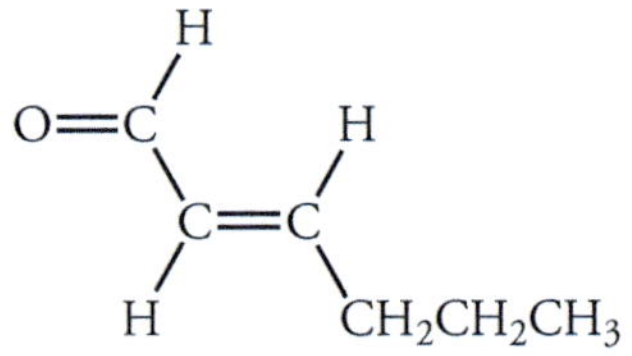

Solution
(E)-2-hexenal. Note that the aldehyde carbon is C-1 by default and does not therefore require a number.

10.4 Oxidation-Reduction Reactions of Carbonyl Compounds

Problem 10.4

Explain how the following two isomeric compounds could be distinguished using Fehling's solution.

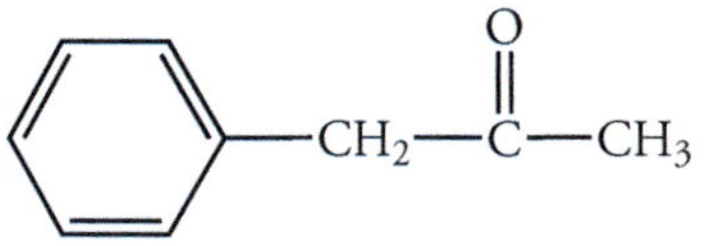

(1) 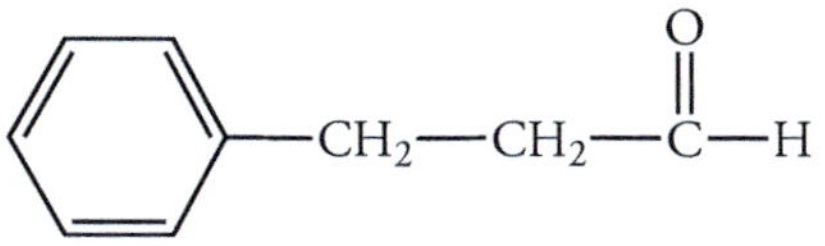   (2)

Solution

Fehling's solution will not react with ketones such as compound (1), but will convert aldehydes such as compound (1) to carboxylic acids.

Problem 10.6

Draw the structure of the product of the reaction of the following compound with each of the following reducing agents: (a) hydrazine and base, (b) palladium and hydrogen at l atm, and (c) Raney nickel and hydrogen at 100 atm.

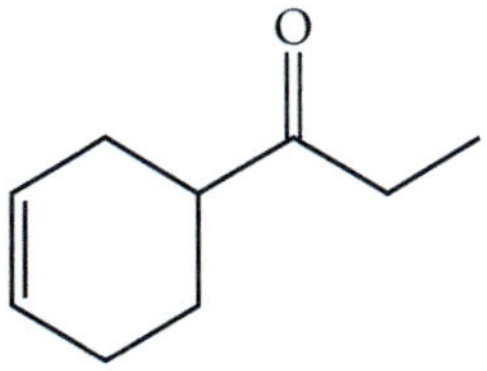

Solution

(a)   (b) 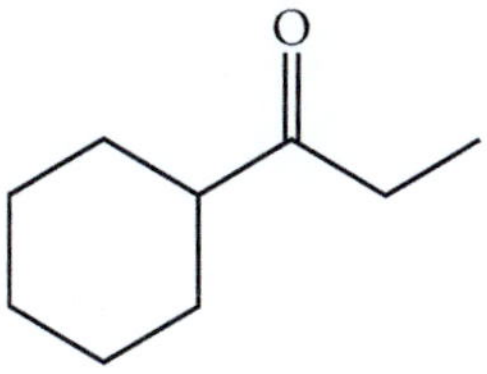   (c)

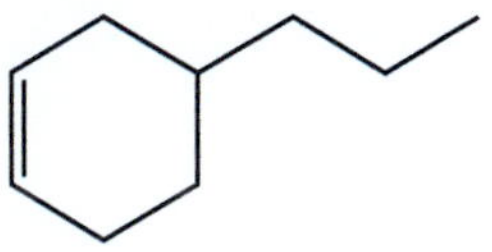

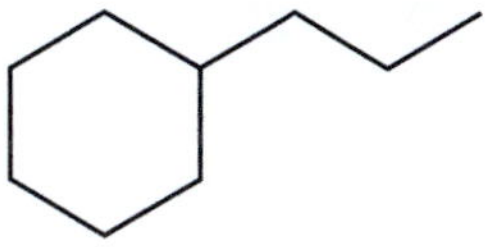

10.6 Synthesis of Alcohols From Carbonyl Compounds

Problem 10.8

The methyl Grignard reagent reacts with 4-*tert*-butylcyclohexanone to give a mixture of two isomeric products. Draw their structures.

Solution

The Grignard reagent adds to the carbonyl group from the bottom to give the most stable isomer in which the 4-*tert*-butyl is equatorial.

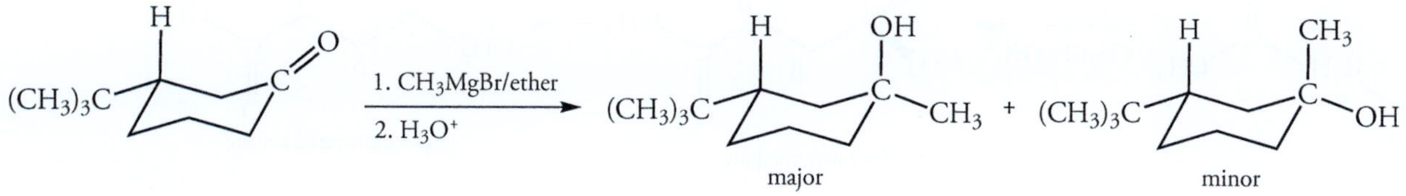

$$\text{1. CH}_3\text{MgBr/ether} \quad \text{2. H}_3\text{O}^+$$

major   minor

10.8 Formation of Acetals and Ketals

Problem 10.10
Identify the class to which each of the following compounds belongs.

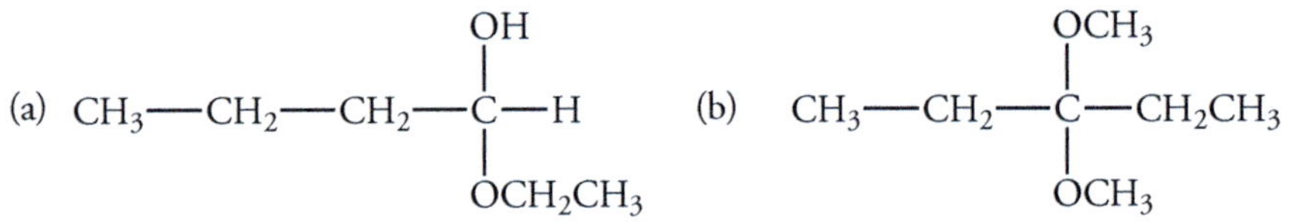

(a) $CH_3-CH_2-CH_2-\overset{\overset{\displaystyle OH}{|}}{\underset{\underset{\displaystyle OCH_2CH_3}{|}}{C}}-H$
(b) $CH_3-CH_2-\overset{\overset{\displaystyle OCH_3}{|}}{\underset{\underset{\displaystyle OCH_3}{|}}{C}}-CH_2CH_3$

Solution
Compound (a) is a hemiacetal. Compound (b) is a ketal.

10.9 Addition of Nitrogen Compounds

Problem 10.12
Write the structure of the reactants required to produce the following compound.

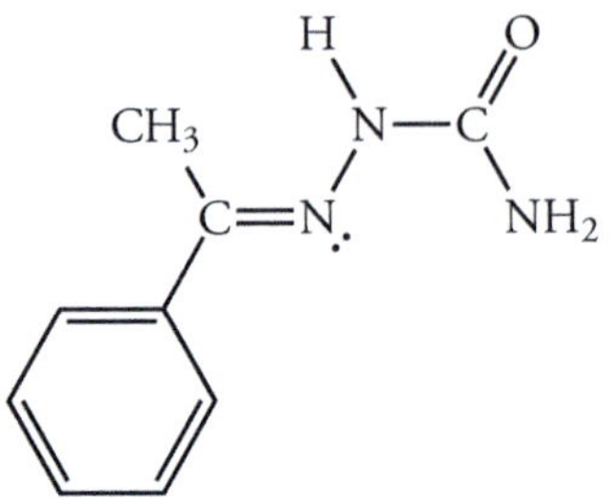

Solution

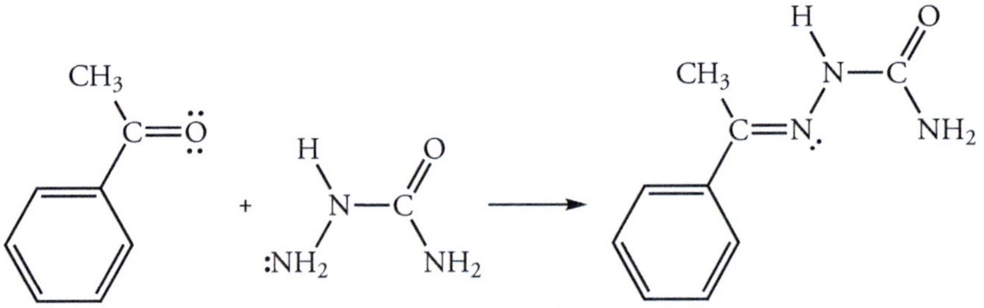

10.11 The Aldol Condensation

Problem 10.14
What compound is required to form the following unsaturated compound using an aldol condensation?

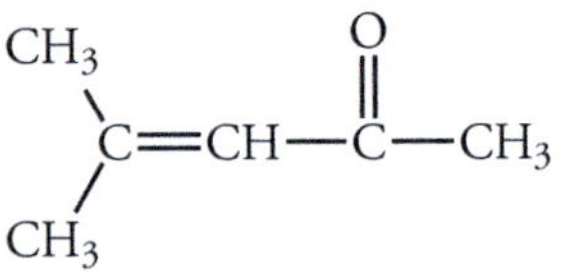

Solution
This compounds results from the aldol condensation of acetone followed by loss of water.

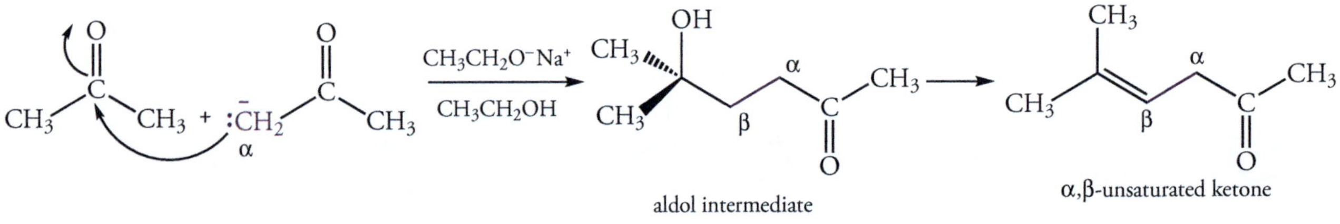

CHAPTER 11

11.2 Nomenclature of Carboxylic Acids

Problem 11.2

Mevalonic acid is required to form isopentenyl pyrophosphate, an intermediate in terpene synthesis. It has the following structure. What is its IUPAC name?

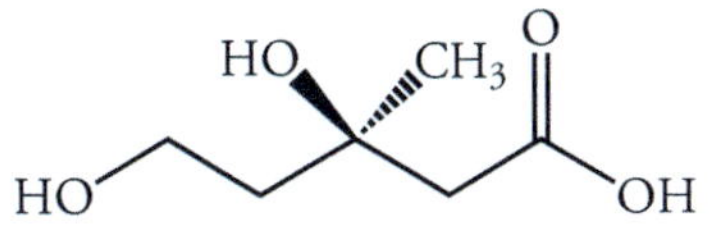

Solution

The IUPAC name of mevalonic acid is (*3R*)-3,5-dihydroxy-3-methylpentanoic acid.

Problem 11.4

Isobutyl formate has the odor of raspberries. Based on this common name, draw its structural formula. What is its IUPAC name?

Solution

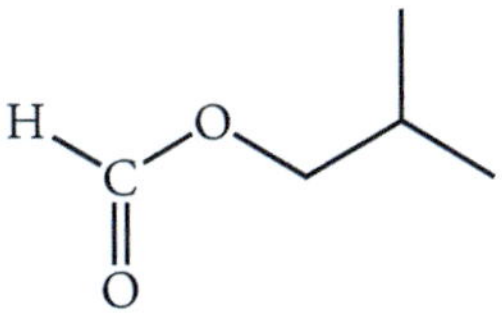

2-methylpropyl methanoate
(isobutyl formate)

11.4 Acidities of Carboxylic Acids

Problem 11.6

The pK_a values for the dissociation of the first of the two carboxyl groups of malic acid and oxaloacetic acid are 3.41 and 1.70, respectively. (a) Which compound is the stronger acid? (b) Which of the two carboxyl groups in each compound is more acidic?

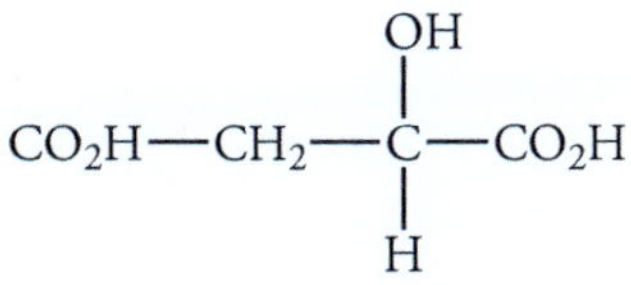

malic acid 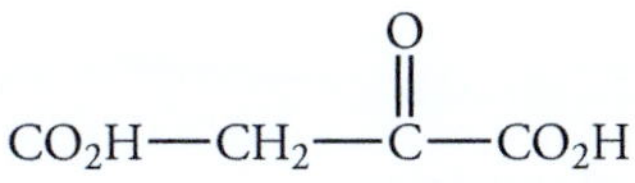   oxaloacetic acid

Solution

(a) Oxaloacetic acid is the more acidic compound because a carbonyl group is more electron-withdrawing than a hydroxyl group. (b) The carboxyl group to the right of the oxygen-containing substituent is the more acidic because a carbonyl group is more electron-withdrawing than a hydrogen atom.

11.5 Synthesis of Carboxylic Acids

Problem 11.8

Outline a series of steps to prepare 2,2-dimethylheptanoic acid starting from 2-methyl-1-heptene.

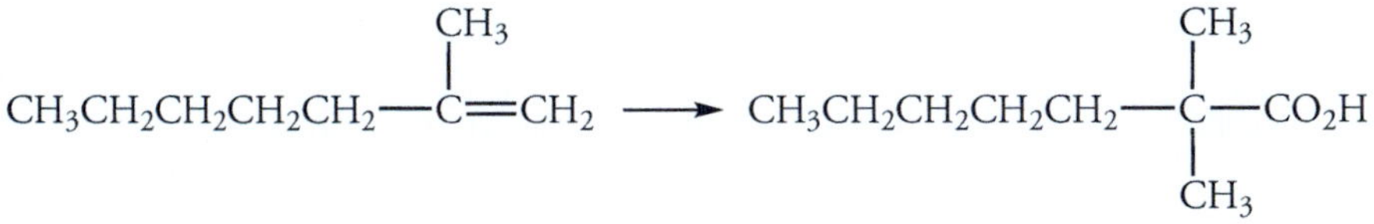

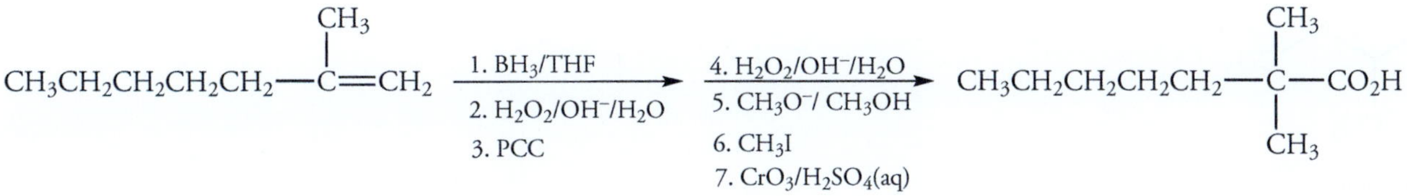

11.8 Esters and Anhydrides of Phosphoric Acid

Problem 11.10

Hydrolysis of diloxanide furanoate in the body is required for it to be effective against intestinal amebiasis. What is the acid component of the drug? Based on the name of the drug, what is the name of the acid?

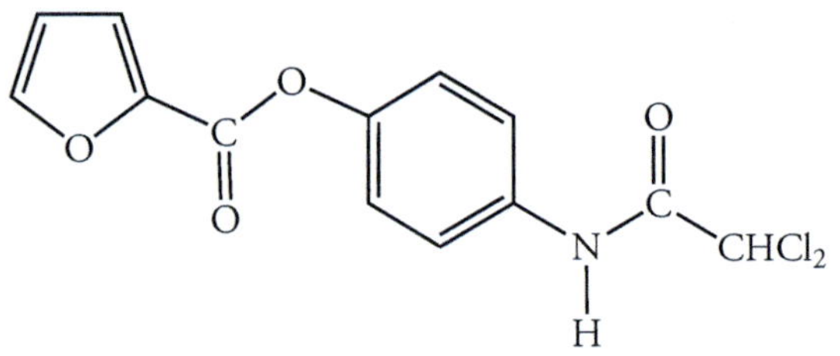

diloxanide

Solution

The acid component of diloxanide furanoate is furanoic acid, as shown below.

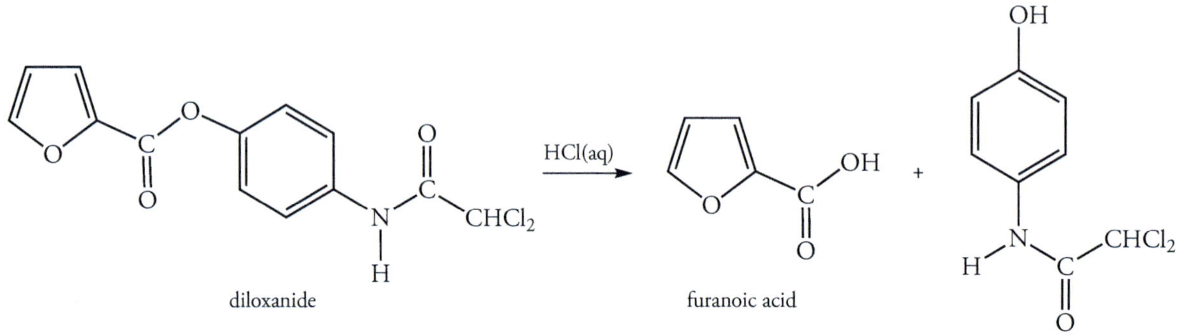

diloxanide   furanoic acid

CHAPTER 12

12.3 Classification and Structure of Amines and Amides

Problem 12.2

Classify the nitrogen-containing functional groups in Mepivacaine®, a local anesthetic.

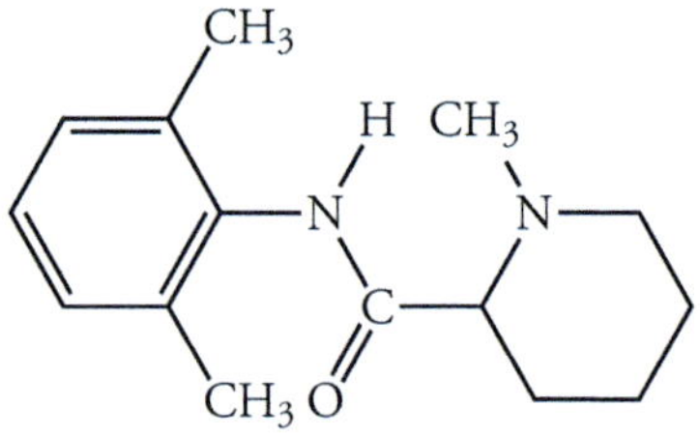

Mepivacaine ®

Solution

The nitrogen atom of the six-membered ring (pyrrolidine) is bonded to three carbon atoms in a tertiary amine. The nitrogen atom bonded to the benzene ring and the carbonyl group is part of a secondary amide.

Problem 12.4

Classify the amine in the illicit drug methamphetamine (also colloquially known as "speed," and infamously as "crystal meth").

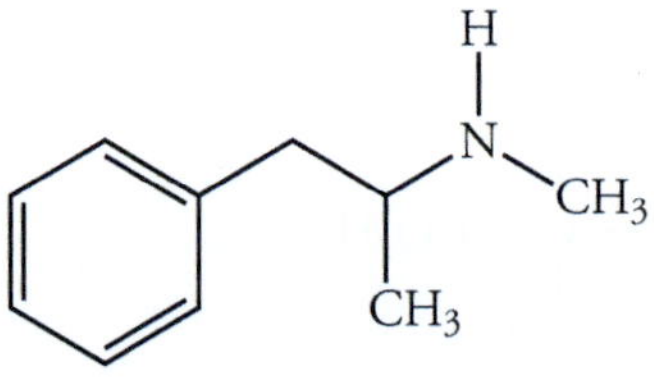

methamphetamine

Solution

The nitrogen atom is bonded to three carbons in a tertiary amine.

12.4 Nomenclature of Amines and Amides

Problem 12.6

2-(3,4,5-Trimethoxyphenyl)ethanamine is the systematic name of mescaline, a hallucinogen. Write its structure.

Solution

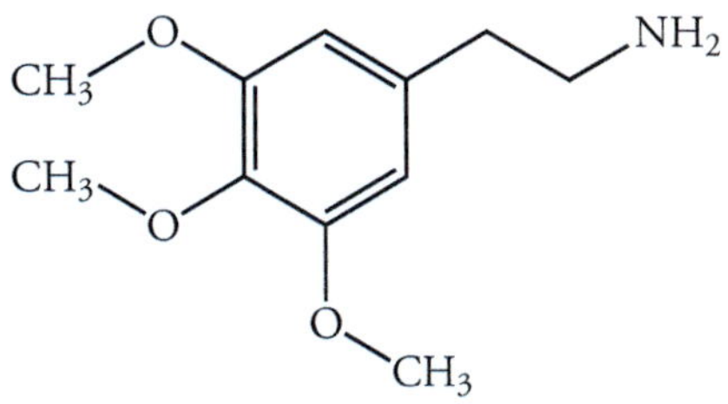

2-(3,4,5-trimethoxyphenyl)ethananime
(mescaline)

Problem 12.8

Assign the IUPAC name of DEET, an insect repellent.

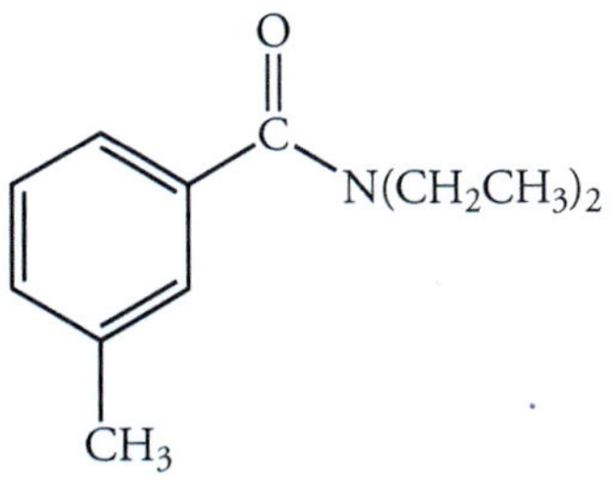

Solution

The IUPAC name of DEET is N,N-diethyl-3-methylbenzamide.

12.6 Basicity of Nitrogen Compounds

Problem 12.10

Which of the two nitrogen atoms of chlorpromazine (Thorazine®), an antipsychotic drug, is the more basic?

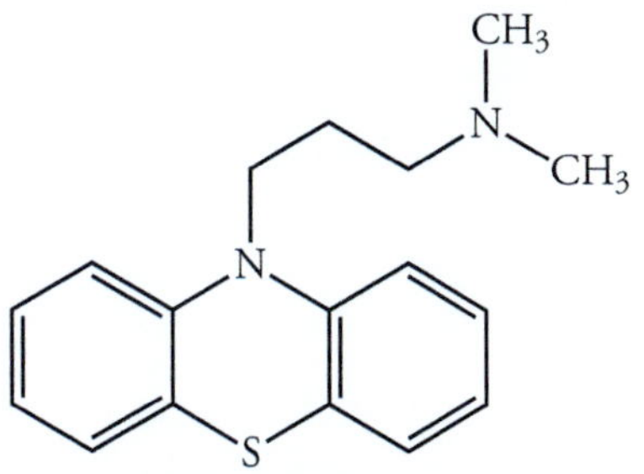

chlorpromazine

Solution

The nonbonded electron pair nitrogen atom that is part of the ring system can be delocalized into the ring, and is therefore less basic than the nitrogen in the tertiary amine.

12.8 Nucleophilic Reactions of Amines

Problem 12.12

Acetaminophen, the analgesic in many drugs, is an amide. Draw the structures of the compounds that could be used to produce it. What possible complications might occur with this combination of reactants?

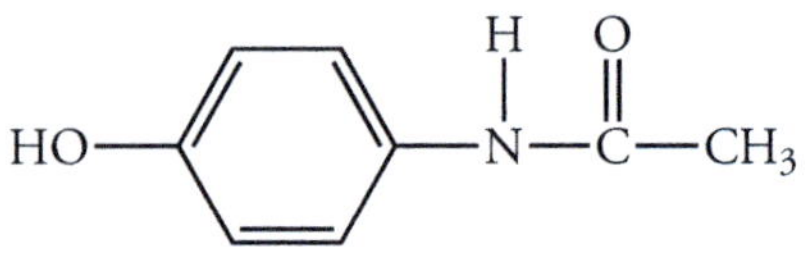

Solution

Acetaminophen could be made by acetylation of *p*-hydroxyaniline (*p*-aminophenol). A competing reaction would also result in acetylation of the hydroxyl group, which is also nucleophilic.

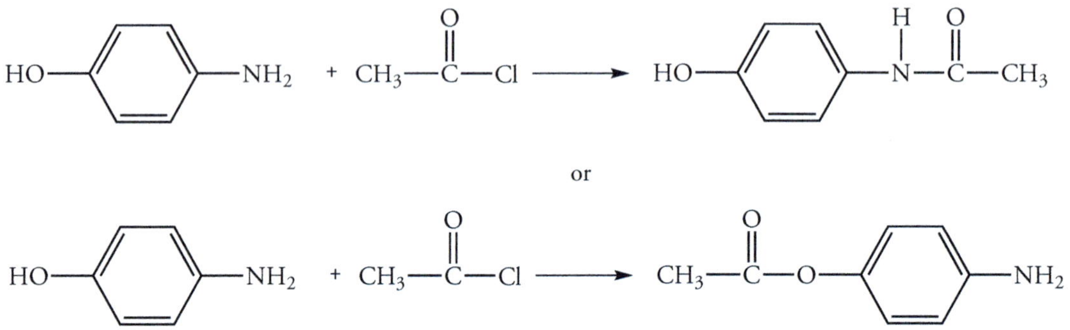

or

12.10 Hydrolysis of Amides

Problem 12.14

What are the products of the hydrolysis of nubucaine by an acid? Nubucaine is a local anesthetic.

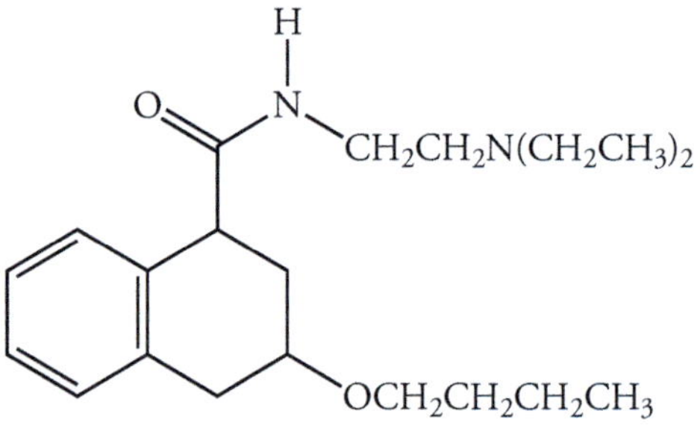

Solution

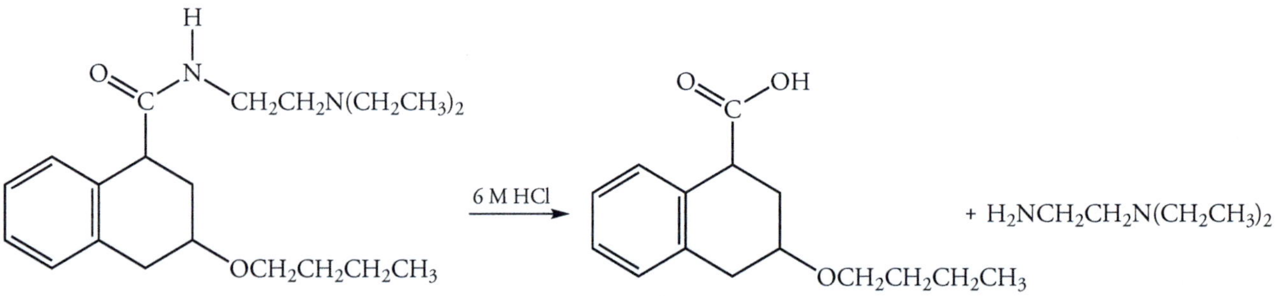

CHAPTER 13

13.2 Chirality of Carbohydrates

Problem 13.2
What relationship exists between D-allose and D-talose (see Figure 13.1)?

Solution
D-allose and D-talose differ at C-2 and C-3. Therefore, they are diastereomers.

Problem 13.4
Draw the structure of the C-3 epimer of D-ribulose (see Figure 13.2).

Solution

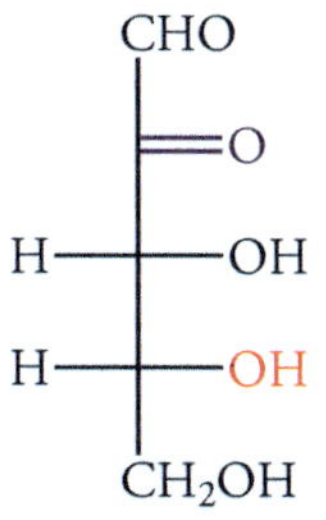

D-ribulose 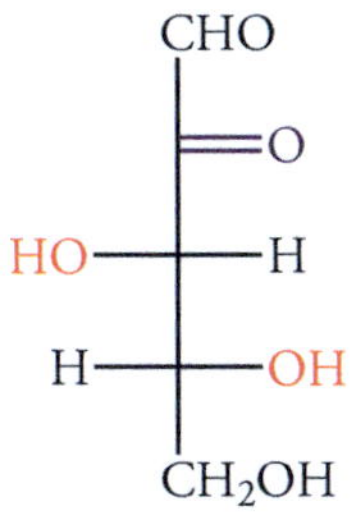   D-xylulose

13.3 Hemiacetals and Hemiketals

Problem 13.7
Draw the Haworth projection formula of the pyranose form of D-mannose with the β configuration; that is β-D-mannopyranose.

Solution

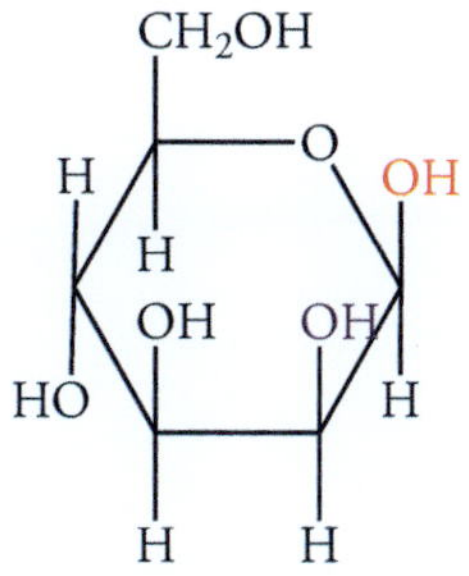

β-D-mannopyranose

13.5 Reduction of Monosaccharides

Problem 13.9
Reduction of ribulose (see Figure 13.2) by sodium borohydride gives a mixture of two isomers with different physical properties. Explain why the isomers have different properties.

Solution
The products of the reduction are diastereomers. Therefore, they have different properties.

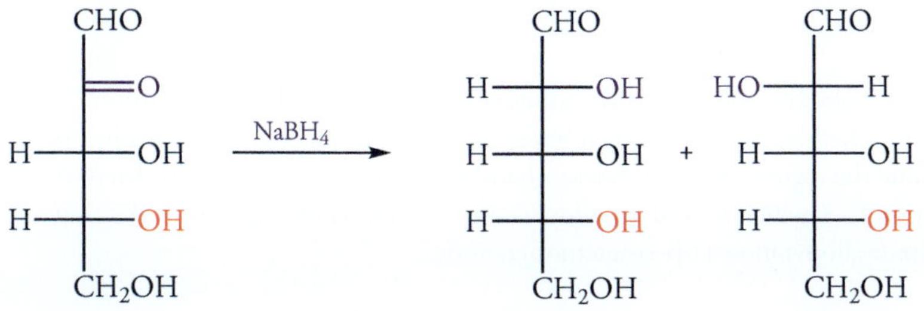

D-ribulose   D-ribose   D-arabinose

13.7 Glycosides

Problem 13.11

Examine the following molecule to determine its component functional groups. From what compounds can the substance be formed?

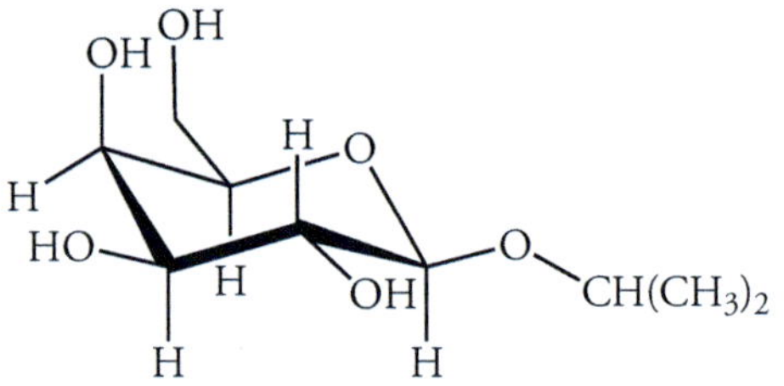

Solution

The molecule shown above is a β-glycoside that can be made from β-D-galactopyranose and isopropyl alcohol.

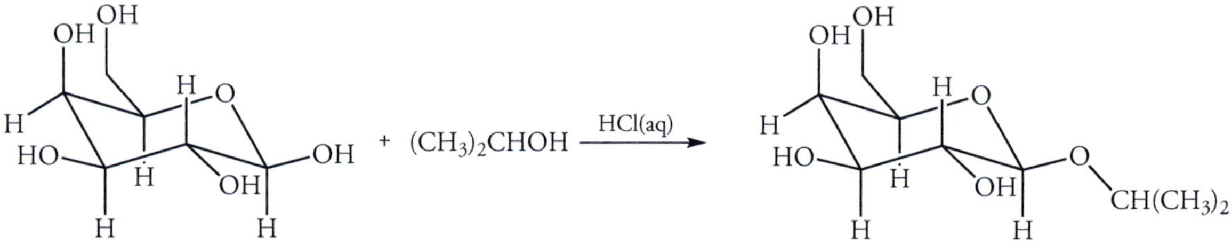

β-D-galactopyranose   4-O-isopropyl-β-D-galactopyranoside

13.8 Disaccharides

Problem 13.13

Describe the structure of the following disaccharide.

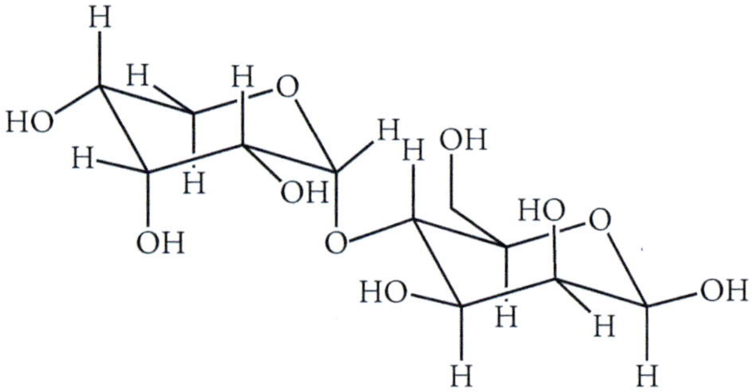

Solution

The hemiacetal center located on the aglycone ring (at the right) has a hydroxyl group in the β-configuration. The glycosidic bond is from C-1 of the acetal ring (on the left) to C-4 of the aglycone ring (on the right). Furthermore, the oxygen bridge is formed through an α-glycosidic bond. Thus, the bridge is α-1,4′. Next, we examine both rings to determine the identity of the monosaccharides. The ring on the left is 6-deoxy-α-D-allopyranose: all of its hydroxyl groups are equatorial except the one at C-3, which is axial. (D-allose is the C-3 epimer of D-glucose.) The ring on the right is β-D-mannopyranose. The compound is 6-deoxy-4-O-(α-D-allopyranosyl)-β-D-mannopyranoside.

CHAPTER 14

14.3 Acid-Base Properties of α-Amino Acids

Problem 14.2

Write the structure of the zwitterion and the conjugate acid of serine (see serine in Table 14.1).

Solution

The conjugate acid of serine, the zwitterion, has protonated carboxyl and amino groups.

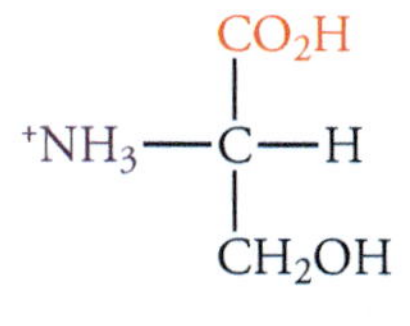

conjugate acid of serine

Problem 14.4

In what ionic form does alanine exist in 0.0 1 M NaOH (see Table 14.1)?

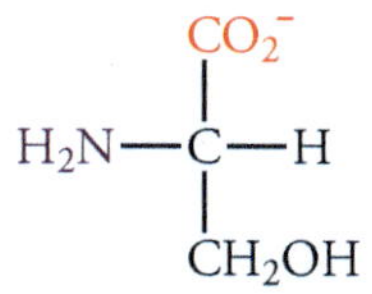

conjugate base of alanine in 1M NaOH

Solution

In 1 M NaOH, the carboxyl group of alanine is ionized and the amino group is unprotonated.

14.5 Peptides

Problem 14.5

(a) Determine the number of isomeric tripeptides containing one alanine and two glycine residues. (b) Write representations of the isomers using three-letter abbreviations.

Solution

(a) There are three isomeric tripeptides containing one alanine and two glycine residues. (b) The three peptides are:

Ala–Ala–Gly   Ala–Gly–Ala   Gly–Ala–Ala

Problem 14.6

(a) Identify the terminal amino acids of tuftsin, a tetrapeptide that stimulates phagocytosis and promotes the destruction of tumor cells. Write the amino acid sequence using three-letter abbreviations for the amino acids. Also write the complete name without abbreviations.

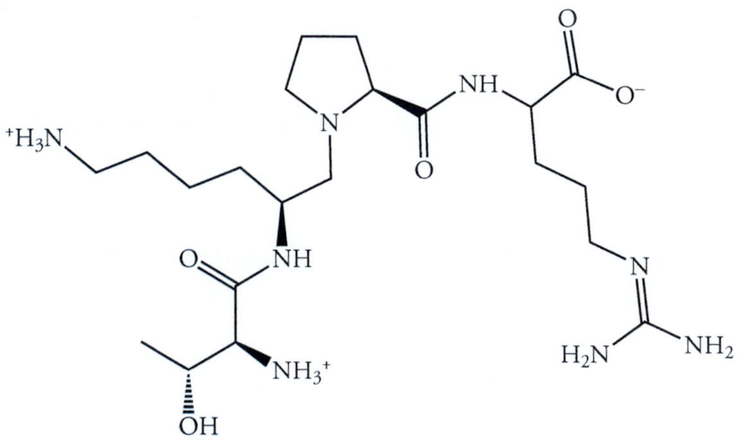

tuftsin

Solution

(a) The N-terminal amino acid of tuftsin is threonine and the C-terminal amino acid is arginine. (b) The abbreviated sequence is Thr-Lys-Pro-Arg, and the complete name is threonyllysylprolylarginine. It is easy to see why it is just called "tuftsin."

14.7 Determination of Protein Structure

Problem 14.8
Predict the products of the trypsin-catalyzed hydrolysis of the following pentapeptide.

Ala-Lys-Gly-Arg-Leu

Solution
Trypsin cleaves proteins on the C-terminal side of basic amino acids. So the trypsin-catalyzed hydrolysis products of the above tetrapeptide are Ala-Lys, Gly-Arg, and Leu. (Partial hydrolysis would also produce some Ala-Lys-Gly-Arg.)

Problem 14.10
Draw the structure of the phenylthiohydantoin obtained by reaction of the following tetrapeptide with phenyl isothiocyanate.

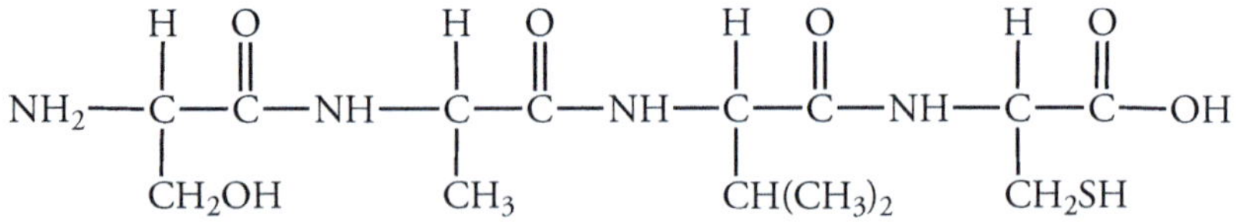

Solution
The N-terminal amino acid of the above peptide is serine. Edman degradation gives the isothiocyanate shown below.

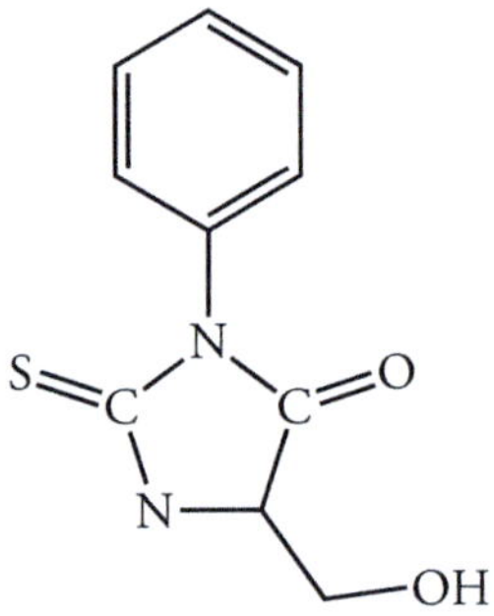

CHAPTER 15

15.2 Structure and Properties of Polymers

Problem 15.2
There are three isomeric benzenedicarboxylic acids. Each reacts with ethylene glycol to produce a polyester. Which one should produce the polyester that packs most efficiently and hence has the largest intermolecular attractive forces?

Solution
The para isomer, p-terephthalic acid, produces a polymer with the smallest number of "bends," so it will pack more tightly and have the largest intermolecular attractive forces.

15.3 Classification of Polymers

Problem 15.4
(a) What type of plastic is best suited to make the handles for cooking utensils for the home? (b) What type of plastic is most likely to be used for the frames of eyeglasses?

Solution
(a) Thermosetting plastic is used for the handles of cooking utensils. (b) Thermoplastics are used for the frames of eyeglasses.

15.8 Stereochemistry of Addition Polymerization

Problem 15.6
Draw the structure of the product of ozonolysis of natural rubber. Would this structure differ from the ozonolysis product of gutta-percha?

Solution
Ozonolysis of natural rubber and of gutta-percha give the same product, as shown below,

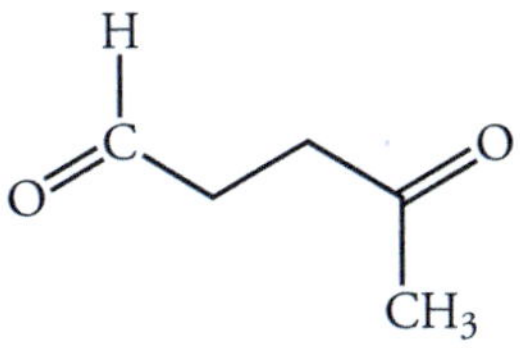

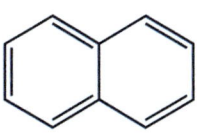

CHAPTER 16

16.3 Ultraviolet Spectroscopy

Problem 16.2
Naphthalene and azulene are isomeric compounds that have extensively conjugated π systems. Naphthalene is a colorless compound, but azulene is blue. Deduce information about the absorption of electromagnetic radiation by these two compounds.

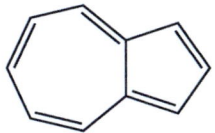

naphthalene   azulene

Solution
Since naphthalene is colorless and azulene is blue, azulene must have the more extensively delocalized conjugated π system.

16.4 Infrared Spectroscopy

Problem 16.4
Explain how you could distinguish between the following two compounds by infrared spectroscopy.

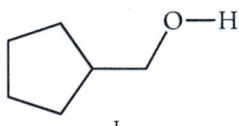

Solution
Compound I has an intense absorption O—H in the 3400-3600 cm^{-1} region of the spectrum. Compound II does not.

Problem 16.6
The carbonyl stretching vibration of ketones is at a longer wavelength than the carbonyl stretching vibration of aldehydes. Suggest a reason for this difference.

Solution
The polar resonance contributing structure to the resonance hybrid of the ketone by the two alkyl groups increases the single bond character of the carbonyl group in a ketone. As a result, it takes less energy to stretch the carbonyl bond.

16.5 Nuclear Magnetic Resonance Spectroscopy

Problem 16.8
How many sets of nonequivalent hydrogen atoms are contained in each of the following ketones?

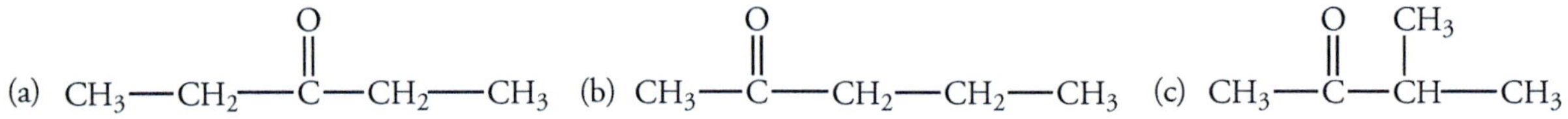

(a) $CH_3-CH_2-\overset{\overset{\displaystyle O}{\|}}{C}-CH_2-CH_3$ (b) $CH_3-\overset{\overset{\displaystyle O}{\|}}{C}-CH_2-CH_2-CH_3$ (c) $CH_3-\overset{\overset{\displaystyle O}{\|}}{C}-\overset{\overset{\displaystyle CH_3}{|}}{CH}-CH_3$

Solution
(a) 2 (b) 4 (c) 3

16.6 Spin-Spin Splitting

Problem 16.10
Describe the NMR spectrum of 1,3-dichloropropane citing the number of resonances, their δ values, and their multiplicities.

Solution
The spectrum of 1,3-dichloropropane is shown below. Protons labeled H_a and H_c are identical, and they have two neighbors (H_b) the result is a triplet that integrates to four hydrogens. Protons labeled H_b have four neighbors and appear as a quintet that integrates to two hydrogens.

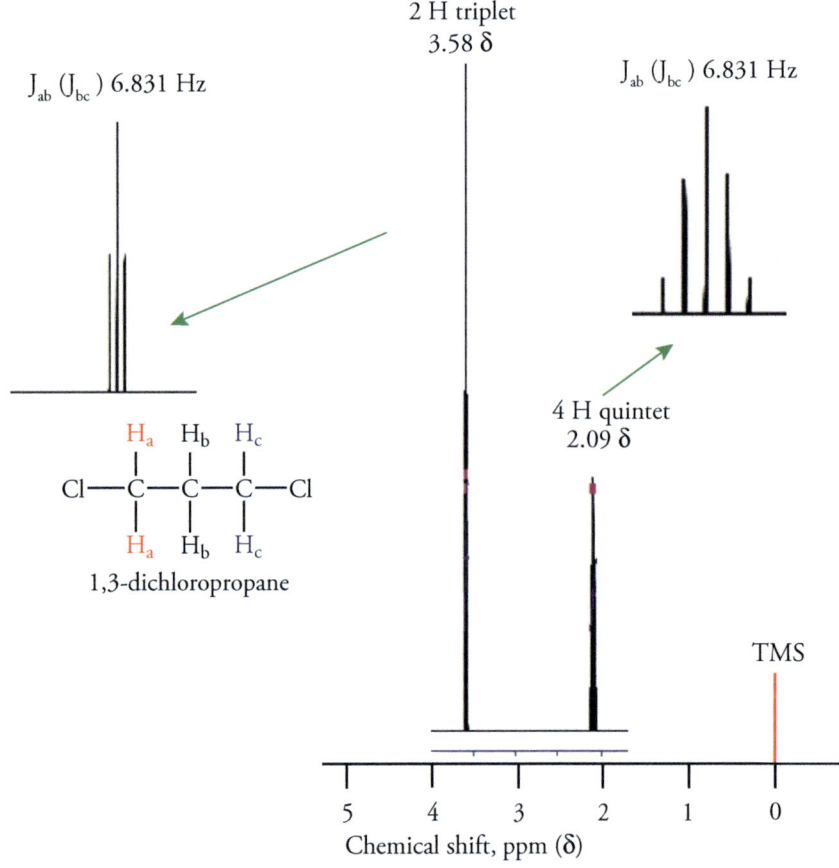

16.7 ¹³C NMR Spectroscopy

Problem 16.12
How can a compound of molecular formula $C_4H_{10}O$ be established as an ether or an alcohol using ^{13}C NMR spectroscopy?

Solution
The compound that is an ether has two low field resonances due to the presence of the ether C—O—C bond. The alcohol has only one.

INDEX

A

Acetal formation, 283
    mechanism of, 275
    reactivity of, 274, 275
Acetaminophen, 470
Acetic acid
    acid ionization constant, 294
    hydrogen-bonded dimer of, 293
Acetyl coenzyme A, 303, 304, 309
Acetylenic alcohols, 270–271
Achiral diastereomers, 174–175
Acid/acyl chloride, 288
Acid anhydrides, 288, 291, 302–303
Acid-base reactions
    of alcohols, 214–215
    Brønsted-Lowry acids and bases, 39–40
    equilibria in, 47–49
    Lewis acids and bases, 40–41
Acid-catalyzed addition reaction, 267–268, 273
Acid-catalyzed ring-opening reactions, 247
Acid chlorides
    nucleophilic acyl substitution, 302
    reduction of, 305
Acidic amino acids, 372
Acidity
    of alcohols, 214
    of carboxylic acids, 294–297
    effect of structure on, 49–51
Activation energy, 55
    S_N1 reaction mechanism, 199
    S_N2 reaction mechanism, 197, 198
Acyl derivatives, reactivity of, 301
Acyl group transfer reaction, 300
Acyl halides, 18
    reactions of amines with, 329, 334
    synthesis of, 310
Adamantane, 76–77
Addition-elimination reaction, 275
Addition polymerization
    chain branching, 405
    chain-transfer agents, 404
    cross-links in, 407
    dimerization reaction, 404
    disproportionation reaction, 404
    inhibitors, 404, 405
    regulation of chain length, 404
    stereochemistry of, 408–410
Addition polymers, 401
Addition reactions, 44
    of alcohols, 273
    of carbonyl compounds, 267–269
    of nitrogen compounds, 275–277, 283
    of oxygen compounds, 272–273
    of water, 272–273
1,4-Addition reactions, 119
Adenosine diphosphate (ADP), 306

Adenosine monophosphate (AMP), 306
Adenosine triphosphate (ATP), 306
Aglycones, 356
Alanylglycine structure, 377, 378
Alcohols, 236–237
    acid-base reactions of, 214–215, 234
    acidity of, 214
    carbonyl compounds, reduction of, 221–223, 232
    classification of, 233–234
    common names of, 209–210
    dehydration of, 216–218, 231, 235–236
    and ethers, 15
    haloalkane synthesis, 231
    nomenclature of, 232–233
    oxidation of, 218–219, 231, 236
    physical properties of, 212–214, 234
    as solvents, 214
    substitution reactions of, 215–216
    synthesis of, 221–226
    toxicity of, 219–220
Aldaric acids, 355
Aldehydes, 284, 285
    general formulas for, 260
    IUPAC names of, 261–262
    naturally occurring, 260, 261
    nomenclature of, 261–263
    oxidation of, 265, 282
    physical properties of, 263–264
    reactions with Grignard reagents, 269–270
    reactivities of, 268, 269
    reduction of, 265, 282, 283
    solubility in water, 264
Alditol, 354
Aldol condensation, 279–281, 283, 466
Aldonic acid, 354
Aldoses, 343
D-Aldoses, 345, 346
Alkadienes, 95, 119–120
Alkanes
    alkyl group, 66
    branched alkanes, 66
    branching point, 66
    carbon atoms classification, 67–68
    conformations of, 72–74, 92
    halogenation of, 84–87
    names and condensed structural formulas, 66
    nomenclature of, 68–72, 91–92
    normal, 65
    oxidation of, 83
    physical properties of, 81–83
    saturated hydrocarbons, 65
Alkenes
    acidity of, 106–107
    addition reactions, 111–115
    bromination, 180–182
    classification of, 96

Alkenes *(Continued)*
    copolymerization of, 405–406
    ethene and ethyne, 95–96
    ethylene, 95
    geometric isomers of, 99–100
    hydration of, 115–116, 125
    hydrogenation of, 107–110, 124
    indirect hydration of, 223–225, 232
    naturally occurring, 98–99
    nomenclature of, 103–106
    oxidation of, 110–111, 124
    physical properties of, 97
    preparation of, 116–118
Alkoxyalkanes, 240
Alkyl alkyl ethers, 240
Alkylation, of amines, 331
Alkyl groups, 66, 90–91
Alkyl halides, amine reaction with, 329–330, 334
Alkynes
    acidity of, 106–107
    addition reactions, 111–115
    classification of, 96
    hydration of, 115–116, 124
    hydrogenation of, 107–110, 124
    naturally occurring, 98–99
    nomenclature of, 103–106
    oxidation of, 110–111
    physical properties of, 97
    preparation of, 116–118
Alkynide ions, 193
Allosteric effects, 392
Allylic carbocation, 119
Allylic oxidation, 121–123
Alternating copolymers, 406
Amides, 16, 291
    bonding in, 318
    classification of, 288
    hydrolysis of, 333–335
    reduction of, 332
    synthesis of, 334, 335, 342
Amines, 16
    alkylation of, 331
    basicity of, 339
    boiling points of, 322, 323
    bonding and structure of, 316
    isomers of, 338
    nomenclature of, 319–322, 338
    nucleophilic reactions of, 328–331
    odor and toxicity of, 325
    physical properties of, 322–325, 339
    solubility in water, 324
    structure and classification of, 317–319
    synthesis of, 331–332, 335, 340–341
Amino acids, 392, 393
    classification of, 372
    ionic form of, 372–374
    R groups in, 372
α-Amino acids
    acid-base properties of, 372–375
    Fischer projection formula for, 371, 372
    pK_a values of, 374–376
    structures of, 372, 373

Ammonium salts, solubility of, 328
Amphetamines, 320–321
Amylopectin, 362–363
Amylose, 362
Angiotensin II, 378
Anionic polymerization, 403–404
Anomeric carbon atom, 351
Anomers, 351
Antibodies, 371
Anti-Markovnikov addition, 224
Antiparallel β-pleated sheet, hydrogen bonding in, 389, 390
Aprotic solvents, 201
L-Arabinose, 348
Arene oxides, 249, 251
Arenes, 138
Aromatic compounds
    acylation, 157
    acyl side chains, 157
    adenine, 136–137
    alkylation, 157
    anthracene *vs.* phenanthrene, 136
    1,2-benzanthracene, 136
    electrophilic aromatic substitution, 139–145
    functional group modification, 152–154
    halogenation, 156
    histamine, 136
    Hückle rule, 135
    interpretation of directing effects, 148–150
    interpretation of rate effects, 145–148
    Kekulé's concept of benzene, 134
    nitration, 156
    nomenclature of, 137–139
    reactions of side chains, 150–152
    resonance theory and benzene, 134–135
    side-chain oxidation, 157
    six-carbon unit, 133
    solid, 133
    sulfonation, 157
    synthesis of, 154–156
Aromatic hydrocarbons, 95
Atactic polymer, 408, 409
Atomic number, 1
Atomic orbitals, 2–3
Atomic structure, organic compounds, 1–4
Automatic sequenator, 383

B

Baclofen, 321, 322
Base-catalyzed addition reaction, 267
Basic amino acids, 372
Basicity, 194
Batzelline, 191
1,2-Benzanthracene, 136
Benzoquinone inhibitors, 404, 405
Benzyl tert-butyl ether synthesis, 463
β-pleated sheet, 389, 390
Bicyclic compounds, 75
Biochemical Claisen condensation, of thioesters, 309
Biochemical reactions, of epoxides, 249–252
Biochemical redox reactions, 42–44
Boiling points

of alcohols, 212, 213
of alkanes, 212, 213
of amines, 322, 323
of carboxylic acids, 292
of ethers, 242
Bombykol, 100
Bond cleavage and formation, types of, 52–53
Bonding electrons, 5
Bond-line structures, 19–20
  Branched alkanes, 66
Branching point, 66
Brevicomin, 77
Bridged ring compounds, 75
1-Bromdecane, 461
Bromination, 140
Bromochlorofluoromethane, 164–165
(Z)-3-Bromo-2-chloro-2-hexene, 105–106
Bromochloromethane, 165
(E)-8-Bromo-3,7-dichloro-2,6-dimethyl-1,5-octadiene, 190–191
4-Bromo-1-pentanol, 464
Brønsted basicity, of nucleophile, 194
Brønsted-Lowry acids and bases, 39–40
Butane, conformations of, 74
2-Butanol, ^{13}C NMR spectrum of, 439, 440
1-Butanol, infrared spectrum of, 429, 430
2-Butanone, 266

C

Cahn-Ingold-Prelog System, 170
Capillin, 263
Carbamic acids, 415
Carbanion, 106
Carbocyclic compounds, 65
Carbohydrates
  chirality of, 344–349
  classification of, 343, 344
α-Carbon atom, reactivity of, 278
Carbonyl compounds
  addition reactions of, 267–269, 286
  alcohol synthesis from, 269–272
  oxidation-reduction reactions of, 265–267, 285–286
  reactions of amines with, 328, 329, 334
  reduction of, 221–223, 232
Carbonyl group, 15, 259–263, 266
Carbonyl oxygen atom, 16
Carboxamide, 291
Carboxylate anion, 290
Carboxylic acids
  acidity of, 294–297, 313
  and acyl group, 288–289
  boiling points of, 292, 293
  bonding in, 287, 288
  derivatives, 290–291
  equivalent representations of, 287
  IUPAC names of, 290
  nomenclature of, 289–292
  physical properties of, 292–294
  pK$_a$ values of, 295
  reduction of, 305, 310
  salts of, 296
  solubility of, 293
  synthesis of, 297–300, 309, 310

Carboxylic acids and esters, 16
β-Carotene, 97, 277
Carvone, 22–23
Caryophyllene, 97
Catalysts, function of, 57
Cationic polymerization, 402–403
Cellobiose, 359–360
Cellulose, 362
Chain growth polymers. See Addition polymers
Chain-transfer agents, 404
Chelates, 252
Chemical equilibrium and equilibrium constants, 45–47
Chemical reactions, 38–39
Chemical shifts
  of ^{13}C atoms, 439, 440
  of hydrogen atoms, 432, 433
Chiral image, 164
Chloramphenicol, 210, 211, 307
Chlorination, 140
2-Chloro-1,3-butadiene, polymerization of, 410
4-Chloro-3,5-dimethylphenol, 228, 229
(E)-1-Chloro-3-ethyl-1-penten-4-yn-3-ol. See Ethchlorvynol
Chlorophene, 228, 229
Chymotrypsin, 384, 385
Chymotrypsin-catalyzed hydrolysis, 385, 386
cis-1,2-dichloroethene, 100
11-cis-retinal, 277
Citronellol, 212, 461
Claisen condensation, 308–309, 311
Clemmensen reduction, 142, 266
^{13}C NMR spectroscopy, 445–446
  advantage of, 439
  characteristics of, 439–441
Coenzyme, 43
Coenzyme Q, 227–228
Concerted reactions, 51
Condensation polymers, 401, 403, 410–411
Condensation reaction, 45, 410–411
Conformers, 73
Conjugated dienes, 95
Copolymerization, of alkenes, 405–406
Copolymers, 405–406
Covalent bonds, 5
Cross-linked polymers, 406–408
Crown ethers, 252
C-terminal amino acid residue, 377
Cyclic ethers, 240–241
Cycloalkanes
  adamantane, 76–77
  bicyclic compounds, 75
  brevicomin, 77
  bridged ring compounds, 75
  cis-1,3-dichlorocyclohexane, 78
  conformations of, 78–81
  fused ring compounds, 75
  geometric isomerism, 75–76
  nomenclature of, 76
  oxidation of, 83
  spirocyclic compounds, 75
Cyclohexane, 78–80, 93
Cyclopropane, 78
Cytochrome c, evolutionary family tree for, 384, 385

D

Dehydration, 115
  of alcohols, 216–218
  of aldols, 280–281
  of citric acid, 216, 217
Dehydrohalogenation, 118, 202
Delocalization, 135
Delta scale, 432
Demerol, 318, 319
Deoxyhemoglobin, structure of, 392, 393
Diastereomers, 174, 182–183, 346
Diazepam (Valium'), 316
Diazotization, 153
Dicarboxylic acids, 290
(Z)-2,3-Dichloro-3-hexene, 191, 460
Dichloromethane, 165
1,3-Dichloropropane, NMR spectrum of, 438–439
Dicyclohexylcarbodiimide (DCCI), 381, 382
Diene polymers, 409–410
Dienes, 95
Dietary protein
  biological value of, 377
  and essential amino acids, 376, 377
2-(Diethylamino)-N-(2,6-dimethylphenyl)ethanamide, 322
Dihydroxyaceton, 347
Dihydroxyacetone, 347
(3R)-3,5-Dihydroxy-3-methylpentanoic acid. See Mevalonic acid
3,3-Dimethyl-1-butene
  hydroboration-oxidation of, 463
  oxymercuration-demercuration of, 462
Dimethyl ether structure, 239
2,4-Dimethyl-1,3-pentadiene, 425
3,7-Dimethy-6-octen-1-ol. See Citronellol
Dioxin, 148
Dipeptide, 377
Dipole-dipole forces, 33–34
Dipole moment, 7
Disaccharides, 343, 370
  cellobiose, 359–360
  lactose, 360–361
  maltose, 358–359
  structure, 472
  sucrose, 361–362
Disubstituted alkene, 96
Diterpenes, 120
p-Divinylbenzene, 406, 407
Dopamine, 315
Doublets, 436

E

Edman degradation, 383–384, 386
Effective collisions, 55
Elastomers, 399–401
Electromagnetic radiation, 422
Electromagnetic spectrum, 422, 423
Electronegativity, 4
Electrophiles, 53, 113
Electrophiles and ring substituents, 161
Electrophilic aromatic substitution, 139–145
Elimination reactions, 44, 201, 208
  dehydrohalogenation, 202
  E1 mechanism, 202

E2 mechanism, 202
  of primary haloalkanes, 204
  of secondary haloalkanes, 204–205
  of tertiary haloalkanes, 203
Enantiomers, 165
β-Endorphin, 386
Enkephalins, 378
Enol, 115
Enolate anion, 278
Epimers, 346–347
Epinephrine, 173, 315, 320
Epoxides, 257, 258
  reactions of, 246–253
  ring cleavage of, 254
  synthesis of, 246, 254
Erythromycin, 173
D-Erythrose, 345
L-Erythrose, 345
Esters, 288
  and anhydrides of phosphoric acid, 305–307
  boiling points of, 293, 294
  flavoring agents, 293, 294
  odors of, 293
  reduction of, 304, 305
  solubility of, 293, 294
  synthesis of, 310
Estradiol, 271–272
Estrogens, 271
1,2-Ethanediol, 37
Ethchlorvynol, 211–212
Ethene structure, 96
Ethers, 255–257
  boiling points, 242
  cleavage of, 254
  dipole moments, 242
  and Grignard reagents, 242–244
  nomenclature of, 240–241
  physical properties of, 241–242
  reactions of, 245–246
  as solvents, 242
  structure of, 239, 240
  synthesis of, 244–245, 254
Ethoxybenzene, 241
2-Ethoxynapthalene synthesis, 244–245
Ethylene glycol, 220, 221
Ethyl 2-(p-chlorophenoxy)-2-methylpropanoate, 292
Ethyne, structure of, 96
E,Z system of nomenclature, 101–103, 127–129

F

Fat-soluble vitamins, 37–38
Fehling's solution, 265, 355, 465
Fibers, 401
Filaments, of thermoplastics, 401
Fischer projection formulas, 168–170, 344, 345
Flavin adenine dinucleotide, 43
Flecainide, 330, 331
Formal charge, 7–8, 28
Formaldehyde (CH_2O) structure, 259, 260
Formalin, 272
Free radicals, 86–87
  bromination, 89

chlorination, 89
 polymerization, 410
 substitution reactions, 53
Freons, 86–87
Friedel-Crafts acylation, 142
Friedel-Crafts alkylation, 141
α-L-Fucose, 363
Functional group isomers, 23
Functional groups, 15–18, 31
Furanoic acid, 468
Furanoses, 350
Fused ring compounds, 75

G

α-D-Galactopyranose, 352, 353
β-D-Galactopyranose, 352, 353
Gasoline, 83
1GB1 structure, 391
Geometric isomerism
 alkenes, 99–100
 cycloalkanes, 75–76
Geometric isomers, 99, 101–103
Geraniol, 100
Germicides, 228, 229
D-Glucitol, 354
Glucose, 25
Glutathione, 196
Glyceraldehyde, 169
D-Glyceraldehyde, 170, 344
L-Glyceraldehyde, 170, 344
Glycine, titration curve of, 376
Glycogen, 363
Glycophorin, 363
Glycosides, 356–358, 366, 369
Glycosidic bonds, 343, 356
Glycylalanine, structure of, 377, 378
Glyptal, 412, 413
Grignard reagents, 242–244, 269–272, 465
Group vibrations, 426
Guanine nucleoside coupled protein receptors (GCPRs), 260
Gutta-percha
 ozonolysis of, 475
 properties of, 409

H

Haloalkanes, 18
 carboxylic acid synthesis, 310
 dehydrohalogenation of, 206
 nomenclature of, 87–89, 190, 206
 nucleophilic substitution of, 206
 reactivity of, 189–190
Halogen compounds, in ocean organisms, 191–192
Haworth projection formulas, 350–351, 353
α-Helix structure, 389
Heme, structure of, 392
Hemiacetals, 349–353, 466
Hemiketals, 349–353
Heptane, 36
3-Heptanol, 440, 441
4-Heptanol, 440, 441
2-Heptanone, infrared spectrum of, 429
Heteroatoms, 65

Heterocyclic aromatic amines, 320
Heterocyclic compounds, 65
Heterolytic cleavage process, 52
Heteropolysaccharides, 343
Hexachlorophene, 228, 229
Hexamethylenediamine, 22
(E)-2-Hexenal, 464
Hexylresorcinol, 229
High-density polyethylene (HDPE), 398
Histamine, 136
Histidine, 377
Homolytic cleavage process, 52
Homopolysaccharides, 343
Hückle rule, 135
Human blood groups, 363–365
Human lysozyme, amino acid composition of, 382, 383
Hybridization on bond length, effect of, 14–15, 29
Hydrates, 272
Hydration, 115
Hydrocarbons
 classes of, 65
 infrared spectrum of, 426–427
Hydrogenation, 107
Hydrogen bond, 35
Hydrogen bonding
 in kevlar, 398, 399
 in nylon 66, 398
Hydrogen-bonding forces, 35–36
α-Hydrogens, acidity of, 278
Hydrolysis
 of amides, 333–335
 reaction, 44
Hydrophilic amino acids, 372
Hydrophobic amino acids, 372
Hydrophobic effect, 391
Hydroxide ion with chloromethane, S_N2 reaction of, 197, 198
Hydroxylation, 110
Hydroxyl group (–OH), 15, 209–212

I

Imines, 16, 275, 276, 331, 332
Indoleacetic acid, 23
Induced dipole, 34
Inductive effect, 50
Inductive withdrawal of electrons, 51
Infrared spectroscopy, 421, 442–443
 1-butanol, 429, 430
 characteristic group frequencies, 427
 group vibrations, 426
 2-heptanone, 429
 hydrocarbons, 426–427
 1-methylcyclopentene, 425, 426
 n-octane, 427, 428
 1-octene, 427, 428
 1-octyne, 427, 428
 oxygen-containing compound identification, 427–431
International Union of Pure and Applied Chemistry (IUPAC)
  names, 25
 of aldehydes, 261–262
 carboxylic acids, 290
 ethers, 240
 of ketones, 262–263

Invert soaps, 330
Ionic bonds, 4–5
Isoionic point, 376, 394
Isomers, 23–25, 32, 92
Isoprene, 120, 424
Isotactic polymer, 408, 409

J

Jones reagent, 219

K

Kekulé's concept of benzene, 134
Ketal formation, 283, 466
    mechanism of, 275
    reactivity of, 274, 275
Keto-enol equilibria, 278–279
Ketones, 284, 285
    general formulas for, 260
    IUPAC names of, 262–263
    naturally occurring, 260, 261
    nomenclature of, 261–263
    physical properties of, 263–264
    reactivities of, 268, 269
    reduction of, 265, 282, 283
    solubility in water, 264
Ketoses, 343, 347–349
D-2-Ketoses, 348, 349
Kevlar, hydrogen bonding in, 398, 399
Kwashiorkor, 377

L

α-Lactalbumin, 382
Lactams, 289
Lactones, 289
Lactose, 360–361
Large chain branching, 405
Leaving groups, 192
Le Châtelier's principle, 46–47, 302
Lewis acids and bases, 40–41
Lewis structure, 5, 27
Lexan, 413–414
Lipid soluble compounds, 43
Lithium aluminum hydride, 305
Liver enzyme alcohol dehydrogenase (LADH), 220, 221
London forces, 34–35
    in polyethylene, 397, 398
Lone pair electrons, 5
Low-density polyethylene (LDPE), 398

M

Malic acid, 297
Maltose, 358–359
β-D-Mannopyranose, 471
Markovnikov addition, 180
Markovnikov addition product, 223
Markovnikov's rule, 113
Mepivacaine®, 317, 468
Mercaptans. See Thiols
Meso compounds, 174
Meta directors, 144
Methamphetamine, 321

Methanol, 299
Methoxyamphetamine, 321
Methylamine structure, 317
1-Methylcyclohexanol, 218
1-Methylcyclopentene, infrared spectrum of, 425, 426
Methylene group, 19
4-Methyl-2-pentanol, dehydration products of, 462
Methylphenidate (Ritalin®), 321
Mevalonic acid, 467
Milk sugar, 343, 360
Molecular models, organic compounds, 20, 29–30
Monosaccharides
    classification, 343, 366–367
    conformations of, 353, 368
    D-and L-series of, 344–345
    epimers, 346–347
    Fischer projection formula, 344, 367–368
    isomerization of, 366, 369
    oxidation of, 354–356, 366, 369
    reduction of, 354, 365, 369
Monosubstituted alkene, 96
Monoterpenes, 120
Monsanto process, 299
Multiple covalent bonds, 5–6
Multiplicity, 436
Mutarotation, 351, 352

N

Native state/conformation, proteins, 386
Natural and synthetic rubbers, 407
Naturally occurring aldehydes and ketones, structures of, 260, 261
Naturally occurring alkenes, 98–99
Naturally occurring alkynes, 98–99
Natural rubber, 400
    ozonolysis of, 475
    properties of, 409
N-butylmethyl ether, 240
Negatively charged carbanion, 52
Neoprene®, 410
N-ethylpropanamide, 320
Neurotransmitter structure, 315, 316
Neutral amino acids, 372
Newman projection formulas, 73–74
Nicotinamide adenine dinucleotide, 43
Nitration, 141
Nitriles, 16, 332
Nitro compounds, reduction of, 332
Nitrogen compounds
    addition reactions of, 275–277
    basicity of, 325–327
    hydrogen bonding in, 323–324
N,N-diethyl-3-methylbutanamide, 320, 469
N-octane, infrared spectrum of, 427, 428
Nomenclature, organic compounds, 25–26
Nonactin, 253
Nonbonding electrons, 5
Nonpolar compounds, 214
Nonsuperimposable mirror image molecules, 165
Norepinephrine, 320
Normal alkanes, 65
N-terminal amino acid residue, 377
Nubucaine, 334

Nubucaine hydrolysis, 470
Nuclear magnetic resonance (NMR) spectroscopy, 421–422
    chemical shift, 431–433
    of 1,3-dichloropropane, 476
    electromagnetic radiation, absorption of, 431
    integrated intensities of, 433, 435
Nuclear spin, 431
Nucleophile, 53, 113
Nucleophilic acyl substitution, 288, 313–314
    acid anhydrides, 302–303
    acid chlorides, 302
    acyl derivatives, reactivity of, 301
    mechanism of, 300
    stoichiometry of, 300
    thioesters, 303–304
Nucleophilic addition reactions, 268
Nucleophilicity
    charge effects on, 195
    definition of, 194
    reaction rates with iodomethane, 194, 195
    of thiolate and alkoxide ions, 195
Nucleophilic reactions, of amines, 328–331
Nucleophilic substitution reactions, 53–54, 192–193, 207, 208
Nylon 6, 414
Nylon 66, hydrogen bonding in, 398
Nylon salt, 414

O

Ocean organisms, halogen compounds in, 191–192
1-Octene
    vs. cyclooctane, 426
    infrared spectrum of, 427, 428
1-Octyne, infrared spectrum of, 427, 428
Oleic acid, 291
Oligopeptides, 377
Oligosaccharides, 343
Oligosaccharides, hydrolysis of, 343
Opiate receptors, 378
Optical isomers, 168
Optically active chiral molecules, 167
Optically inactive achiral molecules, 167
Oral contraceptives, 271–272
Orbital hybridization, 12
Orbitals and molecular shapes, 11–15
Organic and inorganic compounds, 1
Organic nitrogen compounds, 315–316
Organic reactions, classification of, 44–45, 62–63
Oxaloacetic acid, 297, 467
Oxidation-reduction reactions, 41–44, 60–62
Oxidation-reduction reactions, of carbonyl compounds, 265–267
Oxidizing agent, 41
Oxonium ion, 214
Oxygen compounds, addition reactions of, 272–273
Oxytocin, 378
Ozone layer, 86–87
Ozonolysis, 110–111

P

Parallel β-pleated sheet, hydrogen bonding in, 389, 390
2-(*P*-chlorophenoxy)-2-methylpropanoic acid, 292
2-Pentyne, 460
Peptide bonds, 371, 387

Peptides, 377, 394–395
    biological functions of, 378–380
    hormones, 378, 379
    synthesis of, 380–382
Phenacetin, 333
Phenindamine, 327
Phenol coefficient (PC), 228
Phenols, 237
    acidity of, 226–227
    germicides, 228–229
    oxidation of, 227–228
1-Phenyl-2,4-hexadiyn-1-one. *See* Capillin
Phenylpropanolamine, 321
Phenytoin, 166
Pheromones, 21–22
pK_a values of, carboxylic acids, 295
pK values and acid strength, 60
Plane-polarized light, 167
Plastics, 400–401
Polar compounds, 214
Polar covalent bonds, 6–7
Polarimeter, 167
Polarizability, 34
Poly(ethylene terephthalate) (PET), 411–412
Polyamides, 414–415, 419
Polycarbonates, 413–414
Polycyclic and heterocyclic aromatic compounds, 159
Polyesters, 411–413, 418–419
Polyethers, 252–253
Polyethylene, London forces in, 397, 398
Polyisoprene, 400, 407, 408
Polymerization methods, 401–404
Polymers, 416–417
    classification of, 399–401
    condensation polymers, 410–411, 417–418
    cross-linked polymers, 406–408, 417
    diene polymers, 409–410
    hydrogen bonding in, 398, 399
    London forces in, 397–398
Polypeptide chains, enzymatic cleavage of, 384
Polypeptides, 371
Polysaccharides, 343, 362–365
Polyunsaturated compounds, 98
Polyurethanes, 415, 419
Positional isomers, 23–24
Positively charged carbocation, 52
Primary alcohols, 269
Primary carbon atom, 67
Primary haloalkanes, substitution and elimination reactions of, 204
Principal quantum numbers, 2
Propanal, dipole moment for, 263, 264
Prosthetic group, 392
Proteins, 371, 396
    amino acid composition, 382, 383
    Edman degradation, 383–384, 386
    isoionic points of, 376
    primary structure of, 384–388
    quaternary structure of, 392–393
    secondary structure of, 388–391
    tertiary structure of, 391
Protic solvents, 200
Proton counting, 433

Pyranoses, 350
Pyridinium chlorochromate (PCC), 219
Pyruvic *vs.* propanoic acid, 296

Q

Quartets, 436
Quaternary ammonium salts, 330
Quaternary carbon atom, 67
Quinones, 227–228

R

Racemic mixture, 178
Reaction coordinate diagrams, 55–56
Reaction mechanisms, 51–54
Reaction rates, 54–57
Rearrangement reaction, 45
Reducing agent, 41
Reducing sugars, 355
Reductive amination, 331
Resonance stabilization, 50
Resonance-stabilized allylic carbocation, 201
Resonance structures, 8–10
Resonance theory, 134–135
Resorcinols, 229
Rhodopsin, 277
Riboflavin, 211
D-Ribose, 345
Ring-opening reactions
    acid-catalyzed, 247
    direction of, 248
    nucleophiles, 247
Rotamers, 73
Rubber, 399, 400

S

S-adenosylmethionine (SAM), 196, 197
Salt bridge, 391
Salts, of carboxylic acids, 296
Sandmeyer reaction, 153
Saturated hydrocarbons, 65
Secondary alcohols, 270
Secondary carbon atom, 67
Secondary haloalkanes, substitution and elimination reactions
    of, 204–205
Serotonin, 315
Sesquiterpenes, 120
Short chain branching, 405
Skeletal isomers, 23
S_N1 reaction mechanism, 200–201
S_N2 reaction mechanism
    activation energy, 197, 198
    hydroxide ion with chloromethane, 197, 198
    vs. S_N1 reaction, 200–201
    steric effects in, 197, 198
Solubility, 37–38
    of alcohols, 213
    of aldehydes and ketones, 264
    of amines, 324
    of ammonium salts, 328
    of carboxylic acids, 293
    of esters, 293, 294

Sorbitol, 354
Spectroscopy
    principles, 422–423
    structure determination, 421–422
sp hybridization of carbon, 14
sp^2 hybridization of carbon, 12–13
sp^3 hybridization of carbon, 12
Spin-spin splitting, 435–439
Spirocyclic compounds, 75
Starch, 362
Step-growth polymers. *See* Condensation polymers
Stereochemical effects, in S_N1 reaction, 199, 200
Stereochemistry
    absolute configuration, 170–173
    of addition polymerization, 408–410
    configuration of molecules, 163
    diastereomers, 182–183, 187
    Fischer projection formulas, 168–170
    mirror images and chirality, 163–167
    multiple stereogenic centers, 173–177
    plane-polarized light, 167
    specific rotation, 167–168
    stereogenic centers, 179–182
    synthesis of stereoisomers, 178–179
Stereoselective reaction, 182
Steric effects
    on addition reactions, 268, 269
    in S_N2 reaction mechanism, 197, 198
Steric hindrance, 117, 197
Steroids, 80–81
Structural features, organic compunds, 20–21
Structural formulas, organic compounds, 18–23, 30–31
Structure and physical properties, organic compounds, 33–38
Substitution reactions, 44. *See also* Nucleophilic substitution reactions
    of alcohols, 215–216
Sucrose, 361–362
Sulfhydryl group, 229
Sulfides. *See* Thioethers
Sulfonation, 141
Superimposable objects, 163
Symmetrical ether, 239
Symmetrical reagents, 111
Syn addition, 224
Syndiotactic polymer, 408, 409
Synthesis gas, 225, 299
Synthetic macromolecules, physical properties of, 397
Synthetic polymers, 397

T

Tamoxifen, 103
Tartaric acids, 175
Tautomerism, 279
Tautomerization, 279
Tautomers, 279
Terminal alkyne, 96
Terpenes, 120–123, 132
Tertiary alcohols, 270
Tertiary haloalkanes, substitution and elimination reactions of, 203
Tetrahydrofuran (THF), 240
Tetrahydropyran (THP), 239, 240
Tetrasubstituted alkene, 96
Tetraterpenes, 120

Thermoplastics, 400–401, 474. *See also* Polyethylene, London forces in
Thermosetting plastics, 474
Thermosetting polymers, 400, 401
Thioesters, 289
   biochemical Claisen condensation of, 309
   nucleophilic acyl substitution, 303–304
Thioethers, 16
Thiols, 16, 404
   properties of, 229–230
   reactions of, 230
Three-dimensional structures, organic compunds, 20
Titration, of amino acids, 376
Tollens reagent, 265, 266, 355
Toxicity, of alcohols, 219–220
trans-1,2-dichloroethene, 100
Tremorine, 97–98
1,1,2-Tribromo-3,3-dimethylbutane, NMR spectrum of, 435, 436
1,1,2-Trichloroethane, NMR spectrum of, 437, 438
1,2,2-Trichloropropane, NMR Spectrum of, 432–434
2-(3,4,5-Trimethoxyphenyl)ethanamine, 469
Trimethylene oxide, 240
Tripeptide, 377
Triplets, 436
Trisaccharides, 343
Triterpenes, 120
Trypsin, 384
Trypsin-catalyzed hydrolysis, 474
Tuftsin, 380, 473

U

Ultraviolet spectroscopy, 421, 442
   of isoprene, 424
   structural effects on, 424

Universal donors, 363
Unsaturated hydrocarbons, 95–99
Unshared electron pairs, 5
Unsymmetrical ether, 239
Unsymmetrical reagents, 111
Uronic acid, 355

V

Valence-shell electron-pair repulsion (VSEPR) theory, 10–11
Valence shell electrons, 3
Vasopressin, 378–379, 388
Vicinal dihalide, 118
Vulcanization process, 407
Vulcanized polyisoprene, cross-links in, 407, 408

W

Wacker process, 299
Warfarin, 172
Water gas, 299
Water-soluble and vitamins, 37–38
Wavenumber, 422
Williamson ether synthesis, 244
Wolff-Kishner reduction, 266

X

D-Xylitol, 354

Z

Zaitsev product, 216
Zaitsev's rule, 117
Ziegler-Natta catalysts, 409, 410
Zwitterion, 372